The Maize Handbook

Springer
New York
Berlin
Heidelberg
Barcelona
Budapest
Hong Kong
London
Milan
Paris
Santa Clara
Singapore
Tokyo

Michael Freeling Virginia Walbot
Editors

The Maize Handbook

With 120 Illustrations

Springer

Michael Freeling
Department of Plant Biology
University of California
Berkeley, CA 94720 USA

Virginia Walbot
Department of Biological Sciences
Stanford University
Stanford, CA 94305 USA

Front Cover Art: Corn-Stalk Vessel. The Metropolitan Museum of Art, Purchase, Judith S. Randal Foundation Gift, 1989. (1989.62.1) © 1989 The Metropolitan Museum of Art.

Library of Congress Catalog-in-Publication Data
The Maize handbook / Michael Freeling, Virginia Walbot, editors.
p. cm.
Includes bibliographical references and index.
ISBN 0-387-97826-7 (New York).—ISBN 3-540-97826-7 (Berlin)
1. Corn—Handbooks, manuals, etc. 2. Corn—Biotechnology—Handbooks, manuals, etc. I. Freeling, Michael. II. Walbot, Virgiania.
SB191.M2M3275 1993
633.1'5—dc20 92-32462

Printed on acid-free paper.

First softcover printing, 1996.

Acquiring Editor: Robert Garber
Production managed by Henry Krell; manufacturing supervised by Vincent Scelta.
Typeset using the editors' Microsoft Word files by Impressions, Inc., Ann Arbor, MI.
Printed and bound by Edwards Brothers, Inc., Ann Arbor, MI.
Printed in the United States of America.

9 8 7 6 5 4 3 2 1

ISBN 0-387-97826-7 Springer-Verlag New York Berlin Heidelberg (hard)
ISBN 3-540-97826-7 Springer-Verlag Berlin Heidelberg New York
ISBN 0-387-94735-3 Springer-Verlag New York Berlin Heidelberg (soft) SPIN 10534051

Preface

The Maize Handbook represents the collective efforts of the maize research community to enumerate the key steps of standard procedures and to disseminate these protocols for the common good. Although the material in this volume is drawn from experience with maize, many of the procedures, protocols, and descriptions are applicable to other higher plants, particularly to other grasses.

The power and resolution of experiments with maize depend on the wide range of specialized genetic techniques and marked stocks; these materials are available today as the culmination of nearly 100 years of genetic research. A major goal of this volume is to introduce this genetical legacy and to highlight current stock construction programs that will soon benefit our work, e.g. high-density RFLP maps, deletion stocks, etc. Both stock construction and maintenance are relatively straightforward in maize as a result of the ease of crossing and the longevity of stored seeds. Crossing is facilitated by the separate staminate (tassel) and pistillate (ear) flowers, a feature almost unique to maize. On the other hand, many of the genetic methodologies utilized with maize, including the precision of record keeping, can be adapted to other plants.

Facile communication and a spirit of co-operation have characterized the maize genetics community since its earliest days. Starting in the 1930s, institutions such as annual Maize Genetics Cooperation Newsletter, the Maize Genetics Stock Center, and the annual maize genetics meeting provide continuity to the field. Each of these institutions allows geneticists to exchange information, to critique each others work, and to benefit from the enthusiasm, experience and expertise of others in their field. *The Maize Handbook* is clearly an extension of this desire on the part of maize geneticists, and embraced by nongenetical maize researchers as well, to share and to aid the progress of others; in modern times, the view that other scientists are cooperators rather than competitors is unusual, but it is a key part of the maize community of scholars.

The Maize Handbook covers five areas: cell biology, developmental morphology, genetic manipulations, tissue culture techniques, and molecular biology techniques. In consultation with expert section editors—Professors Jim Birchler, Ron Phillips, R. Scott Poethig, Anne Sylvester, and Sue Wessler—we developed a chapter outline and suggested authors. Without exception, each potential author contacted agreed to provide the materials requested.

We have tried to make *The Maize Handbook* comprehensive in its coverage. There are several reference books that complement this volume. Chief among these is the annual edition of the Maize Genetics Cooperation Newsletter (MNL) in which new genetic map information and reports on gene isolation, gene action and genetic methodologies are reported in a "from the notebook" format. Maize research is a dynamic field, and it is impossible to be well informed without access to the Newsletter. For subscription information, contact Dr. E. H. Coe, Editor, Maize Genetics Cooperation Newsletter, USDA-ARS, Curtis Hall, University of Missouri, Columbia MO 65211.

A group of inexpensive books (total cost about $400) would also be part of a robust personal or laboratory library, but should certainly be in an institutional library for easy reference.

1. The new edition of *Mutants of Maize*, by M. G. Neuffer, E. H. Coe, and S. Wessler due from Cold Spring Harbor Press in 1994. Color illustrations of visible mutants (naked eye polymorphisms) and cytogenetic stocks; molecular maps of cloned genes are also included.

2. *Corn and Corn Improvement*, 1988, third edition, G. F. Sprague and J. W. Dudley, eds. American Society of Agronomy, Madison WI. This volume contains comprehensive genetic and cytogenetic information as well as more practical chapters on corn diseases, pests and agronomic practice.

3. *Discussions in Cytogenetics*, second edition, 1992, by C. R. Burnham, Alpha Editions, Edina MN, 55435. Classical text with a dateline of about 1959. Maize research is highlighted, and presented in detail, with complete citations.

4. *Modern Corn Production* by S. R. Aldrich, W. O. Scott and E. R. Leng, 1975, second edition. A & L Publications, Champaign, Illinois. Useful reference on field practices with color illustrations.

5. *Maize Diseases* by D. C. McGee, 1988. APS Press, St. Paul, Minnesota. Paperback book with color illustrations on disease and nutritional deficiencies.

6. *The Structure and Reproduction of Corn*, 1949, by T. A. Kiesselbach. Nebraska Agric. Exp. Station Research Bulletin. Observations of the plant and its developmental stages are illustrated with light micrographs and camera lucida drawings. A short but classic paperback.

7. *Anatomy of Seed Plants*, second edition, 1977, by K. Esau, John Wiley & Sons, New York. Maize is among the featured species.

8. *Compilation of North American Maize Breeding Germplasm*, 1993. Authors: J. T. Gerdes, C. F. Behr, J. G. Coors, and W. F. Tracy. Editors: W. F. Tracy, J. G. Coors, and J. L. Geadelmann. Published by Crop Science Society of America, Madison, Wisconsin. This paperback lists the history of lines, and is useful in determining the likelihood of shared parentage in genetic stocks.

9. *The Discovery and Characterization of Transposable Elements*, 1987, by B. McClintock. The collected papers of Barbara McClintock. Garland Publishing, Inc., New York.

Virginia Walbot and Michael Freeling

Acknowledgments. The editors give heartfelt thanks to Melissa Quilter of the NSF Center of Plant Developmental Biology at U.C. Berkeley for her tireless efforts to assemble the initial draft of the manuscript and to complete the final text. We also thank Jill Otto of the Falconer Biology Library at Stanford University for her assistance with the verification of references. *The Maize Handbook* would not have been possible without the coordination and careful husbandry of the individual contributions by the section editors. Pioneer Hi-Bred International, Inc., generously financed an up-to-date genetic map prepared by Ed Coe and distributed to the maize community. The credit for the idea for *The Maize Handbook* rests entirely with Dr. Robert C. Garber, Life Sciences Editor of Springer-Verlag. No one in their right mind would spontaneously undertake such a large task, but we have found the 18 months of planning and editing highly rewarding and thank Rob for his persistence, his appeal to our love of maize, and his persuasive scientific arguments for initiating this project.

Contents

Preface v

Contributors xix

I. Development and Morphology 1

1. Fertilization and Embryogeny in Maize 3
   William F. Sheridan and Janice K. Clark
2. The Maize Shoot 11
   R. S. Poethig
3. The Maize Leaf 17
   Michael Freeling and Barbara Lane
4. The Maize Root 29
   L. Feldman
5. Morphology and Development of the Tassel and Ear 37
   Ping-Chin Cheng and Dayaker R. Pareddy
6. Gametogenesis in Maize 48
   Patricia Bedinger and Scott D. Russell
7. The Patterns of Plant Structures in Maize 61
   Walton C. Galinat
8. Genetics and the Morphological Evolution of Maize 66
   John Doebley
9. Overview of Key Steps in Aleurone Development 78
   Virginia Walbot

II. Cell Biology 81

10. Light Microscopy I: Dissection and Microtechnique 83
    Anne W. Sylvester and Steven E. Ruzin
11. Light Microscopy II: Observation, Photomicrography, and Image Analysis 95
    Steven E. Ruzin and Anne W. Sylvester
12. Scanning Electron Microscopy 108
    Michelle H. Williams and Anne W. Sylvester

13. Transmission Electron Microscopy: Chemical Fixation, Freezing Methods, and Immunolocalization 118
M. V. PARTHASARATHY
14. Techniques for Histology of Maize Megaspores and Embryo Sacs 135
S. D. RUSSELL AND D. P. WEST
15. Indirect Immunofluorescence: Localization of the Cytoskeleton 140
C. J. STAIGER
16. Immunocytochemistry for Light and Electron Microscopy .. 149
M. J. VARAGONA AND N. V. RAIKHEL
17. Immunolocalization of Nuclear Proteins 158
LAURIE G. SMITH
18. In situ Hybridization 165
JANE A. LANGDALE
19. Photography of Maize 180
PHILIP W. BECRAFT

III. Genetics Protocols 187

20. Genetic Experiments and Mapping 189
E. H. COE
21. Growing Maize for Genetic Studies 197
M. G. NEUFFER
22. A Nine-Step Way to Characterize a Morphological Mutant 209
MICHAEL FREELING AND JOHN FOWLER
23. Mutagenesis 212
M. G. NEUFFER
24. Gene Tagging with *Ac/Ds* Elements in Maize 219
STEPHEN L. DELLAPORTA AND MARIA A. MORENO
25. Using Cytogenetics to Enhance Transposon Tagging with *Ac* Throughout the Maize Genome 234
DONALD AUGER AND WILLIAM F. SHERIDAN
26. Transposon Tagging with *Spm* 240
KAREN CONE
27. Transposon Tagging with *Mutator* 243
PAUL S. CHOMET
28. Mapping Genes with Recombinant Inbreds 249
BENJAMIN BURR, FRANCES A. BURR AND EILEEN C. MATZ
29. The Placement of Genes Using *waxy*-Marked Reciprocal Translocations 255
JOHN R. LAUGHNAN AND SUSAN GABAY-LAUGHNAN

30. Chimeras for Genetic Analysis 258
M. G. NEUFFER

31. The Use of Clonal Sectors for Lineage and Mutant Analysis 262
SARAH HAKE AND NEELIMA SINHA

32. Use of Segmental Aneuploids for Mutant Analysis 270
BEN GREENE AND SARAH HAKE

33. Biased Transmission of Genes and Chromosomes 274
WAYNE R. CARLSON

34. Anthocyanin Genetics 279
E. H. COE

35. Cloned Anthocyanin Genes and Their Regulation 282
KAREN CONE

36. Maize and *Puccinia sorghi*: A System for the Study of the Genetic and Molecular Basis of Host-Pathogen Interactions 286
TONY PRYOR

37. Disease Lesion Mutants 291
M. G. NEUFFER

38. Classification of Pollen Abortion in the Field 297
R. L. PHILLIPS

39. Genetic Fine Structure as Revealed in Pollen Assays 298
OLIVER E. NELSON

40. Genetic Fine Structure from Testcross Progeny Analysis ... 303
HUGO K. DOONER

41. Trisomic Manipulation 307
JAMES A. BIRCHLER

42. B-A Translocation Manipulation 308
WAYNE R. CARLSON

43. Locating Recessive Genes to Chromosome Arm with B-A Translocations 315
J. B. BECKETT

44. Dosage Analysis Using B-A Translocations 328
JAMES A. BIRCHLER

45. Marker Systems for B-A Translocations 330
JAMES A. BIRCHLER

46. Construction of Compound B-A Translocations 332
JAMES A. BIRCHLER

47. A-B-A Compound Chromosomes 334
JAMES A. BIRCHLER

48. Comprehensive List of B-A Translocations in Maize 336
J. B. BECKETT

49. Chromosomal Translocations Involving the Nucleolus Organizer Region or Satellite of Chromosome 6 342
R. L. PHILLIPS
50. Inversions and List of Inversions Available 346
G. G. DOYLE
51. Use of Maize Monosomics for Gene Localization and Dosage Studies 350
DAVID F. WEBER
52. Marker Systems for ***r-x1*** 359
JAMES A. BIRCHLER AND E. H. COE
53. Translocations as Genetic Markers 361
E. B. PATTERSON
54. A-A Translocations: Breakpoints and Stocks 364
E. H. COE
55. Segmental Aneuploid Analysis 377
JAMES A. BIRCHLER
56. Directed Synthesis of Segmental Transpositions 383
JAMES A. BIRCHLER
57. Practical Aspects of Haploid Production 386
JAMES A. BIRCHLER
58. Indeterminate Gametophyte (*ig*): Biology and Use 388
JERRY L. KERMICLE
59. Production of a Ploidy Series 394
JAMES A. BIRCHLER
60. Absorption Cytophotometry of Nuclear DNA 396
RICHARD V. KOWLES, GEORGIA L. YERK AND RONALD L. PHILLIPS
61. Flow Cytometry for Endosperm Nuclear DNA 400
RICHARD V. KOWLES, GEORGIA L. YERK, LIANG SCHWEITZER, FREIDRICH SRIENC AND RONALD L. PHILLIPS
62. Allelism Testing of Lethal Mutations 407
JANICE K. CLARK AND WILLIAM K. SHERIDAN
63. Analysis of Cytoplasmically Inherited Mutants 413
KATHLEEN J. NEWTON
64. Male Sterility and Restorer Genes in Maize 418
SUSAN GABAY-LAUGHNAN AND JOHN R. LAUGHNAN
65. Inbred Lines of Maize and Their Molecular Markers 423
MICHAEL LEE
66. Traditional Analysis of Maize Pachytene Chromosomes 432
ELLEN DEMPSEY
67. Techniques for Preparing Whole-Mount Spreads of Maize Pachytene Chromosome Complements for Electron-

Microscopic Visualization of Synaptonemal Complex Structures 442
Marjorie P. Maguire

68. A Smear Technique for the Study of Meiosis in Pollen Mother Cells of Maize 447
Inna Golubovskaya

69. Protocol for Preparing Maize Macrospore Mother Cells for the Study of Female Meiosis and Embryo-sac Development 450
Inna Golubovskaya and N. A. Avalkina

70. Preparing a Suspension of Microsporocytes for Spreading and Electron Microscopy 454
Inna Golubovskaya and Z. K. Gzebennikova

71. Three-dimensional Fluorescence Microscopy of Maize Chromosomes 457
R. Kelly Dawe

72. Chromosomal Behavior During Microsporogenesis 460
Ming T. Chang and M. Gerald Neuffer

73. A Staining Procedure for Pollen Grain Chromosomes of Maize 476
Bryan Kindiger

74. A Technique for the Preparation of Somatic Chromosomes of Maize 481
Bryan Kindiger

75. A Technique for Somatic Chromosome Preparation and C-banding of Maize 484
David C. Jewell and Nurul Islam-Faridi

76. Duplications 493
James A. Birchler

77. Deficiency Analysis 494
James A. Birchler

78. The Gametophyte Factors of Maize 496
Oliver E. Nelson

79. Ring Chromosomes 503
James A. Birchler

80. In situ Hybridization of DNA and RNA Probes to Maize Chromosomes 504
S. M. Livingston and R. L. Phillips

81. Analysis of Traits with Complex Inheritance in Maize Using Molecular Markers 509
Tim Helentjaris and Manfred Heun

82. Heterofertilization 514
JAMES A. BIRCHLER

IV. Molecular Biology Protocols 517

83. Isolation of Small Nuclear RNAs 519
TAMÁS KISS AND WITOLD FILIPOWICZ

84. Plant DNA Miniprep and Microprep: Versions 2.1–2.3 522
STEPHEN L. DELLAPORTA

85. Urea-based Plant DNA Miniprep 526
JYCHIAN CHEN AND STEPHEN L. DELLAPORTA

86. High-Molecular-Weight Plant DNA Preparation for CHEF Gel Analysis 528
ELSBETH WALKER

87. Preparation of High-Molecular-Weight Maize DNA and Analysis by Pulsed-Field Gel Electrophoresis 530
AVRAHAM A. LEVY

88. Isolation of Genomic DNA from Calli 534
CHRISTINE A. WARREN

89. Isolation of DNA from Immature Cobs 536
CHRISTINE A. WARREN

90. Preparation of Nucleic Acids from Maize Microspores and Pollen 538
ANNE H. BROADWATER AND PATRICIA BEDINGER

91. Preparation of DNA and RNA from Leaves: Expanded Blades and Separated Bundle Sheath and Mesophyll Cells 541
TIMOTHY NELSON

92. Isolation of RNA from *Wx* and *wx* Endosperms 545
SUSAN R. WESSLER

93. RNA Isolation from Electroporated Protoplasts 547
KENNETH R. LUEHRSEN AND JANE HERSHBERGER

94. Procedures for Isolating Mitochondria and Mitochondrial DNA and RNA 549
KATHLEEN J. NEWTON

95. Isolation of Maize Chloroplasts and Chloroplast DNA 556
STEVEN RODERMEL

96. In vitro Capping of Maize Mitochondrial RNA and Transcription Initiation Site Characterization by RNase Protection 559
R. MICHAEL MULLIGAN

97. Editors' Note 565
VIRGINIA WALBOT AND MICHAEL FREELING
98. Southern Blots of Maize Genomic DNA 566
CHRISTINE A. WARREN AND JANE HERSHBERGER
99. Southern Blot Hybridization 569
STEPHEN L. DELLAPORTA AND MARIA A. MORENO
100. Northern Blotting 572
KENNETH B. LUEHRSEN
101. RNase Protection Assay 575
KENNETH R. LUEHRSEN
102. Genomic Sequencing in Maize 579
ANNA-LISA PAUL AND ROBERT J. FERL
103. The Polymerase Chain Reaction: Applications to Maize Transposable Elements 586
ANNE B. BRITT AND DAVID J. EARP
104. In vitro Synthesis of Capped and Polyadenylated mRNA for Translation Studies in vitro and in vivo 592
DANIEL R. GALLIE
105. Construction of a Genomic Library in Lambda Phage 595
CLIFFORD F. WEIL AND THOMAS E. BUREAU
106. Agroinfection 599
JESÚS ESCUDERO AND BARBARA HOHN
107. Polyethylene Glycol-mediated DNA Uptake into Maize Protoplasts 603
LESZEK A. LYZNIK AND THOMAS K. HODGES
108. DNA Delivery into Maize Cell Cultures Using Silicon Carbide Fibers 610
H.F. KAEPPLER AND D.A. SOMERS
109. Transient Gene Expression Assay by Electroporation of Maize Protoplasts 613
KENNETH R. LUEHRSEN AND JEFFREY R. DE WET
110. Assay of Bacterial Chloramphenicol Acetyl Transferase in Transformed Maize Tissues 616
STEVE GOFF
111. Assay of Firefly Luciferase in Transformed Maize Tissues 619
STEVE GOFF
112. Storage of Frozen Maize Tissue 622
SUSAN R. WESSLER
113. Maize Methods—Starch Biosynthetic Genes 624
L. CURTIS HANNAH, EDWIN R. DUKE, KAREN E. KOCH AND B. GREGORY COBB

114. Identifying and Characterizing the TATA Box Promoter Sequence Element in a Maize Nuclear Gene 630
JULIE M. VOGEL
115. Promoters 633
KENNETH R. LUEHRSEN
116. Introns 636
KENNETH R. LUEHRSEN
117. Characterization of Zein Genes and Their Regulation in Maize Endosperms 639
GARY A. THOMPSON AND BRIAN A. LARKINS
118. Overview of Cloning Genes Using Transposon Tagging 647
VICKI CHANDLER
119. How RFLP Loci Can Be Used to Assist Transposon-Tagging Efforts 653
STEVEN P. BRIGGS AND WILLIAM D. BEAVIS

V. Cell Culture Protocols 661

120. Regeneration of Plants from Somatic Cell Cultures: Applications for in vitro Genetic Manipulation 663
CHARLES L. ARMSTRONG
121. Initiation, Maintenance, and Plant Regeneration of Type II Callus and Suspension Cells 671
J. C. SELLMER, S.W. RITCHIE, I.S. KIM AND T.K. HODGES
122. Production of Transgenic Maize Plants via Microprojectile-Mediated Gene Transfer 677
MICHAEL FROMM
123. In vitro Selection 685
D. A. SOMERS AND K. A. HIBBERD
124. Selection of Stable Transformants from Black Mexican Sweet Maize Suspension Cultures 690
JULIE ANDERSON KIRIHARA
125. Maize Protoplast Culture 695
R. D. SHILLITO, G.K. CARSWELL AND C. KRAMER
126. Anther and Microspore Culture 701
J. F. PETOLINO AND A. D. GENOVESI
127. In vitro Culture of Maize Kernels 705
BURLE G. GENGENBACH AND ROBERT J. JONES
128. In vitro Ear Culture System 709
RICHARD V. KOWLES
129. Maize Inflorescence Culture 712
R. I. GREYSON

130. Shoot Meristem Culture 715
ERIN E. IRISH
131. Establishment and Culture of Maize Endosperm 719
JACK C. SHANNON
132. In vitro Pollen Germination 723
D. B. WALDEN
133. Axillary Bud in vitro Culture: Asexual Propagation of Maize 725
R. I. GREYSON AND D. B. WALDEN

Index **727**

Contributors

Charles L. Armstrong
Monsanto Agricultural Company
St. Louis, MO 63198 USA

Donald Auger
University of North Dakota
Biology Department
Grand Forks, ND 58202 USA

N.A. Avalkina
N.I. Vavilov Research Institute
Plant Industry
Department of Genetics
Sankt-Peterburg, 190000 Russia

William D. Beavis
Pioneer Hi-Bred International, Inc.
Johnston, IA 50131 USA

Jack B. Beckett
University of Missouri
Department of Agronomy
Columbia, MO 65211 USA

Philip W. Becraft
University of California
Department of Plant Biology
Berkeley, CA 94720 USA

Patricia Bedinger
University of North Carolina
Department of Biology
Chapel Hill, NC 27599-3280 USA

James A. Birchler
University of Missouri
Division of Biological Sciences
Columbia, MO 65211 USA

Steven P. Briggs
Pioneer Hi-Bred International, Inc.
Johnston, IA 50131 USA

Anne B. Britt
University of California
Department of Botany
Davis, CA 95616 USA

Anne H. Broadwater
University of North Carolina
Department of Biology
Chapel Hill, NC 27599-3280 USA

Thomas E. Bureau
University of Georgia
Botany Department
Athens, GA 30602 USA

Benjamin Burr
Brookhaven National Lab
Biology Department
Upton, NY 11973 USA

Frances A. Burr
Brookhaven National Lab
Biology Department
Upton, NY 11973 USA

Wayne R. Carlson
University of Iowa
Botany Department
Iowa City, IA 52242 USA

G.K. Carswell
CIBA-Geigy Corporation
Agricultural Biotechnology
Research Unit
Research Triangle Park, NC 27709-2257
USA

Vicki Chandler
University of Oregon
Department of Biology
Molecular Biology Institute
Eugene, OR 97403 USA

Ming T. Chang
Garst/ICI Seeds
Slater, IA 50244 USA

Jychian Chen
Yale University
Department of Biology
New Haven, CT 06511 USA

Ping-Chin Cheng
State University of New York
Department of Electrical and Computer Engineering
Buffalo, NY 14260 USA

Paul S. Chomet
DeKalb Plant Genetics
Mystic, CT 06355-1958 USA

Jancie K. Clark
University of North Dakota
Biology Department
Grand Forks, ND 58202 USA

B. Gregory Cobb
University of Florida
Department of Vegetable Crops/IFAS
Gainesville, FL 32611 USA

Ed H. Coe
United States Department of Agriculture —ARS
University of Missouri
Department of Agronomy
Columbia, MO 65211 USA

Karen Cone
University of Missouri
Columbia, MO 65211 USA

R. Kelley Dawe
University of California
Department of Molecular and Cell Biology
Berkeley, CA 94720

Stephen L. Dellaporta
Yale University
Department of Biology
New Haven, CT 06511 USA

Ellen Dempsey
Indiana University
Biology Department
Bloomington, IN 47401 USA

Jeffrey R. de Wet
Pfizer Central Research
Groton, CT 06340 USA

John Doebley
University of Minnesota
Department of Plant Biology
St. Paul, MN 55108 USA

Hugo K. Dooner
DNA Plant Technology Corporation
Oakland, CA 94608-1239 USA

Gregory G. Doyle
United States Department of Agriculture —ARS
University of Missouri
Columbia, MO 65211 USA

Edwin R. Duke
University of Florida
Department of Vegetable Crops/IFAS
Gainesville, FL 32611 USA

David J. Earp
Plant Gene Expression Center
Albany, CA 94710 USA

Jesús Escudero
Friedrich Miescher-Institut
CH-4002, Basel, Switzerland

Lewis Feldman
University of California
Department of Plant Biology
Berkeley, CA 94720 USA

Robert J. Ferl
University of Florida
Department of Vegetable Crops
Gainesville, FL 32611 USA

Witold Filipowicz
Friedrich Miescher-Institut
CH-4002 Basel, Switzerland

John Fowler
University of California
Department of Plant Biology
Berkeley, CA 94720 USA

Michael Freeling
University of California
Department of Plant Biology
Berkeley, CA 94720 USA

Michael Fromm
Monsanto Company
St. Louis, MO 63198 USA

Susan Gabay-Laughnan
University of Illinois
Urbana, IL 61801 USA

Walton C. Galinat
University of Massachusetts, Amherst
Eastern Agricultural Center
Waltham, MA 02154-8096 USA

Daniel R. Gallie
University of California
Department of Biochemistry
Riverside, CA 92521-0129 USA

Burle G. Gengenbach
University of Minnesota
Department of Agronomy and Plant Genetics
St. Paul, MN 55108 USA

A.D. Genovesi
DeKalb Plant Genetics
DeKalb, IL 60115 USA

Steve Goff
CIBA/Geigy Biotechnology
Research Triangle Park, NC 27709-2257 USA

Inna Golubovskaya
N.I. Vavilov Research Institute Plant Industry
Department of Genetics
St. Petersburg, 190000 Russia

Ben Greene
Plant Gene Expression Center USDA—ARS
Albany, CA 94710 USA

Richard I. Greyson
University of Western Ontario
Department of Plant Sciences
London, Ontario N6A 5B7 Canada

Z.K. Gzebennikova
N.I. Vavilov Research Institute Plant Industry
Department of Genetics
St. Petersburg, 190000 Russia

Sarah Hake
Plant Gene Expression Center USDA—ARS
Albany, CA 94710 USA

L. Curtis Hannah
University of Florida
Department of Vegetable Crops/IFAS
Gainesville, FL 32611 USA

Christian Harms
CIBA-Geigy Corporation
Research Triangle Park, NC 27709-2257 USA

Tim Helentjaris
University of Arizona
Plant Science Department
Tucson, AZ 85721 USA

Jane Hershberger
Stanford University
Department of Biological Science
Stanford, CA 94305-2040 USA

Manfred Heun
University of Arizona
Plant Science Department
Tucson, AZ 85721 USA

Ken A. Hibberd
Plant Science Research, Inc.
Minnetonka, MN 55343 USA

Thomas K. Hodges
Purdue University
Department of Botany and Plant Pathology
West Lafayette, IN 47907 USA

Barbara Hohn
Friedrich Miescher-Institut
CH-4002, Basel, Switzerland

Erin E. Irish
University of Iowa
Departments of Botany and Biology
Iowa City, IA 52242 USA

Nurul Islam-Faridi
CIMMYT
International Maize and Wheat Improvement Center
06600 Mexico, D.F. Mexico

David C. Jewell
CIMMYT
International Maize and Wheat Improvement Center
06600 Mexico, D.F. Mexico

Robert J. Jones
University of Minnesota
Department of Agronomy and Plant Genetics
St. Paul, MN 55108 USA

Heidi F. Kaeppler
University of Minnesota
Department of Agronomy and Plant Genetics
Plant Molecular Genetics Institute
St. Paul, MN 55108

Jerry L. Kermicle
University of Wisconsin
Laboratory of Genetics
Madison, WI 53706 USA

I.S. Kim
Purdue University
Department of Botany and Plant Pathology
West Lafayette, IN 47907 USA

Bryan Kindiger
USDA—ARS
Southern Plains Research Station
Woodward, OK 73801 USA

Julie Anderson Kirihara
Plant Science Research, Inc.
Minnetonka, MN 55343 USA

Tamás Kiss
Friedrich Miescher-Institut
CH-4002 Basel, Switzerland

Ted Klein
E.I. DuPont
Medical Products Division
Glasgow, DE 19702 USA

Karen E. Koch
University of Florida
Department of Vegetable Crops/IFAS
Gainesville, FL 32611 USA

Richard V. Kowles
St. Mary's College of Minnesota
Biology Department
Winona, MN 55987 USA

C. Kramer
CIBA-Geigy Corporation
Agricultural Biotechnology Research Unit
Research Triangle Park, NC 27709-2257 USA

Barbara Lane
University of California
Department of Plant Biology
Berkeley, CA 94720 USA

Jane A. Langdale
University of Oxford
Department of Plant Sciences
Oxford OX1 3RB United Kingdom

Brian A. Larkins
University of Arizona
Department of Plant Sciences
Tucson, AZ 85721 USA

John R. Laughnan
University of Illinois
Urbana, IL 61801 USA

Michael Lee
Iowa State University
Department of Agronomy
Ames, IA 50011 USA

Avraham A. Levy
The Weizmann Institute of Sciences
Plant Genetics Department
Rehovot, 76100 Israel

S.M. Livingston
University of Minnesota
Plant Molecular Genetics Institute
Department of Agronomy and Plant Genetics
St. Paul, MN 55108 USA

Kenneth R. Luehrsen
Stanford University
Department of Biological Science
Stanford, CA 94305-2040 USA

Leszek A. Lyznik
Purdue University
Department of Botany and Plant Pathology
West Lafayette, IN 47904 USA

Marjorie P. Maguire
University of Texas
Zoology Department
Austin, TX 78712 USA

Eileen C. Matz
Brookhaven National Lab
Biology Department
Upton, NY 11973 USA

Albrecht Melchinger
Universität Hohenheim
Institut für Pflanzenzüchtung
W-7000 Stuttgart 70
Federal Republic of Germany

Maria A. Moreno
Yale University
Department of Biology
New Haven, CT 06511 USA

R. Michael Mulligan
University of California
Department of Developmental and Cell Biology
Irvine, CA 92717 USA

Oliver Nelson
University of Wisconsin
Department of Genetics
Madison, WI 53706 USA

Tim Nelson
Yale University
Department of Biology
New Haven, CT 06511-8112 USA

M.G. Neuffer
University of Missouri
Plant Sciences Unit
Columbia, MO 65211 USA

Kathleen J. Newton
University of Missouri
Division of Biological Sciences
Columbia, MO 65211 USA

Dayakar R. Pareddy
Dow Elanco
Agricultural Biotechnology Division
Champaign, IL 61820 USA

M.V. Parthasarathy
Cornell University
Division of Biological Sciences
Section of Plant Biology
Ithaca, NY 14853 USA

Earl B. Patterson
University of Illinois
Department of Agronomy
Urbana, IL 61801 USA

Anna-Lisa Paul
University of Florida
Department of Vegetable Crops
Gainesville, FL 32611 USA

J.F. Petolino
Dow/Elanco
Champaign, IL 61824-4011 USA

Ronald L. Phillips
University of Minnesota
Department of Agronomy and Plant Genetics
Plant Molecular Genetics Institute
St. Paul, MN 55108 USA

R. Scott Poethig
The University of Pennsylvania
Department of Biology
Plant Science Institute
Philadelphia, PA 19104-6018 USA

A. J. Pryor
CSIRO Division of Plant Industry
Canberra ACT 2601 Australia

Natasha V. Raikhel
Michigan State University
Plant Research Laboratory
East Lansing, MI 48824-1312 USA

S.W. Ritchie
Purdue University
Department of Botany and Plant Pathology
West Lafayette, IN 47907 USA

Steven Rodermel
Iowa State University
Department of Botany
Ames, IA 50011 USA

Scott D. Russell
University of Oklahoma
Department of Botany and Microbiology
Norman, OK 73019-0245 USA

Steven E. Ruzin
University of California
Center of Plant Developmental Biology
Department of Plant Biology
Berkeley, CA 94720 USA

Liang Schweitzer
University of Minnesota
Institute for Advanced Study in Biological Process Technology
Department of Chemical Engineering and Materials Science
St. Paul, MN 55108 USA

J.C. Sellmer
Purdue University
Department of Botany and Plant Pathology
West Lafayette, IN 47907 USA

Jack C. Shannon
The Pennsylvania State University
Department of Horticulture
University Park, PA 16802 USA

William F. Sheridan
University of North Dakota
Biology Department
Grand Forks, ND 58202 USA

R.D. Shillito
CIBA-Geigy Corporation
Agricultural Biotechnology Research Unit
Research Triangle Park, NC 27709-2257 USA

Neelima Sinha
Plant Gene Expression Center USDA—ARS
Albany, CA 94710 USA

Laurie G. Smith
Plant Gene Expression Center USDS—ARS
Albany, CA 94710 USA

David A. Somers
University of Minnesota
Department of Agronomy and Plant Genetics
Plant Molecular Genetics Institute
St. Paul, MN 55108 USA

Friedrich Srienc
University of Minnesota
Institute for Advanced Study in Biological Process Technology
Department of Chemical Engineering and Materials Science
St. Paul, MN 55108 USA

Chris J. Staiger
John Innes Institute
Department of Cell Biology
Norwich NR4 7UH
United Kingdom

Anne W. Sylvester
University of Idaho
Department of Biological Sciences
Moscow, ID 83843 USA

Gary A. Thompson
University of Arizona
Department of Plant Sciences
Tucson, AZ 85721 USA

Marguerite J. Varagona
Michigan State University
Plant Research Laboratory
East Lansing, MI 48824-1312 USA

Julie M. Vogel
University of California
Department of Plant Biology
Berkeley, CA 94720 USA

Virginia Walbot
Stanford University
Department of Biological Sciences
Stanford, CA 94305-2040 USA

D.B. Walden
University of Western Ontario
Department of Plant Sciences
London, Ontario N6A 5B7 Canada

Elsbeth Walker
Yale University
Department of Biology
New Haven, CT 06511 USA

Christine A. Warren
Stanford University
Department of Biological Sciences
Stanford, CA 94305-2040 USA

David F. Weber
Illinois State University
Genetics Group
Department of Biological Sciences
Normal, IL 61761 USA

Clifford F. Weil
University of Idaho
Department of Biological Sciences
Moscow, ID 83843 USA

Susan R. Wessler
University of Georgia
Department of Botany
Athens, GA 30602 USA

David P. West
Jacques Seed Company
Research Department
Prescott, WI 54021 USA

Michelle H. Williams
University of Florida
Department of Environmental Horticulture
Institute of Food and Agriculture Services
Gainesville, FL 32611-0670 USA

Georgia L. Yerk
University of Minnesota
Department of Agronomy and Plant Genetics
Plant Molecular Genetics Institute
St. Paul, MN 55108 USA

I
Development and Morphology

1

Fertilization and Embryogeny in Maize

WILLIAM F. SHERIDAN AND JANICE K. CLARK

The processes of fertilization and embryogeny in maize and other flowering plants have been examined extensively. For a general review, the classic book by Maheshwari (1950) is especially valuable for its scope and clarity. In addition, the works of Wardlaw (1955) and Johansen (1950) provide a broad historical perspective. More recent reviews include Johri (1984), Raghavan (1986), and Meinke (1991). For an introduction to maize and other grasses the work of Arber (1934) is recommended. The events of double fertilization in maize were first described by Guignard (1901); more accessible descriptions are those of Miller (1919) and Weatherwax (1919). For an understanding of the events of maize embryogeny the reader is referred to the studies of Randolph (1936), Kiesselbach (1949), and Abbe and Stein (1954). More recent works of interest are those of Van Lammeren (1986, 1987), Neuffer et al (1986), Sheridan and Clark (1987, 1993), and Clark and Sheridan (1991).

FERTILIZATION

The embryo sac of maize is of the monosporic 8-nucleate (*Polygonum*) type. The single megasporocyte in the ovule undergoes meiosis to produce four megaspores three of which degenerate. The chalazal-most megaspore undergoes three successive mitotic divisions to produce an embryo sac containing eight haploid nuclei or cells (Maheshwari 1950). Occasionally tetrasporic origin of the embryo sac has been observed to occur in sectors of ears of stocks carrying appropriate genetic markers to detect nonconcordance of the female contribution to the embryo and endosperm (Neuffer 1964). See Maheshwari (1950), Diboll and Larson (1966), Van Lammeren (1986), and Bedinger and Russell (this volume) for a more detailed consideration of embryo sac formation. Depending on temperature and silk length, fertilization occurs between 16 and 24 hours after pollination. Fusion of one sperm cell with the egg cell and of the other sperm cell with the two polar nuclei in the central cell occurs simultaneously (Randolph 1936). Fertilization of the egg results in formation of the diploid zygote while fertilization of the central cell results in for-

The Maize Handbook—M. Freeling, V. Walbot, eds.

mation of the triploid primary endosperm cell. Whereas the primary endosperm cell nucleus divides within 3–5 hours after fertilization (Kiesselbach 1949) and the endosperm nuclei divide rapidly thereafter, the zygote does not divide until about 10 or 12 hours after fertilization (Randolph 1936).

In the large majority of cases the two sperm that participate in double fertilization are genetically identical. But there are some exceptions to this general rule of concordance. One such exception can occur when there is heterofertilization. In such a case the two sperm are contributed by two different pollen grains that may be genetically nonidentical. The frequency of kernels resulting from heterofertilization can range from 1 to 10% depending on the genotype of the pollen parent (Sprague 1932; Coe et al. 1988). The two sperm nuclei participating in double fertilization may also be nonidentical when the generative nucleus of a pollen grain contains a B-A and an A-B reciprocal translocation. Because of the frequent nondisjunction of the B-A chromosome at the mitotic division resulting in sperm formation, one sperm is commonly hypoploid and lacks the B-A chromosome whereas the other sperm is hyperploid and contains two B-A chromosomes. In such cases the hyperploid sperm preferentially fertilizes the egg cell. (See Beckett 1991 and in this volume for additional details and a review of B-A translocation behavior.) Nonconcordance can also occur if transposon insertion or excision occurs after the last mitotic division yielding two sperm of different genotype.

An additional feature of fertilization in maize that merits mention is the difference in expression of the same allele depending on whether it is transmitted through the female (ovule) or through the male (pollen). It has been shown by Kermicle (1970, 1978) in his studies on *R* gene expression in the aleurone of the endosperm that the genes of the sperm have experienced "imprinting" (Kermicle and Alleman 1990).

Studies on the process of fertilization in vitro should become feasible with the improvement and application of techniques for the isolation of viable maize sperm cells (Dupuis et al. 1987) and embryo sacs (Wagner et al. 1989).

EMBRYOGENY

The embryogeny of maize can be considered to occur in three phases. During the first phase the basal-apical asymmetry of the embryo is established, and the embryo is regionalized into suspensor and embryo proper. During the second phase, radial asymmetry appears and the embryonic axis and meristems are established; during the third, vegetative structures are elaborated.

The rate of embryo development is temperature-dependent. In this review the comparison of age and stage of embryo development is largely based on the observations made in Ithaca, New York, by Randolph (1936) and in Lincoln, Nebraska, by Kiesselbach (1949). Observations made in Missouri (Sheridan and Neuffer 1982; Clark and Sheridan 1988) indicate a more rapid rate while those in Minnesota (Abbe and Stein 1954) and North Dakota (Clark and Sheridan 1986, 1988) indicate a slower rate. The scheme of classification of embryo stages developed by Abbe and Stein (1954) is the one used in this review.

Phase 1: Setting Apart of the Embryo Proper and Suspensor—Proembryo Through Midtransition Stage

The first division of the zygote establishes the initial asymmetry of the embryo; it occurs in the transverse plane so as to form a large vacuolated basal cell and a nonvacuolated apical cell. Both Randolph (1936) and Kiesselbach (1949) emphasized that the early stages of embryogeny are characterized by irregular patterns of cell division, not by the orderly patterns that have been so extensively documented for many dicot species. [See Johansen (1950, pp 93–126) for a consideration of his and Soueges's systems of classification of the principal types of embryogeny in the dicots; Maheshwari (1950) may also be consulted.] By 100 hours after pollination (about 75–80 hours after fertilization) the proembryo consists of about 12–24 cells; its shape is that of an apical hemisphere (the embryo proper) consisting of about three-fourths of the total cells of the proembryo all of which are derived from the initial apical cell. At the base of the embryo is the suspensor region comprised of much larger but fewer cells derived from the initial basal cell. The cells of the embryo proper are more densely filled with cytoplasm while the basal cells are vacuolated (Randolph 1936; Van Lammeren 1986). The distribution of vacuole and cytoplasm between the basal and apical region of the proembryo maintains the structural polarity present in the unfertilized egg and zygote (Diboll 1968; Diboll and Larson 1966; Van Lammeren 1986).

Throughout its development the ovoid proembryo is radially symmetrical along an axis extending from its base to its apex (Figure 1.1). As the embryo enlarges and increases in cell number and size it becomes club-shaped. The initial basal-apical asymmetry is maintained; the basal cells comprising the suspensor are large and vacuolated while the apical region of the embryo proper consists of many small cells. This club-shaped embryo elongates into the transition stage embryo (Figure 1.1). Up to the midtransition stage the embryo remains radially symmetrical—that is to say, there is no localized region of growth within the embryo proper; both anticlinal and periclinal cell divisions are common and no tissue differentiation is evident.

Phase 2: Establishment of the Meristems and Embryonic Axis—Late Transition Stage Through Stage 1

On about the eighth or ninth day after pollination (Randolph 1936; Kiesselbach 1949) an important morphogenetic change occurs within the embryo proper, a change that underlies and enables the subsequent morphogenesis of the embryo. At this time a localized region of greater mitotic activity appears under the anterior face of the embryo proper. This region is occupied by a wedge-shaped mass of cells of greater cytoplasmic density than the adjacent cells (Figure 1.1); this region is soon conjoined by an identical-appearing region more centrally and basally located within the embryo proper. This ex-

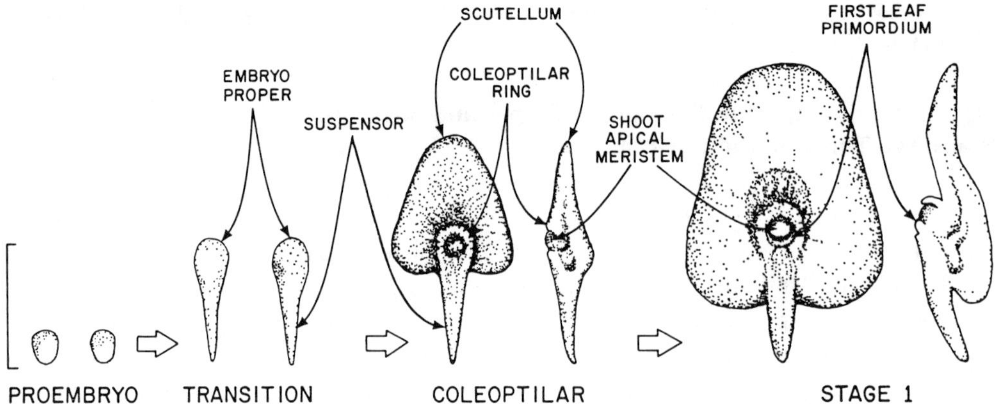

FIGURE 1.1 Early embryogenesis in maize. For each stage the drawing on the left shows the external features of the embryo in face view while the drawing on the right shows the embryo in longitudinal radial section with the face of the embryo at the left. Note that the proembryo is produced by an asymmetrical division of the zygote but that it is radially symmetrical. During the transition stage radial asymmetry is introduced by the formation of an internal wedge-shaped meristematic region in the upper part of the embryo. During the coleoptilar stage this region gives rise to the shoot apical meristem and the surrounding coleoptilar ring. Stage 1 is marked by the appearance of the first leaf primordium which grows upward from the lower border of the face of the shoot apical meristem. The scale bar equals 0.5 mm.

tended region of greater meristematic activity establishes an obliquely angled meristematic axis extending from the anterior upper surface to the lower interior central region of the embryo proper (Figure 1.1). This axis becomes the shoot apical meristem–root apical meristem axis of the embryo over the period from about 11 through 13 days after pollination (Figure 1.1). The transition stage can, therefore, be viewed as the stage of embryogeny wherein the embryo progresses in its development from an ovoid, radially symmetrical proembryo through an elongation process to form a cone-shaped embryo. Subsequently the apical meristems originate at sites of increased mitotic activity. This stage is appropriately named as it encompasses the developmental transition of the radially symmetrical embryo of the first phase to the radially asymmetric embryo of the second phase of embryogeny.

Simultaneous with the bulging out of the newly forming shoot apical meristem a semicircular ridge of cells appears above the meristem on the anterior face of the late-transition-stage embryo. This ridge is progressively extended to enclose the shoot apical meristem within the now-recognizable coleoptilar ring (Figure 1.1). At the same time the embryo proper broadens, particularly on the posterior side, to form the spade-shaped beginnings of the scutellum. The embryo, now possessing a prominent shoot apical meristem surrounded by the coleoptilar ring on its anterior surface and backed by a scutellum, is termed the coleoptilar stage embryo (Figure 1.1). This stage was observed by

Randolph (1936) at about 14 days and by Kiesselbach (1949) at 15 days after pollination.

Beginning with the formation of the proembryo, the embryo steadily increases in size. By the coleoptilar stage the embryo is about 1 mm or more in length; the embryo proper and the suspensor each comprise about equal parts. The first leaf primordium arises on the lower side of the shoot apical meristem as a crescent-shaped bulge. This occurs about 16 days after pollination when the embryo proper is 1 mm or more in length (Randolph 1936), and is referred to as stage 1 (Figure 1.1). In many tissue culture experiments, embryos of this stage have been utilized to initiate embryonic callus. (See chapter 121 on embryo culture, this volume.)

Phase 3: Elaboration of Embryonic Structures—Stages 2–6

During the 30–40-day period following the appearance of the first leaf primordium, additional leaf and root primordia are initiated and elaborated and the scutellum enlarges and surrounds these structures. At maturity, the embryonic axis consists of a plumule with five or six leaf primordia enclosed within the coleoptile and of a primary root primordium with one or more secondary root primordia; the axis is nearly completely enclosed by the enshrouding scutellum with only the tips of the coleoptile and coleorhiza protruding at the top and bottom, respectively. During the third phase of embryogeny storage products accumulate within the embryo, particularly within the scutellum. The scutellum increases greatly in size and becomes the major portion of the embryo. The fresh weight of the embryo as a whole increases from approximately 1 mg at stage 1 to 50 mg or more depending on the genetic stock and growth conditions. During phase 3 of embryogeny the capacity to withstand desiccation is acquired by the embryo. Ultimately, dormancy occurs after massive dehydration of the embryo cells.

THE ROLE OF GENES IN EMBRYOGENY

Two types of maize embryo lethal mutations have been isolated and subjected to developmental and genetic analysis with the goal of obtaining insight into the role of genes in embryo morphogenesis. The first type consists of *defective kernel* (*dek*) mutations; these affect the development of both the endosperm and the embryo. While conducting a large-scale screen of about 200 *dek* mutations by genetic (Neuffer and Sheridan 1980) and embryo culture techniques (Sheridan and Neuffer 1980) in a search for auxotrophic mutations, a group of mutants was identified that appeared to be developmental mutants (Sheridan and Neuffer 1981, 1982). The more detailed analysis of eight of these mutations (Clark and Sheridan 1986, 1988; Sheridan and Clark 1987; Sheridan and Thorstenson 1986) resulted in our drawing five conclusions from these studies with regard to the role of genes in maize embryogeny:

1. The morphogenesis of the embryo and of the endosperm requires many genes in common.

2. The genetic regulation of embryo morphogenesis is distinct and separable from the regulation of other processes including embryo enlargement by cell division and cell expansion. This is evident from the fact that, in seven of the eight mutants altered in morphogenesis, the mutation does not block all of these processes but acts selectively.
3. The control of morphogenesis of the shoot apical meristem is separable from that of the root meristem; three of the mutants are able to form a normal root meristem but not a shoot meristem.
4. Normal gene action is required not only for the morphogenesis of shoot meristems and leaf primordia but also for their maintenance as organized structures throughout embryogeny.
5. The pleiotropic expression of these mutations can extend beyond the endosperm and embryo to include the gametophyte generation. This indicates that the organism employs the same genes to control developmental processes in different tissues and generations.

The second type of mutation that has yielded insight into the role of genes in embryo morphogenesis is the *embryo-specific* (*emb*) class. In mutants of this type the kernel contains a mutant embryo but a normal-appearing endosperm. Mutations of this type were first studied by Demerec (1923), who termed them *germless* (*gm*); their abundance was noted by Wentz (1930). Two characteristics of the *emb* mutations that make them difficult to study are their lethality and the fact that they can only be recognized when the germinal (adaxial) face of the kernel is examined. It is probably for these reasons and not because of a lack of *emb* mutations that there was a gap of 60 years in the description of maize *emb* mutations (Clark and Sheridan 1991).

The set of *dek* mutations described above were all ethyl methane sulfonate (EMS)-induced mutations (Neuffer 1978), but the 51 *emb* mutations isolated and characterized by us (Clark and Sheridan 1991, Sheridan and Clark 1993) arose in active Robertson's Mutator stocks. These 51 mutations represent at least 45 independent mutation events. So far 25 of the mutations have been localized by the use of B-A translocations to six chromosome arms of the eight tested; there appears to be a clustering of mutations on chromosome arms *1S*, *3L*, and *4L*. The embryo morphology of all 51 of the mutations has been examined by dissection at kernel maturity and all are altered in their morphogenesis. About one-third of the mutations affect each of the three phases of embryonic development. Based on the genetic and developmental data, we conclude that many gene loci can mutate to give the *emb* phenotype and that these genes are crucial to the morphogenesis of the embryo.

The developmental data on the 51 *emb* mutants examined so far indicate that:

1. A mutation can block or reduce the development of the embryo proper while the suspensor remains normal appearing or may even enlarge beyond its normal size.
2. A mutation can block the formation of the shoot apical meristem while the remaining parts of the embryo continue their growth and development.

3. The scutellum can be altered in its development so that it is either retarded relative to the embryonic axis or proliferates excessively, but in either case, the normal proportionality in size between the embryonic axis and the scutellum is lost.
4. Because several of the mutations show a reduced transmission frequency, it appears likely that expression of some of the *emb* loci may be important in the gametophyte generation.

We propose that the *dek* and *emb* mutations define genes that play fundamental roles in the morphogenesis of the maize embryo and that characterization of these mutations will provide new insights into the regulation of this interesting developmental phenomenon.

REFERENCES

Abbe EC, Stein OL (1954) The growth of the shoot apex in maize: embryogeny. Am J Bot 41: 285–293

Arber A (1934) The Gramineae: A Study of Cereal, Bamboo, and Grass. Cambridge

Beckett JB (1991) Cytogenetic, genetic, and breeding applications of B-A translocations in maize. In Tsuchiya T, Gupta PK (eds) Chromosome Engineering in Plants, Elsevier, Amsterdam, pp 491–527

Clark JK, Sheridan WF (1986) Developmental profiles of the maize embryo-lethal mutants *dek22* and *dek23*. J Hered 77: 83–92

Clark JK, Sheridan WF (1988) Characterization of the two embryo-lethal defective kernel mutants *rgh*-1210* and *fl*-1253B*: effects on embryo and gametophyte development. Genetics 120: 279–290

Clark JK, Sheridan WF (1991) Isolation and characterization of 51 *embryo-specific* mutations of maize. Plant Cell 3: 935–951

Coe EH, Jr, Neuffer, MG, Hoisington DA (1988) The genetics of corn. In Sprague GF, Dudley JW (eds) Corn and Corn Improvement, American Soc of Agronomy, Crop Sci Soc of America, and Soil Sci of America, Madison, WI, pp 83–258

Demerec M (1923) Heritable characters of maize. XV. Germless seed. J Hered 114: 297–300

Diboll AG (1968) Fine structural development of the megagametophyte of *Zea mays* following fertilization. Am J Bot 55: 797–806

Diboll AG, Larson DA (1966) An electron microscope study of the mature megametophyte in *Zea mays*. Am J Bot 53: 391–402

Dupuis I, Roeckel P, Matthys-Rochon E, Dumas C (1987) Procedure to isolate viable sperm cells from the corn (*Zea mays*) pollen grain. Plant Physiol 85: 876–878

Guignard L (1901) La double fecundation dans le mais. Jour Botanique 15: 37–50

Johansen DA (1950) Plant Embryology. Chronica Botanica Company, Waltham, MA

Johri BM (ed) (1984) Embryology of Angiosperms. Springer, Berlin

Kermicle JL (1970) Dependence of the *R*-mottled phenotype in maize on mode of sexual transmission. Genetics 66: 69–85

Kermicle JL (1978) Imprinting of gene action in maize endosperm. In Walden DB (ed) Maize Breeding and Genetics, John Wiley and Sons, New York, pp 357–371

Kermicle JL, Alleman M (1990) Gametic imprinting in maize in relation to the angiosperm life cycle. Development 1990 Supplement: 9–14

Kiesselbach TA (1949) The Structure and Reproduction of Corn. Univ Nebr Agric Exp Stn Bull 161, 96 pp

Maheshwari P (1950) An Introduction to the Embryology of Angiosperms. McGraw-Hill, New York

Meinke DW (1991) Perspectives on genetic analysis of plant embryogenesis. Plant Cell 3: 857–866

Miller EC (1919) Development of the pistillate spikelet and fertilization in Zea mays. J Agr Research 18: 255–266

Neuffer MG (1964) Tetrasporic embryo-sac formation in trisomic sectors of maize. Science 144: 874–876

Neuffer MG (1978) Induction of genetic variability. In Walden DB (ed) Maize Breeding and Genetics, Wiley-Interscience, New York, pp 579–600

Neuffer MG, Sheridan WF (1980) Defective kernel mutants of maize. I. Genetic and lethality studies. Genetics 95: 929–944

Neuffer MG, Chang MT, Clark JK, Sheridan WF (1986) The genetic control of maize kernel development. In Shannon JC, Knievel DP, Boyers CD (eds) Regulation of Carbon and Nitrogen Reduction and Utilization in Maize, Am Soc Plant Physiol, Rockville, MD, pp 35–50

Randolph LF (1936) Developmental morphology of the caryopsis in maize. J Agr Research 53: 881–916

Raghavan V (1986) Embryogenesis in Angiosperms. Cambridge University Press, Cambridge

Sheridan WF, Clark JK (1987) Maize embryogeny: a promising experimental system. Trends Genet 3: 3–6

Sheridan WF and Clark JK (1993) Mutational analysis of morphogenesis of the maize embryo. Plant Journal 3: 347–358

Sheridan WF, Thorstenson YR (1986) Developmental profiles of three embryo-lethal maize mutants lacking leaf primordia: *ptd*-1130*, *cp*-1418*, and *bno*-747B*. Dev Genet 7: 35–49

Sheridan WF, Neuffer MG (1980) Defective kernel mutants of maize II. Morphological and embryo culture studies. Genetics 95: 945–960

Sheridan WF, Neuffer MG (1981) Maize mutants altered in embryo development. In Subtelney S, Abbot U (eds) Levels of Genetic Control in Development, Alan R Liss, New York, pp 137–156

Sheridan WF, Neuffer MG (1982) Maize developmental mutants: embryos unable to form leaf primordia. J Hered 73: 318–329

Sprague GF (1932) The nature and extent of heterofertilization in maize. Genetics 17: 358–368

Van Lammeren AA (1986) Developmental morphology and cytology of the young maize embryo (*Zea mays* L.) Acta Bot Neerl 35: 169–188

Van Lammeren AA (1987) Embryogenesis in Zea mays L. A structural approach to maize caryopsis development in vivo and in vitro. Doctoral dissertation. University of Wageningen, Wageningen

Wagner VT, Song YC, Matthys-Rochon E, Dumas C (1989) Observations on the isolated embryo sac of *Zea mays* L. Plant Science 59: 127–132

Wardlaw CW (1955) Embryogenesis in Plants. Methuen & Co, London, Wiley & Sons, New York

Weatherwax P (1919) Gametogenesis and fecundation in Zea mays as the basis of xenia and heredity in the endosperm. Bull Torrey Bot Club 46: 73–90

Wentz JB (1930) The inheritance of germless seeds in maize. Iowa Exp Sta Res Bull 121: 347–379

2

The Maize Shoot

R. S. Poethig

The maize shoot is a determinate structure. During its growth, it produces a limited number of structures in a regular and highly predictable pattern. Although the morphology of the shoot plays a major role in the productivity of maize and has been the subject of intensive conscious and unconscious selection, the genetic regulation of shoot morphology in maize is still poorly understood. The following brief review will focus on some major features of shoot development and genes that affect these features.

THE STRUCTURE OF THE SHOOT

The primary axis of the maize shoot is generated by a single apical meristem that is initiated during embryogenesis. This apical meristem produces a reiterated series of four basic structures—the internode, leaf, prophyll, and bud—that are considered the basic structural unit of the grass shoot (Galinat 1959). It is convenient to describe the structure of the shoot in terms of this unit, the *phytomer*, because these four structures or their homologues always occur together in the same spatial relationship. Variation in the morphology of phytomers along the length of the shoot axis results from changes in the degree of development of various parts of the phytomer (e.g., the growth of leaf primordia is suppressed in the tassel and ear) or from differences in their pattern of differentiation (e.g., juvenile leaves vs. adult leaves; leaves vs. glumes). Galinat (1993) has illustrated homologous parts of the phytomer in different parts of the maize shoot. (See Figure 7.1 in Galinat, this volume.) Unfortunately, the developmental relationship between the components of the phytomer is unknown. For this reason, the phytomer has been variously described as consisting of, from top to bottom, (1) a leaf, internode, and bud (Cutler and Cutler 1948), (2) an internode, prophyll, bud, and leaf (Galinat 1959), and (3) a prophyll, bud, leaf, and internode (Poethig 1988).

The shoot apical meristem produces 5 or 6 vegetative phytomers before the seed matures and another 10–20 vegetative phytomers after germination and then produces a male inflorescence known as the tassel. The female inflorescence, the ear, is the terminal

The Maize Handbook—M. Freeling, V. Walbot, eds.

inflorescence of a lateral branch that is usually located 5–7 phytomers below the tassel. In the vegetative part of the shoot, lateral organs are initiated distichously, that is, alternately on opposite sides of the shoot meristem. As a result, leaves and branches fall within a single plane. The morphology of vegetative phytomers varies continuously along the length of the shoot (Greyson et al. 1982). Leaves and internodes increase in size from the base of the shoot to the ear node and then become progressively smaller. Axillary buds in the first 5–7 phytomers form branches that are identical in structure to the primary axis (tillers) but become progressively more earlike at distal nodes (Raman et al. 1980). One of the most conspicuous differences between different lines of maize is in the extent to which tillers and ears develop. Commercial varieties of dent maize produce only one or two mature ears and rarely have tillers. In contrast, most exotic varieties, as well as many commercial varieties of sweet corn and popcorn, produce several tillers and two or more ears. Two mutations that increase tiller production are *teosinte branched* and *grassy tillers* (Coe et al. 1988).

Although each vegetative phytomer is morphologically unique, the vegetative axis of the shoot can be divided into distinct domains on the basis of traits whose expression changes abruptly at specific nodes. These domains are usually described in temporal terms because they are produced at different times in shoot development. The basal 5–7 phytomers (which are produced early in shoot development) are termed the *juvenile* part of the shoot, while more distal phytomers are considered to be part of the *adult* phase of development. Traits that are characteristic of the phytomers in these domains are listed in Table 2.1. The adult region of the shoot can be further divided into a domain below the ear and a domain above the ear because axillary buds and prophylls are present below the ear but are absent above the ear.

The extent of these domains varies in different inbred backgrounds (Poethig 1988)

Table 2.1. Summary of traits that distinguish juvenile and adult regions of the shoot: E = epidermal cells (data from Sylvester et al. 1990; Lyons and Nicholson 1989; Bongard-Pierce and Poethig, unpublished)

Traits	Juvenile	Adult
Epicuticular wax	present	absent
Epidermal hairs	absent	present
Wall crenulation (E)	moderate	extreme
Toluidine blue (E)	Purple	Aqua
Cell shape (cross section) (E)	circular	rectangular
Cuticle thickness (E)	<1 μm	>1 μm
Roots	present	absent
Lateral buds	tillers	ears or absent
Anthracnose resistance	low	high

and is affected by a number of mutations. The expression of juvenile traits, for example, is prolonged by several semidominant mutations (*Teopod1*, *Tp2*, *Tp3*, *Corngrass*) (Galinat 1966; Poethig 1988). These mutations have dramatic effects on shoot morphology that appear to result from the simultaneous expression of traits from different phases of development; phytomers in what is normally the adult part of the shoot possess both juvenile and adult traits, and both the tassel and the ear have a combination of reproductive structures and vegetative structures. The dominant *Hairy-sheath-frayed* mutation also prolongs the expression of some juvenile vegetative traits (Betrand-Garcia and Freeling 1991). In contrast, recessive alleles of *gl15* truncate the expression of some juvenile traits and coincidentally accelerate the expression of adult traits (Evans and Poethig 1991). *Leafy* affects the timing of reproductive development but does not have a major effect on the differentiation of vegetative structures (Shaver 1983). This dominant mutation delays tassel and ear initiation and increases the total number of phytomers by from two to 17 in different inbred backgrounds. In general, *Lfy* increases the number of phytomers above the ear more than phytomers below the ear.

THE CELL LINEAGE OF THE SHOOT

The cell lineage of the shoot has been reconstructed from an analysis of radiation- and transposon-induced somatic sectors (Poethig et al. 1990; Dawe and Freeling 1991). In maize, these studies are facilitated by the existence numerous mutations that can be used as cell markers. Dominant alleles of the *B* and *Pl* loci are particularly useful because these genes cause anthocyanin to be produced by virtually every tissue in the shoot (Coe et al. 1988). Mutations affecting chlorophyll production, such as *wd* (Coe and Neuffer 1978; Poethig et al. 1986), *lw* (Sinha and Hake 1990), and *al* (Becraft et al. 1990), also have proven to be good cell lineage markers, although they are technically more difficult to work with than anthocyanin genes because they must be maintained as heterozygotes.

In maize, as in many other plants, patterns of cell lineage revealed by clonal analysis are quite variable and it is impossible to predict the fate of specific regions of the shoot meristem with any certainty (Poethig et al. 1990). Nevertheless, certain general features of the cell lineage of the shoot have been established. Clones induced at various times during embryogenesis suggest that the shoot apical meristem arises from approximately 200 cells within two or more cells layers of the transition-stage embryo (Poethig et al. 1986). Most of the cells in the meristem at this stage are destined to form the basal 5–6 phytomers of the shoot. A central group of 20–30 cells in at least 2 cell layers (10–15 cells in each layer) of the embryo produces the meristem of the mature seed, from which all the remaining phytomers (number 7 through the tassel) of the shoot arise. Fate maps of the shoot meristem of a mature embryo have been constructed by several investigators (Steffenson 1968; Coe and Neuffer 1978; Johri and Coe 1983; McDaniel and Poethig 1988). Although all the cells in the shoot meristem make a significant contribution to the growth of the shoot, cells at the base of the meristem contribute less to the growth of the shoot than cells near the apex of the meristem. This is not surprising, of course, given that the differentiation of new organs occurs at the base of the meristem. Clones almost always encompass the leaf and the internode below the leaf, indicating that these structures arise from a common primordium (a "disc of insertion"; Sharman 1942) and that this primordium is produced by a small number of progenitor cells. Axillary buds are clonally related to the internode above the bud, rather than to the subtending leaf. The tassel is derived from the apical two tiers of cells in the meristem of a mature embryo. The most apical cells in the meristem produce the central spike of the tassel, while the next ring of cells produces the base of the tassel and several vegetative phytomers below the tassel. The pattern of cell division in the L1 layer of the meristem differs during vegetative growth and the development of the tassel. During vegetative growth it is common for L1 cells to divide periclinally and contribute to internal tissue of the shoot (Poethig et al. 1986; McDaniel and Poethig 1988). In contrast, there is no evidence that periclinal divisions occur in the L1 during the development of the tassel (Dawe and Freeling 1990).

THE GROWTH OF THE SHOOT

An excellent pictorial representation of stages in the vegetative and reproductive development of the maize shoot is available (Hanway 1966). The vegetative phase of shoot growth lasts 3–4 weeks under normal field conditions. During this phase, internodes elongate slowly and the entire aboveground part of the shoot consists solely of leaves. With the initiation of the tassel, all but the basal five or six internodes begin to elongate rapidly and push through the enclosing leaf sheaths. The basal 5–6 internodes remain below ground, where they give rise to the root system. (See Feldman, this volume.)

The number of phytomers produced during the vegetative phase of development is regulated by genetic and environmental factors. In one survey of inbred lines, total leaf number ranged from 10 to 29 (Russell and Stuber 1983). Because the duration of shoot development (as measured by the time to anthesis) in different genotypes correlates well with total leaf number (Chase and Nanda 1967; Hesketh et al. 1969; Shaver 1976; Russell and Stuber 1983), it is reasonable to conclude that variation in total leaf number results primarily from variation in the timing of tassel initiation rather than from variation in the rate of leaf initiation. Variation in flowering time in commercial inbreds appears to be regulated by numerous loci with minor effects (Shaver 1983). The reproductive behavior of extremely early lines, such as Gaspe Flint, may be regulated by no more than two or three major genes (Shaver 1976). Mutations that have a major effect on leaf number and reproductive maturity include *indeterminate* (*id*) and *Leafy* (*Lfy*). *id* is a recessive mutation that blocks ear and tassel initiation under normal field conditions and consequently prolongs the vegetative growth of the shoot. This mutation is thought to affect the photosensitivity of the shoot, making tassel initiation dependent on short day conditions (Galinat and Naylor 1951). As mentioned above, *Lfy* delays tassel and ear initiation and increases leaf number to varying extents in different genetic backgrounds (Shaver 1983).

The vegetative and reproductive development of the shoot are acutely sensitive to both temperature and photoperiod (Hunter et al. 1977; Warrington and Kanemasu 1983; Hanway and Ritchie 1985). Cool temperatures (e.g., 15–20°C) reduce the rate of leaf initiation and often also reduce total leaf number. Photoperiod has little effect on the rate of leaf initiation but may dramatically affect leaf number by virtue of its effect on tassel initiation (Warrington and Kanemasu 1983; Hunter et al. 1977). Most varieties of maize flower earlier under short photoperiods than under long photoperiods. Intermediate- and late-flowering varieties tend to be particularly responsive to photoperiodic conditions, whereas early flowering varieties (total leaf number ≤ 15) are either insensitive to short days or produce a tassel only 1–2 plastochrons earlier under short day conditions (Francis et al. 1969; Russell and Stuber 1983).

Tassel initiation is maximally sensitive to these environmental cues during a relative brief period. In the hybrid GX122, tassel initiation is only accelerated by low temperature or a short photoperiod during the time leaves 5–7 are emerging from the apical whorl (Tollenaar and Hunter 1983). Tassel initiation in this hybrid normally occurs around the emergence of leaf 7. A similar result has been obtained in the case of an A632/W23 hybrid, where the photosensitive phase occurs sometime during the initiation of leaves 15 and 16 (leaf 6 emerged) (Bassiri, Irish, and Poethig, 1992). In this hybrid, tassel initiation usually occurs after the initiation of leaf 18 (leaf 7 emerged). Experiments in which shoot tips were excised and rerooted at various stages of shoot growth have shown that the shoot meristem can be "reprogrammed" to produce a normal number of vegetative leaves even after it has entered the initial stage of tassel differentiation (Irish and Nelson 1988, 1991). Thus, however the transition from vegetative to reproductive growth is regulated, commitment to tassel initiation occurs late in shoot growth and is initially quite weak.

REFERENCES

Bassiri A, Irish EE, Poethig RS (1992) Heterochronic effects of *Teopod2* on the growth and photosensitivity of the maize shoot. Plant Cell 4: 497–504

Becraft PW, Bongard-Pierce DK, Sylvester AW, Poethig RS, Freeling M (1990) The liguleless-1 gene acts tissue specifically in maize leaf development. Devel Biol 141: 220–232

Betrand-Garcia R, Freeling M (1991) *Hairy-Sheath Frayed 1–0*: a systemic, heterochronic mutant of maize that specifies slow developmental stage transitions. Amer J Bot 78: 747–765

Chase SS, Nanda DK (1967) Number of leaves and maturity classification in *Zea mays* L. Crop Sci 7: 431–432

Coe EH Jr, Neuffer MG (1978) Embryo cells and their destinies in the corn plant. In Subtelny S, Sussex IM (eds) The Clonal Basis of Development, Academic Press, New York, pp 113–129

Coe EH Jr, Neuffer MG, Hoisington DA (1988) The genetics of corn. In Sprague GF, Dudley JW (eds) Corn and Corn Improvement, Third Edition, American Society of Agronomy, Madison, WI, pp 81–258

Cutler HC, Cutler MC (1948) Studies on the structure of the maize plant. Ann Mo Bot Gard 35: 301–316

Dawe RK, Freeling M (1990) Clonal analysis of the cell lineages in the male flower of maize. Devel Biol 142: 233–245

Dawe RK, Freeling M (1991) Cell lineage and its consequence in higher plants. The Plant Journal 1: 3–8

Evans M, Poethig RS (1991) *gl15* is a heterochronic mutation. Maize Genet Coop News Lett 65: 91

Francis C A, Grogan CO, Sperling DW (1969) Identification of photoperiod insensitive strains of maize. Crop Sci 9: 675–677

Galinat WC (1959) The phytomer in relation to the floral homologies in the American Maydeae. Bot Mus Leaflets, Harvard U 19: 1–32

Galinat WC (1966) The corn grass and teopod loci involve phase change. Maize Genet Coop News Lett 40: 102–103

Galinat WC (1993) The morphology and homology of plant structures in maize. In Freeling M, Walbot V (eds) The Maize Handbook, Springer-Verlag, New York

Galinat WC, Naylor AW (1951) Relation of photoperiod to inflorescence proliferation in *Zea mays* L. Amer J Bot 38: 38–47

Greyson RI, Walden DB, Smith WJ (1982) Leaf and stem heteroblasty in *Zea*. Bot Gaz 143: 73–78

Hanway JJ (1966) How a corn plant develops. Iowa State Univ Coop Ext Ser Spec Rep 48, Ames, IA

Hanway JJ, Ritchie SW (1985) *Zea mays*. In Halevy AH (ed) Handbook of Flowering, Vol 4, CRC Press, Boca Raton, FL, pp 525–541

Hesketh JD, Chase SS, Nanda DK (1969) Environmental and genetic modification of leaf number in maize, sorghum and Hungarian millet. Crop Sci 9: 460–463

Hunter RB, Tollenaar M, Breuer CM (1977) Effects of photoperiod and temperature on vegetative and reproductive growth of a maize (*Zea mays*) hybrid. Can J Plant Sci 57: 1127–1133

Irish EE, Nelson TM (1988) Development of maize plants from cultured shoot apices. Planta 175: 9–12

Irish EE, Nelson TM (1991) Identification of multiple stages in the conversion of maize meristems from vegetative to floral development. Development 112: 891–898

Johri MM, Coe EH Jr (1983) Clonal analysis of corn plant development 1. The development of the tassel and ear shoot. Devel Biol 97: 154–172

Lyons PC, Nicholson RL (1989) Evidence that cyclic hydroximate concentrations are not related to resistance of corn leaves to anthracnose. Can J Plant Pathol 11: 215–220

McDaniel CN, Poethig RS (1988) Cell lineage patterns in the shoot apical meristem of the germinating maize embryo. Planta 175: 13–22

Poethig RS (1988) Heterochronic mutations affecting shoot development in maize. Genetics 119: 959–973

Poethig RS, Coe EH Jr, Johri MM (1986) Cell lineage patterns in maize embryogenesis: a clonal analysis. Devel Biol 117: 392–404

Poethig RS, McDaniel CN, Coe EH Jr (1990) The cell lineage of the maize shoot. In Mahowald AP (ed) Genetics of Pattern Formation and Growth Control, 48 Symposium of the Society for Developmental Biology, Wiley-Liss, NY, pp 197–208

Raman K, Walden DB, Greyson RI (1980) Propagation of *Zea mays* L. by shoot tip culture: a feasibility study. Ann Bot 45: 183–189

Russell WK, Stuber CW (1983) Effects of photoperiod and temperatures on the duration of vegetative growth in maize. Crop Sci 23: 847–850

Sharman BC (1942) Developmental anatomy of the shoot of *Zea mays* L. Ann Bot NS 6: 245–282

Shaver DL (1976) Conversions for earliness in maize. Maize Genet Coop News Lett 50: 20–23

Shaver DL (1983) Genetics and breeding of maize with extra leaves above the ear. Proc Ann Corn and Sorghum Res Conf 38: 161–180

Sinha N, Hake S (1990) Mutant characters of *Knotted* maize leaves are determined in the innermost tissue layers. Devel Biol 141: 203–210

Steffenson DM (1968) A reconstruction of cell development in the shoot apex of maize. Amer J Bot 55: 354–369

Sylvester AW, Cande WZ, Freeling M (1990) Division and differentiation during normal and *liguleless-1* maize leaf development. Development 110: 985–1000

Tollenaar M, Hunter RB (1983) A photoperiod and temperature sensitive period for leaf number in maize. Crop Sci 23: 457–460

Warrington IJ, Kanemasu ET (1983) Corn growth response to temperature and photoperiod. II. Leaf initiation and leaf appearance rates. Agron J 75: 755–761

3

The Maize Leaf

MICHAEL FREELING AND BARBARA LANE

THE VEGETATIVE LEAF

The leaf is the most conspicuous organ of the repeating shoot segment. A typical elite inbred, such as W23, has approximately 20 leaves. Because leaves are founded sequentially within the meristem over weeks of time, the older, more juvenile leaves arise earlier and in a more basal position than the younger, more adult leaves. Two numbers designate any particular leaf: the leaf number (L) counting the nonleaf coleoptile as zero, and the plastochron number (P), counting the established but predivision meristematic founder cells as zero. By this convention, L10, P3 is the tenth leaf above the coleoptile of the plant at an approximately 4-mm primordial developmental stage, when the leaf is the third leaf from the meristem. Such a leaf might be called a young, adult leaf. L2, P15 is a juvenile, old leaf. A plastochron is actually a unit of time—the time between the initiation of one leaf and the next leaf from the meristem, as if it were a constant; thus, P4 means four units of time have passed since being founded in the meristem. The actual developmental meaning of any particular L, P designation is different in different genotypes and at different stages of shoot development. Even within any inbred line, each leaf is unique, and the plastochron of each may be different. (See Sylvester et al. 1990, and references therein.) Therefore we use both L and P designations when they apply and accurate descriptions of the genotype and growth conditions whenever a leaf is mentioned.

The Maize Handbook—M. Freeling, V. Walbot, eds.

The three dimensions of the leaf have subdivisions. These have names that are specific to grasses, and, in some cases, to maize (Figure 3.1).

The *longitudinal* dimension is divided into the basal portion of the leaf, called the *sheath*, and the distal or tip portion, called the *blade*. At the boundary between sheath and blade is the *ligular region*, consisting of a pair of wedge-shaped, lighter-green *auricles*, and the unique epidermal line of apparently "fused hairs," the *ligule* (Figure 3.1A). Leaves that differ by either L or P differ in dimensions and relative dimensions of the blade, ligular, and sheath parts of the organ. Mutants at several genes alter the presence/absence of the ligule/auricle (Becraft et al. 1990) or the shape of the boundary between blade and sheath (Freeling 1992 and references therein). The ligular boundaries of the leaf may be the *domain* boundaries of some genes.

The *transverse* dimension of the leaf is highly polarized. The upper surface of the blade and the inner surface of the sheath that is pressed against the stalk is *adaxial* epidermis; the outer-lower surface is *abaxial*. These epidermal surfaces and the vascular bundles within the leaf are asymmetrical both with respect to cellular pattern and in the type of cell. This asymmetry is expressed differently in different parts of the organ; for example, the ligule (Figure 3.1D) is an adaxial, epidermal product.

The *lateral* dimension displays bilateral symmetry about the central vein, the first vein that differentiates into the P0-P1 leaf (Sharman 1942). *Regions* along this dimension are defined morphologically. In the blade, the *midvein* (Figure 3.1J) lies at the middle of

---

FIGURE 3.1 Regional and cellular features of the adult leaf, seen in adaxial view. A. An adaxial view of the mature leaf near the ligule. Note how the auricles are narrow at the midrib and expand marginally like a wedge. B. The interlocking, crenulated cell walls and absence of visible wax particles in this replica SEM identify this as the mature blade of an adult leaf. Bar = 28 μm. C. Leaf blade similar to B, but material is fresh whole-mounted in water. Image is of autofluoresence of cell-wall components excited by near UV light. Image has been digitized and recorded on film using an LFR raster scanning image processor. Bar = 37 μm. D. A young (ca. leaf 10, plastochron 8), developing ligule seen from the sheath side in an SEM. Bar = 29 μm. E. As previously, with the ligule ripped away, disclosing the small cells that will become auricle epidermis. Bar = 29 μm. F. This SEM is a view over the ligule looking from the sheath just on the border of a somatic sector that specifies normal auricle (the small, "square" cells) to the right of the arrow, and blade (the larger cells with hairs) to the left. Auricle epidermis transforms to blade distal to the ligule in the absence of the *Lg1* product. Bar = 67 μm. G. A transverse hand-section of sheath stained with toluidine blue O. The inner surface is abaxial. Note that the midvein is no larger than a lateral vein, and there is no midrib region. Also note that there is usually only one intermediate vein between laterals. H. A transverse section of fixed and embedded mature, adult blade stained with toluidine blue O. Note the polarity of epidermal types of cells (bulliform cells and macrohairs—see 1B,C—are found only on the adaxial epidermis) and the polar arrangement of xylem and phloem. Air spaces are located beneath the stomata. I. A replica-SEM view of the mature sheath adaxial surface. Note the stomatal complexes in the rows of wider epidermal cells and the absence of hairs. Bar = 49 μm. J. A freehand transverse section of blade in the region of midrib (between arrows). The live leaf was incubated in methylene blue prior to being sectioned. The majority of the leaf's width has been cut off.

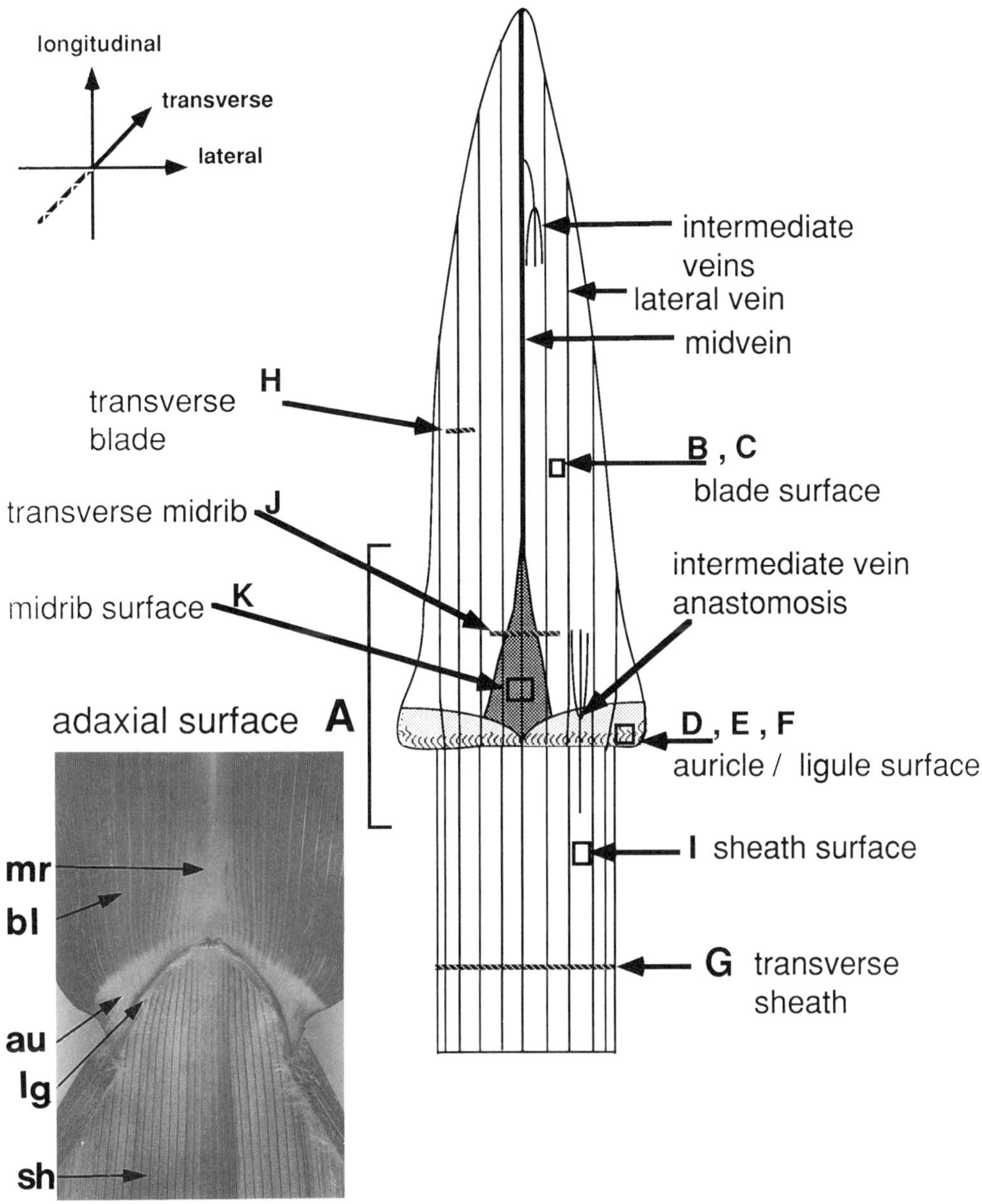

FIGURE 3.1 (Continued) Note the large area of parenchymatous cells. K. A replica-SEM view of the adaxial surface of the midrib of a mature, adult leaf. Although the shiny appearance is sheathlike, the pointy shape of the cells is unique. Bar = 51 μm. Photographs are labeled for type of cell or region of leaf: au, auricle; bc, basal cell; bf, bulliform cell; bl, blade; bs, bundle sheath cell; iv, intermediate vein; lc, long cell; lg, ligule; lv, lateral vein; m, mesophyll cell; mh, macrohair; mih, microhair; mr, midrib region of blade; mv, the midvein; p, parenchyma cells; ph, prickle hairs; pl, phloem; shc, short cell; s, sclerenchyma cells of the bundle sheath extension; sc, stomatal complex; sh, sheath; xy, xylem. B, D, E, and I are reprinted with permission from negatives published in Sylvester et al. (1990). F is reprinted by permission from Becraft et al. (1990).

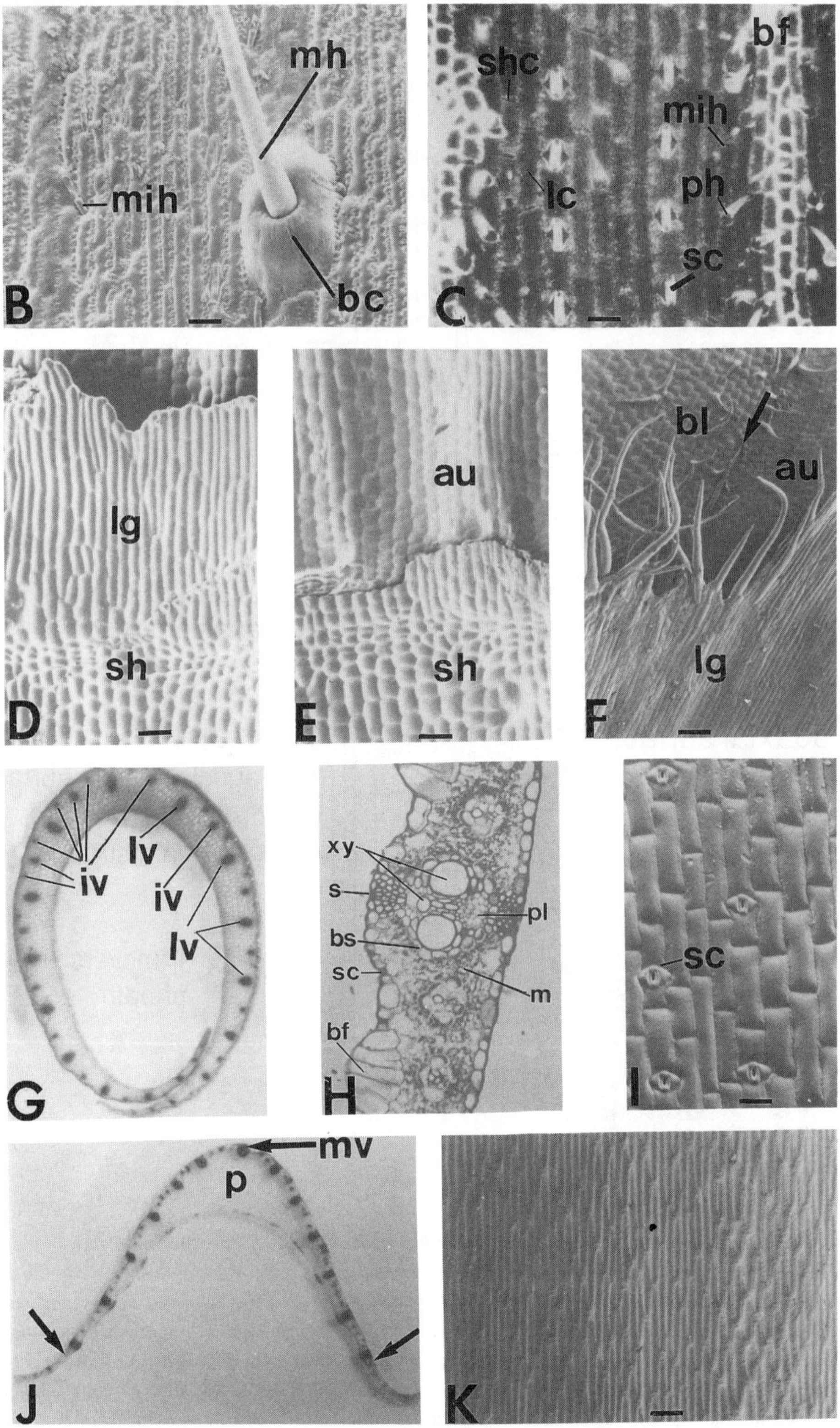

Figure 3.1 (Continued)

a highly lignified triangular region, the *midrib* (Figure 3.1J,K). The phenotype of some mutations affecting the leaf defines the developmental significance of this midrib/blade boundary, while the phenotype of others identifies a morphologically invisible boundary between the *middle blade* and *margin domains* of the blade (Sylvester unpublished; Freeling et al. 1992); *domains* define regions of allele expression. The *auricles* (Figure 3.1A,E) begin as points at the midvein and get larger toward the margins of the leaf.

THE TISSUES AND CELLS OF THE MATURE, ADULT LEAF

There are three *tissues* in the leaf: the outer layer or *epidermis*, which is mainly derived from L1 of the meristem; the *ground tissue*, and the *vascular tissue*, both of which are derived largely from L2 of the meristem (Poethig 1987 and references within). The vascular tissue, bundle sheath cells, and middle mesophyll are derived from the central layer of cells in the leaf primordium, which are, themselves, preferentially derived from the adaxial subepidermal ground cells in the primordium (Langdale et al. 1989). Thus, vascular cells do not represent a unique cell lineage. Our hypothesis that middle mesophyll cells are actually provascular stem cells (Langdale et al. 1989; Dawe and Freeling 1991), if correct, would emphasize the arbitrary nature of tissue designations. Each leaf tissue generates different types of cells, as illustrated and named in the various scanning electron microscope (SEM) and light micrographs of Figure 3.1.

Observation of the leaf as a whole organ discloses unifying patterns in the positions of different cell types. The relationship between venation pattern and cell patterns of the epidermis implies a whole-organ plan of some sort; rows of particular types of epidermal cells (Figure 3.1B,C) differentiate in relation to the position of underlying veins. Figure 3.2 illustrates this pattern relationship. Because veins differentiate first, and certain leaf pattern mutants act only in the middle layers of the leaf but induce surrounding tissue (Sinha and Hake 1990, and references therein), we hypothesize that provascular cells are *organizers* in the developmental sense. However, this proposition remains to be tested experimentally.

JUVENILE AND ADULT LEAVES

Juvenile and adult leaves differ in shape and size. The cells of the juvenile blade epidermis have slightly crenulated noninterlocking cell walls and epicuticular waxes are present (Figure 3.3). Only the production of epicuticular wax is known to be cell autonomous (Salamini and Borghi 1966). Sylvester (1990) and Poethig (1990) show that wax production changes over a few leaves. Usually, but not always, epicuticular wax disappears first from the base of the leaf blade. L4 is entirely covered with wax, L5 has the boundary near the base of the leaf, L6 near the middle, L7 near the tip, and L8 would be entirely waxless. Therefore, the waxy component of juvenility applies to the *apex* as a whole, not the meristem, segment, or leaf. See Poethig (1990) for a discussion of ju-

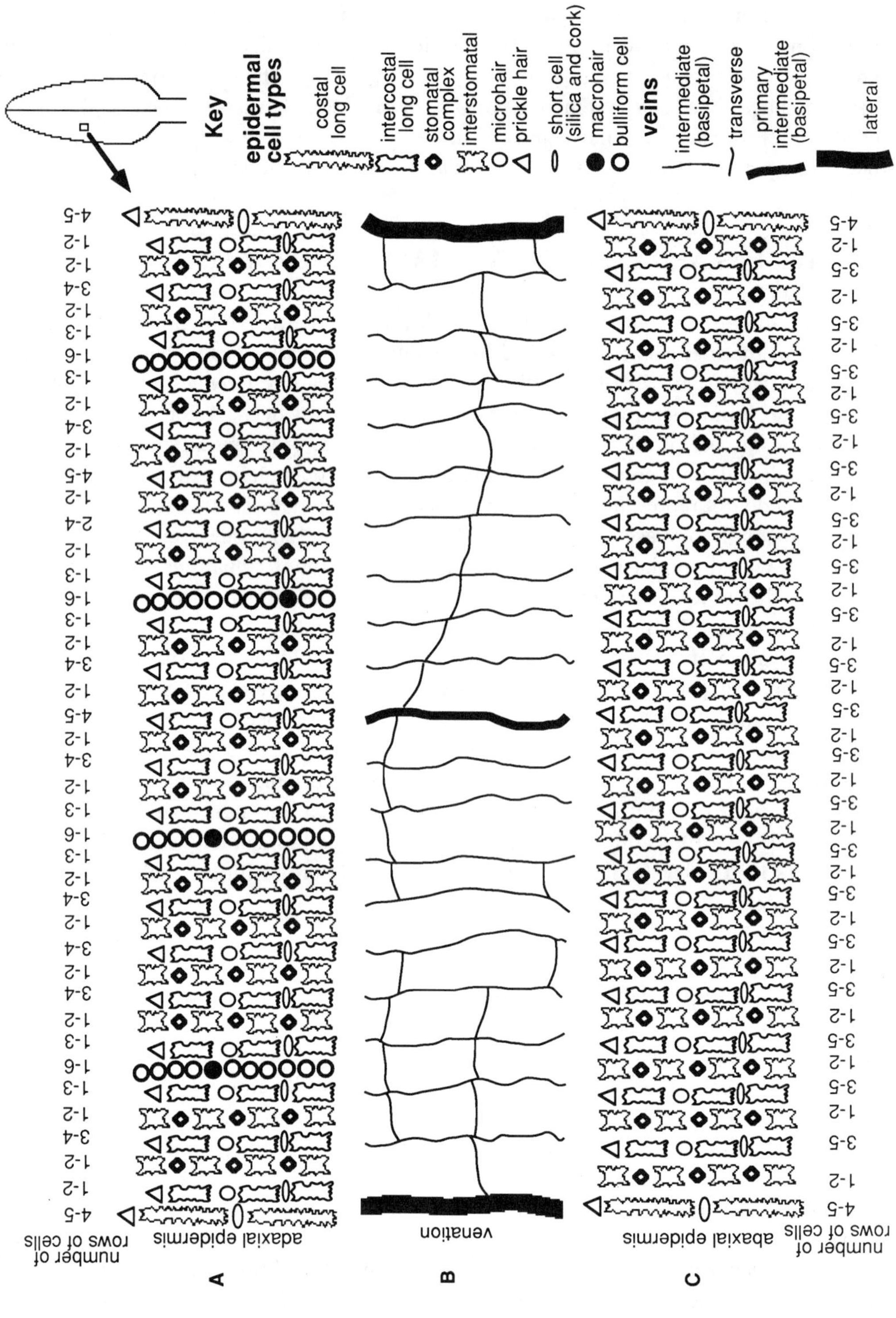
Key
epidermal cell types
costal long cell
intercostal long cell
stomatal complex
interstomatal
microhair
prickle hair
short cell (silica and cork)
macrohair
bulliform cell
veins
intermediate (basipetal)
transverse
primary intermediate (basipetal)
lateral
A
adaxial epidermis
number of rows of cells
B
venation
C
abaxial epidermis
number of rows of cells

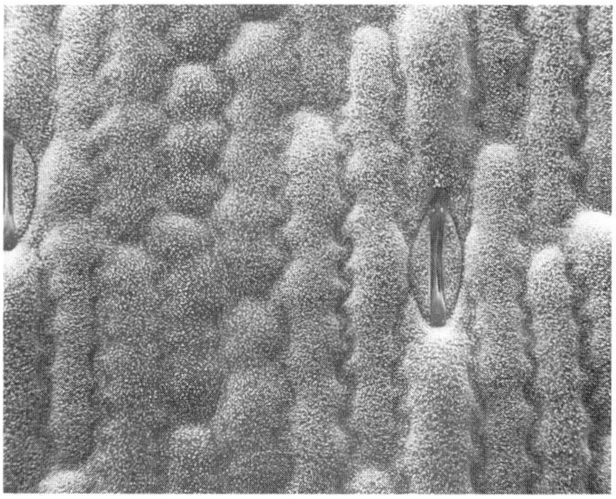

FIGURE 3.3 Wax particle deposition, visualized as a rough coating on the surface of the noninterlocking slightly crenulated epidermal cells of a mature, juvenile blade. Reprinted by permission from Sylvester et al. (1990).

venility in general, and Freeling and co-workers (1992) for the concept of juvenility applied speculatively to the maize leaf.

HOW THE LEAF DEVELOPS FROM THE MERISTEM FOUNDER CELLS TO THE PRIMORDIUM

On the basis of carefully executed genetic mosaic studies and published histological observations, Poethig (1984 and references therein) and McDaniel and Poethig (1988) concluded that an L7 adult leaf originates from about 250 cells in at least two layers forming a ring of *founder cells* within the apical meristem. Further, these researchers concluded that the leaf did not actually grow from the base only, and that there were no division foci at the margins either. (We suggest that the term *meristem* not be used for a division focus within determinate organs, although it is traditional to do so.)

Sylvester and co-workers (1990), using a powerful technique by which replicas of living surfaces are observed with SEM (Williams et al. 1987), found that the depth and

FIGURE 3.2 Cartoon juxtaposing the general arrangement of epidermal cells in surface view and the arrangement of veins. Adaxial (A) and abaxial (C) surfaces are shown in relation to each other and to the underlying vascular tissue (B) between two lateral veins. These lateral veins are in the middle of the blade of a mature adult leaf. Only one row of each type of epidermal cell is depicted and the actual number of rows of cells of that type in the lateral dimension is shown at the top of the figure for the adaxial cell pattern and at the bottom of the figure for the abaxial cell pattern. Within those rows containing microhairs and prickle hairs, both of those kinds of cells may be present or one or the other may predominate. The actual distribution of these types of hair changes along the longitudinal dimension of the mature adult leaf. Refer to Figure 3.1B and 3.1C for views of the actual cells.

direction of recent crosswalls could be used to estimate a mitotic index and the prevalent direction of growth; leaves of representative L and P values were described. Sylvester and co-workers found that each founder cell divides about the same number of times to form the P3–P4 6-mm square-shaped leaf primordium. This mitotically uniform interval, encompassing the primordial stages of development, ends when the lateral veins are complete, and the periclinal divisions of the ligule are in progress (called the *primordium*, which is P3–P4). These workers found that a *preligular region* was visible along the flanks of a plastochron 2 leaf; Figure 3.4B identifies this region with its diagnostic "square" cells. By P3, cells within this region divide anticlinally, giving rise to the preligular band. These divisions occur at about the same time across the lateral dimension of the leaf (Figure 3.4). Subsequently, the now-columnar epidermal cells divide periclinally beginning on either side of the midrib to produce the ligule (Sharman 1942; Becraft et al. 1990), and a wave of division moves out to the margin and in to the midvein. This band of ligular divisions during the primordial stages of leaf development is the only intercalary focus of cell division discovered, and is fully removed by certain mutants (Sylvester et al. 1990). Auricle development progresses along with ligule development. When auricle development is blocked, such as with a *liguless1-O* mutant sector, the auricles begin again as points in the blade and get larger toward the leaf margin (Becraft and Freeling 1991).

POSTPRIMORDIAL LEAF DEVELOPMENT

In contrast to *primordial* leaf development, the postprimordial leaf matures basipetally (Sylvester et al. 1990, and references therein). The cell divisions required to bring the leaf to its mature size are completed first at the tip of the leaf and later at its base. Terminal differentiation sometimes follows a strict pattern of cell divisions, as with the stomatal complex (Stebbins and Shah 1960).

VEINS DURING LEAF DEVELOPMENT

The three stages of leaf development—meristem founder, primordial, and postprimordial—are correlated with three distinct stages of venation. According to Sharman (1942), the *midvein* preexists in the subapical cortex of the shoot meristem and differentiates into the very young leaf (P0–P1). It is unclear whether or not the *lateral veins* arise in the shoot meristem or the very young leaf. In any case, as with the midvein, they arise independently, and differentiate acropetally to the leaf margin, and basipetally into the stem, where they anastomose with each other (Kumazawa 1961; Esau 1943). A large leaf has about 20 lateral veins on each side of the midvein. These laterals are complete in the primordium. Thus the midvein appears to precede the initiation of the leaf, and the laterals arise during the primordial stage of leaf development.

During postprimordial growth, laterals grow via cell division and cell expansion and

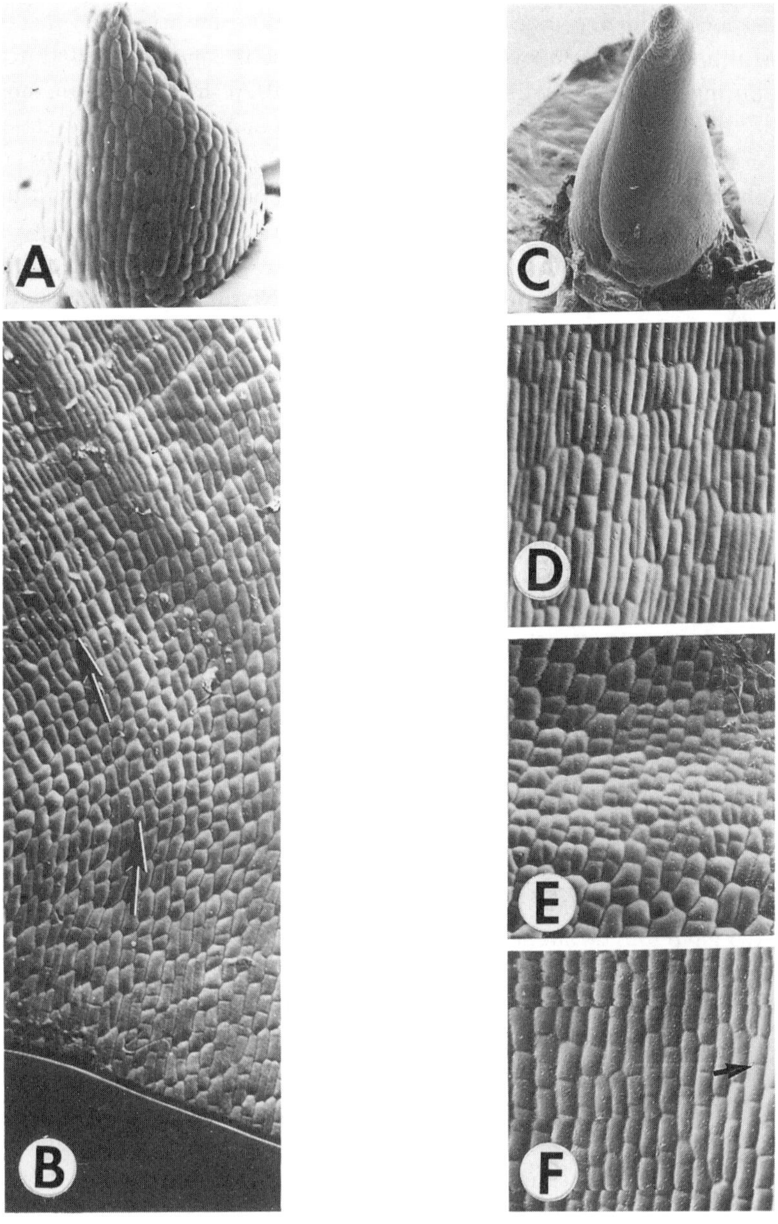

FIGURE 3.4 Replica SEM analyses of the primordial cells along leaf 11/plastochron 2 (left) and leaf 10/ plastochron 3 (right) adult leaf. At P2 the preligular region is visible as the squarish cells between the arrows of B. It is within these cells that anticlinal divisions give rise to the preligular band. E shows the band of P3 anticlinal divisions that occurs within the preligular region. F, the primordial presheath, displays cells where some of the crosswalls are distinctively shallow, as the two daughter cells at the tip of the arrow. Crosswall depth is used to estimate mitotic index and the plane of cell division. Reprinted by permission from Sylvester et al. (1990).

may recruit cells from the surrounding mesophyll. At present we do not have appropriate genetic markers for the vascular cell types to determine the lineage of the cells within the bundle, although it is known that bundle sheath cells can be recruited from adjacent mesophyll during vascular growth (Langdale et al. 1989). At the leaf margin, laterals branch into *primary intermediate veins*, which develop basipetally. As the leaf gets wider, and cell division and expansion progress basipetally, the *primary intermediate veins* branch further to elaborate the intermediate veins. Collectively, these are referred to as the *basipetal veins*. At the widest part of a large leaf there might be 30 intermediate veins between two laterals. In the blade two middle mesophyll cells separate each intermediate vein from the next (Russell and Evert 1985). This pattern has been explained by a stem-cell lineage rule, "the half-vein rule" (Langdale et al. 1989), and the concept of structural templates (Dawe and Freeling 1991) has been suggested to operate during the vein extension phase of postprimordial growth.

Transverse, or *cross veins*, connect all the longitudinal veins in the leaf, and these differentiate last. As the leaf gets narrower, intermediates *anastomose*, and as the auricle is reached, all intermediate veins anastomose, usually into one intermediate that continues differentiation into the sheath. Kumazawa (1961) concludes that the intermediate veins that make it into the sheath enter the stem, where they anastomose with one another into a peripheral network which remains entirely separate from the network formed by the primordial midvein and laterals.

HOW TO MARK TRANSFORMATIONS OF ONE PART OF THE LEAF INTO ANOTHER

SEM observation of surfaces is usually sufficient to show that, for example, a region of blade is transformed to a region of sheath (Becraft and Freeling 1989; Bertrand-Garcia et al. 1991). We have found that some mutants transform a region to a combination of cell or tissue types or to a "new" identity. Then it is necessary to use cell-autonomous alleles that express only in specific leaf areas or regions. Anthocyanin, wax, and hairiness genes all have mutant or variant alleles that are part specific for the epidermis. Their use requires much knowledge of genetic background effects. Figure 3.5 is an abaxial view of a *Liguleless 4-O* mutant heterozygote in a genetic background that colors sheath and auricle red, but not blade; the transformed blade is visualized as red spikes (arrow in Figure 3.5) in the region of blade between the midrib and the margin.

WHAT ARE LEAF HOMOLOGS?

The leaflike glumes and lemmas of the floret, both with a single midvein, are clearly homologous to the vegetative leaf; single mutant alleles transform these organs into obvious leaves with ligules. The bikeeled, leaflike organs that subtend new buds—the co-

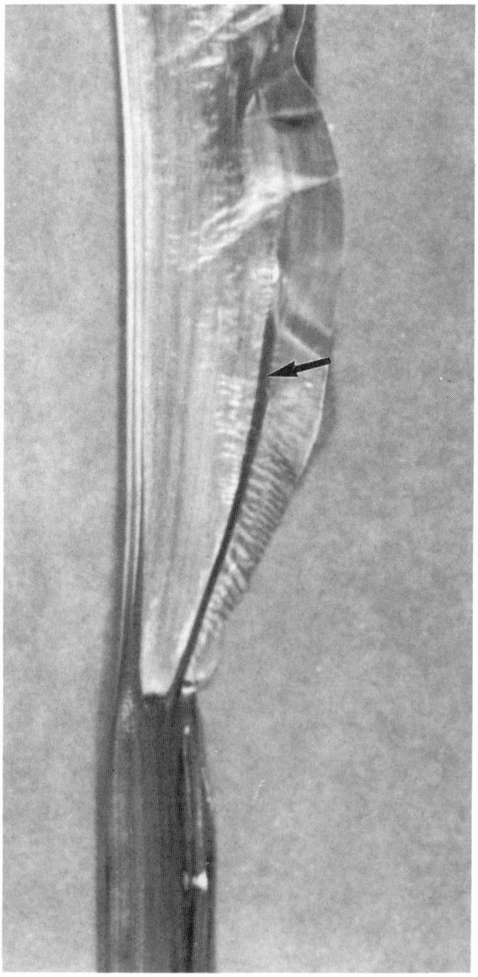

FIGURE 3.5 An abaxial view of a leaf from a plant that is genetically conditioned to express anthocyanin pigment in sheath, and is also heterozygous for the dominant mutant *Lg4-O*, one of several mutants of seven genes that all interfere with leaf cell identity, and push the blade/sheath boundary up the leaf. Note the spiky region of red sheath tissue (arrow) extending well up into the blade. Data provided by John Fowler (unpublished).

leoptile, prophyll, and palea—have unknown relationships to the leaf. The homologies of stamen and carpel programs to the leaf genetic program are also unknown because the appropriate homeotic or heterochronic mutants have not been described. It is not wise to ascribe homologies without a mutant analysis since morphology itself is but the reflection of genetic programs, programs that might merge, submerge, overlap, or interact during evolution.

METHODS

All methods are cited in the legends and they usually are covered in the cell-biology methods sections of this handbook.

ACKNOWLEDGMENTS

The leaf development studies in our laboratory are funded by the National Institute of Health, USA, and the California Agricultural Experiment Station. We thank Assunta Chytry and the NSF Center of Plant Developmental Biology for help with graphics and photomicroscopy, and especially thank present and past laboratory members for contributing photographs and comments.

REFERENCES

Becraft P, Freeling M (1989) Use of the scanning electron microscope to ascribe leaf regional identities even when normal anatomy is disrupted. Maize Genetics Cooperation News Letter 63: 37

Becraft P, Bongard-Pierce DK, Sylvester AW, Poethig RS, Freeling M (1990) The *liguleless-1* gene acts tissue-specifically in maize leaf development. Dev Biol 141: 220–232

Bertrand-Garcia R, Freeling M (1991) *Hairy-sheath frayed #1-O*: A systemic, heterochronic mutant of maize that specifies slow developmental stage transitions. Am J Bot 78(6): 747–765

Dawe RK, Freeling M (1991) Cell lineage and its consequences in higher plants. Plant J 1(1): 3–8

Esau K (1943) Ontogeny of the vascular bundle in *Zea mays*. Hilgardia 15: 327–368

Freeling M (1992) A conceptual framework for maize leaf development. Develop. Biol. 153: 44–58

opmental time and their heterochronic interactions. Bio Essays, 14: 227–236

Kumazawa M (1961) Studies on the vascular course in maize plant. Phytomorphology 11: 128–139

Langdale JA, Lane B, Freeling M, Nelson T (1989) Cell lineage analysis of maize bundle sheath and mesophyll cells. Dev Biol 133: 128–139

McDaniel CN, Poethig RS (1988) Cell lineage patterns of the shoot apical meristem in the germinating corn embryo. Planta 175: 13–22

Poethig S (1984) Cellular parameters of leaf morphogenesis in maize and tobacco. In White RA, Dickison WC (eds) Contemporary Problems in Plant Anatomy, Academic Press, New York, pp 235–259

Poethig RS (1987) Clonal analysis of cell lineage patterns in plant development. Am J Bot 74 (4): 581–594

Poethig RS (1990) Phase change and the regulation of shoot morphogenesis in plants. Science 250: 923–930

Russell SH, Evert RF (1985) Leaf vasculature in *Zea mays* L. Planta 164: 448–458

Salamini F, Borghi B (1966) Analisi genetica di mutanti *Glossy* di mais. II. Frequenze di reversione al locus *gl*. Estratto da "Genetica Agraria" Vol. XX, pp 239–248

Sharman BC (1942) Developmental anatomy of the shoot of *Zea mays* L. Ann Bot 6: 245–284

Sinha N, Hake S (1990) Mutant characters of Knotted maize leaves are determined in the innermost tissue layers. Dev Biol 141: 203–210

Stebbins GL, Shah SS (1960) Developmental studies of cell differentiation in the epidermis of monocotyledons. II. Cytological features of stomatal development in the Gramineae. Dev Biol 2: 477–500

Sylvester AW, Cande WZ, Freeling M (1990) Division and differentiation during normal and *liguleless-1* maize leaf development. Development 110: 985–1000

Williams MH, Vesk M, Mullins MG (1987) Tissue preparation for scanning electron microscopy of fruit surfaces: comparison of fresh and cryopreserved specimens and replicas of banana peel. Micron Microsc Acta 18: 27–31

4

The Maize Root

L. Feldman

ROOT HABIT

During an average period of growth, the root system of a single maize plant may exploit over 200 cubic feet of soil and may absorb 35–50 gallons of water. Depending on soil texture, the lateral spread of a mature root system may be 3–4 feet on all sides of the plant and typically penetrates to depths of 5–6 feet (Figure 4.1), although depths of 8 feet are not unusual (Weaver 1926). The depth of planting appears to bear no relationship to depth of rooting.

As in other herbaceous monocotyledons, the root system of maize is derived from several distinctive types of roots:

1. The primary or tap root and associated lateral roots.
2. Seminal roots (that is, roots other than the primary root preformed and present in the ungerminated seed).
3. Shoot-borne ("adventitious")[1] roots that originate at nodes on portions of the germinated seedling axis below the soil level.
4. Shoot-borne ("adventitious"—see footnote 1) roots that form at nodes on portions of the stem above the soil, usually at the lowermost 2–3 nodes; these are usually referred to as "prop" or "brace" roots.

The primary root originates early during embryogenesis and becomes defined as a distinctive region 10–15 days after pollination. Vascular development is evident 12–15 days after pollination (Avery 1930). The primary root is enclosed by a protective ensheathing structure, the coleorhiza (Figure 4.2). At germination the coleorhiza grows or

[1]Technically, adventitious roots are roots formed at a cut surface or on portions of the plant as a result of injury. Because "adventitious" roots in maize are part of the normal development of the plant, they are therefore not truly adventitious. However, because these roots are usually described as adventitious, I will continue to use this terminology.

The Maize Handbook—M. Freeling, V. Walbot, eds.

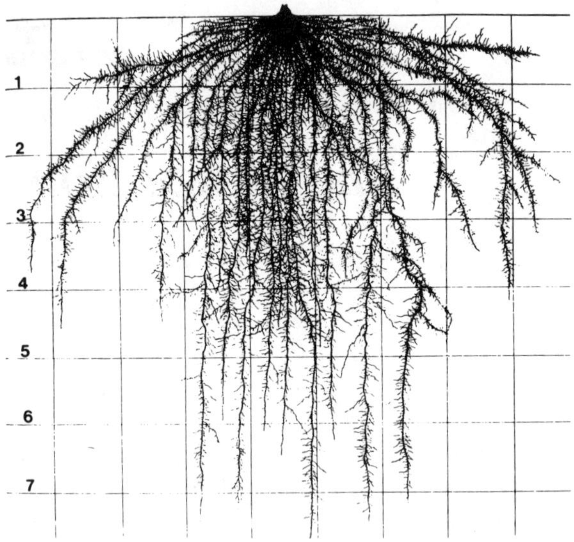

FIGURE 4.1 Root system of a mature corn plant (depth indicated in feet) (from Weaver© 1926 McGraw Hill, Inc.).

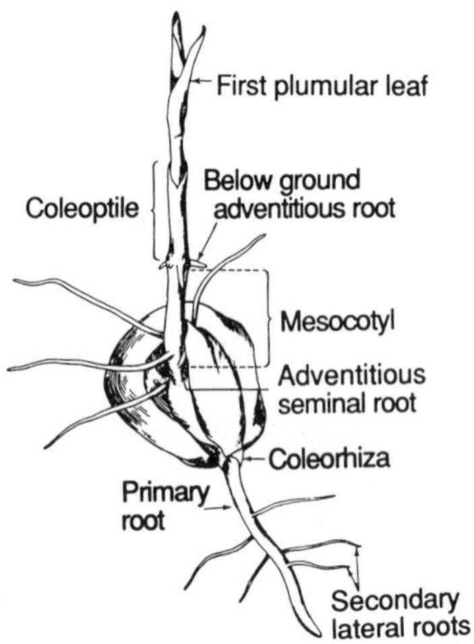

FIGURE 4.2 Young corn seedling showing early stages in the development of various types of roots (from Wallace and Bressman 1949).

is pushed through the seed coat and is itself then pierced by the primary root (the radicle). Although the radicle is reportedly capable of extending at rates of up to 11 cm per day, rates of 1–4 cm/day are most common.

At about the time the coleoptile reaches the soil surface, the adventitious seminal roots emerge between the scutellum and the first internode. Mature maize kernels typically contain from 3 to 7 seminal root primordia (Figure 4.2); these originate 30–40 days postanthesis. At germination these roots initially grow upward or horizontally, depending on the original orientation of the seminal root primordia, but soon bend downward.

For the first 2–3 weeks following germination the primary root and the seminal roots comprise the bulk of the seedling root system. In some varieties of maize, seminal roots never grow or are retarded in their development; the initial root system is therefore derived almost exclusively from the primary root. Seminal roots are usually evident, however. Both the seminal and tap roots are important during early seedling development, but they assume less importance and frequently decay following development of the permanent root system.

Permanent, adventitious roots originate from nodes on the belowground portion of the stem. Usually many whorls of roots arise from these nodes. Because underground internodes fail to elongate, and hence are very short, close inspection is often necessary to ascertain the node of origin of each root. Collectively, the whorls comprise the root crown. Permanent roots typically grow horizontally for considerable distances before eventually angling and growing downward. Foth et al. (1960) report that early root development is characterized by prominent lateral growth in the upper foot of the soil after which vertically growing roots become more prominent. There is some evidence that in many varieties light influences downward growth of horizontally oriented roots. In some types of soil, light may penetrate several inches, where it can interact with the photomorphogenetic pigment phytochrome, thereby triggering downward growth. In roots requiring light for downward growth, this may be a mechanism allowing roots to exploit upper horizons of the soil while at the same time ensuring that the root does not wander too close to the soil surface where it risks desiccation. As a consequence of the initial horizontal (lateral) growth the large, permanent, adventitious roots heavily exploit the upper 1–2 feet of the soil where they are most concentrated.

The prop, or brace, roots are also adventitious in origin, forming on the lowermost 2–3 nodes of the aboveground portion of the stem (Figure 4.3). Although it is quite evident that the prop roots function in keeping a plant upright, or in preventing a plant from lodging completely, only recently have workers appreciated that prop roots also function in nutrient and water uptake. Until the prop root reaches the soil, it remains unbranched and relatively rigid, and is usually covered by a mucilaginous substance that protects the root from drying. Following contact with the soil the root branches and numerous laterals are produced (Figure 4.4). This allows the plant to exploit a region of the soil that even late in the growing season may have few roots active in mineral or water uptake. Once they contact the soil, prop roots from the first node or two above the ground are important in maintaining a healthy upright plant. Adventitious roots initiated at higher nodes probably only function to prevent a plant that has fallen sideways from completely lodging.

Weaver (1926) reports that maize root systems develop best in deep, well-drained soil that has an abundant supply of water throughout the growing season. He also emphasizes the detrimental effects of even shallow cultivation after the establishment of the permanent root system. In one study, "cultivating to a depth of 4 inches during a period of 9 years gave a decreased yield in every season but one as compared with similar cultivation to a depth of 1.5 inches." He concludes, "for highest yields, cultivation should be deep enough to kill the weeds, but shallow enough to reduce root injury to a mini-

FIGURE 4.3 Adventitious "prop" or "brace" roots.

mum." To avoid damaging these roots, all cultivation should have been completed before this growth occurs (usually 3–4 weeks after germination). Generally only the smaller lateral roots reach depths greater than 4 feet (estimated to be about 2% of the total root mass).

Hilling during cultivation can encourage the development of more prop roots and thus can provide the plant additional support. But if the hills are later washed away or are otherwise removed, so that roots are exposed to the atmosphere, desiccation and injury occur, usually leading to the death of the root.

As in other plants, maize shows marked varietal (heritable) differences in root habit, including root mass per plant, number of branches per unit length, lateral root spread, and number of seminal roots. These varietal differences are further influenced by differences in soil types and irrigation practices, both of which can have pronounced effects on root habit. In well-irrigated soils containing adequate supplies of water and nutrients, the roots of 6-week-old plants were almost all within 1 foot of the soil surface and grew parallel to the soil surface. Only a few roots angled downward. In contrast, in relatively dry soils only a few roots grow horizontally; most grow downward 1–2 feet (Foth et al. 1960). Thus, it appears that moisture content of the soil affects the gravity response of the root. While the number of permanent roots was the same in both irrigated and dry

FIGURE 4.4 Aboveground unbranched prop roots bearing numerous laterals; these are produced once the prop root enters the soil.

soils, irrigated roots had fewer laterals than roots growing in dry soils. In well-irrigated soils, laterals were considerably shorter than on roots in dry soils. Hence, a controlling factor in variation of root habit is soil water content.

For experimental studies, maize roots may be grown and manipulated in sterile liquid culture. A relatively simple medium (White 1954) supplemented with 2–4% sucrose is sufficient. To obtain roots for culture, kernels are surface-sterilized for 8 minutes in half-strength commercial bleach (approximately 5% sodium hypochlorite) and rinsed three times in sterile water. The seeds are then germinated aseptically on filter paper. When primary roots are about 2–4 cm long they are excised and placed in liquid culture on a shaker (50–60 rpm) in darkness at 23–25° C. A characteristic of roots in culture is that as they grow they become quite thin but are otherwise normal in anatomy and morphology.

ROOT ANATOMY

The anatomy of the maize root has been reviewed in detail several times (Avery 1930; Hayward 1938; Sass 1977). In cross section, the mature primary root is comprised of a central pith encircled by a ring of vascular elements. The largest of the vascular elements, the metaxylem, is composed of vessels that may be broader than they are long (Figure 4.5). The number of metaxylem elements may vary from the typical 6–10 in a primary root to upward of 48 in thick adventitious roots. Peripheral to and encircling the metaxylem is the protoxylem (Figure 4.5). The number of protoxylem points varies somewhat, but there are commonly 2–3 protoxylem "strands" for each metaxylem element. Pockets of primary phloem alternate with strands of the protoxylem (Figure 4.5). Surrounding the vascular tissues is a single layer of thin-walled cells, the pericycle, which is itself bounded by the endodermis, in which a Casparian strip is often evident. In older portions of the root, in regions in which root hairs have ceased functioning, walls of the endodermis are thickened as a consequence of suberin deposition. The cortex beyond the endodermis typically consists of several layers of parenchymatous cells surrounded by a short-lived epidermis. In older roots, the epidermis is usually lost and its protective functions are taken over by a lignified, suberized exodermis that develops from the outer cells of the cortex. The types and distribution of cells in seminal and adventitious roots are essentially similar to those of the primary root, the main difference being in the number of primary xylem points. In addition, in aerial roots the epidermis does not disintegrate but instead is retained and develops a waxy cuticle.

In longitudinal section, the most distinctive region of the root is the terminal 2–3 mm consisting of the root cap, the apical meristem, and a portion of the elongation zone (Figure 4.6). The root cap originates early during embryogenesis. In maize, the cap may consist of upward of 10,000 cells whose origins are traceable to the activities of a specific

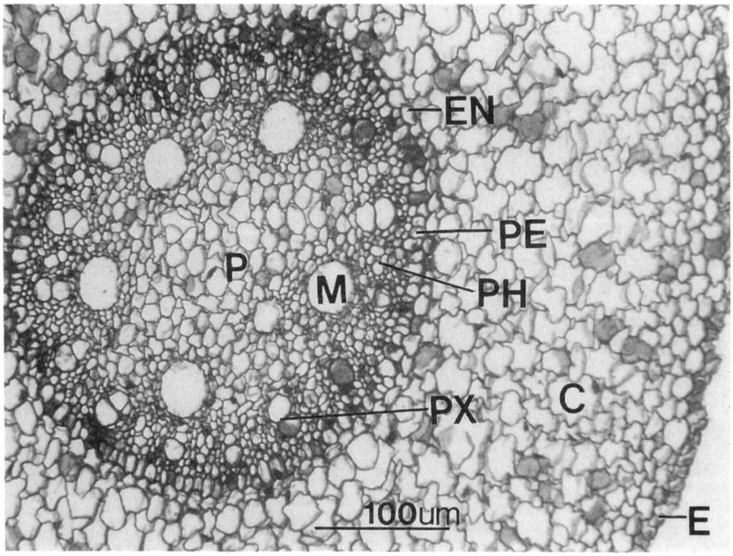

FIGURE 4.5 Cross section through the primary root of maize. Note central pith (P), metaxylem (M), protoxylem (PX), pericycle (PE), endodermis (EN), cortex (C), epidermis (E), and phloem (PH).

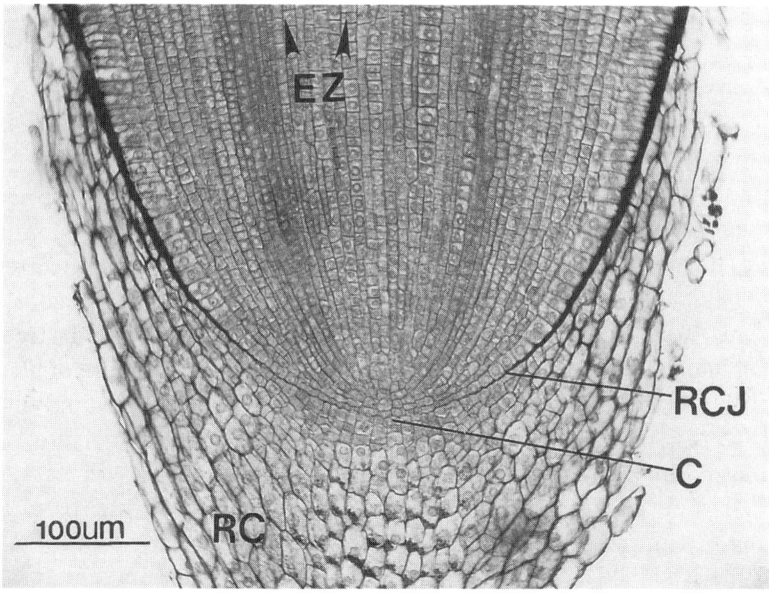

FIGURE 4.6 Longitudinal section through a maize root apex. Note the root cap (RC) extending downward, the calytrogen layer (C), and the cell files converging to a region just basal to the root/root cap junction (RCJ).

root cap meristem layer, the calyptrogen, which functions only in the production of new cap cells (Figure 4.6). Cells comprising the calyptrogen have the highest rate of cell division in the root, with a cell cycle time of about 12 hours. After a new cell is produced at the calyptrogen it is moved (displaced) through the root cap, ultimately to be shed by the root into the soil. Passage of a single cell through the cap usually takes between 5 and 8 days. Upon reaching the surface of the cap, the walls of the cap cells disintegrate, the protoplasts lyse, and the contents become part of the "slime" or mucilage covering root tips. There is some evidence that the slime influences the mitotic activity of the calyptrogen. In a root growing in the soil, slime is normally continually removed as a result of physical contact between the root tip and the soil. However, if the tip of the root is kept from contacting soil, and slime is allowed to accumulate, the calyptrogen ceases dividing. Removing the slime from the tip restores the mitotic activity of the calyptrogen. Aside from the production of slime, which may be a lubricant for root passage through the soil, the root cap appears to have many other functions, including sensing gravity, light, and temperature and moisture gradients in the soil.

In many cultivars of maize the cap can be excised from the root, facilitating studies of its physiological functions. Such dissections are best done under a dissecting microscope using microdissecting holders fitted with pieces of slightly dulled razor blades tapered to a point. The slime is first wiped off using tissue paper and then the blade is placed at about a 45° angle at the junction of the cap and the root proper (most easily seen by illuminating the root from beneath). Slight pressure is exerted (this takes some practice) and an intact cap can be readily isolated. The tear occurs through the calyptrogen, destroying these cells in the process. Not all cultivars of maize are suitable, but two varieties that are particularly useful are Merit and Kelvedon 33.

Just basal (proximal) to the root cap is the subterminal root apex. Files, or lineages, of cells radiate from a point adjacent to the root cap/root boundary, leading earlier workers to conclude that cell production occurs as a consequence of continued mitoses at the focus of the cell lineages (Figure 4.6). We now know this view to be incorrect. Rather, the region of cells where cell files converge is relatively inactive mitotically, with cells dividing on average of about once every 200 hours. This relatively inactive region of about 1,000–1,500 cells is known as the quiescent center. It is a feature of all maize and indeed of all angiosperm root apices and is perhaps best revealed by supplying roots ^{3}H-thymidine (1–5 μCi/ml of liquid culture medium [White's medium: White 1954] for 4–8 hours), followed by autoradiographic analysis. The actual source of most new cells in a growing root is an arc of meristematic cells that is adjacent to the basal face of the quiescent center. Upward of 10,000–20,000 new cells arise daily from this meristem. Because this meristem is basal or proximal to the root/root cap junction, it is called the proximal meristem. Depending on the rate of root growth, the proximal meristem extends 1–3 mm from the root cap boundary (Figure 4.7).

Overlapping and extending basally from the proximal meristem is the region of cell elongation (Figure 4.6). In actively growing roots maintained at 25° C, the region of elongation is about 7–8 mm in length. The pattern of elongation within the elongation zone was studied by Erickson and Sax (1956), who coated maize roots with fine carbon particles and characterized the displacement of particles from each other and from the tip of the root. They found that the rate of elongation varies within this region and is maximal 3–4 mm from the cap boundary. As a cell moves through the elongation zone, its rate of elongation increases and then decreases to zero. The final length of various cell types varies because cells in the same region of the root stop dividing at different times. Epidermal and cortical cells elongate to about 40–60× their original length (attaining an

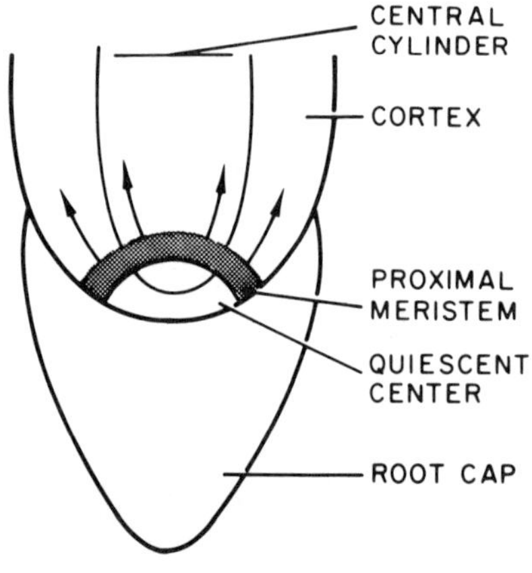

FIGURE 4.7 Diagram indicating the location of the proximal meristem and the quiescent center. Arrows denote the direction of production of new cells by the proximal meristem.

average length of about 190 μm), whereas metaxylem elements elongate about 150× their original length (attaining a final length of about 1,000 μm) (Luxova 1981).

REFERENCES

Avery GS (1930) Comparative anatomy and morphology of embryos and seedlings of maize, oats and wheat. Botanical Gazette 89: 1–39

Erickson RO, Sax KB (1956) Elemental growth rate of the primary root of *Zea mays*. Proc Am Philosoph Soc 100: 487–514

Foth HD, Kinra KL, Pratt JN (1960) Corn root development. Michigan Agr Exp Quart Bull 43: 2–13

Hayward HE (1938) The Structure of Economic Plants, The Macmillan Co., New York

Luxova M (1981) Growth region of the primary root of maize (*Zea mays* L.). In Brouwer R, Gasparikova O, Kolek J, Loughman BC (eds) Structure and Function of Plant Roots, Martinus Nijhoff/Dr W. Junk Publishers, The Hague, 9–14

Sass JE (1977) Morphology. In Sprague GF (ed) Corn and Corn Improvement, Am Soc Agronomy, Madison, WI

Wallace HA, Bressman EN (1949) Corn and Corn Growing, John Wiley and Son, New York

Weaver JE (1926) Root Development of Field Crops, McGraw-Hill Book Co., New York

White PR (1954) The Cultivation of Animal and Plant Cells, The Ronald Press, New York

5

Morphology and Development of the Tassel and Ear

Ping-chin Cheng and Dayakar R. Pareddy

As a monoecious plant, maize develops unisexual male and female flowers in physically separated parts of the plant. The tassel (staminate or male inflorescence) arises from the shoot apical meristem, while the ears (pistillate or female inflorescences) originate from the axillary bud apices. Initially, both inflorescences contain bisexual flowers. During the course of development, however, they become unisexual through abortion of gynoecia in the tassel flowers and stamens in the ear flowers. As a result, the tassels develop male

The Maize Handbook—M. Freeling, V. Walbot, eds.

flowers and the ears develop female flowers. (See descriptions of Weatherwax 1916; Bonnett 1948, 1953, 1966; Kiesselbach 1949; Cheng et al. 1983; Stevens et al. 1986; Hanway and Ritchie 1987.)

MORPHOLOGY

The Tassel

The tassel is a branched inflorescence located at the tip of the main stem. It consists of a central spike (rachis) and about 10–50 lateral branches (Figure 5.1). The paired spikelets (pedicellate and sessile; Figure 5.2) occur in many ranks around the central spike, but are arranged in only two rows on the lower (adaxial) surface of the lateral branches (Figure 5.2). Each spikelet contains two florets, the upper and lower floret. The development of the upper floret is about 2–3 days ahead of the lower floret measured at anthesis.

Each spikelet has a pair of leaflike glumes, which encase two florets (Figures 5.3 and 5.4). Within the glumes, each floret is further enclosed with a pair of thin scales, a lemma (located adjacent to the glume), and a palea (located opposite to the lemma, between the two florets). Two of the three anthers present in each floret are located adjacent to the palea; the third is located adjacent to the lemma and is flanked by two lodicules (Figure 5.4). These lodicules swell at anthesis allowing extrusion of anthers by elongation of the filaments. Following extrusion, anthers dangle downward and shed pollen from openings at the tip (Figure 5.3).

The Ear

The top one or several axillary buds terminate in an ear capable of producing mature kernels. Each of these axillary buds is covered with about 8–14 modified leaves called husks, and a prophyll (Figure 5.5). The ear branch, or shank, consists of nodes and short internodes (Figure 5.6). The ear does not have lateral branches but its thick axis, called the cob, is similar to the central spike of the tassel in that it produces multiple rows of paired spikelets (polystichous phyllotaxy).

Each spikelet is enclosed by a pair of glumes and contains only one functional floret. Although two florets are initiated, only the upper one develops; the lower floret aborts at an early stage of development (Figure 5.7). Both florets produce a lemma, but usually only the upper floret produces a palea. The palea of the lower floret either does not initiate or remains rudimentary. Each functional floret produces an ovary with an elongated silk (style) covered with hairs (trichomes).

FIGURE 5.1 Mature tassel inflorescence at anthesis. Note the central rachis (CR) and branches (BR). **FIGURE 5.2** Placement of spikelets on the lateral branch (A) and central rachis (B). Note the sessile spikelet (SS) and pedicellate spikelet (SP). **FIGURE 5.3** portion of the lateral branch showing extrusion of anthers. Note elongated filaments (F). **FIGURE 5.4** Scanning electron microscopic image of a mature tassel spikelet. The glumes, lemma, and part of the palea (P) as well as two anthers (An) from each floret have been removed. Go, remnant of the outer glume; Gi, remnant of the inner glume; Lo, remnant of outer lemma; asterisk, lodicules; arrowhead, degenerated gynoecium.

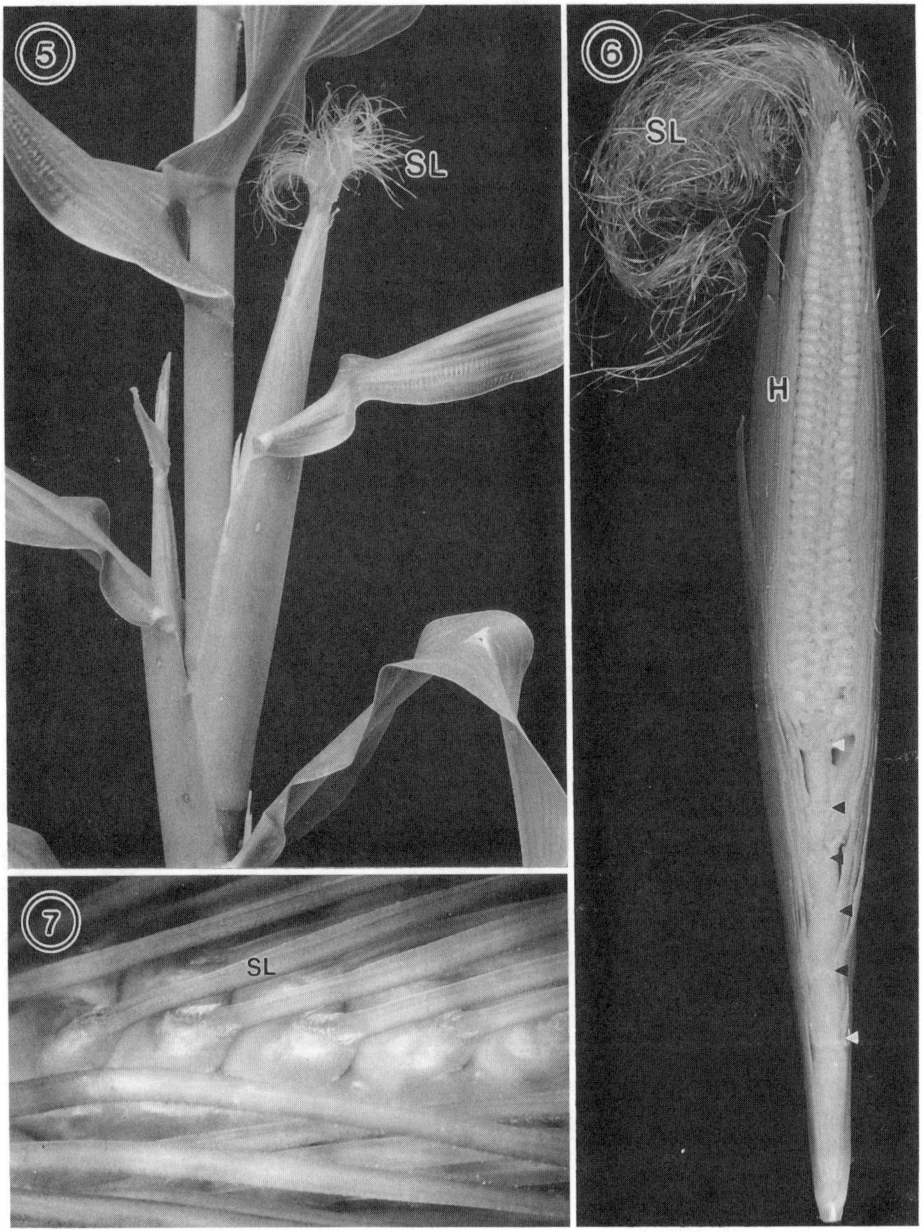

FIGURE 5.5. Ear inflorescence showing silk (SL) emergence. FIGURE 5.6 Longitudinal section of an ear inflorescence. Note the short internode of the lateral branch (arrowheads) and the placement of husks (H). SL, silks. FIGURE 5.7 Ear spikelets showing gynoecium with long styles. SL, silks.

DIFFERENTIATION AND DEVELOPMENT

The Tassel

The development of the tassel precedes that of the ear. Following the initiation of all leaf primordia, the shoot apical meristem elongates and is transformed into a reproductive (tassel) meristem. The tassel differentiates in an acropetal sequence. The first primordia to arise develop into major lateral branches. This is followed by the initiation of rows of spikelet-pair primordia, which form first on the rachis and subsequently on the lateral branches (Figure 5.8).

Each spikelet-pair primordium differentiates into a pedicellate and sessile spikelet (Figures 5.9 and 5.10). About 500–1,000 spikelets are produced on each tassel, depending on the genotype and growth conditions. Initially, the development of the pedicellate spikelet precedes that of the sessile spikelet. However, this difference seems to disappear when the spikelets reach maturity (Cheng et al. 1983; Stevens et al. 1986).

Spikelet development begins with the initiation of the outer glume; this is followed by the initiation of the inner glume (Figure 5.10) and then of the outer lemma and inner lemma. Subsequently, basally and abaxial to the outer lemma, the lower floret is initiated (Figure 5.11). The development of the upper floret precedes that of the lower floret. The extent of this developmental difference varies between genotypes and with the stage of development, and is maintained through to anthesis (Hsu et al. 1988; Cheng et al. 1983). A pair of laterally positioned stamen primordia and a palea are then inititiated in the upper floret. This is followed by the initiation of the third stamen on the abaxial side of the floral meristem. Gynoecial development begins with the initiation of an adaxial ridge on the remainder of the floral meristem distal to the stamens (Figure 5.12), but soon afterward the gynoecium begins to abort and subsequently degenerates (Figure 5.13). Only the stamens develop further, resulting in imperfect staminate flowers (Figure 5.14). A similar sequence of organ initiation occurs in the lower floret (Bonnett 1948, 1953, 1966; Cheng et al. 1983). As a result of the order in which they are initiated, spikelets in the middle portion of the rachis reach anthesis first (Stevens et al. 1986). A similar pattern of maturation is observed on the lateral branches as well.

The stamen primordia in the spikelets enlarge and differentiate into microsporangia. The sporogenous tissue in the anther, which produces pollen mother cells (microsporocytes), is enclosed by the anther wall, which consists of four cell layers (epidermis, endothecium, middle layer, and tapetum). Pollen mother cells undergo meiosis to produce quartets (tetrads) containing four haploid cells. Soon after meiosis, the haploid cells in the quartet break apart and form haploid microspores (Cheng et al. 1979). These microspores undergo two further mitotic divisions, producing trinucleate pollen grains that contain numerous starch granules and thick walls (exine and intine) at maturity.

At anthesis, the lodicules swell and pry apart the glume and lemma (Figure 5.4), allowing the anthers to extrude (Figure 5.3). Soon after extrusion, pores at the tip of the anthers break open and pollen is liberated. The extrusion and subsequent shedding of pollen occurs first in the spikelets located a little above the middle of the rachis and

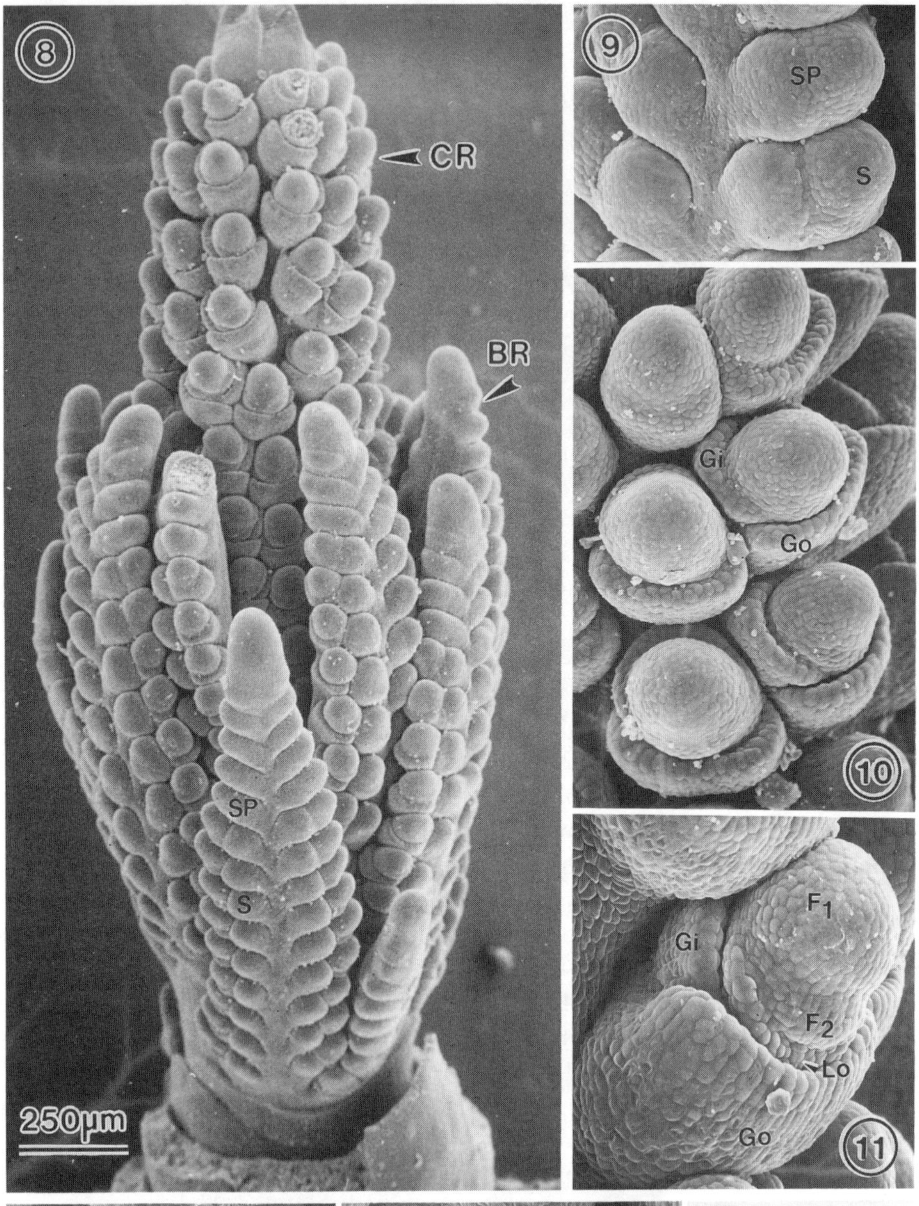

FIGURE 5.8–5.14

then proceeds on subsequent days toward the tip and base. The extrusion of spikelets occurs somewhat later on the lateral branches, following a sequence similar to that of the rachis. As a result of the asynchronous maturation of spikelets within the tassel, and the developmental difference between the upper and lower florets, pollen is shed over a period of a week or more.

The Ear

Ears develop from axillary buds, and are initiated after the tassel. First, the axillary bud meristem initiates a prophyll and many leaflike husks; then it produces rows of spikelet-pair primordia in acropetal sequence along the inflorescence meristem (Figure 5.15). The ear inflorescence is unbranched, but it shares many developmental features with the tassel inflorescence. As in the tassel, spikelets originate in pairs from a single primordium. Spikelet development begins with the initiation of the outer and inner glumes, followed by the initiation of the outer and inner lemmas, and then the lower floret (Figures 5.15 and 5.16). The upper floret produces a palea, but it develops into a much smaller structure than the palea of the male floret. The palea of the lower floret either does not initiate or remains rudimentary (Cheng et al. 1983).

Following the initiation of glumes, the differentiation of the two florets proceeds. The lower floret arises following the initiation of the glumes and lemmas. As in the tassel, the development of this floret lags behind that of the upper floret (Figure 5.17). Also, the sequence of stamen initiation is similar to that in the tassel florets (Cheng et al. 1983).

Following the initiation of stamens, gynoecium development in the upper floret begins with the production of a ridge on the abaxial surface of the apical meristem, distal to the third stamen (Figure 5.18). This ridge (ovary wall) encompasses the meristem as a ring (Figure 5.18) and initiates the style at its tip (Figure 5.19). Continued overgrowth of the shoot apex (which becomes the ovule primordium) by the ring of tissue leads to the formation of the stylar canal (Figure 5.19), which is detected as a slight protruberance on the mature ovary. A similar process occurs in the lower floret, but does not proceed beyond the early ridge stage (Bonnett 1948, 1953; Cheng et al. 1983).

FIGURE 5.8 Immature tassel showing the differentiation of spikelet-pair primordia. Note the double-row placement of spikelet-pair primordia on the lateral branches (BR) and multirows on the central rachis (CR). **FIGURE 5.9** Spikelet-pair (SP) and spikelet (S) initiation on tassel. **FIGURE 5.10** Initiation of outer (Go) and inner (Gi) glumes. **FIGURE 5.11** Initiation of outer lemma (Lo) and lower floret (F). F: upper floret; Go: outer glume; Gi: inner glume. **FIGURE 5.12** Initiation of gynoecial ridge on upper floret and stamen (StL) initiation on lower floret. Ld, lodicule; Lo, outer lemma; Go and Gi, outer and inner glume. **FIGURE 5.13** A tassel spikelet prior to anthesis. Degenerating gynoecium in the center of the floret is evident (arrow-head). Ld, lodicule; F, filament (anther are removed); An, anther. **FIGURE 5.14** Diagrammatic representation of a tassel spikelet. Ax, axis of inflorescence; Go, outer glume; Gi, inner glume; Lo, outer lemma; Gi, inner lemma; P, palea; Ld, lodicule; An, anther; shaded structure, aborted gynoecium.

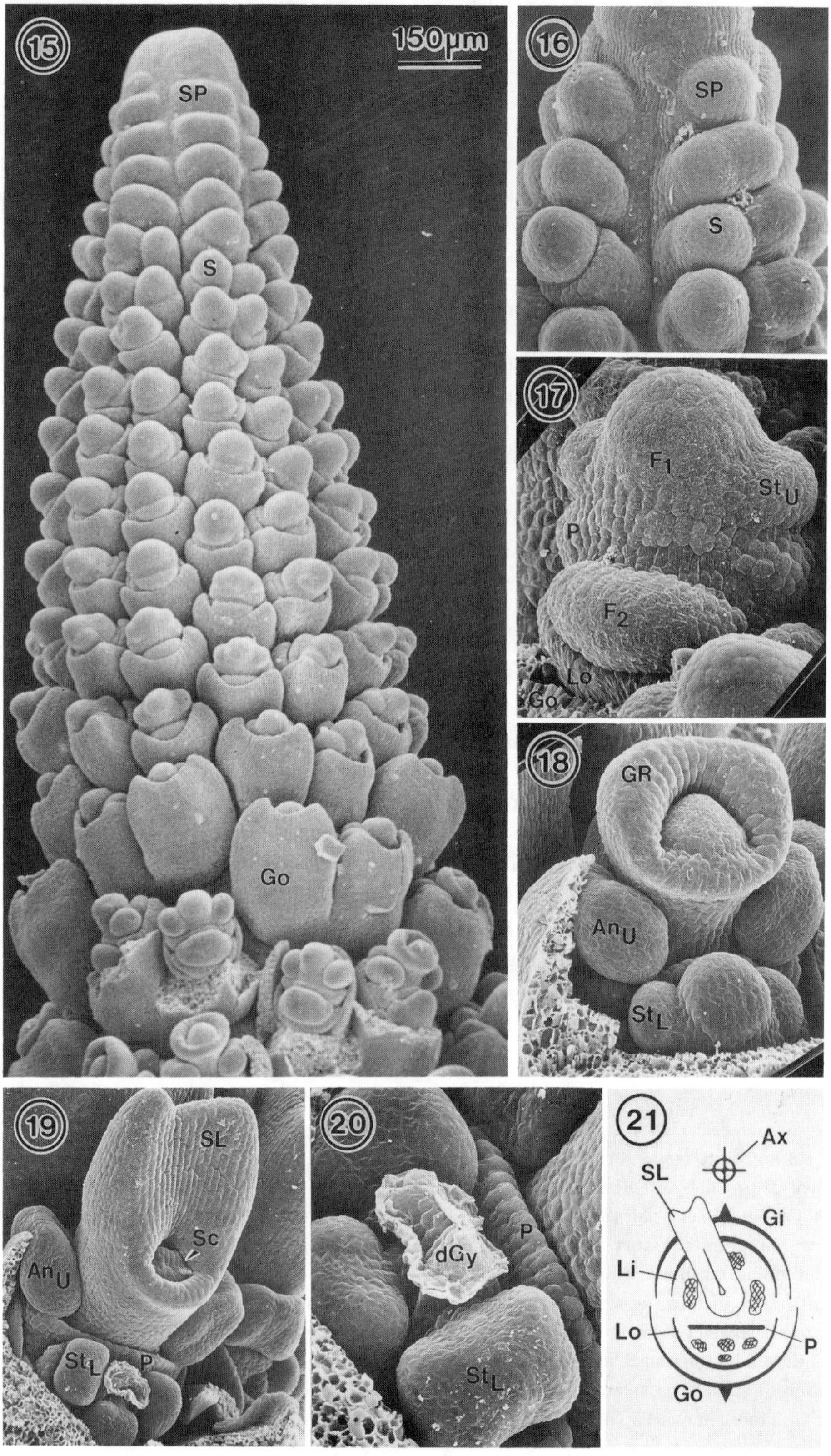

FIGURE 5.15–5.21

During the period when the style of the upper floret is developing, the anthers with distinct lobes begin to abort. At this stage, the lower floret also begins to degenerate, first in the region of the gynoecium (Figure 5.20) and subsequently in the region of the developing stamens. This results in a single functional female floret in each ear spikelet (Figure 5.21).

In the functional florets, a single ovule develops from the meristem enclosed by the gynoecial ridge. The archesporial cell differentiates from a hypodermal cell in the nucellus and enlarges to become the megaspore mother cell. Meiosis in this cell results in a linear tetrad of megaspores. Three megaspores degenerate, while the fourth develops into an embryo sac with eight haploid nuclei, produced by three successive mitotic divisions (Kiesselbach 1949).

During the development of the ovary, the silk grows to a length of about 1 foot and becomes covered with numerous hairs (trichomes) (Bonnett 1948, 1953; Kiesselbach 1949). At anthesis, the silks emerge from the husks at the tip of the ear. Due to the acropetal differentiation of the ear, the spikelets at the base of the ear mature before more apical spikelets. As a result, receptive silks appear over a period of about 4–5 days.

MUTANTS OF THE TASSEL AND EAR

Many genetic mutations alter tassel and ear development in interesting ways (Coe et al. 1988). Some mutations affect the sexual determination in the tassel and ear. For example, the tassel seed (*ts1*, *ts2*, *Ts3*, *ts4*, *Ts5*, *Ts6*) and *terminal ear* (*te*) mutants produce pistillate flowers in the male tassel (Figures 5.22 and 5.23). A reverse condition, i.e., the development of staminate flowers in the ear, is produced in andromonoecious dwarf (*d1*, *d2*, *d3*, *d5*, and *D8*), *anther-ear-1* (*an1*) (Figure 5.24; Karpoff 1983), and *teosinte branched* (*tb*) mutants. Some other mutants transform the floral parts into leaflike structures as in the case of Teopod (*Tp1*, *Tp2*) and *Corngrass* (*Cg*) mutations, while other mutationss

---

FIGURE 5.15 SEM view of the tip of an ear inflorescence. The sequence of development from spikelet-pair primordia to spikelets is evident. Some of the outer glumes of older spikelets were removed. From PC Cheng et al. (1983). **FIGURE 5.16** Initiation of spikelet-pair (SP) and spikelet (S) primordium in ear inflorescence. **FIGURE 5.17** Initiation of upper (F) and lower (F) florets in ear spikelet. Note the initiation of stamens (StU) and palea (P) in the upper floret. Lo, outer lemma; Go, outer glume. **FIGURE 5.18** Early stage of silk development from gynoecial ridge (GR) in the upper floret. In the lower floret, the two lateral stamens (StL) are visible. AnU, anther of the upper floret. **FIGURE 5.19** Onset of abortion of gynoecium degeneration in the lower floret. Note the developing silk (SL); the anthers of the upper floret (AnU). StL, stamen of the lower floret; Sc, stylar canal. **FIGURE 5.20** The lower floret of an ear spikelet showing the degenerating gynoecial ridge (dGy). **FIGURE 5.21** Diagrammatic representation of an ear spikelet. Ax, axis of inflorescence; Go, outer glume; Gi, inner glume; Lo, outer lemma; Li, inner lemma, P, palea; SL, silk; shaded structures, aborted stamens and gynoecium.

Figure 5.22 Tassel of *tassel seed-4* (*ts4*) mutant homozygote. **Figure 5.23** Tassel of *terminal ear* (*te*) mutant homozygote showing male spikelets on the upper portion, few rows of bisexual spikelets and female spikelets on the lower portion. **Figure 5.24** An ear of anther ear (an1) mutant homozygote. Female, bisexual, male, and sterile spikelets can be found on the ear inflorescence in acropetal sequence (photograph courtesy of Dr. A. Karpoff, University of Louisville, KY).

affect the size or proliferation of specific floral parts of the tassel and ear. For example, alterations in branching are conditioned by the *ramosa-1*, *ramosa-2*, and *ramosa-3* (*ra1*, *ra2*, and *ra3*), *Polytypic* (*Pt*), and *branched silkless* (*bd*) mutations. Silk formation is affected by the *silky* (*si*) and *silkless* (*sk*) mutations, and glume development by *Tunicate* (*Tu*), *Vestigial glumes* (*Vg*), and *Papyrescent* (*Pn*) mutations. Several other mutations also affect the tassel and ear development in a variety of ways. (See Coe et al. 1988 for a complete list and description.)

These genetic mutations produce broad variation in the morphology of the tassel and ear. The use of these mutations together with the capability to culture both tassels and ears in vitro (Greyson et al. 1992) should be useful for genetic, physiological, and biochemical studies of sex determination, floral differentiation, and other aspects of tassel and ear development in maize.

REFERENCES

Bonnett OT (1948) Ear and tassel development in maize. Ann Mo Bot Gard 35: 269–287

Bonnett OT (1953) Developmental morphology of the vegetative and floral shoots of maize. Univ Ill Agric Exp Stn Bull No. 568

Bonnett OT (1966) Inflorescences of maize, wheat, rye, barley and oats: their initiation and development. Univ Ill Agric Exp Stn Bull No. 721

Cheng PC, Greyson RI, Walden DB (1979) Comparison of anther development in genic male-sterile (*ms10*) and in male-fertile corn (*Zea mays*) from light microscopy and scanning electron microscopy. Can J Bot 57: 578–596

Cheng PC, Greyson RI, Walden DB (1983) Organ initiation and the development of unisexual flowers in the tassel and ear of *Zea mays*. Am J Bot 70: 450–462

Coe EH Jr, Neuffer MG, Hoisington DA (1988) The genetics of corn. In Sprague GF, Dudley JW (eds) Corn and Corn Improvement, Am Soc Agronomy, Madison, WI

Greyson RI, Bommineni VR, Pareddy DR, Polowick PL (1992) In vitro culture of maize inflorescences. In Bajaj YPS (ed) Biotechnology in Agriculture and Forestry: Maize, Springer-Verlag, New York, in press

Hanway JJ, Ritchie SW (1987) *Zea mays*. In Halvey H (ed) Handbook of Flowering, Vol IV, CRC Press Inc, Boca Raton, FL

Hsu SY, Huang Y-C, Peterson PA (1988) Development pattern of microspores in *Zea mays* L. The maturation of upper and lower florets of spikelets among an assortment of genotypes. Maydica 33: 77–98

Karpoff A (1983) A closer look at the gibberellin effect and sex expression in the reversible dwarf, Anther ear (an-1). Maize Genetics Cooperation News Letter 57: 78–80

Kiesselbach TA (1949) The structure and reproduction of corn. Univ Nebraska Agric Exp Stn Res Bull No. 161

Stevens SJ, Stevens EJ, Lee KW, Flowerday AD, Gardner CO (1986) Organogenesis of the staminate and pistillate inflorescences of pop and dent corns: relationship to leaf stages. Crop Sci 26: 712–718

Weatherwax P (1916) Morphology of the flowers of *Zea mays*. Torrey Bot Club Bull 43: 127–144

6

Gametogenesis in Maize

Patricia Bedinger and Scott D. Russell

MICROSPOROGENESIS

The process of microsporogenesis, or pollen development, encompasses all stages from microsporocyte meiosis to maturation of the male gametophyte. Meiosis is described in some detail in another chapter of this volume, and therefore the focus of this chapter will be on postmeiotic events. For previous reviews of microsporogenesis in maize, readers are referred to Cheng et al. (1979), Albertsen and Phillips (1981), and Chang and Neuffer (1989). The numerous sporophytic male-sterility mutations (Beadle 1932b; Albertsen and Phillips 1981) provide genetic evidence supporting the notion that both gametophytic and sporophytic cells play crucial roles during microsporogenesis. Therefore, notable developmental events in both gametophytic and other anther cells will be discussed in this section.

The anatomy of the male flower of maize is well described in Kiesselbach (1949). To briefly summarize, the male flowers are contained in spikelets on the tassel of the plant (Figure 6.1, and see previous section). Each spikelet contains two florets that differ in developmental age; the upper floret is the more mature of the two. Each flower has three four-lobed anthers. The anther wall consists of four cell layers. The outer anther wall layer is the epidermis, the second layer is the endothecium, and the third layer is the transient middle layer. The innermost anther wall layer is the tapetum, which plays a very active role in microsporogenesis prior to undergoing a programmed cell death. The anther locule contains the microsporocytes, or pollen mother cells. After meiosis, the haploid microspores develop into pollen within this fluid-filled locule (Figure 6.2).

Within the anther locule, the microsporocytes are encased in an impermeable β-1,3-glucan (callose) wall (Knox and Heslop-Harrrison 1970). Meiosis produces a tetrad of four haploid cells called microspores, still encased within the callose wall. During meiosis, the surrounding tapetal cells become binucleate, lose their primary cellulosic cell wall, and fill with endoplasmic reticulum and secretory vesicles, which are transported in a polar fashion to the locular surface of the cells. One of the first products secreted by

The Maize Handbook—M. Freeling, V. Walbot, eds.

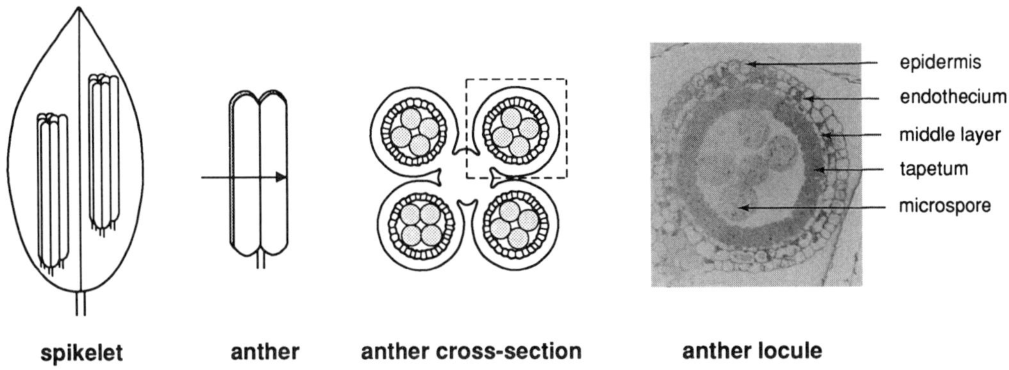

FIGURE 6.1 Schematic depiction of a maize spikelet, anther, anther cross section, and single anther locule with all cell layers identified. The stage shown is postmeiotic, with free young microspores in the anther locule.

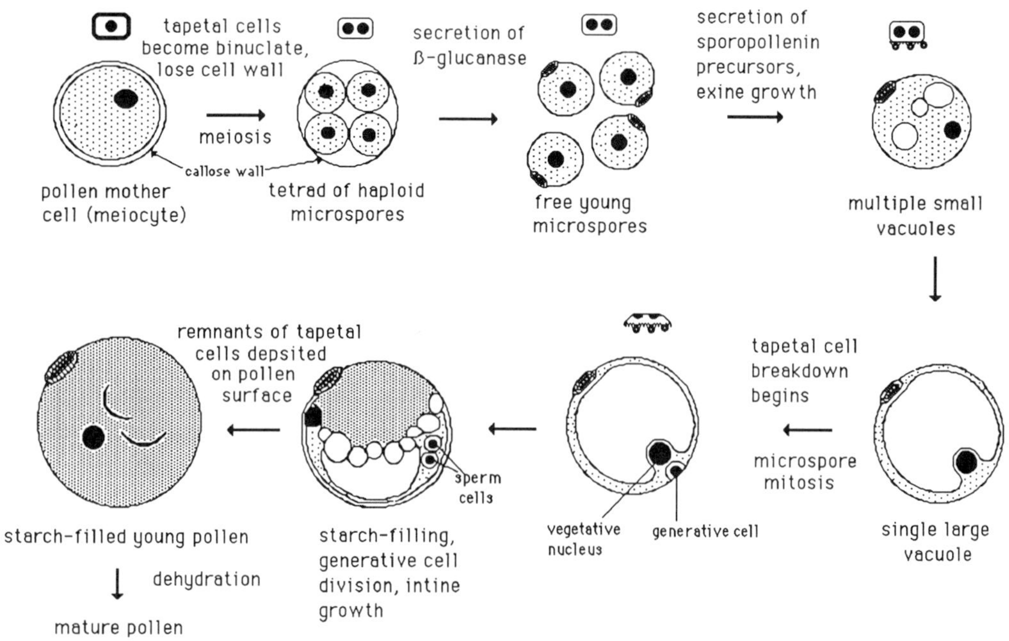

FIGURE 6.2 Diagram indicating notable steps in pollen development, including morphological changes in microspores and tapetal cells.

the differentiated tapetal cells is a β-glucanase, which degrades the callose wall and releases the individual young haploid microspores into the anther locule (Steiglitz 1977). The young microspores undergo a period of rapid growth after release from the tetrad.

The distinctive extracellular matrix that forms the outer pollen wall, or exine, is synthesized at this time (Skvarla and Larson 1966). The major structural material of the exine is sporopollenin, a highly resistant biopolymer. Even though sporopollenin has been studied for nearly 200 years (John 1814), its molecular structure is not currently understood (Prahl et al. 1985). It is thought that sporopollenin precursors are secreted from the tapetal cells and polymerized onto a template system, known as the primexine or glycocalyx, on the microspore surface. As the exine is forming on the microspore surface, a sporopollenin-based "orbicular wall" and associated orbicules (also called Ubisch bodies) form on the locular face of the tapetal cells. The pollen pore, the future site of pollen tube extrusion, is a polar feature visible very early in exine biosynthesis. A second polar event is the migration of the centrally located nucleus to a position opposite the pollen pore.

The young microspore fills with multiple small vacuoles that coalesce into a single large vacuole, compressing the cytoplasm to a small region opposite the pollen pore. Yellow flavonoids accumulate in and pigment the pollen wall. At this time, the asymmetric nuclear division called microspore mitosis occurs, producing two nuclei with very different fates. The nucleus nearest to the plasma membrane becomes cellularized to form the generative cell. The generative cell slowly separates from the pollen wall and becomes enclosed within a double membrane—one lipid bilayer derived from its own membrane and the other from the plasma membrane of the vegetative cell. In this way, the generative cell becomes entirely contained within the plasma membrane of the vegetative cell. The chromatin of the generative cell is highly condensed. The inner, vegetative nucleus has a higher nuclear volume, a larger number of nuclear pores, and decondensed chromatin, indicating that it is the transcriptionally active nucleus (LaFountain and LaFountain 1973; Wagner et al. 1990a).

Microspore mitosis appears to be a critical point in the pollen developmental pathway. Anther and microspore culture experiments suggest that late uninucleate microspores can be more easily induced than young binucleate pollen to enter a sporophytic developmental program (Gaillard et al. 1991). Studies of RNA and protein populations (Bedinger and Edgerton 1990) and of protein synthesis in isolated microspores (Mandaron et al. 1990) indicate that a major developmental switch in gene expression occurs at this time. Studies of pollen-specific gene expression are also consistent with new gene expression after microspore mitosis (Stinson et al. 1987). At this time, the tapetal cells also begin to undergo a programmed cell death.

The bicellular products of microspore mitosis are called pollen. The generative cell within the young pollen divides to form two sperm cells, which elongate and become sickle-shaped. In the case of maize, the two sperm cells are not physically interconnected with each other or with the vegetative nucleus. Thus mature maize pollen does not possess the "male germ unit" found in many plant species (McConchie et al. 1987). However, it is quite possible that such connections are made upon germination of the pollen, as has been detected in other grasses (Mogensen and Wagner 1987). The pollen grain accumulates starch granules starting from the pore region, until the grain is entirely filled. The pollen prepares for release into the environment with the synthesis of the inner pollen wall, or intine, which is cellulosic and pectic in nature. The pollen then dehydrates, such that pollen grains contain 40–58% H_2O at the time of dehiscence (Barnabas 1985). Tapetal cell remnants, called tryphine (proteinaceous) and pollenkitt (lipoidal), are deposited on the maturing pollen surface. The walls of the endothecial cells undergo thickening, particularly near the anther tips. Cuticular ridges form on the epidermal cells. Elongation of the anther filament occurs to accomplish dehiscence, and the anther tip opens to release mature pollen. Once released into the environment, further dehydration of the pollen grain causes domains of the vegetative cell plasma membrane to enter a gel/liquid crystal state (Kerhoas et al. 1987).

Once the pollen interacts with the silk and begins to hydrate, a pollen tube is extruded through the pollen pore. The pollen tube grows rapidly through the transmitting tissue to the ovule. The remarkable growth rate of the pollen tube (on the order of 0.5 cm/hour) is rivaled by few other biological systems. The transport of the pollen cytoplasm, vegetative nucleus, and sperm cells through the pollen tube represents a rare example of cell migration in plant biology.

Gametophytic factors affecting pollen development are thought to be numerous, because even small deficiencies are only rarely transmitted through the pollen (Carlson 1977) and chemical mutagenesis of seeds leads to a high level of pollen abortion in the treated plants (Birchler and Schwartz 1979). Gene expression in sporophytic tissues, as well as in the developing gametophyte, is required for pollen development, as evidenced by the many sporophytic mutations that cause male sterility (Beadle 1932; Albertsen and Phillips 1981). The point of pollen abortion of many of these sporophytic male-sterile mutations is summarized in Table 72.1 of chapter 72 of this volume. The large numbers of both sporophytic and gametophytic genes required for microsporogenesis are perhaps not so surprising, given that the mature pollen grain is a highly specialized three-celled organism that once released from the anther must survive outside of the sporophytic plant, interact with the female silk, and germinate a rapidly growing pollen tube to accomplish fertilization.

MEGASPOROGENESIS AND MEGAGAMETOGENESIS

The development of the female gametophyte, or embryo sac, of maize is divided into two connected developmental phases: *megasporogenesis*, which begins with the formation and maturation of the initial products of meiosis (termed megaspores), followed by *megagametogenesis*, which begins with the mitotic division of the meiotic products and continues through the cellularization and maturation of the embryo sac. As described in the previous section, the "ear" is a spike composed of spikelets, the apical of a pair of florets maturing to form the kernel (after fertilization) and the lower typically aborting before meiosis. The carpel and contained ovule are formed from periclinal and anticlinal divisions occurring in superficial and subsuperficial layers of the shoot apex. The ovule develops as a mass of cells that initiates annular integument primordia (inner and outer) on the outside of the chalazal end of the ovule. The integuments ultimately enclose the ovule in two layers of tissue, interrupted at the apex by the micropyle (formed by the inner integument), which provides the pathway for pollen tube entry.

Megasporogenesis

Within the ovule, one centrally located cell differentiates from the others to form the archesporial cell, which is characterized by its larger size, prominent nucleus, and large nucleoli. Ultrastructurally, this is an unpolarized cell with a full complement of organelles. In maize, the archespore directly becomes the megaspore mother cell (or megasporocyte) as this cell becomes committed to meiosis (Figure 6.3). Initially, this cell is isodiametric, but becomes teardrop-shaped and later elongates as the cell approaches meiosis. In maize, prophase I occurs over a protracted period. (For techniques and figures, see Russell and West, this volume.) Cytokinesis accompanies meiosis, forming first

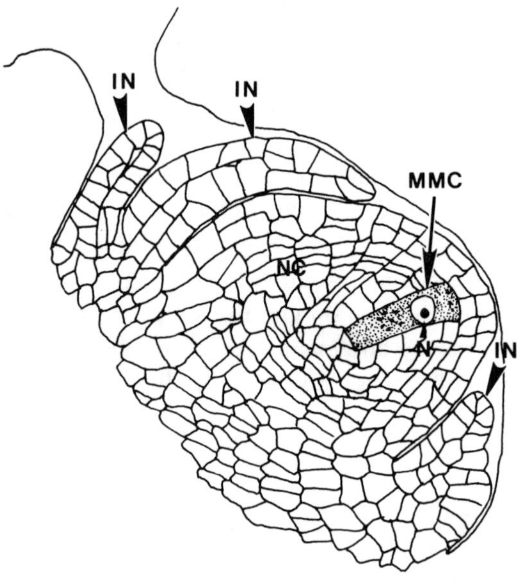

FIGURE 6.3

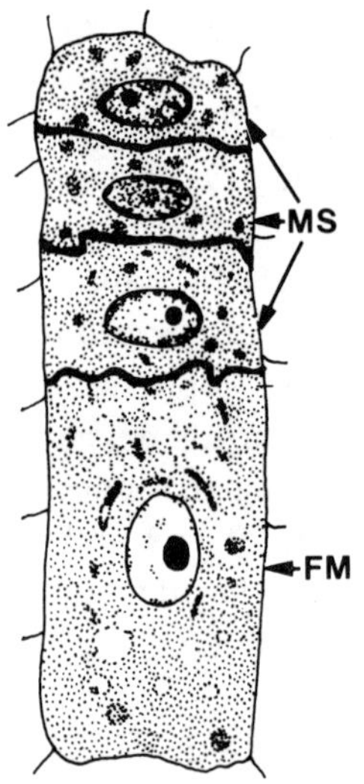

FIGURE 6.4

FIGURE 6.3 Megaspore mother cell (MMC) or megasporocyte, surrounded by nucellar cells (NC) and forming integuments (IN). The nucleus (N) is in prophase I. FIGURE 6.4 Tetrad of megaspores (MS). The production of four megaspores is variable; the division of the micropylar dyad cell (toward the top) may result in triads, incomplete divisions, T-shaped tetrads, and linear tetrads, as shown here. The chalazal megaspore is the functional megaspore (FM) that forms the embryo sac.

two dyad cells and then the megaspores, which classically form a linear tetrad (Figure 6.4).[1]

Variability in form of the tetrad occurs because (1) the micropylar dyad cell may arrest without completing meiosis, forming a "triad"; (2) wall formation in the micropylar tetrad cells may be incomplete and/or variable, forming transverse, oblique, or longitudinal cell walls (Weatherwax 1919; Russell 1979); and (3) the apportionment of cytoplasm may be nearly equal (Kiesselbach 1949), or noticeably unequal (Weatherwax 1919; Cooper 1937; Russell 1979), with the largest volume inherited by the functional megaspore. Variability in tetrad formation occurs in different strains and in some strains, even within the same ear.

The factors determining megaspore selection are not yet well understood; however, the functional megaspore may be determined by a combination of (1) a biophysical "null stress" point, which may initiate developmental divergence (Russell 1979); (2) the formation of a callose wall in the megasporocyte at future sites of megaspore abortion (Rodkiewicz 1970; see Russell and West, this volume, for figures); (3) the severance of plasmodesmatal connections except at the functional pole of the megasporocyte (Russell 1979); (4) cytoplasmic polarization (review: Willemse 1981); and (5) other positional factors not yet appreciated. In theory, kin conflict in monosporic species is resolved by the ovule and is exerted externally (Haig, 1986). The chalazal-most cell, termed the *functional megaspore*, forms the embryo sac.

The onset of megaspore abortion is indicated by dense areas of hydrolytic activity, evident in ultrastructure (Russell 1979) and cytochemistry of peroxidases, esterases (Willemse and Bednara 1979), and acid phosphatases in other monosporic angiosperms (Schulz and Jensen 1981, 1986). The degenerating cells are initially similar to the functional megaspore except for a slight impoverishment of organelles, but eventually vital intracellular membrane systems are broken down and the breakdown of materials becomes general. Hydrolytic enzymes, when present in the functional megaspore, are compartmentalized in lytic vacuoles. Presumably, the degraded products of the abortive megaspores are ultimately resorbed, as the embryo sac later expands to occupy this site.

The condition of the heritable cytoplasmic organelles (plastids and mitochondria) appears to fluctuate during megasporogenesis, as does the concentration of ribosomes. At the earliest stages, the organelles appear similar to those of meristematic cells; mitochondria have distinct, well-developed cristae and proplastids have few lamellae. During meiosis, mitochondria and proplastids undergo dramatic structural changes and dedifferentiate, reflecting to some degree the dramatic changes occurring in the nucleus at this stage. Although early cytologists questioned whether heritable organelles persisted throughout meiosis, subsequent ultrastructural and biochemical studies have confirmed

[1]This pattern of development, known as the monosporic Polygonum pattern, has been described in maize and other grasses studied to date. Miller (1919) has described tetrasporic development in corn, but to date, this has only been confirmed in specific trisomic sectors of certain lines of maize plants.

that DNA-containing organelles remain present. The significance of buds generated from the nuclear envelope at this stage, which later appear to enter the cytoplasm, is unknown. The concentration of ribosomes reaches the lowest levels during prophase I.

Autophagic vacuoles are frequently found in the cytoplasm generated from endoplasmic reticulum and preexisting vacuoles. These occasionally contain membranes or organelle-like objects (Russell 1979). The presence of such nonspecific hydrolases as acid phosphatase and esterase (Willemse and Bednara 1979; Schulz and Jensen 1981) suggests that selected cellular materials are degraded into their precursor components. Autolytic activity is also present in the surrounding nucellus and presumably hastens the degeneration of these cells as the female cells expand and encroach upon these cells.

The functional megaspore is characterized by the redifferentiation of mitochondria and plastids and an ultrastructural appearance of increased synthetic activity. This includes an increased accumulation of cytoplasmic ribosomes, the multiplication of organelles, and cellular enlargement.

Megagametogenesis

Megagametogenesis begins with the maturation of the megaspore and is completed when the embryo sac reaches maturity. In corn, the nucleus undergoes three cycles of free nuclear division (without intervening cytokinesis), eventually resulting in eight free nuclei (Figures 6.5, 6.6); after the eight nuclei are produced, cell wall formation occurs simultaneously between the nuclei (Figure 6.7), presumably from the interzonal microtubules present at the completion of telophase (review: Huang and Russell 1992).

The embryo sac expands dramatically during this phase, accompanied by the degeneration of surrounding nucellar cells. The organization of the cytoplasm changes dramatically during megagametogenesis: (1) the prominent central vacuole is formed beginning at the two-nucleate stage; (2) synthetic activity is evidenced by significant increases in cytoplasmic volume and the concentration of organelles, including ribosomes, mitochondria, plastids, endoplasmic reticulum, and Golgi bodies; and (3) one final cycle of organellar dedifferentiation occurs (Russell 1979). Central vacuole formation typically involves the production of small vacuoles, de novo or from endoplasmic reticulum, that coalesce to produce the prominent vacuole characteristic of the central cell at maturity. In corn, the dedifferentiation and redifferentiation of mitochondria during the four-nucleate megagametophyte stage coincides with the completion of intense lytic activity in the central vacuole.

Once free-nuclear division is completed, cell walls simultaneously form around the nuclei (Figure 6.7). Three nuclei toward the micropyle form the egg apparatus, which is composed of the egg cell and two synergid cells; three nuclei toward the chalaza form three antipodal cells; and two remaining nuclei, termed *polar nuclei*, migrate toward the egg cell and partially fuse within the so-called *central cell*. The cell plate arising between the egg and the central cell forms a transverse *embryo sac top wall*, with the cells subdivided by further longitudinal plate formation. The egg apparatus cells subsequently expand, causing the cell walls to become locally discontinuous, appearing beaded at the

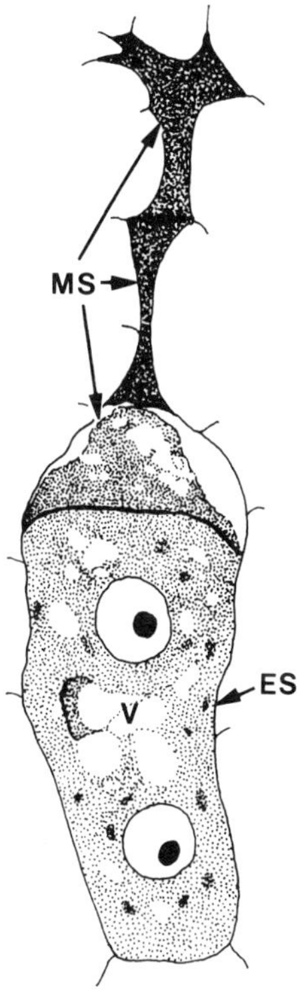

FIGURE 6.5

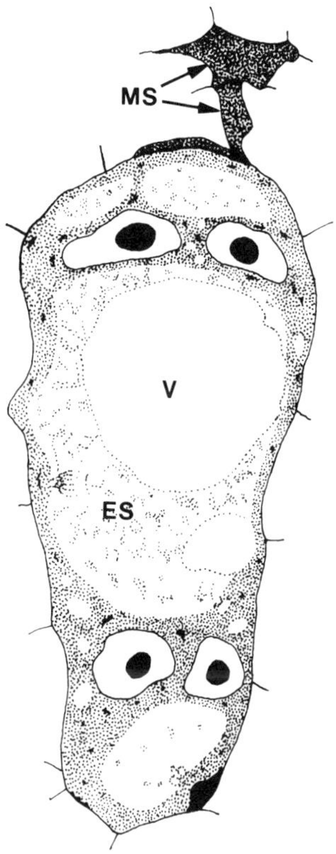

FIGURE 6.6

FIGURE 6.5 Two-nucleate embryo sac (ES). At this stage, the nuclei separate and a large vacuole (V) begins to form between them. MS, degenerating megaspores. FIGURE 6.6 Four-nucleate embryo sac (ES). The central vacuole (V) continues to expand, displacing the nuclei to the periphery. Degenerate megaspores (MS) and nucellar cells are crushed by the continued expansion of the embryo sac.

chalazal end of the cells; this is a structural modification that later facilitates the deposition of sperm and occurrence of gametic fusion, as in other angiosperms. The development of the so-called "hooks" of the synergids is the result of the chalazal expansion of the synergids relative to their initial attachment point.

Mature Embryo Sac

At maturity (Figure 6.8), the embryo sac achieves and the individual cells develop their mature form. The synergid is modified to attract and accept the pollen tube (Diboll and Larson 1966). In other grasses, the cytoplasm has been reported to contain vacuoles with large concentrations of calcium: this ion is known to attract pollen tubes (Chaubal and Reger 1992). There is a modified secretory cell wall at its micropylar end known as the filiform apparatus, which receives the pollen tube. Based on the frequency of mitochondria and density of the cytoplasm, the synergid is believed to possess a high level of physiological activity. The egg cell appears to be relatively quiescent prior to fertilization, possessing a large chalazal nucleus with a prominent nucleolus and a vacuolated micropylar cytoplasm. The central cell is largely vacuolated, with most of the cytoplasm present near the polar nuclei and toward the periphery of the cell. The polar nuclei partially

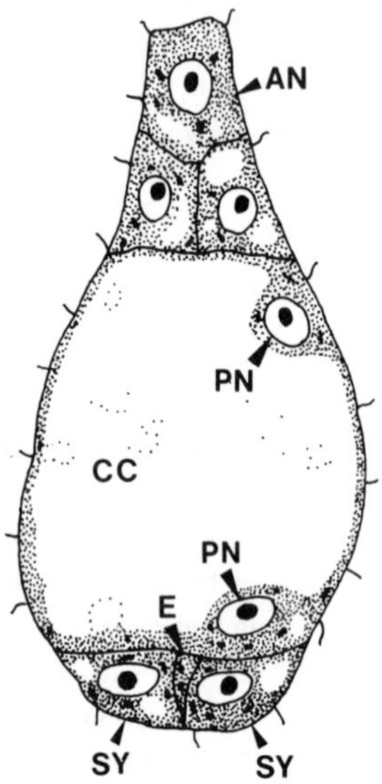

FIGURE 6.7

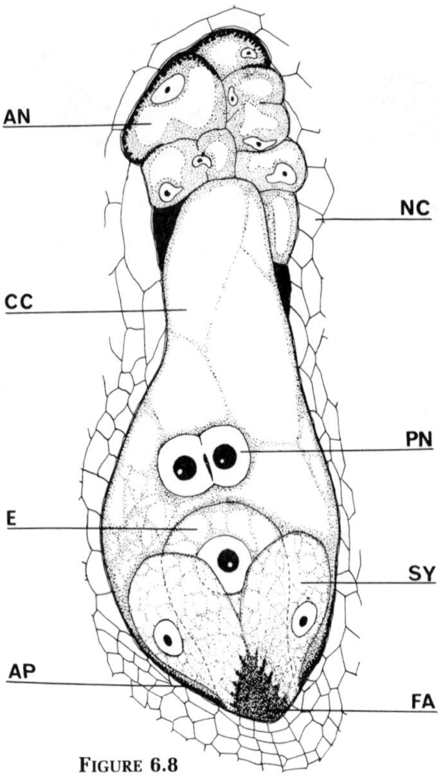

FIGURE 6.8

FIGURE 6.7 Immature eight-nucleate embryo sac. Cellularization coincides with the completion of the third mitotic cycle. This results in seven cells, including an egg cell (E), two synergids (SY), three antipodal cells (AN), and a binucleate central cell (CC), containing the two polar nuclei (PN). Micropyle is down. **FIGURE 6.8** Mature embryo sac. The synergids contain prominent cell wall ingrowths, called the filiform apparatus (FA), at the micropylar end of the embryo sac. The antipodals multiply to nearly 100 prior to maturity and because of incomplete division may be polyploid and coenocytic. Cell wall ingrowths also occur at the periphery of the embryo sac in the antipodals and in the apical pocket (AP), formed by the so-called synergid "hooks" (Figures 6.3–7, original; Figure 6.8, after Diboll and Larson 1966).

fuse prior to fertilization, but await the arrival of the sperm nucleus before completing fusion. The antipodals are densely cytoplasmic, possessing cell wall ingrowths next to the chalazal nucellus. Their nuclei are frequently polyploid and the cells often have incomplete closure of some of the cell walls and are thus, to some degree, coenocytic (Diboll and Larson 1966). In maize, the three original antipodal cells multiply to accumulate a mature complement of as many as 100 cells. Several different approaches are currently being developed to introduce genetic material into the corn embryo sac. The fate of these and the other cells are the subject of reviews on fertilization and embryogenesis elsewhere in this volume.

Table 6.1 Maize meiotic/gametophytic mutants[a]

Mutant	Male[b]	Female[b]	Behavior	Name and reference
am1	sterile	sterile	Mother cells enter mitosis, degenerate	*ameiotic* (Rhoades 1956; Palmer 1971)
am2	sterile	sterile	Similar to *am1*, but not allelic	*ameiotic* (Doyle, personal communication)
as1	variably sterile	variably sterile	Variable synaptic failure, univalents result	*asynaptic* (Beadle 1930)
dsy1	usually sterile	usually sterile	Bivalents separate before diakinesis, numerous univalents are produced	*desynaptic* (Golubovskaya and Mashnenkov 1976)
dys2	usually sterile	usually sterile	Bivalents separate before diakinesis, numerous univalents are produced	*desynaptic* (Golubovskaya and Mashnenkov 1976)
dv1	usually sterile	little sterility	Spindles diverge instead of converge, resulting in multiple nuclei, spores; possibly caused by UV	*divergent spindle* (Clark 1940)
dy1	low sterility	no effect	Bivalents separate before diakinesis	*desynaptic* (Nelson and Clary 1952)
el1	abnormal	abnormal	Uncondensed chromosomes, frequently form unreduced gametes, with variable results based on ploidy of endosperm	*elongate* (Rhoades 1956)

(Continued)

Table 6.1 (Continued)

Mutant	Male[b]	Female[b]	Behavior	Name and reference
ig1	some male sterility	variable	Number and type of cells in embryo sac are variable; twinning, heterofertilization, haploids common; can produce male nuclear, female cytoplasmic seeds and other products	*indeterminate gametophyte* (Kermicle 1969)
lo2	no effect	sterile	Ovules with *lo2* embryo sacs abort	*lethal ovule* (Nelson and Clary 1952)
lp1	poor competitor	no effect	Mutant pollen fails in competition with *Lp1* pollen	*lethal pollen* (Carangal M.S. thesis, Purdue Univ. 1958)
po1	sterile	partial sterility	Reduction division occurs repeatedly, losing chromosome material to microcytes and megacytes	*polymitotic* (Beadle 1932b)
st1	sterile	partial sterility	Chromosomes sticky in metaphase I; *st1-e* mutant has heightened response to high temperatures	*sticky chromosome* (Beadle 1932a)

[a]Authors are indebted to Dr. M. G. Neuffer for providing the information synopsized in this table.

[b]Manifestation of mutant in male (pollen) and female (ovule, embryo sac). Blank entry = characteristic not given.

Two new developments are now being exploited for manipulation of the maize embryo sac. Kranz et al. (1991) have developed methods to isolate, combine, and culture specific gametes to form artificial zygotes. Wagner et al. (1990b) have described a structural method to determine the exact position of the embryo sac within the mature floret for purposes of microinjecting genetic material and then culturing segments of the ear in vitro.

A summary of selected mutants affecting meiosis and sporogenesis is included in tabular form (Table 6.1). Three are particularly noteworthy. The lethal ovule mutation (*lo2*) is expressed only in ovules in which the functional megaspore contains the *lo2* gene; this has no effect on male reproduction. The polymitotic gene (*po1*) is a late meiotic mutant in which reduction division appears to continue to occur in meiosis II and beyond, resulting in fractional chromosome complements in the resulting in cells termed *microcytes* and *megacytes* (D.P. West, personal communication). This results in complete male sterility but only partial female sterility, making it useful for breeding. The indeterminate gametophyte mutant (*ig1*) mutant produces variable numbers and types of embryo sac cells, with highly variable results. In some cases, supernumerary eggs form haploid embryos, but interestingly in a few (1–2%) of some strains, haploid embryos can be produced that have male nuclei and female cytoplasm. These progeny, when pollinated by other progeny, may infrequently result in autodiploids, making this also a useful mutant.

ACKNOWLEDGMENTS

We thank M. G. Neuffer and David P. West for helpful discussions about female-expressed mutants of maize and for providing unpublished data for this review. We also thank Vince Wagner for helpful discussions.

REFERENCES

Albertsen MC, Phillips RL (1981) Developmental cytology of 13 genetic male sterile loci in maize. Can J Genet Cytol 23: 195–208

Barnabas B (1985) Effect of water loss on germination ability of maize (*Zea mays*) pollen. Ann Bot 48: 861–864

Beadle GW (1930) Genetical and cytological studies of Mendelian asynapsis in *Zea mays*. Cornell Univ Agr Exp Station Mem 129: 1–23

Beadle GW (1932a) A gene for sticky chromosomes in *Zea mays*. Ztsch ind Abstr u Vererb 63: 195–217

Beadle GW (1932b) Genes in maize for pollen sterility. Genetics 17: 413–431

Birchler JA, Schwartz D (1979) Mutational study of the alcohol dehydrogenase-1 FCm duplication in maize. Biochem Genet 17: 1173–1180

Bedinger P, Edgerton MD (1990) Developmental staging of maize microspores reveals a transition in developing microspore proteins. Plant Physiol 92: 474–479

Carlson WR (1977) The cytogenetics of corn. In Sprague GF (eds) Corn and Corn Improvement, American Society of Agronomy, Madison, WI, pp 225–304

Chang MT, Neuffer MG (1989) Maize microsporogenesis. Genome 32: 232–244

Chaubal R, Reger BJ (1992) Calcium in the synergid cells and other regions of pearl millet ovaries. Sex Plant Reprod 5: 34–46

Cheng PC, Greyson RI, Walden DB (1979) Comparison of anther development in genic male-sterile (*ms10*) and in male-fertile corn (*Zea mays*) from light microscopy and scanning electron microscopy. Can J Bot 57: 578–596

Clark FJ (1940) Cytogenetic studies of divergent meiotic spindle formation in *Zea mays*. Amer J Bot 27: 547–559

Cooper DC (1937) Macrosporogenesis and embryo-sac development in *Euchlaena mexicana* and *Zea mays*. J Agric Res 55: 539–551

Diboll A G, Larson DA (1966) An electron microscopic study of the mature megagametophyte in *Zea mays*. Am J Bot 53: 391–402

Gaillard A, Vergne P, Beckert M (1991) Optimization of maize microspore isolation and culture conditions for reliable plant regeneration. Plant Cell Reports 10: 55–58

Golubovskaya In, Mashnenkov As (1976) Genetic control of meiosis. 2. desynaptic mutant in maize induced by N-nitroso-N-methyl urea. Genetika 12 7–14

Haig D (1986) Conflicts among megaspores. J Theor Biol 123: 471–480

Huang B-Q, Russell SD (1992) Female germ unit: organization, isolation and function. Int Rev Cytol 140: 233–293

John JJ (1814) Uber befruchtenstrasse nebst eine analyse des tulipen pollens. J Chemie Physik 12: 244–261

Kerhoas C, Gay G, Dumas C (1987) A multidisciplinary approach to the study of the plasma membrane of *Zea mays* pollen during controlled dehydration. Planta 171: 1–10

Kermicle JL (1969) Androgenesis conditioned by a mutation in maize. Science 166: 1422–1424

Kiesselbach TA (1949) The structure and reproduction of corn. Univ Nebraska Coll Agric, Agric Exp Station Res Bull 161: 1–96

Knox RB, Heslop-Harrison J (1970) Direct demonstration of the low permeability of the angiosperm meiotic tetrad using a fluorogenic ester. Z Pflanzenphysiol Bd 62: 451–459

Kranz E, Bautor J, Lörz H (1991) In vitro fertilization of single, isolated gametes of maize mediated by electrofusion. Sex Plant Reprod 4: 12–16

LaFountain JR, LaFountain KL (1973) Comparison of density of nuclear pores on vegetative and generative nuclei in pollen of *Tradescantia*. Exp Cell Res 78: 472–476

Mandaron PM, Niogret F, Mache R, Moneger F (1990) In vitro protein synthesis in isolated microspores of *Zea mays* at several stages of development. Theor Appl Genet 80: 134–138

McConchie CA, HoughT, Knox RB (1987) Ultrastructural analysis of the sperm cells of mature pollen of maize, *Zea mays*. Protoplasma 139: 9–19

Miller EC (1919) Development of the pistillate spikelet and fertilization in *Zea mays*. J Agric Res 18: 255–267

Mogensen HL, Wagner VT (1987) Associations among components of the male germ unit following in vivo pollination in barley. Protoplasma 138: 161–172

Nelson OE, Clary GB (1952) Genic control of semisterility in maize. J Heredity 43: 205–210

Palmer RG (1971) Cytological studies of ameiotic and normal maize with reference to premeiotic pairing. Chromosoma 35: 233–246

Prahl AK, Springstubbe H, Grumbach K, Wiermann R (1985) Studies on sporopollenin biosynthesis: the effect of inhibitors of carotenoid biosynthesis on sporopollenin accumulation. Z Naturforsch 40: 621–626

Rhoades MM (1956) Genic control of chromosomal behavior. Maize Genet Coop News Lett 30: 38–42

Rodkiewicz, B. (1970) Callose in cell walls during megasporogenesis in angiosperms. Planta 93: 39–47

Russell SD (1979) Fine structure of megagametophyte development in *Zea mays*. Can J Bot 57: 1093–1110

Schulz P, Jensen WA (1981) Pre-fertilization ovule development in *Capsella*: ultrastructure and ultracytochemical localization of acid phosphatase in the meiocyte. Protoplasma 107: 27–45

Schulz P, Jensen WA (1986) Prefertilization ovule development in *Capsella*: the dyad, tetrad, developing megaspore and two-nucleate megagametophyte. Can J Bot 64: 875–884

Skvarla JJ, Larson DA (1966) Fine structural studies of *Zea mays* pollen I: cell membranes and exine ontogeny. Am J Bot 53: 1112–1125

Steiglitz H (1977) Role of β-1,3-glucanase in postmeiotic microspore release. Dev Biol 57: 87–97

Stinson JR, Eisenbery AR, Willing RP, Pe ME, Hanson DD, Mascarenhas JP (1987) Genes expressed in the male gametophyte of flowering plants and their isolation. Plant Physiol 83: 442–447

Wagner VT, Cresti M, Salvatici P, Tiezzi A (1990a) Changes in volume, surface area, and frequency of nuclear pores on the vegetative nucleus of tobacco pollen in fresh, hydrated and activated conditions. Planta 181: 304–309

Wagner, VT, Dumas C, Mogensen HL (1990b) Quantitative three-dimensional study on the position of the female gametophyte and its constituent cells as a prerequisite for corn (*Zea mays*) transformation. Theor Appl Genet 79: 72–76

Weatherwax P (1919) Gametogenesis and fecundation in *Zea mays* as the basis of xenia and heredity in the endosperm. Bull Torrey Bot Club 46: 73–90

Willemse MTM (1981) Polarity during megasporogenesis and megagametogenesis. Phytomorphology 31: 124–134

Willemse MTM, Bednara J (1979) Polarity during megasporogenesis in *Gasteria verrucosa*. Phytomorphology 29: 156–165

7

The Patterns of Plant Structures in Maize

WALTON C. GALINAT

PHASE CHANGE

The most spectacular transformations of morphogenesis in maize occur at phase change when the vegetative phenotypes suddenly switch into floral ones. Less dramatic changes develop as secondary sex traits when the male or tassel and female or ear phenotypes diverge from a common primordial type of inflorescence. The differences between the juvenile and adult vegetative stages are even less striking in appearance but they also

The Maize Handbook—M. Freeling, V. Walbot, eds.

Table 7.1. Manifestation of the phytomer in different regions of the maize plant.

six different phytometric cycles		*i*	*l*	*p*	*ab*
	A	internode	leaf	prophyll	axillary bud(ear)
	B	rachis segment	glume cushion	cupule-lining	spikelet pair
	C	rachilla segment	lemma	palea	pistil +
	D	rachis segment	leaf rudiment	pulvinus	tassel-branch
	E	rachid* segment	glume cushion	rachid* scab	spikelet pair
	F	rachilla segment	lemma	palea	3 anthers +

*axis of a tassel branch

depend on some system of molecular switching. Recognition of the vegetative-floral homologies is essential to understanding how a framework of differential genetic activities regulates and synchronizes repetitious and reversible patterns of growth that result in structures of incredible beauty that delicately adapt to carry out the functions of intercepting solar radiation in the vegetative phase and the functions of sexual reproduction in the floral phase—all within a certain external environment. Once the structures of the inflorescenses are completed, the controlling genes must reverse epimutate at, or soon after, sexual reproduction so that the developmental clock can be reset for the cycle of the next generation.

Most of the literature on developmental genetics in maize ignores the central problem of somatic changes in gene action during phase change and structure elaboration and how this is reflected in homologies between vegetative and floral structures as adaptations in form to serve different functions. Research has concentrated on hormonal and biophysical mechanisms as regulators of patterns of division, enlargement, and specialization of cells during development (reviewed by Sheridan 1988).

THE PHYTOMER AND ORGAN HOMOLOGIES

Figure 7.1 is a drawing of mine (Galinat 1959 and 1963) that was first published 33 years ago. It depicted for the first time how the components of the maize phytomer are mod-

FIGURE 7.1 The explanation of the illustration must be made in conjunction with the details of Table 7.1 that identifies the modified expressions of four components of the phytomer (*i*-internode, *l*-leaf, *p*-prophyll, *ab*-axillary bud) in six different regions of the plant. They range from complete manifestation in the vegetative phase of position A to their various modified phenotypes under female (positions B, C) and male (positions D, E, F) influences during the floral phase.

F
E
D
C
B
A
ab
p
i
l
W.C. Galinat

ified in expression under different phases and sexes (Table 7.1). I should point out that our concept of the phytomer as constituting a genetic pattern of activity controlling growth and form is itself always evolving. It is not helpful to go back to some obsolete definition of the phytomer as a discrete individual and then create a straw man for the purpose of blowing it down. The expression of the phytomer in the language of DNA is not always discrete in that the movement away from DNA may be difficult, leading to the developmental counterpart of stuttering—which may, in turn, lead to repeats of a morphological component of the phytomer.

PHYLLOTAXY

In maize, as in the rest of the grasses, there is usually only one leaf component per phytomer; but in more primitive plants, and in the dicots in general, each phytomer may have two or more leaves, perhaps selected to function like a battery of solar panels designed to track the sun from east to west as the world rotates eastward once a day. With radial phyllotaxy, some of the leaves are always at right angles to the solar rays—directed toward the sun like a dish antenna. However, when the phyllotaxy is bilateral, a new system for solar ray abutment (right-angle interception) has evolved in which the bilateral orientation of each branch axis is always at right angles to that of its parent axis. Thus, in a profusely branched grass plant, the overall phyllotaxy of the whole plant is radial, with some leaves always in a solar abutment position.

In U.S. corn-belt maize we have bred for unbranched single stalk plants adapted to high density stands and grown under conditions of technological agriculture. Because the leaf orientation of each plant is at random in comparison with the other plants, it is the maize field as a whole rather than an individual plant that has the display for optimum solar gain and highest yield in modern maize. The broad leaves of modern maize with their undulating margins that constantly ripple in the wind may have greater solar interception than their solitary bilateral phyllotaxy would predict.

In contrast with the two-ranked phyllotaxy in the vegetative phase, with one leaf subtending one axillary bud within each phytomer, the floral phase of maize is many ranked with whorls of leaf primordia each subtending paired spikelets. The many-ranked ears and central spike of the tassel in maize are a result of the domestic selection for an epimutant gene, (*mr* or *Tr*), linked to *lg* on the short arm of chromosome 2, during the domestication of teosinte some 8,000 years ago. This floral phyllotaxy of maize has yield advantages over its two-ranked teosinte ancestor, and it might be transferable to the two-ranked small grains. Normally regulation of phyllotaxy behaves as an epimutation in that it always seems to be turned back to two-ranking for the vegetative phase of the next generation. However, there have been several reports of instances where the *mr*, (multiranking) gene was not turned back based on the floral-type phyllotaxy appearing in the vegetative phase of the next generation—as if the epigenic state had become transformed into the standard mutant state. It was demonstrated by Greyson et al. (1978) that the embryos in the seed carrying abnormal vegetative phyllotaxy had a thicker meristem than normal, as if this were the first manifestation of multiranking. This is supported by the fact that the multiranked central spike of the normal tassel is derived from a thicker apical meristem than the lateral meristems that produce two-ranking on the same tassel. In an independently occurring mutation to vegetative multiranking in my material but allelic to that of Greyson et al. (1978), I found a higher rate of *mr* transmission from *mr* pollen collected from the central spike in comparison with that from the lateral branches of the same tassel (Galinat 1990). Each pollen source was applied to a number of twin ears. It appears that the differences in ranking between the central spike and lateral branches of the normal tassel are not just a mechanical by-product of meristem size, but do have an inherited difference on an epigenic basis. Subsequently I have introduced the *ub* (unbranched tassel) gene, which produces single-spike tassels, into my vegetative multiranked stocks in order to simplify the collection of pollen. Partly as a result of this genetic technique, I now have a few stocks that are homozygous for vegetative multiranking.

POSTSCRIPT

Finally, the analogy of plant form to good music (Galinat 1959), or even to the meter and rhyme of poetry, is that they all depend on a sort of heartbeat or rhythmic pattern marked by the repetition of a related element that is important to the construction and organization of the whole. The recurring pattern in maize and other plants starts with the tempo of the plastochron beat upon which the phytomer framework with its various elaborations and variations must be constructed. In the terminology from recent levels of understanding, it is the internal environment of the host that controls the transposon movement involved in the genetics of morphogenesis.

REFERENCES

Galinat WC (1959) The phytomer in relation to floral homologies in the American *Maydeae*. Bot Mus Leafl, Harvard University 19: 1–32

Galinat WC (1963) Form and function of plant structures in the American *Maydeae* and their significance for breeding. Econ Bot 17: 51–59

Galinat WC (1990) Increased transmission of full vegetative multiranking through pollen from the central spike of the tassel in comparison with pollen from the lateral branches. Maize Genetics Cooperation Newsletter 64: 120

Greyson RI, Walden DB, Hume JA (1978) The ABPHYL syndrome in *Zea mays*. II. Patterns of leaf initiation and the shape of the shoot meristem. Can J Bot 56: 1545–1550

Sheridan WF (1988) Maize developmental genetics: genes of morphogenesis. Annu Rev Genet 22: 353–385

8

Genetics and the Morphological Evolution of Maize

JOHN DOEBLEY

Maize (*Zea mays* L. ssp. *mays*) and its nearest wild relatives, the teosintes (*Zea* spp.), differ profoundly in both vegetative and inflorescence architecture. Nevertheless, maize and the Mexican annual teosintes (*Z. mays* ssp. *parviglumis* and ssp. *mexicana*) are members of the same biological species (Iltis and Doebley 1980). These teosintes and maize form fully fertile hybrids with normal meiosis and chromosome pairing (Beadle 1932; Wilkes 1967). Differences in chromosome morphology of maize and Mexican annual teosinte are no greater than those within maize itself (Kato 1976; Smith et al. 1982). Moreover, isozymes of maize and Mexican annual teosinte are encoded by the same suite of genes, and allelic constitutions at these genes show no greater differentiation between maize and Mexican annual teosinte (especially ssp. *parviglumis*) than exists among the races of maize themselves (Doebley et al. 1984, 1987a). Finally, the chloroplast genomes of maize and some Mexican annual teosintes are identical in their restriction endonuclease maps (Timothy et al. 1979; Doebley et al. 1987b).

This paradox—profound morphological differentiation in the absence of substantial genetic differences—led Beadle (1939) to propose that maize is simply a domesticated form of teosinte and that mutations in as few as five major genes selected by the prehistoric peoples of Mexico may have accomplished the evolutionary feat. Since 1939, Beadle's hypothesis that maize is a domesticated teosinte has received strong support from diverse sources (see Doebley 1990 for a review), and an alternative hypothesis that maize evolved from an extinct wild maize (Mangelsdorf and Reeves 1939) has faced increased opposition. (But see Goodman 1988; Wilkes 1989.) Nevertheless, many questions remain unanswered including: (1) Is Beadle's specific hypothesis of five major genes correct? and (2) If so, then what are these genes, their genomic locations, and their effects on inflorescence development? In this chapter, I review the current state of knowledge bearing on these issues.

The Maize Handbook—M. Freeling, V. Walbot, eds.

MORPHOLOGY OF MAIZE AND TEOSINTE

Both maize and Mexican annual teosinte plants have a main culm that is terminated by a tassel (Figure 8.1). Both plants also bear lateral branches at one or more nodes on the main culm. In annual teosinte plants grown in their native habitat, the primary lateral branches are normally elongate, and the inflorescences terminating these branches (primary lateral inflorescences) are normally male and branched, i.e., tassels (Figures 8.1–8.2). In some teosinte plants, the primary lateral branch does not fully elongate, and the primary lateral inflorescence can be mixed male-female or complete female. It is not known whether this results from environmental or genetic factors or both. In typical maize, the lateral branches are short, and the primary lateral inflorescences are female and unbranched, i.e., ears (Figure 8.1). In teosinte, there are secondary (and higher order) lateral branches that are terminated by female inflorescences or ears. In maize, secondary lateral branches terminated by female inflorescences may also be present, although they are most frequent in more primitive landraces (Figure 8.1).

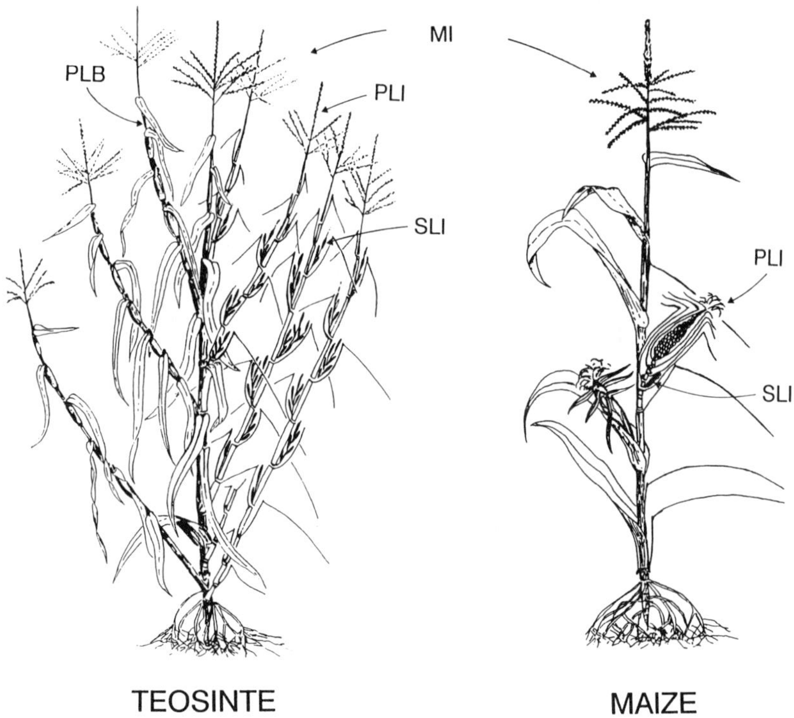

FIGURE 8.1 Mexican annual teosinte and maize plant architectures. MI is the main inflorescence, PLI is the primary lateral inflorescence, SLI is the secondary lateral inflorescence, and PLB is the primary lateral branch. Adapted from Iltis (1983)

FIGURE 8.2 Annual teosinte (*Zea mays* ssp. *parviglumis* race Balsas) grown under short days in Hawaii. The primary lateral branches are elongated. The upper primary lateral branches end in branched male inflorescences (tassels). Subspecies *parviglumis* has occasionally been characterized as having a profusely tillered or grassy-tillered habit. It and other annual teosintes show profuse tillering when grown at northern latitudes but not when grown at tropical latitudes to which they are adapted.

The most dramatic differences between maize and teosinte involve the architecture of their female inflorescences (Figure 8.3). The teosinte ear is composed of 5–10 (or more) distichously (in two ranks) arranged cupulate fruitcases (Figure 8.3A, C). The cupule of the cupulate fruitcase is formed from the invaginated rachis internode. The cupule contains a single sessile spikelet that is oriented parallel to the axis of the rachis (Figure 8.3C, E). The outer glume of this sessile spikelet seals the opening of the cupule, thus obscuring the kernel from view. Both the rachis internode and its outer glume are highly indurated in teosinte. The cupulate fruitcases are separated from one another by abscission layers, thus enabling the fruitcases to separate (disarticulate) at maturity for dispersal.

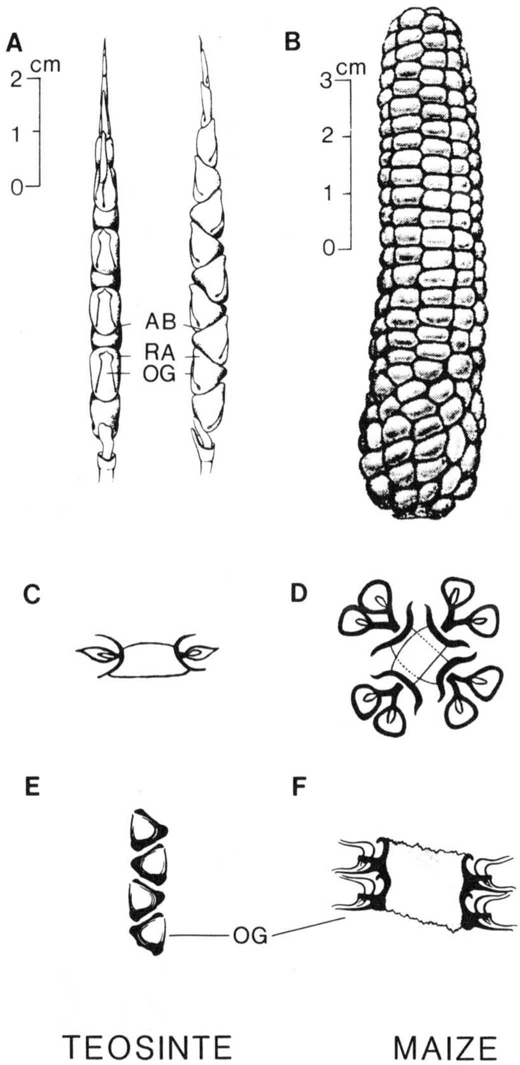

FIGURE 8.3 Architecture of Mexican annual teosinte and maize ears (female inflorescences), adapted from Iltis (1983) and Galinat (1969). A. Teosinte ear: AB is the abscission layer, OG is the outer glume, and RA is the rachis internode. B. Maize ear. C,D. Schematic transverse cross sections. C. Teosinte, showing two ranks of cupules with one spikelet per cupule. D. Maize, showing four ranks of cupules with two spikelets per cupule. E,F. Longitudinal cross sections. E. Teosinte, showing the outer glume oriented upward, parallel to the axis of the ear. F. Maize, showing the outer glume oriented outward, perpendicular to the axis of the ear.

The cob (rachis) of the maize ear, like that of its teosinte counterpart, is composed of invaginated internodes or cupules (Figure 8.3D). Maize cupules are arranged polystichously (in four or more ranks) with usually 100 or more cupules in a single ear (Figure 8.3B, D). Unlike teosinte, the cupules of maize are shallow, often collapsed, and do not envelop the kernels (Figure 8.3F). Maize cupules may be indurate, but the outer glumes are softer than the highly indurated glumes of the teosinte ear. In contrast to teosinte, there are two spikelets associated with each cupule, one pedicellate and the other sessile. Thus, an ear with four ranks of cupules will have eight rows of kernels (Figure 8.3D). The female spikelets of maize also differ from those of teosinte in that they are oriented perpendicular and not parallel to the axis of the ear (Figure 8.3F). Finally, maize ears lack abscission layers as found in teosinte, so the ear remains intact at maturity.

GENETIC CONTROL OF THE KEY MORPHOLOGICAL TRAITS

Reports on genetic control of the key traits distinguishing maize and teosinte are often conflicting. The reasons for these conflicts include the following: (1) Different authors used different maize and teosinte parents. The genes differentiating a 10-rowed maize from teosinte may not be the same as those for a 20-rowed variety. (2) Most authors used elite agronomic lines as their maize parents. These are not optimal for mapping genes involved in the origin of maize. (3) Most authors grew their plants at northern latitudes where, because teosinte is a short-day plant, its development is severely affected by improper day-length and some genotypes fail to mature. (4) While some authors (Collins and Kempton 1920; Langham 1940) clearly understood the morphology of the plant, others did not. For example, some never recognized the distinction between the primary lateral and secondary lateral inflorescences in which some traits (e.g., inflorescence phyllotaxy, floral sex) are expressed differently on the same plant. (5) Some authors categorized plants as maizelike or teosintelike so that "Mendelian" ratios could be calculated, despite the fact that variation for the traits is continuous (in some cases bimodally continuous) and not Mendelian (Collins and Kempton 1920; Mangelsdorf 1947; Rogers 1950; Doebley et al. 1990). Forcing such variation into a Mendelian pattern is apt to introduce a bias.

Ear Disarticulation

Factors affecting ear disarticulation have been mapped to chromosomes *1*, *2*, *3*, *4*, *5*, *6*, and *8* (Mangelsdorf 1947; Doebley et al. 1990). Based on unpublished results, Galinat (1978) reported that two linked genes in *4S* (*Ph*, pith abscission; *Ri*, rind abscission) control the formation of abscission layers in the ear. Later, Galinat (1985) reported that a single gene, *Ab* (abscission), controls this trait, although he presented no data to demonstrate the existence of *Ab*. The inheritance of this trait is in need of further study. The concept that one or two major genes control this trait is not well supported at present.

Ear Phyllotaxy

Langham (1940) concluded that the switch from distichous to polystichous phyllotaxy is controlled by a single locus that he designated *tr* (two-ranked). He reported that the maize allele (*Tr*) is dominant. Other authors have reported that phyllotaxy does not segregate in a Mendelian fashion in maize-teosinte F_2 and backcross populations (Collins and Kempton 1920; Mangelsdorf 1947; Rogers 1950). My colleagues and I (Doebley et al. 1990) also found ear phyllotaxy to vary continuously, although bimodally, with many ears being mixed distichous-polystichous. Thus, this trait could not be readily scored as a simple Mendelian trait in our F_2 population.

Langham (1940) found no linkage of phyllotaxy with markers in chromosomes *1–6*. Mangelsdorf (1947) and Rogers (1950) reported significant linkage with markers in chromosomes *1*, *2*, *3*, *4*, *6*, *8*, *9*, and *10*. My colleagues and I (Doebley et al. 1990) found effects on this trait in chromosomes *2*, *3*, *4*, *5*, *8*, *9*, and *10*. We were also able to quantify the relative contributions of these different regions and showed that *2S* accounts for most of the difference. The *2S* locality was confirmed in a second primitive maize-teosinte F_2 population (Doebley and Stec, unpublished). The chromosome *2S* locality agrees with a report by Galinat (1973a) placing a gene affecting phyllotaxy in *2S*. The switch from two-ranked to four-ranked may have involved a major locus in *2S* and minor loci elsewhere.

Glume Induration

There is a suite of traits necessary for the formation of the teosinte cupulate fruitcase including (1) indurate rachis internodes and glumes, (2) an invaginated rachis internode, and (3) a spikelet oriented parallel to the axis of the ear. These traits are highly correlated in maize-teosinte hybrid populations (Rogers 1950; personal observation). Factors affecting glume induration and spikelet orientation have been reported in all chromosomes (Mangelsdorf 1947; Rogers 1950; Doebley et al. 1990). However, one location (*4S*, linked to *su1*) has been consistently implicated in several studies and accounts for most of the difference between maize and teosinte in this regard (Rogers 1950; Doebley et al. 1990).

One could hypothesize that the origin of maize involved a mutation in a regulatory locus in *4S* that normally activates a series of genes necessary for the formation of the teosinte cupulate fruitcase. If the action of such a regulatory locus were blocked or modified, then the default program could result in a more male-like softer glumes and a rachis internode without a strong invagination. This model is consistent with the view that maize differs from teosinte in that it manifests male secondary sex traits (soft glume and un-invaginated rachis internodes) on a female background as proposed by Galinat (1983).

Inflorescence Sex and Plant Architecture

Several authors (Collins and Kempton 1920; Kempton 1924; Langham 1940) noted that the length of the primary lateral branch and the sex of the primary lateral inflorescence segregate in maize-teosinte populations; however, these authors did not attempt to analyze the inheritance of these traits nor comment on the importance of these traits in the evolution of maize. In contrast, Iltis (1983) drew attention to these differences in plant architecture and argued that they are some of the key morphological differences between maize and teosinte. Doebley and Stec (1991) studied the inheritance of these traits in a maize-teosinte F_2 population. These traits were highly correlated (R = 0.75) in their F_2 population. Both traits are controlled by a region in *1L* with a large effect and by regions in *3*, *6* and *8* with smaller effects. These traits appear to represent different phenotypic effects of the same genetic-developmental program.

Paired-Spikelets

In analyzing spikelet pairing, Langham (1940) found that many plants in his maize-teosinte populations were intermediate between the maize (paired) and teosinte (single) phenotypes. Nevertheless, he classified those that tended toward maize as paired-spikelets and those that tended toward teosinte as single-spikelet and obtained 3:1 segregation. Based on these data, Langham (1940) concluded that paired-spikelets is controlled by a single Mendelian locus that he named *Pd*. Mangelsdorf (1947) and Rogers (1950) disagreed with Langham's conclusion, suggesting that genetic control of paired-spikelets is more complex. Galinat (1971, 1988), referring to unpublished data, suggested that there may be an additional locus involved, which he named *Pd2*. More recent results suggested complex inheritance (more than two loci) of this trait (Doebley and Stec 1991).

Attempts to map the genes controlling paired-spikelets have yielded variable results. Kempton (1924) shows linkage with chromosomes *1* and *3*; Langham (1940) with 3; Mangelsdorf (1947) with *4* and *8*; Rogers (1950) *3* and *7*; and Doebley et al. (1990) with *1*, *2*, *3–4*. These variable results probably manifest an interaction between inflorescence phyllotaxy and spikelet pairing (Galinat 1988). Paired-spikelets result from the bifurcation of the spikelet-pair primordia, a process that may also be involved in the evolution of polystichy (Sundberg and Doebley 1990, Maize Genetics Newsletter 64: 21–22). Thus, spikelet pairing may show different inheritance when maize lines with different kernel row numbers are used as the maize parent. More detailed study of paired-spikelets on two- and four-ranked backgrounds is needed to better understand its inheritance. Nevertheless, based on evidence from several sources (Langham 1940, Mangelsdorf 1952, Doebley et al. 1990), it appears that one of the principal factors for this trait is on *3L*.

THE ROLE OF KNOWN MAIZE GENES

Several genes of maize produce teosintelike phenotypes. There has been some discussion concerning the potential role of these genes in the origin of maize. The *Tunicate (Tu)*, *Corngrass (Cg1)*, *Teopod (Tp1, Tp2)*, and *teosinte* branched (*tb1*) genes have been of particular interest.

Tunicate (*Tu1*)

This locus causes the glumes of the ear to become long and papery, forming a "pod" that envelopes the kernel. Mangelsdorf and Reeves (1939) suggested a role for *Tu1* in the origin of maize, believing as they did that maize evolved from an extinct wild pod-corn. In 1972, Beadle first suggested that *Tu1* might be the gene that converted the hard glume of teosinte into the soft glume of maize. He also suggested that it may pleiotropically retard ear disarticulation and make the cob softer. Beadle's hypothesis has been elaborated upon by Galinat (1983, 1988).

My colleagues and I (Doebley et al. 1990) found the principal gene(s) controlling the difference in glume induration and spikelet orientation to be located in *4S*. We found no significant effects on these traits in *4L* where *Tu1* is located. The only published evidence supporting Beadle's hypothesis comes from the archaeological specimens with a *Tunicate*-like phenotype (Mangelsdorf et al. 1967). These specimens could equally as well be explained by the gene(s) in *4S*. While it remains possible that *Tu1* played some role in the origin of maize, there is no compelling evidence to support this view. *Tu1* is not one of the key genes differentiating maize and teosinte.

Teosinte Branched (*tb1*)

tb1 causes the maize plant to produce numerous tillers and elongate primary lateral branches terminated by tassels. These are some of the same differences in plant architecture that distinguish teosinte from maize. Doebley and Stec (1990) analyzed the inheritance of these traits in a maize-teosinte F_2 population and found that a gene(s) near the molecular marker *UMC107* in *1L* (ca. 7 map units proximal to *Adh1*) controls much of the variance for these traits. This is close to the known location of *tb1*. It seems an attractive hypothesis that an allele of *tb1* has a principal role in governing the differences in plant architecture between maize and teosinte.

Corngrass and *Teopod* (*Cg1*, *Tp1*, *Tp2*)

The *Corngrass* and *Teopod* genes have similar effects on plant form, producing many tillered, narrow-leaved, and small-eared plants. These genes appear to cause prolonged expression of the juvenile phase of plant development (Poethig 1988). Because these genes cause maize to superficially resemble teosinte, it is natural to consider their potential role in the origin of maize. Galinat (1954) rejected any central role for *Cg1* in origin of maize. Recent efforts at mapping the genes distinguishing maize and teosinte (Doebley et al. 1990) are consistent with Galinat's view in that *3S* (where *Cg1* is located) has only a small effect on the traits differentiating teosinte and primitive maize. Based on the phenotype of the *Teopod* mutants, Poethig (1988) broached the idea that the evolution of modern maize involved "the progressive truncation of a juvenile developmental phase." This is an appealing hypothesis that could be further examined through a comparison of modern maize with more primitive, tropical varieties of maize.

HOW MANY GENES?

Beadle (1939) hypothesized that if maize was derived from teosinte by intense selection under domestication, then a small number of major genes would have been involved. The first evidence bearing on the number of genes controlling the key traits differentiating maize and teosinte was provided by Mangelsdorf and Reeves (1939). Through analysis of crosses between teosinte and maize marker stocks, these authors inferred the existence of four independent chromosomal segments or genes with large effects on ear morphology. Because Mangelsdorf and Reeves (1939) did not analyze single vs. paired spikelets, Beadle (1939) postulated that an additional gene for this trait must be involved, giving a total of five major genes.

Beadle (1972, 1980) published the results of an experiment intended to test his 1939 hypothesis of five major genes. He grew 50,000 F_2 progeny of a cross of a primitive maize (Race Chapalote) and Mexican annual teosinte (ssp. *mexicana* Race Chalco). Beadle used the frequency of recovery of the two parental types (1:500) to infer that "five major and independently inherited gene differences" distinguish primitive maize and teosinte. Beadle recognized that linkage and dominance complicated his interpretation. Nevertheless, he believed that he had established that genetic control of the key morphological differences between maize and teosinte was not so complex as to render untenable his hypothesis that maize was derived from teosinte by human selection under domestication.

Recently, my colleagues and I (Doebley et al. 1990; Doebley and Stec 1991) repeated Beadle's experiment with the addition of molecular analyses that enabled us to follow segregation of specific chromosomal segments in our F_2 population. We employed the same cross as Beadle (Race Chapalote maize × Race Chalco teosinte), recognizing as Beadle did that one can not study the genes involved in the origin of maize by using elite strains with 14 or more rows of grain and other advanced traits. We also were able to take advantage of recent advances in quantitative genetics (Edwards et al. 1987; Landers and Botstein 1989) and measure the key traits quantitatively rather than attempting to categorize each plant arbitrarily as teosinte-like or maizelike so that "Mendelian" ratios could be calculated. The results of our analyses are congruent with Beadle's interpretations, although they do not provide conclusive proof for his hypothesis. We found five restricted regions of the genome that account for most of the differences in inflorescence morphology between maize and teosinte. These regions are in *1L*, *2S*, *3L*, *4S*, and *5S* (Doebley and Stec 1991). We also detected smaller effects in all other chromosomes.

BLOCK INHERITANCE OF THE KEY GENES

Mangelsdorf and Reeves (1939) hypothesized that maize and teosinte are differentiated by four or five chromosomal segments (blocks) each containing many linked genes. These authors presented no compelling evidence for this view, but simply stated that isogenic lines containing the hypothesized blocks produce intermediate phenotypes at low frequency. They concluded that these intermediates resulted from crossovers within the blocks, breaking up a complex of genes that confers a teosintelike phenotype to the isogenic line. Mangelsdorf (1947, 1952) repeated this interpretation, but provided no additional evidence to support it.

Mangelsdorf (1974) reported some evidence regarding the existence of two of the hypothesized four or five blocks. He reported that, in backcross populations between a maize line containing a portion of teosinte chromosome *3L* and the maize recurrent parent, no plants with intermediate phenotypes were found. Thus, his data were consistent with a single Mendelian locus and provided no evidence for a block of genes in *3L*. However, in a backcross population for an isogenic line with a portion of teosinte *4S*, Mangelsdorf (1974) reported that 21% of the progeny had intermediate phenotypes. He concluded that the intermediates represented crossovers within a block of genes in *4S* controlling the phenotype. This evidence is not conclusive for two reasons. First, Mangelsdorf did not have flanking markers that would have enabled him to show that the intermediate phenotypes were associated with actual crossover events. Second, the isogenic lines used were the product of only three generations of backcrossing. Thus, there were potentially other unlinked genes segregating in the populations that could influence the phenotype.

Galinat (1971, 1973b, 1977, 1978, 1985, 1988) provides further discussion of block inheritance of the genes differentiating maize and teosinte. This author refers to unpublished data for two loci (*Ph* and *Ri*) that are linked to the glume factor in *4S* identified by Rogers (1950). Galinat refers to this block as the "chromosome *4* complex." This block is remarkable in that it covers nearly all of *4S* from *Ph1* at position 0 to *Su1* at position 66.

The concept that maize and teosinte are differentiated by many genes located in four or five blocks is an interesting one, although it is not well supported at present. To evaluate the importance of block inheritance in maize evolution, it may first be necessary to map many or most of the genes differentiating maize and teosinte. At that point, it should be possible to discern whether the amount of linkage is greater than that expected by chance alone. These data are necessary because some linkage of the genes differentiating maize and teosinte should be expected by chance if Beadle's (1939) model of five major plus many minor genes is correct. For example, if (1) the size of a block is 60 cM (based on the "chromosome *4* complex"), (2) there are five major and 50 minor genes distinguishing maize and teosinte, and (3) the maize genome is 1,500 cM in size, then 20% of the minor loci (or 10 of 50 minor loci) will fall within a "block" near one of the major loci by chance alone.

CONCLUDING REMARKS

Understanding the genetic basis of the morphological differences between maize and teosinte is central to the elucidation of the origin of maize. At present, Beadle's (1939) hypothesis that maize was derived from teosinte by as few as five mutations of large effect plus a greater number of minor mutations is generally consistent with the available data.

There are now candidates for major loci controlling three of the key traits. First, a locus in *1L* may prevent the elongation of the primary lateral branch during development and concurrently enable the primary lateral inflorescence to develop into a female rather than a male structure. *Tb1* could be this locus. Second, a locus in *2S* may alter inflorescence phyllotaxy from two- to four-ranked. Third, a locus in *4S* may inhibit the expression of a suite of secondary-sex traits (including glume induration and spikelet orientation) that are involved in the formation of the teosinte cupulate fruitcase. The inheritance of two other key traits (ear disarticulation and paired spikelets) is less clear and may not involve single major loci plus modifiers.

Considering the available evidence, one might argue that the origin of maize involved mutations at several major loci and then the movement of these mutant genes into favorable genetic backgrounds controlled by the effects of more numerous minor loci (Beadle 1939; Galinat 1988). Allelic variation at such minor loci may exist in teosinte populations as neutral polygenes, so that a large number of new mutations would not have been required. An understanding of the interactions between major and minor loci as well as between the putative major loci themselves is crucial to understanding the steps involved in the origin of maize.

ACKNOWLEDGMENT

This work was supported by a National Science Foundation grants (BSR-88–06889 and BSR-91-07175) and by Pioneer Hi-Bred, Intl. of Johnston, Iowa.

REFERENCES

Beadle GW (1932) Studies of *Euchlaena* and its hybrids with *Zea* I. Chromosome behavior in *Euchlaena mexicana* and hybrids with *Zea mays*. Z Indukt Abstamm Vererbungsl 62: 291–304

Beadle GW (1939) Teosinte and the origin of maize. J Hered 30: 245–247

Beadle GW (1972) The mystery of maize. Field Mus Natl Hist Bull 43: 2–11

Beadle GW (1980) The ancestry of corn. Sci Am 242: 112–119, 162

Collins GN, Kempton JH (1920) A teosinte-maize hybrid. J Agric Res 19: 1–37

Doebley J (1990) Molecular evidence and the evolution of maize. Econ Bot 44 (3 supplement): 6–27

Doebley J, Stec A, Wendel J, Edwards M (1990) Genetic and morphological analysis of a maize-teosinte F_2 population: implications for the origin of maize. Proc Natl Acad Sci USA 87: 9888–9892

Doebley J, Stec A (1991) Genetic analysis of the morphological differences between maize and teosinte. Genetics 129: 285–295

Doebley J, Goodman M, Stuber C (1984) Isoenzymatic variation in *Zea* (Gramineae). Syst Bot 93: 203–218

Doebley J, Goodman M, Stuber C (1987a) Patterns of isozyme variation between maize and Mexican annual teosinte. Econ Bot 41: 234–246

Doebley J, Renfroe W, Blanton, A (1987b) Restriction site variation in the *Zea* chloroplast genome. Genetics 117: 139–147

Edwards MD, Stuber CW, Wendel JF (1987) Molecular-marker-facilitated investigations of quantitative-trait loci in maize. I. Numbers, genomic distribution and types of gene action. Genetics 116: 113–125

Galinat WC (1954) Corn grass: I. Corn grass as a prototype or a false progenitor of maize. Am Nat 88: 101–103

Galinat WC (1969) The evolution under domestication of the maize ear: string cob maize. Mass Agric Exper Sta Bull 577: 1–19

Galinat WC (1971) The origin of maize. Annu Rev Genet 5: 447–478

Galinat WC (1973a) Intergenomic mapping of maize, teosinte and *Tripsacum*. Evolution 27: 644–655

Galinat WC (1973b) Preserve Guatemalan teosinte, a relic link in corn's evolution. Science 180: 323

Galinat WC (1977) In Sprague GF (ed) Corn and Corn Improvement, Am Soc Agron, Madison, WI, pp 1–47

Galinat WC (1978) The inheritance of some traits essential to maize and teosinte. In Walden DB (ed) Maize Breeding and Genetics, John Wiley and Sons, New York, pp 93–111

Galinat WC (1983) The origin of maize as shown by key morphological traits of its ancestor, teosinte. Maydica 28: 121–138

Galinat WC (1985) The missing links between teosinte and maize: a review. Maydica 30: 137–160

Galinat WC (1988) The origin of corn. In Sprague GF, Dudley JW (eds) Corn and Corn Improvement, Am Soc Agron, Madison, WI, pp 1–31

Goodman MM (1988) The history and evolution of maize. CRC Crit Rev Plant Sci 7: 197–220

Iltis HH, Doebley J (1980) Taxonomy of *Zea* (Gramineae). II. Subspecific categories in the *Zea mays* complex and a generic synopsis. Am J Bot 67: 994–1004

Iltis HH (1983) From teosinte to maize: the catastrophic sexual transmutation. Science 222: 886–894

Kato TA (1976) Cytological studies of maize. Mass Agric Exper Station Res Bull No 635

Kempton JH (1924) Inheritance of the crinkly, ramose, and brachytic characters of maize in hybrids with teosinte. J Agric Res 27: 537–596

Lander ES, Botstein D (1989) Mapping Mendelian factors underlying quantitative traits using RFLP linkage maps. Genetics 121: 185–199

Langham DG (1940) The inheritance of intergeneric differences in *Zea-Euchlaena* hybrids. Genetics 25: 88–107

Mangelsdorf PC (1947) The origin and evolution of maize. Adv Genet 1: 161–207

Mangelsdorf PC (1952) Hybridization in the evolution of maize. In Gowen JW (ed) Heterosis, State College Press, Ames, IA, pp 175–198

Mangelsdorf PC (1974) Corn: Its Origin, Evolution and Improvement,. Harvard University Press, Cambridge, MA

Mangelsdorf PC, MacNeish RS, Galinat WC (1967) Prehistoric wild and cultivated maize. In Byers DS (ed) The Prehistory of the Tehuacan Valley, Vol. I, University of Texas Press, Austin, TX, pp 178–200

Mangelsdorf PC, Reeves RG (1939) The origin of indian corn and its relatives. Texas Agric Exp Sta Bull 574

Poethig RS (1988) Heterochronic mutations affecting shoot development in maize. Genetics 119: 959–973

Rogers JS (1950) The inheritance of inflorescence characters in maize-teosinte hybrids. Genetics 35: 541–558

Smith JSC, Goodman MM, Kato TA (1982) Variation within teosinte. II. Numerical analysis of chromosome knob data. Econ Bot 36: 100–112

Timothy D, Levings CS, Pring DR, Conde MF, Kermicle JL. (1979) Organelle DNA variation and systematic relationships in the genus *Zea*: teosinte. Proc Natl Acad Sci USA 76: 4220–4224

Wilkes HG (1967) Teosinte: The Closest Relative of Maize, Bussey Institute, Harvard University, Cambridge, MA

Wilkes HG (1989) Maize: domestication, racial evolution, and spread. In Harris DR, Hillman GC (eds) Foraging and Farming, Unwin Hyman, London, pp 440–454

9

Overview of Key Steps in Aleurone Development

VIRGINIA WALBOT

THE aleurone is the epidermis of the triploid endosperm. During germination, the aleurone secretes hydrolases that aid in the digestion of stored reserves in the interior cells of the starchy endosperm. Aleurone cells are readily distinguished from inner endosperm cells by their small size, lack of prominent protein bodies, and potential for anthocyanin pigmentation. Unlike most cereals, the aleurone of maize is composed of only a single layer of cells. Moreover, the tissue has a simple developmental history involving mostly synchronous nuclear division cycles. Consequently, aleurone development can be readily modeled using powers of two, and this feature allows assignment of the time (cell division number) at which events leading to a visible phenotype occur, i.e., anthocyanin deposition following transposable element excision from a reporter allele or activation of *R* gene expression (Coe 1978; Levy and Walbot 1990).

STAGES OF ALEURONE DEVELOPMENT

Fertilization

At fertilization one haploid sperm nucleus and the diploid polar nucleus are combined, resulting in the triploid primary endosperm cell. This is a large cell occupying most of the volume of the embryo sac, and the single nucleus is centrally located.

Syncitial Development

1. Division cycles 1–3. The primary endosperm nucleus undergoes three mitotic divisions while centrally located. These divisions are in relatively predictable "planes" such that there is a regular "fate map" of the endosperm. Each of the initial pair of

The Maize Handbook—M. Freeling, V. Walbot, eds.

nuclei (division cycle 1) will form approximately half of the endosperm; viewed from the top of the kernel the vast majority of half-sectors bisect the kernel through the embryo face (Coe 1978). Quarter kernel events divide the endosperm in the alternative plane, creating a top center point on the endosperm; this point is displaced slightly back from the silk scar, away from the embryo. One-eighth sectors are similar to segments of an orange, extending from the top center of the endosperm around to the bottom of the kernel. The nuclei are paired after mitotic cell cycle 1; after division cycle 2, four nuclei are found in the same plane, in a precise orientation (McClintock 1978); after division cycle 3, there are two sets of four nuclei indicating that a new plane of nuclear separation was used. These initial nuclear cycles are synchronous; the information programming nuclear location and division planes has not been investigated.

2. Division cycles 4–9. At the eight nuclear stage, the centrally placed nuclei migrate to the surface of the primary endosperm cell. Each nucleus migrates to a different area. There must be a more-or-less fixed pattern to this migration to account for the regularity of sector shape described above; however, the control of nuclear migration has not been investigated. At the periphery, the nuclei continue to proliferate, progressing through the 16 (2^4 = fourth division cycle), 32, 64, 128, 256, and 512 (2^9) stages. The peripheral nuclei are thought to cycle synchronously; this feature is not all that surprising considering that the nuclei are in the same cytoplasm. All of the daughter nuclei lie along the plasma membrane, approximately evenly spaced.

Cellularization

At the 256–512 nuclear stage (depending on genotype), cellularization begins. To accomplish this, for the first time, the nuclei divide so that the daughter nucleus is pushed toward the inside of the endosperm. After this nuclear separation a plasma membrane forms to enclose each peripheral cell; there is also a partial wall, anticlinal to the endosperm surface, but there is no wall (or significantly less wall) between the peripheral cell and its daughter, interior nucleus. The net outcome of the initial step of cellularization is the formation of 512 peripheral cells, and of a layer of 512 syncitial nuclei just interior to these newly formed cells.

When the 512 interior nuclei subsequently divide, a wall separating the peripheral and interior cells is visible. The subepidermal nuclei become cellularized within a box with a bottom and sides, but no top; and just interior to these partially walled cells is a layer of 512 daughter nuclei still sharing a single cell. The "top wall" forms when the interior daughter nuclei subsequently divide in parallel to what occurred at the periphery at the preceding division. This pattern of a two-step cellularization continues until the cytoplasm of the entire primary endosperm cell is contained within discrete cells. Further information on the clonal organization of the inner endosperm is found in McClintock (1978).

Throughout the syncitial stage of development, the nuclei present should be considered a meristem because they are responsible for producing cells of more than one tissue (aleurone and inner endosperm). The distinction between these two cell types minimally involves a single division that separates the peripheral nucleus and an interior daughter; this step is coincident with the cellularization of the peripheral (future aleurone) cell. With cellularization the aleurone emerges as a discrete tissue. Subsequent cell divisions in this tissue will be exclusively anticlinal to enlarge the epidermal layer.

This description of aleurone development is somewhat idealized as some peripheral cells divide more than once to contribute interior daughter cells. In addition to cell division in the aleurone, the subepidermal layer is mitotically active during the phase of rapid endosperm growth during the middle of seed maturation.

Growth as an Epidermis

After cellularization, aleurone proliferation continues until there are approximately 250,000 (2^{18}) cells. These divisions are also relatively synchronous; i.e., it can be assumed that most cells make the same clonal contribution to the aleurone. Working backward from the final stage, single-cell sectors correspond to events that occur after the last division (18th), 2-cell sectors correspond to cell division cycle 17, 4-cell sectors correspond to cycle 16, etc. These terminal divisions in the aleurone are highly regular, based on the distribution of sector types. For example, most four-cell sectors are a square, suggesting a specific proliferation pattern (alternate planes of anticlinal division). Eight cell sectors are usually a 2×4 rectangle, while 16-cell sectors are usually a square (Coe 1978). Despite the simplicity and elegance of the terminal stages of aleurone development, there is not yet a genetic analysis of such issues as the regulation of alternative planes of anticlinal cell division.

REFERENCES

Coe EH (1978) The aleurone tissue of maize as a genetic tool. In Walden DB (ed) Maize Breeding and Genetics, John Wiley and Sons, New York, pp 447–459

Levy AA, Walbot V (1990) Regulation of the timing of transposable element excision during maize development. Science 248: 1534–1537

McClintock B (1978) Development of maize endosperm as revealed by clones. In Subtelny S, Sussex IM (eds) The Clonal Basis of Development, Academic Press, New York, pp 217–237

II
Cell Biology

10

Light Microscopy I: Dissection and Microtechnique

ANNE W. SYLVESTER and STEVEN E. RUZIN

THE microscope is one of the primary tools of the cell biologist. In this first section we set forth a number of approaches that can be used to study maize using the stereo, compound, light, and epifluorescence microscopes. Successful cellular analysis is dependent on three factors: proper specimen preparation, careful use of the instrument, and appropriate observation techniques. Furthermore, any cellular analysis must be conducted within the context of the whole organism; hence, we begin in Part I by describing methods of dissection and observation using the whole plant. Then we present several methods of specimen preparation. Standard methods of observation using transmitted light and epifluorescence microscopy and for recording and analyzing the image are presented in part II.

DISSECTION

Observations of whole plants are generally essential before beginning a more detailed cellular analysis. Many of the important features of the shoot system are obscured by the enclosing leaf sheaths of maize. Careful dissections are therefore often required in order to record plant structure in a systematic fashion. Two methods of dissection of the shoot system are outlined below.

Destructive Dissection

1. Rinse soil from the root system.
2. Wrap the cleaned root system with Saran Wrap or foil (to be used as a handle).
3. Remove each leaf sequentially: a curved edged scalpel is useful to rip down the length of the leaf. (Also: see Irish, this volume, for other methods.) If the outer leaves are

The Maize Handbook—M. Freeling, V. Walbot, eds.

not of interest the shoot may be severed 1 cm above the position of the shoot apex (which is in a predictable position related to the plant age as determined empirically).

4. Tightly rolled leaves may be slit open and attached to the slide using double stick tape or Spot O'Glue from Avery. They may also be "relaxed" in 1% borax before adhesion.
5. Stabilize the plant when ready for fine dissections (e.g., of a young leaf or shoot primordium). The enclosed root system may be tethered to the bench top or microscope stand using wads of clay, thereby freeing both hands for dissection. A useful device is also described in Green (1988).
6. Bathe young leaves and the meristem with water regularly to maintain hydration.
7. Very young leaves and the meristem are unpigmented and hard to see. Adjust the stereomicroscope with oblique lighting if possible; use a black background or use a water soluble ink to highlight the cell surfaces. Mix the ink to desired concentration and brush on with a moist fine paintbrush; the ink will evaporate, leaving sufficient residue to produce a shadowing affect.

Nondestructive Dissection

Two methods are available if you wish to observe young leaves or the meristem without killing the plant: shoot meristem culture (see Irish, this volume) or windowing (outlined here).

1. Carefully prepare the root system as above, but avoid damaging the roots.
2. Determine where the shoot apex is.
3. Cut a window large enough to insert various tools and for observation. (A rectangle of approximately 2 × 1 cm is sufficient for a 6-leaf seedling.) The base of the rectangle should be just below the level of the meristem. The window itself consists of only the outermost heavily lignified leaf sheaths.
4. Using a straight scalpel, slice through the enclosing leaves at the upper edge of the window. If the window was big enough, this will not cut the meristem or youngest leaves.
5. Carefully cut off each leaf layer in the window. This can be accomplished by snipping the leaf attachment at the base with reverse-action scissors (e.g., Vannas scissors from Fine Science Tools). Peel cut leaves away with forceps.
6. Rinse out debris regularly and maintain hydration in the chamber.
7. Proper lighting is essential; use the ink dye for visual contrast.
8. After arriving at your target (e.g., young leaf or meristem) and accomplishing your goal (e.g., manipulation of tissue, photography, surface impression), the window may be sealed and the plant will continue to grow. First, thoroughly hydrate the opening

and then wad the window loosely with *wet* cotton. Close the window by wrapping the entire area with Saran Wrap. Seal with tape or glue. A splint using a slit piece of tygon tubing of appropriate diameter and length is useful. Replant the windowed plant in either soil or, preferably, in a hydroponic tank in a greenhouse.

9. The plant will regrow quickly if it has survived. After 2–3 days of acclimation, check the plant under the microscope. The regrowing primordium must be encouraged to grow at an angle out of the window: The regrowing plant will become distorted if forced to continue growing within the cylinder of the parent plant sheath.

TISSUE CLEARING

Plant cells contain a variety of pigments and inclusions (ergastic substances) that render them opaque, or at best translucent. A number of techniques can be used to clear the cytoplasm from tissues of interest without altering their cellular patterns. These methods are particularly useful for studying vascular anatomy in maize and for identifying other three-dimensional relationships among cells and tissues. There are essentially two approaches to tissue clearing (Herr 1972): the first relies on harsh chemicals (e.g., NaOH) to remove all cytoplasmic components. The second approach does not remove cytoplasm but somewhat equalizes refractive indices among cytoplasmic components, thereby rendering the cells transparent. The following are two examples of clearing techniques. Maize leaves are heavily lignified and therefore need extended times in each solution. The length of treatment must be determined empirically for each developmental stage.

Clearing Tissues With NaOH and Chloral Hydrate (Arnott 1959)

1. Treat tissue in 5% NaOH at room temperature (RT) to 37 °C for 1 to several days or until the tissue clears.
2. Rinse briefly in deionized (DI) water.
3. If dark areas persist, bleach 2–5 minutes in full strength commercial bleach (Clorox). Longer times are detrimental to tissue. Stockwell's solution is acceptable (90 ml water, 10 ml glacial acetic acid, 1 g potassium bichromate, 1 g chromic acid) but takes much longer. (Schmid, 1977) Rinse with three changes in DI water. (See following notes regarding bleach and staining properties.)
4. Treat tissue with concentrated chloral hydrate (250 g/100 ml DI water) at RT to 37 °C until material is transparent (several hours to days). Young, growing leaf tissue will clear faster than older, nongrowing tissue. Tissue may be stored in chloral hydrate, but take care not to let it dry. Note that chloral hydrate is toxic and must be handled with care; wear gloves.
5. Wash 2× in DI water 15 minutes each, then overnight in DI water.

6. Stain 20–30 minutes in Johansen's 1% safranin solution (1940, 2:1:1 methyl cellosolve : 95% EtOH : DI water). Avoid overstaining.
7. If material is overstained, destain 2–3 minutes in acidic alcohol (95% EtOH + 0.5% picric acid).
8. Dehydrate in 100% EtOH.
9. Transfer to 1:1 absolute EtOH:xylene and rinse for a few minutes.
10. Transfer to xylene for a few minutes to remove precipitated stain.
11. Transfer to fresh xylene. Examine the amount of staining. Then either destain further (steps 7 and 8), or restain (step 6), or continue.
12. Transfer to xylene and store until mounting.
13. Mount with Permount, Piccolyte, etc., on conventional glass microscope slides or between thin glass plates (whole leaves). It may be necessary to place a small lead weight on the coverslip or top glass plate during drying.

Notes

- For delicate materials use 2.5% NaOH.
- Prior to mounting, tissues must be soaked in 100% EtOH for enough time to ensure complete dehydration *before* placing in xylene.
- To restore the staining properties of tissues treated with bleach, soak for 15 minutes in 10% benzoyl peroxide in acetone and wash in 2:3 xylene:acetone followed by 100% EtOH.
- Fragile material should be transferred through a graded series between major steps—for example, at the transfer from DI water to full-strength Clorox.
- Material for paraffin sectioning can be carried through steps 1–5 but omit Clorox.
- Dried tissues (herbarium specimens, pressed leaves) clear better than fresh tissues.

Clearing Without Removing Cytoplasmic Components (Herr 1972)

1. Fix tissue in either FAA or CRAF. (See below.)
2. Dehydrate tissues to 70% EtOH. Carry out necessary dissections in 70% EtOH.
3. Optional treatment for large or dense tissues: Pretreat tissues in a solution of lactic acid saturated with chloral hydrate (approximately 250 g/100 ml) at RT for 12–24 hours. A higher temperature may be required for sturdier tissues.
4. Transfer to BB–4 1/2 clearing solution for 12–24 hours or until cleared [a 1:9 ratio of benzyl benzoate : 4 1/2 clearing solution (2:2:2:2:1, lactic acid : chloral hydrate : phenol crystals : clove oil : xylene)].
5. Examine cleared tissue using DIC optics or continue with staining.

6. To differentially stain tissues use any of a number of acid stains. (See notes that follow.)

Notes

- Chloral hydrate is toxic and should be handled with great care. Use gloves.
- Lignified tissues may be stained with basic fuchsin (0.05 g in 2 ml 95% EtOH, diluted to 100 ml DI water; Johansen 1940; Fuchs 1963) or phloroglucinol (1% phloroglucin in 95 % EtOH acidified with HCl); note, however, phloroglucinol fades rapidly.
- Phloem may be stained using aniline blue, Peterson and Fletcher 1973.
- Toluidine blue O (0.05% aqueous) is a good general polychromatic stain.

HAND-SECTIONING

Hand sectioning of fresh, dried and rehydrated, or even prefixed material is an often-ignored but excellent method of cellular analysis. Hand-sections are surprisingly versatile: they can be produced rapidly and abundantly; they can be disposable or can be preserved indefinitely; they can be observed in the fresh, unadulterated state or can be stained with a number of histochemical dyes. Most importantly, three-dimensional relationships of cells and tissues are readily apparent in hand-sectioned material. Free-hand sectioning with a razor blade is surprisingly easy with many maize tissues because the heavily lignified walls and vascular tissue provide a stable medium. With some practice, skilled sectioning produces uniform sections of between 20 and 40 μm in thickness. Suggestions are outlined below.

Live Tissue (Viewed Live or Fixed *After* Sectioning)

1. Store the plant organ or tissue in the refrigerator until it is well chilled. Submerging the tissue in water is ideal: the goal is to maximize turgidity. Separate portions of the tissue into separate containers so that small areas can be sectioned while the rest remains chilled.
2. Get organized so that sectioning will proceed quickly. This is especially important for young maize tissue, which relies entirely on cell turgidity for stability.
3. Sectioning may be done against either a slide or a firm surface such as a carrot. (See O'Brien and McCully 1981.)
4. Arrange the tissue in the appropriate orientation. Transverse sections are easiest but all orientations are possible. Even young maize seedlings are remarkably easy to section: the rigid enclosing sheaths provide good support even if the tissue of interest is the meristem. The unwanted leaf sheaths can then be teased away and discarded.
5. *The keys to sectioning are*: (a) always use a fresh region of the razor blade for each

cut; (b) cut from a small piece of tissue; (c) always use maximally turgid tissue that is chilled and fully hydrated; and (d) section in a pool of cold water. *Using a dull, dry razor blade on wilted tissue will needlessly frustrate your attempts at sectioning.*

6. Place the sections in water (or fixative, preferably an aldehyde; see below) on a slide and arrange them with a probe or paintbrush. Placing all sections in the same orientation and keeping track of the section order eases serial reconstruction of the tissue.
7. Initially, until skill is developed and sections are of uniform thickness, it is a good idea to place only a few sections under a single coverslip for viewing purposes. If sections are thick by default or by design, make tiny clay feet on the coverslip by scraping the coverslip corner in the clay to raise the slip so that it is parallel with the slide.
8. Unfixed sections may be viewed fresh or may be stained. (See below.) Toluidine blue is especially useful, but it will not stain the interior of a thick, unfixed section.
9. Sections may be made permanent by fixing, staining, and mounting in a permanent medium. Alcohol-based stains are better for permanent slides: sections must be transferred by hand through EtOH to xylene or toluene before mounting in a permanent medium such as Permount. Alternatively, slides can be mounted in glycerine and sealed with nail polish. (see O'Brien and McCully 1981.)

Fixed Material

Tissue fixed in a standard fixative (see below) can also be sectioned freehand, but the tissue will be considerably more difficult to work with because the cells are no longer turgid. In general, fixing sections *after* they have been cut is recommended.

Dried Material

Dried, dehydrated maize plants can be rehydrated and sectioned. It is best to dehydrate them intentionally, either by pressing the tissue or by removing the source of water abruptly. Even plants that have senesced naturally are useful, providing they have not begun to decay. Plants left to slowly rot and die are the least desirable. Suggestions follow:

1. Dry a plant from the field or greenhouse completely.
2. Soak reasonable sizes of tissue in 1% borax in water overnight or until the tissue is rehydrated and also relaxed.
3. Freehand section following suggestions above, except chilling is not necessary.
4. Remember tissue will be fairly flaccid so this requires a careful hand and *MANY* razor blades to get uniform thin sections.
5. These sections may be viewed stained (using protocols below) or unstained. (Remember there will be shifts in wavelengths, if you are looking at autofluorescence.)

Microtechnique

If thinner, more uniform sections are required, it is best to fix, embed, and microtome-section maize tissue (Johansen 1940; Jensen, 1962, Feder and O'Brien 1968; O'Brien and McCully 1981). There are several options for semithin sectioning. The tissue can embedded in either wax or plastic. Useful protocols for plastic-embedded sections of maize are described in Russel and Evert (1985), Van Lammeren (1986), and Becraft et al. (1990). See also Langdale (this volume) for applications of paraffin sections to in situ hybridization and Parthasarathy (this volume) and Varagona and Raikhel (this volume) for the use of LR White for immunolocalization. Any fixation and embedding procedure used for electron microscopy is also usually excellent for light microscopy. Spurr's is an excellent resin of choice for this purpose but see Parthasarathy (this volume) for protocols. Paraffin and glycol methacrylate embedding methods are outlined below. Useful figures that show the results of paraffin embedding may be found in Smith, Langdale and various chapters in the section on Development and Morphology (this volume).

Fixation

Tissue may be chemically fixed using either acid fixatives (FAA, CRAF) or aldehyde fixatives (paraformaldehyde, glutaraldehyde, acrolein). Glutaraldehyde is a finer fixative and can yield much better cellular detail, but only when coupled with plastic embedding. Tissues fixed in glutaraldehyde or CRAF will harden beyond use in a relatively short time and must be removed from the fixative within 24–48 hours. Place a penciled label inside the specimen vial with the specimen. The label will follow all steps. (The solvent will dissolve ink labels.) Glutaraldehyde alone may be too harsh a fixative for immunolocalization or in situ hybridization, but a combination of 2.5% paraformaldeyde + 0.25% glutaraldehyde in a phosphate buffer (see below) will yield excellent results. (See relevant chapters for protocols.) The fixatives below are for standard light microscopy.

1. FAA: 50 ml EtOH, 10 ml formalin (37%), 5 ml glacial acetic acid in 35 ml DI water. Fix small, soft tissues for 4–6 hours; and hard or dense tissues for more than 24 hours. Tissues can be stored in FAA. Important: vacuum infiltrate the fixative for 3–5 minutes until the tissue sinks.
2. CRAF III: Mix 30 ml of 1% CrO_3 (w/v), 20 ml of 10% glacial acetic acid (v/v), 10 ml formalin (37%), and 40 ml DI water. Fix tissues 12–24 hours. Do not let CRAF solution turn purple. Remove tissues within 48 hours and proceed with dehydration.
3. 4% glutaraldehyde: Mix 8 ml of 25% glutaraldehyde in 42 ml Sorenson's buffer at pH 7.0 (made by adding 39 ml of 0.2 M NaH_2PO_4 + 61 ml of 0.2 M Na_2HPO_4 + 100 ml DI water). Fix soft small (<5 mm) tissues for 4–12 hours at 4 °C and rinse 2× with buffer (10–30 minutes each).

Dehydration

Both ethanol and Methyl Cellosolve (ethylene glycol monomethyl ether) yield acceptable results. Methyl Cellosolve is a more rapid dehydration and is somewhat easier for cytological studies, but use ethanol for in situ hybridization studies.

1. EtOH dehydration: Dehydrate glutaraldehyde-fixed tissue through graded ethanol series starting with 30% EtOH (then 50, 70, 90, 95, 100 × 2 for 30 minute to 1 hour each). Increase the time for hard or large samples. The final EtOH step should include Safranin (1% w/v) to stain tissues. The intermediate steps between the EtOH and paraffin treatments must include the paraffin solvent (a) xylene or (b) tert-butyl alcohol (TBA). The use of TBA may preserve some tissue softness.
   a. Transfer into paraffin via xylene: Replace the EtOH with xylene in graded steps (3:1 EtOH : xylene, then 1:1, 1:3, 100% xylene × 2) for 1 hour each. After the final xylene step, add several small pieces of solid paraffin to the xylene, cap the vial, and store at 42° C for a few hours. Repeat until the paraffin no longer dissolves (a few days); then transfer the vial (uncapped) to a 58° C paraffin oven and proceed with infiltration.
   b. Transfer into paraffin via TBA: Replace the EtOH with TBA in graded steps (5:1 EtOH : TBA, then 5:2, 5:3, 1:1, 1:3, absolute TBA × 3) for 1 hour each. TBA is solid at room temperature but can be stored at 42° C to be maintained as a liquid. Pure TBA steps should be at 42–58° C.
2. Methyl Cellosolve (MC) dehydration. This method utilizes the chemical reaction between MC and water to form methanol and acetone. The treatments are:
   a. Replace aqueous fixative with pure MC.
   b. Incubate at 4° C for 2 hours.
   c. Replace with fresh MC, incubating 3× at 2 hours each.
   d. Add absolute EtOH.
   e. Let stand 20 minutes at 4° C.
   f. Go back to step d and repeat twice.
   g. Store at 4° C or go to the standard EtOH or TBA dehydration (above).

Infiltration

This step will replace the solvent with liquid paraffin or plastic monomer. It is very important to allow enough time during the steps for the solutions to completely penetrate the tissues. Times listed are only starting points and should be determined empirically for each tissue type. Developmental stage should be taken into account: older tissues will be harder to infiltrate.

1. Paraffin.
   a. Carefully pour one-third volume melted paraffin into the vial and leave off the vial cap. (The paraffin will solidify on top of the solvent.) Leave room in the vial

for additional wax. Incubate at 58° C until the wax melts. After 4 hours add 1 volume additional liquid paraffin.

b. After 4 hours (or longer) pour off half of the liquid and replace with an equal volume of melted paraffin. Repeat 2–4 times to slowly replace the solvent with liquid paraffin.
c. Pour off all of the paraffin/TBA (or paraffin/xylene) mixture and add pure liquid paraffin. Redo 3× at 1–4-hour intervals (overnight for the last step).
d. Proceed with "Embedding" step.

2. Glycol methacrylate (JB-4; GMA). The tissues can be dehydrated to 95% EtOH since GMA is hydrophilic. Mix GMA monomer with peroxide according to the manufacturer's instructions, and infiltrate (3 EtOH: 1 GMA; 1:1; 1:3; 100% GMA × 2) for a minimum of 1 hour each (overnight for the final step).

Embedding

1. Paraffin. On the hot side of the embedding tray, pour the paraffin + sample into a "boat" (make aluminum or paper boats about 2 × 1 × 0.5 inches), arrange the specimen appropriately (i.e., upright if you want to make cross sections), and move the boat to the cool end of the tray to solidify the specimen within the paraffin. Move the tray to an intermediate temperature first. The paraffin at the bottom of the boat will solidify first, causing the specimen to "stick" so that you can manipulate it into an upright position. Place small, penciled labels in the wax, facing out, so that the specimen blocks can be identified later. Pour the paraffin at a temperature no greater than 60° C. Hotter paraffin tends to shrink, causing tissue damage.

2. GMA. GMA embedding is done at room temperature. Mix monomer + accelerator + catalyst according to the manufacturer's instructions. The highest ratio (slowest polymerization) is recommended. Pour into a suitable mold and place in the specimen. Molds can be film-can tops or special JB-4 molds. Arrange the specimens appropriately. Finally, cover the mold with Parafilm to exclude oxygen. (O_2 inhibits polymerization.)

Sectioning

1. Paraffin. Use a standard rotary microtome for paraffin sections of a thickness greater than 5 μm. Sections are cut using a metal knife and come off in ribbons that can then be mounted directly on slides. For short-term storage, place the ribbons into a suitable dust-free box. (A photographic paper box works well.) Sections cut at 8–12-μm thickness give the best results. During sectioning, store the next paraffin blocks on ice.

2. GMA. Use a retracting microtome (Sorval JB-4, Zeiss Microm) for GMA sections of a thickness less than 5 μm. To get ribbons, coat the top and bottom surfaces of the

block with Barge or Pliobond cement (obtainable at any hardware store). This yields a sticky edge that (with luck) will allow the sections to stick together as a continuous ribbon. Glass knives are used to cut plastic sections. The knives are made from microscope slides and attached to an aluminum knife holder with dental wax. D-face tungsten-carbide knives or standard triangular glass knives made on an LKB knife maker may also be used to cut these plastics.

Mounting the Sections

Put a drop of Haupt's solution (1 g gelatin, 0.5 g sodium benzoate dissolved in 100 ml DI water at 30–35° C) on a clean glass slide and smear it into a thin film. Let air-dry and store desiccated. Place the slide on the warming tray and flood it with 4% formalin. Then float the sections onto the liquid. Allow for about 25% expansion as the sections will heat up and stretch out. (See procedure below.) After drying, the slides can be stored in slide boxes. Be careful however, because the mounted sections will be fragile and could be scratched easily. After mounting, proceed with paraffin removal and staining.

Staining

There are innumerable staining methods ("schedules") for histological investigation. See O'Brien and McCully (1981), Johansen (1940), and *Staining Procedures, Fourth Edition* (Clark 1981) for details. Following are examples of a general-purpose multiple staining schedule (Safranin/fast green), a cell-component-specific stain (orange gold), and a general-purpose aqueous stain (Toluidine Blue).

Safranin and Fast Green (Johansen 1940)

This method yields a brilliant staining of plant tissues. Safranin tends to overstain tissues and thus requires acid differentiation followed by a destaining step. Safranin appears brilliant red in chromosomes and nuclei and in lignified, suberized, and cutinized cell walls. Fast green appears brilliant green in cytoplasm and cellulosic cell walls. Fast green turns blue in basic solutions.

1. Deparaffinize in xylene and bring slides to 70% EtOH (i.e., xylene, absolute EtOH, 95%, 70%).
2. Stain 2–24 hours in 1% Safranin in a mixture of 1:1 Methyl Cellosolve:50% EtOH, plus 1%(w/v) sodium acetate (intensifies stain) and 2% (v/v) formalin solution (mordant).
3. Wash out excess stain for a few moments with DI water.
4. Dehydrate 10 seconds in 95% EtOH/0.5% picric acid for Safranin differentiation. Be aware that picric acid is explosive in the crystalline form.

5. Wash 1–2 minutes (no longer) in 95% EtOH + 4 drops ammonia/100 ml to stop picric acid action.
6. Dip briefly (10 seconds to 10 minutes) into 100% EtOH to finish dehydration.
7. Counterstain <15 *seconds* in fast green FCF solution (0.5% w/v in Methyl Cellosolve:Absolute EtOH:clove oil, (1:1:1). Methyl salicylate may be substituted for clove oil. As the fast green dye solution evaporates with use, add additional solvent *minus* fast green.
8. Rinse excess fast green with used clove oil (or methyl salicylate). You can gently apply clove oil with a pipet, taking care not to squirt it directly on the sections. Save clove oil rinse, dilute 50% with a 1:1 mixture of absolute EtOH:xylene, and reuse this solution for rinsing slides.
9. Clear by dipping the sections for a few moments (<10 seconds in a mixture of clove oil:absolute EtOH:xylene, 2:1:1).
10. Remove clearing solution by dipping for a few moments into xylene plus 2–3 drops absolute EtOH to remove residual water.
11. Rinse twice in absolute xylene.
12. Mount coverslip.

Safranin and Orange Gold (Sharman 1943)

This staining series is a good general-purpose stain for meristematic tissues. The following is a partial list of results: Cell walls—blue-black; nuclei—yellow to orange; starch grains—black; lignified walls—brilliant red. Tissues may be preserved with any fixative.

1. Deparaffinize and bring tissues to water through xylene, xylene/EtOH, and EtOH.
2. Transfer to 2% $ZnCl_2$ (filtered) for 1 minute.
3. Wash in DI water for 5 seconds.
4. Transfer to safranin diluted 1:25,000—2 ml of a 2% stock in 1 liter of water, filtered. Stain for 5 minutes.
5. Wash in DI water for 5 seconds.
6. Transfer to orange G–tannic acid and stain for 1 minute. Make this stain by mixing 2 g orange G; 5 g tannic acid; 0.3 ml 1:5 HCl; a few crystals thymol; 100 ml DI water, filtered.
7. Wash in DI water for 5 sec.
8. Transfer to filtered tannic acid for 5 minutes. Make this solution by mixing 5 g tannic acid and a few crystals thymol in 100 ml DI water.
9. Wash in DI water for 5 seconds.
10. Transfer to $NH_4Fe(SO_4)_2$ (1% aqueous, filtered) for 2 minutes.

11. Wash in DI water for 5 seconds.
12. Dehydrate through 45%, 90%, 100% EtOH, about 10 seconds each step.
13. Transfer to methyl salicylate/xylene for >1 minute.
14. Transfer to pure xylene >1 minute and mount coverslip.

Toluidine Blue (TBO)

TBO is useful because it exhibits metachromasia: at low pH, acidic components of the cell appear red and more basic components appear blue. Lignin and other polyphenols appear blue to blue-green. Use a 0.1% solution TBO in benzoate or acetate buffer at pH 4.4 to bring out metachromasia. Paraffin sections must be deparaffinized but proceed directly to step 2 for plastic or freehand sections.

1. Deparaffinize and bring to water following above directions.
2. Flood the sections with 0.1% aqueous toluidine blue and incubate 1–10 minutes. If the stain is too intense, decrease the concentration to 0.01%.
3. Wash with DI water and mount the coverslip.

Mounting Coverslip

Slides should be brought to xylene through a graded EtOH series if they were last stained with an aqueous stain (i.e., 30%, EtOH, 50, 75, 90, 95, 100, 1:1 xylene: absolute EtOH, 100% xylene). Put coverslips on slides one at a time, keeping the remainder in xylene to avoid tissue damage by drying. Put a drop or 2 of mounting medium (Permount) on a slide, place the coverslip on the slide, putting one edge down first, and then lower the other edge until the adhesive touches the coverslip. Continue lowering the slip, taking care that no bubbles form. Dry on the warming tray for at least 24 hours. Label the slide and store upright in a slide box.

REFERENCES

Arnott HJ (1959) Leaf clearings. Turtox News 37(8): 192–194

Becraft P, Bongard-Pierce DK, Sylveser AW, Poethig RS, Freeling M (1990) The *liguleless-1* gene acts tissue specifically in maize leaf development. Dev Biol 141: 220–232

Clark G (ed) (1981) Staining Procedures, Fourth Edition. Williams & Wilkins, Baltimore/London

Feder N, O'Brien TP (1968) Plant microtechnique: some principles and new methods. Am J Bot 55: 123–142

Fuchs C (1963) Fuchsin staining with NaOH clearing for lignified elements of whole plants or plant organs. Stain Tech 38: 141–144

Green PB (1988) A theory for inflorescence development and flower formation based on morphological and biophysical analysis in *Echeveria*. Planta 175: 153–169

Herr JM (1972) Application of a new clearing technique for the investigation of vascular plant morphology. J Elisha Mitchell Sci Soc 88(3): 137–143

Jensen WA (1962) Botanical Histochemistry, WH Freeman, San Francisco

Johansen DA (1940) Plant Microtechnique, McGraw-Hill, New York, 523 pp

O'Brien TO, McCully ME (1981) The Study of Plant Structure: Principles and Selected Methods, Termarcarphi, Melbourne, Australia

Peterson CA, Fletcher RA (1973) Lactic acid clearing and fluorescent staining for demonstration of sieve tubes. Stain Tech 48(1): 23–27

Russel SH, Evert RF (1985) Leaf vasculature in *Zea mays* L. Planta 164: 448–458

Schmid R (1977) Stockwell's bleach, an effective remover of tannin from plant tissues. Bot Jahrb Syst 98: 278–287

Van Lammeren AAM (1986) Developmental morphology and cytology of the young maize embryo Acta Bot Neerl 35: 169–188

11

Light Microscopy II: Observation, Photomicrography, and Image Analysis

STEVEN E. RUZIN and ANNE W. SYLVESTER

A number of methods to prepare maize tissue for microscopical investigation were outlined in the previous section. Successful microscopy also depends on appropriate observation skill and proper use of the instrument. This next section considers some ways in which to observe the specimen, record an image, and also analyze and print the image using computer technology.

METHODS OF OBSERVATION

The compound light microscope is an instrument of great versatility. It is useful to have a working understanding of the general physics and construction of a compound microscope. A good summary of the subject may be found in a Kodak data book entitled *Photography Through the Microscope, Ninth Edition* (1988; available from Eastman Kodak Co., Rochester, NY). Standard methods of illumination and tips for optimizing the image are outlined below:

The Maize Handbook—M. Freeling, V. Walbot, eds.

Desirable Features of the Microscope

1. Access to photography
2. A full complement of lenses that optimize the resolving power of the microscope *for your purposes* (see notes below)
3. A clean dust-free location
4. A stable surface with minimal vibration
5. A room that may be easily darkened

Notes on Resolving Power and Lenses

Resolution depends on several factors, an important one being the numerical aperture of the objective lens. There are many different types and grades of objective lenses. Each lens is engraved with the Numerical Aperture (NA), the magnification, and the type of optical correction.

Numerical aperture (NA) The NA is related to the refractive index of the medium the lens is immersed in (air, oil, etc.), and to the angular aperture of the lens. It refers to the light gathering ability and the resolving power of a particular lens. Lenses of high numerical aperture (NA) are desirable because they have higher resolving power and are brighter for fluorescence microscopy. Any objective with an NA over 1 must be immersed in a medium that has a refractive index over 1 (e.g., immersion oil). Resolving power also depends on the NA of the substage condenser lens, so a high-NA objective lens is best combined with a high NA condenser lens (preferably oiled, but this can be and usually is air). The specimen is then sandwiched between two layers of oil and topped by two high-NA lenses.

Magnification Final magnification of the object will depend on the magnification of the objective times that of the eyepiece times any intermediate magnification (usually 1.25×). It is possible to produce "empty magnification" when magnification of an image is beyond the resolving ability of the microscope. Multiply the NA by 1,000 to get a rough estimate of the magnification range of a particular lens. Going beyond this range will not increase resolution.

Degree of optical correction A number of inherent optical aberrations may be corrected in a given lens. (See below.) Note that the most completely corrected lens is a Plan Apochromat. Combined with a high NA, this lens is the most expensive and yields a superior image.

Achromat: chromatic aberrations corrected for two wavelengths (blue and red) and spherical aberrations corrected for one wavelength (green).

Apochromat: chromatic aberrations corrected for three wavelengths (red, green, and blue) and spherical aberrations corrected for two wavelengths (blue and green); hence they are best for color photography.

Flatfield: Curvature of image field is corrected in these lenses. Most modern objectives are flatfield objectives and are thus designated with the prefix 'plan-' or 'plano-' as in planapo chromat.

Adjusting the microscope for Köhler illumination Köhler illumination provides even, fairly high intensity illumination. When a microscope is properly adjusted for Köhler illumination, the illumination source is centered and focused on the object plane, resulting in the objective aperture being illuminated with the cone of light of the same diameter (on the focus plane) as the field of view.

The simple steps outlined below assume the lamp filament is centered. See instrument instructions for details if you must change a bulb or suspect the filament is out of alignment. These steps should become second nature when you sit down at the microscope.

1. Focus on a specimen at low magnification (eg. 10×)
2. Close down the field diaphragm (located nearest the light source).
3. Bring the edges of the field diaphragm into focus by raising or lowering the substage condenser.
4. Center the image of the field diaphragm using the condenser-centering knobs (usually facing you on the condenser).
5. Open the centered and focused field diaphragm so the light just fills the field of view. This must be repeated each time you change lenses in order to reduce glare.
6. Adjust the aperture (substage) diaphragm on the condenser for image contrast. The optimal position depends on the specimen and the optical system present. It must be determined empirically. Closing down this diaphragm will increase contrast at the expense of resolution. Opening the diaphragm does the reverse: it optimizes resolution but decreases contrast and depth of field. It is always helpful to adjust this aperture until you get the combination you need. Never use this aperture to control light intensity. (See photography section that follows.)

Other Optical Techniques

The standard brightfield microscope can be outfitted with a variety of attachments that will enhance contrast or change the image appearance. Their use should be determined by the specimen in question. See the Kodak data book for more details.

Darkfield illumination utilizes only scattered or reflected light to make the specimen appear bright against a black background. This is useful for looking at single cells, silica inclusions in maize or other crystals, and autoradiograph silver grains. A special substage condenser with a darkfield stop, which blocks the cone of light from entering the objective lens, may be used. A standard microscope may be converted by inserting an opaque filter of the appropriate diameter (determined empirically) in a filter carrier below the substage diaphragm. Glare can be a problem with darkfield so be sure the lenses, slide, and coverslip are dust- and scratch-free. The condenser NA must be greater than the objective NA. An oiled condenser further improves the image but is not required.

Phase contrast is accomplished by recombining light rays so that phase shifts imparted by the specimen cause wave interference. When out of phase by one-half wavelength, maximum reduction of intensity occurs. An annular aperture located at the condenser diaphragm illuminates the specimen in such a way that diffracted and direct beams of light are separated. A phase plate located in the objective lens retards the diffracted light from the direct beam by one-quarter wavelength. Light passing through the specimen may be shifted an additional one-quarter wavelength. Interference is generated when these out-of phase beams are recombined in the image plane. This greatly increases contrast of the specimen and permits visualization of otherwise transparent objects. It is particularly useful for observing single cells or thin slices of living tissue. Required microscope attachments include a phase condenser, with an annular aperture for each objective, and special objectives, each with a built in phase plate. Important steps to setting up phase contrast follow:

1. Be sure all components are dust-free.
2. Set up proper Köhler illumination. (See above.)
3. Align the annular diaphragm with the phase plate. Remove the eyepiece and look at the back focal plane of the objective lens with a phase telescope or Bertrand lens. Center the light ring with the annulus using the centering knobs on the condenser.

Notes

Oiling the condenser is always optimal, but it is messy and not essential, especially if you tend to not clean after each use. Dried-on oil is worse than no oil.

A thin specimen is required: a thick specimen will destroy the phase relations because of multiple phase changes.

Differential Interference Contrast (*DIC, Nomarsky optics*) separates incident light into reference and direct beams before it interacts with the specimen. The phase of the reference beam may then be changed by interaction with the specimen before it is recombined with the direct beam. Specimen contrast is created by constructive or destructive interference of the two beams. The DIC microscope is based on polarization principles to accomplish interference. See listed references for detailed discussion. The image produced by DIC enhances contrast just as phase contrast does, but it eliminates the "halo effect". The prisms used may also introduce color differences and highlight the specimen with a three-dimensional relief. DIC works best for nonstained, living tissue. Microscope attachments are required, but the position of the attachments varies considerably with each microscope. In order of passage, the light proceeds through a plane polarizer, a Wollaston prism, a condenser, the specimen, the objective, another (inverted) Wollaston prism, another polarizing filter (analyzer) set at right angles to the first, the ocular, and the eye or film plane. Simplified steps for setting up DIC follow:

1. Follow steps 1–3 for phase-contrast microscopy above.
2. Set the polarizer at maximum extinction, if the microscope permits; i.e., move it slightly until the image is as dark as possible. Many new microscopes do not have moveable polarizers: they are set to extinction (in relation to the analyzer) by the manufacturer.

An older microscope will have a polarizing filter below the condenser; it is usually marked to indicate the proper position but it may need fine tuning.

3. Insert the Wollaston prism (located in each objective in modern Zeiss microscopes but above the objective turret in all other microscopes) and translate the prism using the adjusting knob. If the full order of interference colors is available, you will see the image pass through a full spectrum as you adjust the prism through its full range. Keep adjusting the prism until you find the position at which the background is a uniform gray, but at which the specimen appears to be illuminated from one side. This is the optimal position for the prism. You will notice the specimen appears three-dimensional and has the finest resolution.

Polarization microscopy relies on using cross-polarizers to detect optical anisotropy in the specimen. Anisotropy refers to the degree of molecular orientation of crystals or compounds such as cellulose in the cell wall. Maize cell walls are highly birefringent and can be studied effectively using polarized light. The light is first polarized, using a polarizing filter located somewhere between the light source and the specimen. Another polarizing filter, the analyzer, is located between the specimen and the eye or film plane. The goal is to orient the polarizer and analyzer so that their polarization planes are crossed and no light is transmitted. When an anisotropic specimen is oriented such that its crystalline structure is at 45° to the crossed polarizers, light will pass through the analyzer. By rotating the specimen through all angles relative to the crossed polarizers, specimen birefringence can be detected. Simple polarization can be done by using sheets of polarizing filters in the appropriate places. Special polarization microscopes are available also. These are equipped with a compensator or retardation plate and a fully rotatable stage, necessary for detecting weak birefringence and to measure thickness based on interference colors. Steps for setting up a standard microscope for polarization follow.

1. Follow steps 1–3 for phase contrast above.
2. Place a polarizer between the light and the condenser, usually on the field diaphragm. Tape an analyzer to the ocular (now you are limited to only using that eye) or else insert an analyzer into any available slot between the objective and the ocular.
3. Rotate the polarizer only until there is no light available (maximum extinction).
4. Rotate the specimen. Any bright areas indicate the presence of birefringent material. Note: You must have a rotatable stage for this. To orient yourself to the specimen, shift back and forth between brightfield and polarization by removing the polarizer.

Epifluorescence microscopy is usually accomplished using epiillumination from a high pressure mercury lamp. Xenon arc or tungsten-halogen lamps are also used. Fluorescence microscopy using transmitted light is not discussed here but is described in O'Brien and McCully (1981) and elsewhere. Object fluorescence is generated because light of short wavelength (e.g., ultraviolet) is absorbed by the specimen or by a specific fluorochrome and then is reemitted at a longer wavelength. An epifluorescence microscope is set up with a series of filters that are customized to the particular use. First, an excitation filter provides light over a selected narrow band of wavelengths. Then a dichroic mirror (chromatic beam splitter) reflects light coming through the excitation filter to the specimen and selectively cuts off specific wavelengths. Light then impinges on the specimen and is reirradiated and collected by the objective lens. Emitted fluorescence (having a longer wavelength) then passes through the dichroic mirror, through a barrier filter (to block excitation light), and finally to the eye or camera. Spectral transmission characteristics are available from the microscope manufacture. Each microscope can be customized with a particular complement of filters to suit individual needs. A good standard array of filters includes (1) a UV filter that excites around 360–380 nm (Zeiss UV-H 365), (2) a blue filter (Zeiss Blue 450–490), and (3) a green filter (Zeiss Green H 546). Two types of investigations using epifluorescence microscopy are commonly used:

Autofluorescence Maize and other plant cells contain a vast array of inherently fluorescent compounds, including chlorophyll, lignin, ferulic acid, *p*-coumaric acid, suberin, sporopollenin, etc. These can be used effectively to study genetic mosaics (Hake and Freeling 1986), wall development, and tissue stages. Each compound is excited by and emits at a characteristic wavelength. These features will determine the particular complement of filters for the microscope (specified later in this chapter). Note that some autofluorescence characteristics may be quenched by certain dyes. For example, toluidine blue quenches the UV-induced autofluorescence of lignin in the cell walls of maize.

Fluorochromes A number of fluorochromes are commercially available for a variety of applications. Fluorescent dyes are used to label DNA, to couple with antibodies in indirect immunofluorescence, to trace movements of aqueous solutions, to stain cell walls with optical brighteners, etc. See Staiger (this volume) for the use of fluorochrome-labeled antibodies in the study of maize cytoskeleton using indirect immunofluorescence. The labels discussed there include FITC (using the blue filters), rhodamine (using green filters) and DAPI (using UV filters). Other specific fluorochromes are given later in this chapter. Note that a number of histochemical stains, used for standard light microscopy, also exhibit fluorescence. These include aniline blue and Schiff's reagent.

The steps to set up epifluorescence illumination are outlined below.

1. Turn on the light source according to the manufacturer's instructions.
2. Be sure the lamp is centered, following the manufacturer's instructions. (See notes below.) An uncentered lamp will greatly reduce intensity and evenness of illumination.
3. Examine the specimen first with bright field to orient position and magnification.
4. Insert appropriate filters for the application but be certain that the blocking filter is in place until you are ready for viewing.
5. View the specimen by covering the bright field light (or turn it off) and by sliding

open the blocking filter. It is a good habit to slide the blocking filter in frequently, for example while changing objectives, scanning the specimen, etc., because many fluorochromes bleach rapidly and should receive minimal light exposure.

6. The objective is also the condenser in epi-illumination, so focusing the object also focuses the exciting light on the specimen. A diaphragm in the light path may be used to reduce the amount of light entering the filters.

Notes Mercury bulbs should be allowed to warm up and cool down before they are turned off or on. Instructions vary, but one guideline is to leave the lamp on for at least 15–30 minutes before turning it off. The lamp should not be turned on again until it is cool. If the microscope is used heavily, lamp life can be extended by leaving the bulb on rather than turning it on and off. A conservative estimate for lamp life is approximately 200 hours. The bulb can generally be used somewhat longer, until it is dimming or flickering. It is best not to push the life too long because the bulb may explode.

Centering the mercury lamp should be done at the time the bulb is changed and then fine-tuned once the filament is in place and the lamp is lit. One approach (there are many) is to place a blank piece of paper (such as an index card) on the stage, focus on the cellulose fibers in the card, remove the objective, and observe the light pools on the white card, which represent the real and mirror images of the filament. Move the position of the filament controls on the lamp housing to align them so they are next to each other. *Do not superimpose the two light spots.* The resulting image when the objective is replaced should be even, fairly intense illumination. The paper card is useful because it is an even field itself and it will readily show shadows where the bulb is out of alignment.

RECORDING THE IMAGE: PHOTOMICROGRAPHY

Recording microscope images on film is, of course, a primary goal of microtechnique. Each microscope and/or camera has its own particular method of operation and the instruction manuals should be studied carefully. A camera may be an integral part of the microscope system, a proprietary attachment with exposure meter and shutter, or a standard 35-mm single-lens reflex (SLR) coupled to the microscope using an adapter tube and T-mount. SLRs have the advantage of having a focal-plane shutter that can achieve shutter speeds of 1/4,000th second or faster (compared to the 1/100th second maximum speed of the Copal shutter).

Color Photography With Brightfield Illumination

The light source of microscopes is usually a tungsten filament light bulb and thus the color temperature (emission spectrum) of the bulb is different from sunlight. (It's more red.) When using color film, it is important to match the color balance of the film to the spectral emission of the light source. Film comes in two types: daylight (5500° K) and tungsten (3200° K). Daylight-balanced film such as Kodachrome 64 slide film and most print films require a daylight balancing color compensation filter (blue: 80B) between the microscope objective lens and light source. The emission spectrum of a tungsten bulb is correlated to the temperature of the filament; more voltage yields a hotter filament. It is necessary to turn the light control to maximum to get the correct color temperature. Some microscope exposure meters are coupled to the light source to automatically adjust it to the appropriate color temperature. The intensity of the lamp at 3200° K may cause the exposure meter of the microscope or camera to register overexposure. To correct this but maintain the correct color spectrum, interpose one (or more) neutral density filters. See Blaker (1976) for additional information regarding color photography. For an in-depth discussion of B/W photography see Schaefer (1992).

The recommended slide film for tungsten-illuminated microscopes is Kodak Ektachrome 160 (ET 135–24, 36).

Fluorescence Photomicrography

Color photomicrography of fluorescent objects is best accomplished using a high-ASA (ISO) film to decrease the exposure time to a reasonable level. Although the concept of color balance does not apply, the long exposure times frequently associated with fluorescence microscopy necessitate cognizance of the reciprocity characteristics of film.

The exposure characteristic of film (exposure = light intensity × time; i.e., reciprocity) applies only to a normal range of light and departs from that relationship at very long or very short exposures. At low light intensities, film becomes less sensitive to light than the ASA rating predicts, requiring an increased exposure time to yield an acceptable film contrast. Some microscope exposure meters (e.g., Zeiss) have predefined reciprocity settings. The appropriate setting for your particular film should be programmed into the exposure meter so as to obtain properly exposed pictures, especially of low intensity fluorescent objects. Increased exposure times due to reciprocity departure may also be determined empirically.

Focusing

For the photograph to be in focus, the cross hairs on the focusing screen *must* be focused to sharp parallel lines. Relax your eyes, cover one eye (do not squint), and focus the eyepieces until the cross hairs of the frame come into focus. It's best to not have a specimen in place so as to reduce the natural tendency for your eye to focus between the microscope slide and the cross hairs.

Contrast

Image contrast is best controlled using the condenser diaphragm (see above), however with black and white photography there are two alternate methods available.

1. Interposing a color filter between the light source and specimen. The color of the filter should be opposite on the color wheel from the primary color of the specimen. eg: Use a green filter for safranin-stained or other red-stained tissue.
2. Underexpose 1–2 stops and overdevelop the film 50–100%. This technique exploits the properties of B/W film whereby dark areas (silver grains) "develop" at a faster rate than light areas artificially increasing contrast. [Of course, in a rudimentary form, this is the basis of the White/Adams technique of the Zone System for standard macrophotography.]

Photographing Very-High-Contrast Objects

Standard center-weighted exposure metering is not useful for fluorescence or darkfield microscopy because these specimens frequently consist of a small number of bright objects on a dark background. Center-weighted metering tends to overexpose the specimens as it averages the entire field of view, which is mostly black. In this situation, spot metering is required. Most microscope photography systems have spot metering capability as do a number of modern SLR cameras (Nikon, Contax, Canon).

1. Center the bright object directly under the cross hairs or the center circle in the focusing screen of the SLR and determine the exposure of the specimen.
2. Store the exposure time.
3. If necessary reposition the specimen to view the maximum number of objects in the frame or to get the most aesthetic arrangement.
4. Take the photograph.
5. Reset the exposure meter before taking another photograph.

CONFOCAL LASER SCANNING MICROSCOPY

A confocal microscope is a type of fluorescence microscope that visualizes thin optical sections of (transparent) objects by blocking from detection all returning nonaxial light. Nonaxial light is light that emanates from below and above the plane of focus. A conventional light microscope transmits this nonaxial light and thus limits resolution at the focal plane itself. Epifluorescence microscopy is particularly susceptible to out-of-focus light since all fluorescent particles in the illuminated sample field emit light equally. This tends to obscure resolution and produce the typical fluorescent "fuzzy" image. A confocal laser scanning microscope (CLSM) illuminates the sample only by a small spot that traverses the object in a zigzag or raster pattern. In this way, any one small sample area is illuminated for only a very brief time. As the spot of excitation light scans the sample, an electronic photodetector (or detectors for more than one color detection) records the resulting fluorescence from that focal plane as a level of light intensity and the scanner controls record the X and Y coordinates of the spot. A computer then reconstructs the fluorescence image spot by spot and paints onto the computer screen a representation of the fluorescent sample plane.

The CLSM may record many thin serial optical sections of a given sample starting at "the bottom" and scanning to "the top," and thus a three-dimensional view of the original biological specimen can be reconstructed. Present computer software can render internal details of the sample as well as the surface because all of the spots (pixels) of all of the optical sections are retained in the computer image files.

Manipulating the three-dimensional CLSM image is a computational-intensive process that requires a large amount of computer power. CLSM manufacturers have approached this problem by either (1) utilizing a powerful UNIX workstation (Sarastro/Molecular Dynamics uses a Silicon Graphics workstation), (2) adding an additional, RISC-based, graphics processor board to a personal computer (BioRad adds an optional RISC board to their DOS-based machine), or by (3) developing a proprietary computer hardware/software combination (Leitz).

Preparing specimens for CLSM observation is the same as for conventional fluorescence microscopy. (See Staiger, this volume and Pawley (1990).) Because of limitations and/or peculiarities of CLSMs, certain specimen preparation techniques must be considered.

1. Three-dimensional image reconstruction. For 3-D imaging the specimen must be transparent (or nearly so) for the excitation beam (laser) to penetrate and for the fluorescence emission to be collected without distortion by overlying tissues. Small tissues or individual cells are best. Large tissues and organs must be cleared. (See previous chapter.)

2. Chlorophyll fluorescence from green tissues can obscure probe signal, especially rhodamine-conjugated probes. Ethanol extraction (2–4 washes, 70–90%) will remove most of the pigment. An alternate (or complementary) technique is to place a 565-nm short pass filter in front of the detector to block most of the chlorophyll fluorescence. Rhodamine fluorescence (peak 565 nm) will pass through the filter. UV lasers are available for excitation of different fluorochromes, but presently they are expensive, somewhat unstable, and have a relatively short lifetime.

3. Common fluorochromes. Most commercial CLSMs come equipped with argon-ion lasers that emit three major lines: 454 nm (violet), 488 nm (blue), and 514 nm (green). Commonly used fluorochromes are fluorescein, rhodamine, Texas red, bodipy, Lucifer yellow (a membrane dye), calcium green and fluo 3 (Ca^{+2}-responsive dyes). Consult the catalog of fluorescent dyes from Molecular Probes Inc. for additional and newly developed probes.

4. DAPI cannot be used. The chromatin dye DAPI excites only in the UV and thus cannot be utilized in conventional CLSMs. The alternate dye propidium iodide (PI) can be substituted. Use the same protocol as with DAPI (1 μg/ml). PI is a potent carcinogen so use appropriate precautions.

5. Observation time. Since most CLSMs are raster imaging devices, true "real-time" confocal microscopy is not possible. Scanning a single focal plane, even at large pixel size (low resolution), takes 1 to several seconds.
6. Surface features. Most CLSMs have the ability to record the surface of objects by imaging reflected light.
7. Transmitted light. Most CLSMs have an additional transmitted light detector. Be aware, however, that the transmitted light is not confocal.

VIDEO MICROSCOPY AND COMPUTER-MEDIATED IMAGE MANIPULATION

Computers may be used to perform a number of tasks that are frequently avoided because they are difficult or tedious. Particle counting, object size, and area (morphometrics) are just a few of the manipulations available to the microscopist who has access to powerful desktop computers. Image analysis software is available on a number of computer platforms including DOS, Macintosh, and UNIX workstations. We will describe a system based on the Apple Macintosh because it is a relatively inexpensive, yet powerful computer. Furthermore, the Macintosh's ease of use and uniform interface across applications make it accessible to a larger portion of the scientific community than either DOS or UNIX. Finally, there are a number of image analysis programs available for the Macintosh including one of the strongest analysis programs: NIH Image. A powerful incentive for using NIH Image is that it is obtained free from common bulletin board services. (see below.)

Microcomputer image analysis systems are usually 8-bit, gray-scale systems. They are capable of displaying up to 256 shades of gray or different colors (for false color images) or both gray scale and color. Images obtained and manipulated on the computer are stored on disk and recalled at any time for further processing and/or printing. Printing the image may be through any compatible printer and at from 72 to over 4,000 DPI. Images printed at 4,000 DPI or greater retain near photographic quality contrast and resolution.

In the following section we describe a system based on the Apple Macintosh computer running the public domain software NIH Image. Consult Inoué (1981) and Russ (1990) for further, detailed information on video microscopy and image processing.

Hardware (Figure 11.1)

1. Any Macintosh computer with at least 4 Mb RAM, >40 Mb (HD) hard disk, an 8-bit monitor and at least one NUBUS slot.
2. Digitizing board (frame grabber): Data Translation (Marlboro, MA) Quick Capture DT2255 or Scion (Frederick, MD) LG-3 NUBUS boards. This electronic component

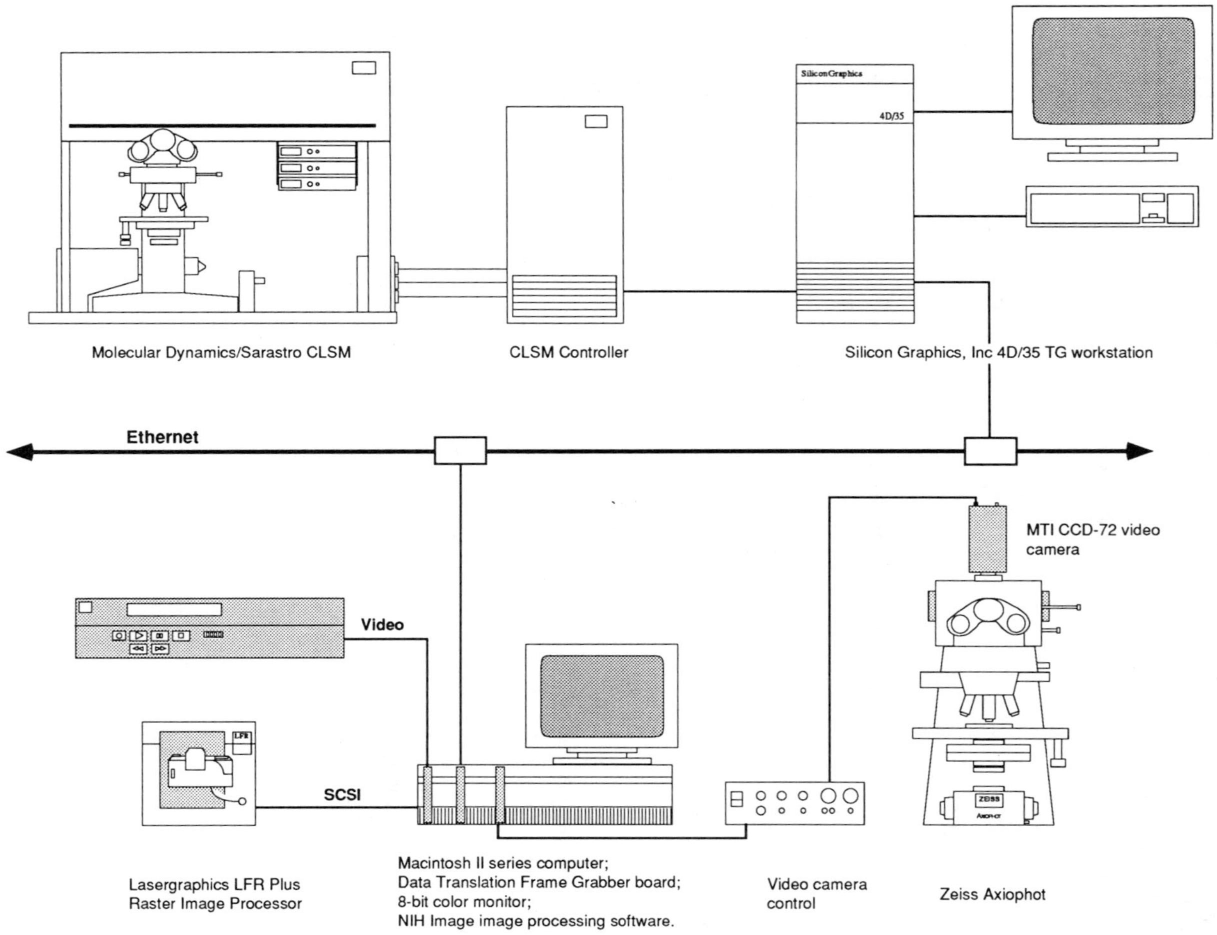

FIGURE 11.1 Network configuration linking confocal microscope, Macintosh computer, and image printing hardware.

digitizes and stores the analog video signal from the video camera. These are 8-bit boards and are capable of displaying up to 256 shades of gray.

3. Video camera: Any high resolution camera will do. Because the pixel density of the camera is a main factor in determining the resolution of the final image, choose the best you can afford. Recommendation: Dage/MTI CCD-72.
4. Software: NIH Image and documentation, available from anonymous ftp on Internet (IP address: 128.231.128.7) and other public sources. Image is the creation of W. Rasband of the National Institutes of Health.
5. System software: Mac OS 6.0.5 or higher (including System 7).
6. Confocal images can be transferred to the Macintosh via Ethernet cabling. Most UNIX workstations are equipped with built in Ethernet adaptors. To connect to Macintosh computers install an Ethernet board (Asanté, TechWorks, Farallon, etc.) Images can be transferred as raw data (binary) or TIFF format. NIH Image can interpret both formats.

Possible Image Manipulations

1. Gray-scale manipulations: lightening, darkening, shading, copying and pasting to make plates (montages), retouching, etc.
2. Pseudo (false) coloring. Assigning color to indicate gray value is useful for image data display and interpretation.
3. Image mathematics. Manipulations with two images (+, −, ×, ÷) can elucidate positional changes of objects in two different images.
4. Pixel convolutions. Mathematically comparing a pixel value to its local planar (e.g., 3 × 3) or volume (e.g., 3 × 3 × 3) neighborhood allows certain image characteristics to be enhanced or subdued. Examples are despeckling, smoothing, and edge detection.
5. Morphometrics: length, width, area, perimeter, particle counting.
6. Densitometry. Rudimentary densitometry may be performed within the dynamic range of the sample (autoradiographic or fluorescent spot) and the electronic components.
7. Printing. Images may be printed at low resolution (600 DPI) on a laser printer (e.g., Apple LaserWriter IIg) or at high resolution (up to 4,000 DPI) to film on a raster image processor (e.g., Mirus Montage, Lasergraphics LFR Plus). Images printed on a RIP device are of publication quality.

REFERENCES

Blaker AA (1976) Field Photography: Beginning and Advanced Techniques, W Freeman, New York

Hake S, Freeling M (1986) Analysis of genetic mosaics shows that the extra epidermal cell di-

visions in Knotted mutant maize plants are induced by adjacent mesophyll cells. Nature 320: 621–623
Inoué S (1989) Video Microscopy, Plenum, New York
O'Brien TO, McCully ME (1981) The Study of Plant Structure: Principles and Selected Methods, Termarcarphi, Melbourne, Australia
Pawley JB (1990) Handbook of Biological Confocal Microscopy, Plenum, New York
Russ JC (1990) Computer-Assisted Microscopy: The Measurement and Analysis of Images, Plenum, New York
Schaefer JP (1992) Basic Techniques of Photography: An Ansel Adams Guide, Little, Brown, Boston, Toronto, London

12

Scanning Electron Microscopy

MICHELLE H. WILLIAMS and ANNE W. SYLVESTER

Scanning electron microscopy (SEM) reveals details of the surface of a specimen that are unavailable by any other method. Essentially a beam of electrons is focused into a small spot and is scanned over the specimen in a regular fashion. The electrons in the beam hit the specimen surface and the resulting radiation from the specimen is detected and monitored onto a screen, forming an image. The SEM has been used extensively in the study of normal maize embryos, leaves, meristems and flowers as well as mutants (Van Lammeren 1987; Becraft et al. 1990; Bertrand-Garcia and Freeling 1991; Sylvester et al. 1990; Veit et al. 1991). All of the necessary background and detailed information on the SEM, the theory, its applications, and specimen preparation are provided in a series of books edited by Hayat (1974–1978).

There are many options for specimen preparation depending on the type and level of sophistication of the SEM available. SEMs currently used are:

1. The traditional, high-vacuum SEMs found in most facilities.
2. SEMs with cryostages; these are particularly useful for preservation of difficult materials (Echlin 1992).
3. Environmental SEMs; these have the advantage of a low vacuum and so do not require fixation or coating of the specimen.

The Maize Handbook—M. Freeling, V. Walbot, eds.

4. High-resolution (field emission) SEMs.
5. Low voltage SEMs.

Some of the new technologies available have many advantages, but at present most of these microscopes (numbers 2–5 above) are costly and their availability is limited. The specimen preparation methods for these new microscopes involve freezing tissue, fracturing, freeze substitution and critical point drying (Goldstein et al. 1981; Robards and Sleytr 1985; Echlin 1992). These methods require costly and precise preparative instruments; however, recent studies have demonstrated that they reveal details of both the internal organelles and of the surface of specimens impossible to observe in any other way.

At present, the traditional high vacuum SEMs are the most commonly used instruments. Here, we discuss a variety of methods available for observing the surface of a specimen by SEM, but we focus on a technique that involves making a replica of the specimen surface. The replica technique is simple and effective; a major advantage is that it can be used on living tissue because it permits nondestructive sampling (Williams et al. 1987; Williams and Green 1988).

PREPARATION OF SPECIMEN

Conventional Techniques

Tissue must be dehydrated to avoid collapse of the specimen under the SEM vacuum. It is possible to observe fresh tissue using high-vacuum SEM, especially at low kilovolts. Tissue collapse is a limitation of this technique and therefore rapid observation and photographing are required. The following methods describe preparative techniques for fixed and dehydrated specimens.

Preserving Tissue

1. Fix tissue in fixative, e.g., 3% glutaraldehyde in 0.1 M cacodylate or phosphate buffer pH 7.2–7.4 for 12 hours.
2. Rinse in 0.1 M buffer.
3. Dehydrate in a graded solvent series, e.g., 30%, 50%, 70%, 100% ethanol or acetone for 15 minutes each.

Drying Tissue

1. Air drying is a simple method but can cause collapse and distortion of the specimen unless special precautions are taken.
2. Critical-point drying involves heating a specialized pressure chamber in which the specimen is immersed in liquid CO_2 so that the liquid around the specimen is converted to a supersaturated vapor (Boyde 1980).

3. Freeze drying cools the specimen to a temperature at which point water can be removed by sublimation. The specimen must be cooled rapidly to avoid the formation of large ice crystals (Robards and Sleytr 1985; Echlin 1992).

Note: Fast freezing of specimens may be accomplished by methods such as plunge freezing, slam freezing, propane-jet freezing, and high-pressure freezing, sometimes followed by fracturing and then critical point drying (Robards and Sleytr 1985). These methods avoid any fixation prior to drying.

Advantage of Preserving and Drying Tissue

Techniques can be used to observe the internal structures of cells when specimens are frozen and then fractured.

Disadvantages of Preserving and Drying Tissue

1. Dehydration steps involve the use of an organic solvent, such as acetone or ethanol, which may dissolve and/or distort the surface waxes of some plant materials. Waxes of maize leaves are well preserved by this method because they are not generally soluble in these solvents.
2. Sampling is destructive.
3. Specimens need to be carefully stored, as they are often brittle and can reabsorb moisture.
4. Processing times may be prolonged during either fixation and dehydration or freezing, fracturing, and substitution, followed by drying of the specimen.
5. Artifacts may be introduced when tissues are distorted or collapsed during treatment.

Replica Technique

An alternative to both preserving and drying tissue is to make a replica of the specimen surface. The replica is made by first taking a negative impression or mold of the specimen. A positive resin replica is then made, and this is identical to the specimen's surface. The reader is referred to the scanning electron micrographs in the sections by Freeling and Lane (this volume) and in Sylvester et al. (1990) for examples of the use of the replica technique for maize. Figure 12.1 is an example of the resolution accomplished by the replica technique when applied to maize leaves.

Materials

- Specimens
- Dental impression kits of various viscosity and colors. Check availability with dental supply companies, such as Patterson Dental Supply, Sunnyvale, CA. One example is G.C. Exaflex, a low viscosity vinyl silicon impression material produced by G-C Dental Industries Corp. Japan. Note: Impression kits have a shelf life. Their ability to poly-

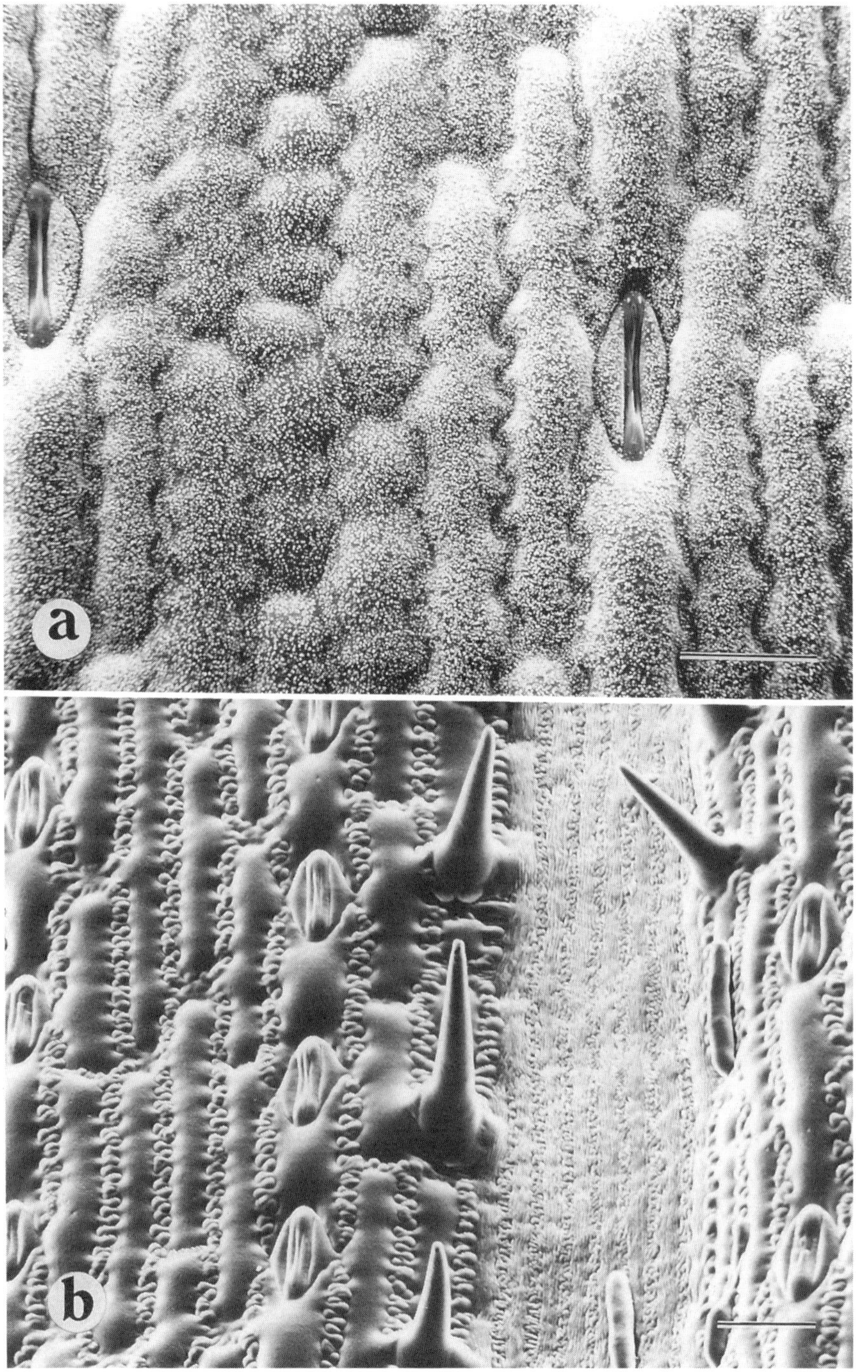

FIGURE 12.1 Scanning electron micrographs of the adaxial surface of maize leaves using the replica technique of specimen preparation. a. Epicuticular waxes are preserved in the juvenile leaves using the method. b. Adult leaves lack epicuticular waxes; hairs and wall surface features are preserved using the method. See micrographs in Freeling & Lane for additional examples of the surface resolution using the replica technique. Scale bars = 40 μm.

merize properly appears to diminish with time. Check the date printed on the crimp of the tube to estimate the shelf-life of a particular batch.

- Modeling clay or plasticene (available from art supply companies).
- Fine spatula.
- Glass slides.
- Tissues.
- Needles.
- A stereo microscope for detailed work.
- Forceps.

Methods

1. Secure the specimen with modeling clay so that the region of tissue to be replicated is easily accessible. It is preferable to work under a stereo microscope.
2. The impression kit has two components—a tube of base and a tube of catalyst. On a glass slide mix together equal quantities from each tube with a fine spatula. With a fine needle transfer small quantities of the impression material from the glass slide to the tissue surface until a region of the specimen is coated.

Note: Gently agitate the impression material onto the surface of the tissue so that the replicating material flows into any contours on the specimen surface. This agitation also tends to reduce the chances of air bubbles being trapped at the interface between the specimen and impression material. Take care not to press on the specimen to avoid damaging surface structures.

3. Polymerize the impression material at room temperature. This takes approximately 10 minutes depending on air temperature.
4. Using forceps, gently lever the edge of the impression material mold away from the specimen and remove/peel off the negative impression.

Note: Determine when to peel the mold from the specimen by testing the remaining impression material left on the glass slide. When it develops a solid rubberlike consistency, the mold is ready to be removed from the surface of the specimen. Do *not* remove the mold before polymerization is complete. The surface of the mold may change upon removal from the specimen, if polymerization is not complete.

5. A negative mold that is small and flat may need support. Again mix equal quantities of an impression kit and spread this onto a glass slide. The negative mold from the specimen should be placed or pushed into the support-impression material on the glass slide, with the impression of the specimen surface facing upwards, ready for the next step.

Note: The impression kit used for support can be either the same sort of low viscosity kit used for making impressions of the specimen surface or it can be any medium viscosity kit. It is helpful to use an impression kit with a different color from that used for making the molds of the specimen.

6. Fill the negative mold with a fluid resin that can be polymerized to form a hard resin replica, e.g., Spurr's resin kit (Spurr 1969) or Araldite.

Notes: The final resolution of the specimen will be limited by the quality of the impression mold and the resin used. Araldite can be used to provide a quick replica of the specimen; however, Araldite tends to provide lower resolution images than those observed using Spurr's resin. The Spurr's resin kit has four components. These are usually added to each other in a plastic container while weighing on a balance, but may also be measured volumetrically.

ERL 4206 (VCD)	11.5 g
NSA	31.0 g
DER 736	7.0 g
DMAE	0.5 g

Mix the first three components together well using a disposable pipette, add the DMAE last, and stir thoroughly. Care should be taken; use gloves (preferably polyvinyl, not latex gloves) and avoid both contact with the skin and breathing in fumes. Once finished, place all used pipets, tissues, gloves, and containers in a 70° C oven overnight to polymerize the resin before discarding. Disposable test tubes with 50 ml of Spurr's resin can be stored in a freezer in a desiccator/glass jar containing silica gel. Remove the desiccator from the freezer and bring to room temperature before opening the Spurr's resin container. Water condensation will ruin the resin.

The Spurr's resin can be placed into the negative mold by droplets from a pasteur pipet or from the end of forceps. Avoid air bubbles in corners and at the interface between the resin and impression material. Bubbles may be removed by gently vibrating the mold from the outside of the replica through the support-impression material using forceps or a needle. If this fails, place the forceps/needle into the droplet of resin and move them around; care must be taken to avoid damaging the inside surface of the impression mold, however, this encourages air bubbles to rise to the surface of the resin. Alternatively, the resin filled mold may be placed in a vacuum oven to remove air bubbles.

7. Polymerize the glass slide with the support material, negative impression mold and Spurr's resin in an oven at 70° C overnight. Use a vacuum oven if bubbles are a problem.

8. Remove from the oven and cool. Lever the hardened resin replica from the impres-

sion mold. The resin replica now represents a positive replica of the specimen surface.

Note: The resin replica can be trimmed with a razor blade to reveal any occluded areas on the specimen surface. Also, the negative impression mold can be refilled with Spurr's resin several times and multiple copies of any single specimen can be made provided care is taken when removing the resin replicas from the mold.

Advantages of Making Replicas of Tissue

1. Quick and easy relative to preserving and drying methods. Making replicas takes approximately 15–30 minutes and polymerization of Spurr's resin is overnight. If Araldite is used, polymerization time is approximately 10 minutes.
2. Specimen impressions can be made in the field or greenhouse and negative molds can easily be stored, labeled, and returned to the laboratory.
3. Resin replicas remain durable in response to the vacuum in the coating chamber and to the electron beam.
4. Resin replicas are easily stored and can be observed repeatedly.
5. Multiple positives can be made (and trimmed) from any single impression.
6. The method is nondestructive; therefore it can be used for sequential developmental studies of any single tissue (Williams 1991). In addition, the plant surface can be replicated, to observe by SEM, and then it can be analyzed by some other destructive method, e.g., transmission electron microscopy, biochemical analysis, etc.

GROUNDING AND STABILIZING THE SPECIMEN

Grounding

All specimens, regardless of preparation method, must be grounded to avoid charging artifacts in the SEM. Securely attach the specimen to a microscope stub, e.g., using carbon or silver paint, double-sided adhesive tape, silver putty, or epoxy glue.

Note for the replica technique: Use only water-soluble paints. Solvent paint will solubilize the polymerized resin and may alter the surface detail of the replica; preferably use double-sided tape and silver putty.

Coating

Coat specimens with a thin film of gold, gold/palladium, or platinum using a sputter coater.

Note for replica technique: A coating of approximately 20 nm is sufficient.

OBSERVATIONS

The coated specimen is inserted into the column of the SEM and the chamber is evacuated. An accelerating voltage is selected. A good starting voltage for specimens prepared as described here is 10–15 kV. The image is observed and photographed on a display cathode ray tube (CRT). Image quality on the CRT can vary dramatically depending on the operation and alignment procedures. The reader is referred to one of the many texts on SEM usage (Chapman 1981). Improper imaging will negate all of the careful procedures used during specimen preparation. Assuming the microscope is properly aligned, a few tips may increase success with the SEM:

Optimizing the Image

Shadows on the specimen depend upon the position of the specimen in relation to the electron detector within the microscope chamber. Any adjustments to visual shadowing can be made by tilting the specimen stage. Tilting will cause some image distortion; and therefore, it is not useful for quantitative analysis.

Appropriate Magnification

A contact print of a negative will give the best resolution. Select a scan speed that will provide the necessary detail in the negative for the final magnification that may be required.

Record Keeping

Draw sketches of the specimen with the important regions highlighted so that it is possible to relocate an area when further photography is required.

Optimizing Photography

Stereopairs permit three-dimensional viewing of the image; this will provide a more accurate perspective of the topography of the specimen surface. Stereopairs can be made by photographing the same surface at approximately 3° of tilt. Focus the image carefully by fine-focusing at a magnification *higher* than that required for photography. Optimize contrast and brightness at the time of photography. It is easier to print a balanced negative than to print one that is over- or underexposed. (See Becraft, this volume, for discussion of techniques in printing.) Remember that the Polaroid negative will have more contrast than the polaroid print. Lower contrast paper may be used in the darkroom to print overly high contrast negatives.

Quantitative Analysis

All microscopes have an inherent distortion factor. Reduce this effect by keeping the specimen surface perpendicular to the beam; adjust the stage to compensate for any angles in the specimen surface.

Recognize artifacts, a few of which are listed below for the replica technique:

1. Round, smooth balls on the specimen surface usually result from air bubbles that were trapped between the specimen surface and the impression material at the time of making the replica.
2. If the specimen surface appears featureless with small holes, the impression material was already partially polymerized at the time of application. Be sure to apply the material *immediately* after mixing the two components.
3. Excess water will either disperse across the specimen surface or form puddles, both of which will obscure details. Be sure the specimen surface is dry. Carefully wick the moisture away but do not dehydrate the specimen. (See #4 below.)
4. Tissue will become distorted if the specimen wilts or is *too* dry prior to taking an impression. (Tissue distortion and collapse are also major artifacts encountered with conventional fixing and drying techniques, as discussed above.)
5. Be aware that artifacts may result during any stage of SEM work including specimen preparation, operation of the microscope, and interpretation of the photographs. See Crang and Klomparens (1988) for discussion of the various pitfalls to avoid in electron microscopy.

REFERENCES

Becraft PW, Bongard-Pierce DK, Sylvester AW, Poethig RS, Freeling M (1990) The *liguleless-1* gene acts tissue specifically in maize leaf development. Dev Biol 141: 220–232

Bertrand-Garcia R, Freeling M (1991) Hairy-sheath frayed 1-O: a systemic, heterochronic mutant of maize that specifies slow developmental stage transitions. Am J Bot 78(6): 747–765

Boyde A (1980) Review of basic preparation techniques for biological scanning electron microscopy. Electron Microscopy (EUREM) 2: 768–777

Chapman SK (1981) Working With a Scanning Electron Microscope, Lodgemark Press LTD, Kent, England, 113 pp

Crang RFE, Klomparens KL (1988) Artifacts in Biological Electron Microscopy, Plenum Press, New York, 233 pp

Echlin PE (1992) Low-Temperature Microscopy and Analysis, Plenum Press, New York, 539 pp

Goldstein JI, Newbury DE, Echlin P, Joy DC, Fiori CE, Lifshin E (1981) Scanning Electron Microscopy and X-Ray Microanalysis: A Text for Biologists, Materials Scientists and Geologists, Plenum Press, New York, 673 pp

Hayat MA (1974–1978). Principles and Techniques of Scanning Electron Microscopy, Vols. 1–16, Van Nostrand Reinhold, New York

Robards AW, Sleytr VB (1985) Low Temperature Methods in Biological Electron Microscopy, Elsevier, Amsterdam, 551 pp

Spurr, AR (1969). A low-viscosity resin embedding medium for electron microscopy. J Ultrastruct Res 26: 31–43

Sylvester AW, Cande WZ, Freeling M (1990) Division and differentiation during normal and *liguless-1* maize leaf development. Development 110: 985–1000

Van Lammeren AAM (1987) Developmental morphology and cytology of the young maize embryo (*Zea mays* L.). Acta Bot Neerl 35: 169–188

Veit B, Greene B, Lowe B, Mathern J, Sinha N, Vollbrecht E, Walko R, Hake S (1991) Genetic approaches to inflorescence and leaf development in maize. Development Suppl 1: 105–111

Williams MH (1991) A sequential study of cell divisions and expansion patterns on a single developing shoot apex of *Vinca major.* Ann Bot 68: 541–546

Williams MH, Green PB (1988) Sequential scanning electron microscopy of a growing plant meristem. Protoplasma 147: 77–79

Williams MH, Vesk M, Mullins MG (1987) Tissue preparation for scanning electron microscopy of fruit surfaces: comparison of fresh and cryopreserved specimens and replicas of banana peel. Micron Micr Acta 18: 27–31

13

Transmission Electron Microscopy: Chemical Fixation, Freezing Methods, and Immunolocalization

M. V. PARTHASARATHY

The two main techniques for studying maize and other plant material with the transmission electron microscope (TEM) are ultrathin sectioning of embedded material and freeze-fracture. Conventional or chemical fixation is the most commonly used method for preserving plant specimens for TEM. The majority of information that we have on plant cell structure was obtained from ultrathin sections of chemically fixed material (e.g., Figure 13.1). Only the protocols related to specimen preparation are presented here. All the micrographs are of maize material. Superscript numbers within parentheses in the text refer to comments related to the particular subject under "Notes." See also Varagona and Raikhel (this volume) for an immunolabeling procedure on chemically fixed root tissue.

CONVENTIONAL (CHEMICAL) FIXATION, DEHYDRATION AND EMBEDDING

Materials for Primary Fixation

25% glutaraldehyde or 8% glutaraldehyde in sealed glass ampules (both commercially available)

- 0.2 M sodium cacodylate buffer, pH 6.8–7.0
- 0.1 M sodium cacodylate buffer, pH 6.8–7.0[(3)]
- Specimen vials
- Petri dishes
- Scalpels

The Maize Handbook—M. Freeling, V. Walbot, eds.

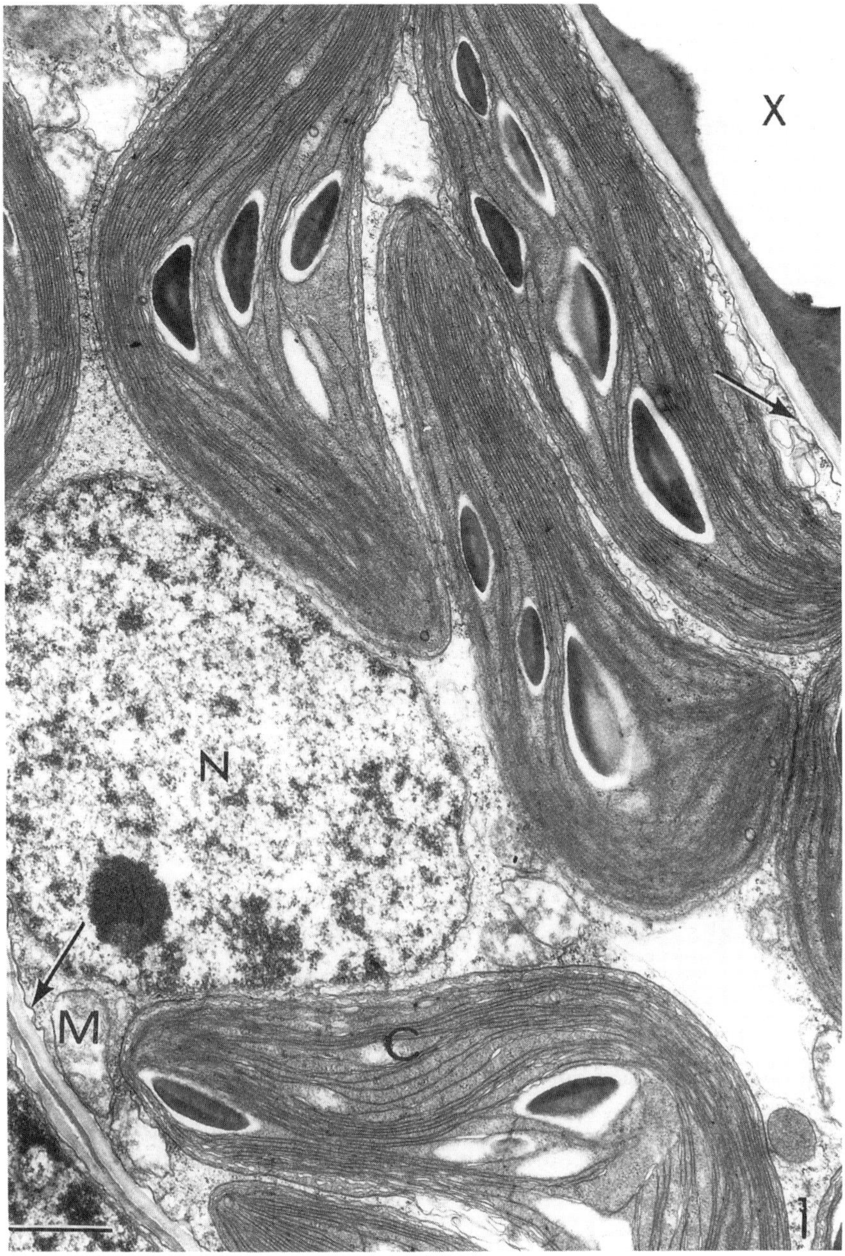

FIGURE 13.1 Longitudinal section of a bundle sheath cell of a seedling leaf. Conventionally fixed with glutaraldehyde/osmium tetroxide. Note the wavy appearance of plasma membrane (unlabeled arrow), which is characteristic of most conventionally fixed cells. C, chloroplast with starch grains; M, mitochondrion; N, nucleus; X, mature xylem vessel element. Scale = 2 μm.

Forceps
Pasteur pipets
Disposable polyethylene gloves
Refrigerator or crushed ice

Materials for Secondary Fixation

1 or 2% osmium tetroxide (OsO_4) in 0.1 M sodium cacodylate buffer; to be prepared just before use
2% aqueous uranyl acetate

Materials for Dehydration

Ethanol concentrations in stoppered bottles: 15%, 30%, 50%, 70%, 90%, 100%
Refrigerator or crushed ice

Materials for Embedding

Spurr's embedding medium
Silicone or silicone-rubber flat embedding molds (commercially available)
Glass beakers
Glass stirring rods
Top-loading balance
Slow rotating/tilting shaker
70° C oven

Method

Fixation

1. Fixation must be carried out in a fume hood and gloves should be worn when handling fixatives.
2. Prior to fixation, mix the appropriate concentration of glutaraldehyde with an equal volume of 0.2 M sodium cacodylate buffer to make a final fixative concentration of 2.5 or 5% glutaraldehyde in 0.1 M buffer. Adjust pH to 6.8 to 7.0, if necessary.[1,2,4,5]
3. Cut 0.5 cm pieces of the plant material with a scalpel and immediately transfer them to the glutaraldehyde fixative in a petri dish.
4. Cut the 0.5 cm pieces to smaller 1-mm^2 bits while holding the larger piece under the fixative with forceps.
5. Transfer the 1-mm^2 pieces of material to fresh glutaraldehyde fixative in a vial and allow to fix for 2 hours at room temperature.

6. Transfer the vial containing specimen/glutaraldehyde to 4° C and leave it for 30 minutes.
7. Replace glutaraldehyde with 0.1 M sodium cacodylate buffer cooled to 4° C and change the buffer every 10 minutes, three times.
8. After the final buffer change, replace it with 1% or 2% OSO_4 at 4° C and allow to fix for 2–4 hours.
9. Replace OsO_4 with 0.1 M sodium cacodylate buffer at 4° C and proceed with buffer wash as in step 7.[6]
10. Rinse three times with distilled water at 4° C.
11. Replace distilled water with 2% aqueous uranyl acetate at 4° C and stain en bloc for 1 hour.

Dehydration

1. Dehydrate by a graded series of solutions of increasing concentration of ethyl alcohol (EtOH) at 4° C allowing 10 minutes for each concentration step (15%, 30%, 50%, 70%, 90%).
2. Let the vial containing the specimen/90% EtOH come to room temperature and replace 90% EtOH with absolute EtOH three times, once every 10 minutes.

Plastic Infiltration

1. Epoxy resins must be handled with disposable plastic gloves.
2. Infiltration can be carried out in a rotary shaker as follows:

Infiltration of Spurr's embedding resin

Recommended media (in the shaker):	*Time*
1 part embedding medium: 3 parts EtOH	30 minutes
1 part embedding medium: 2 parts EtOH	30 minutes
1 part embedding medium: 1 part EtOH	1 hour
2 part embedding medium: 1 part EtOH	2 hours
3 part embedding medium: 1 part EtOH	4 hours
4 part embedding medium: 1 part EtOH	12 hours (overnight)

3. Transfer material to pure embedding medium and leave it 12–24 hours in the shaker.
4. Transfer material to fresh embedding medium and embed in flat molds to facilitate desired orientation of specimens.
5. Ensure that there are no air bubbles close to the embedded material and transfer the molds to a 70° C oven for 24 hours to cure.
6. Remove cured blocks from the mold after they reach room temperature. Embedded material can be stored indefinitely at room temperature. Ultrathin sections of embedded material can be obtained with an ultramicrotome, and the sections can be picked

up on a Formvar-coated copper grid and stained before observing with a TEM. See Varagona and Raikhel (this volume) for coating methods. See Reid (1975) for details on ultramicrotomy and Hayat (1989) for thin-section staining procedure.

Notes

1. 2.5% glutaraldehyde can be used when young and soft material such as shoot and root tips are to be fixed. 5% glutaraldehyde is recommended if woody or mature tissues are to be fixed. Karnovsky's (1965) formaldehyde/glutaraldehyde is recommended where deep penetration of fixative is required.
2. Addition of 0.1% tannic acid to the primary fixative can help preserve the cytoskeleton better.
3. 0.1 M sodium phosphate or 0.03 M PIPES can also be used instead of sodium cacodylate buffer.
4. Fixation of mature leaf material might require 75–100 mM sucrose in the buffer.
5. Thick cuticle and dense epidermal hairs will retard rapid penetration of the primary fixative. A minimal amount of Triton X-100 added to the fixative prior to fixation will usually solve this problem.
6. 1% OsO_4 is usually adequate for shoot and root tips while 2% OsO_4 may be needed for mature tissues.
7. A modified chemical fixation procedure is reported to preserve microfilament bundles in maize root cap cells (Vaughn and Vaughn 1987).

CRYOFIXATION/FREEZE-SUBSTITUTION

Although the conventional chemical fixation is the routine method of preserving plant material for TEM, the fixatives penetrate cells slowly, often fail to preserve labile structures and can cause morphological changes in cellular structures (Mersey and McCully 1978). Cryofixation can physically stabilize cellular components within a few milliseconds and is therefore superior to chemical fixation in preserving cell structures in their native state as well as in freezing dynamic events (Gilkey and Staehelin 1986). Cryofixation has not been used widely for fixing corn tissues because of the difficulty in routinely achieving good freezing. Nevertheless, depending on the freezing method/instrument used, good freezing can be achieved to a depth that ranges from 10 to 250 μm. Freeze-substitution involves dissolving the ice from frozen specimens with an organic solvent that normally contains chemical fixatives. To prevent secondary ice-crystal growth, freeze-substitution is carried out at temperatures lower than −78° C.

Materials

Materials for Propane Plunge-Freezing

0.5-liter dewar, flask, or Styrofoam container

2.0-cm (0.75 inch)-diameter copper pipe with a copper well (about 1.5-cm diameter and 2 cm deep welded on to one end). The copper pipe with the well should be about 0.5–0.75 cm taller than the dewar or flask container in which it will be placed. Holes should be drilled at the sides of the copper pipe near the bottom and below the well to facilitate free flow of liquid and gas through the pipe. Retaining rods radiating from the copper pipe to the wall of the container will help keep the pipe standing vertical during freezing.
Propane (commercial variety in a cylinder)[1]
Regulator for propane cylinder
Flexible tubing
Liquid nitrogen
Insulated or welders' forceps

Materials for Freeze-Substitution

Specimen carriers: 1–2-ml eppendorf tube that is modified by heat-fusing stainless steel or nylon mesh (e.g., Cell Microsieves-BioDesign Inc. or Nicro) of appropriate mesh size to the cutaway tip and the cap
Molecular sieves
Acetone
Uranyl acetate
Osmium tetroxide (OsO_4)
Screw-cap vials (30 ml)
Rack to hold vials in the freezer
Insulated or welders' forceps
Thermal gloves
Ultracold freezer (−80° to −90° C)[2]

Methods

Cryofixation

1. Specimens to be frozen must be 1 mm^3 in size or smaller. In general, the smaller the better. It is best if the material is frozen without any pretreatment. But it is often difficult to obtain satisfactory freezing of good depth in plant tissues without any cryoprotectant.[3,4]
2. Pretreatment.
   a. Sucrose of appropriate concentration in PIPES, HEPES, or MES buffer is very good if a penetrating cryoprotectant is to be used. For plant leaf material, 20 mM MES buffer (pH 5.5) containing 2 mM $CaCl_2$, 2 mM KCl, and 0.15–0.2 M sucrose works well (Ding et al. 1991). It is important to make sure that the sucrose concentration used does not induce plasmolysis, shrinking, or other artifacts. It is best to test the pretreatment carefully with living cells or tissue under a phase-contrast or Nomarski DIC light microscope to monitor plasmolysis before a suit-

able sucrose concentration is chosen. Depending on the material, 15–30 minutes of incubation in the buffer/sucrose solution should be sufficient. If the tissue has intracellular spaces (e.g., leaf), gentle aspiration may be required to fill the spaces with the buffer/sucrose.

b. Among nonpenetrating cryoprotectants, 15% polydextran (35,000–40,000 MW) and 1-hexadecene are good. Hexadecene can cause severe artifacts in some cells and therefore should be carefully tested on living cells before use. Polydextran seems less harmful. An incubation time of 15–30 minutes in polydextran is usually sufficient. Unlike hexadecene, polydextran often persists as a heavy coating over the frozen specimen. This can make it difficult to orient the specimen during embedding.

3. Freezing: Place the copper pipe (with the well on top) inside the dewar (or similar container) and fill the container with liquid N_2, taking care not to splash the liquid into the well. Wait for the boiling of the liquid to stop and add more liquid N_2 if necessary. The well should protrude about 1 cm above the liquid N_2. NOTE THAT YOU MUST HAVE THE WHOLE SETUP IN A FUME HOOD IF LIQUID PROPANE IS USED FOR FREEZING AND THERE SHOULD BE NO OPEN FLAME IN THE LAB! Commercially available propane is quite adequate and high-purity propane is not required. Propane can be dispensed into the well through a pasteur pipet connected to a plastic tubing. The tubing is connected to any commercially available regulator screwed to the propane cylinder. It takes almost 1 hour for the propane to form a slush. A clean glass slide can be placed on the receptacle to prevent condensation when the propane in the well is being cooled by the surrounding liquid N_2. Material to be frozen should be picked up with an insulated forceps for freezing. Electron microscope (EM) grids are suitable carriers for freezing single cells and cell cultures.
4. Hold the specimen with the insulated forceps and quickly plunge it into the propane with a swift motion without hitting the bottom of the well.
5. After holding the specimen for about 2 seconds in the propane, transfer the frozen specimen quickly to the specimen carrier placed under liquid N_2 in a nearby separate container (e.g., dewar or styrofoam cup). Depending on the specimen and pretreatment, good freezing up to a depth of 10–15 μm can usually be achieved (Figure 13.3).

OTHER FREEZING METHODS

Cold-metal Block Freezing (CMBF)

This method of freezing is achieved by impacting (slamming) the biological material against a highly polished surface of a copper block that is kept at liquid N_2 or helium temperatures. There are several commercial CMBF freezing devices that are good for freezing surface cell layers of relatively large tissues (up to 2 mm^2, Figure 13.2). A common artifact in material frozen with some of the commercially available CMBF is crushed cells. This

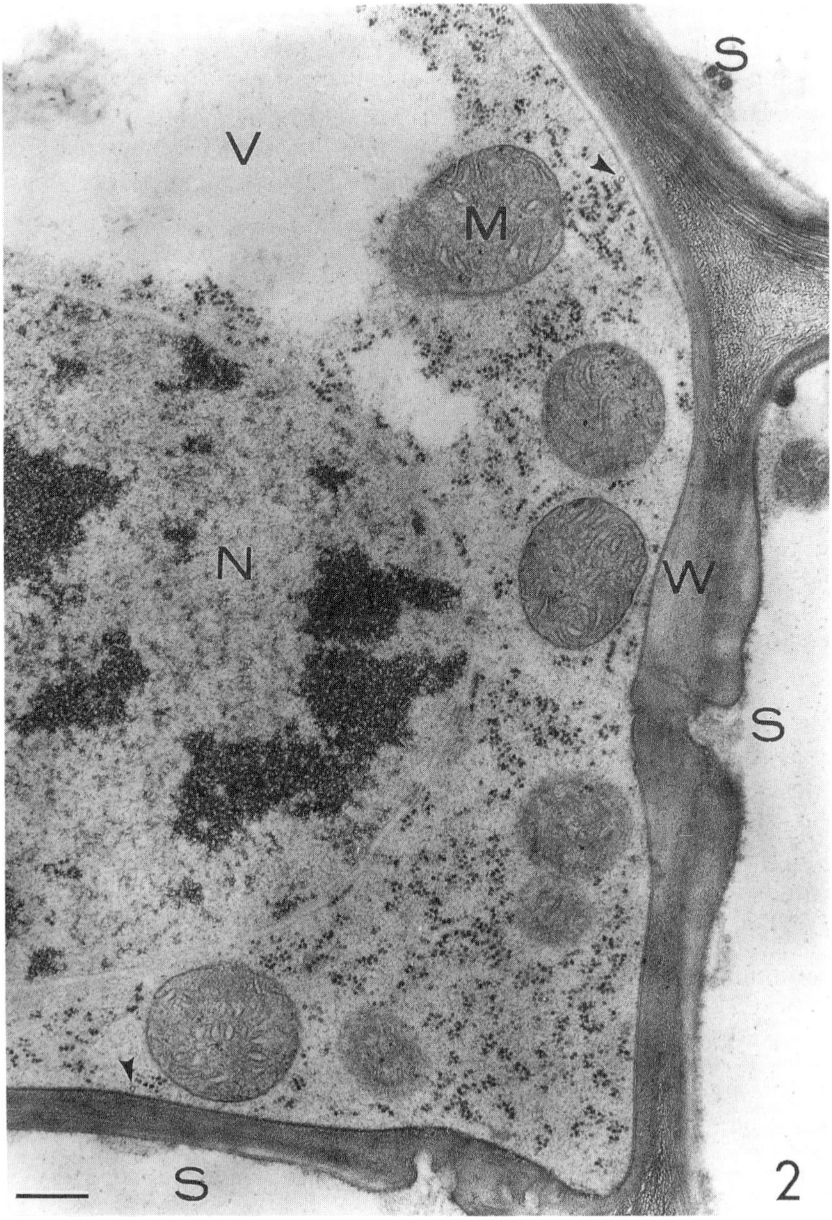

FIGURE 13.2 Transverse section of a cryofixed phloem parenchyma cell in coleoptile showing good freezing presumably due to the high amount of sugar normally found in the phloem tissue. Cryofixed without any pretreatment by quickly clamping the coleoptile with LN_2-cooled copper blocks attached to regular pliers. Freeze-substituted with osmium tetroxide/uranyl acetate in acetone. Note the smooth profile of the plasma membrane and compare it with that in Figure 13.1. W, cell wall; N, nucleus; M, mitochondrion; S, mature sieve element; V, vacuole; darts indicate microtubules. Scale = 0.2 μm.

can often be minimized by using the right kind and combinations of "shock absorbers" behind the specimen (e.g., agar, foam). The frozen specimen can then be quickly transferred to liquid N_2 for storage. A simple clamping device such as one with copper blocks attached to regular pliers can also work relatively well when the copper blocks are cooled to liquid N_2 temperature and the material is quickly clamped by the cold blocks (e.g., Figure 13.2). Depending on the specimen, good freezing up to a depth of 15 μm can be usually achieved.

Propane Jet Freezing (PJF)

In this method, two high velocity jets of propane hit the opposite sides of a specimen that is sandwiched between two thin copper hats. Two commercial PJFs are available—one from RMC, Tucson, Arizona, and a recently designed one from Baltec/Balzers, Liechtenstein. The general principle of PJF and technical details have been discussed well by Gilkey and Staehelin (1986). With penetrating cryoprotectants such as sucrose, good freezing up to depth of 100 μm can be achieved (Ding et a1. 1991).

High Pressure Freezing (HPF)

This method utilizes high pressure during freezing with liquid N_2 as a means to cryofix relatively large specimens. The only commercial HPF available is made by Balzers/Baltec. The theory and practice of HPF have been discussed by Moor (1987). HPF is the only method by which good freezing up to a depth of 250 μm or more can be achieved (Studer et al. 1989). However, the possible effects of high pressure on labile components of some cells should be of some concern (Ding et al. 1992).

Freeze-Substitution

1. Transfer the frozen specimens (with the carrier) quickly into the vial containing the freeze-substitution fluid that has been precooled to −90°C. The fluid is a mixture of 2% OsO_4, 2% uranyl acetate, and molecular sieves in acetone.[5,6]
2. Keep in freezer for 2 days at −90° C to achieve freeze-substitution.[7]
3. Transfer to a regular freezer at −20° C for 8–12 hours.
4. Transfer to 4° C for 2–4 hours.
5. Rinse in pure acetone 3×5 minutes at 4° C.
6. Bring the material to room temperature.
7. After one more rinse with pure acetone at room temperature, proceed with plastic infiltration and embedding.

Embedding

Similar to chemical fixation/embedding.[8]

Thin sections need be stained the usual way (Hayat 1989) before viewing under a TEM.[9]

Figure 13.5 summarizes various freeze-substitution methods recommended for maize.

Notes

1. In labs without a fumehood, Freon-22 can be tried instead of liquid propane although Freon-22 is not as good a fluid for freezing.
2. If an ultracold freezer is unavailable, a container with a mixture of acetone and dry ice that is placed in a freezer at −20° C will usually bring down the temperature in the container to about −78° C.
3. Tissue-culture cells generally freeze well because the medium usually contains sugar. The same is true for pollen in germinating medium that contains sugar and for nonhydrated pollen grains.
4. Cells that normally contain high sucrose such as those in the phloem freeze better than other cells (e.g., Figure 13.2).
5. Molecular sieves (e.g., Davison Chemical—grade 514) will have to be baked for 2–3 hours in an oven at the self-cleaning setting and brought to room temperature before use. A screw-cap glass vial (30 ml) with about one-half filled with molecular sieves is normally enough. There should be sufficient space left at the top of the vial to let the specimen container submerge under the fluid.
6. Most of the uranyl acetate salt will dissolve in acetone if left in a properly covered beaker over a stir-plate for 1–2 hours. Uranyl acetate/acetone is first prepared and then the osmium crystals are added to the solution. The solution is unstable at room temperature and has to be transferred to a freezer immediately after osmium crystals are dissolved.
7. The freeze-substitution time will vary depending on the type of tissue frozen.
8. Freeze-substituted cells generally require longer plastic infiltration time than those chemically fixed and therefore low viscosity embedding media such as Spurr's or LR White is preferable. But if infiltration is gradual and prolonged, other plastics such as Epon/Araldite or Quetol may be used. For low-temperature embedding with Lowicryl, the procedure of Humbel and Muller (1986) or Carelmalm et al. (1986) can be tried.
9. Sections from material freeze-substituted with osmium/uranyl acetate still need to be stained before viewing because the contrast of images is inherently lower than in conventionally fixed material.
10. Protocols I & II indicated in Figure 13.5 normally yield images of poor contrast, but these methods are recommended for X-ray microanalysis and immunogold locali-

zation studies. The most commonly used protocols are III, IV and V (Figure 13.5), because they yield consistently good results for ultrastructural studies (e.g., Figures 13.2, 13.3). Protocol V (Figure 13.5) is especially recommended for cytoskeletal studies (e.g., Figure 13.4). Protocol VI (Figure 13.5) is recommended if clarity of the endomembrane system is important.

IMMUNOGOLD LOCALIZATION

The advantages and limitations of immunogold localization at the ultrastructural level in plant cells have been well reviewed by Herman (1988). Although the antigenicity of some proteins can be lost during chemical fixation, others do retain their antigenicity (e.g., Hurley and Taiz 1989; Ludevid et al. 1990). Protocols for chemically fixed as well as freeze-substituted material are included here. See also Varagona and Raikhel (this volume) for immunolabeling procedure with root tissue.

Materials

Chemical Fixation

8% paraformaldehyde in sealed ampules (commercially available)
8% glutaraldehyde in sealed ampules (commercially available)
0.2 M and 0.1 M sodium phosphate buffer solutions
Rest of the materials are the same as in chemical fixation describe before

Cryofixation

Same materials as indicated earlier under "Cryofixation"

Freeze-substitution

Same as described before

Embedding

Acetone or Ethanol
LR White acrylic resin
Gelatin capsules
60° C oven

Immunogold Labeling

Forceps
Formvar-coated nickel grids
Paraffin film (Paraplast)
Saturated solution of sodium periodate
0.1 N HCl
Glass distilled water

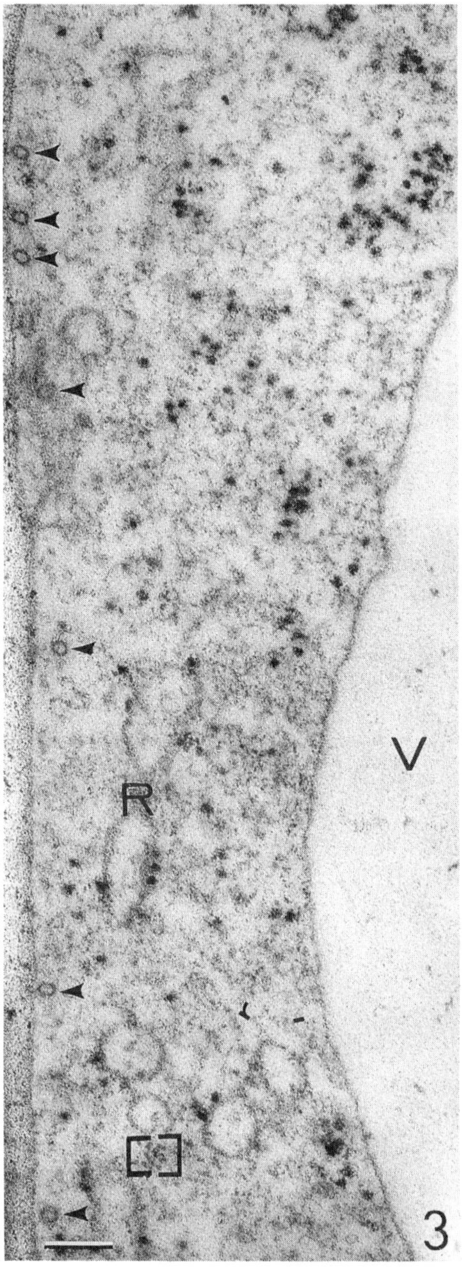

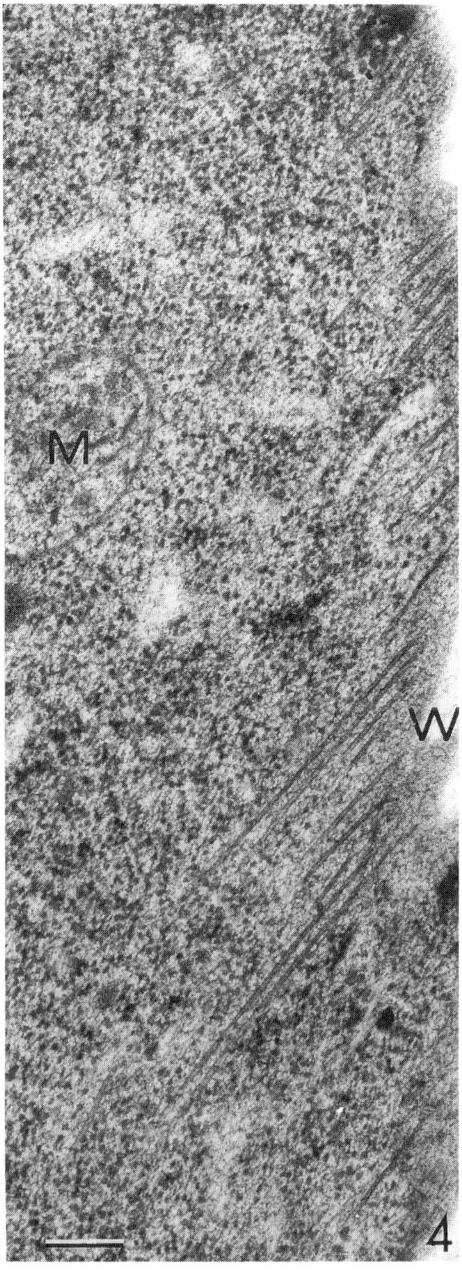

FIGURE 13.3

FIGURE 13.4

FIGURE 13.3 Transverse section of an inner epidermal cell of coleoptile cryofixed by plunging a small piece of peeled epidermis, without any pretreatment, into liquid propane. Freeze-substituted as in Figure 13.2. Unlabeled arrows indicate microtubules close to the plasma membrane and the region delimited by brackets shows a microfilament bundle in transverse view. R, rough ER; V, vacuole. Scale = 0.1 μm. **FIGURE 13.4** Array of microtubules in a cortical cell of root cryofixed with a high-pressure freezer and freeze-substituted with tannic acid/osmium tetroxide/uranyl acetate in acetone. The presence of a large number of ribosomes is obvious. W, cell wall; M, mitochondrion. Scale= 0.2 μm.

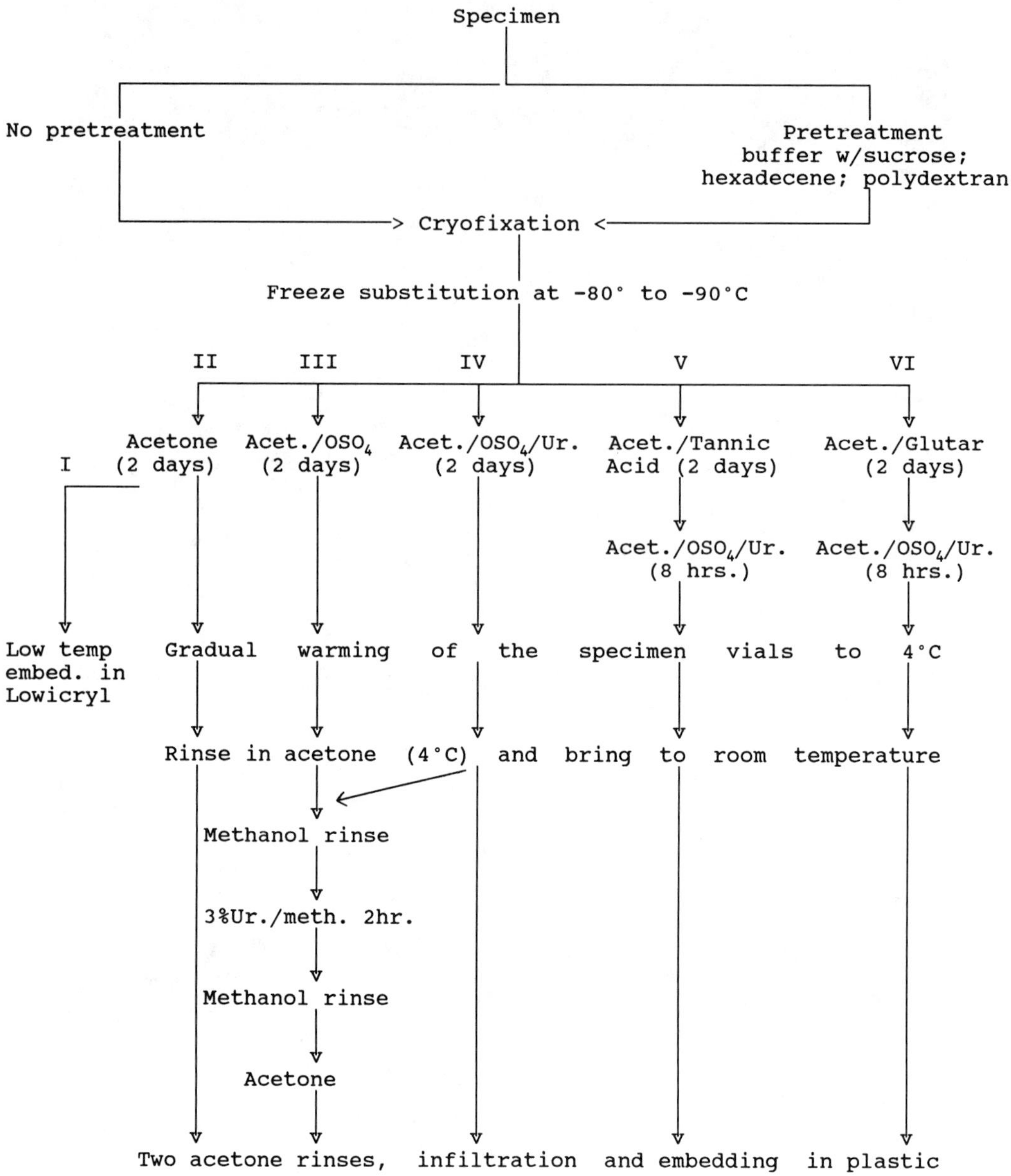

FIGURE 13.5 Flow chart showing freeze-substitution protocol options for cryofixed maize material.

TBST-BSA (50 mM Tris-HCl, pH 7.4, 150 mM NaCl, 0.05% v/v Tween 20, and 0.5% w/v bovine serum albumin)
Primary antibody
Secondary antibody conjugated to 10-nm or 15-nm colloidal gold

Method

Chemical Fixation

1. Fix material for 2 hours at 4° C or room temperature in a fixative that has a final concentration of 0.5% glutaraldehyde and 4% paraformaldehyde in 0.1 M sodium phosphate buffer, pH 7.2.
2. Proceed with the buffer wash, secondary fixation with 1% OsO_4, and dehydration as previously described for chemical fixation.

Cryofixation/Freeze-substitution[1,2]

1. Cryofix specimen as described before.
2. Transfer frozen material into 0.5% uranyl acetate in acetone at −90° C and leave it there for 2 days.
3. Transfer to 1% OsO_4 in acetone at −90° C and bring the specimen gradually to room temperature as previously described for cryofixation/freeze-substitution.

Embedding

1. Infiltrate the plastic for both chemically fixed and cryofixed material as previously described.
2. Embed material in LR White using gelatin capsules.[3]
3. Cure the plastic in a 60° C oven for 24 hours.

Immunogold Labeling

1. Thin sections obtained from both chemically fixed and cryofixed material are processed in the same manner.
2. Pick up thin sections on Formvar-coated, nickel grids.
3. Place the section, sitting on a grid, on a drop of saturated aqueous sodium periodate solution that is in turn placed on a piece of paraffin film, for 15 minutes.
4. Place the sections/grid similarly on a drop of 0.1 N HCl for 10 minutes.
5. Wash thoroughly with distilled water.
6. Place the sections/grid on a drop of TBST-BSA for 15 minutes.
7. Incubate the sections/grid on a drop of primary antibody, suitably diluted with TBST-BSA (usually diluted 50–100 times) for 1 hour.
8. Wash in TBST-BSA 3×5 minutes.

9. Incubate the sections/grid as before in the secondary antibody/colloidal gold conjugate diluted 1:100 with TBST-BSA for 1 hour.
10. Wash in distilled water for 5 minutes.
11. Stain the sections/grid the usual way (Hayat 1989) before observing with a TEM.[4–6]

Notes

1. The cryofixation/freeze substitution method described here has been shown to significantly improve the antigenic preservation in some plant tissues (Kandasamy et al. 1991).
2. Although freeze-substitution with acetone alone followed by the usual embedding procedure can be used for immunogold localization in plants (e.g., Lancelle and Hepler 1989), the method often yields images of poor contrast lacking structural details. Freeze-substitution in acetone alone followed by low temperature embedding (e.g., Lowicryl K4M, K11M, HM20, HM23, LR White, or LR Gold) is probably the best method for preserving antigenicity of plastic-embedded material (Humbell and Muller 1986; Carlemalm et al. 1986). But in plant material the method can pose problems with infiltration, insufficient contrast, and poor preservation, and hence fail to reveal ultrastructural details.
3. Assuming that the antibody concentrations were optimal and the washes were thorough, high background labeling, if present, can usually be suppressed by increasing the concentration of BSA in the primary antibody incubation solution and by the addition of 0.5–2% fish gelatin in the secondary antibody incubation solution.
4. For controls omit the primary antibody and replace with buffer only, replace primary antibody with nonimmune serum, and/or replace primary antibody with an antibody raised against an antigen known to be absent in the tissue.
5. Refer to Varndell and Polak (1987) for a detailed description of immunolabeling methods for electron microscopy.

FREEZE-FRACTURE

Freeze-fracture technique employs cryofixation and metal film replication methods (Rash and Hudson 1979). This method is invaluable for studying membrane morphology (e.g., Zuo et al. 1989; and Vom-Dorp et al. 1990).

Materials

If the specimen is not chemically fixed before being frozen, the materials indicated earlier under "Cryofixation" should suffice. The following additional materials will be needed if the specimen is to be chemically fixed before freezing:

2% glutaraldehyde in 0.1 M sodium phosphate buffer, pH 6.8–7.0
0.1 M sodium phosphate buffer, pH 6.8–7.0
Glycerol
Small petri dishes

Method

Unfixed Specimens

1. Incubate the specimen in the phosphate buffer or any appropriate physiological buffer for 10 minutes.
2. Freeze the specimen by one of the methods indicated earlier under "Cryofixation" and transfer it to liquid N_2 for storage and subsequent freeze-fracture.

Fixed Specimens[1,2]

1. Fix the specimen in 2% glutaraldehyde in 0.1 M sodium phosphate buffer for 10 minutes.
2. Wash 2×5 minutes in the buffer.
3. Gradually add glycerol to the specimen in buffer until a final concentration of 25% is reached in about 10 minutes.
4. Incubate an additional 10–20 minutes in the glycerol/buffer.
5. Freeze the specimen as described before and transfer to storage in liquid N_2 for subsequent freeze-fracture.

Notes

1. It is obviously best to cryofix specimens without any pretreatment, but it is often difficult to obtain satisfactory freezing of plant material without a pretreatment.
2. Although the glutaraldehyde/glycerol pretreatment is known to produce some artifacts, useful information on membrane structure can often be obtained with the procedure (e.g., Robards et al. 1980, 1981).
3. Refer to Rash and Hudson 1979 for a detailed description of the freeze-fracture procedure and to Severs (1991) for a review of freeze-fracture cytochemistry.

REFERENCES

Carlemalm E, Villigier W, Acetarin J-D, Kellenberger E (1986) Low temperature embedding. In Müller M, Becker RP, Boyde A, Wolosewick JJ (eds) The Science of Biological Specimen Preparation for Microscopy and Microanalysis, SEM, AMF, O'Hare, IL, pp 147–154

Ding B, Turgeon R, Parthasarathy MV (1991) Routine cryofixation of plant tissues by propane jet freezing for freeze substitution. J Electron Microsc Techn 19: 107–117

Ding B, Turgeon R, Parthasarathy MV (1992) Effect of high pressure freezing on plant microfilament bundles. J Microsc (165: 367–376)

Gilkey JC, Staehelin LA (1986) Advance in ultra

rapid freezing for the preservation of cellular ultrastructure. J Electron Microsc Techn 3: 177–210

Hayat MA (1989) Principles and Techniques of Electron Microscopy, Third Edition, CRC Press, Boca Raton, FL, 469 pp

Herman EM (1988) Immunocytochemical localization of macromolecules with the electron microscope. Annu Rev Plant Physiol Plant Mol Biol 39: 139–155

Humbel BM, Müller M (1986) Freeze-substitution and low temperature embedding. In Müller M, Becker RP, Boyde A, Wolosewick JJ (eds) The Science of Biological Specimen Preparation for Microscopy and Microanalysis, SEM, AMF, O'Hare, IL, pp 175–183

Hurley D, Taiz, L (1989) Immunocytochemical localization of the vacuolar H^+-ATPase in maize root tip cells. Plant Physiol 89: 391–395

Kandasamy MK, Parthasarathy MV, Nasarallah ME (1991) High pressure freezing and freeze substitution improve immunolabeling of S-locus specific glycoproteins in the stigma papillae of Brassica. Protoplasma 162: 187–191

Karnovsky MJ (1965) A formaldehyde-glutaraldehyde fixative of high osmolarity for use in electron microscopy. J Cell Biol 27: 137A

Lancelle SA, Hepler PK (1989) Imunogold labelling of actin on sections of freeze substituted cells. Protoplasma 150: 72–74

Ludevid MD, Ruiz-Avila L, Vallés MP, Stiefel V, Torrent M, Torné JM, Puigdomènech P (1990) Expression of genes for cell-wall proteins in dividing and wounded tissues of *Zea mays* L. Planta 180: 524–529

Mersey B, McCully ME (1978) Monitoring of the course of fixation of plant cells. J Microsc 114: 49–56

Moor H (1987) Theory and practice of high pressure freezing. In Steinbrecht RA, Zierold K (eds) Cryotechniques in Biological Electron Microscopy, Springer-Verlag Berlin, pp 175–191

Rash JE, Hudson CS (eds) (1979) Freeze-fracture: Methods, Artifacts and Interpretations, CRC Press, Boca Raton, FL, 204 pp

Reid N (1975) Ultramicrotomy. Practical Methods in Microscopy. In Vol 3, Part 2, Elsevier/North Holland, Amsterdam, pp 213–353

Robards AW, Newman TM, Clarkson DT (1980) Demonstration of the distinctive nature of the plasma membrane of the endodermis in roots using freeze-fracture electron microscopy. In Lucas WJ, Dainty J (eds) Plant Membrane Transport: Current Conceptual issues, Elsevier/North Holland, Amsterdam pp 395–396

Robards AW, Bullock GR, Goodall MA, Sibbons PD (1981) Computer assisted analysis of freeze-fractured membranes following exposures to different temperatures. In Morris GJ, Clarke A (eds) Effects of Low Temperatures on Biological Membranes, Academic Press, New York, pp 219–238

Severs NJ (1991) Freeze-fracture cytochemistry: a simplified guide and update on developments. J Microsc 161: 109–134

Studer D, Michel M, Müller M (1989) High pressure freezing comes of age. Scanning Microsc. Suppl. 3: 253–269

Vaughn MA, Vaughn KC (1978) Effects of microfilament disrupters on microfilament distribution and morphology in maize root cells. Histochemistry 87: 129–137

Varndell IM, Polak JM (1987) EM immunolabeling. In Sommerville J, Scheer U (eds) Electron Microscopy in Molecular Biology. IRL Press, Oxford, Washington, DC

Vom-Dorp B, Scherer GFE, Canut H, Brightman AO, Liedtke C, Morre DJ (1990) Identification of tonoplast fractions resolved from the plasma membrane by free-flow electrophoresis using fillipin labeling and antibody to the tonoplast ATPase. Protoplasma 156: 57–66

Zuo BY, Tang CQ, Jiang GZ, Kuang TY (1989) Changes of freeze-fracture ultrastructure and light harvesting chlorophyll protein complex in thylakoid membranes during ontogeny of maize. Acta Botanica Sinica 31: 903–908

Techniques for Histology of Maize Megaspores and Embryo Sacs

S. D. RUSSELL AND D. P. WEST

DEVELOPMENTAL STAGING

In corn, many of the stages of megasporogenesis and megagametogenesis may be harvested from the same ear; an external morphological gradient can be used to find specific stages in the progression. A 2-cm ear will contain the stages of meiosis and megasporogenesis, and those longer than 2 cm will contain later stages of megagametogenesis; developmental stages in the ears follow those in the tassels by about 10 days (D. P. West, unpublished). Silk length will differ between cultivars and plantings, but the following should provide a rough guide: 1–2 mm = megaspore mother cells, 2.5 mm = early tetrad of megaspores, 3.0 mm = late tetrad and early embryo sac, 4.5 mm = 2-nucleate (N) embryo sac, 7.0 mm = 4-N embryo sac, 8.5 mm = 8-N embryo sac, 10 mm to maturity = maturing embryo sac (Russell, unpublished). Fixing a morphological progression (as is present in each ear) will facilitate staging. The normal sequence of development has been reported by Weatherwax (1919), Cooper (1937), and Russell (1979). Earlier stages of floral organ initiation and staging are reported by Cheng et al. (1983).

PREPARING MATERIAL

Material may be prepared using one of four methods: paraffin histology, plastic histology, ovule clearing, and isolation of megaspores and embryo sacs. The first and third methods provide limited information about cytoplasmic organelles, whereas the second and fourth require more delicate manipulation but may be adaptable to refined techniques such as electron microscopy (Russell 1979) or the isolation, characterization (Wagner et al. 1988, 1989), fusion (Kranz et al. 1991a), and culture of gametes (Kranz et al. 1991b), and to determine the internal position of the embryo sac for microinjection or transformation (Wagner et al. 1990). For all fixed material, air bubbles should be knocked off the surface

The Maize Handbook—M. Freeling, V. Walbot, eds.

of the material by rapping the vial against the table or subjecting the material to a mild vacuum (not recommended for EM). First-time histologists should work in close consultation with a lab that routinely uses histological techniques or the experience may be unduly frustrating or dangerous. Many of the routine chemicals for histology and ovule clearing are toxic, some are carcinogenic, and many produce vapors that require the use of a fumehood for safety.

PARAFFIN HISTOLOGY

Material is fixed in FAA (1 part formalin:1 part acetic acid:18 parts 70% ethanol) for 6 hours to overnight (substitute propionic acid for acetic acid to make FPA), dehydrated with 50%, 70%, 80%, 90% and absolute ethanol, transferred to tertiary butyl alcohol, and infiltrated in a 60–65° C oven with paraffin. Alternatively, CRAF III may be used to fix the material (3 parts 1% chromic acid:2 parts 10% acetic acid:1 part formalin:4 parts water—mix immediately before fixation); followed by an extended wash, the material is dehydrated and infiltrated as before. The material is embedded, cooled, microtomed, and mounted using standard methods (Berlyn and Miksche 1976). Material stained with Sass's hemalum is well suited for meiotic detail (as in Figures 14.1–3); greater cytoplasmic detail is provided by a combination of safranin and fast green (Figure 14.4) (general stain reference, see Clark 1981). Callose present in the cell walls of megasporocytes and degenerate megaspores can be localized using 0.005% aniline blue in 0.1 M KH_2PO_4, pH 8.2, filtered through decolorizing charcoal and viewed (Figures 14.5, 14.6) using fluorescence microscopy at 366 nm (Jensen 1962); be aware that fluorescence yield varies greatly according to manufacturer and batch. The paraffin histology techniques preserve the nucleus well, but cause significant extraction of material in the cytoplasm (Figures 14-1–14.4).

PLASTIC HISTOLOGY

Material is fixed in 3% glutaraldehyde in a suitable buffer (e.g., 0.1 M cacodylate, pH 7.4—caution: contains arsenic) for 6 hours to overnight, rinsed in four 10 minute changes of buffer. For electron microscopy, the material is trimmed to less than 0.5 mm to be fixed and postfixed in 2% buffered osmium tetroxide at 4° C. (Caution: OsO_4 is a potent vapor fixative, particularly dangerous to mucosal surfaces, e.g., eyes, lungs! Use fumehood.) Material is dehydrated in 15–20-minute changes in an ethanol series of 20%, 40%, 50%, 60%, 70%, 80%, 90%, and absolute ethanol (three changes). Material is transferred to propylene oxide (through a 1:1 change) and then transferred to one-third, two-thirds, and pure Spurr's resin or to another light microscopy resin (e.g., JB-4, methacrylate, etc.). Material is then sectioned (1 to as thick as 4 μm) using a glass knife and mounted according to standard techniques. Material is stained using heated toluidine or methylene blue for general histology; for general histochemical localizations, periodic acid-Schiff's

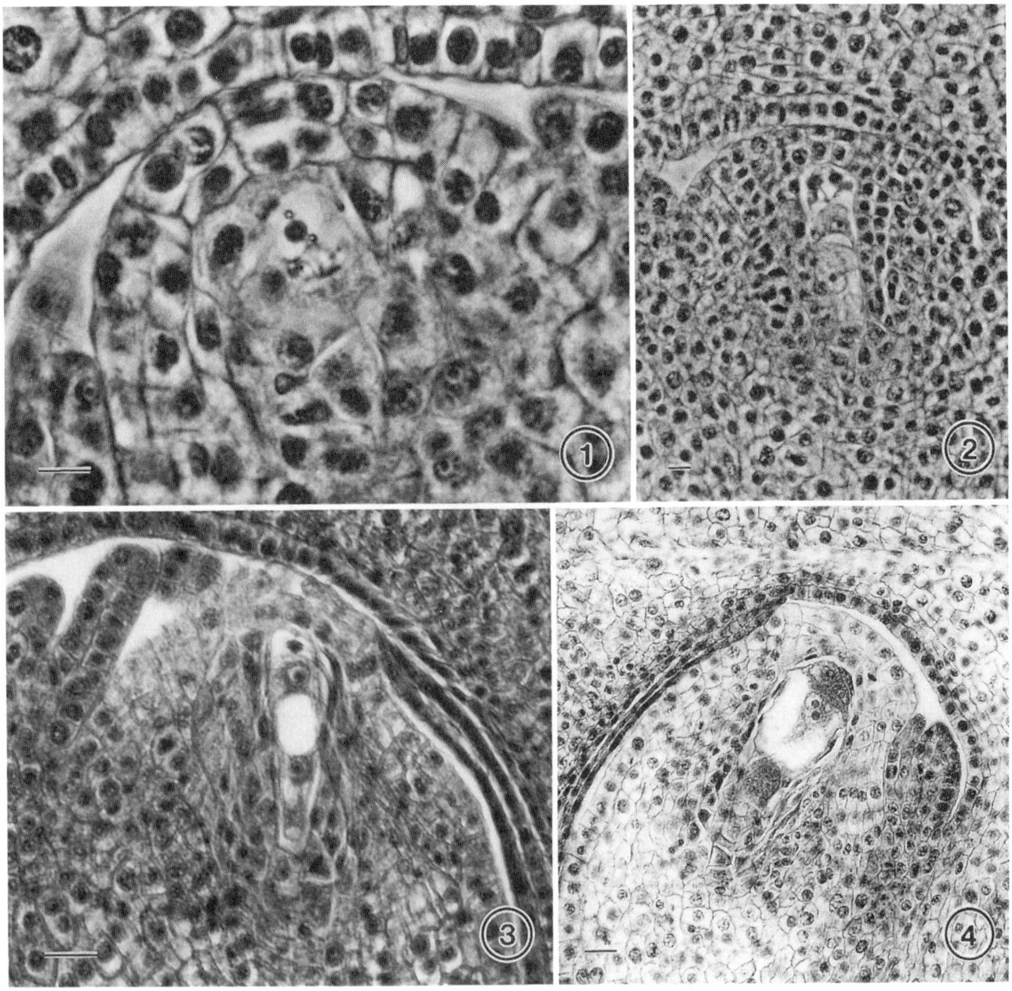

FIGURES 14.1–14.4 Light microscopy of material embedded in paraffin. Bar = 10 μm. **FIGURE 14.1** Prophase I in megaspore mother cell. Stain = Sass's hemalum, 2 hours. **FIGURE 14.2** Telophase II. Micropylar megaspores arrested in anticlinal anaphase. Stain = Sass's hemalum, 2 hours. **FIGURE 14.3** Embryo sac. Stains = safranin, 25 minutes; fast green, 30 seconds; Sass's hemalum, 1 hour. **FIGURE 14.4** First postmeiotic mitosis. Stains = Safranin, 25 minutes; fast green, 30 seconds; Sass's hemalum, 1 hour.

reaction is used for insoluble polysaccharides (Feder and O'Brien 1968) and 1% aniline blue-black in 7% acetic acid for proteins. Suitably fixed material may also be prepared for transmission or scanning electron microscopy. This technique preserves the nucleus and the cytoplasm well (Figures 14.7, 14.8) and provides the highest potential resolution at the light or electron microscopic level.

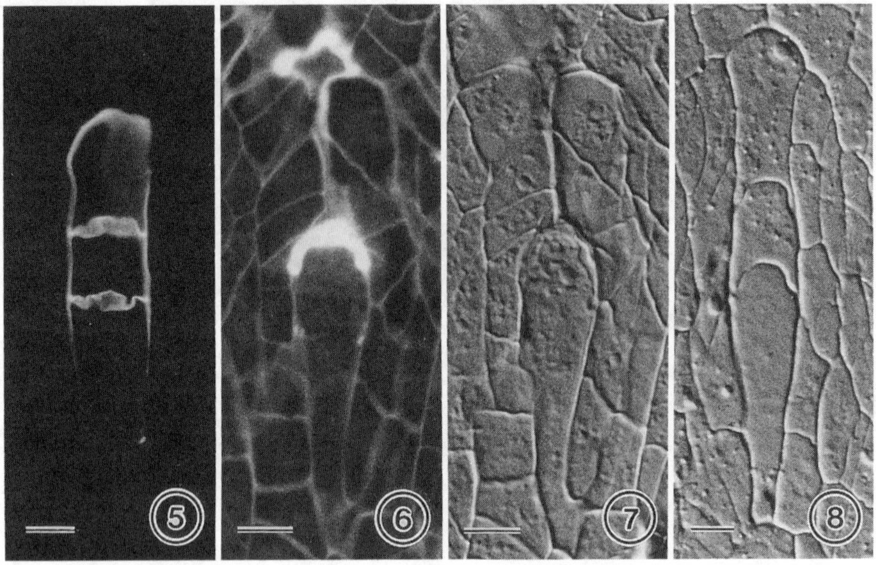

FIGURES 14.5–14.6 Fluorescence microscopy of plastic sections stained with aniline blue. Bar = 10 μm. **FIGURE 14.5** Calloselike substances in wall of triad of megaspores. **FIGURE 14.6** Calloselike substances of degenerating megaspores. **FIGURES 14.7–14.8** Interference contrast microscopy of unstained plastic sections. Bar = 10 μm. **FIGURE 14.7** Triad of megaspores with enlarging functional (chalaza) megaspore **FIGURE 14.8** Two-nucleate embryo sac with degenerating megaspores from Fig. 14.6. (Figures 14.4–4, courtesy of D.P. West; Figures 14.5–8, courtesy of S.D. Russell).

OVULE CLEARING

Material fixed as for paraffin histology is placed in a clearing solution composed of Herr's 4–1/2 solution (2 parts lactic acid: 2 parts chloral hydrate: 2 parts phenol: 2 parts clove oil: 1 part xylene: 1 part benzyl benzoate, by weight; Herr 1971) or 100% methyl salicylate and incubated until it is translucent. The material may then be examined using phase contrast, differential-interference contrast, or confocal scanning microscopy. This technique permits a rapid view of cytoplasmic structures and nuclei, but resolution is limited. Considerable cytoplasmic structure is extracted during the clearing process.

ISOLATION OF MEGASPORES, EMBRYO SACS, AND GAMETES

Fixed (FAA, FPA, or glutaraldehyde) or fresh material is incubated in an enzyme solution of 1% cellulase and 1% pectinase for 30 minutes or longer (Wagner et al. 1988, 1989). Fixed material may be digested for longer, but in fresh material, viability decreases with increased enzyme incubation. If short incubations are used, a minimal amount of micromanipulation may be needed to dislodge the embryo sac from the integuments and surrounding nucellus. Suitably prepared living material may be maintained using breeder cells for several weeks (Kranz et al. 1991a,b); however, techniques are still developing for manipulating gametophytes, conducting fusions using electroporation, and culturing recombinants with the goal of eventual regeneration. This technique provides high-quality fixed and fresh preparations, suitable for fluorescent probes such as DAPI or Hoechst 33258 for DNA (1 μg/ml) or for immunochemistry using FITC, TRITC, or rhodamine; this is an emerging method that requires some manual skills to be developed, but is not difficult.

REFERENCES

Berlyn GP, Miksche JP (1976) Botanical Microtechnique and Cytochemistry, Iowa State University, Ames, IA, 326 pp

Cheng PC, Greyson RI, Walden DB (1983) Organ initiation and the development of unisexual flowers in the tassel and ear of *Zea mays.* Am J Bot 70: 450–462

Clark G (1981) Staining Procedures, Willliams & Wilkins, publishers, Waverly Press, Baltimore, MD, 512 pp

Cooper DC (1937) Macrosporogenesis and embryosac development in *Euchlaena mexicana* and *Zea mays.* J Agric Res 55: 539–551

Feder N, O'Brien TP (1968) Plant microtechnique: some principles and new methods. Am J Bot 55: 123–142

Herr JM, Jr (1971) A new clearing squash technique for the study of ovule development in angiosperms. Am J Bot 58: 785–790

Jensen WA (1962) Botanical histochemistry: principles and practice. W. H. Freeman, San Francisco, CA, 468 pp

Kranz E, Bautor J, Lörz H (199la) *In vitro* fertilization of single, isolated gametes of maize mediated by electrofusion. Sex Plant Reprod 4: 12–16

Kranz E, Bautor J, Lörz H (199lb) Electrofusion-mediated transmission of cytoplasmic organelles through the *in vitro* fertilization process, fusion of sperm cells with synergids and central cells, and cell reconstitution in maize. Sex Plant Reprod 4: 17–21

Russell SD (1979) Fine structure of megagametophyte development in *Zea mays.* Can J Bot 57: 1093–1110

Weatherwax P (1919) Gametogenesis and fecundation in *Zea mays* as the basis of xenia and heredity in the endosperm. Bull Torrey Bot Club 46: 73–90

Wagner VT, Dumas C, Mogensen HL (1990) Quantitative three-dimensional study on the position of the female gametophyte and its constituent cells as a prerequisite for corn (*Zea mays*) transformation. Theor Appl Genet 79: 72–76

Wagner VT, Song Y, Matthys-Rochon E, Dumas C (1988) The isolated embryo sac of *Zea mays.* Structural and ultrastructural observations. In Cresti M, Gori P, Pacini E (eds) Sexual Reproduction in Higher Plants, Springer-Verlag, Berlin, pp 125–130

Wagner VT, Song Y, Matthys-Rochon E, Dumas C (1989) Observations on the isolated embryo sac of *Zea mays* L. Plant Sci 59: 127–132

15

Indirect Immunofluorescence: Localization of the Cytoskeleton

C. J. Staiger

The plant cytoskeleton is composed of two structural elements known as microfilaments and microtubules. Microtubules are 24-nm-diameter hollow fibers constructed from α/β heterodimers of the protein tubulin. Microfilaments (or F-actin) are 5–7-nm-diameter homopolymers of 42 kD actin subunits. These two cytoskeletal components play a major role in a wide variety of cellular processes. For example, cytoplasmic streaming is driven by the mechanochemical enzyme myosin moving along actin microfilaments. The complex events of mitosis, including chromosome segregation and cell plate deposition, utilize two poorly understood microtubule-based structures, the spindle and phragmoplast. Intracellular positioning of nuclei and certain organelles is also dependent on microtubules and microfilaments. For a more detailed analysis of cytoskeletal function, the reader is referred to several excellent reviews and a monograph on this topic (Baskin and Cande 1990; Seagull 1989; Lloyd 1987, 1991).

Our understanding of the plant cytoskeleton has been greatly assisted by immunological techniques (Lloyd 1987). In particular, the use of highly specific antibodies coupled with fluorescent dyes allows great accuracy in determining the distribution of a wide range of protein antigens within cells. Using immunofluorescence microscopy, it is now a routine matter to characterize the three-dimensional arrangement of microtubules and microfilaments in a variety of cell types. By examining populations of dividing or differentiating cells, we know that the cytoskeleton is capable of dynamic organizational changes. In this chapter, I describe methods for fixing and staining microtubules in maize microsporocytes, root meristematic cells, and Black Mexican Sweet corn (BMS) suspension cells. Procedures for labeling actin microfilaments with fluorescent phallotoxins are also discussed.

MATERIALS

Poly-L-lysine, MW 250,000 (Sigma); 1 mg/ml stock in water

The Maize Handbook—M. Freeling, V. Walbot, eds.

Paraformaldehyde solution, 16% (Electron Microscopy Sciences, Fort Washington, PA)
4,6-diamidine-2-phenylindole dihydrochloride (DAPI, Sigma); 1 mg/ml stock in water
1,4-diazobicyclo-(2,2,2)-octane (DABCO, Ernest F. Fullam, Schenectady, NY); 100 mg/ml stock in 1 part phosphate-buffered saline (PBS) and 9 parts glycerol
m-maleimidobenzoyl *N*-hydroxysuccinimide ester (MBS, Pierce); 100 mM stock in DMSO
Rhodamine-phalloidin (Molecular Probes, Eugene, OR); 3.3 μM stock in methanol

Cell-Wall Digesting Enzymes

NovoZym 234 (NovoBio Labs, Danbury, CT)
Cellulase "Onozuka" R-10 (Yakult Honsha Co., Tokyo)
Pectolyase Y23 (Yakult Honsha Co., Tokyo)
Cellulysin (Calbiochem, La Jolla CA)

Solutions

PHEMS, microtubule stabilizing buffer I:
60 mM PIPES (pH 6.8), 25 mM HEPES, 10 mM EGTA, 2 mM $MgCl_2$, 0.35 M sorbitol
PME, microtubule stabilizing buffer II:
100 mM PIPES (pH 6.9), 5 mM EGTA, 5 mM $MgCl_2$, 0.35 M sorbitol
PBS, phosphate buffered saline:
137 mM NaCl, 2.7 mM KCl, 8 mM $Na_2HPO_4 \cdot 7H_2O$, 1.47 mM KH_2PO_4, pH 7.4
Protease inhibitor cocktail
2 mg/liter each: leupeptin and pepstatin A
20 mg/liter each: benzyl arginyl methyl ester, *p*-tosyl-1-arginine methyl ester, *L*-1-tosylamide-2-phenylethyl chloromethyl ketone, and soybean trypsin inhibitor
1 mM phenylmethylsulfonyl fluoride (PMSF)
MES wash:
0.5% MES (pH 5.8), 80 mM $CaCl_2$, 0.25 M mannitol
MFSB, microfilament stabilizing buffer:
50 mM PIPES (pH 7.0), 5 mM $MgCl_2$, 5 mM EGTA, 0.35 M sucrose

METHODS

Localization of Microtubule Arrays in Meiotic Cells

In this section I set forth the techniques that have proven successful for localizing microtubules during meiosis in maize microsporocytes. For full details of the results obtained, the reader is referred to an article by Staiger and Cande (1990). Descriptions of the temporal and spatial changes in microtubule distribution during wild type meiosis served as the foundation for comparing microtubule organization in several meiotic mutants (Staiger and Cande 1990, 1991).

The keys to successful fluorescence microscopy lie in preserving the antigen by proper fixation and in digesting the plant cell wall to allow sufficient antibody penetration. Fixation of plant microtubules is best achieved using paraformaldehyde in the presence of microtubule-stabilizing buffers. Certain modifications of published techniques were necessary in order to stain microtubules in pollen mother cells. (See Note 1.) First, it was essential to extrude the microsporocytes from the anther locule to allow rapid fixation. Second, an enzyme capable of digesting the predominantly callose wall of sporocytes had to be found. Finally, detergent treatment was required to reduce the cytoplasmic autofluorescence observed in certain maize lines. Steps summarizing the procedure are outlined below:

1. Anthers containing meiotic cells are isolated and staged according to the methods of Burnham (1982).
2. Coverslips are prepared before collecting fresh plant material. To enhance sticking of single cells to the coverslip, a solution of 1 mg/ml poly-*l*-lysine is applied to acetone-cleaned coverslips, and allowed to soak for 1 hour. Coverslips are then rinsed with distilled water and air-dried.
3. Three to five anthers are placed on each coverslip in a drop of fixative solution containing 4% (v/v) paraformaldehyde and 0.2% (v/v) Triton X-100 in a microtubule stabilizing buffer such as PHEMS or PME.
4. To extrude meiotic cells, the distal tip of each anther is held by a pair of fine forceps while the proximal end is cut with a sharp scalpel blade. The scalpel blade or a blunt probe is then used to gently stroke the anther from the distal end toward the cut. By this method, meiotic cells from zygotene and pachytene stages onward are readily teased from fresh anthers. Very early prophase and premeiotic cells are generally resistant to isolation and suffer too much damage to provide useful cytological images.
5. Anther wall debris is removed from the coverslip, and the cells are allowed to settle.
6. A 1-hour fixation is generally sufficient to give good preservation of microtubules. Care should be taken to ensure that the coverslip does not dry during fixation.
7. Excess fixative solution is carefully aspirated from the coverslip and replaced with microtubule stabilizing buffer without fixative or detergent. At this stage it is sometimes useful to briefly allow the coverslip to air dry. This increases adhesion of meiotic cells, but may cause distortion of microtubule arrays. In all subsequent steps, the coverslip is used to move the cells from one solution to the next. Great care must be taken when transferring the coverslip between solutions so as not to lose cells.
8. Following several rinses in microtubule stabilizing buffer, the callose meiocyte wall is partially digested to allow antibody penetration. NovoZym 234 is an enzyme mix-

ture enriched for β-1,3-glucanase that works well on maize meiocytes. A solution of 1 mg/ml NovoZym in microtubule stabilizing buffer is allowed to digest the cell walls for 10–15 minutes. A protease cocktail (1/100 dilution) should be added to the digestion solution to reduce protease action. After wall digestion, the wash solution can be changed to PBS and several washes of 10 minutes each should be used before application of the primary antibody.

9. For a discussion on the choice of primary antibodies, see Note 2 below. In practice, a small amount of diluted primary antibody against tubulin is placed on each coverslip so that the entire surface is covered. For a 22-mm coverslip, 25 μl is adequate. Primary antibody incubation is done at 37° C for 1 hour in a humid chamber to prevent drying.
10. This is followed by three 10-minute washes with PBS to remove nonspecific antibody.
11. A fluorescently labeled secondary antibody is applied and incubated for 1 hour at 37° C. See Note 3 below concerning the choice of second antibody.
12. Excess secondary antibody is removed by washing three times, 10 minutes each, with PBS.
13. To determine the meiotic stage of microsporocytes, chromatin can be stained with a DNA-specific dye included in one of the final washes. I use DAPI at a final concentration of 1 μg/ml. Both propidium iodide and Hoechst 33662 also work well on sporocytes.
14. Most of the liquid should be removed and the coverslips should be placed cell-side down onto a drop of antifade solution on a slide. Antioxidants that retard bleaching of the fluorochrome include 10% DABCO and 2% n-propyl gallate made up in PBS and glycerol. Slides can be made semipermanent by sealing the edges with clear nail polish. Slides can then be stored at −20° C for more than 1 year.

Note 1. Unfortunately, the techniques described above are not directly applicable to study of the very interesting processes of pollen mitosis and pollen maturation. Maize pollen grains have proven quite resistant to a cytoskeletal analysis, perhaps because of penetration problems associated with the impervious pollen grain wall. The protocol for staining meiocytes can be adapted for staining maize root tip cells and BMS suspension cells. (See below.)

Note 2. Choosing the primary antibody against tubulin can be somewhat complicated. A number of polyclonal and monoclonal antibodies have been raised against purified plant tubulins, and all of those tested recognized maize microtubules. However, none of these are currently commercially available. Antibodies raised against animal and fungal tubulins can be purchased from companies, including Amersham, Sigma, and Polysciences. In my experience, many of these are perfectly adequate for staining plant microtubules. I routinely use a polyclonal rabbit antibody raised against sea urchin tubulin (#17870, Polysciences, Inc, Warrington, PA). Therefore, you should either purchase commercial antibodies or find a generous researcher to give you some of his/her own antitubulin. The titer of every antibody will obviously be different; therefore the useful concentration must be empirically determined. Fortunately, most manufacturers offer recommendations for diluting their product. Alternatively, start by testing of dilution series of 1/10, 1/100, and 1/1,000. It is best to choose the highest-possible dilution that still gives fibrillar microtubule staining while eliminating nonspecific background. Antibodies should be diluted in PBS with 0.01% (w/v) azide as a preservative. Diluted antibodies can be stored for months at 4° C. The stock solution should be frozen in small aliquots at −80° C.

Note 3. The secondary antibody should be chosen to recognize the species in which the primary antibody was raised. For example, to complement the rabbit polyclonal against tubulin, I use a goat antirabbit Ig (Sigma). Again, proper dilution of the secondary antibody is determined by trial and error. A good starting point is to try 0.1 mg/ml total immunoglobins. Whenever a new secondary antibody is purchased a control must be performed in which the first antibody is replaced with PBS. This reaction should give practically no cellular staining. If bright cytoplasmic fluorescence is observed, try greater dilutions or a different supplier.

Localization of Microtubule Arrays in Root Tip Cells

Root tip cell microtubules are easily stained with a modification of techniques first described by Wick and co-workers (1981). Steps are summarized here:

1. Four-day-old primary roots are excised 2–3 mm from the tip and placed in 4% paraformaldehyde in PHEMS buffer for 1 hour.
2. Extensive buffer washes are required to remove fixative from the intact tissue (i.e., a minimum of 30 minutes and three buffer changes).
3. Root tips are then digested for 30 minutes with 1% (w/v) Cellulysin in a 0.3 M sucrose solution. Enzyme solution is removed by washing with PBS.
4. Cells are released from the root tip by teasing the tissue apart with fine dissecting needles.
5. Large debris are removed and the isolated cells are air-dried.
6. All subsequent processing follows the guidelines for microsporocytes.

Localization of Microtubule Arrays in BMS Cells

To stain BMS cell microtubules it is necessary to first digest the cell walls and then fix the cytoplasm. (See also Wang et al. 1989; Goodbody et al. 1989).

1. Gravity-settled BMS cells are resuspended in an enzyme solution containing: 1.5% (w/v) Cellulase R10 and 0.1% (w/v) pectolyase Y23 in MES wash.

2. Cells are incubated at 25° C with shaking for 2–3 hours in a small petri dish until protoplasts are formed.
3. Protoplasts are washed two times with MES wash by collecting on a 40-μm nylon filter or by gentle centrifugation (5 minutes, 100*g*).
4. A small drop of this solution is placed onto polylysine-coated coverslips and the protoplasts are allowed to settle for 10–15 minutes.
5. The excess liquid is then removed by gentle aspiration.
6. Protoplasts are fixed for 30 minutes in 4% paraformaldehyde in PME buffer with the addition of 1% Triton X-100.
7. Antibody staining follows the protocol for microsporocytes.

Microtubule Localization in Other Maize Cell Types

Staining of microtubule arrays within the cells of intact tissues has proven to be far more difficult than working with isolated cells. A few groups have approached this problem by fixing intact tissues and then cryosectioning or embedding in polyethylene glycol before sectioning and processing for immunofluorescence. The reader is referred to one such publication for guidance (Hogetsu 1989). Another approach is to fix and stain hand-sectioned tissue or epidermal peels. This technique works well for mature maize leaves, but is less successful on young primordia (A. Sylvester, personal communication).

Actin Microfilament Staining

The techniques used to visualize F-actin differ significantly from the protocol for observing microtubules in maize microsporocytes. Several researchers have reported difficulties in preserving plant microfilaments with aldehyde fixatives (Parthasarathy et al. 1985; Seagull et al. 1987), and consequently new methods have been developed. The most common alternative technique is one pioneered by Traas and co-workers (Traas et al. 1987, 1989), that has since been applied to maize sporocytes (Staiger and Cande 1991). This method involves treating unfixed, detergent-permeabilized cells with rhodamine-phalloidin. Another successful method uses a mild cross-linking reagent to stabilize microfilaments before a short aldehyde fixation. Microfilament distribution in cells stained with these two techniques is virtually indistinguishable.

Microsporocyte F-actin arrays are most easily stained using the techniques of Traas et al. (1987, 1989), as outlined below.

1. Sporocytes from cut anthers are extruded on a glass slide into a drop of permeabilization/stabilization medium composed of MFSB plus 5% (v/v) DMSO and 0.1% (v/v) Nonidet P-40.
2. A coverslip is gently placed over the isolated cells, and extraction is allowed to continue for 10–15 minutes at room temperature.

3. Permeabilization solution is replaced with staining solution by applying a small drop at one edge of the coverslip and drawing the liquid through by wicking with a piece of bibulous paper placed at the opposite side.
4. The staining solution includes 0.33 μM rhodamine-phalloidin and 1 μg/ml DAPI to stain microfilaments and chromatin. Staining is allowed to progress for 10 minutes.
5. An antifade agent is added either by perfusion or by removing the coverslip and adding a small drop of solution before replacing. Rhodamine-phalloidin stained sporocytes prepared by this method are best viewed immediately or within a couple hours if the slide is stored at 4° C.

A second technique adequate for staining microsporocytes, but requiring more effort, is a modification of the methods published by Sonobe and Shibaoka (1989).

1. Isolated sporocytes are initially treated with MFSB containing the mild protein cross-linking reagent MBS at a concentration of 100 μM and 0.05% Triton X-100.
2. After 30 minutes of MBS treatment, cells are postfixed in 4% paraformaldehyde for 10 minutes in MFSB with 0.05% Triton.
3. Following several rinses of buffer solution without fixative, rhodamine-phalloidin and DAPI are applied as for unfixed, permeabilized cells.
4. This method has the advantage that slides can be stored for long periods in the refrigerator before viewing.

Observation of Stained Cells

Microtubules, microfilaments, and chromatin are visualized using a microscope equipped for epifluorescence illumination (Figure 15.1). The excitation and emission properties will depend on the specific fluorescent tag used. For example, FITC is excited by light between the wavelengths 450 and 500 nm (blue light), with a maximum absorption at 490 nm. The molecule then emits a yellow-green fluorescence (500–550 nm). Rhodamine conjugates absorb green wavelengths (520–560 nm) and emit red fluorescence (550–600 nm). The DNA-specific dye DAPI is excited by UV light (maximum 365 nm) and emits blue light (470 nm). Because the excitation/emission spectra for these three molecules overlap, some precautions must be taken to avoid confusion when using more than one label on a single slide or coverslip. It is possible to get "bleed-through" of signal when viewing in a given channel. The most frequent problem occurs when both fluorescein and rhodamine antibodies are used. The green light emission is often strong enough to excite the rhodamine, allowing one to visualize both elements simultaneously. When using multiple fluorescent tags, it is important to look first at the molecule requiring the lowest energy for excitation (i.e., longest wavelength). After determining the staining pattern (and usually taking a photograph), switch to a shorter wavelength.

Microtubules and microfilaments are fibrous components and are found in several different cytoplasmic locations depending on the stage of the cell cycle. Microtubules

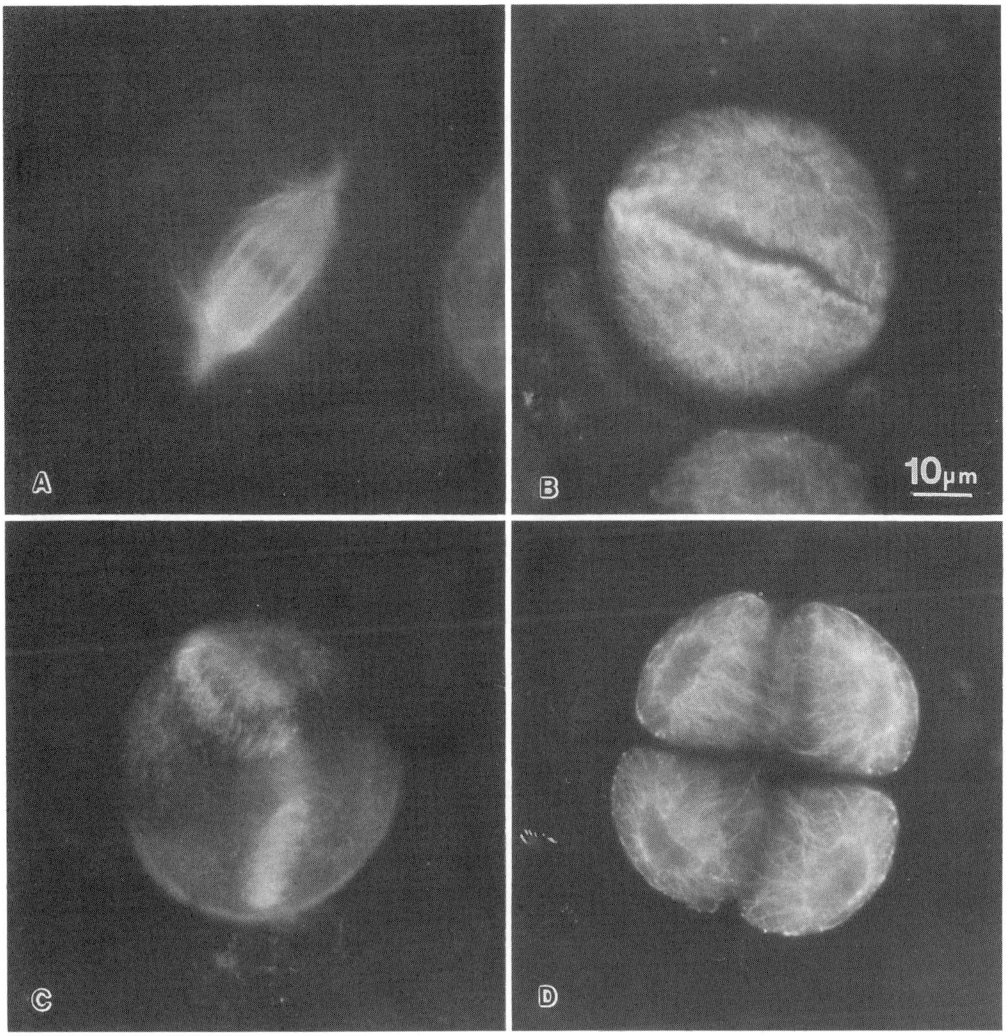

FIGURE 15.1 Indirect immunofluorescence visualization of microtubule arrays in maize meiotic cells. A. Metaphase I spindle in a primary microsporocyte. B. A pair of secondary microsporocytes with cytoplasmic microtubules radiating from the nucleus toward the cortex. C. Phragmoplast microtubule arrays during the second division cytokinesis appear as a ringlike structure. D. A tetrad of microspores with characteristic radial cytoplasmic microtubules. Bar = 10 μm.

can be found as highly organized cortical arrays in interphase somatic cells or as a complex reticulate network in meiocytes. They are also the primary components of the meiotic and mitotic spindles and the cytokinetic phragmoplast. The staining pattern obtained using the methods described above should be distinctly fibrillar. If the cytoskeletal staining appears fragmented or punctate, fixation is usually to blame. Try making some fresh fixative. Alternatively, but less likely, the primary antibody concentration is too low and should be increased. Another common problem is high background fluorescence in the cytoplasm. This problem has two main sources: autofluorescence of unknown cytoplasmic constituents and excessive or poor quality secondary antibody. The first problem is sometimes remedied by increasing the detergent concentration during or after fixation. Occasionally I have even used 1% Triton X-100. The second problem is eliminated by testing the dilution of secondary antibody as recommended in Note 3 above. If this fails to solve the problem, try a different supplier. A few commercial antibodies, especially rabbit sera, have nonspecific affinity for cytoplasmic components.

REFERENCES

Baskin TI, Cande WZ (1990) The structure and function of the mitotic spindle in flowering plants. Annu Rev Plant Physiol Plant Mol Biol 41: 277–315

Burnham, CR (1982) Details of the smear technique for studying chromosomes in maize. In Sheridan WF (ed) Maize for Biological Research, University Press, Grand Forks, ND, pp 107–118

Goodbody KC, Hargreaves AJ, Lloyd CW (1989) On the distribution of microtubule-associated intermediate filament antigens in plant suspension cells. J Cell Sci 93: 427–438

Hogetsu T (1989) The arrangement of microtubules in leaves of monocotyledonous and dicotyledonous plants. Can J Bot 67: 3506–3512

Lloyd CW (1987) The plant cytoskeleton: the impact of fluorescence microscopy. Ann Rev Plant Physiol 38: 119–139

Lloyd CW (1991) The Cytoskeletal Basis of Plant Growth and Form, Academic Press, London, 322 pp

Parthasarathy MV, Perdue TD, Witztum A, Alvernaz J (1985) Actin network as a normal component of the cytoskeleton in many vascular plant cells. Am J Bot 72: 1318–1323

Seagull RW (1989) The plant cytoskeleton. CRC Crit Rev Plant Sci 8: 131–167

Seagull RW, Falconer MM, Weerdenburg CA (1987) Microfilaments: dynamic arrays in higher plant cells. J Cell Biol 104: 995–1004

Sonobe S, Shibaoka H (1989) Cortical fine filaments in higher plant cells visualized by rhodamine-phalloidin after pretreatment with m-maleimidobenzoyl N-hydroxysuccinimide ester. Protoplasma 148: 80–86

Staiger CJ, Cande WZ (1990) Microtubule distribution in *dv*, a maize meiotic mutant defective in the prophase to metaphase transition. Dev Biol 138: 231–242

Staiger CJ, Cande WZ (1991) Microfilament distribution in maize meiotic mutants correlates with microtubule organization. Plant Cell 3: 637–644

Traas JA, Burgain S, Dumas de Vaulx R (1989) The organization of the cytoskeleton during meiosis in eggplant (*Solanum melongena* L.): microtubules and F-actin are both necessary for coordinated meiotic division. J Cell Sci 92: 541–550

Traas JA, Doonan JH, Rawlins DJ, Shaw PJ, Watts J, Lloyd CW (1987) An actin network is present in the cytoplasm throughout the cell cycle of carrot cells and associates with the dividing nucleus. J Cell Biol 105: 387–395

Wang H, Cutler AJ, Saleem M, Fowke LC (1989) Microtubules in maize protoplasts derived from cell suspension cultures: effect of calcium and magnesium ions. Eur J Cell Biol 49: 80–86

Wick SM, Seagull RW, Osborn M, Weber K, Gunning BES (1981) Immunofluorescence microscopy of organized microtubule arrays in structurally stabilized meristematic plant cells. J Cell Biol 89: 685–690

16

Immunocytochemistry for Light and Electron Microscopy

M. J. Varagona and N. V. Raikhel

Several immunocytochemical protocols have been optimized for the localization of particular proteins in maize. (See references.) Due in large part to multiple steps in the procedure and to structural differences between tissues, all protocols should be viewed as guidelines that may need to be modified for specific objectives. Immunocytochemistry adapted for the plant tissue has been reviewed in detail (Jeffree et al. 1982; Mishkind et al. 1987). In this chapter we present an immunocytochemical protocol that has been successfully used for the detection of the Opaque 2 (O2) protein in sections from maize kernels at the light microscope level (Varagona et al. 1991). In addition, we are offering a protocol for localization of barley lectin protein in the roots of barley seedlings. This protocol was successfully used for the detection of proteins in roots of several monocotyledonous plants. See also Parthasarathy (this volume) for additional protocols for immunoelectron microscopy.

The Maize Handbook—M. Freeling, V. Walbot, eds.

IMMUNOLABELLING PROCEDURE FOR MAIZE KERNELS AT THE LIGHT MICROSCOPE LEVEL

Materials

NaCl
KCl
Na_2HPO_4
KH_2PO_4
Na-azide
Paraformaldehyde (Polysciences, Inc.)
Chromerge (Fisher)
Sucrose (Baker Analyzed)
Triton X-100
Glycine (Sigma)
Glycerol (Aldrich Chem. Co., Inc.)
Goat serum (Sigma)
Poly-D-lysine hydrobromide (Sigma)
Tris-HCl (pH 8.0)
HistoPrep (Fisher)
HistoFreeze (Fisher)
Specific antibody (assumed to be derived from a rabbit)
Preimmune serum
Rhodamine-conjugated goat antirabbit IgG (Pierce)
Liquid N_2
Cryostat
Hot plate
Slides
Fluorescence microscope

Method

Preparation of Poly-D-Lysine Coated Slides

1. Etch new slides with identifying numbers, then soak in Chromerge overnight. Rinse slides for 3 hours in tap H_2O, then 1 hour in double-distilled (dd) H_2O. Rinse each slide individually (held with forceps) under running dd H_2O. Dry slides at room temperature (RT) overnight or at 80° C for a few hours.
2. Soak dried slides in 50 μg/ml poly-*d*-lysine hydrobromide in 10 mM Tris-HCl (pH 8.0) for 30 minutes at RT (250 ml Tris-HCl + 12.5 mg poly-*d*-lysine). Air-dry slides in a dust-free area. Slides can be stored in boxes at RT for several weeks before use.

Tissue Fixation

1. Cut maize kernels (10 to 22 DAP, Figure 16.1) longitudinally, bisecting the embryo.
2. Cut each half kernel both longitudinally and crosswise, yielding four equal quarters.
3. Fix immediately in fixative containing 4% paraformaldehyde in 1× PBS[a] with 0.1 M sucrose. Incubate 24 hours on a shaker at 4° C.
4. Wash tissues three times, 10 minutes each in 1× PBS containing 0.5 M sucrose. Store (for up to 3 months) in 1× PBS containing 0.5 M sucrose.
5. Put a drop of HistoPrep compound into the bottom of a beam capsule.
6. Orient the tissue in this drop; then add HistoPrep to fill the beam capsule about half full.
7. Immerse the capsule with the specimen in a tube containing hexane (hexane holder) and submerge into liquid N_2 until the HistoPrep compound completely freezes (30–60 seconds).

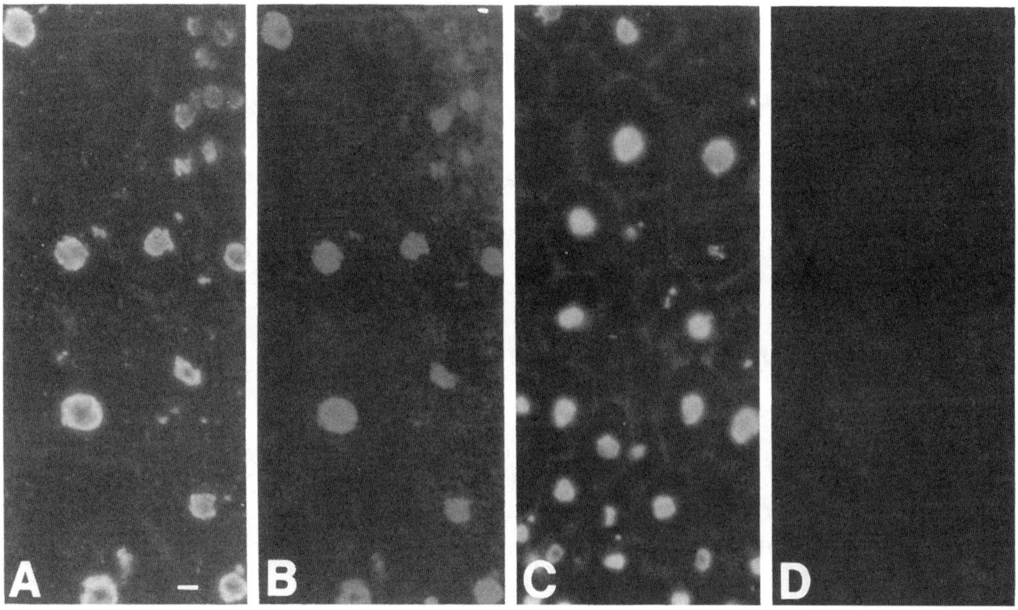

FIGURE 16.1 Localization of O2 protein in maize tissue (Varagona et al. 1991; figure reprinted with permission from *The Plant Cell*). Photomicrographs are shown in pairs with tissues treated with the nuclear-specific stain DAPI (A and C) and with immunolocalization, using polyclonal anti-O2 sera and IgG-conjugated rhodamine (B and D). A, B. 18-DAP O2 endosperm. C, D. 22-DAP O2 embryo. The O2 protein is localized in nuclei of endosperm cells of O2 kernels (B). No specific staining was observed in sections from O2 embryos (D). Bar = 10 μm.

8. Remove the capsule from the hexane holder with forceps, and quickly cut away the beam capsule with a fresh razor blade or scalpel.
9. Place the frozen block onto a drop of HistoPrep on the specimen support and freeze the drop with HistoFreeze aerosol.
10. Equilibrate the specimen in the cryostat chamber for 20 minutes.
11. Mount the specimen onto the microtome and section the block down to the tissue.
12. Collect sections, 8–12 μm thick (either individually or in ribbons), onto slides pretreated with poly-*d*-lysine.
13. After enough sections have been collected (to fit under one coverslip), place the slide on a hot plate for 20 minutes at 40–50° C. Slides may be used immediately or within 2 days.

Immunocytochemistry[b]

1. Treat sections with a few drops of 0.05% Triton X-100 for 20 minutes in a petri dish.
2. Wash in 1× PBS[a] containing 10 mM glycine and 0.02% Na-azide, three times, 10 minutes each.
3. Block with goat serum (undiluted) for 15 minutes.
4. Incubate overnight at 4° C or 1.5 hours at RT with specific primary antibody.[c]
5. Wash in 1× PBS containing 10 mM glycine and 0.02% Na-azide, three times, 10 minutes each.
6. Incubate in goat serum (undiluted) for 15 minutes.
7. Incubate in rhodamine-conjugate secondary antibody[d] for 1.5 hours at RT.
8. Wash in 1× PBS containing 10 mM glycine and 0.02% Na-azide, three times, 10 minutes each.
9. Mount[e] the sections in 90% glycerol in 1× PBS.

Notes

a. A 10× stock of PBS, pH 7.2, contains 1.37 M NaCl, 30 mM KCl, 80 mM Na_2HPO_4, 15 mM KH_2PO_4.

b. All steps should be done in a humid chamber (petri dish containing toothpicks over wet Kimwipes) at RT unless indicated otherwise.

c. In this protocol, polyclonal antibodies raised in rabbit against an O2-β-galactosidase fusion protein (Varagona et al. 1991) were used to localize O2 protein (Figure 16.1). Primary antibodies were diluted 1:500 in 10% goat serum, 90% 1× PBS containing 0.02% Triton X-100. The same concentration of preimmune serum was used in the control experiments.

d. The secondary antibodies used in Figure 16.1 were rhodamine-conjugated goat antirabbit IgG diluted 1:3000 in the same solution as primary antibodies.

e. Mount solution used in Figure 16.1 contains 1× PBS, 90% glycerol, 0.02% Na azide, and 20 μg/mL DAPI (Sigma) for visualization of nuclei.

IMMUNOLABELING PROCEDURE FOR ROOT TISSUE AT THE ELECTRON MICROSCOPE LEVEL

Materials

Formvar (Ladd Res. Ind., Inc.) (0.25% in ethylene dichloride)
Paraformaldehyde (Polysciences, Inc.)
Glutaraldehyde (Polysciences, Inc.)
Na-phosphate buffer (pH 7.2)
OsO_4 (Polysciences, Inc.)
Ethanol
LR White (Polysciences, Inc.)
Sodium meta-periodate (Aldrich Chem. Co., Inc.)
HCl
Dried milk
Na-phosphate buffer (pH 7.2)
NaCl
Tween-20
Aqueous uranyl acetate (Ladd Res. Ind., Inc.)
NaOH
Reynolds lead citrate (Reynolds 1963)
Specific antibodies
Protein A colloidal gold (E Y Labs, Inc.)
Beam capsules (Polysciences, Inc.)
Vacuum oven
Rotator
300 mesh Ni grids (Ted Pella, Inc.)
Microtome

Method

Preparation of Formvar-Coated Grids

1. Dip dust-free slides in Formvar solution and air dry.
2. Trim edges with a razor.
3. Slowly and gently dip the slide into a deep glass container filled with ddH_2O. The Formvar film should float to the surface.

4. Place clean grids on the floating Formvar film and pick up film with grids using a clean piece of Parafilm or paper that slips under the film.
5. Air dry and then store in a dust-free container. Formvar-coated grids are good for 3–6 months.

Tissue Fixation

1. Fix tissue (roots of 3-day-old barley seedlings, Figure 16.2) in 2% paraformaldehyde and 3% glutaraldehyde in 0.05 M Na-phosphate buffer (pH 7.2) for 2 hours at RT; then at 4° C overnight.[a]
2. Rinse in 0.05 M Na-phosphate buffer three times, 10 minutes each.
3. Postfix with 2% OsO_4[b] in 0.05 M Na-phosphate buffer for 1 hour.
4. Rinse in ddH_2O three times, 5 minutes each.
5. Dehydrate in an ethanol series: 30%, 50%, 70%, 90%, 100% two times, 10 minutes each.

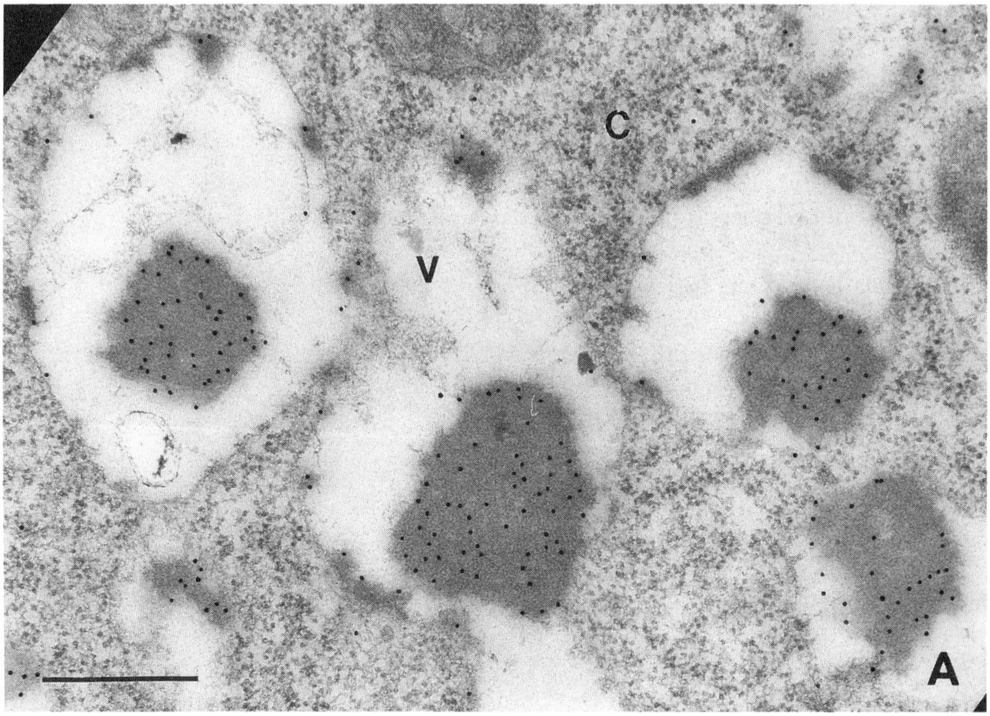

FIGURE 16.2 Electron micrographs of tissue obtained from roots of 3-day-old barley seedlings. Sections were treated with anti-WGA (A) and preimmune serum (B) followed by protein A–colloidal gold. Barley lectin, a WGA homolog, is localized in the vacuoles. Bar = 0.5 μm, C = cytoplasm, V = vacuole.

6. Infiltrate in LR White (Polysciences Inc.).[c]
   2:1 ethanol = LR White 1 hour
   1:2 ethanol = LR White 2 hours
   100% LR White 4 hours at RT and then overnight at 4° C
7. Embed in beam capsules and polymerize at 60° C under vacuum for 24 hours. Embedded samples will last for up to 2 years.
8. Mount specimen on the microtome and section block down to the tissue.
9. Collect 0.08-μm (silver or gold) sections in ribbons onto Formvar-coated grids. Use samples within 2 weeks.

Immunocytochemistry[a]

1. Float grids with sections on freshly made 5% Na-meta-periodate, for 30 minutes.[b]
2. Wash with ddH_2O three times, 2 minutes each.
3. Treat with 0.1 N HCl for 10 minutes.
4. Wash with ddH_2O three times, 2 minutes each.

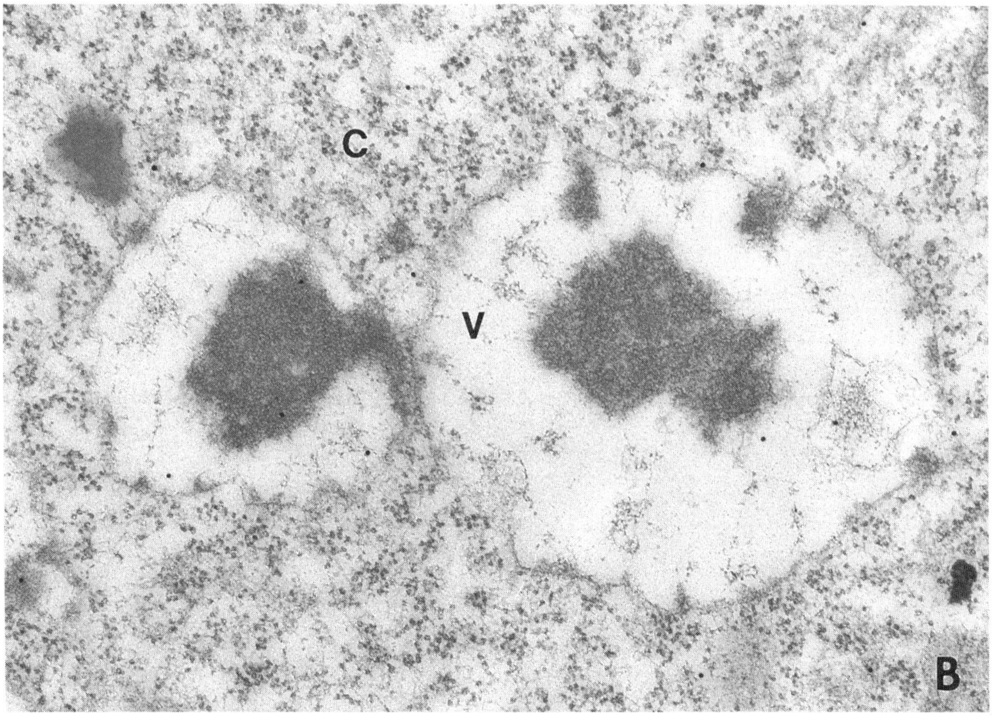

FIGURE 16.2 (Continued).

5. Immerse grids in blocking solution (PBST[d] containing 5% low-fat milk) for a total of 30 minutes, two times, 15 minutes each.
6. Wash grids with PBST wash solution,[e] three times, 2 minutes each.
7. Incubate grids in specific antibody for 1 hour.[f]
8. Wash grids with a continuous stream of PBST wash solution[e] for 30 seconds.
9. Incubate grids in protein A colloidal gold, diluted 1:50 in PBST for 30 minutes.[f]
10. Wash grids in PBST washing solution,[e] three times, 2 minutes each.
11. Wash with ddH_2O three times, 2 minutes each.
12. Stain with 5% aqueous uranyl acetate for 45 minutes.
13. Thoroughly rinse with ddH_2O.
14. Stain grids with Reynolds lead citrate (Dawes 1971) for 3 minutes and thoroughly rinse first with 0.02 N NaOH and then with ddH_2O.

Notes

a. All steps are carried out at RT unless otherwise indicated.

b. OsO_4 is used as a postfixative to aid visualization of membranes; however it is thought that it might interfere with antigenicity. Treatment with Na-meta-periodate before immunolabeling (steps 1–4) releases the bound osmium, thus recovering the antigenicity.

c. Use a rotator in step 6 of the "Tissue Fixation" section.

d. PBST contains: 0.01 M Na-phosphate buffer, 500 mM NaCl, and 0.1% Tween 20.

e. PBST wash solution contains 0.01 M Na-phosphate buffer, 500 mM NaCl, and 0.5% Tween.

f. In this protocol, anti-WGA antibodies raised in rabbits (Raikhel and Quatrano 1986) were used to localize barley lectin, antibodies diluted 1:50 in PBST (0.01 M Na-phosphate, 500 mM NaCl, and 0.1% Tween 20). The same concentration of preimmune serum was used in the control experiments (Figure 16.2).

CONTROLS

Interpretable immunocytochemical data can only be generated with antibodies that have been thoroughly characterized at the biochemical level (Mishkind et al. 1987). A further requirement is that a series of standard controls be performed to verify the specificity of immunochemical reaction products. One type of control is performed by omitting particular steps in the labeling protocol. Among the artifacts detected by this process are the binding of secondary antibodies to particular sites and the presence of endogenous enzymes and particles that may be confused with the exogenously added labels. A second type of control replaces the primary antibody with either a preimmune serum or a serum obtained from a different animal or an adsorbed antiserum that had been depleted of a particular antibody by affinity chromatography.

Misleading localization patterns may be generated by antibodies directed to sugar residues present in the carbohydrate determinants of glycoproteins and other antigens. If an antiserum reacts specifically with tissues from a wide diversity of species, it should be questioned whether the antiserum contains immunoglobulins directed to ubiquitous determinants such as common carbohydrates. To control for these artifacts, sugars that are components of the carbohydrate portion of an antigen can be included in buffers. Alternately, antibodies directed to carbohydrates can be removed from an antiserum on a matrix of insolubilized sugars or glycoproteins.

REFERENCES

Augeri MI, Angelini R, Federico R (1990) Sub-cellular localization and tissue distribution of polyamine oxidase in maize (*Zea mays* L.) seedlings. J Plant Physiol 136: 690–695

Felker FC, Muhitch MJ (1990) Immunohistochemical localization of glutamine synthetase in developing maize kernels. Can J Bot 68: 1916–1920

Geetha KB, Lending CR, Lopes MA, Wallace JC, Larkins BA (1991) Opaque-2 modifiers increase γ zein synthesis and alter its spatial distribution in maize endosperm. Plant Cell 3: 1207–1219

Hack E, Lin C, Yang H, Horner HT (1991) T-URF13 protein from mitochondria of Texas male-sterile maize (*Zea mays* L.). Plant Physiol 95: 861–870

Hurley D, Taiz L (1989) Immunocytochemical localization of the vacuolar H^+-ATPase in maize root tip cells. Plant Physiol 89: 391–395

Jeffree CE, Yeoman MM, Kilpatrick DC (1982) Immunofluorescence studies on plant cells. Intl Rev Cytol 80: 231–265

Langdale JA, Zelitch I, Miller E, Nelson T (1988) Cell position and light influence C4 versus C3 patterns of photosynthetic gene expression in maize. EMBO J 7: 3643–3651

Lending CR, Larkins BA (1989) Changes in the zein composition of protein bodies during maize endosperm development. Plant Cell 1: 1011–1023

Mishkind ML, Plumley FG, Raikhel NV (1987) Immunochemical analysis of plant tissue. In Vaughn KC (ed) CRC Handbook of Plant Cytological Methods, CRC Press, Boca Raton, FL, pp 65–119

Raikhel NV, Quatrano RS (1986) Localization of wheat-germ agglutinin in developing wheat embryos and those cultured in abscisic acid. Planta 168: 433–440

Reynolds ES (1963) The use of lead citrate at high pH as an electron opaque stain in electron microscopy. J Cell Biol 17: 208–212

Rowland LJ, Chen Y-C, Chourey PS (1989) Anaerobic treatment alters the cell specific expression of *Adh-1*, *Sh*, and *Sus* genes in roots of maize seedlings. Mol Gen Genet 218: 33–40

Slocum RD, Furey MJ, III (1991) Electron-microscopic cytochemical localization of diamine and polyamine oxidases in pea and maize tissues. Planta 183: 443–450

Varagona MJ, Schmidt RJ, Raikhel NV (1991) Monocot regulatory protein Opaque-2 is localized in the nucleus of maize endosperm and transformed tobacco plants. Plant Cell 3: 105–113

17

Immunolocalization of Nuclear Proteins

LAURIE G. SMITH

Immunohistochemistry permits the localization at the tissue and subcellular level of antigens for which specific antisera or monoclonal antibodies are available. A wide variety of methods are used for this purpose, and the best method for any particular application must unfortunately be determined empirically. Described here are two methods that have been used successfully to localize the protein encoded by *Knotted-1* to nuclei (Smith et al. 1992); another method for the localization of nuclear proteins is described by Varagona and Rahikel in this volume.

In the methods described here, tissues are fixed and then infiltrated and embedded in either paraffin or Steedman's Wax. Paraffin-embedded tissue is easier to section and generally yields better sections of mature, lignified tissues, but antibody reactivity is often lost in paraffin sections. In some cases antibody reactivity can be restored by treating paraffin sections with protease. KNOTTED-1 protein was successfully localized in protease-treated paraffin sections and also in Steedman's Wax sections without protease treatment. Antibody binding to tissue sections is visualized with silver-enhanced immunogold (IGSS) labeling. IGSS provides an attractive alternative to immunofluorescence for plant immunolocalization studies because plant tissues tend to autofluoresce. Unlike fluorescently labeled sections, IGSS labeled sections can be mounted permanently and analyzed without a fluorescence microscope. By counterstaining the tissue sections with a histological stain, the pattern of antibody labeling can be observed superimposed over the counterstained tissue section in a single image. This is an advantage over immunofluorescence, where two images (fluorescence vs. brightfield) must always be compared to place the labeling signal in the context of the tissue section.

SLIDE PREPARATION

Materials

Frosted-end slides
Gelatin
Chromium potassium sulfate

The Maize Handbook—M. Freeling, V. Walbot, eds.

Method

The most effective adhesive that does not cause backgound problems with IGSS labeling is gelatin-chrome alum. Clean slides before treatment by soaking in 70% ethanol containing a few drops of soapless detergent, rinsing in 70% ethanol, and drying. For slide coating, dissolve gelatin to a final concentration of 0.5% by heating the solution to 30–35 degrees with constant stirring. Then add chromium potassium sulfate to a final concentration of 0.05% and continue stirring until dissolved. Dip clean slides into this solution and allow them to dry (do not rinse).

TISSUE FIXATION

Materials

Paraformaldehyde
Dimethylsulfoxide
1 M $NaPO_4$ pH 7.2
Freshly harvested tissue
Small capped vials for fixation

Method

On the day the tissue will be fixed, prepare a 4% paraformaldehyde solution in water. This requires heating the mixture to 60° C, adding a few drops of 1 N NaOH, and stirring for several minutes. After cooling to 4° C, add DMSO to 1% and $NaPO_4$ to 100 mM. Pass through a 45-μm filter. Cut small pieces of tissue (less than 8 mm^3) and add them to fixative on ice. If tissues do not sink immediately, apply vacuum until they do (10–30 minutes). Rotate end over end at 4° C for 8–16 hours (depending on the size of tissue pieces; use a longer time for larger tissue pieces). After fixation, rinse tissues twice for 1 hour in 100 mM $NaPO_4$ with end-over-end rotation at 4° C.

TISSUE INFILTRATION, EMBEDDING, AND SECTIONING

Paraffin Wax

Materials

Jars with lids or caps capable of holding 200–500 ml
Ethanol
Tertiary butyl alcohol (TBA)
Paraffin (Paraplast Plus, Monoject Scientific)

Tissue cassettes (Fisher Scientific)
Molds for casting wax blocks (Polysciences Inc.)
Microtome mounting blocks
Gelatin-chrome alum-coated slides

Method

Place fixed, rinsed tissues into tissue cassettes for easy transfer through alcohol and wax mixtures. Transfer tissues through the following series (in the capped jars) at room temperature: 30% EtOH in water, 50% EtOH in water, 70% EtOH in water (30–60 minutes each). Proceed with paraffin infiltration and embedding as described by Langdale in this volume, beginning with the 80% EtOH step.

Using a razor blade, trim paraffin-embedded tissue down to a small rectangle or square. Mount on to a microtome mounting block by melting a small amount of wax onto the block and then pressing the trimmed paraffin square onto the melted wax and allowing it to harden. Section paraffin-embedded tissue on a rotary microtome with a sharp stainless steel blade or disposable blade at 8 μm and mount onto gelatin-chrome alum-coated slides by floating ribbons on water-flooded slides warmed to 37° C on a slide warmer. After sections have completely expanded, drain or pipette off excess water and allow to dry on the slide warmer for several hours or overnight at 37° C. Dried sections can be stored if necessary in a dust-free slide box at room temperature.

Steedman's Wax

Materials

Jars with lids or caps capable of holding 200–500 ml
Ethanol
PEG 400 distearate (Aldrich Chemicals)
1-hexadecanol (Aldrich Chemicals)
Tissue cassettes (Fisher Scientific)
Molds for casting wax blocks (Polysciences, Inc.)
Cryostat mounting blocks
Gelatin-chrome alum-coated slides

Method

Infiltration and embedding in Steedman's Wax (SW) is described by Brown et al. (1989). SW is made by mixing, at 60° C, 9 parts melted PEG 400 distearate (molten at 40° C) with one part melted 1-hexadecanol (molten at 60° C). The mixture is then resolidified for storage. Place fixed, rinsed tissues into tissue cassettes and leave for 30–60 minutes in jars containing the following mixtures at room temperature: 30% EtOH in water, 50% EtOH in water, 70% EtOH in water, 80% EtOH in water, 90% EtOH in water, 100% EtOH, fresh 100% EtOH. Incubate tissues for a minimum of 4 hours in each of the following mixtures: 25% SW in EtOH (40° C), 50% SW in EtOH (40° C), 75% SW in EtOH (40° C), 100% SW (60° C, uncapped to allow ethanol evaporation), fresh 100% SW (60° C, uncapped), fresh 100% SW

(60° C, uncapped). If a 60° C vacuum oven is available, do the final SW incubation under vacuum to remove the last traces of ethanol. Remove tissues from cassettes and transfer to mold chambers filled with fresh, molten SW; orient. Allow to harden at room temperature. As a general rule, do not heat wax and wax-alcohol mixtures any longer than necessary; they can be re-used several times.

Steedman's Wax is harder at 4° C than at room temperature, and better sections are obtained using a cryostat set at 4° C than with a rotary microtome at room temperature. Trim SW-embedded tissue and mount on cryostat mounting block as described above for paraffin sectioning. Allow to cool at 4° C for several minutes before beginning sectioning. Float ribbons of 8 μm sections on gelatin-chrome alum-coated slides flooded with water for a few minutes at room temperature. Drain or pipette off excess water and allow to dry at room temperature overnight. If necessary, store in a dust-free slide box at room temperature.

ANTIBODY LABELING

Materials

Paraffin or Steedman's Wax sections
Histoclear (National Diagnostics)
Ethanol
Proteinase K (Sigma)
Phosphate-buffered saline (PBS; 150 mM NaCl, 20 mM $NaPO_4$, pH 7.2)
Bovine serum albumin (Sigma)
Glass slide staining jars (Thomas Scientific)
PBS/BSA (PBS containing 1 mg/ml BSA)
Antibodies, including appropriate negative control antibodies
Gold-conjugated secondary antibody (5 nm gold conjugates from Amersham are good)
Silver enhancement reagents for light microscopy (Caltag, South San Francisco or BioCell, available from Ted Pella, Redding, CA)

Method

Sections must be dewaxed prior to antibody incubation. Paraffin sections are dewaxed by incubating for 30 minutes in Histoclear, then removing Histoclear in 100% ethanol (5 minutes, twice). Rehydrate by incubating sections for 5 minutes in each: 95% ethanol in water, 80% ethanol in water, 60% ethanol in water, 30% ethanol in water, water, PBS. Sections are now ready for antibody incubation but may require pretreatment with protease to "reveal" antibody binding sites. To determine whether this is necessary and what concentration of protease will give the best results, incubate sections for 10 minutes at room temperature with each of the following dilutions of proteinase K in PBS (proteinase K can be stored at −20° C as a 1 mg/ml stock in water): 500 μg/ml, 100 μg/ml, 20 μg/ml, 4 μg/ml, PBS only. Briefly rinse twice in fresh PBS to remove protease, then incubate sections for at least 30 minutes in PBS containing 1 mg/ml BSA (PBS/BSA) to saturate sections with protein. Higher protease concentrations typically enhance antibody binding, but also degrade the tissue more; try to find the best compromise. When labeling the sections with antibodies, be very careful to compare positive (immune serum) and negative (preimmune or nonimmune serum) samples for *each* protease concentration, because higher protease concentrations may also increase nonspecific antibody binding.

Steedman's Wax sections are dewaxed for 30 minutes in 100% ethanol and then rehydrated by incubating for 5 minutes in each: 95% ethanol in water, 80% ethanol in water, 60% ethanol in water, 30% ethanol in water, water, PBS. Incubate slides for at least 30 minutes in PBS/BSA to saturate sections with protein.

The best antibody concentration to use for immunolocalization must be determined empirically by trying different concentrations, and is typically at least tenfold higher than the optimal concentration for Western blotting. For crude antisera, dilutions ranging from 1:100 to 1:2,000 could be tried. The results for each dilution *must* be directly compared with the same dilution of preimmune or nonimmune serum to control for nonspecific immunoglobulin binding. Monoclonal antibody culture supernatants can be tested undiluted-1:50, and compared with the same dilutions of a negative control monoclonal antibody. Incubate sections with antibodies diluted into PBS/BSA for at least 2 hours at room temperature. 100 μl of diluted antibody should be sufficient to cover the sections if the slides are placed onto flattened, wet paper towels on the bench top. During the incubation, minimize evaporation by inverting a shallow container lined with wet paper towels over the sections. After the incubation, drain off antibody and wash by incubating twice for 15 minutes in fresh PBS/BSA. Incubate sections as before for at least 1 hour with gold-conjugated secondary antibody diluted into PBS/BSA. As for primary antibody, the optimal dilution of gold-conjugate is determined empirically, but the manufacturer usually recommends a dilution for each lot of conjugate. Following the secondary antibody incubation, drain it off and wash by incubating twice for 15 minutes in fresh PBS/BSA.

Throughout the silver enhancement steps that follow, do not use metal forceps, metal racks, or metal rack handles to transfer slides, or allow metal to come into contact with the slides or the water rinses (metal ions nucleate the precipitation of silver grains from the IGSS reagents). Wash the sections three times for 5 minutes in the highest grade distilled water available. Just prior to use, mix the silver enhancement reagents according the manufacturer's instructions and apply a few drops of the mixture with a pasteur pipette to each slide to cover the sections. The silver enhancement reaction can be monitored under a light microscope, but avoid prolonged, high-intensity illumination of the slides during silver enhancement. The time required for optimal silver enhancement is usually between 10 and 20 minutes. Prolonged silver enhancement times tend to produce high background, so stop the reaction as soon as a positive signal is clearly visible. Stop the reaction by draining off the enhancement reagents and rinsing briefly in several changes of distilled water.

COUNTERSTAINING AND PERMANENT MOUNTING OF SECTIONS

To highlight nuclei and other cellular structures in a color that contrasts well with black silver grains, the sections can be counterstained with basic fuschin. A variety of other tissue stains can also be employed for counterstaining; many are described in *Staining Procedures, Fourth Edition* (1981). The choice of a counterstain depends on what features of the tissue section are to be highlighted, but if immunogold/IGSS is used, dark colors should be avoided, because they will tend to obscure the pattern of silver grains.

Materials

Basic fuschin (Eastman Kodak)
Ethanol
Histoclear (National Diagnostics) or xylene
Merckoglas (E. Merck, available from EM Science) or other permanent histological mounting medium

Method

Prepare a 1% stock of basic fuschin in 50% ethanol. Dilute 1:20 in water for tissue staining. Submerge slides in diluted staining solution for 10 minutes. Rinse out excess stain briefly in distilled water; then dehydrate tissues by transferring through the following series (at least 2 minutes in each): 30% ethanol in water, 60% ethanol in water, 80% ethanol in water, 95% ethanol in water, 100% ethanol, 100% ethanol. Basic fuschin is very ethanol soluble and will gradually diffuse out during this process, particularly during the 95% and 100% ethanol steps. The tissue must therefore be overstained initially and can be left in 100% ethanol until the desired staining intensity is obtained. If tissues are destained too much, they can be rehydrated by passing them back through the ethanol/water series, repeating the staining, and dehydrating again. Staining times can be reduced or lengthened as necessary to achieve the desired staining intensity after dehydration. Meristematic cells, for example, have a much higher affinity for basic fuschin than other cells and may only require 30–60 seconds of staining. After the second 100% ethanol step, incubate slides twice for 15 minutes in Histoclear or xylene. In a fumehood, drain Histoclear or xylene from slides one at a time, apply a few drops of mounting compound, and gently lay down a coverslip. Allow mounting compound to harden several hours.

EVALUATING THE RESULTS

Slides can be examined by brightfield microscopy. In examining the slides, keep in mind that two types of unwanted signals can be present. The antibody may bind specifically but to the wrong antigen, or to the wrong antigen in addition to the right antigen. Biochemical characterization of the antibody (by Western blotting and/or immunoprecipitation studies on plant tissue extracts) is the best way to evaluate the potential for reactivity to more than one antigen. Antibodies can also bind nonspecifically, and this nonspecific binding will tend to increase with higher antibody concentrations. Nonspecific antibody binding can sometimes be deceptively "tissue-specific," labeling certain tissues or cells and not others. Therefore, it is desirable to use the lowest antibody concentration that still produces a specific labeling signal, and it is *essential* to compare each positive sample with a negative sample labeled with the same concentration of preimmune or nonimmune serum (for a polyclonal antibody) or the same concentration of a negative control monoclonal antibody. No labeling is meaningful except by direct comparison to the appropriate negative control.

To produce color slides, the results can be photographed with Kodak Ektachrome 160 tungsten film. These slides can be used to generate color prints of the results. In my experience, better (and less expensive) color prints are obtained if the results are photographed with color print film. Because color print film balanced for a tungsten light source is not available, daylight color print film must be used with a blue color correction filter. Kodak Ektar 25 color print film gives excellent high-resolution prints. For both types of film, set the light source on 12 V to achieve good color balance; use a neutral density filter to control light intensity.

REFERENCES

Brown RC, Lemmon BE, Mullinax JB (1989) Immunofluorescent staining of microtubules in plant tissues: improved embedding and sectioning techniques using polyethylene glycol (PEG) and Steedman's Wax. Bot Acta 102: 54–61

Smith LG, Greene B, Veit B, Hake S (1992) A dominant mutation in the maize homeobox gene, *Knotted-1*, causes its ectopic expression in leaf cells with altered fates. Development 116: 21–30.

Staining Procedures, Fourth Edition (1981) Clark G (ed) Williams & Wilkins, Baltimore, 512 pp

18

In situ Hybridization

Jane A. Langdale

In situ hybridization techniques enable gene expression to be monitored in individual cells (Figure 18.1). Methods for use with the light microscope are as sensitive as northern blot analysis for detecting mRNAs that accumulate equally in all cells, and more sensitive than northern analysis for detecting mRNAs that are present in only a small population of cells within an organ (Langdale et al. 1988a,b; Martineau and Taylor 1986). The following protocol has been used most frequently to monitor gene expression in maize leaf tissue. By varying tissue fixation conditions, all organs of the maize plant have been analyzed. The method uses paraffin-embedded sections and ^{35}S-radiolabeled or digoxigenin nonradiolabeled riboprobes. The use of nonradiolabeled probes is a fairly recent advance and at this stage they appear to be less sensitive than radiolabeled probes. However, the method can undoubtedly be improved with further use and optimization. Tissue that is fixed and embedded as described below may also be used for immunocytochemistry. Standard methods of microscopy and photomicroscopy are used for signal detection. See appropriate chapters in this volume for details.

PREPARATION OF TISSUE SECTIONS

Preparation of Poly-D-Lysine-Coated Slides

Material

Microscope slides with a frosted edge (buy them "washed and cleaned")
Poly-D-lysine hydrobromide (Sigma P1149) made up as a 10× stock solution (500 μg ml^{-1} in 100 mM Tris-HCl pH 8.0) and stored frozen at −20° C

Method

1. Dip slides in 50 μg ml^{-1} poly-D-lysine in 10 mM Tris-HCl pH 8.0 for 30 minutes at room temperature.

The Maize Handbook—M. Freeling, V. Walbot, eds.

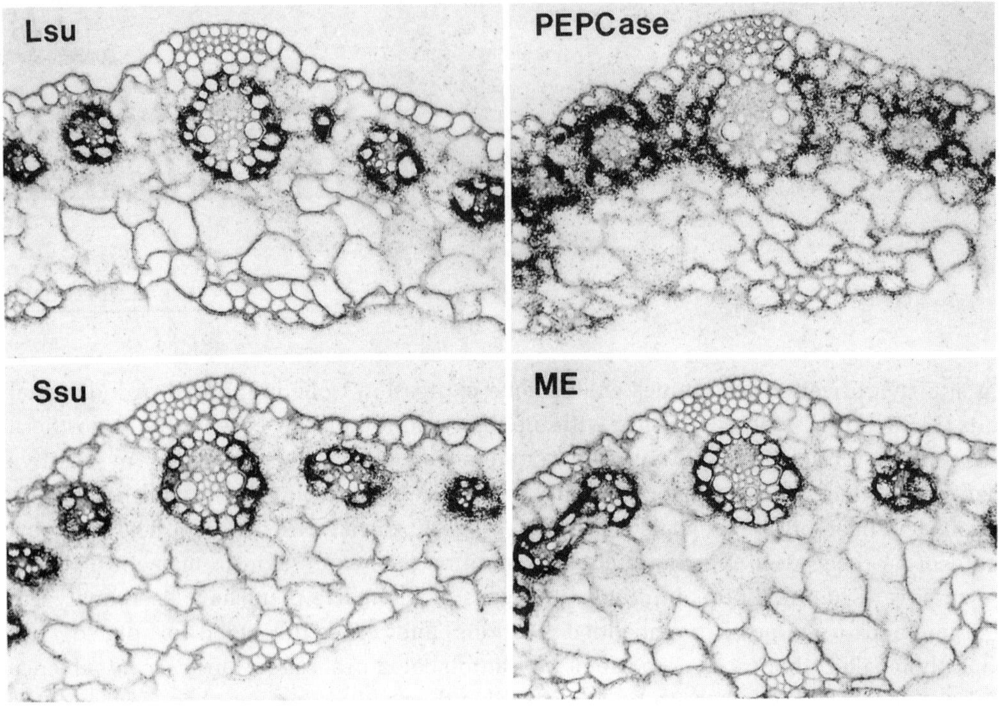

Figure 18.1 In situ localization of photosynthetic mRNAs in the midrib region of light-grown leaf blade sections. Malic enzyme (ME) and ribulose biphosphate carboxylase large subunit (Lsu) and small subunit (Ssu) mRNAs are detected in the bundle sheath cells that surround the vascular tissue. In contrast, phosphoenolpyruvate carboxylase (PEPCase) mRNA is detected in the mesophyll cells. Autoradiographs were exposed to approximately equal intensities and do not reflect relative levels of mRNAs in the leaf. Magnification: ×140. (Langdale et al., 1988a EMBO Journal 7, by permission of Oxford University Press.)

2. Air dry in a dust-free place.

Notes

1. Coated slides must be used to ensure sections remain fixed to slides throughout the many treatments they will receive.
2. Treated slides can be kept at room temperature for several months.

Fixation of Leaf Sections

Material

Freshly harvested tissue
Glass vials (scintillation vials work well)

Freshly made ethanol: acetic acid (3:1)
FAA—50% EtOH, 10% formalin, 5% acetic acid; keep at 4° C
50% and 70% EtOH

Method

1. Razor cut 2–3mm leaf sections.
2. Fix in 3:1 ethanol/acetic acid (a few milliliters in a glass vial) for 30 minutes at room temperature.
3. Remove fixative and add 2 ml 70% EtOH. Leave 30 minutes at room temperature.
4. Replace 70% EtOH with fresh 70% EtOH and store indefinitely at 4° C.

Notes

1. If fixing tissues other than leaves, use FAA as a fixative. Overnight fixation at room temperature is usually adequate. However, larger or tougher samples (e.g., whole kernels or mature husk leaves) may need longer. Samples can be left indefinitely in FAA at 4° C or be transferred to 50% and then 70% EtOH prior to long-term storage.
2. With particularly tough tissue, infiltration can be improved by placing the sample (in fixative) under vacuum for a few minutes. Vacuum infiltration is essential for tissues, such as vegetative apices, that do not readily sink into the fixative (D. Jackson, personal communication).
3. An alternative fixative is 4% paraformaldehyde. See Jackson (1992) for suggested protocols.

Paraffin Embedding Tissue

Material

Fixed tissue in 70% EtOH
Dri-Block or oven at 60° C
30 ml disposable centrifuge tubes (Sarstedt)
Plastic scintillation vial inserts
Nylon net (greater than 100 μm mesh)
Peel-a-way disposable plastic molds (Polysciences Inc.)
Paraplast Plus
70%, 80%, 90%, and 100% EtOH
25%, 50%, and 100% tertiary butyl alcohol (TBA) (made up in 100% EtOH)

Method

1. Half-fill 30 ml disposable centrifuge tubes with Paraplast Plus pellets and place in a Dri-Block (or oven) at 60° C. It will take 1–2 hours for the Paraplast to melt, during which time samples should be dehydrated as follows.

2. Transfer fixed sections through the following EtOH/tertiary-butyl alcohol series. (See Note 1.)

30 minutes	80% EtOH
30 minutes	90% EtOH
30 minutes	100% EtOH
30 minutes	100% EtOH
30 minutes	25% TBA
30 minutes	50% TBA
30 minutes	100% TBA
30 minutes	100% TBA

3. Transfer sections to a basket made from a plastic scintillation minivial. (Cut bottom of vial off and remove all but the sides of the lid. Use the remainder of the lid to secure a piece of nylon mesh over the top of the vial).
4. Layer about 2 ml 100% TBA onto the molten Paraplast Plus.
5. Lower basket containing the sections into the TBA/Paraplast mix. (See Note 2.) Incubate at 60° C overnight.
6. Transfer basket into 5-ml fresh Paraplast. Incubate 60° C overnight.
7. Repeat transfer, incubate 2 hours at 60° C.
8. Prewarm disposable plastic mold on 60° C Dri-Block. Transfer fresh Paraplast and section to mold using a cut-off prewarmed pasteur pipet. (Heat pipet in burner and cool with spare molten Paraplast immediately before use—see Note 3.) Make sure section is entirely covered in Paraplast.
9. Align section in mold using warmed forceps. Once surface begins to solidify, float block on ice cooled water (about 18° C). After about 1 minute, push the block below the surface of the water. This ensures that the paraffin solidifies evenly but it takes some practice—the block will crack if chilled too soon.

Notes

1. Dehydration times should be increased up to 2–3 hours per step for large or tough samples.
2. Basket will initially float at the TBA/Paraplast interface but will gradually sink overnight.
3. Sectioning is impaired if Paraplast is heated over 62° C at any stage.

Sectioning Paraffin-Embedded Tissue

Material

Paraplast-embedded tissue
Polylysine coated slides
Rotary microtome

Microtome mounting blocks
Slide warming tray

Method

1. Remove Paraplast-embedded sample from Peel-a-way mold. Trim away excess Paraplast and cut around section to form a trapezoid with the smallest face presenting the surface for sectioning.
2. Heat a spatula and melt spare Paraplast onto microtome mounting block. Press trimmed sample onto block. Once cool, use a heated spatula and spare Paraplast to secure sample to block.
3. Use a razor blade to trim the sample to the smallest possible size.
4. Lay polylysine coated slides on a slide warmer at 42° C and cover each with a few milliliters of degassed distilled H_2O.
5. Mount block in microtome and cut 8 μm sections. (See Note 2.)
6. Lift 4–5 section ribbons with forceps and float shiny side down on water-covered slides.
7. Once paraffin has expanded, use a paper tissue to remove surplus H_2O and then dry slides at 42° C overnight.
8. Store away from dust, heat, and moisture. A slide box in a desiccator at 4° C is best. (See Note 3.)

Notes

1. If air conditions are humid, sectioning is improved by chilling both the block and knife. Ribbons will coil from static if the air is too dry.
2. Central sections from a smooth ribbon are usually the best. The easiest way to compare samples is to place a number of different ones on a single slide.
3. Samples keep for at least a month. After this time, sections have an increased tendency to fall off the slides during subsequent manipulations (and usually not until the very last step).

Staining Sectioned Tissue

Material

Sections mounted on slides
Coplin jars
Coverslips
Xylenes or cnp30
70% and 100% EtOH
1% Safranin O in 50% EtOH

0.5% fast green FCF in 95% EtOH
Permount

Method

1. Transfer slides through the following solutions (50 ml in Coplin jars):

5 minutes	cnp30
5 minutes	cnp30
2 minutes	100% EtOH
2 minutes	100% EtOH
2 minutes	70% EtOH
30 minutes	Safranin O
2 minutes	70% EtOH
30 seconds	Fast Green
2 minutes	100% EtOH
2 minutes	100% EtOH
5 minutes	cnp 30

2. Drain slide and place a drop of Permount on top of sections. Seal with a coverslip.

Notes: A few sections of each sample should be stained to check that the tissue is intact. This also provides a useful record of morphological detail. (Often at the end of the hybridization procedure tissue morphology is not as clear.)

PREPARATION OF RIBOPROBES

cDNAs should be cloned into an appropriate vector containing both T7 and T3 (or SP6) RNA polymerase promoters. Template DNA must be linearized prior to RNA synthesis so that run-off transcripts of discrete content are synthesized. Both sense (control) and antisense transcripts should be synthesized. After digestion with the appropriate restriction enzyme, extract twice with phenol/chloroform/isoamylalcohol (25:24:1) and ethanol precipitate and resuspend at a concentration of $1 mg ml^{-1}$ in DEPC-treated water (see Molecular Biology section for details of these methods); 10 mM DTT should be present in all solutions when making ^{35}S labeled probes. EVERYTHING MUST BE KEPT RNase FREE!

^{35}S labeling

Materials

Linearized template DNA (see above)
Microfuge
Dri-block or incubators at 30° C, 37° C, and 60° C
Scintillation counter
Scintillation vials and scintillant

DE81 paper
Triton X-100
RNAguard (Pharmacia) or equivalent RNase inhibitors
T7 polymerase
T3 or SP6 polymerase
DNase I (RNase free)
^{35}S-UTP
95% and 100% EtOH
phenol/chloroform/isoamyl alcohol (25:24:1)
0.5 M Na_2HPO_4
DEPC-treated distilled H_2O (see Molecular Biology section)
The following solutions need to be made up in DEPC dH_2O:
1 mg ml^{-1} BSA
1 M DTT
100 mM DTT
10 mM ATP
10 mM CTP
10 mM GTP
50 mg ml^{-1} tRNA
7.5 M NH_4OAc
3 M NaOAc pH 4.5
5% acetic acid (v/v)
5x polymerase buffer (200 mM Tris-HCl pH 8.0, 30 mM $MgCl_2$, 20 mM spermidine)
10× DNase I buffer (400 mM Tris-HCl pH 7.5, 60 mM $MgCl_2$)
0.1 M $NaHCO_3$ pH 10.2, 10 mM DTT
50% deionized formamide, 10 mM DTT

Method

1. Make up the following mix for five reactions. Add components in order given:

	μl
water	5
1 mg ml^{-1} BSA	2.5
5× polymerase buffer	5
Triton X-100	0.25
1 M DTT	0.5
10 mM ATP	1.25
10 mM CTP	1.25
10 mM GTP	1.25
T7 or T3/SP6 polymerase	1
RNAguard	2

Mix carefully—RNAguard is very unstable.

2. Dry down 0.1 nm ^{35}S-UTP for each reaction. This is easily done in Eppendorf tubes in a Speed Vac concentrator.
3. To each tube of label, add 4 μl of reaction mix.
4. Add 1 μl DNA template (500–1,000 ng μl^{-1}) to start the reaction.
5. Incubate at 30° C for 3 h.
6. Dilute into 50 μl by adding:

water	32 μl
RNAguard	2 μl
10× DNase buffer	5 μl
100 mM DTT	5 μl

7. Remove 2 μl of mix to quantify incorporation. (See Note 1).
8. Add 1 unit of DNase I and incubate for 15 minutes at 37° C.
9. Extract once with phenol/chloroform/isoamyalcohol (25:24:1).
10. Add 25 μl 7.5 M NH_4OAc, 1 μl 50 mg ml^{-1} tRNA, and 188 μl 100% EtOH.
11. Precipitate at −70° C for 1 hour or in dry ice/EtOH for 10 minutes.
12. Pellet RNA for 10 minutes in microfuge.
13. Resuspend in 0.1 M $NaHCO_3$ pH 10.2, 10 mM DTT (50 μl).
14. Incubate at 60° C to hydrolyze RNA. Appropriate piece size is about 150 bases, mass average. (See Note 2 to calculate hydrolysis time.)
15. Add 0.1-volume 5% acetic acid.
16. Add 0.1-volume 3 M sodium acetate pH 4.5 and 2.5 volumes 100% EtOH. Precipitate as above (step 11) or overnight at −20° C.
17. Pellet for 10 minutes in microfuge, dry off in air, and then resuspend in 50% deionized formamide, 10 mM DTT, at 5× concentration of probe (1× probe = 0.2–0.3 μg ml^{-1} kb^{-1}).
18. Store at −70° C. Probes are stable in this form for at least 2 months.

Notes

1. To quantitate incorporation, spot 2 × 1 μl of reaction mix onto 2 x 1 cm^2 DE81 paper.

   Take one filter and rinse: 5 x 5 minutes in 0.5 M Na_2HPO_4
   2 x 1 minute in dH_2O
   2 x 1 minute in 95% EtOH

   Count both in scintillant. Unwashed sample represents total counts, while the washed sample represents incorporated counts.

$$\text{RNA synthesized (ng)} = 0.1\ \text{nM} \times \%\ \text{incorporation} \times \underset{(300\ \text{nM} = 1\ \text{ng})}{300} \times \underset{(\#\ \text{of NTPs})}{4}$$

2. Hydrolysis time obeys the following relationship:

$t = \dfrac{Lo - Lf}{kLoLf}$ where t= time in minutes

Lo = initial length in kb
Lf = desired length in kb
k is approximately 0.11 cuts/kb/minutes

[from Cox et al. (1984) Dev Biol 101: 485–502]

3. It is advisable to check hydrolyzed RNA prior to hybridization. Run 1 μl of each reaction on a 6% polyacrylamide/urea gel alongside radiolabeled markers. (See Molecular Biology section for details.) Dry down gel and autoradiograph overnight. Mean probe length should be about 150–350 bases.

Digoxigenin Labeling

Material

Linearized DNA template
Dri-Block or incubators at 37° C and 60° C
Digoxigenin RNA labeling kit (Boehringer catalog # 1175 025)
T3 polymerase (only if vector has T3 promoter—kit comes with SP6 polymerase)
RNAguard
DEPC-treated dH_2O
10× DNase buffer (400 mM Tris-HCl pH 7.5, 60 mM $MgCl_2$)
0.1 M $NaHCO_3$ pH 10.2

Method

1. Set up transcription reaction as directed by manufacturer.
2. Incubate at 37° C for 2 hours.
3. Dilute into 50 μl by adding:

RNAguard	2 μl
10× DNase buffer	5μl
water	23 μl

4. Follow steps 8–16 as for ^{35}S labeling but omit DTT from buffer in step 13 and omit phenol extraction step 9.

5. Resuspend in 200 µl 50% formamide (this will be used as 5x probe as with ^{35}S) and store indefinitely at −70° C.

6. Synthesis can be checked by running 1 µl on a 2% agarose gel. (Everything should run around 150 bp).

PRETREATMENT OF SECTIONS

Pretreat slides immediately before hybridization. Pretreatment is identical for ^{35}S and DIG probes. Solutions do not need to be made RNase-free.

Materials

Slides of sections
Metal slide racks that hold approximately 20 slides
Plastic sandwich boxes that racks fit into (should hold 250 ml buffer)
Glass dish that holds two slide racks (one on top of the other)
Xylenes or cnp30
Acetic anhydride
100%, 95%, 80%, 60%, 30% EtOH
0.2 M HCl
2× SSC (0.3 M NaCl, 0.03 M sodium citrate, pH 7.6)
1 µg ml^{-1} proteinase K, 100 mM Tris-HCl pH 8.0, 50 mM EDTA
PBS (10 mM Na-phosphate pH 7.5, 150 mM NaCl)
2 mg ml^{-1} glycine in PBS
4% formaldehyde in PBS (needs to be heated at 60° C to dissolve)
0.1 M triethanolamine pH 8.0 (6.5 ml triethanolamine, 2 ml HCl, 491.5 ml H_2O)

Method

1. Deparaffinize sections by incubating in xylenes or cnp30 (the latter is less carcinogenic) for 2 × 10 minutes in a fumehood.
2. Rehydrate through an ethanol series. (100% × 2, 95%, 80%, 60%, 30%, distilled H_2O × 2—2 minutes each).
3. Incubate in 0.2 M HCl for 20 minutes.
4. Rinse in distilled H_2O and transfer to 2 × SSC (150 mM NaCl, 15 mM Na citrate)for 30 minutes.
5. Rinse in distilled H_2O, blot excess H_2O with a paper towel, and incubate in proteinase K for 30 minutes at 37° C (See Note 2).
6. Rinse briefly in PBS and then block protease with 2 mg ml^{-1} glycine in PBS (2 minutes).
7. Rinse in PBS—2 × 30 seconds.

8. Fix for 20 minutes in freshly prepared 4% formaldehyde. (See Note 3.)
9. Rinse in PBS for 2 × 5 minutes.
10. Put 500 ml 0.1 M triethanolamine buffer in a glass dish. Put an empty slide rack and stir bar in the bottom. Place on stirrer and stir vigorously. Add 500 μl acetic anhydride and quickly place slide rack containing samples in the dish. (See Note 4.) Stir for 10 minutes.
11. Rinse in PBS for 2 × 1 minute and then dehydrate back through the ethanol series used in step 2.
12. Air-dry slides wrapped in paper towels. Slides are now ready to hybridize.

Notes

1. With the exception of step 5, everything is done at room temperature.
2. Proteinase digestion can be varied via proteinase concentration or time in order to obtain good hybridization signals without loss of morphology. These parameters must be optimized empirically for different tissues and different proteinases.
3. Steps 8–10 should be carried out in a fumehood.
4. Acetic anhydride is very unstable ($T_{1/2}$ is under 2 minutes) in aqueous solution, and therefore slides really do have to be added immediately.

HYBRIDIZATION

Apart from the fact that DTT is required for ^{35}S probes, hybridization conditions are identical for ^{35}S and DIG probes. Wash conditions vary slightly. Set up three antisense and two sense strand hybridizations for each set of sections to be examined.

Materials

Pretreated slides
Dri-block at 80° C
Microfuge
Oven at 50° C
Humid chamber (sealed sandwich boxes containing wet paper towels)
Metal slide racks
Sigmacoted coverslips
Rubber cement (Cow gum in UK)
^{35}S or DIG-labeled probe
10 mg ml^{-1} RNase A (boiled for 10 minutes)
10× salts (3 M NaCl, 10 mM Tris-HCl pH 6.8, 100 mM Na phosphate pH 6.8, 50 mM EDTA)

SPB (everything made with DEPC-H_2O)	100 μl 10× salts
	400 μl deionized formamide
	200 μl 50% dextran sulphate
	20 μl 50 mg ml^{-1} tRNA
	10 μl 1 M DTT
	50 μl 10 mg ml^{-1} poly A
	20 μl H_2O

Wash buffer (1× salts, 50% formamide, 10 mM DTT)—not DEPC treated
NTE (500 mM NaCl, 10 mM Tris-HCl pH 8.0, 1 mM EDTA, 10 mM DTT)—not DEPC treated
PBS (10 mM Na phosphate pH 7.5, 150 mM NaCl)—not DEPC treated
100%, 95%, 80%, 60%, 30% EtOH

Method

1. Heat probe at 80° C for 30 seconds.
2. Spin tube 30 seconds in microfuge and then mix 1:4 with SPB (make 12 μl per 22 mm^2 coverslip).
3. Apply probe to sections, cover with a Sigmacoted coverslip and seal with rubber cement.
4. Incubate at 50° C overnight in a humid chamber.
5. Remove rubber cement with a pair of tweezers and place slides into a slide rack. Place rack in wash buffer and allow coverslips to float off. This is helped by raising the rack enough to allow the coverslips to fall below it, and by stirring the buffer slowly.
6. Transfer rack to fresh wash buffer and incubate at 50° C for 4–5 hours. Agitation is not necessary.
7. Incubate at 37° C for 30 minutes in 20 μg ml^{-1} RNAase A in NTE. (See Note 3.)
8. Wash 5 × in NTE at 37° C for a total of 1 hour.

^{35}S

9. Transfer back into fresh wash buffer and incubate 1 hour to overnight at 50° .
10. Dehydrate slides through an ethanol series.
11. Air dry in a paper towel.

DIG

9. Transfer back into fresh wash buffer and incubate 1 hour at 50° C.
10. Transfer into PBS and place at 4° C overnight

Notes

1. DTT can be omitted from SPB, wash buffer, and NTE when using DIG probes.

2. Use deionized formamide for the hybridization buffer and Analar grade (Fluka) straight out of the bottle for the washes.
3. Under these conditions, RNase A only digests single-stranded RNA. RNA-RNA hybrids are left intact while background is removed.

DETECTION OF HYBRIDS

^{35}S hybrids are detected by autoradiography and DIG by immunoreaction with an alkaline phosphatase conjugated anti-DIG antibody.

^{35}S

Materials

Double door dark room with red safelight
Disposable double slide mailers (Sarstedt)
Waterbath at 45° C (take bulb out of on/off indicator or it will fog the emulsion)
Metal tray
Test tube rack large enough to stand slides upright
Light-tight slide boxes
Kodak NTB2 emulsion
Kodak D-19 developer
Kodak Rapid Fix (360 ml H_2O, 125 ml solution A, 14 ml solution B)
100%, 95%, 80%, 60%, 30% EtOH
0.5% fast green FCF in 95% EtOH
Xylenes or cnp30

Method

1. Melt aliquots of KODAK NTB2 emulsion in a waterbath at 45° C in the darkroom. (See Note 1.) Use red safelight.
2. Mix emulsion 1:1 with H_2O in a slide mailer at 45° C. Allow bubbles to rise to top (about 15 minutes) before dipping slides.
3. Dip slides in emulsion.
4. Drain excess emulsion on a paper towel, wipe the back of slide with a tissue, and then place the slide on metal tray supported on ice.
5. Allow emulsion to gel on an ice tray for 1 minute.
6. Transfer slide to test tube rack to dry.
7. Dry for 4 hours in total darkness.
8. Put at 4° C in a light-tight slide box with desiccant. Wrap the box in foil to be double

sure light cannot penetrate. Leave to expose for the appropriate amount of time (overnight for abundant mRNAs).

9. Leave slides at room temperature for 30 minutes before developing.
10. Cool developer, water, and fixer to 15° C (See Note 2.)
11. Develop for 5 minutes in Kodak D-19 developer.
12. Rinse in distilled H_2O.
13. Fix for 5 minutes in Kodak rapid fix.
14. Wash in distilled H_2O for 10 minutes.
15. Dehydrate through EtOH to 95%.
16. Counterstain with 0.5% fast green FCF in 95% EtOH for 30 seconds. (See Note 3.)
17. Wash in 100% EtOH.
18. Dip in xylenes or cnp30 for 2 minutes.
19. Mount in Permount without allowing the slides to dry. (See Note 4.)

Notes

1. NTB2 emulsion has a shelf life of approximately 6 months. On receipt, aliquot emulsion in scintillation vial inserts and store in the dark at 4° C. Agitate as little as possible, because a lot of mixing will increase the number of background grains.
2. Cooling developing solutions reduces background.
3. Choice of counterstain may vary depending upon tissues used. In some cases it may prove to be unnecessary.
4. If a slide dries out at all, emulsion will remain cloudy on the slide. If this happens you cannot use darkfield microscopy to look at the sections.

DIG

Materials

Hybridized sections in PBS
DIG detection kit (Boehringer catalog # 1175 041)
Metal slide rack
Sandwich boxes
Humid chamber
PBS (10 mM Na-phosphate pH 7.5, 150 mM NaCl)
BSA/Triton/PBS (1% BSA, 0.3% Triton X-100 in PBS)
AP buffer (100 mM Tris-HCl pH 9.5, 100 mM NaCl, 50 mM $MgCl_2$)
Substrate solution (45 μl NBT solution, 35 μl X-phos solution in 10 ml AP buffer)
100%, 95%, 80%, 60%, 30% EtOH

Method

1. Incubate slides for 45 minutes in 0.5% blocking agent in PBS.
2. Transfer to BSA/Triton/PBS for 45 minutes.
3. Drain off excess fluid and wipe the back of the slide with a tissue. Place 100–200 μl of 1/2,000 dilution of anti-DIG antibody in BSA/Triton/PBS on top of the sections. Place slide in a humid chamber and leave for 1 hour at room temperature.
4. Wash 3 × 20 minutes in BSA/Triton/PBS.
5. Wash 5 minutes in AP buffer, drain, and place slides back in humid chamber.
6. Put 100–200 μl of substrate solution on each slide and leave at room temperature overnight (or longer) to develop.
7. When signal is clearly visible, dehydrate through ethanol (without counterstaining) and mount in Permount.

Analysis of Results

^{35}S-labeled hybrids can be viewed using darkfield or brightfield microscopy. In general, brightfield is better for strong signals and darkfield for weak signals. Photographs should be taken either using low contrast black and white film or using tungsten color slide film. (Ektachrome 160 works well.) DIG-labeled hybrids should be viewed using brightfield microscopy. If background noise is low, high contrast black and white film will work best. Ektachrome 160 also gives good results.

Data interpretation is absolutely dependent upon comparisons between signals generated by sense and antisense probes. In some cases sense and antisense probes will adhere to epidermal cells. This appears to be a function of probe size (generally seen when inserts greater than 1 kb are used to generate riboprobes). The quickest solution to this problem is to subclone smaller fragments.

The most frequently encountered problems are:

1. Sections fall off the slide.
   Make sure all water is removed after sections are floated on slides. However, some sections will still fall off so it is necessary to hybridize more than you need. You can minimize loss by being particularly gentle when removing coverslips after hybridization.
2. Morphology is bad after hybridization (but okay prior to).
   Optimize proteinase K digestion and, again, be careful removing coverslips.
3. Background noise is high.
   This is the most common problem, particularly with DIG probes. In my experience, this usually occurs with a particular probe preparation. Often making new probe solves the problem. Also check the RNase digestion postwashing.

REFERENCES

Cox, KH, Deleon, DV, Angerer, LM and Angerer, RC (1984) Detection of mRNAS in sea urchin embryos by in situ hybridization using asymmetric RNA probes. Dev. Biol. *101* 485–502

Jackson D (1992) *In situ* hybridization in plants. In Gurr SJ, Bowles DJ, McPherson MJ (eds) Molecular Plant Pathology: A Practical Approach, Volume 1 IRC Press, Oxford, pp 163–174

Langdale JA, Zelitch I, Miller E, Nelson T (1988a) Cell position and light influence C4 versus C3 patterns of photosynthetic gene expression in maize. EMBO J 7: 3643–3651

Langdale JA, Rothermel BA, Nelson T (1988b) Cellular patterns of photosynthetic gene expression in developing maize leaves. Genes Dev 2: 106–115

Martineau B, Taylor WC (1986) Cell-specific photosynthetic gene expression in maize determined using cell separation techniques and hybridization in situ. Plant Physiol 82: 613–618

19

Photography of Maize

PHILIP W. BECRAFT

This is a guide to help the neophyte successfully photograph maize. The scope of this guide is limited to the use of the 35 mm camera, macro lenses, extension tubes, and dissecting microscopes. Microphotography is discussed in Ruzin and Sylvester (this volume). For successful photography, the film, camera (including optics), subject, and lighting must all be matched for the particular image you wish to capture. The film, camera and subject will be dealt with separately below. Lighting will be discussed in each section as it relates to that subject.

The most important advice is to experiment and to bracket exposures. This will teach you more than anything you read and in the long run will save film and time. Take a shot with several camera settings; alter the lighting; photograph the subject from several angles and distances. Bracketing is especially important with perishable and irreplaceable

The Maize Handbook—M. Freeling, V. Walbot, eds.

subjects. Record the conditions and settings for each photograph. As you photograph, be alert to idiosyncrasies of your camera. For example, the field of view on your photographs may be slightly larger or smaller than that through your viewfinder. Pay attention to details such as centering the subject in the frame and keeping feet, fingers or unnecessary shadows out of the field of view. Always be conscious of *light*, and use it to your advantage. The human eye interprets the lightest part of a photograph as the top, so try to arrange the photograph so that the lightest part will coincide with the top of the subject.

CAMERA

Camera Parts

The camera consists of the camera back and the optics. The camera back holds the film and advances it, controls the shutter, and usually contains a built-in light meter. If you are buying a camera back, get one with an automatic exposure setting. The optics (lens) focuses the image on the film and controls the magnification and depth of field. A 50–100-mm zoom macro lens is a versatile lens for general use. For close-up photography a lens with an image ratio of at least 1:4 is recommended. This produces an image on the film one-fourth the size of the actual image. For more magnification, extension tubes can create an image ratio of 1:1 or greater. These fit between the lens and the camera back and are much less expensive than a higher-power lens. The longer the extension tube, the greater the degree of magnification. For yet-greater magnification, use a dissecting microscope with a photo tube for mounting a camera.

Camera Settings

Important settings and features on the camera are listed below along with a brief description of their purpose.

ASA or *ISO* refers to the film speed. The camera setting must match the ASA or ISO rating of the film. The exposure times given by the light meter are adjusted to match the film speed setting. Some cameras automatically read the ASA from the film.

The *light meter* measures the amount of light coming through the lens. A needled scale inside the viewfinder indicates exposure time in relation to the set ASA and the available light.

Exposure is the length of time the shutter stays open, exposing the film to light. The automatic setting exposes for the time indicated by the light meter. The numbered settings give the respective length exposure regardless of the light meter reading.

f-stop controls the lens aperture size. This controls the depth of field and the amount of light coming through the lens. High f-stop settings (e.g., 22) have small apertures, large depth of field, and require high light or a long exposure. Low f-stops (e.g., 4) have wide apertures, narrow depth of field, and require a short exposure.

The choice of settings depends on the circumstances, but keep the following in mind:

(1) At exposure times of about 1/60th of a second, movement will cause blurring; adjust the f-stop to give a shorter exposure time or use a tripod and mechanical cable release. (2) The light meter averages the available light to determine the exposure time. Therefore, if the subject is darker than the rest of the field of view, it will be underexposed in the photograph unless you compensate by overexposing. Conversely, underexpose the photograph if the subject is lighter than the rest of the field. (3) Adjust the depth of field to emphasize the subject. For a particular plant in the field, a shallow depth of field will focus the subject against a blurry background. If you want objects at different distances to be clear, use a high f-stop. If you cannot adjust the exposure acceptably, use a tripod or take your subject to a camera stand (light table, copy stand). A cable release will help avoid moving a mounted camera during long exposures.

Close-up photography poses a set of problems. The closer the subject is to the camera, the less light is reflected into the lens. Also, less depth of focus is available requiring a high f-stop if the subject has any relief. Both of these factors contribute to long exposures, so mount the camera stably on a tripod or copy stand. Also, light coming through the viewfinder can influence the light meter, particularly when light coming through the lens is low. In these circumstances, cover the viewfinder.

Examples of useful equipment are listed below, but many manufacturers produce comparable equipment.

Camera back:	Nikon N2000
Lens:	Nikon AF Micro Nikkor, 60 mm, 1:2.8 image ratio
Extension tubes:	Vivitar 12 mm, 20 mm, 36 mm
Dissecting microscope:	Wild M5
Phototube:	Nikon PFX with Nikon adapter

SUBJECT

Arranging the subject is a crucial part of good photography. Foremost, emphasize the features of interest by paying attention to lighting.

Outdoor Photography

For field shooting, overcast days are better than bright sun. Clouds diffuse the light evenly, whereas sun creates stark shadows. Avoid shooting toward the sun because backlit subjects are difficult to expose properly. Use a background that highlights the subject. If the particular plant is dispensible, dig it up and move it to an appropriately lit place with a suitable background. Leaving roots attached will prevent rapid wilting.

Indoor Photography

Indoor shooting is often preferable to outdoor because you control lighting and background. Blue background contrasts well with most plant parts and permits a good exposure. Black backgrounds can produce striking results but pose more difficulty for proper exposure. Try underexposing slightly. Velvet is an ideal background material but any solid textile or paper works. Avoid material with textured designs, and keep the material free from folds or wrinkles. For upright plants, hang the background material, making sure it fills the whole frame and is placed slightly out of the plane of focus. For spreads of whole plants or large leaves, arrange the subjects on the background material spread on the floor and photograph from as nearly a vertical position as possible. Camera stands or tripods are ideal for this type of photography because bracketing is routine once the subject is arranged.

For *artificial lighting* use at least two light sources. Arrange them at 45° on either side of the camera for even lighting. To reduce glare, shift the lights or diffuse them with tissue or gauze (don't place it directly on the bulbs), or use a white or tan umbrella. To emphasize three-dimensional structure with shadows, dim the light on one side or shift the lamps. Try to keep the shadows soft or the shaded areas may lose detail. For large subjects such as whole plants or leaves, inexpensive studio lights work well. For smaller subjects, camera stands are equipped with good light sources, and for dissecting microscopes a fiber optic light source is best. All these light sources use incandescent bulbs with *tungsten* filaments, so use the appropriate film. (See "Film" section.)

Arranging the subject properly can be difficult. Tape or glue portions down to prevent shifting. Slightly wilted leaves will lie flatter than turgid ones. It is easier to unroll the sheath on young leaves than old. A small piece of modeling clay is ideal for positioning a kernel. Freshly harvested kernels are more pleasing than older ones, which become dull. To make kernels or other small subjects appear to float in space, place them on a piece of clean glass and shine light below the glass onto a sheet of paper several centimeters below.

Take photographs at several *magnifications*. It is often tempting to use the highest magnification possible, but meaning may be lost because the photograph contains no context. Lack of context may not be apparent until later; therefore, photograph at lower magnifications as well.

FILM

Choosing a Film

The film to use depends on what you want. For publications, black and white (B/W) prints are generally best, although color prints may be required to emphasize certain details. Color slides are desired for presentations. The products listed in the following discussion are all Kodak brand for the sake of simplicity and availability. Other brands may work equally well.

Black and white film relies on variation in tone as opposed to color; thus contrast is very important. For most plant subjects, soft contrast or continuous tone film is preferable. For prints, the film in the camera must produce a negative. Good films of this type are T-max (ASA 100) and Plus-X Pan (ASA125). The exposed film can be easily developed according to the manufacturer's specifications. For B/W prints it is best to photograph in B/W; however, a B/W print can be obtained from a color slide by having an internegative made by a professional photographer. Specify that you want a black and white internegative.

The film choice for color photography is Ektachrome slide film. This is cheaper than using print film, and excellent prints can be made from slides. The only thing to account for is light source. Use Ektachrome Daylight film under daylight or fluorescent lights, and Ektachrome Tungsten film with a tungsten light source (any incandescent lamp such as studio lights, a light stand, or dissecting microscope). Turn out fluorescent room lights when using a tungsten film and light source or the photographs may appear blue.

Film speed is indicated by the ASA (or ISO) number. The larger the ASA number, the faster the film (the less light is required to expose the film). The grain size for a given film type is related to film speed such that faster films have coarser grain. The best ASA depends on the circumstances. For enlarged prints, fine grain is important; however, for photographing outdoors on a dark, breezy day, a fast speed prevents blurring. For general purposes, an ASA from 100 to 200 is good. Above 400 films become grainy, and at ASA 32 long exposure times may be required.

PRINTING B/W NEGATIVES

Printing is done in a light-tight darkroom. Place the negative in an enlarger with the emulsion side (least shiny surface) away from the light source. Remember "emulsion to emulsion": emulsion of the film faces the emulsion of the paper. Adjust the enlarger so that the image is the desired size and focused. Use an enlarging microscope to focus the silver grains in the image. Working under a safelight, place a test strip of photographic paper under the enlarger, emulsion side (shiny surface) up. Expose it for different times by setting the enlarger timer for 1 or 2 seconds, covering most of the strip, and giving consecutively longer exposure times by uncovering more of the strip with each exposure. Develop the strip and decide which exposure is best. Adjust the exposure by controlling (1) the brightness with the f-stop on the enlarger lens and (2) the duration with the enlarger timer. If the negative is not evenly exposed, compensate during printing by shadowing overexposed portions of the image (a process called "dodging") or by "burning in" underexposed areas by shading the rest. The object used for shading must be moved rapidly to prevent visible edges from appearing on the print. This is easier with long exposure times.

There are many types of paper available for printing. Kodabrome II is a good standard photographic paper for prints. An automatic processor requires the use of "resin-coated" paper (Kodabrome II RC). For developing with liquid chemicals it does not matter whether the paper is RC. The contrast is determined by an F rating; F1 is soft contrast while F5 is hard. For most plant subjects, F2–3 is best, but if there is stark contrast in the subject, F1 may be needed to bring out details in both the dark and light areas. The contrast is controlled either by the paper itself or by a set of filters used on the enlarger. To use filters you must use Polycontrast paper. Otherwise the contrast is designated by the F# rating of the paper.

An automatic processor produces excellent quality prints and is convenient, but affords less flexibility in the printing process. To process the prints with an automatic processor, turn it on, make sure all the chemical reservoirs are full, and slide the exposed paper in one side; the developed print comes out the other. With liquid chemicals, immerse the paper in trays containing the solutions according to the schedule specified for the paper you are using. Follow standard developing, stopping, fixing, and washing procedures.

III
Genetics Protocols

20

Genetic Experiments and Mapping

E. H. Coe

ESTABLISHMENT OF INHERITANCE PATTERN

When a variation is found in experimental materials, genetic analysis requires simple determination of its heritability and of its conformance to one of the several alternative patterns of inheritance (e.g., dominance-recessiveness vs. codominance-additivity; unit factorial vs. multifactorial; biparental vs. uniparental vs. cytoplasmic). A unit-factor variation normally should follow the conventional rules of 3:1 or 1:2:1 in self-fertilization and 1:1 in test crosses, ratios that are evaluated by commonplace statistical tests.

Variants crossed with standard strains should reappear in the progeny (i.e., they should not "disappear"), etc. It is commonplace for accidents, injuries, chance variations, diseases, physiological anomalies, and developmental errors to give rise to discrete-appearing variations that do not show inheritance. In addition, spontaneous or induced aneuploidy (trisomy or monosomy), chromosome breakage, haploidy, and other ploidies are by no means rare in maize and represent occasional variant plants in progenies. In other words, a gene-controlled variation cannot be concluded before inheritance studies are carried out. Concordantly, cytogenetic variations are among the most significant and valuable resources for analysis of biological systems and of the genome; such exceptions should be treasured and explored when found.

ANALYSIS OF SEGREGATION AND PLANNING OF EXPERIMENTS

The size of progeny needed to include at least one of the recessive class in a family segregating in a ratio of p:q (e.g., 3/4:1/4) can be calculated from the first term of the binomial expansion. The p^n term represents the frequency with which all n members of the progeny will be dominant (p class). The frequency of failing to see a recessive in a progeny is desired to be small, e.g., 1%. Thus, if $p^n = 0.01$ and $p = 3/4$, n is 16, which means that a progeny of 16 will contain one or more of the recessive class in 99% of

The Maize Handbook—M. Freeling, V. Walbot, eds.

such progenies. Similar calculations for a 1:1 ratio yield $n = 7$. Table 20.1 gives progeny sizes for several frequencies and probabilities.

A shorthand calculation uses the equation:

$n \log(1 - \text{fraction}) = \log P$, where
n = # of plants required
$1 - \text{fraction} = 1 -$ fraction of the "expected" type
P = probability

For example, with 3 : 1 segregation, how many plants are required to be 95% sure of finding one example of the one-fourth class?

$$n \log(3/4) = \log 0.05$$

$$n = \frac{\log 0.05}{\log 3 - \log 4} = \frac{1.30103}{0.12494} = 10.4 = 11 \text{ plants}$$

EXAMPLE

Suppose that, in a case where the segregation is 3:1, a progeny large enough to include at least one of the recessive (aa) class is to be obtained. The class desired occurs with a

Table 20.1. Minimum progeny size needed in order to include at least one of the desired class, with a chosen probability level. After Mather (1957) and Hanson (1959).

	Probability level			
Class desired	90%	95%	99%	99.9%
1/2	3.3	4.3	6.6	10.0
1/3	5.7	7.4	11.4	17.0
1/4	8.0	10.4	16.0	24.0
1/7	14.9	19.4	29.9	44.8
1/8	17.2	22.4	34.5	51.7
1/9	19.5	25.4	39.1	58.6
1/16	35.7	46.4	71.4	107.0
1/27	61.0	79.4	122.0	183.0
1/32	72.5	94.4	145.1	217.6
1/64	146.2	190.2	292.4	438.6
1/100	229.1	298.1	458.2	687.3
1/1000	2301.4	2994.2	4602.9	6904.3
2/3	2.1	2.7	4.2	6.3
3/4	1.7	2.2	3.3	5.0

frequency of one-fourth (third row of the Table 20.1), and we choose a probability level of 99%. In the body of the table we find that 16.0 individuals are sufficient—i.e., that about 99% of progenies of 16 individuals will contain at least one of the recessive class. While the progeny size on purely mathematical grounds should be rounded upward, these numbers are suitable to use as approximate guides to adequate progeny sizes. Note, for example, that for a 15:1 ratio (1/16 of the double recessive aabb class) the 99% level requires over 70 individuals (average expectation about 4.5 double recessives) to have this level of confidence that one or more of the double recessive class will be included in the progeny.

A convenient table developed by W. L. Stevens (1942; also in Fisher and Yates' tables, 1963) provides for estimation of limits and fit for binomial and Poisson distributions, for low expectations in small to infinite progeny sizes; this table can be used also (in reverse) to help in planning progeny sizes. For progeny sizes to distinguish between two different proportions (e.g., 3:1 vs. 9:7), and for several other planning calculations, see Hanson (1959) or Mather (1957).

TESTING FOR ALLELISM WITH EXISTING MUTANTS

Many prior mutations are known in maize, so new mutations that resemble previously established ones should be tested for allelism (complementation tested) by intercrossing. When crosses between recessives show the trait, allelism is assumed; when complementation is found (i.e., normal F_1) between recessive variants this is considered proof of nonallelism. Ordinarily this means crossing homozygotes with homozygotes, though it is equally valid (and sometimes necessary) to cross heterozygotes *inter se*, or even putative heterozygotes *inter se*. Testing allelism among two recessives, both of which are homozygous inviable, requires either confirmation of each parent (typically one by selfing, the other by testcrossing onto several sibs) or confirmation that a pooled pollen sample and a pool of tested ear parents both contain recessive gametes. Dominants with recessives or dominants with dominants are not easily proven to be allelic, and biochemical or mutational evidence may be necessary.

Because the volume of old and new mutations is large and is increasing rapidly, alternate strategies are needed and are beginning to be employed. Mapping is the primary alternate strategy that is coming into use. Current mapping techniques now permit localization of recessive mutations with the help of B-A translocations to a specific chromosome arm (i.e., to a specific 5% of the genome) in one generation. (See chapter by Beckett in this volume.) Mapping with molecular markers [Restriction Fragment Length Polymorphisms (RFLPs) and Randomly Amplified Primed DNAs (RAPDs)] promises to increase the resolution further. The compromise in choosing to map before testing allelism directly is between prompt answers vs. a sacrifice of time and patience for one or two generations; if the number of candidates to be tested inter se is modest (e.g., five or so), allelism tests may be justifiable, but larger numbers soon become prohibitive in time, expense, and feasibility.

Selection of candidate mutations for allelism should be done by (1) including all degrees of expression of the affected trait and (2) maintaining suspicion toward "physiologically different" variations in the same system or process.

CHARACTERIZATION OF EXPRESSIONS

If a mutation is to be studied, or is to be used as a marker, its expression (whether biochemical or visible) must be examined to determine its stability, reliability, and diversity. One objective should be to determine how it can be recognized consistently: At what *stages*, in what *tissues or organs*, under what *growth conditions*, under what *observational conditions*? *Viability*, *conditions of viability*, and *transmission* information are needed. A second objective should be to evaluate genetic diversity in expression: With what *other genetic constitutions*, and in what *genotypic backgrounds*, is its expression altered? Each of these characterizations increases potential utility of the mutation and increases knowledge of the developmental, biochemical and molecular implications; concurrently the probability of serendipitous discovery, a major driving force in research advances in genetics, is enhanced.

Crosses with defined normal strains that have been chosen for their informativeness can be recommended emphatically. The choice of strains depends upon the goals of the study, but for almost any purpose crosses to uniform, inbred strains will prove most valuable, if not critical. Isogenic lines are not necessarily the goal of such crosses, though they can be. In addition to the obvious value of genetic comparisons and uniformity, current mapping strategies with RAPDS and RFLPs are dramatically more efficient if the molecular markers near the target locus, in one of the parents (i.e., the normal parent), can be determined unambiguously. Elite inbred lines (lines in widespread commercial use in hybrid production) have advantages that include prior characterization, vigor, stability, and uniformity along with the potential for research information and impact on quantitative traits and production efficiency. Nonetheless, for many characteristics the genotypes of elite inbreds, per se, may be obstructive for the analysis of certain mutants: Anthocyanin genetics is an example in which essential complementary factors are lacking in elite inbred lines of the U.S. corn belt (all of which are recessive *r1*, usually *c1*, and often carry dominant inhibitors of pigmentation), and multifactorial segregations confound the expressions. Another example is the shrunken-kernel expression of *sh1*, which is excellent and unequivocal in selected genetic strains grown under 'good conditions' but is unreliable in the dent-corn backgrounds of most inbred lines. Similarly, the expressions of many morphological traits (e.g., *Cg1*, *ts2*, *Kn1*) vary greatly in different backgrounds.

Variability of modifying genetic factors in the source from which a mutation has been derived may greatly influence its expression, and specific interactions in one background may show effects that are not predictive for other backgrounds. Thus, generalizations about the expression are not warranted until deliberate comparative studies have been carried out.

What does a skillful observer note? Observations at each stage of development often reveal valuable properties. Observation under cloudy vs. sunny conditions, with spectral filters, under ultraviolet light, after rapid growth, or during intense sunlight or drying conditions can be informative. Whenever form, size, proportions, etc., are affected morphological and anatomical explorations or quantifications, sectioning, staining, and biochemical parameters should be considered. Cytological studies that examine cell types and orientations, chromosome counts and constitutions, and organelle modifications are often more instructive than any other form of information. On more than one occasion, aroma, reflectivity, wilting, necrosis, water repellency, transient chlorosis, or relative proportions of plant parts have opened avenues to analysis of a mutation's cause.

Applications of test treatments, whether corrective (such as hormones) or analytical of the expression (such as an action spectrum; for inhibitory compounds), will be most effective when the physiological or biochemical basis of an expression is carefully considered and tests are devised that are targeted to specific processes.

MAPPING TECHNIQUES

A new factor that has not yet been placed to chromosome has classically been located and mapped by use of gene marker stocks across the genome. This method can be tedious. Efficient techniques of crossing with B-A translocations, or with reciprocal translocations involving chromosome 9 marked with *wx*, both described in other chapters, are far superior. Several other techniques for determining linkage and mapping are described in Coe et al. (1988).

Mapping is relatively straightforward, once a factor has been located to chromosome, because the linkage groups with gene loci are extensively established. Mapping, per se, is conventionally conducted by preparing a hybrid with three or more heterozygously marked points and testcrossing whenever possible, preferably with the homozygous recessive as pollen parent in order to avoid problems arising from heterofertilization.

Tabulations of recombination data should be presented as follows (see Emerson, et al. 1935, for examples):

For two-point data:

		AB	Ab	aB	ab	T	Recombination
Ab/aB	F2	#	#	#	#	#	%

For a three-point testcross: AbC/aBc × abc/abc:

Parentals		Recombinants						
		Region 1		Region 2		Regions 1 & 2		
AbC	aBc	ABc	abC	Abc	aBC	ABC	abc	Total

AbC/aBc	#	#	#	#	#	#	#	#	#
Subtotals	#		#		#		#		#
Percentages	%		%		%		%		

The complete, systematic tabulation of all classes (as above) is essential to evaluation of the data by others and to reanalysis as new information develops.

For recombination data from a testcross (e.g., ++/ab × ab/ab), the percentage of recombination, p, is calculated by adding the two recombinant classes and dividing by the total observed, n. With three point data, the recombinant classes for the region are summed (recombinants in region 1 plus recombinants in regions 1 and 2) and divided by the total, or their percentages can be summed. The standard error, s, for a testcross is the square root of (pq/n). The percent recombination in a region should be presented with the standard error, e.g., 20 ± 5%.

For planning purposes the progeny size, n, required for a certain precision (i.e., confidence limits around p) can be calculated for an assumed p and standard error, s, from the relationship s = the square root of (pq/n). For example, n required for 20 ± 5% is pq divided by s squared, e.g., $(0.20)(0.80)/(0.05)(0.05) = 64$. Following are examples calculated by this means:

$p = 10 \pm 1\%$ $n = 900$
$p = 10 \pm 5\%$ $n = 36$
$p = 20 \pm 5\%$ $n = 64$
$p = 40 \pm 5\%$ $n = 96$

These calculations are valid for *testcrosses in coupling* (cis) or in *repulsion* (trans). The 95% confidence interval for p can be determined by multiplying the standard error by 1.96 (i.e., by approximately 2); for 99%, multiply by 2.58.

To plan progeny size to distinguish independence (50% recombination) from linkage, consider the last example above. Ninety-six, i.e., approximately 100 individuals, will be sufficient to show significant association between two loci that undergo 40% recombination (1.96 times the standard error defines the 95% confidence interval; thus 40 plus or minus about 10 defines an interval between 30 and 50%).

Formulas of relationships between p and s for data other than testcrosses are quite different from the above, and are given in Allard (1956). Examples for an *F_2 in coupling* (A dominant to a, B dominant to b):

$p = 10 \pm 5\%$ $n = 41$
$p = 20 \pm 5\%$ $n = 83$
$p = 40 \pm 5\%$ $n = 176$

For an *F_2 in repulsion*, however, larger numbers are required:

$p = 10 \pm 5\%$ $n = 390$
$p = 20 \pm 5\%$ $n = 363$
$p = 40 \pm 5\%$ $n = 275$

For an F_2 with two codominant factors, when p is small, the value of s approaches the square root of $(pq/2n)$, in coupling or in repulsion, approximately:

$p = 1 \pm 0.5\%$ $n = 198$
$p = 2 \pm 1\%$ $n = 98$
$p = 5 \pm 2\%$ $n = 59$
$p = 10 \pm 3\%$ $n = 50$

These examples display the resolution of mapping experiments with multiple, closely linked codominant isozymes and RFLP markers analyzed in F_2 progenies. For example, an interval with 10% recombination requires roughly 50 individuals to generate a 95% confidence interval between 4% and 16% recombination.

Recombination data from an F_2 progeny or other sources are only treated with precision, accounting for all the proportions observed, by the mathematically exact method of maximum likelihood. Tables, formulas, examples, and evaluations for a wide range of data sources are given in Allard (1956). A simple method that has been shown to be equally exact mathematically for normal purposes is the product method. Immer (1930) and Immer and Henderson (1943) describe and give tables and formulas for the product method in F_2 for 3:1 A:a with 3:1 B:b, several interaction ratios (e.g., 9:7 with 3:1), and partial backcross (3:1 A:a with 1:1 B:b, from AaBb × Aabb). Stevens (1939) gives a convenient table for the 3:1 with 3:1. Standard errors of p values for the F_2 and other ratios, which are second derivatives of the appropriate function, can be calculated with the aid of the same tables, or from formulas and tables in Allard (1956).

The advent of molecular markers, having the combined efficiency of codominant expression (i.e., heterozygotes and homozygotes are each fully classifiable), unlimited numbers of markers, and absence of effects on phenotype, has set a new paradigm and new protocols for mapping. Genes now are being located more often in relation to molecular markers than to other genes.

The strategy of *interval mapping* (Lander and Botstein 1987; Hoisington and Coe 1990) exploits the fact that, in a genetic interval, segregating markers nearest to a target gene remain linked with the target and tend to become homozygous when it becomes homozygous. The stock materials required for interval mapping are segregating families in which multiple-point polymorphism is present; in fact the extent of polymorphism is such that virtually any segregating family inherently has sufficient interval marking. Interval mapping is typically done individual by individual, by determining associations of markers and the target gene in 25–30 homozygous individuals, directly giving a measure of frequency of recombination. Matz et al. (unpublished data in MNL65:104–105, 1991)present a demonstration example of the method. A simplified strategy that has proven effective in recent studies in other species is to pool samples (e.g., 10 individuals) of the recessive class and score for disappearance of a marker band that is associated with the dominant alleles (See Coe and Chao, unpublished data in MNL65:53–54, 1991). A suggested protocol for viable recessives is as follows:

Cross mutants with inbred lines (preferably lines that have been well characterized for molecular markers—e.g., A619, A632, B73, Mo17).

Self the F1, classify F2 individuals and sample tissues in pools from 10 individuals (e.g., 2 pools of the dominant class and 3 pools of the recessives).
Prepare Southern blots or RAPD gels from the pooled samples.
Seek disappearance of mapped bands characteristic of the dominant (i.e., inbred-line) parent in the recessive pools.
Strategies for dominants, and for inviable mutations, require appropriate modifications in design.

Mapping of molecular sequences (e.g., of cloned genes) is efficiently and promptly done by probing for polymorphisms of the sequence in ready-made populations (or blots of them) that have been comprehensively mapped already. Efficient and precise results can be achieved with either recombinant inbred (RI) lines, described elsewhere in this volume, or with immortalized F_2 (IF_2) populations (Gardiner et al., Genetics, in press, 1993). Mathematical treatment for maximum-likelihood mapping of multipoint data in F_2 or IF_2 populations has been developed (Lander et al. 1987; MAPMAKER: Lincoln et al. 1990) and is in widespread application. A particular value of MAPMAKER is in the reflection of statistical precision for alternative maps from the data. Comparable treatments of RI data are becoming available (RI Plant Manager: Manly 1987) for assessment and application. Programs for combining of mixed families of data that are under development and assessment include GMendel, developed by S. Knapp (Echt et al. unpublished data in MNL66:27–29, 1992).

Mapping of quantitative trait loci (QTLs) is developing rapidly. In principle this is similar to unit-factor mapping and is conducted with extensions and enhancements of the same mathematical treatments. The principal differences, other than the involvement of multiple factors, are in the collection of measurement data rather than "this vs. that" and in the requirements for replicated and randomized designs that are essential to quantitative-genetic studies.

Maps presented in this volume should not be treated as immutable or final, in view of the imprecision and of the rapid pace of change in mapping information. Annual updates of the working linkage maps for maize may be found in the *Maize Genetics Cooperation News Letter* and biennially in *Genetic Maps* (Cold Spring Harbor Laboratory). The developing database effort (see chapter in this volume) is being planned to incorporate maintenance of current mapping information.

REFERENCES

Allard RW (1956) Formulas and tables to facilitate the calculation of recombination values in heredity. Hilgardia 24: 235–278

Coe EH, Hoisington DA, Neuffer MG (1988) The genetics of corn. In Sprague, GF and Dudley, JW (eds) Corn and Corn Improvement, Am Soc Agron, Madison, WI, pp 81–259

Emerson RA, Beadle GW, Fraser AC (1935) A summary of linkage studies in maize. Cornell Univ Agric Exper Sta Mem 180, 83 pp

Fisher RA, Yates F (1963) Statistical Tables for Biological, Agricultural, and Medical Research, Sixth Edition, Hafner Publishing, New York, 146 pp

Hanson WD (1959) Minimum family sizes for the planning of genetic experiments. Agron J 51: 711–715

Hoisington DA, Coe EH (1990) Mapping in maize using RFLPs. In Gustafson JP (ed) Gene Manipulation in Plant Improvement II, Plenum, New York, pp 331–352

Immer FR (1930) Formulae and tables for calculating linkage intensities. Genetics 15: 81- 98

Immer FR, Henderson MT (1943) Linkage studies in barley. Genetics 28: 419–440.

Lander ES, Botstein D (1987) Homozygosity mapping: a way to map human recessive traits with the DNA of inbred children. Science 236: 1567–1570

Lander ES, Green P, Abrahamson J, Barlow A, Daly M, Lincoln S, Newburg L (1987) MAPMAKER: an interactive computer program for constructing genetic linkage maps of experimental and natural populations. Genomics 1: 174–181

Lincoln SE, Daly MJ, Lander ES (1990) Constructing Genetic Linkage Maps With MAPMAKER: A Tutorial and Reference Manual, Whitehead Institute, Cambridge, MA

Manly K (1987) RI Plant Manager: A Program for Genetic Mapping With Selfed Recombinant Inbred Strains. Health Research, Roswell Park Cancer Institute, Buffalo, New York

Mather K (1957) The Measurement of Linkage in Heredity, Reprint of Second Edition, Methuen, London, 149 pp

Stevens WL (1939) Tables of the recombination fraction estimated from the product ratio. J Genetics 39: 171–180

Stevens WL (1942) Accuracy of mutation rates. Genetics 43: 301–307

21

Growing Maize for Genetic Studies

M. G. Neuffer

CORN is a large kerneled, highly domesticated, vigorous annual plant of tropical origin. Because it is a natural cross-pollinator it is highly heterogeneous and responsive to selection pressure. Corn has been taken by humans into all but the harshest of agricultural environments. Strains of corn are known that grow as far north as southern Canada while others range to the extremes of the tropical forest and desert oasis. Some grow at sea level and others at 11,000 feet of elevation. Those lines and strains most used for breeding and genetic purposes are much more restricted in adaptation and therefore require narrower limits on growth conditions in order to produce a useful crop. There is no doubt that strains could be found or developed that would be satisfactory for use under most plant

The Maize Handbook—M. Freeling, V. Walbot, eds.

growing conditions. For the purpose of this chapter the conditions of choice will be described while keeping in mind that, by proper manipulation and by appropriate choice of strains, there is considerable flexibility in the conditions leading to a successful corn crop.

FIELD CULTIVATION

Choice of Field

Corn grows best and is most easily handled on fairly flat, well-drained, fertile land. The field should be large enough to permit some crop rotation. Continuous cropping with corn is feasible but requires closer attention to the buildup of injurious pathogens and insect pests. Proximity to forests (raccoons), marshes (blackbirds), rodent populations, college residence houses, and dense human populations should be avoided. The area should be appropriately fenced; a wandering four-wheel-drive vehicle can mean disaster.

Soil Preparation

In most temperate zones, fall ploughing is best. When soil is dry enough to work in spring it should be tilled to a medium texture with a disc harrow. Avoid working when the soil is too wet as large, hard clods will form that prevent good seed-soil contact or lead to excessive compaction. Best germination and weed control are obtained by working the soil the final time just before planting. When final soil preparation is complete mark the rows using a two row corn planter (strings in small plots). The rows should be 36 inches (91.4 cm) apart or at distances that match and are in synchrony with cultivating equipment that will be used later in the season. At the time of marking a band of starter fertilizer (NPK balanced for area soil) may be applied through the planter. Following the marking a preemergence herbicide spray may be applied.

Planting

Planting should not begin until the soil temperature has reached 50° F (10° C) but after that any time that allows a 90-day growth period before the first frost is suitable. Early spring plantings usually perform better than late ones because this allows the crop to flower before summer stress periods and to avoid diseases and insects. Early planting also allows plants to make use of long days and relatively warm nights. (Note: tropical strains will not flower in the long days of the temperate zone.) Seed should be prepared in envelopes labeled with family or culture numbers. Stakes with corresponding family numbers (and an optional notation of the seed number) should be prepared and placed in the field in the marked rows; seeds are planted with an "Allan Jab" hand planter at a spacing of 1 foot (30 cm) apart in rows 3 feet (91.4 cm) apart. Up to one-third closer planting is feasible on fertile land and under ideal conditions. Wider spacing is of little advantage except under dry conditions. Wide spacing does allow high-tillering lines to spread out and produce several stalks if such are needed. The seed should be planted not more than 1 inch (2.5 cm) deep and be pressed firmly into the moisture zone of the soil. If the soil is dry, irrigation immediately after planting will be required to achieve uniform germination. Some workers prefer to plant 2 kernels per hill and thin at the 4– to 5–leaf stage to 1 plant per hill. This should be done after all seedlings are up and growing and before any tillers are produced. See Wallace and Bressman (1937) and Larson and Hanway (1977) for additional details on cultivation practices.

Pest Control

Corn can be grown in most areas without herbicides, insecticides, or fungicides and using only organic fertilizers, but to do so the grower must pay careful attention to many critical agronomic practices such as crop timing, rotation, isolation, residue removal, and cultivation. The crop must be watched daily and those invaders that will periodically attack this crop must be physically removed. Barring such dedication a successful crop may be obtained through the prudent use of whichever of the following practices apply in the planned planting locale. See your local extension agent, farmers group, or government agricultural representative for advice.

Weed Control

Control may be obtained either by frequent cultivation and hoeing or by combining herbicide treatments and judicious supplementation of hand cultivation. In cultivation avoid cutting roots by refraining from deep, close cultivation. Herbicide control may be attained by using a preemergence spray on the soil just before or just after planting. Always follow the manufacturer's instructions for applying any type of chemical. Use a combination of broad-leaf and grass herbicides according to the recommendation of the local farm experts and the supplier of the chemicals. Materials used and methods of application vary from time to time and place to place. A combination product of Lasso and Atrazine applied in aqueous solution at the recommended rate depending on soil type and sprayed over the soil surface has been successful. For good control the soil must not be disturbed for some weeks after application. Later control may be obtained by hoeing out surviving weeds and/or applying a contact herbicide such as Roundup. Do not allow it to touch corn leaves or drift to nearby plants.

Insect Control

In areas infested by cutworm and armyworms an application of Lorsban insecticide should be made before or immediately after planting. Later earworm control can be obtained by timely sprayings of Dipel (*Bacillus thuringensis*). In the greenhouse, red spider mites and aphids are a serious problem, and are best controlled by a combination of sanitary conditions and the use of the insecticide Avid (Avermectin) for mites and Enstar 5E (Kinoprene) for aphids. See Dicke (1977) for additional information. Note: *extreme care* should be taken to follow manufacturer's directions in the use of modern insecticides.

Control of Other Pests

In areas where fungal and bacterial diseases are prevalent, fair control can be obtained by regular leaf surface application of a spray containing Dithane. Use according to directions given by the supplier. Insect-borne diseases are best controlled by clean agronomic practices and by insect control. See Ullstrup (1977) and Shurtleff (1981) for additional information on diseases and their control.

Irrigation

In many localities adapted corn strains and hybrids produce a satisfactory crop in most years without irrigation; however, the investigator will no doubt want greater assurance of a better-than-average crop or will be required by the constraints of the experiment to use genetic stocks that are often poorly adapted to the locality. In any case supplemental water is important in most areas of the world. Water is most easily applied by an overhead sprinkler system. The system should be relatively easy to turn on or off.

Water should be applied whenever the crop shows wilting or other signs of drought stress. Flowering is the most critical period in which to avoid drought stress. Irrigation during the pollination period should be done in the afternoon and evenings so as not to disrupt other operations. Good seed set requires adequate moisture and moderate temperature during pollination and early kernel growth. Excessive moisture due to irrigation followed by unexpected rains may cause yellowing of plants. This may be overcome by hand application of a nitrogen fertilizer (ammonium nitrate) at the rate of 50 pounds (22.5 kg) per acre spread over the root area around the plant. Excess water may also cause the plants to fall over (root lodging). In case of lodging do not lift plants upright but let them recover their upright position naturally; this will occur in a few days without help. Except in the driest of climates irrigation should cease 10–14 days after the last pollination. This helps to speed maturity.

CONTROLLED POLLINATIONS

Controlled pollinations require several careful steps:

Ear Shoot Bagging

Bagging is a fundamental requirement and yet, of all the pollinating operations, it is the most difficult to do properly. The ear shoots must be covered before the silks emerge so they will be protected from all but the desired pollen. The bag used (Lawson 217) measures 2 × 1 × 7 inches (5 × 25 × 18 cm) and is made of semitransparent treated 36-pound wet strength paper with waterproof glue. The daily shoot-bagging operation should begin when the first tassels appear. At that time the tip of the first ear shoot may be visible in the axil of the sixth or seventh leaf from the top. This shoot may be covered by placing the bag over the tip of the shoot with the longer lip next to the culm toward the stem so the shorter lip slides between the tip of the ear shoot and the leaf sheath. The edges of the bag should be pulled around to conform to the shape of the culm with an extra sharp downward pull to keep it in place. The plants should be examined every day to catch new ears before the silks emerge. Bags already on the plants may be resecured if they have been loosened by wind or field activity. Bagging too early is not very satisfactory because the tissues will not hold the bag firmly in place. Bagging too late risks silk exposure and contamination. Ears should not be bagged after silks appear. Complete honesty is required of all helpers in this regard, otherwise, contamination will get completely out of hand.

Some workers use a larger bag 2 1/2 × 1 × 8 1/2 inches (6.4 × 2.5 × 22 cm) (Lawson 218), bend back the seventh leaf, cover the whole auricle and ligule area at the top of the leaf sheath, and staple the bag around the culm even before the first shoot appears. This allows one-trip shoot bagging but sometimes the wrong leaf axil is taken and the shoot comes out elsewhere and is therefore exposed. In any shoot bagging practice care must be taken in handling the plant so that the support the leaf sheath provides to the culm of the next node is not weakened; otherwise the top of the plant will break off and the whole plant will be lost.

Under natural conditions receptive silks appear over a 4– to 5–day period and are pollinated by the new pollen shed each day, resulting in a full set ear. When hand pollinated, the silks receive pollen only once in a very short period of time. A partially filled ear results unless additional steps are taken. The standard practice to ensure a full ear is called *cutting back.*

Cutting Back

When the first day's silks are visible, use a knife to cut off the tip of the husks and silks. Immediately recover the ear to prevent exposure. The cut should be made squarely and cleanly across the ear and as far down the husks as possible without cutting off the tip of the cob inside. Contamination is avoided because the wet ends of freshly cut silks are not receptive to stray pollen grains that may fall on them during the operation. Mark the shoot bag by folding the corner to indicate which ears were cut back. The next day the silks will have grown to form a thick brush; all silks will be the same length (about 2–5 cm) and all will be ready for pollination. Pollen applied at this time will reach all silks and result in a fully set ear. Failure of unpollinated silks to grow or irregular growth indicates that the silks were already past the most receptive stage or that conditions were not good for pollen germination. Unpollinated receptive silks will grow continually to a length of 6–8 inches (15–20 cm) or more. Pollinated silks stop growing in an hour or so and become darker in color. Under midwestern U.S. conditions silks are usually receptive for about 3 days. After that the chance of successful pollination diminishes rapidly.

Collecting Pollen

Pollen is shed from mature anthers beginning on the upper third of the main spike of the tassel, usually 1 or 2 days before silks appear on the ear shoots. Pollen sheds in daily increments that spread from tip to base of each tassel branch and upward toward the tip of the main spike. The shedding moves along the tassel in two waves about 2 days apart. The first wave begins from the primary anthers in each floret and the second wave from the secondary anthers. A healthy tassel should shed pollen for about 1 week. On a warm, sunny Missouri day fresh anthers begin to extrude from the florets by filament elongation at about 7 A.M. but this can occur much later on cool or cloudy mornings. These anthers will open up (dehisce) about 30 minutes later and pollen will fall out and be scattered by air currents. Anthers will continue to appear until about 10 A.M. Then no new anthers will appear until the next morning. If the weather is cool or cloudy and humid the extrusion of anthers will be delayed until later, sometimes until evening. Some workers report successful pollinations made in the evening but the author finds this undependable if not impossible in most conditions. As long as humidity is near 100% the anthers will not open. If the weather is hot and dry the process will proceed more rapidly.

Pollination

To make controlled pollinations it is necessary to collect viable pollen. This can be done by covering shedding tassels, before the time the pollen is shed, with a brown paper tassel bag and then collecting and carrying the pollen to the desired protected silks. If the plants to be used as male parents are sturdy and heavy winds are not in the forecast, the best time to cover the tassels is in the evening or after pollen has shed on the previous day. If the plants are weak or heavy winds are expected, the tassels may be bagged after the dew is off the tassels but before shedding has ended on the day of the desired pollinations. Moisture from dew trapped in the tassel bag will prevent the anthers from opening.

Cover the tassel with the bag, keeping it as flat as possible; pull it down past the first flag leaf (which may be removed); then fold firmly around the sheath and stem of the tassel and secure it in place with a regular paper clip. If tassels were bagged the day before, pollination can begin as soon as nearby unbagged tassels are shedding. Bees determine the proper time very well and can be a good indicator. If tassels are bagged in the morning at least one-half hour should be allowed for all stray pollen grains to die and for anthers with fresh pollen to emerge inside the bag.

To collect the pollen, carefully bend the plant so that the top, clipped end of the bag is higher than the bottom. Remove the paper clip and shake the bag and tassel sharply. Withdraw the tassel, being careful not to allow the open end of the bag to drop low enough so that the pollen falls out. The pollen can be carried in this manner to make a self-pollination or to cross on silks of one or two nearby plants. One cannot depend upon pollen remaining viable in the removed tassel bag more than 10–30 minutes depending on conditions. The best procedure for standard pollination is to carry the tassel bag, with open end folded, to the plant with intended silks. Place the stalk and tassel of the female parent under the pollinator's arm, thus bending the plant so that the tassel is at the pollinator's back and the ear is immediately in front. (This protects the ear from pollen falling from its own tassel.) Then raise the shoot bag slightly, tear off the top, and squeeze it so that the shoot bag forms a chimney above the silks. (This protects the silks from pollen carried by air currents from the side.) Open the tassel bag in such a way that the pollen can be poured out in small amounts and dust a small amount of pollen on the silks; then quickly fold over the open torn end of the shoot bag to protect the silks and close the tassel bag if it is needed to make additional pollinations. After the pollinations with one pollen source are completed, return to the plants carrying the ears pollinated with that source. Use a tassel bag to note the family or culture and plant number of the female (ear) and also the family and plant number of the male (pollen) along with notes about the pollination, plant types of the parents, date, and so forth. Then pull the tassel bag over the pollinated ear so that the long lip slides between the stalk and the ear and the short lip hangs loosely on the outside. Pull the two edges on the long-lip side of the bag near the bottom around the stalk and staple them together on the side of the stalk opposite the ear, thus securing the bag to the stalk so that the ear is free to enlarge without tearing the bag.

If a large number of pollinations are to be made on a single day from one pollen source (as many as 100 pollinations may be made from one tassel) the best procedure is as follows: Prepare a glassine bag by folding up and sharply creasing the bottom inch of the bag. Then pour in the contents of the tassel bag. Sift the pollen past the crease into the bottom fold, taking care to keep the anthers in the top where they may be discarded. Turn the bag on its side and tear off the upper of the two bottom corners of the bag. Carry the pollen in this bag to the waiting silks, where it may be sifted sparingly through the torn corner of the glassine bag into the top of the torn shoot bag covering each ear. Then fold the shoot bag to protect the pollinated silks and move rapidly to the next ear. Speed is of the essence here because the pollen will remain viable only a short time. Fresh pollen is light yellow and will flow freely from the glassine bag. If the pollen becomes deeper yellow or clumps together, these are signs that it is no longer viable (collectively speaking, even though there may be some viable grains mixed among the majority of nonviable ones). After all pollinations from one source are completed there is then time to go back and write information on the tassel bags, to place them over the ears, and to staple them on the plants. It is often more efficient to have two people working together—a pollinator and a bagger.

SOME PRECAUTIONS

Contamination

During the shedding period pollen will be flying about on air currents; it will be on leaves, and on the pollinator's clothing and hands. Most of it will already be nonviable, but enough will be viable to make contamination a serious problem. Precautions should be taken to keep receptive silks covered at all times.

Pollen Viability

The life of a pollen grain is usually less than 20 minutes after it leaves the anther. It is shorter than that in hot, dry weather and much longer than that in cool, humid weather. (Ways to considerably extend pollen viability are known but the author has not found them to be reliable. They require special knowledge and handling.) There is little hope for fertilization if the temperature goes much above 95° F (35° C) during the shedding period or for several hours thereafter. Conversely, pollen may appear to be normal but will not grow on the silks unless the temperature during pollination and some hours thereafter is above 55° F (13° C). Pollen is especially sensitive to drying conditions. Any time pollen grains change in color or begin to collapse, irreversible degeneration may have already occurred. High humidity has different but equally important consequences. Anthers will not open in high humidity, thereby trapping the pollen within. While this does not kill the pollen immediately, it does pose mechanical difficulties in spreading pollen and causes premature loss of viability. Moisture in bags or other containers for carrying pollen will cause clumping. Pollen is promptly killed when immersed in water and most other liquids. Known exceptions are paraffin oil and properly buffered aqueous solutions. See Poehlman (1987) for additional information or pollination practices.

HARVESTING

Most early strains of corn grown under optimum conditions will be ready for harvesting 6 weeks after pollination. Viable seed may be obtained after only 1 month. However, for well-matured seed with high viability, a period of 7–8 weeks may be required. Midseason or long-season lines may take even longer. Kernels of most strains reach physiological maturity 35–60 days after pollination. Actual maturity time is affected by many factors including temperature, day-length, moisture, and soil conditions. A safe time to harvest is when the husks have become dry and the kernel is hard.

Harvesting may be accomplished by breaking off and husking the ears and tagging them directly in the field, or by husking the ear in the field, placing it in the tassel bag, and bringing the ears into the laboratory where they may be tagged under more comfortable conditions. The ears may be labeled with a 1 × 2-inch (2.5 × 5 cm) heavy paper label with a reinforced hole. The label can be attached to the ear by a wire parcel hook. The completed label is strung on the parcel hook which is then inserted into the soft center at the base of the cob. In some stocks the cob is quite hard so a hole needs to be made with a medium-size screwdriver. Ears should be tagged prior to drying while the cob is still soft. After tagging, the ears should be stored in some sort of drying facility for a few days to reduce the moisture content to 12% or less. This is important as ears stored at high moisture content will mold or lose their viability rapidly, and are also more subject to attack by seed storage insects.

After drying is completed, the ears should be laid out consecutively by female parent family number and a careful harvest record should be made. The ears may be stored by placing them tag up in a No. 10 flat-bottom Kraft paper bag that is cut to 6 inches (15 cm) high. Then these bags should be placed in consecutive order in a waterproof cardboard box (approximate dimensions 13 × 24 × 6 inches (33 × 61 × 15 cm) with a close-fitting cardboard lid and a label indicating the included consecutive families. To prevent insect damage about one-fourth cup (225 g) of naphthalene flakes should be sprinkled in the box before closing. Good viability may be maintained in storage by keeping the corn at comfortable room temperature as long as the humidity is kept fairly low. For long-term storage it is best to keep the corn below 60° F (15° C) and at as low a humidity as can be obtained by refrigeration-type cooling. Under these latter conditions acceptable germination is still possible after 20 years.

MAINTAINING PEDIGREES

One of the most important aspects of conducting genetic research is the maintenance of proper pedigrees. This can be done by keeping a system of records in which every individual has an identity and an ancestry and, for selected individuals, a recorded posterity. Every individual should be identified by year, by crop (e.g., winter or summer), by family, by family subdivision, and finally by individual number. All of the information about a particular family or culture planted can be recorded on a pedigree card; these cards are a portable source of information, a place to jot field notes, and may constitute a permanent record of the transactions of that culture (for future reference). Different lineages of maize geneticists use different systems, and they are all adequate. A sample of the card that we use appears in Figure 21.1. We use 7 × 4 inch (18 × 10 cm) sheets of heavy-weight subdued blue paper (colored to reduce glare in the field). The left-hand 1-inch margin of the sheet is punched for insertion in a ring-binder notebook. The card is identified by a pedigree number, including series (crop year), family number, and family subdivisions. It also contains numerical pedigrees of parents, genotypes of parents, date planted, types of seed and subdivisions planted, and handling directions; an area on the card is reserved for field notes and later notations as they are needed for this family's permanent record. Each event in the family is recorded in the appropriate places on the card. For example, we note the pollinations completed in blue ink in the same area of the card as the handling directions. We also use red ink to write in the lower-right-hand corner of the card each future planting of this family as it is made. On the back of the card we indicate in pencil the harvest record showing the actual ears obtained in the harvest. When followed systematically, the combination of typed information and handwritten notations in various colors of ink or pencil allows for a great deal of information to be stored on a single piece of paper. These sheets are eventually cut down to 4 × 6 inch by trimming off the 1-inch margin and are filed in a standard

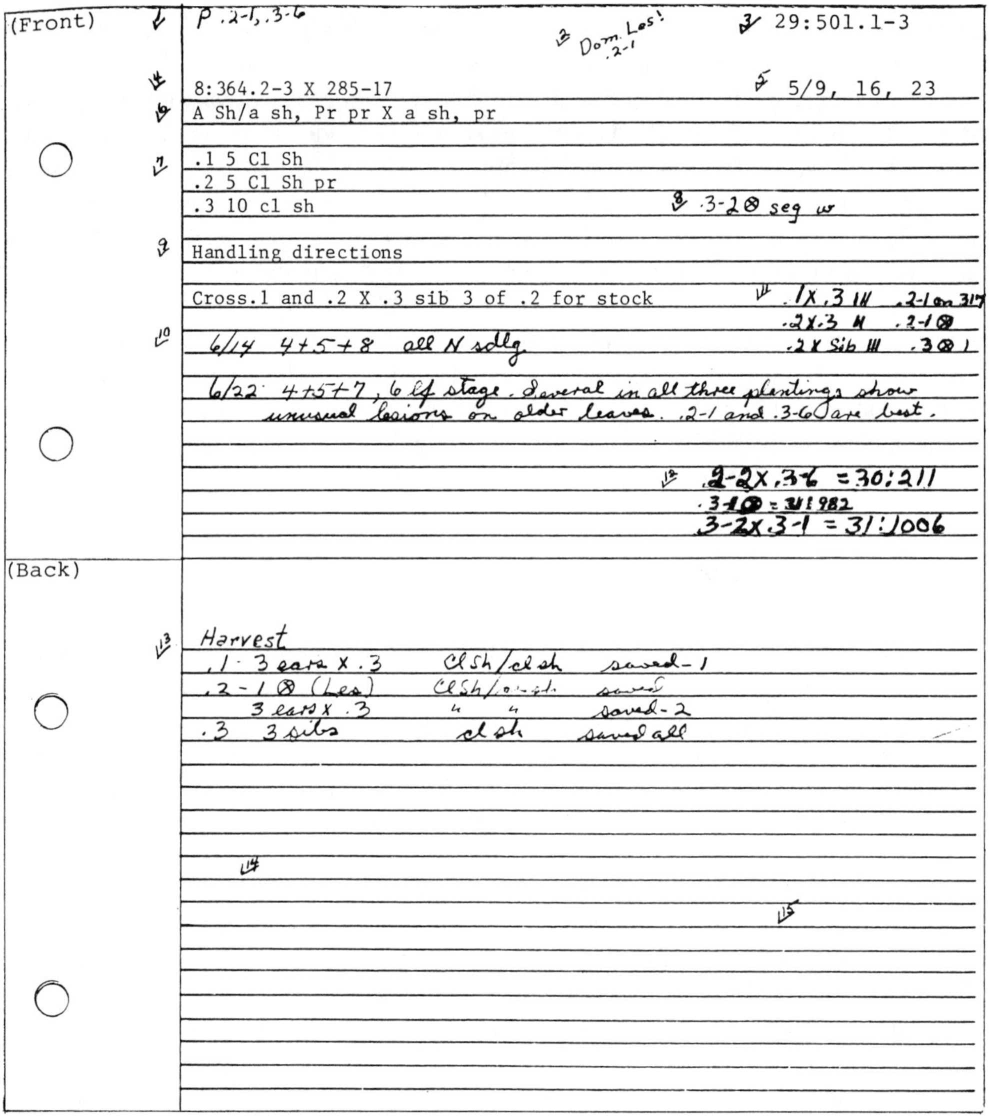
(Front)
P .2-1, .3-6
Dom. Les! .2-1
29:501.1-3
8:364.2-3 X 285-17
5/9, 16, 23
A Sh/a sh, Pr pr X a sh, pr
.1 5 Cl Sh
.2 5 Cl Sh pr
.3 10 cl sh
.3-2⊗ seg w
Handling directions
Cross.1 and .2 X .3 sib 3 of .2 for stock
.1X.3 III
.2X.3 II
.2X Sib III
6/14 4+5+8 all N sdlg
6/22 4+5+7, 6 lf stage. Several in all three plantings show unusual lesions on older leaves. .2-1 and .3-6 are best.
.2-2X.3-6 = 30:211
.3-1⊗ = 31:982
.3-2X.3-1 = 31:1006
(Back)
Harvest
.1 3 ears X .3 Cl Sh/cl sh saved-1
.2-1 ⊗ (Les) Cl Sh/cl sh saved
3 ears X .3 " " saved-2
.3 3 sibs cl sh saved all

FIGURE 21.1 Pedigree card. Use light blue ε40 stock printed with lines and punched to make 4 × 7 3/16-inch field sheets that can be cut to 4 × 6-inch cards for filing. Information is placed on the sheets as follows:

1. Photo record (black ink)
2. Special attention-getting note
3. Family number. Series 29, family 501, subdivisions, 1, 2, & 3
4. Numerical pedigree of parents
5. Planting dates
6. Genotype of parents
7. Subdivision and types of seed planted
8. Seedling progeny test (graphite pencil)
9. Handling directions
10. Field notes (pencil)
11. Pollination record (blue ink)
12. Descendants (red ink)
13. Harvest record (pencil)
14. Field Notes continued (pencil)
15. Descendants continued (red ink)

file box. They are kept up to date as additional information about the family arises. The file boxes are kept in consecutive order and individual crops are indexed for easy access.

Following this system one can go directly to the index and then to the file for a particular genotype and can follow its ancestry to the original source, or follow the descent forward to succeeding generations, and can find detailed information at any point. Every corn plant and every ear of corn has an individual identity and can be tied into this network.

This system of record keeping and data storage does not require computer access. Individual laboratories are developing computer programs for pedigree storage, label preparation, and for storage and access to field notes and harvest records.

The crucial step in this system is the preparation of a pedigree card or record at the time the seed is being prepared. **Without exception,** when seed is being prepared for planting the pertinent information for the pedigree card should be written upon the seed envelope. After the planting is made the information can be copied off the envelope onto a new pedigree card, and a permanent record is initiated.

GREENHOUSE CULTIVATION

Corn can be grown in the greenhouse by following good cultural practice that provides conditions that fall within the crop's natural requirements. Methods include groundbed, sandbench, and pot culture. Groundbeds should be fertile, well-drained soil. When grown in pots, 9–10 inch clay pots or 2-gallon plastic pots are best. With careful management, moderately vigorous corn will produce an ear even in a 5-inch clay pot. A medium of 20% soil and 80% Pro-Mix BX by Premier is recommended. Sandbenches for seedling observation should contain about 6 inches of coarse sand. The greenhouse grower must consider fertility, pH, watering, and lighting. Fertility can be maintained throughout the growing season by using a "time-release" fertilizer such as Osmocote by Sierra, and spot fertilization with Peters Soluble by W. R. Grace. Generally, lowering pH is necessary and iron sulfate is a good acidifier. Drip-irrigation watering systems are convenient; soils with high peat moss content must be kept moist. The light intensity is critical for growing corn in the winter greenhouse in northern latitudes. To grow a suitable crop in those locales requiring greenhouse protection in the winter one should have supplemental lighting. The supplement should be enough to bring the total light level to 3,000 foot-candles during normal daytime hours and extended at 2,000 foot-candles for a total of 12–14 hours; 1,000-w metal halide (metalarc) lamps provide good spectral distribution and intensity. Rectangular fixtures, which distribute light better than round ones, should be suspended by adjustable chains or ropes 5 feet (1.5 m) apart in rows 8 feet (2.4 m) apart and held at a distance of 3–4 feet (0.9–1.2 m) above the canopy of leaves. The lamps should be controlled by an automatic timer.

Lines of corn originating in the southern latitudes of the tropics generally have a short-day requirement for flowering. Lines originating in northern latitudes and the corn belt of the United States generally will flower under most day length situations.

A BRIEF LIST OF SUPPLIERS

This list is provided only as a convenient guideline. Many of the supplies listed below are available from other sources; some supplies suited for one investigator's project may be less suitable for others.

Transparent glassine paper bags for collecting pollen; miscellaneous supplies for field and greenhouse: Midco Enterprises, 145 Grand Ave., Kirkwood, MO 63122; 800/444-0983

Plastic marking tags, plastic pots, fertilizer; miscellaneous supplies for field and greenhouse: Hummert Seed Co., 2746 Chouteau Ave., St. Louis, Mo 63103; 800/325-3055

Coin envelopes for seed storage: K.C. Envelope Co., Inc., 8638 N.E. Underground Dr., Kansas City, MO 64161; 816/455-3980.

Wooden field stakes: Dayton Garden Labels, 1215 Ray Street, Dayton, OH 45404; 513/223-4650

Hand planter: Almaco, Box 296, 99 M Ave., Nevada, AZ 50201

Ear shoot bags and tassel bags: Lawson Bags, 480 Central Ave., POB 8577, Northfield, IL 60093; 800/451-1495

Canvas pollinating apron, for carrying pollinating supplies while in the field: Corn States Hybrid Service, POB 2706, Des Moines, IA 50315; 515/285-3091

Onion bags for gathering harvested ears: Kenneth Fox Supply, POB 2288, McAllen, TX 78502; 512/682-6176

Parcel hooks for attaching label to ear: Tape Products Co., 11630 Deerfield Rd., Cincinnati, OH 45242; 800/543-4930

Tags for labeling ears: Avery/Dennison, 1104 E. 40 Highway, Indeppendence, MO 64055; 816/358-4488

Waxed cardboard boxes for storing ears (freezerboard): Lawrence Paper Co., Lawrence, KS 66044; 913/843-8111

Lights for supplemental lighting in greenhouse (1,000-W metal halide): Energy Technics, POB 3423, York, PA 17402; 717/755-5642

Hand corn sheller: Decker Manufacturing Co., POB 368. Keokuk, IA 52632-0368; 319/524-3304

REFERENCES

Dicke FF (1977) The most important corn insects. In Sprague GF (ed) Corn and Corn Improvement, American Society of Agronomy, Madison, WI, pp 501–590

Larson WE, Hanway JJ (1977) Corn production. In Sprague GF (ed) Corn and Corn Improvement, American Society of Agronomy, Madison, WI, pp 625–669

Poehlman JM (1987) Breeding corn (maize). In Breeding Field Crops. Third Edition, AVI Publishing Co, Westport, CT, pp 451–507

Shurtleff MC (1981) A Compendium of Corn Diseases, The American Phytopathological Society, St. Paul, MN

Ullstrup AJ (1977) Diseases of corn. In Sprague GF (ed) Corn and Corn Improvement, American Society of Agronomy, Madison, WI, pp 391–500

Wallace HA, Bressman EN (1937) Corn and Corn Growing, Fourth Edition, John Wiley and Sons, New York, 436 pp

22

A Nine-Step Way to Characterize a Morphological Mutant

MICHAEL FREELING and JOHN FOWLER

IT is important to decide which of the numerous maize mutants are most worthy of study, and once a specific mutant has been singled out for analysis, it is important to have a logical plan for its characterization. In our laboratory we use the following strategy to analyze new mutants, so that as each mutant is characterized we can hypothesize a biological function for the wild type gene, and as more mutants are analyzed, our understanding of the biological process in which the genes act is enriched.

1. *Analyze expressivity and penetrance in various genetic backgrounds.* Find genetic backgrounds that alter the expression of the mutant allele. Complete penetrance is desirable for mapping, but suppression or altered expression provides important clues as to function and may identify interacting genes. This procedure requires crossing a plant displaying the mutant phenotype to several inbred or tester lines, sometimes for successive generations. Mutant allele dosage effects can now be determined as near-isogenic lines are constructed (i.e., in a given background, is the mutation dominant or recessive?). See if "time-to-flowering" or other quantitative traits affect penetration. If a transposon insertion is suspected, cross the mutation to lines that turn

The Maize Handbook—M. Freeling, V. Walbot, eds.

the transposon system "on" or "off." If a point mutation is suspected, check for temperature sensitivity.

Mutations displaying complex expressivity are potentially very important. For example, one might imagine that altered expression of factors involved in interpreting graded positional information would produce phenotypes sensitive to genetic background.

2. *Observe phenotypes.* Catalog and observe as many components of the mutant phenotype as possible in divergent genetic backgrounds. Observe all plant structures throughout development. Organ region identities, reflected by characteristic cells, can be most easily distinguished on epidermal surfaces using SEM. Imagine different models that could explain the phenotypes.
3. *Cross to other, similar mutant alleles* to discover interactions. Double-mutant phenotypes, particularly those involving two recessives, are crucial to all models postulating a sequence of events. If all similar mutants are introgressed into a few inbreds, problems of expressivity can be controlled.
4. *Map* by an efficient method: B-A translocations, *wx*-marked reciprocal translocations, or *r-x1*-induced monosomy. If there are penetrance problems, consider using RFLP techniques rather than traditional genetic methods in order to preserve genetic backgrounds. Mapping using recombinant inbred lines is also possible if linkage to a unique sequence usable as a probe can be established. Follow with a three-point cross. One should begin this step and the previous three steps in the first year.
5. *Use aneuploids* for dosage analysis. If the mutant is dominant, decide which type (neomorph, antimorph, etc.) using aneuploids generated with B-A translocations or multiple chromosomal interchanges. This same approach can be used to test whether a recessive allele is a hypomorph or an amorph. Aneuploids may also suggest the presence of interacting genes on a given chromosomal segment whose dosage affects the mutant phenotype.
6. *Identify the site of action* of any dominant allele's function using precise tissue-layer genetic mosaics. X-rays, ring chromosomes, or chromosome-breaker transposons (i.e., double *Ds*) may be used to induce hemizygous sectors that concomitantly lose the region of chromosome carrying the dominant allele and uncover a linked marker on the homolog. This procedure also tests for organ, organ region, tissue, and cell autonomy. In situ localization of gene products is not an adequate substitute.

It is important to carefully choose the recessive, cell-autonomous marker allele that will indicate hemizygosity for a particular region of chromosome in the organ desired. It is best that the marker lie close to and proximal to the mutant allele being studied, so all breaks that uncover the marker will also uncover the recessive allele of the gene being analyzed. Inversions and translocations can be used to move a particular locus closer to or distal to the marker.

7. *Attempt Experimental Manipulations* on developing plants. Techniques such as incisions, laser ablations, and physical or chemical blocks or inductions should be at-

tempted in order to interrupt or change possible signal-receptor interactions or to create a phenocopy of the mutant. Somatic sectors generated in the mosaic analysis (step 6) may also interrupt cell-cell communication, resulting in phenotypes that may provide clues about whole tissue or organ communication processes.

8. *Clone the Gene* and characterize its normal and mutant products at the molecular level. Transposon tagging is tried and true, and our laboratory has had particular success with the *Mutator* transposons. In addition, it is possible to clone genes based on their homology to a conserved structural motif (e.g., a homeobox.) Molecular walking may also prove useful should functional genes be clustered and as the RFLP maps continue to improve. The identity of cloned sequences must be proved genetically using revertants of the mutation or additional alleles of the gene *or* by complementation of the phenotype by transformation.
9. *Generate complete null alleles.* In diploid plants, it is difficult to prove that any allele is completely null, as deletions are hard to obtain. The monoploid gametophytes filter out all but the smallest deletions, because larger deletions tend to lack genes necessary for gametophyte function or development. One practical method to circumvent this problem is to obtain transposon insertions into an allele as soon as possible; later, transposon excision events will generate many nulls from imprecise events that remove parts of the gene. Such derivatives can be confirmed as nulls by molecular analysis (step 8).

Keep in mind that the knockout phenotype may be either germless (embryo defective) or a defective kernel. In cases where the homozygous null phenotype is lethal, the recessive allele can be uncovered in hemizygous sectors to reveal the phenotype in a specific organ. If the null cannot be passed through either gametophyte, it is possible to construct segmentally disomic stocks to cover the lethal mutation in the haploid cells after the mutagenesis occurs.

At this point the gene's product can be localized within the plant using specific nucleic acid and antibody probes. Interactions with other gene products and mutants can be analyzed similarly. In addition, screens for revertants, suppressors, and modifiers of the original mutation can be initiated to identify other genes that are potentially relevant to the process being studied. These nine steps provide a thorough strategy that integrates phenotypic observation with molecular description. This builds a strong foundation of data from which one can begin to understand the biological meaning of a gene and its product.

23

Mutagenesis

M. G. NEUFFER

THE production of heritable changes in the maize genome is an important part of maize genetic research. Chromosome breaks, deficiencies, duplications, translocations, inversions, gene mutations, and transposon insertions all have their place. There are unique and efficient ways to produce each, as described in Stadler and Roman (1948), Sax (1957), Nuffer (1957), Sparrow (1961), and McClintock (1951). Special attention should be given Stadler's (1946) descriptions of methods of studying mutation; the implications of seed vs. pollen treatment in relation to the ontogeny of the germ cells are discussed by several authors (Coe and Neuffer 1978; Johri and Coe 1983; Coe et al. 1988). Reviews of mutagenic agents, techniques, applications, and consequences are presented in Bird and Neuffer (1985) and Neuffer and Chang (1989).

Potential uses, methodology, and problems of mutagenesis are too diverse and sophisticated to detail here; therefore, only the rationale and protocols of chemical mutagenesis, the most efficient current method of producing gene mutations in maize, are described. For brief protocols for X-ray, UV treatment, and transposon insertion, see Neuffer (1982). A more current description of transposon insertion is discussed elsewhere in this volume.

When conducting mutagenesis experiments it is important to consider the characteristics of the organism being treated. Seed treatment is generally not efficient for corn because the mature kernel has separate male and female germ cell primordia. Because of this a recessive mutant produced in the embryo would not be expressed until the M_3, at which time it will have been replicated many times, thus adding to the difficulty of analysis. On the other hand, pollen treatment allows gametes to be treated individually, and new recessive mutants may be observed to segregate unambiguously in the M_2. Protocols for both pollen and seed treatment are presented below, however, because under certain conditions seed treatment may be useful.

The Maize Handbook—M. Freeling, V. Walbot, eds.

PARAFFIN OIL POLLEN TREATMENT USING ETHYL METHANESULFONATE (EMS)

The chemical ethyl methanesulfonate is the most efficient mutagen known for maize, and the treatment of pollen in paraffin oil with EMS is the most efficient method. The following are step-by-step considerations to follow in paraffin oil treatment. (See Figure 23.1).

STOCKS AND GENERAL HANDLING

If the experiment calls for improving an inbred line or the production of mutants in a carefully controlled background, use a good inbred line. Use seed from a selfed ear that has been observed to be free from variations. If background variation is not critical, much better material can be obtained by using one inbred (a good ear producer) for ear stock and another (a good pollen shedder) for pollen stock. The M_1 will have the advantage of hybrid vigor, so more mutants will survive for observation. Again, both parents should be from selfed seed.

In handling, have an excess of plants available for treatment so only fresh material is used. One hundred ears from treatment should produce 3,000–10,000 M_1 kernels. Cut back fresh silks the night before pollination and at the same time select tassels for pollen. Strip off all old anthers, and bag tassels. Remove all pollen contaminant sources.

TREATMENT SOLUTION

In a fume hood, prepare a stock solution of 1 ml EMS (Eastman #7830) in 100 ml of light domestic paraffin oil (Fisher 0121-1 or equivalent). Keep at room temperature—stir vigorously for at least 1 hour. This stock solution can be kept indefinitely. The morning of treatment stir the stock solution vigorously (1 hour) and prepare a treatment solution of 1 part stock solution and 15 parts paraffin oil. Stir this vigorously also.

TREATMENT

At the proper time collect fresh pollen free of anthers and mix less than 1 part pollen to 10 parts treatment solution in a Nalgene bottle with a close-fitting cap. Shake immediately before and after mixing and at 3–5-minute intervals for at least 45 minutes. Spread pollen in oil suspension on good silks with a #10 camel-hair brush. Use a manageable amount of mixture for each ear, being sure to stir pollen for each brushfull for each ear. Pollination may continue for up to 75 minutes.

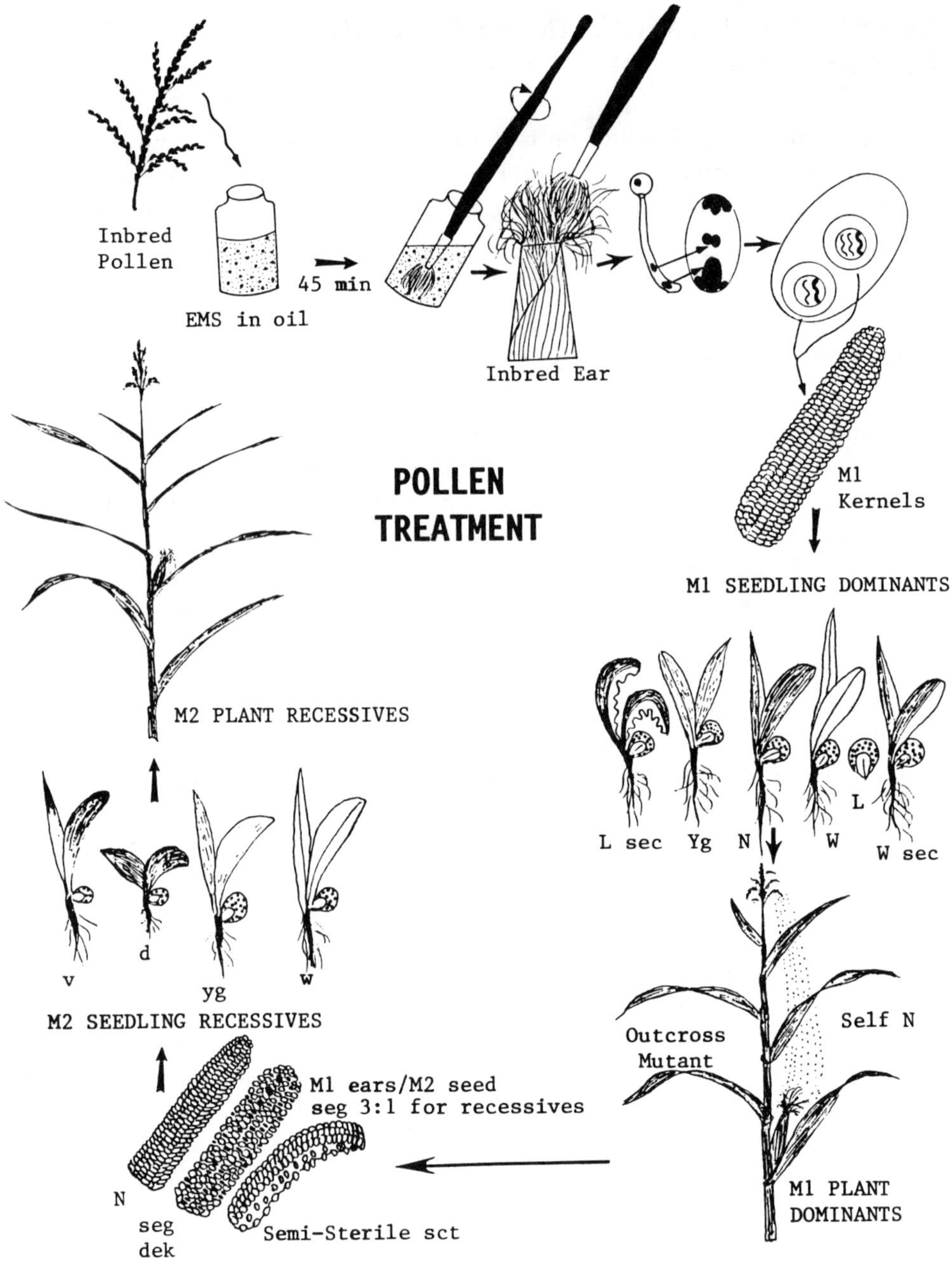

FIGURE 23.1 A representation of the essential features of the paraffin oil method for treating maize pollen with ethyl methanesulfonate, showing consecutively the handling of treatment solutions and subjects, the gametophyte nuclei involved, the appearance of dominant mutants (both whole and chimeric) in the M_1, the handling of the M_1, and the appearance of segregating recessive kernels, seedlings, and whole plants in the M_2. (Drawing by M.T. Chang.)

PRECAUTIONS

Use special precautions (protective clothing, glasses, etc.) to protect handlers, and keep untrained persons away from the field. EMS is a dangerous carcinogen.

HARVEST

Ears and kernels are safe for handling after 30 days in field weather and sunlight. Harvest using standard procedures.

HANDLING M_1

M_1 kernels will have one treated (from male) and one untreated (from female) genome. Examine ears for dominant kernel mutants; these will not transmit the mutant because the embryo and the endosperm are nonconcordant; however, note their type and frequency. Plant M_1 kernels and examine for dominant seedling mutants. These will be whole seedling cases such as: failure to germinate (dominant lethals); loss of chlorophyll (yellow-green, white); and morphology and stature changes. Also look for half-seedling sectored cases for the same traits; these will occur at the same frequency as whole seedling cases. Observe plants at regular intervals for dominant traits—both whole plant and half-plant sectored cases. Tag all unusual plants for later identification. For dominant cases, self if possible and outcross on to a normal standard. If selfing is not feasible, outcross by normal. Use the mutant part of the tassel of sectored plants for crosses. Self all normal plants to produce M_1 ears with M_2 seed.

HANDLING M_2

Examine ears for kernel mutants; these should segregate 3 normal to 1 mutant (except for rare dominant kernel mutants that will segregate 3 mutants to 1 normal). Also look for sectored ears for mutants; these may have a much-reduced frequency of transmission, depending upon the size and distribution of the sector in the reproductive organs. These ears will also segregate for gamete and kernel abortion (semisterile ears, both whole ear and sectored; the latter ears are usually bent toward the sectored side). Plant a 20–40-kernel M_2 (depending on how rigorous the test should be) from each ear in a sandbench and observe for seedling mutants. These should segregate 3 normal to 1 mutant (except for sectorals that will segregate $>3:1$). Plant a 20-kernel sample from the dominant outcrosses to confirm the mutant. Valid dominant mutants should appear as one normal to one mutant segregant in the outcross. If mature plant traits are desired, plant 20 normal M_2 kernels from each ear in the field. Cross mutant plants on a standard stock for F_1 and subsequent F_2 to prove heritability. For an M_2 segregating for lethal mutants, self three phenotypically normal sibs, two-thirds of which should be heterozygotes, to test transmission to the next generation.

EXPECTED FREQUENCIES

With successful treatment, the frequency of recessives should exceed 1 recessive mutant per locus per 1,000 M_1 produced for dominants, and 1 dominant per 200 recessive mutants (0.5×10^{-6} with a range of 1×10^{-3} to 0×10^{-6}). A population of 3,000–5,000 M_1 tested in the M_2 should produce all the recessive mutants conceivably needed while 100,000–300,000 M_1 plants are needed to get most of the dominant mutants.

RATIONALE FOR SEED TREATMENT

Under some circumstances seed treatment may be advantageous for maize. When seeking dominant kernel and seedling mutants or when screening for recessive mutants that can be subject to mass screening techniques, a minimal sample method (Redei 1974; Redei et al. 1984) may make seed treatment practical. If dry seed is treated with a chemical mutagen before germination the following logic applies.

Each kernel contains approximately 4 separate meristem cells for the tassel and 2–4 cells for the ear. Assuming 4 each, then these 8 cells will carry 16 genomes. Treatment of 10,000 kernels will affect 160,000 genomes and, assuming a 50% mutagenesis rate, will produce 80,000 recessive and 400 dominant mutants. Planting and selfing the resulting M_1 plants will produce ca. 10,000 ears with the above-stated mutants present in heterozygotes but unexpressed except for dominant kernel mutants.

Dominant mutants may be identified by observation (kernel mutants) and by planting and screening as seedlings or plants. Since each kernel will represent 2 genomes and one-eighth of the genomes available, a 107-kernel sample will be required from each ear to save 99.9% of the mutants (22 kernels for 95%). In practice, screening a 100-kernel sample from 10,000 ears will save almost all mutants produced. This number (1×10^6) is feasible for some seedling screens but probably not for mature plant traits.

Recessive mutants may be identified using the minimum sample method. A one-kernel sample each from 10,000 selfed ears, when planted and selfed, will test 20,000 genomes and should express 10,000 recessive mutants or roughly 10 mutants for each known locus in the genome, each of which will be a completely independent event. A repeated sample of the M_1 would produce an additional large number of new mutants, only one-eighth of which should be duplicates of the previous sample. These would be at the cost of 10,000 M_1 selfs and 10,000 single sample M_2s for 20,000 total plants to produce 10,000 mutants for a mutant-producing efficiency of 0.50; this is comparable to pollen treatment.

A labor-saving procedure would be to plant the treated seed. Allow the plants pro-

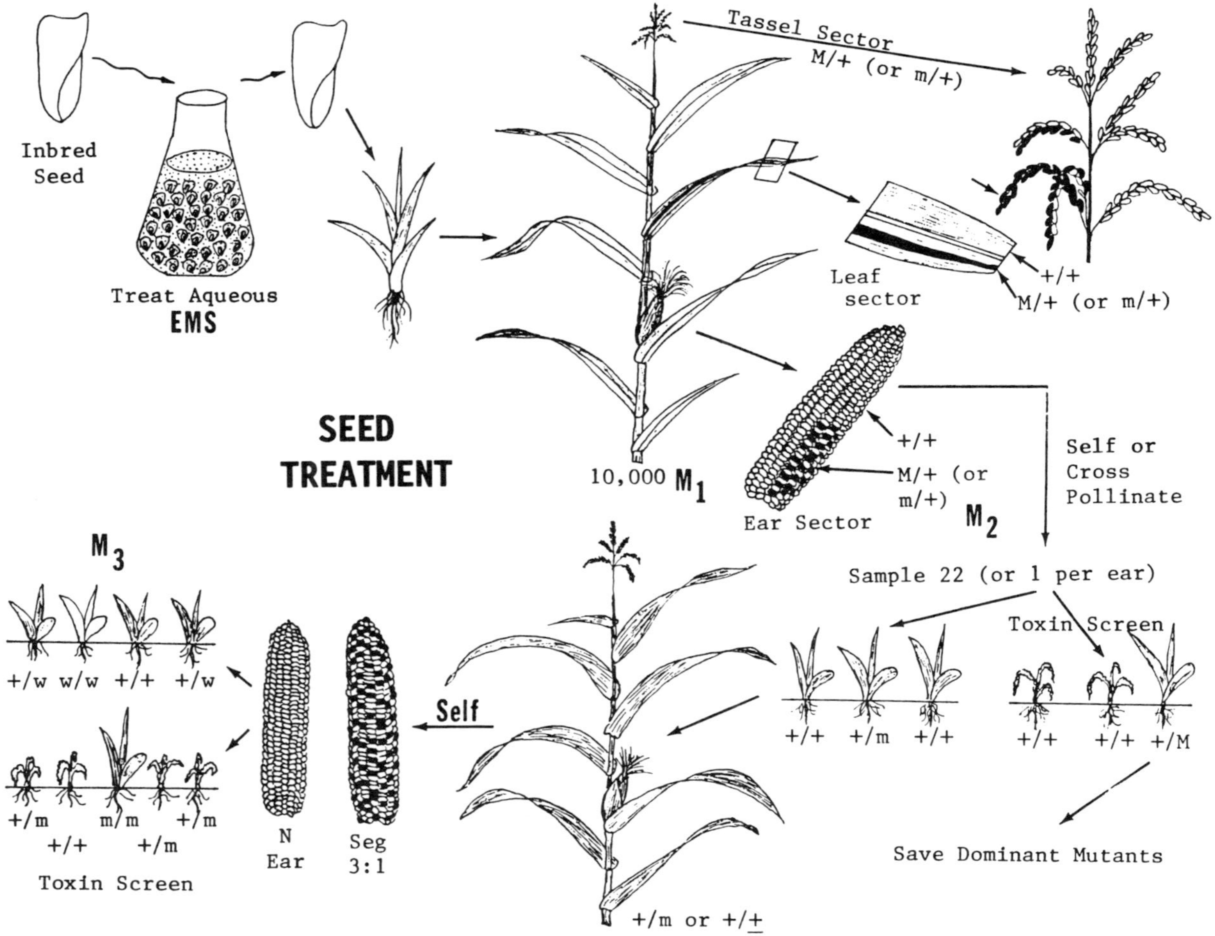

FIGURE 23.2 A representation of the essential features of the seed method for treating maize with ethyl methanesulfonate, showing consecutively the handling of treatment, the appearance of dominant chimeras on the M_1 plant (also reflects the presence of the nonvisible recessive mutants produced), the handling of the M_2 to most effectively screen, sample, and save both recessive and dominant mutants, and the handling of the M_3 to effectively identify useful mutants. (Drawing by M.T. Chang.)

duced to open pollinate; harvest and shell by machine; mix to randomize sampling; and take repeated single-seed equivalent samples to screen for the dominant mutants; or, use a single-kernel sample to grow, self, and test for recessive lethals.

DRY SEED TREATMENT USING EMS

The following protocol (see Figure 23.2) should produce the results indicated.

1. Select a good inbred stock. Use dry seed. Soak for 4–8 hours in an 0.075 M aqueous solution of EMS at 20° C. Wash 4 hours with fresh running water. Plant kernels directly into soil using gloves and protective clothing.
2. Observe seedlings through the five-leaf stage for effectiveness of treatment and for dominant chimeras, which will appear after the six-leaf stage.
3. Self M_1 plants for maximum number of tested gametes, or cross by normal to produce half the number of tested gametes but in a cleaner background.
4. Plant the M_2 seed to screen seedlings for dominant mutants; expect frequency of 2×10^{-5} per treated germ cell genome.
5. Self M_2 and M_3 seed; examine directly for kernel mutants (alternatively, open pollinate and machine harvest).
6. Plant 20-seed M_3 samples of each M_2 ear to screen for recessive seedling and plant mutants. Mutants will segregate 3:1 and will occur at a frequency of 1/1,000 per locus (alternatively, single seed samples from randomized bulk M_3 at this rate = approximately 1 kernel per ear).

REFERENCES

Bird RM, Neuffer MG (1985) Odd new dominant mutations affecting the development of the maize leaf. In Freeling M (ed) Plant Genetics, Alan R Liss, New York, pp 818–822

Coe EH, Jr, Neuffer MG (1978) Embryo cells and their destinies in the corn plant. In Subtelny S, Sussex IM (eds) The Clonal Basis of Development, Academic Press, New York, pp 113–129

Coe EH Jr, Neuffer MG, Hoisington DA (1988) The genetics of corn. In Sprague GF (ed) Corn and Corn Improvement, Third Edition, Am Soc Agronomy, Madison, WI, pp 81–258

Johri MM, Coe EH, Jr (1983) Clonal analysis of corn plant development. I. The development of the tassel and the ear shoot. Dev Biol 97: 154–172

McClintock B (1951) Chromosome organization and genic expression. Cold Spring Harbor Symp Quant Biol 16: 13–47

Neuffer MG (1982) Mutant induction in maize. In Sheridan WF (ed) Maize for Biological Research, Plant Mol Biol Assoc, Charlottesville, VA, pp 61–64

Neuffer MG, Chang MT (1989) Induced mutations in biological and agronomic research. Science for Plant Breeding. Proc. of the XII Congress of EUCARPIA, Gottingen, Germany 16: 165–178

Nuffer MG (1957) Additional evidence on the effect of X-ray and ultraviolet radiation on mutation in maize. Genetics 42: 273–282

Rédei GP (1974) Economy in mutation experiments. Z. Pflanzenzüchtg. 73: 87-96

Rédei GP, Acedo GN, Sandhu SS (1984) Mutation induction and detection in *Arabidopsis*. In Chu Ehy, Generoso WM (eds) Mutation, Cancer and Malformation, Plenum Pub. Corp., pp 285-313

Sax K (1957) The effect of ionizing radiation on chromosomes. Quart Rev Biol 32: 15–26

Sparrow AH (1961) Types of ionizing radiation and their cytogenetic effects. Symp on Mutation and Plant Breeding No. 891: 55–112. Natl Res Council, Washington, DC

Stadler LJ (1946) Spontaneous mutation at the *R* locus in maize. I. The aleurone-color and plant-color effects. Genetics 31: 377–394

Stadler LJ, Roman H (1948) The effect of X-rays upon mutation of the gene *A* in maize. Genetics 33: 273–303

24

Gene Tagging with *Ac*/*Ds* Elements in Maize

STEPHEN L. DELLAPORTA AND MARIA A. MORENO

Activator (*Ac*) and *Ds* (*Dissociation*) comprise a family of maize transposable elements. *Ac* is the autonomous member of the family, capable of producing a transposase factor needed for mobility. *Ds* elements are nonautonomous and capable of transposition only when transactivated by *Ac* (reviewed by Fedoroff 1989). The mobility of these elements provides a useful molecular genetic tool for gene tagging in maize.

Two strategies for gene tagging using the *Ac*/*Ds* transposable element system of maize are discussed. These strategies are nontargeted methods and rely on the ability to select kernel progeny carrying independent transposed *Ac* or *Ds* elements. The first strategy employs an *excision assay*—the donor *Ac* or *Ds* element, resident at a known locus, excises and is subsequently recovered elsewhere in the genome. Transposition events are detected as somatic sectors or germinal events whereby gene action at the donor locus is restored. The second strategy is based on a unique feature of *Ac*—transposition is developmentally delayed as the number of *Ac* elements increases in the maize genome. Transpositions are detected by an increase in *Ac* copy number using this method. With

The Maize Handbook—M. Freeling, V. Walbot, eds.

both methods, the ability to tag genes is based on recovery of single transposition events of *Ac* or *Ds* elements. Transposed elements are recovered in heterozygous condition in T_1 kernel progeny. T_1 plants are field grown, screened for dominant mutations, and self-pollinated to uncover recessive mutations in T_2 families.

Both tagging methods are essentially based on single element mutagenesis strategies. The advantage of using a single element for mutagenesis is that induced mutations can be readily cloned because the element causing the mutation is known and it may be the only element of its kind in the genome. (See also the chapter by Chandler on the distribution of *Ac* and *Ds* in the maize genome.) Its disadvantage is the low frequency of mutations since, in theory, only one element is participating in the mutational process. Also, because the mutation frequency is low, the background of mutations not caused by the donor element is relatively high, thus interfering with tagging efforts. Yet, as discussed below, these disadvantages can be partially offset by simple genetic tests and by a mutagenesis program whereby large numbers of transposition events are examined for mutations. In our laboratory we have used these nontargeted strategies to screen over 10,000 chromosomes carrying transposed *Ac* elements and have recovered and cloned many *Ac*-induced mutations.

THE TRANSPOSITION PROCESS AND GENE TAGGING CONSIDERATIONS

Studies on the transposition process of *Ac* show that transposition is a two-step process: *Ac* excises from a donor locus after DNA replication and inserts into a target site (Figure 24.1). The target site may (Figure 24.1B) or may not (Figure 24.1A) be replicated at the time of integration. As shown in Figure 24.1, after cell division twinned lineages are formed—one lineage retains *Ac* at the donor site in addition to the transposed *Ac* element (tr-*Ac*) while its sister lineage loses *Ac* from the donor locus. (see Chen et al. 1987, 1991 for further details). Whether or not the excision lineage receives a tr-*Ac* element will depend on the replication status of the target site. For purposes of gene tagging, only those cell lineages carrying a transposed *Ac* element are important. The two strategies outlined below are designed to recover progeny derived from cell lineages containing a tr-*Ac* using genetic markers. For purposes of simplification, we outline the strategies for recovering transposed *Ac* elements from the *P* locus. In practice, these strategies can be applied to any *Ac* inserted at other chromosomal locations with minor modifications.

STRATEGY I: RECOVERING TRANSPOSED *Ac* OR *Ds* ELEMENTS BY AN EXCISION ASSAY

Transposition involves excision of *Ac* and its reintegration into a new chromosomal target site. Excision of *Ac* from a donor locus is a reliable indication that a transposition

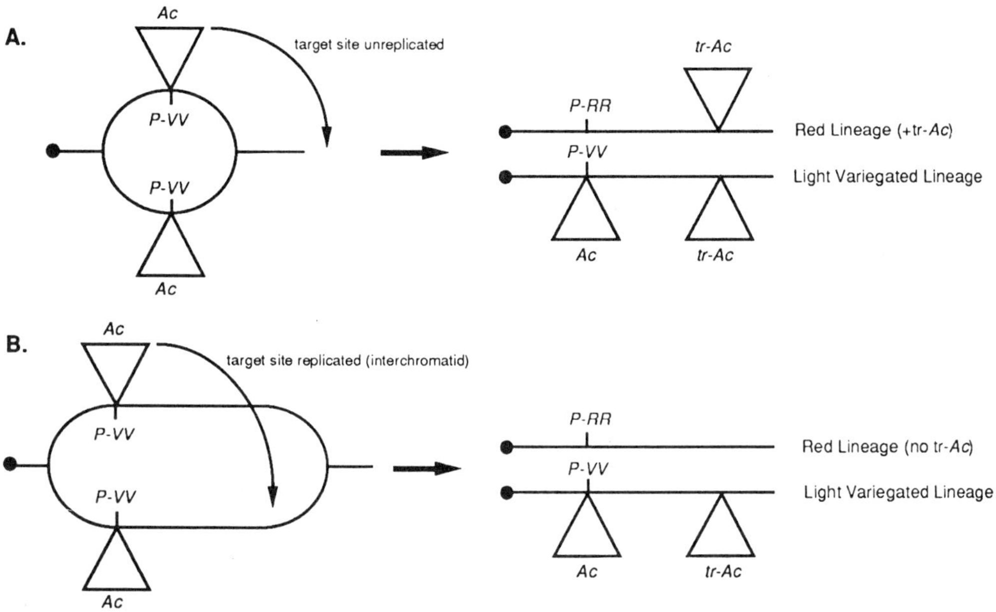

FIGURE 24.1 Mechanism of Ac transposition. **A.** Transposition from replicated donor site to unreplicated target site. **B.** Transposition from replicated donor site to replicated target site.

event has occurred. As shown in Figure 24.1A, however, excision of *Ac* does not mean that a cell lineage will actually carry the transposed element. How often *Ac* is retained in both twinned lineages after transposition has been estimated from genetic studies of *Ac* transposition from the *P-VV* allele. *P-VV* is an *Ac*-induced mutation of the *P* gene, a regulatory gene involved in the production of visible flavonoid pigments in the pericarp. The *P-VV* allele conditions a variegated pericarp phenotype (Figure 24.2A*), consisting of numerous red pericarp stripes on an otherwise colorless pericarp background. Each red stripe in the pericarp is a clonal sector derived from an excision of *Ac* from the *P-VV* allele that restores *P* action. (The *Ac* element is inserted into an intron of the *P* gene so even imprecise excision of *Ac* almost always restores *P* gene action.) Hence, a red pericarp stripe indicates that *Ac* has excised from the *P* gene. If a red sector contains *Ac* activity, a tr-*Ac* element should also be present elsewhere in the genome. Greenblatt and Brink (1962) and Greenblatt (1984) have estimated the frequency of red sectors containing *Ac* activity to be two-thirds of all red sectors. From these data, we can assume a 66% efficiency of recovering a tr-*Ac* element in a lineage using excision as the sole indicator of transposition. These tr-*Ac* elements in red pericarp sectors can be recovered in kernel offspring because the pericarp cells share a common lineage with the underlying megaspore mother cell which, after meiosis and fertilization, produces the egg sac of the kernel. There is a high probability of genetic correspondence between the somatic peri-

*Figure 24.2 is on page 761 in black and white, and on the back cover of the book in color.

carp and the megaspore mother cell when a red pericarp sector covers all or most of the kernel surface. Therefore, choosing kernels with a uniform red pericarp should greatly enrich for transposed elements in kernel progeny. In summary, tr-*Ac* or tr-*Ds* elements can be recovered by selecting kernels in which *Ac* or *Ds* elements have excised from their original location.

The intervening meiosis between transposition and kernel fertilization is a second important factor that will reduce the efficiency of *Ac* transmission to underlying kernel progeny. The pericarp tissue is somatic diploid tissue derived from the carpels. A newly transposed *Ac* element in the sporophyte will be heterozygous. During gamete formation, meiosis will result in only one of two homologs being transmitted to the gametophyte. The efficiency of *Ac* transmission to any kernel will, therefore, be further reduced by 50%. The overall efficiency of recovering tr-*Ac* in any one kernel underlying a red sector is approximately 0.66×0.5 or 33%. Therefore, on average, only one-third of all offspring derived from a somatic excision sector will carry a transposed *Ac* element.

The efficiency of recovering tr-*Ac* can be increased to nearly 100% by incorporating genetic markers that detect the presence or absence of *Ac* activity in kernel progeny. This activity is monitored using *Ds* reporter alleles. A *Ds*-induced mutation is stable in the absence of *Ac* activity and unstable in the presence of *Ac* activity regardless of the location of *Ac* in the genome. The presence of *Ac* can be determined in a single kernel using the appropriate *Ds*-induced mutation.

Consider that the endosperm and embryo share a common genetic history. During embryo sac formation, mitotic divisions of a single haploid spore (the product of meiosis) give rise to both the single egg nucleus and two polar nuclei. Except for rare mutational events during these mitotic divisions, these nuclei are genetic clones of one another. Double fertilization of the proembryo nucleus and two polar nuclei by haploid sperm nuclei leads to formation of the diploid embryo and triploid endosperm. Likewise, the sperm nuclei participating in double fertilization are mitotic descendants of a single haploid spore and therefore also genetic clones. Except for differences in ploidy and occasional heterofertilization, the genotype of the endosperm should accurately reflect the genotype of the embryo. The endosperm fully differentiates during kernel maturation with the outer cell layer forming the aleurone; this tissue is capable of anthocyanin production. For gene tagging purposes, *Ds*-induced mutations in anthocyanin production in the aleurone can be used to monitor *Ac* activity in the kernel. Several *Ds*-induced mutations are available for this purpose. We have employed the *Ds*-induced *R-sc* mutation, *r-sc:m3* , for gene tagging. The *R-sc* allele fully pigments the aleurone. The *Ds*-induced *r-sc:m3* mutation (Kermicle 1984) conditions a completely colorless aleurone in the absence of *Ac* activity and a variegated (spotted) aleurone in the presence of *Ac* activity. In a red pericarp lineage where *Ac* has left the *P* locus, an underlying kernel carrying the *r-sc:m3* allele will show a spotted aleurone only if a tr-*Ac* element is transmitted to the kernel.

In practice, on a medium variegated ear (with one copy of the *Ac*-containing *P* allele in the pericarp: *P-VV/P-WR r-g/r-g*) fertilized with *P-WR r-sc:m3* pollen (Figure 24.2B), red pericarp kernels with spotted aleurones should represent those progeny that received a tr-*Ac* element (Figure 24.2C). In summary, kernels carrying a transposed *Ac* element can be identified by excision of the donor *Ac* (red pericarp) and by detection of its presence elsewhere in the genome using a *Ds* reporter allele. A typical series of crosses for gene tagging using this excision assay is outlined in Figure 24.3.

For gene tagging purposes, a large number of independent transpositions are desirable. In a large clonal red sector, each kernel with a spotted aleurone should carry the same tr-*Ac* element (except for occasional secondary transpositions). Therefore, only a single kernel from each red sector is needed to represent an independent transposition event in the T_1 population. Noncontiguous red sectors represent independent transpositions. As a result of selecting a single kernel with spotted aleurone from noncontiguous red sectors, each kernel selection should carry an independent tr-*Ac* element in its ge-

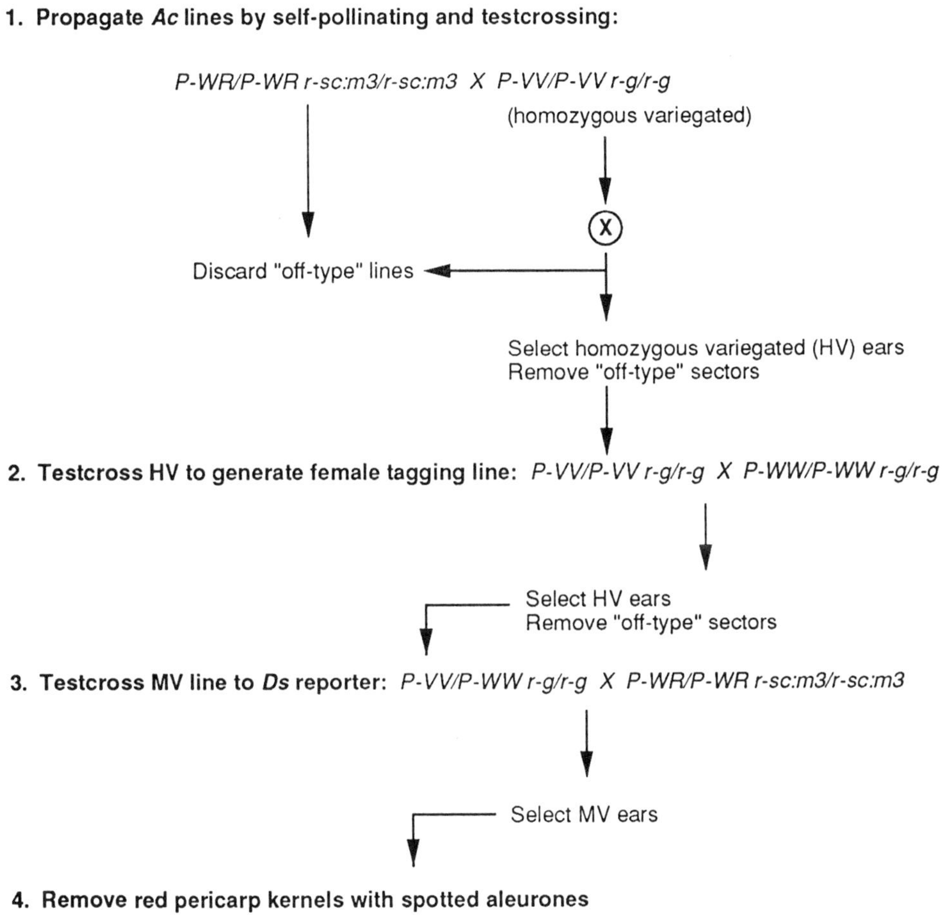

FIGURE 24.3 Strategy I: Selecting tr-*Ac* by Excision and *Ds* Transactivation

nome. In our experience, each medium variegated ear, with about 500 kernels, will carry about 10 independent red sectors that cover the surface of at least one; approximately 66% of the multikernel sectors will have at least one spotted aleurone kernel within them. (Spotted aleurone kernels are less frequent in small red sectors because even though the sector may contain *Ac*, there is only a 50% chance, on average, of any one kernel receiving the element after meiosis.) We find about 3–5 independent red pericarp kernels with spotted aleurones on a typical medium variegated ear. Each one of these kernels potentially contains an *Ac*-induced mutation.

By choosing the appropriate donor *Ac* and *Ds* reporter alleles, gene tagging studies are possible with *Ac* elements on several chromosome arms of maize. This is an important consideration, because transpositions of *Ac* tend to be to linked chromosomal sites. This tendency, however, does not preclude transposition to unlinked sites and to other chromosomes. In addition, when using *Ds* as a gene tag only an excision assay is possible. Because *Ds* possesses no genetic activity, there is no genetic means of identifying a tr-*Ds* element in the genome. Under these circumstances, kernels or plants where *Ds* has excised from a donor locus are selected and self-pollinated. Extrapolating from *Ac* transposition studies, approximately 33% of these selections should contain a tr-*Ds* element.

STRATEGY II: RECOVERING *Ac* TRANSPOSITIONS BY CHANGES IN *Ac* DOSAGE

Strategy II relies on the ability to recover tr-*Ac* elements by an increase in *Ac* copy number. A typical series of crosses for gene tagging using a dosage suppression assay is outlined in Figure 24.4. T_1 kernels are selected from crosses between homozygous variegated females (genotype: *P-VV/P-VV r-g/r-g*) and a male *Ds* tester line (*P-WR/P-WR r-sc:m3/r-sc:m3*). In this cross, the female parent transmits a *P-VV* chromosome to most kernel progeny (endosperm genotype: *P-VV/P-VV/P-WR r-g/r-g/r-sc:m3*) and shows the expected spotted aleurone phenotype (Figure 24.2D). However, transpositions that result in additional copies of *Ac* occur when both the donor and transposed *Ac* elements are transmitted to the kernel offspring (outlined in Figure 24.1). Kernels with both donor and tr-*Ac* elements should possess four copies of *Ac* in the endosperm—two maternal copies of *P-VV* and two maternal copies of the tr-*Ac* (endosperm genotype: *P-VV/P-VV/P-WR tr-Ac/tr-Ac/- r-g/r-g/r-sc:m3*). The increase in *Ac* copy number from two to four copies causes a considerable delay in *Ds* transactivation at *r-sc:m3*. This results in an aleurone phenotype that is nearly colorless or very finely spotted; this fine spotting is usually visible under low magnification. (Kernel selections are shown in Figure 24.2E.) Note that the colorless aleurone phenotype is also expected if the female transmits a *P-RR* homolog lacking *Ac* elsewhere in the genome. Nevertheless, this dosage suppression strategy can be used to recover transposed *Ac* elements from any homozygous *Ac* line. It is not necessary that the donor *Ac* condition a visible phenotype, although this helps considerably to minimize "off-type" progeny, as discussed below.

Strategy II will yield more false positives (kernels lacking a tr-*Ac* element) for several

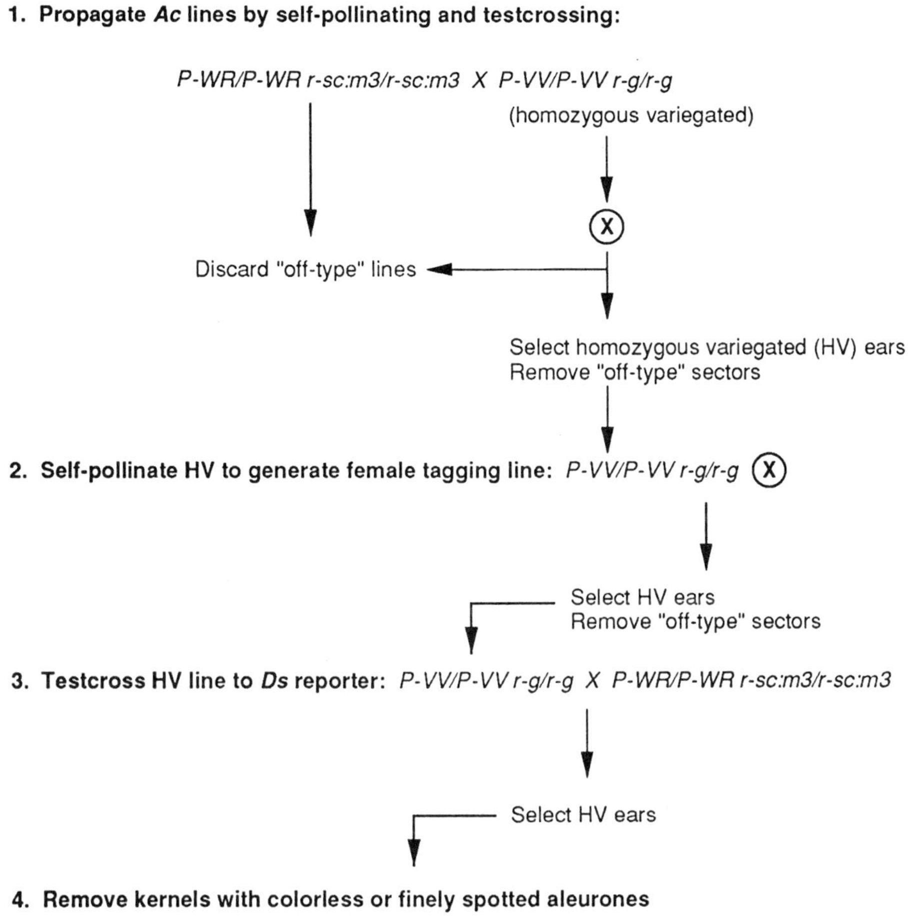

FIGURE 24.4 Strategy II: Selecting tr-*Ac* by Dosage Suppression

reasons. First, unlike strategy I kernel selections, where both the somatic pericarp and the underlying kernel phenotypes are scored, strategy II relies on detecting transpositions based solely on the appearance of the near colorless aleurone phenotype. These kernel types will contain progeny without *Ac* as well as the kernels with a high *Ac* dosage. Moreover, because transposition is frequent during the mitotic divisions leading to embryo sac formation (S. Dellaporta, unpublished results), a significant fraction of strategy II progeny will have an embryo and endosperm that are genetically nonidentical—i.e., the endosperm phenotype indicates a high *Ac* dose but the embryo contains only *P-VV* or *P-RR*. Essentially the endosperm and embryo are twinned lineages! Transposition during this same developmental period is less problematic with strategy I. Strategy I offspring are derived from a red lineage pericarp; i.e., they occurred during ear formation.

Even though secondary transpositions of the tr-*Ac* may occur during this period, there is still a high probability that the embryo will carry the original tr-*Ac* element or the secondary tr-*Ac* or both. In summary, expect a significantly higher frequency of progeny from strategy II that lack a tr-*Ac* element and, therefore, plan experiments accordingly.

Even though strategy II has its disadvantages there are some novel advantages worth mentioning. First, with minor modifications this strategy can be adapted to any donor *Ac* element regardless of its position—even *Ac* elements that do not cause a visible mutation. The only requirement is that *Ac* dosage in the kernel progeny be able to be monitored using a *Ds*-induced reporter allele. This can be especially important when trying to use donor *Ac* elements at various chromosomal sites in order to saturate specific regions of the maize genome. A second advantage is that because a high *Ac* dosage is maintained, somatic and germinal excision is minimal, which should tend to stabilize the mutation. This may be an important factor when the mutation is cell nonautonomous—expression due to somatic excision may result in correction of the phenotype or weakened expressivity and penetrance.

PRECAUTIONS

Logistically, hundreds of independent transpositions can be recovered in a single generation using either gene-tagging strategy. However, several precautions are necessary especially when propagating *Ac* lines and generating crosses for gene tagging. In addition to the standard precautions used in any genetic experiment (e.g., minimizing pollen contamination), several special precautions should be taken when gene tagging with *Ac*. Most importantly, crosses need to be carefully examined to assure the desired genotype is maintained in the stock. This is an especially serious problem with transposon stocks because these lines are inherently unstable and often the *Ac* element will transpose unpredictably. The precautions outlined below will maximize the frequency of the desired final product—an *Ac*-induced mutation—and minimize undesirable results. These precautions will make the difference between success and failure.

Propagating *Ac* Lines

A line containing an *Ac* element is inherently unstable. To maximize the efficiency of recovering a tr-*Ac* element, it is important to minimize transposition events during the propagation of *Ac* lines and the generation of tagging lines. Unwanted transpositions will contribute to off-type progeny and ears. The frequency of these off-types can be minimized but not eliminated. Start by selfing the homozygous *Ac* line to generate stocks. Test the genotype of each plant used for stock by self-pollinating and testcrossing as a male to *r-sc:m3* females (Figures 24.3 and 24.4, line 1). This testcross should yield >90% of the kernels with a coarse spotted aleurone phenotype (endosperm genotype = *P-WR/P-WR/P-VV r-sc:m3/r-sc:m3/r-g*). The occasional colorless or fine spotted aleurone kernels on these ears represent transposition events. For instance, when propagating the *P*-

VV line, retain the self-pollinated ear only when two conditions are met: (1) the pericarp phenotype is homozygous variegated (*P-VV/P-VV*) on the selfed ear and (2) >90% of the testcrossed progeny show the appropriate coarse spotted aleurone. Expect a significant fraction (<10%) of the selfed plants to fail this test and be discarded. This same precaution applies to other *Ac* lines as well—always testcross to determine the *Ac* genotype in stock plants.

Remove Off-type Somatic Sectors

On homozygous variegated ears, all kernels underlying off-type somatic sectors should be removed. The occasional red, light-variegated, and very light variegated sectors on these ears will contribute to off-type progeny and must be eliminated. It takes some practice and experience to distinguish these phenotypes, especially in small sectors, but having a set of standard ears of known genotype for comparison will help considerably. Eliminating the large off-type sectors from homozygous variegated ears will maximize the number of progeny kernels with the desired *P-VV/P-VV* genotype. Even when taking these precautions, a certain frequency of off-type progeny is unavoidable. When propagating the homozygous *P-VV* line, each time the plant is selfed, off-type chromosomes from the male parent cannot be avoided. Moreover, transpositions during megagametophyte formation and early during sporophyte development will contribute to the frequency of off-type progeny. In any case, by taking the proper precautions, the desired product can be maximized.

We recommend the following procedure. To generate the female tagging line for strategy I crosses, select self-pollinated ears that meet the criteria outlined above. Cross this line as a female to *P-WR r-g males*. Eliminate off-type ears and somatic sectors from these crosses. For strategy II crosses, propagate the homozygous *Ac* line as outlined above. To generate the female line, choose ears that meet both criteria—(1) homozygous variegated pericarp and (2) mostly (>90%) kernels with coarse spotted aleurones in the *r-sc:m3* testcross. The kernels from these ears are grown and self-pollinated to bulk up the female tagging line.

Standard Precautions

Pollen contamination can be a significant problem. Follow good genetic practices. For instance a pollen marker (e.g., *y1*, *sh*, or *wx*) can be essential in identifying and eliminating contaminated offspring. Another critical precaution is prescreening stock lines for the presence of unwanted mutations. If such a mutation preexists in these lines, it will greatly interfere with the identification of tagged mutants in T_1 and T_2 lines. For instance, a glossy mutation present in one of our homozygous *P-VV* lines caused this phenotype to show up in a large fraction of T_2 families, preventing us from identifying *Ac*-induced glossy mutations (unpublished results). Therefore, by taking the precaution of checking the integrity of the parental lines, major problems can be avoided at later stages when it is too late to correct the situation.

Before any stock is used to generate gene tagging lines, we recommend that all ears be checked for segregating mutations. Carefully examine each individual ear for segregating kernel mutations. Before bulking up any line, perform a seedling screen on selfed ears to identify segregating mutations by planting 25 kernels in sand benches. (See below.) Even the *Ds* tester line should be carefully checked before bulking up the seed for gene tagging. In the nursery always be alert to the appearance of vegetative or floral abnormalities in the tagging lines. Remember that *Ac* is active in these lines and spontaneous mutations are common. Any off-type plants in the nursery should be eliminated as a source of gene tagging stock. Uniformity is an important factor in the parental lines that minimizes interference with subsequent de novo genetic variability caused by authentic transposition events. We found that the use of inbred lines is an important factor in genetic uniformity in T_2 families. Whenever possible, the use of inbreds should be considered. If these precautions are taken, your chances for success are great.

LOGISTICAL CONSIDERATIONS

A typical gene tagging experiment may follow the following sequence. The homozygous variegated and *Ds* tester lines are propagated by selfing with all the precautions outlined above. Let's assume that approximately 5,000 T_1 kernels carrying independent tr-*Ac* elements are desired. To generate this number of transpositions, we will make the conservative estimate that each ear will yield, on average, 3–4 independent transpositions. Therefore, a minimum of 1,500 testcrossed ears are needed. Remember to account for the frequency of off-type ears and plants and for losses of plants that typically occur under field conditions. A rule of thumb is to plant twice as many kernels as the final number of ears needed. Therefore, 3,000 (strategy I—*P-VV/P-WR r-g/r-g*, or strategy II—*P-VV/P-VV r-g/r-g*) females are planted and crossed to homozygous *P-WR r-sc:m3* males. If larger number of ears are needed, consider performing this cross by detasseling the female plants and allowing open pollination with the pollen parent. This should be done in isolation plots—mutations found in the typical maize nursery will contribute to the appearance of false mutations as a result of pollen drift. A pollen marker is essential when crosses are made by open pollination.

The next step is to examine the pericarp phenotype of these ears. In strategy I crosses, only medium variegated pericarp ears (Figure 24.2B) should be retained for further analysis. Only homozygous variegated ears (Figure 24.2D) are retained for strategy II crosses. Cull out the expected off-type ears. On medium variegated ears, red pericarp sectors are identified and an underlying kernel with spotted aleurone is removed for further analysis (Figure 24.2C). The size of the red sector is unimportant as long as the red sector covers at least most of the surface of the underlying spotted kernel. Smaller red sectors are often not transmitted to the underlying kernel. At minimum the red sector should completely cover the germinal face of the kernel. For large multikernel sectors only a single spotted kernel is retained. On homozygous variegated ears from strategy II crosses, choose all colorless or fine spotted kernels (Figure 24.2E). Expect around 5,000 or more T_1 kernels from these crosses. Keep in mind that the frequency of recovering *Ac*-induced mutations will be proportional to the number of T_1 and T_2 families screened for mutations.

MUTATIONAL SCREENS

Figure 24.5 outlines four types of screens for mutations. All T_1 kernels are planted in the nursery, and the plants are examined for aberrant phenotypes. Gross variation (i.e., dwarfism, early or late flowering, etc.) can usually be readily detected. Examine both dermal (i.e., trichome and cuticular wax formation) and subdermal (i.e., venation patterns, photosynthetic pigments, etc.) features for subtle abnormalities. Inflorescences (i.e., tassel and ear branching and sexuality) and floral structures (i.e., anthers, glumes, etc.) can also be examined for abnormalities. Because the tr-*Ac* is heterozygous in this generation (T_1), only dominant or semidominant mutations are expected to be visible in T_1 plants. The frequency at which *Ac* causes dominant or semidominant mutations is unknown but several semidominant phenotypes have been recovered that are linked to a tr-*Ac* element (unpublished data). Unfortunately, dominant lethals will be lost, but it may be possible to recover DNA from such plants. Pollen production in T_1 plants can also be examined. Gametophytic steriles should show 50% pollen sterility. In general, any plant

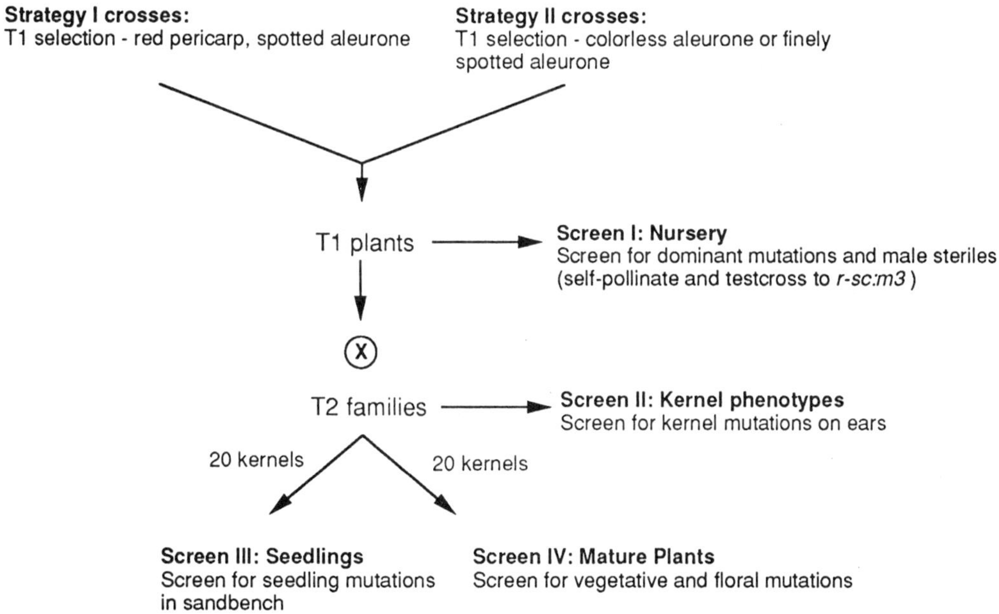

FIGURE 24.5 Mutant screens

showing an aberrant phenotype should be identified and handled separately from the remaining T_1 population.

Normal T_1 plants are self-pollinated to generate T_2 families. All ears from these selfs should be examined and only those families carrying a tr-*Ac* element should be retained for further analysis. Strategy I ears (Figure 24.2F) should have red or colorless pericarp, and many of the kernels on the ears should have a spotted aleurone phenotype. (Because *r-sc:m3* and the tr-*Ac* element are segregating in the T_2 kernel population, the frequency of coarse spotted aleurone phenotype should only be about 12%.) Cull any ears that have variegated pericarp or do not show *Ac* activity. Of the remaining ears, the frequency of red pericarp should exceed the frequency of colorless pericarp. This deviation from a 1:1 Mendelian ratio reflects the tendency of *Ac* to transpose to nearby chromosomal sites. By selecting for tr-*Ac* activity, nearby transpositions will cause an indirect selection for an excess of the *P-RR* (red pericarp) homolog. Similarly, only light-variegated progeny ears should be retained for further analysis from strategy II (Figure 24.2G). The medium variegated and red ears from this population are probably the result of noncorrespondence between the embryo and endosperm as discussed above.

Each individual T_2 family is then subjected to three additional screens (Figure 24.5). First, examine each T_2 family for the segregation of aberrant kernel phenotypes (i.e., defective kernels, shrunken or brittle endosperm, defective embryos, etc.). These ears are retained and analyzed separately from the remaining ears. Next, randomly select two 20 kernel samples from each T_2 family. One 20 kernel sample is germinated in sand benches and examined for aberrant seedling phenotypes (i.e., albino, yellows, pale greens, abnormal shoots, leaves, high chlorophyll fluorescence, etc.). The uniformity of the greenhouse environment is an important factor when scoring germination rates and other characters that are sensitive to the environmental fluctuations found in the field during germination and early seedling growth. Uniform greenhouse conditions allow greater accuracy in recording germination data and seedling phenotypes. For instance, low germination (75% or less) often indicates an abnormality in embryo, shoot, or root development. The second 20 kernel sample is planted in the field nursery, and plants are examined periodically for vegetative and floral abnormalities. Periodic examinations are important because, in some instances, mutations visible at one stage of development are corrected at later stages or undetectable in earlier stages. Moreover, under field conditions we often detect vegetative phenotypes that go undetected in the seedling screens (i.e., wilting, dwarfism, etc.). Examine ears on T_2 plants to detect mutations that affect ear branching and fertility. All abnormalities are carefully recorded, i.e., by a written description and/or a photograph. When possible both normal and mutant plants in the T_2 family are self-pollinated and testcrossed as males to the *Ds* tester line. (See below.)

GENETIC TESTS TO IDENTIFY *Ac*-INDUCED MUTATIONS

From previous experiments we have found that the overall frequency of mutations exceeds the frequency of *Ac*-induced mutations. This indicates that it is imperative to distinguish between *Ac*-induced and non-*Ac*-induced mutations before substantial effort is invested in cloning or analysis. The only distinction available at this time is linkage of the mutation to the tr-*Ac* element. Linkage is a good indication that the mutations are *Ac*-induced. Fortunately, *Ac* possesses two features that allow linkage to be established. First, it provides a convenient dominant phenotype (e.g., *Ds* transactivation) that can be used in standard linkage analysis. For a simple recessive mutation, mutant plants should be homozygous for the tr-*Ac* element; plants heterozygous for the tr-*Ac* element should also be heterozygous for the mutation as well. Second, *Ac* elements can be readily detected in the maize genome by Southern hybridizations. This allows one to perform a molecular genetic segregation analysis—the segregation of the tr-*Ac* fragment on Southern blots can be compared to the segregation of the mutant allele in a T_2 population. Unfortunately, it can be difficult to distinguish between two copies (homozygous) and one-copy (heterozygous) *Ac* elements on Southern blots although the presence or absence of *Ac* is easily ascertained.

There is one major problem with linkage tests involving *Ac*: the element continues to remain active and capable of transposition. This instability may cause misleading conclusions to be drawn. For instance, secondary transposition of *Ac* will interfere with standard linkage tests. We recommend that all T_2 families that show genetic linkage to *Ac*, even if linkage is incomplete, be examined by Southern analysis. If the primary tr-*Ac* element gives a distinctive fragment, its segregation can be more reliably scored in a segregating population. Be aware that imprecise excision of *Ac* from the mutant allele may result in mutant progeny that no longer carry the tr-*Ac*.

MOLECULAR ANALYSIS

Linkage analysis using Southern blots is performed on segregating T_2 seedlings or plants. Internal *Ac* probes (e.g., the 1.6-kb *Hin*dIII fragment) will hybridize to multiple fragments (cryptic elements) in the maize genome that will typically range in copy number from 8 to 16. An active *Ac* element has a tendency to transpose into hypomethylated DNA; this allows its identification among these cryptic fragments using methylation-sensitive restriction enzymes. (See Chen et al. 1987; Chomet et al. 1987 for further details.) Methylation-sensitive restriction enzymes such as *Pst*I, *Apa*I, and *Sal*I are especially useful for identifying tr-*Ac* on Southern blots. If inbred material is used in gene tagging crosses, the cryptic copy number can be low (eight copies) and constant among individual plants. In these circumstances, methylation-sensitive enzymes are often unnecessary, and an *Eco*RI digest can be employed instead. Because *Eco*RI cuts once inside *Ac*, two unique *Ac*-genomic junction fragments are produced. A single genomic blot containing *Eco*RI-digested DNA is sequentially probed with the *Hin*dIII-EcoRI 900-bp and 700-bp internal *Ac* fragments. Two segregating bands linked to the mutation should be identified, and these can be cloned. Several convenient λ vectors are available for cloning *Ac* junction fragments that accept up to 20 kb *Eco*RI fragments. DNA is size-selected by gradient centrifugation or by agarose gel electrophoresis and ligated into the appropriate vector.

For genomic fragments less than 8 kb consider using an insertion type λ vector for cloning junction fragments. For larger DNA fragments consider using a substitution type λ vector.

UTILITY OF *Ac*-INDUCED MUTATIONS

Besides the obvious advantage of cloning genes, *Ac*-induced mutations are useful for subsequent genetic studies. The instability of *Ac*-induced mutations may be a disadvantage during the early stages of analysis but at later stages this proves extremely valuable. For instance, one criterion in judging that a mutation is caused by an insertion is that reversion be associated with excision of the element. Often, reversion is difficult to attain with some transposable elements, such as Robertson's Mutator, and nearly impossible with retrotransposons. Revertant alleles can be especially valuable when only a single allele defines the mutation; this is often the case for novel phenotypes and genetic pathways with large numbers of target genes (e.g., embryogenesis, photosynthesis, etc.). Revertant alleles of *Ac*-induced mutants can usually be obtained by self-pollinating the homozygous mutant line. If a stable reference allele is available, consider testcrossing the mutants to obtain revertants. The frequency of phenotypic reversions depends on the tolerance of the insertion site for imprecise excision. (See Moreno et al. 1992 for further details.) A typical reversion frequency between 0.1 and 2% is usually observed for *Ac*-induced mutations. Phenotypic reversion should be correlated with excision of *Ac* from the gene and detected molecularly if the gene is indeed cloned.

Once a mutation is cloned, the flanking DNA should be used as a molecular probe to examine the progenitor allele (the female tagging line), the putative *Ac*-induced allele, any other mutant alleles available, and revertant alleles recovered by subsequent excision of *Ac*. The pattern of hybridization should be consistent with expected results: (1) the progenitor allele should contain a 4.56-kb insertion of *Ac*; (2) other insertion mutations or rearrangements should be within detectable limits of the probe; and (3) phenotypic revertants should be molecularly detectable as excisions of *Ac* from the locus. Regarding the last criterion, often short-range transpositions of *Ac* can be over short physical distances, especially if the *Ac* element is at the extreme 5′ or 3′ ends of the gene. Hence, reversion may be associated with transpositions over a short interval. These events may make it appear that the *Ac* element is retained in a particular genomic fragment when, in reality, the element has transposed out of the cistron to a nearby flanking chromosomal site within the same restriction fragment. (See Moreno et al. 1992 for further details). These events should only comprise 10% of all transpositions, however. In summary, the ability of *Ac* to excise can be used to generate alleles that can prove valuable.

Genetic and molecular studies of a gene can greatly benefit from the availability of a extensive set of mutant alleles. To generate an allelic series of mutant alleles, *Ac*-induced mutations can be used to perform insertional mutagenesis using a reconstitution strategy. (See Moreno et al. 1992.) Excision of *Ac* to nearby chromosomal sites and its subsequent reinsertion back into the gene can cause a large number of *Ac* insertions to be generated with insertion sites dispersed throughout the locus. From these unstable mutations, stable mutant alleles can be generated by imprecise excision from each insertion site of *Ac*. This process allows a series of amorphic, hypomorphic, and possibly hypermorphic and neomorphic alleles to be generated for each gene. Imprecise excision can also be used to modify the amino acid sequence of proteins for structure-function studies.

Finally, the instability of *Ac*-induced mutations can be exploited for clonal analysis. (See Dawe and Freeling 1990; Dellaporta et al. 1991.) Transpositions in epidermal and subepidermal lineages are independent events, allowing a variety of sectors to be generated during plant development. These sectors may be useful in the study of cell-to-cell interactions and lineage relationships.

In summary, *Ac* possesses unique genetic features that can be exploited for gene tagging purposes. The ability to tag and clone *Ac*-induced mutations should be considered especially when a gene tagging program is nontargeted or *Ac* elements are found in the vicinity of a locus under study. The advantages of *Ac*-induced mutations in genetic and molecular studies enhance the attractiveness of this system for gene tagging.

REFERENCES

Chen J, Greenblatt IM, Dellaporta SL (1987) Transposition of *Ac* from the *P* locus of maize into unreplicated chromosomal sites. Genetics 117: 109–116

Chen J, Greenblatt IM, Dellaporta SL (1991) Molecular analysis of *Ac* transposition and DNA replication. Genetics 130: 665–676

Chomet PS, Wessler SR, Dellaporta SL (1987) Inactivation of the maize transposable element *Activator* (*Ac*) is associated with its DNA modification. EMBO J 6: 295–302

Dawe RK, Freeling M (1990) Clonal analysis of the cell lineages in the male flower of maize. Dev Biol 142: 233–245

Dellaporta SL, Greenblatt IM, Kermicle JL, Hicks JB, Wessler SR (1988) Molecular cloning of the *R-nj* gene by transposon tagging with *Ac*. In Gustafson JP, Appels R (eds) Chromosome Structure and Function: Impact of New Concepts, Plenum Press, New York, pp 263–282

Dellaporta SL, Moreno MA, DeLong A (1991) Cell lineage analysis of the maize gynoecium using *Ac*. Development s1: 141–147

Fedoroff NV (1989). Maize transposable elements. In Berg DE, Howe MM (eds) Mobile DNA, Am Soc Microbiol, Washington DC, pp 377–411

Greenblatt IM (1984) A chromosome replication pattern deduced from pericarp phenotypes resulting from movements of the transposable element *Modulator* in maize. Genetics 108: 471–485

Greenblatt IM, Brink RA (1962) Twin mutations in medium variegated pericarp in maize. Genetics 47: 489–501

Kermicle JL (1984) Recombination between components of a mutable gene system in maize. Genetics 107: 489–500

Moreno MA, Chen J, Greenblatt I, Dellaporta SL (1992) Reconstitutional mutagenesis of the maize *P* gene by short-range *Ac* transpositions. Genetics 131: 939–956

25

Using Cytogenetics to Enhance Transposon Tagging with *Ac* Throughout the Maize Genome

Donald Auger and William F. Sheridan

Our long-term goal is to modify a group of reciprocal translocations and pericentric inversions to bring the transposable element *Ac* into a contiguous relationship with a large portion of the maize genome. This will provide material for local transposition of *Ac* to tag molecularly genes of interest on all 20 chromosome arms. This project relies on many of the same strategies discussed by Dellaporta and Morino (this volume); see that chapter for details on working with *P-vv*.

MAIZE TRANSLOCATION AND INVERSION STOCKS

Immediately after the end of World War II, a cooperative project was established between the Agricultural Research Service of the USDA and the California Institute of Technology. During this project maize kernels were exposed to the radiation emitted by the Bikini and Eniwetak atomic bomb tests. From these materials, and from additional maize materials exposed to X-rays and to gamma rays, a total of 1,003 reciprocal translocations and 60 inversions were isolated; intensive cytological examination allowed calculation of the breakage point positions (Longley 1961). During this period and for several years thereafter E. G. Anderson propagated and prepared these materials for use by other geneticists and deposited them in the Maize Genetics Cooperation Stock Center. About 865 of these translocation stocks and 24 of the inversion stocks have been maintained and are available upon request. These stocks are listed periodically in the *Maize Genetics Cooperation News Letter* (1981, 55: 140–144) and elsewhere in this volume.

We selected 40 reciprocal translocations and five inversions that all involved an exchange between chromosome arm *1S* and another chromosome arm. An additional consideration was to have each of the other 19 arms (other than *1L*) represented in the col-

The Maize Handbook—M. Freeling, V. Walbot, eds.

lection. Furthermore, the stocks were selected so that the maximum amount of the other arm would be in linkage with the region of *1S* containing the *P* locus. *P* is located just distal to cytological position 0.48 on chromosome arm *1S* according to the Cytogenetic Working Map (Maize Genetics Cooperation News Letter 52: 129–145). For instance, as shown in Table 25.1, the two 1S-2L reciprocal translocations selected were 1S-2L 1–2c (1S.77–2L.33) and 1S-2L 036–7 (1S.37–2L.33). The former translocation has the distal 67% of chromosome *2L* attached to chromosome 1 at a site distal to the *P* locus and in coupling with it. The latter translocation has the distal 63% of chromosome arm *1S* (including the *P* locus) attached to chromosome 2 thereby coupling the *P* locus with the proximal 33% of chromosome arm *2L*. The five *1S-1L* inversions were selected so as to bring the greatest extent of chromosome arm *1L* in the closest coupling with the *P* locus. The other arm region in linkage with the *P* locus is shown for each translocation stock and inversion stock in Table 25.1. We estimate that, on a cytological map basis, together the 40 translocation stocks and five inversion stocks bring the *P* locus into coupling with virtually all of the genome. The linkage marker stocks that can be used to map the breakpoints in the inversion and translocation stocks are also listed in Table 25.1.

SOURCE OF *Ac*, THE *P* ALLELES, AND TESTER STOCKS

Most of the translocation and inversion stocks carried the *P-wr* allele but some carried the *P-ww* allele. The source of *Ac* was the *P-vv* allele in a colorless yellow background. For monitoring the presence and dosage of *Ac* in the translocation and inversion stocks we have used the *Ds2*-containing *bronze-2* mutable (*bz2-m*) allele located on chromosome arm *1L*. In addition we have used the *Ds2*-containing *r1-sc-m3* allele to monitor *Ac* dosage in certain stocks.

MODIFICATION STRATEGY

The basic strategy is to make an initial cross of each of the translocation and inversion stocks as female parents by pollen from the *P-vv* stock and then perform recurrent backcrosses using the *bz2-m* stock as a pollen parent. The presence of the translocation or inversion is monitored by scoring the ears for a semisterile pattern of kernel set and also by scoring pollen for semisterility by use of a pocket microscope. The variegated pericarp striping of the *P-vv* phenotype is in maternal tissue; the presence of *Ac* revealed by this striping lags one generation behind the genetic constitution of the embryo in that kernel. But the use of the *bz2-m* or *r1-sc-m3* monitor genes allows for the reading of the presence and dosage of *Ac* in the embryo. The goal is to obtain a recombinant chromosome linking *P-vv* to the breakpoint in each stock.

TABLE 25.1 Reciprocal translocations and inversions between *1S* and the other arms

	Inversion	Break points	Other arm region to be linked to P-vv	Markers
1S-1L	Inv.1m	1S.80 1L.10	1L .10–1.00	*br1 bz2*
	Inv.1g	1S.41 1L.35	1L .35–.00	*br1 bz2*
	Inv.1l	1S.82 1L.46	1L .46–1.00	*br1 bz2*
	Inv.1a	1S.86 1L.50	1L .50–1.00	*br1 bz2*
	Inv.1f	1S.85 1L.56	1L .56–1.00	*br1 bz2*
	Translocation			
1S-2S				
	1–2b	1S.43 2S.36	2S .36–.00	*fl1 wt1*
1S-2L				
	1–2c	1S.77 2L.33	2L .33–1.00	*fl1 v4*
	036–7	1S.37 2L.33	2L .33–.00	*fl1 v4*
1S-3S				
	5883	1S.88 3S.60	3S .60–1.00	*d1 g2*
	013–9	1 ctr 3 ctr	3S or 3L	*gl6 lg2*
1S-3L				
	5597	1S.77 3L.48	3L .48–1.00	*lg2 a1*
	5982	1S.77 3L.66	3L .66–1.00	*lg2 a1*
	1–3k	1S.17 3L.34	3L .34–.00	*gl6 lg2*
	8995	1S.49 3L.06	3L .06–1.00	*gl6 lg2*
1S-4S				
	1–4i	1S.13 4S.42	4S .42–.00	*la1 su1*
1S-4L				
	4308	1S.65 4L.58	4L .58–1.00	*gl4 Tu*
	1–4b	1S.55 4L.83	4L .83–1.00	*c2 dp1*
	064–20	1S.23 4L.19	4L .19–.00	*su gl4*
	8602	1S.41 4L.81	4L .81–.00	*c2 dp1*
1S-5S				
	5525	1S.75 5S.53	5S .53–1.00	*a2 bm1*
1S-5L				
	5045	1S.94 5L.50	5L .50–100	*bt1 pr1*
	6899	1S.40 5L.10	5L .10–.00	*bm1 pr1*
	043–15	1S.10 5L.63	5L .63–.00	*bt1 pr1*

Table 25.1 (Continued)

	Inversion	Break points	Other arm region to be linked to P-vv	Markers
1S-6S				
	5495	1S.25 6S.80	6S .80–.00	*y1 Pl1*
1S-6L				
	7097	1S.46 6L.62	6L .62–.00	*y1 Pl1*
	028–13	1S.56 6L.54	6L .54–1.00	
	7352	1S.40 6L.60	6L .60–.00	*y1 Pl1*
1S-7S				
	4444	1S.65 7S.50	7S .50–1.00	*o2 v5*
	4405	1S.43 7S.46	7S .46–.00	*o2 v5*
	6796	1S.40 7S.39	7S .39–.00	*o2 v5*
1S-7L				
	4837	1S.73 7L.55	7L .55–1.00	*gl1 ij1*
	010–12	1S.35 7L.57	7L .57–.00	*gl1 ij1*
	1–7i	1S.31 7L.26	7L .26–.00	*gl1 ij1*
1S-8S				
	6591	1S.18 8S.43	8S .43–.00	*Bif v16*
	055–23	1 ctr 8 ctr	8S or 8L	*Bif v16*
1S-8L				
	8919	1S.53 8L.44	8L .44–1.00	*Bif v16*
	4307–4	1S.42 8L.61	8L .61–.00	*Bif v16*
	4685	1S.20 8L.21	8L .21–.00	*Bif v16*
1S-9S				
	7535	1S.33 9S.27	9S .26–.00	*c1 wx1*
1S-9L				
	8302	1S.55 9L.29	9L .29–1.00	*wx1 v1*
	8001	1S.51 9L.24	9L .24–1.00	*wx1 v1*
	6762	1S.16 9L.53	9L .53–.00	*wx1 v1*
1S-10S				
	4885	1 ctr 10 ctr	10S or 10L	*zn1 g1*
1S-10L				
	1–10g	1S.80 10L.21	10L .21–1.00	*zn1 g1*
	1–10f	1S.04 10L.30	10L .30–.00	*zn1 g1*

Ac AND THE *P-vv* ALLELE

The transposable element *Activator* (*Ac*) is the autonomous element and *Dissociation* (*Ds*) is the nonautonomous element of the two element *Ac-Ds* family (McClintock 1947). The *Ac* element transposes from its initial insertion site to a new locus where a recessive mutation of the locus may result. Often this new mutation is unstable (mutable) and reverts to normal expression upon subsequent transposition of the *Ac* element away from that locus.

The *P-variegated* allele (*P-vv*) at the *P* locus on chromosome arm *1S* results in a variegated pericarp as reported 75 years ago by Emerson (1917). This condition was shown to be a result of an autonomous transposable element named *Modulator* (of pericarp) (*Mp*) by Brink and Nilan (1952) and shown to be indistinguishable from *Ac* (Barclay and Brink 1954). Molecular cloning confirmed the genetic deduction: the *P-vv* allele contains an *Ac* inserted at that locus and the transposition of *Ac* away from that locus results in reversion of *P-vv* to *P-rr* with the appearance of red pigmentation in the pericarp tissue (Chen et al. 1987; Lechelt et al. 1989; Peterson 1990). When the maternal pericarp tissue contains one dose of *Ac* (*P-vv/P-ww*) the transposition event occurs earlier and therefore produces more red stripes in the pericarp (medium variegation) than when two doses are present (*P-vv/P-vv*). With two doses, transposition is later and fewer red stripes are produced (light variegation).

An important feature of Ac transposition is that the transposed *Mp* (*Ac*) often remains linked to the *P* locus (van Schaik and Brink 1959; Greenblatt and Brink 1962). Therefore, nearby chromosomal sites were the most common sites for insertion of the element following transposition away from the *P* locus.

A frequent event accompanying transposition of *Ac* away from the *P* locus is that the element replicates, then transposes and inserts at a new site of unreplicated DNA. This results in twin spots consisting of a sector of red kernels and a sector of light variegated kernels. In both sets of kernels the embryos contain the transposed *Ac* element at the same new site, while those in the light variegated kernels also contain the *Ac* element still present at the *P* locus. Greenblatt (1984) mapped the location of the transposed *Ac* element by studying 105 light variegated/red twin sectors on medium variegated pericarp ears. He found that 61% of the receptor sites were linked to the *P* locus but there was a region of four map units just proximal to *P* where no insertions occurred. Recently it has been shown that *Ac* transposes from the *bz1* locus into regions very close to the gene on both its proximal and distal sides (Dooner and Belachew 1989; Schwartz 1989). It is evident that *Ac* is biased for insertion into sites near its original location.

Because the presence and dosage of *Ac* can be detected by the mutability of loci containing it as well by test with *Ds* stocks, *Ac* is well suited for transposon mutagenesis. The main limitation of *Ac* is its strong tendency to transpose to a nearby location (Greenblatt 1984; Schwartz 1989; Dooner and Belachew 1989). Although stocks are available that carry *Ac* at other loci only a few chromosome arms are represented. By introducing *Ac* near known chromosomal breakpoints in inversions and translocation stocks, *Ac*-insertion mutations could be induced at many loci. This strategy was successfully employed in the tagging and cloning of the *R* locus (Delaporta et al. 1988). In many cases a mutant allele will already have been obtained by transposon mutagenesis using *Mutator* or some other transposable element. The availability of two mutant alleles, each induced by a different transposable element, greatly simplifies the cloning of a gene (Shepherd et al. 1988). This has been demonstrated by O'Reilly et al. (1985), who cloned the *a1* locus of maize using the transposable elements *En/Spm* and *Mu1*.

REFERENCES

Barclay PC, Brink RA (1954) The relation between *Modulator* and *Activator* in maize. Proc Natl Acad Sci USA 40: 1118–1126

Brink RA, Nilan RA (1952) The relation between light variegated and medium variegated pericarp in maize. Genetics 37: 519–544

Chen J, Greenblatt IM, Dellaporta S (1987) Transposition of *Ac* from the *P* locus of maize into unreplicated chromosomal sites. Genetics 117: 109–116

Delaporta SL, Greenblatt I, Kermicle JL, Hicks JB, Wessler SR (1988) Molecular cloning of the maize *R-nj* allele by transposon tagging with *Ac*. In Gustafson JP (ed) Chromosome structure and function: Impact of new concepts, Plenum Press, New York, pp 263–282.

Dooner HK, Belachew A (1989) Transposition pattern of the maize element *Ac* from the *bz-m2(Ac)* allele. Genetics 122: 447–457

Emerson RA (1917) Genetical studies of variegated pericarp in maize. Genetics 2: 1–35

Greenblatt IM (1984) A chromosomal replication pattern deduced from pericarp phenotypes resulting from movements of the transposable element, *Modulator*, in maize. Genetics 108: 471–485

Greenblatt IM, Brink RA (1962) Twin mutations in medium variegated pericarp maize. Genetics 47: 489–501

Lechelt C, Peterson P, Laurel H, Chen J, Dellaporta S, Dennis E, Peacock WJ, Starlinger P (1989) Isolation and molecular analysis of the maize *P* locus. Mol Gen Genet 219: 225–234

Longley AE (1961) Breakage points for four corn translocations series and other corn chromosome aberrations. ARS Crops Res (34)16: 1–40

McClintock B (1947) Cytogenetic studies of maize and Neurospora. Carnegie Inst Wash Year Book 46: 146–152

O'Reilly C, Shepherd NS, Pereira A, Schwarz-Sommer Z, Bertram I, Robertson DS, Peterson PA, Saedler H (1985) Molecular cloning of the *a1* locus of *Zea mays* using the trasposable elements *En* and *Mul*. EMBO J 4: 877–882

Peterson T (1990) Intragenic transposition of *Ac* generates a new allele of the maize *P* gene. Genetics 126: 469–476

Schwartz D (1989) Pattern of *Ac* transposition in maize, Genetics 121: 125–128

Shepherd NS, Sheridan WF, Mattes MG, Deno G (1988) The use of Mutator for gene tagging: Cross referencing between transposable element systems. In Nelson OE (ed) Plant-Transposable Elements, Plenum Press, New York, pp 137–147

van Schaik NW, Brink RA (1959) Transpositions of Modulator, a component of the variegated pericarp allele in maize. Genetics 44: 725–738

26

Transposon Tagging with Spm

KAREN CONE

This protocol is designed for targeted mutagenesis in which the gene to be tagged has been defined phenotypically by a recessive, viable mutation.

MATERIALS

Female

1,500–2,000 plants carrying a recessive allele (preferably homozygous) of the gene of interest; these plants should be detasseled.

Male

50–100 plants carrying *Spm* and the dominant allele of gene of interest. (To make sure that pollen shed is ample and spans the time required for all female plants to reach silking, it is advisable to make multiple plantings of the male line, staggered by 1–2 weeks. Alternatively, plants located in different parts of the field will frequently mature at slightly different times, thus effectively increasing the number of days when pollen is available.)

METHODS

1. Cross the two stocks by hand-pollination.
2. For recessive kernel traits, putative mutants can be scored directly and then planted. For traits affecting seedlings or mature plants, F_1 kernels should be planted, and scored at the appropriate developmental stage.
3. To generate a segregating population suitable for molecular analysis, putative mu-

The Maize Handbook—M. Freeling, V. Walbot, eds.

tants should be self-pollinated to generate an F_2 and/or crossed back to the parental line carrying the recessive mutation to generate a backcross population.

4. To confirm that mutants are controlled by *Spm* (as opposed to some other transposable element), putative mutants should be crossed to a tester line containing a *dSpm* at another locus and then self-pollinated and/or crossed back to the *dSpm*-containing line. The resulting segregating populations should be scored for cosegregation of the new mutant phenotype with mutability of the independent *dSpm*.

CONSIDERATIONS

Frequency of Transposition of *Spm*

The frequency of transposition of *Spm* from an unlinked (or unknown) location is about $4–5 \times 10^{-6}$. The frequency can be up to 100-fold higher if the *Spm* is linked to the gene of interest. In addition, the frequency of transposition is usually higher if the *Spm*-containing line is used as the male.

Expected Phenotype

An *Spm*-induced mutant should exhibit a typical "mutable" phenotype, i.e., recessive as a result of "suppression" of gene expression with somatic sectors showing the wild-type pattern of expression. The size of the wild-type sectors depends on the timing of *Spm* activity. If *Spm* is active very early in development of the appropriate tissue, the sectors can be very large; however, in cases where *Spm* activity occurs very late in development, wild-type sectors may not be visible at all. In the absence of *Spm*, mutants caused by insertion of defective *Spm* (*dSpm*) elements have a stable phenotype. Usually this phenotype is recessive; but occasionally, depending on the site of insertion within the gene, the stable phenotype associated with a *dSpm* insertion can be wild-type.

Expected Segregation Ratios

If the mutable phenotype of the new mutant is caused by the insertion of an autonomous *Spm* element, then one expects close to a 3:1 ratio of mutable to recessive progeny in an F_2, and close to a 1:1 in backcross progeny. If the mutable phenotype is caused by the insertion of a defective *Spm* element and the controlling autonomous *Spm* is unlinked, then one expects a ratio of 9 mutable:7 recessive in the F_2 and a ratio of 1 mutable:3 recessive in the backcross. If the mutation is caused by the insertion of a *dSpm*, and the controlling *Spm* element is very closely linked, the segregation ratios will be very close to those obtained when the gene of interest is tagged with an autonomous element. These possibilities can usually be distinguished by performing crosses to confirm that the mutation is induced by *Spm*, as discussed below.

Confirmation That the Mutation is *Spm*-controlled

Many maize lines contain more than one family of active transposable elements. Therefore, even if a line is selected with *Spm*, sometimes a new mutable phenotype is caused by insertion of an element from another family. To demonstrate that the gene of interest has indeed been tagged with *Spm*, one must show that mutability at the new locus depends on an active autonomous *Spm* element. For this purpose, F_2 or backcross progeny should be generated by crossing the putative mutant to a tester line containing a *dSpm*, preferably at a gene affecting an easily scored kernel trait such as starch composition (the *waxy* gene) or anthocyanin production (the *c1* or *a1* gene).

Progeny that contain both the tester *dSpm* and the new mutant of interest should be examined for mutability with the following criteria in mind.

1. No matter where the autonomous *Spm* resides in the genome, *if the gene of interest is controlled by Spm,* then any progeny carrying both the tester *dSpm* and the new mutant of interest that show mutability of the tester *dSpm* will also show mutability at the locus of interest.
2. *If the autonomous Spm is at the locus of interest*, then progeny that contain both the tester *dSpm* and the new mutant should show always show mutability at *both* loci.
3. If the gene of interest contains a *dSpm*, but *the controlling Spm is closely linked* but in repulsion, progeny should exhibit mutability at the tester *dSpm*, but not at the gene of interest. In coupling, recombination can occur between the closely linked *Spm* and the gene of interest allowing subsequent segregation of the *Spm* away from the new mutant locus. Because detection of this type of event depends on (probably rare) recombination events, it may be necessary to screen large numbers of progeny, especially if the new mutant locus and the controlling *Spm* are very tightly linked.

REFERENCE FOR MORE COMPLETE DISCUSSION

Cone KC, Schmidt R, Burr B, Burr FA (1988) Advantages and limitations of using *Spm* as a transposon tag. In Nelson OE (ed) Plant Transposable Elements, Plenum Publishing Corporation, New York, pp 149–159

27

Transposon Tagging with Mutator

Paul S. Chomet

Transposable elements have been useful in the identification and cloning of genes that were previously inaccessible by other cloning methods. In maize, three transposable element systems have been utilized for transposon tagging: *Activator/Dissociation*, *Suppressor-mutator* (*Enhancer/Inhibitor*), and Robertson's *Mutator*. Each system has inherent advantages and disadvantages, and the choice of system depends on the tagging approach taken, the genomic position and the expression of the gene sought, as well as the stocks available for tagging.

The *Mutator* transposable element family was originally identified in lines that exhibited an unusually high frequency of forward mutation (Robertson 1978). Extensive genetic and molecular analyses have demonstrated that the increase in mutation frequency is caused by a family of transposable elements, designated *Mutator* (*Mu*) elements. The *Mutator* system consists of more than eight different classes of *Mu* transposable elements each of which can be found in multiple copies. (For review see Chandler and Hardeman 1992.) Each element class is defined by a unique internal sequence flanked by inverted repeats about 200 bp long common to all *Mu* elements. Germinal transposition (and presumably forward mutation) and somatic excision of the nonautonomous *Mu* elements are under the control of the autonomous *MuR1* element (Chomet et al. 1991), now designated *MuDR-1*. Other *Mu* elements have been cloned that are similar or identical to *MuDR-1* in sequence; these were called *MuA2* (Qin et al. 1991) and *Mu9* (Hershberger et al. 1991), and are now called *MuDR* transposons. Lines harboring the autonomous element(s) are referred to as *Active Mutator* lines. The *Mu* system has been useful in cloning a number of genes, including *a1* (O'Reilly et al. 1985), *bz2* (McLaughlin and Walbot 1987), *vp1* (McCarty et al. 1989), *hcf106* (Martienssen et al. 1989), *y1* (Buckner, et al. 1990), and *hm1* (Briggs, personal communication).

THE *MUTATOR* SYSTEM FOR TAGGING

One of the advantages of *Mu* over the *Ac/Ds* and the *Spm* families lies in its non-localized mutagenic action. *Ac* and *Spm* often transpose to linked regions of the chromosome

The Maize Handbook—M. Freeling, V. Walbot, eds.

(Greenblatt and Brink 1962; Dooner and Belachew 1989; Nowick and Peterson 1981). *Mu* lines exhibit an increase in mutation frequency for all loci examined (Robertson 1978, 1983). This does not prove that *Mu* elements readily transpose to unlinked sites although recent evidence suggests this is the case (Chomet, Lisch, and Freeling, unpublished).

TARGETED AND NONTARGETED APPROACHES TO TAGGING

Two approaches to obtain mutations can be pursued dependent on the needs of the investigator. With the targeted (or directed) method the goal is to recover insertions in a previously identified locus. This method presumes a mutant allele of the target locus is available in a tester line. In such a case, homozygous wild-type *Mutator* stocks for the target gene are crossed to homozygous tester lines. The F_1 progeny are then screened for the mutant phenotype. The reported frequency of mutation for such an approach varies from 10^{-3} to 10^{-6} (Robertson 1985; Walbot et al. 1986; Patterson et al. 1991; Brown et al. 1989). Therefore, if possible, at least 10^5 gametes should be screened for a directed tagging experiment. Furthermore, it is important to have prior knowledge of the spontaneous mutation rate for the targeted locus. Some loci can be highly unstable in the absence of a transposable element system (Pryor 1987; Stadler 1946).

In the case of directed tagging of dominant loci, it is possible to "knock out" the dominant mutant allele with *Mu* (Hake 1992; Becraft, unpublished; Fowler, unpublished). In this case, a *Mutator* line homozygous for the dominant allele would be crossed (as a female) to a wild-type tester. The F_1 would then be screened for phenotypically wild type plants. This presumes a deletion heterozygote for the given locus is phenotypically wild-type. Such an assumption may not always be the case and tests should be undertaken to determine the phenotype of the hemizygote before proceeding with the screening of the F_1 population.

The nontargeted approach allows the investigator to identify new, uncharacterized mutations as well as to identify new mutations with previously characterized phenotypes. Active *Mutator* lines should be crossed by a non-*Mutator* line (inbred) or a tester line carrying a *Mu*-induced unstable marker allele. (See *nonautonomous genetic marker*, below.) The F_1 seed then would be selfed to produce F_2 populations; 20–40 seeds from each F_2 ear are then screened for the mutant phenotype(s). In this case, lethal or sterile phenotypes can be identified as homozygotes and recovered as heterozygotes in the population.

MONITORING MUTATOR ACTIVITY

Before making the first cross to generate either an F_1 population or the selfed populations, it is best to determine which plants in the *Mutator* stock carry *Mutator* activity and are mutagenic. This information can be obtained wholly or partially by a number of tests outlined below. Because each assay measures a different aspect of *Mu* transposition, each test does not necessarily match results from another test for *Mutator* activity.

ROBERTSON'S MUTATOR TEST

Robertson devised an assay for *Mutator* activity that gives an estimate of the forward mutation frequency of *Mutator* plants (Robertson 1978). The test is performed by selfing and crossing the *Mutator* plant as a male to a hybrid non-*Mutator* line to produce F_1 seed. The second ear on the non-*Mutator* plant is selfed as well. F_1 seed, from plants which did not segregate for a visible mutant (as determined by the parental selfs), are sown and plants are selfed to produce F_2 ears. Kernels from these F_2 ears are planted in a sand bench and observed for new seedling traits (i.e., albino, yellow, yellow-green, etc.). With active *Mutator* lines, Robertson reported approximately 10% of the F_2 families exhibited a new seedling mutation (1980).

NON-AUTONOMOUS GENETIC MARKER(S)

Nonautonomous *Mu* insertion alleles report on the presence or absence of *MuDR*, the regulator element (Chomet et al. 1991). The presence or absence of spotting of a *Mu*-induced insertion allele reports only on the the somatic reversion of one *Mu* element (such as *Mu1* at *bz1mum9*, or *Mu1* at *a1-mum2*). This is in contrast to the Robertson test, which can measure the forward mutation or insertion of diverse *Mu* into many different loci. Since these two assays do not measure the same event, use of one assay is not always a substitute for another. However, there is a correlation between an increase in forward mutation rate with lines containing multiple regulator elements that exhibit a high frequency of spotting. Plants that segregate for one *MuDR-1* (position 1) usually do not exhibit a high forward mutation rate (Robertson and Stinard 1989; Chomet unpublished). When using a mutable *Mu* allele as a marker for *Mu* activity, it is best to propagate the high spotting pattern. This will likely select for an increased number of *MuDR-1* elements in the stock (Chomet et al. 1991).

MOLECULAR MARKERS

A number of studies have demonstrated a correlation between loss of *Mutator* activity and methylation of *Mu1* at a number of methylation-sensitive restriction enzyme sites including *Hin*f1 (Chandler and Walbot 1986; Bennetzen 1987). *Hin*fI sites are located in the ends of *Mu1* and produce a 1.3-kb fragment upon cleavage. To determine the status of *Mu1* methylation and the associated *Mutator* activity, the DNA from *Mu* lines is cleaved with *Hin*f1 and hybridized to a *Mu1* internal probe by Southern blot analysis. The predominance of a 1.3-kb fragment indicates cleavage, whereas the presence of a ladder of fragments (and a reduced amount or lack of a 1.3-kb fragment) indicates *Hinf*1 sites in the ends of *Mu1* are methylated and *Mu* activity is probably absent. It should be noted that this correlation is not absolute (Bennetzen 1987; Bennetzen et al. 1988).

The generation of unique *Mu1*-homologous fragments, not detected in parent plants, has been correlated with *Mutator* activity (Alleman and Freeling 1986; Bennetzen et al. 1987). To detect new *Mu1* elements, DNA from parent and progeny of a *Mutator* lineage should be cleaved with an enzyme that does not cut within *Mu1* such as *Eco*R1, *Bam*H1, or *Hin*dIII. Southern blot analysis with the *Mu1* probe will reveal a series of *Mu1*-hybridizing fragments that represent *Mu1* at unique genomic locations. *Mu1* fragments not found in the parent most likely represent new insertion events, indicating the presence of *Mutator* activity.

CROSSES WITH NEWLY ARISEN MUTANTS

Once a mutant is identified, it is useful to have crosses to a number of different lines. Therefore, it is wise to plant a number of tester lines along with the population to be screened.

1. If possible, selfing, to recover homozygotes, is important. A homozygous stock can be utilized to screen for germinal reversion events; revertants are a useful tool in molecular identification of a tagged locus.
2. Outcrossing the mutation to a few different non-Mutator (inbred) lines will be useful for subsequent molecular analyses. Identifying the original, tagged locus of interest will be important and can be accomplished because RFLPs associated with the locus can be associated within a given line. Crossing to different lines will also allow the production of F_2 populations segregating for the mutant allele. These populations are necessary for subsequent molecular analyses. Furthermore, introduction of the mutant allele into a number of different lines will also insure expression and subsequent recovery of the mutant in later generations because expressivity or penetrance of the mutant can be affected by genetic background.
3. If the *Mutator* line is recessive for *A1*, *C1*, *Bz1* or *Bz2* (or other anthocyanin loci), it would be advantageous to outcross the mutant to a line lacking *Mutator* activity and homozygous for a *Mu*-induced allele of the same gene (such as *bz-mum9*, or *a1-mum2*). This would allow selection in the F_1 generation of seed that carried or lacked *Mutator* activity. These seed can then be grown, selfed, and screened for the mutant phenotype. Production of populations segregating for the mutation of interest and lacking *Mutator* activity facilitates molecular analyses. New *Mu* fragments are not generated, and existing, unlinked *Mu* fragments segregate out of the line with subsequent outcrosses. It is important to point out that some *Mu*-induced alleles are suppresssible; that is, in the absence of *Mu* activity the phenotype is wild type (Martienssen et al.

1990). In such a case, selection of inactivity would select for phenotypically wild-type plants.

DNA ANALYSIS

Identifying the gene responsible for the mutant phenotype involves screening for a *Mu*-homologous fragment that cosegregates with the mutant phenotype. This is done by examining the DNA of the segregating population(s) (as produced above) by Southern blot analyses. The preliminary screen is expedited by examining a small population first. As many different outcrossed segregating lines should be examined as possible. (See Walbot 1992.) It is also useful to examine the population utilizing a number of different restriction enzymes, since segregating fragments may be obscured by other *Mu* homologous bands. The population should also be screened with probes to all known *Mu* elements. (For review see Chandler and Hardeman 1992.) Inclusion of DNA from the parent lines on these blots is also important. A cosegregating fragment should not be present in the parental plant. Once a cosegregating fragment is identified, additional analyses with different populations and a larger population set should be performed. Furthermore, as noted above, *Mu*-induced suppressible alleles can confound the cosegregaton analysis (Martienssen et al. 1989, 1990). For this reason, it is important to emphasize linkage of a fragment with mutant individuals, whereas phenotypically wild-type individuals may be genotypically mutant.

Once a cosegregating band is identified, cloning or PCR is used to obtain a flanking, unique sequence. This flanking probe is then used to prove the locus is responsible for the mutant phenotype. This can be accomplished in a variety of ways:

1. Identification of DNA rearrangements, insertions, or deletions at the locus of independently generated alleles (O'Reilly et al. 1985) demonstrates the clone is (or is nearby) the locus of interest. Multiple alleles generated within the same tagging experiment can be used for this purpose.
2. The correlation of the loss of the transposon from the locus in germinal revertant alleles or within somatic sectors of revertant tissue also indicates the cloned locus is responsible for the phenotype. The generation of stocks homozygous for the mutation and subsequent analyses (see above) is useful here.
   Mu elements transpose germinally from a locus at a very low frequency as compared to other transposon systems. It may soon be possible to utilize an early reverting line identified by V. Walbot (1991, 1992) to increase the frequency of germinal reversion. It has been shown that a high frequency (4%) of germinal revertants of *bz2-mu2* are recovered in this line, but work with other *Mu*-induced alleles is necessary before its generalized use can be established.
3. Differential RNA hybridization can facilitate identification of the correct clone in special cases where the expression of the locus is well understood, as was done for

Bz2 (McLaughlin and Walbot 1987). Hybridizations were performed with the putative clones to RNA isolated from various tissues or allelic variants that showed a predicted pattern of expression.

REFERENCES

Alleman M, Freeling M (1986) The *Mu* transposable elements of maize: evidence for transposition and copy number regulation during development. Genetic 112: 107–119

Bennetzen JL, Brown WE, Springer PS (1988) The state of DNA modification within and flanking maize transposable elements. In Nelson OE (ed) Plant Transposable Elements, Plenum Press, New York, pp 237–250

Bennetzen JL (1987) Covalent DNA modification and the regulation of *Mutator* element transposition in maize. Mol Gen Genet 208: 45–51

Brown WE, Robertson DS, Bennetzen JL (1989) Molecular analysis of multiple Mutator-derived alleles of the *Bronze* locus of maize. Genetics 122: 439–445

Buckner B, Kelson TL, Robertson DS (1990) Cloning of the *y1* locus of maize, a gene involved in the biosynthesis of carotenoids. Plant Cell 2: 867–876

Chandler VL, Walbot V (1986) DNA modification of a maize transposable element correlates with loss of activity. Proc Natl Acad Sci USA 83: 1767–1771

Chandler VL, Hardeman KJ (1992) The *Mu* elements of *Zea mays*. Adv Genet, in press

Chomet P, Lisch D, Hardeman KJ, Chandler VL, Freeling M (1991) Identification of a regulatory transposon that controls the Mutator transposable element system in maize. Genetics 129: 261–270

Dooner HK, Belachew A (1989) Transposition pattern of the maize element *Ac* from the *bz-m2 (Ac)* allele. Genetics 122: 447–457

Greenblatt IM, Brink RA (1962) Twin mutations in medium variegated pericarp maize. Genetics 47: 489–501

Hake S (1992) Unraveling the knots in plant development. Trends Genet 8: 109–114

Hershberger RJ, Warren CA, Walbot V (1991) Mutator activity in maize correlates with the presence and expression of the *Mu* transposable element *Mu9*. Proc Natl Acad Sci USA 88: 10198–10202

Martienssen R, Barken A, Taylor WC, Freeling M (1990) Somatically heritable switches in the DNA modification of *Mu* transposable elements monitored with a suppressible mutant in maize. Genes Dev 4: 331–343

Martienssen RA, Barken A, Freeling M, Taylor WC (1989) Molecular cloning of a maize gene involved in photosynthetic membrane organization that is regulated by Robertson's *Mutator*. EMBO J 8: 1633–1639

McCarty DR, Carson CB, Stinard PS, Robertson DS (1989) Molecular analysis of *viviparous-1*, an abscisic acid-insensitive mutant of maize. Plant Cell 1: 523–532

McLaughlin M, Walbot V (1987) Cloning of a mutable *bz2* allele of maize by transposon tagging and differential hybridization. Genetics 117: 771–776

Nowick EM, Peterson PA (1981) Transposition of the *Enhancer* controlling element system in maize. Mol Gen Genet 183: 440–448

O'Reilly C, Shepherd NS, Pereira A, Schwarz-Sommer Z, Bertram I, Robertson DS, Peterson PA, Saedler H (1985) Molecular cloning of the *a1* locus of *Zea mays* using the transposable elements En and Mu1. EMBO J 4: 877–882

Patterson GI, Harris LJ, Walbot V, Chandler VL (1991) Genetic analysis of *B-peru*, a regulatory gene in maize. Genetics 127: 205–220

Pryor A (1987) Stability of alleles of Rp. Maize Genetics Cooperation News Letter 61: 37–38

Qin M, Robertson DS, Ellingboe AH (1991) Cloning of the *Mutator* transposable element *MuA2*: a putative regulator of somatic mutability of

the *a1-Mum2* allele in maize. Genetics 129: 845–854
Robertson DS (1978) Characterization of a mutator system in maize. Mut Res 51: 21–28
Robertson DS (1980) The timing of *Mu* activity in maize. Genetics 94: 969–978
Robertson DS (1985) Differential activity of the maize mutator *Mu* at different loci and in different cell lineages. Mol Gen Genet 200: 9–13
Robertson DS, Stinard PS (1989) Genetic analyses of putative two-element systems regulating somatic mutability in Mutator-induced aleurone mutants of maize. Dev Genetics 10: 482–506
Stadler LJ (1946) Spontaneous mutation at the *R* locus in maize I. The aleurone-color and plant-color effects. Genetics 31: 377–394
Walbot V (1991) Early events in Mutator lines. I. Germinal revertants, II. somatic reversion. Maize Genetics Cooperation News Letter 65: 95–96, cited by permission
Walbot V (1992) Strategies for mutagenesis and gene cloning using transposon tagging and T-DNA insertional mutagenesis. Annu Rev Plant Phys Plant Mol Biol, in press
Walbot V (1992) Early sectoring and germinal reversion in a Mutator line with the *bz2-mu2* reporter allele. Maize Genetics Cooperation News Letter 66: 101, cited by permission
Walbot V, Briggs CP, Chandler V (1986) Properties of mutable alleles recovered from mutator stocks of *Zea mays* L. In Gustafson JP, Stebbins GL, Ayala FJ (eds) Genetics, Development and Evolution, Plenum Press, New York, pp 115–142

28

Mapping Genes with Recombinant Inbreds

BENJAMIN BURR, FRANCES A. BURR AND EILEEN C. MATZ

Recombinant inbreds (RIs) are derivatives of an F_2 population in which linked blocks of parental alleles are now essentially fixed. Like other replicated populations—doubled haploids, for example—these populations can be continuously propagated. This means that different investigators can work with the same materials and that all data generated from the same recombinant inbred population(s) contribute to a common database. Because RIs have undergone multiple rounds of recombination before attaining homozygosity, they also permit higher mapping resolution for short linkage distances. RI families were

The Maize Handbook—M. Freeling, V. Walbot, eds.

first employed for gene mapping in mice, where they continue to be extensively used. Taylor (1978) and Bailey (1981) have published detailed, helpful discussions on the theory of their use.

There are at least four maize RI populations in which there has been extensive mapping. We have previously written about the use of RIs for mapping in maize (Burr et al. 1988, 1991). The present paper is meant to be an informal discussion on mapping with RIs in maize, drawn from empirical lessons we have learned.

HOW ARE RECOMBINANT INBREDS DERIVED?

The starting materials for a recombinant inbred family are two distinct inbred lines. This assures that there will be only two alleles for any locus segregating in the subsequent population. An F_2 population is obtained from these two original inbreds. The F_2 individuals and their subsequent progeny are self-pollinated, giving rise to new inbred lines. Effective homozygosity is obtained in less than ten generations of inbreeding beyond the F_2. During the inbreeding process, it is important to avoid selection in order to prevent skewing allele distributions or maintaining heterozygosity. Because inbreeding leads to the lack of vigor in some genotypes, selection cannot be entirely avoided. Propagation is by the ear-to-row method. When RI seed is distributed, only limited amounts are sent to any one laboratory to ensure that all groups will be working from the same individuals in a given generation.

WHAT CAN BE MAPPED?

Any trait for which there is polymorphism between the two original inbred parents of the RI family can be mapped. This can be a morphological trait (e.g., plant color), a physiological reaction (e.g., disease resistance), an isozyme or protein mobility polymorphism, or a DNA polymorphism detected by Southern blotting or by PCR. The trait does not have to exhibit simple inheritance but it cannot be cytoplasmically inherited, because in an RI family all the progeny receive only one cytoplasm. Multigenic traits can be mapped. In fact, these replicated families, if they are large enough, are ideal for mapping segregating factors controlling quantitative traits.

HOW IS MAPPING DONE?

Mapping is straightforward and simple. After a polymorphism that distinguishes the original inbred parents has been identified, each recombinant inbred line in the family is typed to determine which parental allele it carries. This information comprises a strain distribution pattern for the newly mapped gene. A database search locates previously mapped loci having similar strain distribution patterns. The more tightly linked two genes are, the greater the resemblance in their strain distribution patterns. A computer program reports linked loci and also computes the map distance between the new locus and the previously mapped loci.

Seed of the most recent expansion of the T×CM and CO×Tx RI families are available by writing to B. Burr, Biology Department, Brookhaven National Laboratory, Upton, NY 11973. Because these plants do not do well in some environments and because we can send only a limited amount of seed to any one investigator, care should be exercised in planting, in keeping track of the seed, and in preparing DNA from the resulting plants. Use of both RI families serves three functions: (1) Results from one family check the results from a second. (2) As long as the same locus is scored in both families, it is possible to combine the data for a more accurate estimate of map position. (3) When polymorphism cannot be found in one family, it is usually present in the other.

A common use of the recombinant inbreds is in mapping newly cloned genes. DNA is first prepared from the original parental inbreds and from all the recombinant inbreds of one or both families. As already mentioned, the inclusion of both families, if possible, will provide greater accuracy in map position. For best results, efforts should be made to prepare DNAs from all the members of the families. We generally prepare DNA from leaves taken from young plants that are ca. 30–35 cm high. The leaves of several plants of a recombinant inbred are pooled. This minimizes sampling error. If the plants are to be grown to maturity for propagation, DNA may be made from second ears picked when the ears are immature and only about 5 cm long. Young ears are an exceptionally good source of DNA; see the Molecular Biology section of this volume for directions on freezing ears and extracting their DNA.

DNA of the parental inbreds is commonly digested with three restriction enzymes that have six-base pair recognition sites and are not sensitive to cytosine methylation. We generally begin with *Bam*HI, *Bgl*II, and *Eco*RI, but other enzymes have also been useful. Because maize is such a polymorphic species, three enzymes are usually sufficient to reveal a polymorphism between any two inbreds. After probing with the new clone, the enzyme that best shows polymorphism between the parental inbreds is then used to digest the DNAs of the RIs. The autoradiograms of the resultant Southern blots are then scored for the alternative parental bands (Figure 28.1A). We have assigned arbitrary numerical values of 1 or 2 to designate the alleles of each parent in a pair. Thus in the T×CM population, CM37 alleles receive a value of 1 and T232 alleles acquire a value of 2 regardless of the locus being scored. In the other family, Tx303 alleles receive a value of 1 and CO159 alleles a value of 2. In both populations heterozygotes are assigned a value of 3. (The computer actually reads this as 1.5, but we use the integer because it is easier to type.) Zeros are put in when there is no information for that recombinant inbred. The proper format for the data is to put on one line the name of the gene followed after a few spaces with the allele distribution (Figure 28.1B). All of the recombinant inbreds for a given family must be accounted for. If data from both families have been obtained, then the first line is used for T×CM and the second line for CO×Tx. We also keep track of the name of the clone, its origin, the restriction enzyme used to

A

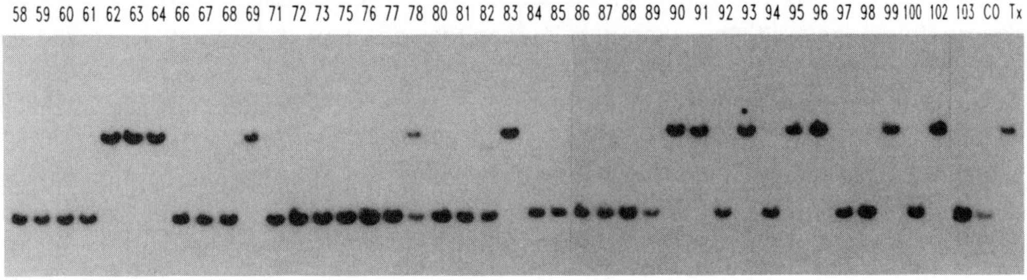

B

npi340A 22221112221222222322212222221121211221212

FIGURE 28.1 **A.** Southern blot showing segregation of the RFLP locus *npi340A* in the CO×Tx RI family whose DNA was digested with *Bgl*II. The RI numbers appear above. The Tx303 allele has a band at 9.4 kb and the CO159 allele has a band at 6.6 kb. **B.** Allele distribution pattern for *npi340A*: 1 = Tx303; 2 = CO159.

detect the polymorphism, the sizes of the allelic bands, and the name of the investigator who contributed the mapping information. The numerical array is compared with the existing database using Keith Thompson's INBRED program. The program reports the most closely linked loci in the database along with their assigned map positions and calculates the distances between these loci and the new gene (Table 28.1). In the example shown, *npi340A* maps at the same location as *umc159* at the end of chromosome *6S*. The program does not assign an actual position for the new marker but this is easily done by visual inspection and simple calculation.

Data obtained with the T×CM or the CO×Tx family can be sent to us preferably by E-mail to burr@bnlux.bnl.gov or by FAX to 516–282–3407, or by diskette to our institutional address. We will run the data through the INBRED program and report the results expeditiously. We hope to have this service available soon through Internet so that geneticists can run the program themselves in an interactive fashion. As of this writing the present consolidated RFLP map of maize has 991 markers, about half of which have been contributed by other investigators.

POTENTIAL PROBLEMS

Because recombinant inbreds undergo multiple rounds of meiosis while they are heterozygous, there have been more chances for recombination between tightly linked mark-

TABLE 28.1. The output from INBRED mapping the allele distribution of *npi340A* given in Fig. 28.1B. This RFLP locus was not separated from *umc159* and maps to the current end of *6S*.

CR	LOCUS	MAP	M[a]	DIFF[b]	R[c]	UNIT[d]
6	*umc*159	6S 00	41	0.0	0.0000	0.0000
6	*nor*	6S 04	40	3.0	0.0750	0.0405
6	*umc*85	6S 04	40	3.0	0.0750	0.0405
6	*bnl*6.29	6S 06	40	4.0	0.1000	0.0556
6	*npi*235	6S 07	41	5.0	0.1220	0.0694

[a]Number of non-zero comparisons.
[b]Total of absolute differences between allele distributions.
[c]Recombination fraction.
[d]Estimated distance in Morgans.

ers. On one hand, this means it is easier to detect nonallelism compared with a similar-sized backcross or F_2 population, but it also means that detection of linkage beyond 20 cM can be unreliable. On the current map, the 991 markers—which include eight sub-telomeric loci—are distributed over 1,680 cM with an average map density of less than 3 cM. Several potential telomeric loci remain unlinked, so there is the remote possibility that we could fail to map a new locus that is distal to one of the present linkage groups.

The most common problems arise from experimental errors and data mishandling. These problems lead to nonlinear results in their mildest form or to the inability to detect reliable linkage at their worst. The database is arranged so that all mapped loci are aligned from the end of *1S* through the end of *10L*. As such, the allele values for each RI can be viewed as a linkage map of parental alleles for a given chromosome. When the newly mapped locus is inserted into its optimal position, there should be few, if any, apparent double crossovers. Of course this is contingent on its proximity to adjacent linked markers, but the presence of two or more double crossovers is usually indicative of problems with the data. Obviously it is imperative that the RI numbers and the DNA samples not be accidently switched. This can be verified, if there is a possibility of confusion, by using a previously mapped clone as a probe. For this purpose we have distributed two clones which each probe two unlinked loci—*bnl5.21A* on *7L*, *bnl5.21B* on *2L*, *bnl17.18A* on *5S*, and *bnl17.18B* on *1L*—and their allele distributions. The allele distributions should match. We would like to emphasize that only unambiguous data should be reported. When faint or obscured blots or poorly resolved bands are misinterpreted, the data are useless both for obtaining a map position and as a contribution to the database. There were about 2% heterozygotes at all loci in the RIs after eight generations beyond the F_2. The presence of a small, but finite, number of heterozygotes is an indication of authentic data. Errors can also creep into data recording. The most frequently encountered mishap is the omission of a data point; this produces a frameshift in the allele distribution. Another common mistake is a reversal in the assignment of parental allele values.

The INBRED program is robust enough to permit a few missing data points, but when only three-fourths or fewer of an RI family are present, only approximate map positions can be assigned. The problem with too few data points is that there are not enough opportunities to observe recombination, so the new locus may appear to be allelic with several adjacent loci.

Mapping is conceptually simple. It must, however, be approached with sufficient rigor to obtain meaningful results. As is the case with other genetic experiments, care must be taken to keep track of strains and DNA samples, good technique should be used in preparing and interpreting Southerns, and every effort should be expended to obtain a full data set.

REFERENCES

Bailey DW (1981) In Foster HJ, Small JD, Fox JG (eds) The Mouse in Biomedical Research, Academic Press, New York, pp 223–239

Burr B, Burr FA, Thompson KH, Albertsen MC, Stuber CW (1988) Gene mapping with recombinant inbreds in maize. Genetics 118: 519–526

Burr B, Burr FA (1991) Recombinant inbreds for molecular mapping in maize: theoretical and practical considerations. Trends Genet 7: 55–60

Taylor BA (1978), In Morse HC (ed) Origins of Inbred Mice, Academic Press, New York, pp 423–438

The Placement of Genes Using waxy-Marked Reciprocal Translocations

JOHN R. LAUGHNAN AND SUSAN GABAY-LAUGHNAN

In maize, translocations with breakpoints closely linked to endosperm marker genes have been employed in the placement of new genes to chromosome. A series of *waxy*-marked reciprocal translocations was developed by E. G. Anderson (1943, 1956). These translocation stocks are homozygous for the *waxy* (*wx*) gene and for a translocation having one of its breakpoints in chromosome 9 not far from the *wx* locus. The other breakpoints are distributed among the other nine chromosomes. Crossing-over is suppressed in the region of the translocation and this increases the efficiency of linkage detection. These translocation strains, along with B-A translocation stocks (Beckett 1978, 1990; this volume), have been widely used by maize geneticists to determine the chromosomal placement of unlocated genes.

PLACEMENT OF GENES EXPRESSED IN THE SPOROPHYTE

The protocol involves crossing plants exhibiting the trait to be located with a *wx* tester strain. The F_1 plants are self-pollinated and the starchy (*Wx*) and *wx* kernels from the F_2 generation are planted separately. If the plants grown from the *Wx* and *wx* kernels exhibit the mutant trait with equal frequency, it can be said that the unplaced gene is not linked to the *wx* locus. Plants exhibiting the unplaced gene are also crossed to the series of *wx*-marked translocations. The F_1 plants are self-pollinated and the *Wx* and *wx* kernels from these ears are planted separately. All but one of these populations should exhibit random distribution of the mutant trait between the *Wx* and *wx* classes. The exceptional line has the chromosome on which the unplaced gene is located translocated to chromosome 9. Genes residing in the distal regions of some chromosome arms might not be placed using the *wx*-translocation series because of high recombination between the breakpoint and very distal loci; other means would be used to locate them. The great majority of genes, however, can be located to chromosome relatively easily.

The Maize Handbook—M. Freeling, V. Walbot, eds.

This technique involves scoring the sporophytes of the F_2 generation. If the mutant trait is exhibited in the endosperm, the test is carried out by scoring kernels on the self-pollinated F_1 ears. If it is exhibited in the seedling, the scoring can be accomplished in a seedling bench. If the unlocated gene is expressed only in the mature plant, this procedure requires classification in a field nursery.

PLACEMENT OF GENES EXPRESSED IN THE MALE GAMETOPHYTE

We have been investigating a series of newly arisen restorer-of-fertility (*Rf*) genes. These genes restore fertility to S-type male-sterile cytoplasm (*cms-S*) and male fertility is a mature plant trait. Restoration of *cms-S* is gametophytic and therefore an *Rf* gene can be scored in the male gametophyte; pollen carrying *Rf* is normal and that carrying *rf* aborts. Reciprocal translocations and *Wx-wx* are also expressed in the male gametophyte. Translocations result in pollen abortion when heterozygous; *Wx* pollen grains stain blue with iodine while *wx* grains stain red. Because the traits involved can all be scored in the male gametophyte, the *wx*-translocation techniques have been adapted to this generation, the pollen of the F_1 plant. This greatly simplifies the linkage analysis of new *Rf* genes.

The protocol involves crossing *cms-S* plants heterozygous for the unplaced *Rf* gene

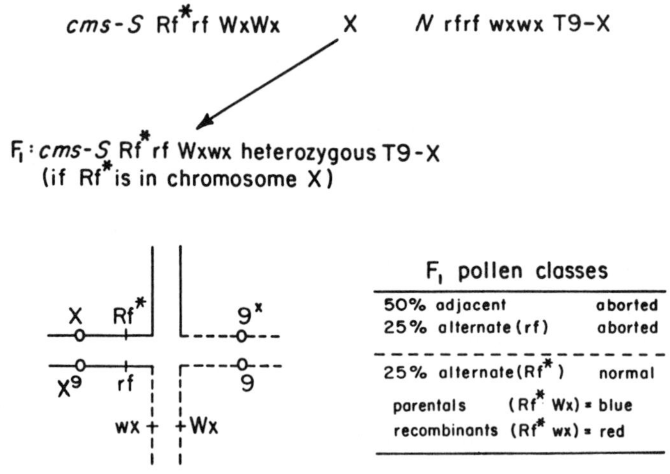

FIGURE 29.1 Pollen method for determination of linkage group of a new restorer, designated *Rf**. The new restorer strain carrying S cytoplasm is crossed as female parent with the *waxy* translocation tester series and iodine-stained pollen samples from F_1 plants exhibiting 75% abortion are analyzed. Linkage of the new *Rf** with one of the chromosomes involved in the interchange is detected on the basis of a discrepant ratio for blue:red-staining pollen grains (from Laughnan and Gabay © 1978 reprinted by permission of John Wiley and Sons Inc.)

as female parents with nonrestoring (*rf*) versions of the *wx*-marked translocations. Samples of tassel branches from fertile F_1 plants exhibiting 75% pollen abortion are collected when the anthers are mature and stored in 70% ethyl alcohol. Restorer gene heterozygosity results in 50% pollen abortion. Translocation heterozygosity is also expressed as 50% pollen abortion. A plant heterozygous for both a reciprocal translocation and an *Rf* gene exhibits 75% pollen abortion. The *waxy* trait is scored among the 25% normal pollen grains. Iodine-stained pollen samples are examined under the dissecting microscope (total magnification 27×). If the restorer gene being tested is located on a chromosome other than the two involved in the reciprocal translocation, there should be an equal frequency of blue- and red-staining pollen grains. If linkage is encountered however, more than 50% of pollen grains stain blue, the proportion of blue- and red-staining grains being a function of the recombination between the *Rf* and *wx* loci (Figure 29.1).

If linkage is indicated from pollen analysis, it can be verified by crossing the F_1 plants whose pollen analyses indicate linkage as male parents onto homozygous *wx* tester plants. Since only pollen grains carrying *Rf* will function, linkage results in an excess of the *Wx* endosperm class. Here, as with the pollen procedure, the proportion of *wx* kernels is a function of the recombination between *Rf* and *wx*. Concordance of testcross data with those from pollen analyses has served to validate the pollen technique (Laughnan and Gabay 1978).

REFERENCES

Anderson EG (1943) Utilization of translocations with endosperm markers in the study of economic traits. Maize Genet Coop News Letter 17: 4–5

Anderson EG (1956) The application of chromosomal techniques to maize improvement. In Genetics in Plant Breeding, Brookhaven Symposia in Biology: No. 9, Upton, New York, pp 23–36

Beckett JB (1978) B-A translocations in maize. J Hered 69: 27–36

Beckett JB (1990) Cytogenetic, genetic and plant breeding applications of B-A translocations in maize. In Gupta PK, Tsuchiya T (eds) Chromosome Engineering in Plants: Genetics, Breeding, Evolution, Part A, Elsevier Science Publishers BV, Amsterdam, pp 493–529

Laughnan JR, Gabay SJ (1978) Nuclear and cytoplasmic mutations to fertility in S male-sterile maize. In Walden DB (ed) Maize Breeding and Genetics, John Wiley and Sons, New York, pp 427–446

Chimeras for Genetic Analysis

M. G. Neuffer

A chimera—defined as an individual (plant) composed of two or more genotypes—usually results from any heritable change that provides different expression by the descendants of two daughter cells from a mitotic cell division. Chimeras can be very useful in revealing the action of mutant genes in plants. McClintock (1938) first demonstrated the feasibility of using deficiencies and unstable ring chromosomes to study the consequences of the absence of functional genes in homozygous mutant tissue. Stadler and Roman (1948) identified three previously unknown genes near the *a1* locus by uncovering different lengths of X-ray induced deficiencies (*a-x1*, *a-x2*, *a-x3*) with an unstable fragment (A-b Frag) carrying functional genes located in the deficiency segments. Similar studies by McClintock (1951, 1965), Steffensen (1968), Coe and Neuffer (1978), and Johri and Coe (1983) demonstrate the utility of this approach in solving problems in biology. An advantage of chimeras in genetic research is that they provide mutant tissue that is of identical age and supported by adjoining normal tissue; this allows expression of lethal mutants that would not normally be observable. Chimeras may arise spontaneously for unspecified reasons or they may be produced by unstable chromosome configurations (inversions, rings, centric fragments, monosomics), by chromosome breaking agents (radiation or chemicals) or by unstable loci resulting from transposon insertion (*Ds*, etc.). Analysis of chimeras can tell us that some genes encode phenotypes that are cell autonomous (anthocyanin and chlorophyll genes) and that some are not (andromonoecious dwarfing). They also (1) show that some lethal mutants may be rescued by supplying missing chemicals, (2) indicate the developmental pattern of the plant parts, and (3) provide clues as to the sequence of events in a biosynthetic pathway.

This presentation deals with the production and use of chromosome breaking *Ds* sites on various chromosome arms as a means of studying defective kernel lethal mutants on those arms. Transposon-induced chromosome breakage has several advantages. Ring chromosomes produce beautiful sectors but are quite unstable and practically impossible to maintain. Radiation-induced breaks are much less frequent and are hard to reproduce, and induced mutations and multiple breaks can confound the analysis. *Ds*-induced chromosome breaking, on the other hand, occurs at a high frequency and at a specific

The Maize Handbook—M. Freeling, V. Walbot, eds.

location, so repeated events can be seen in the same individual. McClintock's (1951) elegant demonstration of these events on the short arm of chromosome 9 showed what can be done.

New locations for the chromosome-breaking *Ds* effect were produced (Neuffer 1986) on all chromosome arms for which there was a reliable marker, by crossing families of 100 ears of each of the marker stocks in an isolated detasseled plot by pollen of a stock carrying a chromosome-breaking *Ds* and its activator *Ac*. The resulting ears were examined for single kernels or for seedlings that showed chimeral losses for the marker in question. The marker stocks used as female lines were carrying one of the following aleurone or endosperm genes; *dek1* (1S), *bz2* (1L), *b* (2S), *dek5* (3S), *a1 sh2* (3L), *bt2* (4S), *c2* (4L), *a2* (5S), *bt1* (5L), *pro1* (8L), *cp*-1381* (9L), or *dek 14* (10S) or the seedling genes *w3* (2L), *cl* (3S), *wl**-217A (6L), *vp9* (7S), or *o5* (7L). The pollen stock, supplied by Dr. Jerry Kermicle, had a chromosome-breaking *Ds* on the long arm of chromosome 10 proximal to the *R-sc* allele at the *r1* locus and *P1-vv* (an active *Ac* at the *p1* locus on chromosome 1). A variant of this stock carrying the aleurone color *B1* allele (*B:Peru*) was used to produce chromosome 2S cases. Good cases of insertion were located on 11 chromosome arms. These are listed below with the number of cases and marker genes indicated.

1S	*Dek1*–(4)	4L	*C2*–(7)
1L	*Bz2*–(1) *bz2-m* (5)	5S	*A2*–(2) *a2-m* (1)
2S	*B:Peru* (2) *b-m* (2)	5L	*Bt1*–(3)
2L	*W3*–(1)	9S	*C1*–(McClintock) (1)
3S	*Cl1*–(3)	10L	*R-sc* (Kermicle) (2)
3L	*A1–Sh2*–(2)		

Tentative *Ds* sites on seven other arms have appeared but have not been analyzed.

An appreciation of chimeral analysis is demonstrated by a study of the colorless floury defective (*dek1*) mutant. This mutant has a colorless aleurone (no anthocyanin) because there is actually no aleurone layer present, a white floury endosperm because there are no carotenoids and no corneous starch, and a nonviable embryo because it fails to develop stem and leaf primordia. The embryo does have normal root primordia and will produce a root. The effect of the mutant on chlorophyll and plant morphology cannot be seen because mutant kernels do not produce plants.

From the cross of *Dek1/dek1* ears by *Dek1/Dek1* pollen that carried a chromosome-breaking *Ds* near *Dek1* and *Ac* at the *pl* locus, kernels with *dek1/Dek1-Ds* embryos were obtained. Because both parents also had the required anthocyanin genes for aleurone color (*AlBzlClRl*), mosaicism for aleurone color in the kernel was observed (Figure 30.1). The colored areas had purple aleurone cells with underlying yellow corneous starch in the endosperm while the colorless areas had no aleurone layer and underlying white floury endosperm tissue. The borders of colorless and floury starch were not identical but roughly parallel: the floury area was smaller than the colorless area. When planted these kernels germinated and produced nearly normal green seedlings that had tiny leaf sectors of irregularly formed tissue (Figure 30.2). These were assumed to result from loss

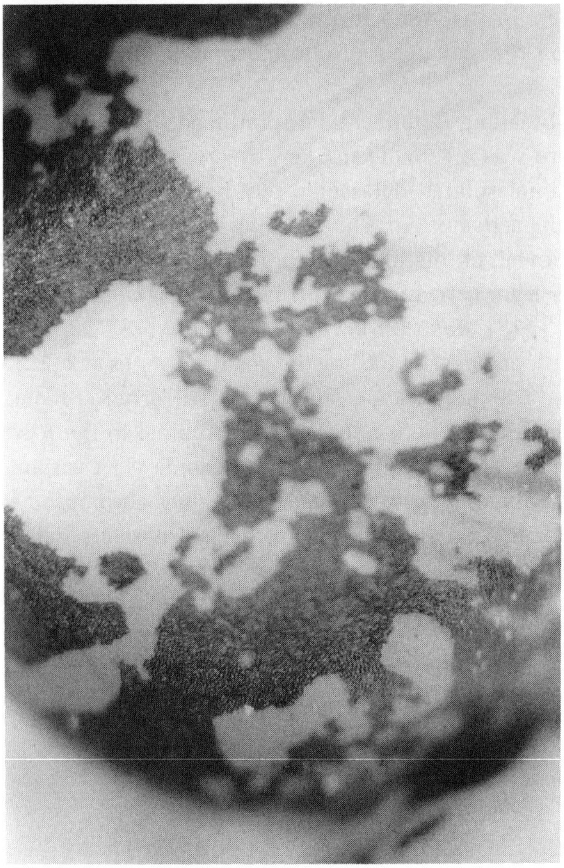

FIGURE 30.1 A heterozygous *Dek1 Ds-4*/*Dek1 Ds-4*/*dek1* kernel from a selfed ear of *Pl-vv Dek1 Ds-4*/*pl-ww dek1*, showing frequent early and late losses of *Dek1*(colorless areas). The absence of color results from the absence of the aleurone where purple pigment is produced. The colorless areas also have floury noncorneous starch in the underlying endosperm. Note that the floury area is smaller than the flinty area, indicating a lack of cell autonomy.

of the *Dek1Ds* segment. Cross sections through an area of the leaf bearing such a sector showed a morphological displacement of tissue such that there was an indentation of the top surface and a protrusion of the underneath surface within the borders of the sector. Occasional larger variants of these sectors were albino, indicating the loss of chlorophyll function.

Thus from chimeral analysis using a chromosome breaking *Ds* we have learned three additional facts about *dek1*—namely, that the floury starch aspect is not cell autonomous, that the mutant causes morphological changes in leaf tissue, and that the mutant leaf tissue is not capable of producing chlorophyll.

One aspect of this material that is not entirely satisfactory is the fact that while chromosome breakage events in the kernel (aleurone and endosperm) are early and therefore produce large chimeras, the embryo and hence the seedling and plant events are very late, so the chimeras are very small. Mutant tissue may be obscured by surrounding or overlapping layers of normal tissue. This is less desirable from an observational point of view than the ring chromosome losses, which are large and easily observed. On the other

FIGURE 30.2 Picture of the underneath surface of a mature leaf of a *Pl-vv Dek1 Ds-4/pl-ww dek1* plant showing small chimeral losses of the normal *Dek1* allele. Note that (1) the sectors of mutant tissue protrude from the leaf surface and (2) the larger sectors are albino, indicating morphological and chlorophyll blocking properties of *dek1*.

hand, smaller sectors are more desirable because of better tissue survival in the case of serious lethals.

REFERENCES

Coe EH Jr, Neuffer MG (1978) Embryo cells and their destinies in the corn plant. In Subtelny S, Sussex IM (eds), The Clonal Basis of Development, Academic Press, New York, pp 113–129

Johri MM, Coe EH Jr (1983) Clonal analysis of corn plant development. I. The development of the tassel and the ear shoot. Dev Biol 97: 154–172

McClintock B (1938) The production of homozygous deficient tissues with mutant characteristics by means of the aberrant mitotic behavior of ring-shaped chromosomes. Genetics 23: 315–376

McClintock B (1951) Chromosome organization and genic expression. Cold Spring Harbor Symp Quant Biol 16: 13–47

McClintock B (1965) Components of action of the

regulators *Spm* and *Ac*. Carnegie Inst Wash Year Book 64: 527–534

Neuffer MG (1986) Transposition of chromosome breaking *Ds* to marked chromosome arms. Maize Genetics Coop News Letter 60: 55–56

Stadler LJ, Roman H (1948) The effect of x-rays upon mutation of the gene *A* in maize. Genetics 33: 273–303

Steffensen DM (1968) A reconstruction of cell development in the shoot apex of maize. Am J Bot 55: 354–369

31

The Use of Clonal Sectors for Lineage and Mutant Analysis

SARAH HAKE and NEELIMA SINHA

Clonal analysis can be used to study the relationship of cell lineages in the formation of the plant and to study the autonomy and action of a gene product. Following a brief review, methodology for clonal analysis of gene expression will be presented.

In clonal analysis, a genetic event visibly marks a cell and its progeny. X-ray-induced chromosome breakage has traditionally marked cells by exposing recessive alleles, but mutable alleles that revert to a dominant phenotype can also be used for cell lineage studies. By examining a large number of cell lineages, or sectors, generalizations can be made as to their contribution in development. Clonal analysis has shown that relatively few cells, two to four, give rise to an ear, whereas many more cells give rise to a leaf (Coe and Neuffer 1978; Johri and Coe 1983). The number of cells that give rise to a tassel has been determined to be three to five (Johri and Coe 1983); however, the number of nodes in the plant may alter the results (McDaniel and Poethig 1988). The relationship between organs can be determined by this technique. For example, it has been shown that the ear is most closely related in cell lineage to the internode and leaf above it (Johri and Coe 1983; McDaniel and Poethig 1988). The relationship of cells within an organ can also be determined. In an anther, the procedure has established that pollen is related in lineage to the inner anther wall, while the outer anther wall is a separate lineage

The Maize Handbook—M. Freeling, V. Walbot, eds.

(Dawe and Freeling 1990). In the same vein, mesophyll and bundle sheath cells have been shown to be clonally related (Langdale et al. 1989). A clonal analysis of the embryonic cell lineage of the shoot meristem was performed by Poethig and co-workers (Poethig et al. 1986; McDaniel and Poethig 1988). While this study provided guidelines as to a meristem cell's fate, enough exceptions can be found that it appears there are no compartments, or developmentally set apart cells. An overriding conclusion from lineage analysis is that the fate of cells is determined by their position and not their lineage.

Clonal analysis of a genetic mutation requires a linked gene with a visible phenotype that is easily scored and cell autonomous. The marker gene allows one to ascertain whether or not the mutant gene is present. Basic questions that can be answered are whether the gene product is diffusible, in what cell type the gene product acts, at what time in development the gene acts, what effect the gene product has on cells that lack the product, or in the case of dominant mutations, what effect the mutant gene product has on wild-type cells. A gene is considered autonomous if the effect of the gene (a particular allele) is only visible in the cells carrying that gene (allele). In a number of cases the distinction between cell autonomous or not depends on the size of the sector in the organ. For example, Harberd and Freeling (1989) found that the tassel branches of the dwarf mutation, *D8*, were short except when the dominant gene was lost on the entire tassel branch, an event marked by uncovering a linked albino mutation. In contrast, sectors of wild-type cells within a dwarf leaf rarely had an effect on the overall morphology of the leaf (Harberd and Freeling 1989). Periclinal chimeras, in which different layers have different genotypes, can also be examined to determine the effect one cell layer has on another. A general finding from examining periclinal chimeras is that the contribution of the epidermis is minimal except when the trait is largely epidermal (Hake and Freeling 1986; Sinha and Hake 1990; Harberd and Freeling 1989; Becraft et al. 1990).

EXPERIMENTAL APPROACHES FOR MUTANT ANALYSIS

1. Determine the map position of gene.
2. Find a marker that is closely linked or at least proximal with respect to the centromere (Figure 31.1). Most X-ray-induced aberrations in maize seem to be terminal deletions or large interstitial deficiencies. Recessive mutations should be on the same chromosome as the recessive marker, while dominant mutations should be in repulsion to the recessive marker (on the homologous chromosome) (Figure 31.2). If a linked marker is not available, then a reciprocal translocation can be used as diagrammed in Figure 31.3. The translocation brings the chromosome arm carrying the gene of interest into linkage with a chromosome that has an appropriate marker. The translocation needs to be crossed to a line carrying the recessive marker and the mutation.

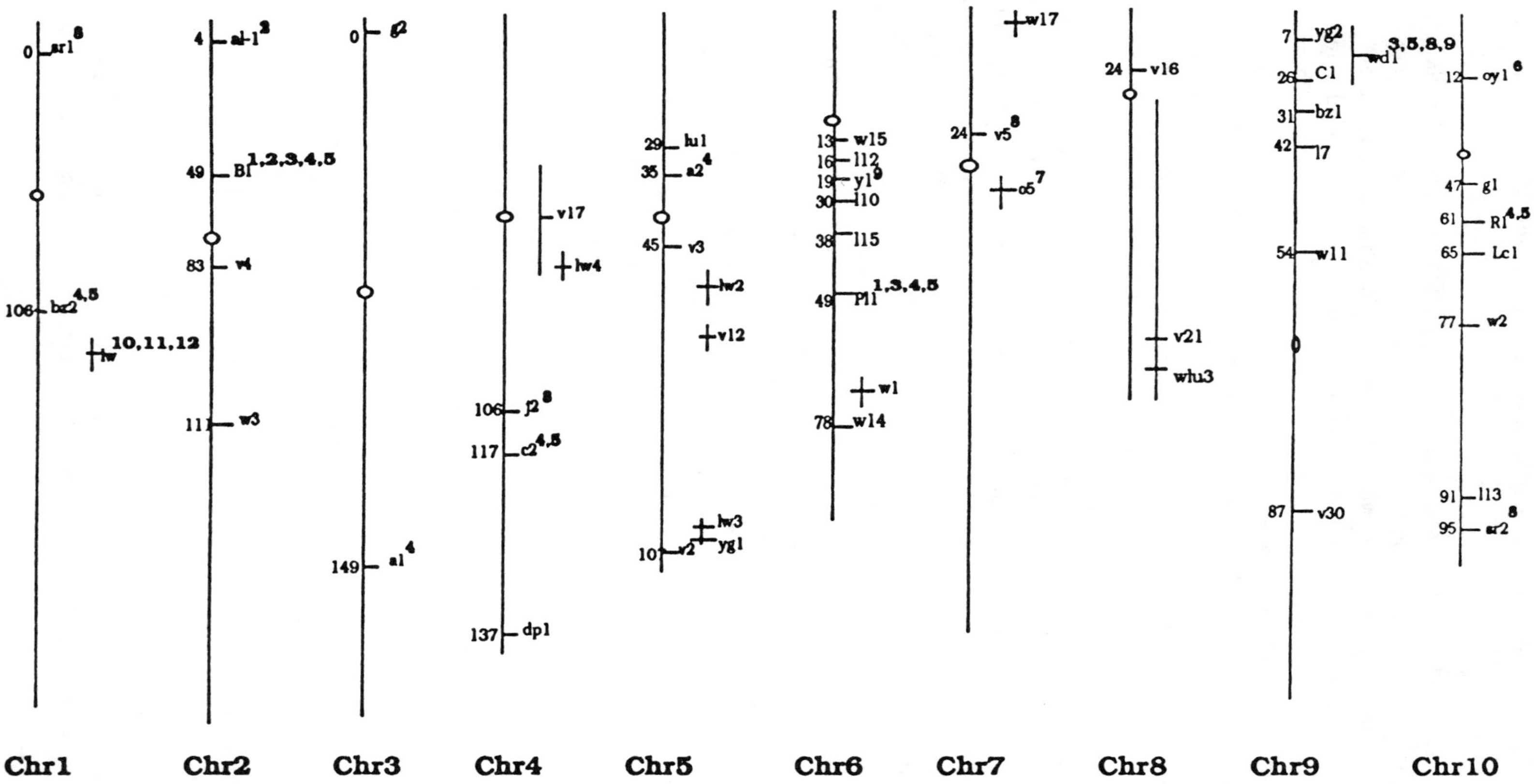

Chr1
Chr2
Chr3
Chr4
Chr5
Chr6
Chr7
Chr8
Chr9
Chr10
10,11,12
1,2,3,4,5
1,3,4,5
3,5,8,9
v17
v12
w15
w11
w14
w17
v16
v21
v30
bz1
dp1
C1
Lc1
w2
w3

FIGURE 31.1

FIGURE 31.1 Maize chromosomes are redrawn from the *Maize Genetics Cooperation Newsletter* (65). Mutants not precisely mapped have approximate map locations on lines to the right of the appropriate chromosome. Markers considered suitable for lineage and mutant analyses are indicated. Superscript numbers next to markers indicate references to use of these markers in published studies: [1]McDaniel and Poethig (1988); [2]Becraft et al. (1990); [3]Poethig et al. (1986); [4]Dawe and Freeling (1990); [5]Johri and Coe (1983); [6]Bennetzen et al. (1988); [7]Poethig (1988); [8]Langdale et al. (1989); [9]Steffensen (1968); [10]Hake and Freeling (1986); [11]Sinha and Hake (1990); [12]Harberd and Freeling (1989).

Consideration of Marker

The most important consideration of a marker is that it be autonomous. A second consideration is the spectrum of cells in which the gene is expressed (Table 31.1). Albino mutations are preferable for a number of reasons: (1) they can be observed in many different tissues and organs; (2) cell layer changes are detectable as variations in pale green hues; (3) sectors on seedlings or juvenile plants can be observed; (4) segregation of the albino gene only needs special consideration. Color genes offer the advantage of not affecting viability when the progeny are homozygous recessive. The requirement that all the anthocyanin genes (*A1*, *A2*, *C2*, *B*, *Pl*, *Bz2*, and *Bz1*) be homozygous dominant, except the marker gene in use, can complicate stock construction. It is also difficult to score sectors on young plants because full coloration is not achieved quickly. Anthocyanin genes are exceptionally good markers for following

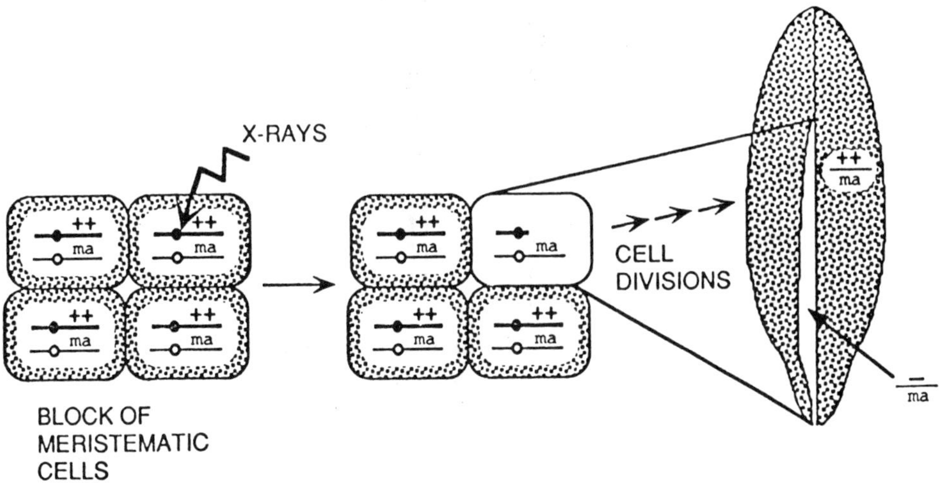

FIGURE 31.2 A graphic illustration of X-ray-induced clonal analysis in leaf tissue showing how a single "marked" cell can give rise to a clone of "marked" tissue.

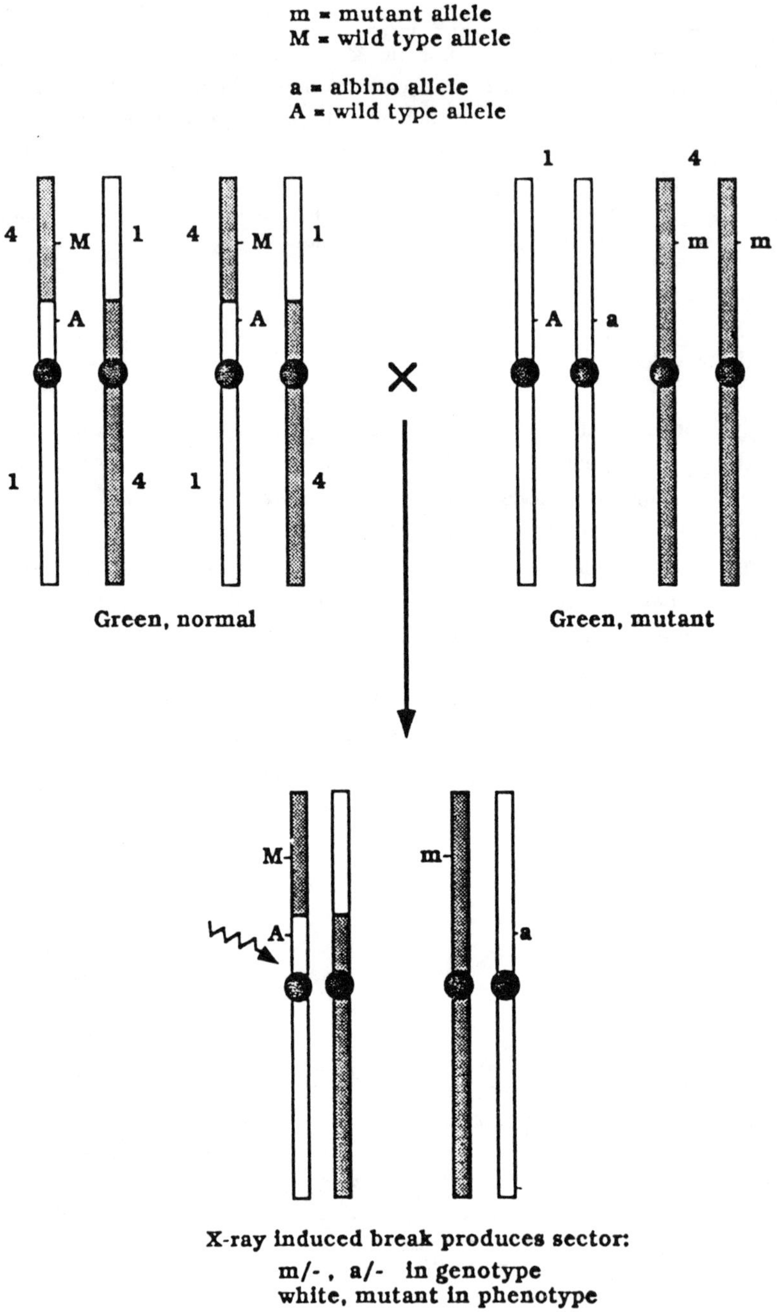

FIGURE 31.3 Translocations can be used to move mutant alleles close to suitable markers on a different chromosome arm to facilitate generation of X-ray-induced clonal sectors.

development in certain organs such as the tassel or stalk in which chlorophyll pigmentation is pale or absent (Table 31.1). Hair phenotypes have been used as successful epidermal markers in tomato species (Goffreda et al. 1990) but have not been exploited in maize.

3. Irradiation Conditions

Dry Seeds

10,000–30,000 rads.

Developing Embryos on the Ear

500 rads.

Seedlings

1,000 rads. 2,000 rads does not cause visible damage to the cells (Stein and Steffensen 1959), but does lead to a lower germination frequency.

Ear Irradiation

If early or large sectors are required for the experiment, irradiation of ears is suggested. The plants need to be grown in pots for easy transport to an X-ray source; or alternatively, a mobile X-ray machine can be taken into the field. Because embryo development varies considerably, a few embryos should be dissected out at the time of irradiation for fixation and subsequent staging.

Seedling Irradiation

Most experimenters have irradiated seedlings—lower radiation dose is required and there is a high frequency of sectors. Once enough seed has been generated (at least 500–1,000 kernels that will have the appropriate genotype), the seeds are imbibed on moist white paper towels for 2–3 days. It is preferable to germinate them on trays that are perforated for good drainage. When the majority of the seeds have germinated, and the radicle has protruded 1–4 cm, the seeds are ready for irradiation. The seeds are kept moist and irradiated directly on the wet paper towels. If possible, a dose-response curve should be obtained with a sample of identical seed to determine a dose that provides 70–80% viability. It is also desirable to have the seeds placed on a rotating platform to give a more uniform dose to all the seeds. Radiation badges must be used; complete safety instructions should be obtained for the particular machine in use.

Table 31.1. Expression of Marker Genes in Maize Tissues

ALBINO MARKER
- coleptile
- leaves
  - blade
    - L1 (epidermis): detected by florescence in guard cells
    - L2 (inner layers): pale green to white
  - sheath
- stem
- cob: husks need to be peeled back[a]
- tassel
  - glumes
  - branches
- husks

ANTHOCYANIN MARKER[b]
- leaves
  - blade: L1 only, poor detection
  - sheath: L1 and L2
- stem
  - L1
  - L2: outer cell layers and scattered vascular bundles
- husks
- cob
- scutellum (need *Rsc* or *B-Peru*)
- tassel
  - anthers (need *R* alleles)
  - glumes

[a](Harberd and Freeling, 1989)

[b]Before attempting to use anthocyanin genes, consideration should be given to the different alleles and interactions, for example, in some tissues, *B* and *R* can substitute for one another (Coe et al., 1988).

4. Planting and Observations

   Seeds can be immediately planted or kept moist until the following day. It is easier to plant them without damage when they are small. If the soil is rocky or has large clods, the seeds can be planted in a bed of vermiculite laid into a trough. Watering should commence immediately following planting. Controls should be planted that are of the same genotype, germinated in a similar way, but without radiation.

Each plant with a sector should be given a number. Each sector can be assumed to arise from one cell. All details of that plant should be noted, especially which leaf the sector is on, the size and placement of the sector, and whether the sector continues on to the next leaf. For chlorophyll-containing marker genes, a piece of tissue containing the sector can be removed, and a thin hand-section can be observed under the fluorescence microscope. Anthocyanin sectors can also be examined by simple hand-sections. It is important to determine in which layer the sector occurs (Table 31.1). In sectors that extend from one leaf to another, the position of the sector within the stem can be observed and used to extrapolate to the cell layer of the leaf, or vice versa. Sectors can change layers by cell invasion, so extrapolation should be done with caution; invasion occurs primarily within the inner layers or from the epidermis into the inner layers (by periclinal division). There are virtually no cases of invasion into the epidermis by interior cells.

5. Complications
   Complications that arise usually concern the marker. If it is not closely linked and proximal, it is possible to lose the gene bearing the visible marker, but not the gene in question. The analysis of such a sector would incorrectly suggest that the gene was not autonomous. Another consideration is whether the mutant phenotype expresses in the background of the marker. A last consideration is whether the expression of the gene is affected by aneuploidy. This possibility can be explored by crossing the mutation to the B-A translocation for the particular arm.

REFERENCES

Becraft PW, Bongard-Pierce DK, Sylvester AW, Poethig RS, Freeling M (1990) The *liguleless-1* gene acts tissue specifically in maize leaf development. Dev Biol 141: 220–232

Bennetzen JL, Blevins WE, Ellingboe AH (1988) Cell-autonomous recognition of the rust pathogen determines *Rp1*-specified resistance in maize. Science 241: 208–210

Coe EH Jr, Neuffer MG (1978) Embryo cells and their destinies in the corn plant. In Sussex IM, Subtelny S (eds) The Clonal Basis of Development, Academic Press, New York, pp 113–129

Dawe RK, Freeling M (1990) Clonal analysis of the cell lineage in the male flower of maize. Dev Biol 142: 233–245

Goffreda JC, Szymkowiak EJ, Sussex IM, Mutschler MA (1990) Chimeric tomato plants show that aphid resistance and tricylglucose production are epidermal autonomous characters. Plant Cell 2: 643–649

Hake S, Freeling M (1986) Analysis of genetic mosaics shows that the extra epidermal cell divisions in *Knotted* mutant maize plants are induced by adjacent mesophyll cells. Nature 320: 621–623

Harberd NP, Freeling M (1989) Genetics of dominant gibberellin-insensitive dwarfism in maize. Genetics 121: 827–838

Johri MM, Coe EH Jr (1983) Clonal analysis of corn plant development I. The development of the tassel and the ear shoot. Dev Biol 97: 154–172

Langdale JA, Lane B, Freeling M, Nelson T (1989)

Cell lineage analysis of maize bundle sheath and mesophyll cells. Dev Biol 133: 128–139
McDaniel CN, Poethig RS (1988) Cell-lineage patterns in the shoot apical meristem of the germinating maize embryo. Planta 175: 13–22
Poethig RS, Coe EH Jr, Johri MM (1986) Cell lineage patterns in maize embryogenesis: a clonal analysis. Dev Biol 117: 392–404
Poethig S (1988) A non-cell-autonomous mutation regulating juvenility in maize. Nature 336: 82–83
Sinha N, Hake S (1990) Mutant characters of *Knotted* maize leaves are determined in the innermost tissue layers. Dev Biol 141: 203–210
Steffensen DM (1968) A reconstruction of cell development in the shoot apex of maize. Am J Bot 55: 354–369
Stein OL, Steffensen D (1959) Radiation-induced genetic markers in the study of leaf growth in *Zea*. Am J Bot 46: 485–489

32

Use of Segmental Aneuploids for Mutant Analysis

BEN GREENE and SARAH HAKE

The utility of reciprocal and B-A translocations for generating segmental aneuploids extends beyond assignment of recessive mutations to a particular chromosome arm. Many aspects of mutant analysis rely upon altering the dosage of either the mutated or wild type gene copy. In addition, segmental aneuploids can be used to screen systematically entire genomes for unlinked modifiers, thereby identifying additional members of a pathway. This brief review will consider such analyses in maize, discussing first the use of aneuploids in studies of dominant mutations, followed by the application of such techniques to uncover interacting loci.

In 1932, Muller outlined criteria for the classification of dominant mutations based upon the expressivity of such mutant character in a background of varying dosage of the wild-type allele. This work initiated analysis of dominant heritable effects. In this classic monograph, he defined four broad classes of dominant lesions (Figure 32.1): (1) A mutation is *hypomorphic* when one dysfunctional gene copy is sufficient to give an altered phenotype [as for haplo-insufficient loci, i.e., the *Notch* mutation in *Drosophila* (Artavanis-Tsakonas

The Maize Handbook—M. Freeling, V. Walbot, eds.

1988)]. The hypomorphic mutant phenotype is lessened upon addition of wild-type copies of the gene. (2) Mutations are *hypermorphic* when the mutant character is caused by the overproduction of a normal gene product. In this case, the mutant phenotype is exacerbated with the addition of wild-type gene copies. An example of an overproducing mutation is *Beadex* in *Drosophila* (Lifschytz and Green 1979). (3) A mutation is *antimorphic* when the mutated gene product competes with or directly interferes with the wild-type gene product (one can think of antisense RNA or competitive inhibition by the production of substrate analogs). The dominant phenotype is lessened by addition of wild-type gene copies. An example in maize of a dominant mutation that formally fits these criteria is *Teopod-1* (Poethig 1988).

(4) *Neomorphic* mutations produce a completely novel product and are unaffected by addition or reduction in the number of wild-type alleles present. For example, the *Knotted-1* mutation in maize is neomorphic (Freeling and Hake 1985). Recent data suggest the mutation results in the production of a normal gene product in novel tissues and/or during a novel developmental time (ectopic expression). Other dominant mutations in maize that show insensitivity to the copy number of the wild-type gene include *Teopod-2* (Poethig 1988) and the gibberellin-insensitive dwarfs, *D8* and *Mpl* (Harberd and Freeling 1989). Most hypermorphic and many neomorphic lesions act semidominantly in that the homozygote is more severely transformed than the heterozygote.

Aneuploid studies can also be used to uncover genes that interact with a particular locus. In *Drosophila*, for example, changes in the levels of glucose-6-phosphate dehydrogenase have been monitored while varying the dosage of small regions of chromosomes. Such experiments have uncovered discrete, unlinked regulatory loci (Rawls and Lucchesi 1974a,b). Similarly, the expression of ADH in *Drosophila* (both mRNA and protein) was found to be reduced in individuals harboring small duplications (interstitial trisomics) of 10 unlinked regions of the genome (Birchler 1985). In maize, the accumulation of glucose-6-phosphate dehydrogenase reaches 200% of the diploid level in individuals segmentally monosomic for a chromosome arm that does not carry the structural gene (Birchler, 1979). This "inverse effect" has been demonstrated in maize with smaller chromosome regions as well as with entire chromosome arms (Birchler 1985). This effect can bring about dosage compensation such as occurs with a dosage series for alcohol

HYPOMORPHIC	dysfunctional + (normal) allele +/− hypoploid shows mutant phenotype
HYPERMORPHIC	overproducer ++/++ shows mutant phenotype
ANTIMORPHIC	competes with function of normal product additional + allele reduces mutant phenotype
NEOMORPHIC	new product a + allele is inconsequential

FIGURE 32.1 Four classes of dominant mutations. The classification is based on the effect that dosage of the wild-type copy has on the mutant phenotype.

dehydrogenase levels. The net compensation of ADH was found to consist of two discrete regions: the structural locus, which, when isolated, *did* exhibit a dosage effect, and a second region, not including the structural gene, that exerts an inverse effect upon ADH expression.

Dosage aneuploids have also been used to identify chromosome regions that affect endosperm development (Birchler and Hart 1987). Kernels resulting from crosses to some, although not all, of the B-A translocated arms exhibit a reduction in size indicative of endosperm hypoploidy. (The endosperm is segmentally disomic, rather than the normal triploid constitution.) Parental imprinting plays a role in the reduction of endosperm size for at least two of the arms studied (1L, 10L) because endosperm size could not be rescued by providing additional copies through the megagametophyte.

Similar to the studies searching for regulatory loci involved in the expression of various enzymes, modifiers of dominant mutations have also been found in *Drosophila* using a set of synthetic duplications that cover >99% of the autosomal complement (Kennison and Russell 1987). In maize, smaller scale studies have identified chromosome regions that suppress or enhance dominant mutant phenotypes. For example, Poethig (1989) crossed lines carrying the *Teopod* (*Tp1* and *Tp2*) mutations that affect phase transition in maize to heterozygotes hyperploid for the B-A translocations. Of eight chromosome arms tested, he found that hyperploidy resulted in modifications of the mutant character for three different chromosome arms. Interestingly, distinct aspects of the mutant phenotype were differentially modified in the various aneuploids. For example, the formation of additional tillers characteristic of *Tp1* was suppressed in *Tp1* individuals hyperploid for TB-1La, whereas another mutant character, that involving a unique tassel morphology, was enhanced when the mutants were hyperploid for TB-1La, TB-3La, or TB-4Lf. It is interesting to note that two of the arms that show a dosage-dependent modification to the *Teopod* character also harbor genes that, in a mutated condition, have previously been shown to interact with the mutant *Teopod*-gene, namely, *teosinte branched* (*tb1*) on 1L and *tunicate* (*Tu1*) on 4L.

Although studies with aneuploid stocks provide a solid framework for analysis of dominant mutations, certain caveats need to be considered. As discussed earlier, alterations in gene expression may accompany alterations in structural gene dosage, but may also include the "inverse effect" or dosage compensation. To address just such concerns, Freeling and Hake (1985) utilized both whole-arm aneuploids, encompassing 80% of the chromosome arm, and a synthetic interstitial aneuploid, varying the dosage of a segment of only 18% of the arm, in their aneuploid analyses of *Kn1* mutations. Another caveat in conducting dosage analysis is the variability produced by different backgrounds. Investigators should cross their mutations into the same stock in which the B-A series is available. In addition to variations of genetic background, one must keep in mind the many environmental contingencies of a given mutant's expressivity. Finally a word of caution if one is to construct segmental aneuploids using sets of reciprocal translocations: (1) breakpoints are not always placed correctly or accurately, and (2) when constructing duplication-bearing progeny as described, gametophytes duplicated for a region between two breakpoints will also be segmentally trisomic for the region between the other two breakpoints! Lines carrying extra copies of appropriate regions as controls should also be constructed. As a final note, never assume that recombination will be suppressed or will not occur in such aneuploids. In fact, Birchler (1988) noted that recombination in such individuals occurs readily.

REFERENCES

Artavanis-Tsakonas S (1988) The molecular biology of the *Notch* locus and the fine tuning of differentiation in *Drosophila*. Trends Genet 4: 95–100

Birchler JA (1988) Chromosome manipulations in maize. In Gupta PK, Tsuchiya T (eds) Chromosome Engineering in Plants—Genetics, Breeding, and Evolution, Part A, Elsevier, Amsterdam, pp 531–559

Birchler JA (1985) The inverse effect in maize and *Drosophila*. In Freeling M (ed) Plant Genetics, Alan R Liss, New York, pp 547–559

Birchler JA (1979) A study of enzyme activities in a dosage series of the long arm of chromosome one in maize. Genetics 92: 1211–1229

Birchler JA, Hart JR (1987) Interaction of endosperm size factors in maize. Genetics 117: 309–317

Freeling M, Hake S (1985) Developmental genetics of mutants that specify *Knotted* leaves in maize. Genetics 111: 617–634

Harberd NP, Freeling M (1989) Genetics of dominant gibberellin-insensitive dwarfism in maize. Genetics 121: 827–838

Kennison JA, Russell MA (1987) Dosage-dependent modifiers of homoeotic mutations in *Drosophila melanogaster*. Genetics 116: 75–86

Lifschytz E, Green MM (1979) Genetic identification of dominant overproducing mutations: the beadex gene. Mol Gen Genet 171: 153–159

Muller HJ (1932) Further studies on the nature and causes of gene mutations. In Jones DF, Menasha WI (eds) Proceedings of the Sixth International Congress of Genetics, Brooklyn Botanic Gardens, pp 213–255

Poethig RS (1988) Heterochronic mutations affecting shoot development in maize. Genetics 119: 959–973

Poethig S (1989) Genetic modifiers of heterochronic mutations in maize. In Goldberg R (ed) The Molecular Basis of Plant Development, UCLA Symposium on Molecular and Cellular Biology, New Series, Vol 92, Alan R Liss, NY, pp 25–35

Rawls JM, Lucchesi JC (1974a) Regulation of enzyme activities in *Drosophila*. I. The detection of regulatory loci by gene dosage responses. Genet Res 24: 59–72

Rawls JM, Lucchesi JC (1974b) Regulation of enzyme activities in *Drosophila*. II. Characterization of enzyme responses in aneuploid flies. Genet Res 24: 73–80

33

Biased Transmission of Genes and Chromosomes

Wayne R. Carlson

TYPES OF BIASED TRANSMISSION IN MAIZE

Most genes and chromosomes in maize are represented in the gametes according to their frequencies in the parent. However, a number of exceptions have been found. The exceptions can usually be attributed to one of the following phenomena:

1. Accumulation mechanisms
2. Lethal or semilethal chromosomal deficiencies
3. Translocation heterozygotes that produce viable adjacent 1 gametes
4. Genes that affect zygotic or gametophytic viability

The linkage of a gene to one of the four factors mentioned above is manifested as a reduced or enhanced transmission of one allele over another. In addition, certain systems have the unique ability to "uncover" the recessive phenotype in crosses of homozygous dominant × homozygous recessive. The latter phenomenon results from transmission of deficient-viable gametes by the dominant parent. Properties of the four systems of biased transmission are discussed below.

Accumulation mechanisms are systems designed to increase the populational frequency of a particular genetic region, without regard to any selective advantage for the individual. They seem to have a parasitic function. For example, a system exists in maize that increases the transmission of heterochromatic "knobs" in the female meiosis. Because knobs are polymorphic, some stocks will carry a particular knob and others will

The classical terminology for B-A translocations is used in this chapter. The chromosome listed first carries the centromere; the second chromosome (an arm donor) is written as a full-sized superscript character, i.e., B^A or A^B. New terminology has been proposed that will probably render superscripts obsolete (B^A becomes B-A). When a final decision is reached, the new "rules" will be published in the *Maize Genetics Cooperation News Letter.* The Editors

The Maize Handbook—M. Freeling, V. Walbot, eds.

not. In a knob heterozygote, such as knob chromosome 3/knobless chromosome 3, the transmission of knobbed vs. knobless chromosomes is usually 1:1. However, when a structurally modified chromosome 10 is present, the knobbed chromosome 3 is recovered at a ratio of about 70:30 over the knobless chromosome (Rhoades and Dempsey 1966). The aberrant ratio occurs only in the female, but it occurs for all heterozygous knobs in a plant. Genes that are linked to the knobs show enhanced transmission through the female.

The key to the system is abnormal chromosome 10. This chromosome differs from normal 10 as a result of the presence of extra chromatin near the terminus of 10L. The extra chromatin includes a heterochromatic knob plus two separate regions of euchromatin (Rhoades and Dempsey 1985: Figure 3). Abnormal chromosome 10 causes preferential recovery of knobs, including its own knob, through an effect on chromosome movement. The system depends on the transformation of knobs into neocentromeres during meiosis. The knobs attach spindle fibers at both divisions of meiosis. At anaphase 2 of the female meiosis, neocentromeres lead knobbed chromosomes to the outer poles of the meiotic quartet. One of the outer cells is the basal megaspore, the cell that gives rise to the embryo sac (Figure 33.1). The system was explained by Rhoades and Dempsey (1966, 1985) and has been reviewed by Carlson (1988).

Another unusual chromosome in maize, the B chromosome, also has an accumulation mechanism. Nondisjunction at the second pollen mitosis allows the B chromosome to increase its frequency in a population. If one B chromosome is present at the second pollen mitosis, nondisjunction produces dissimilar sperm, with one having two B's and the other zero B's. The sperm with two B's tends to fertilize the egg more often than the zero B sperm, thus increasing B chromosome frequency. The system has been reviewed

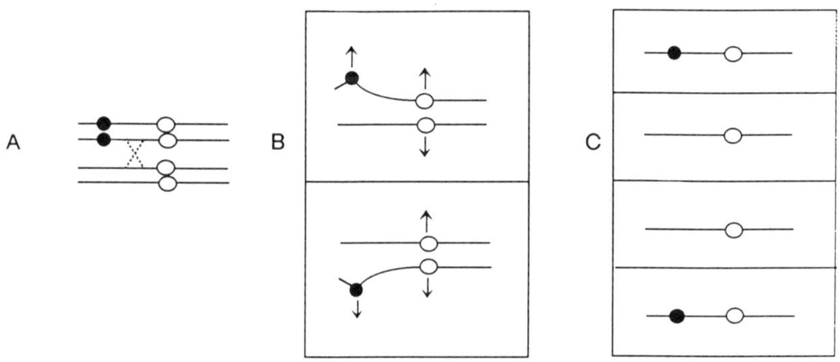

FIGURE 33.1 . Accumulation mechanism of abnormal chromosome 10. A. Pachytene bivalent. A chiasma is present between the knob and the centromere. As a result, each chromosome has one knobbed and one knobless chromatid. B. Anaphase 2. Neocentromeres lead knobbed chromosomes to the outer poles. C. Quartet stage. Knobbed chromosomes occupy the outer cells, with knobless chromosomes in the central cells. The basal cell will give rise to the embryo sac.

by Carlson (1986) and Beckett (1991) and is discussed in detail in other chapters of this volume. The B chromosome can be linked to standard genes through translocations with members of the normal (A) chromosomal complement. In B-A translocations, the nondisjunctional system produces pollen with two B^A's in one sperm (duplicate) and zero B^A's in the other (deficient). If the zero B^A sperm fertilizes the polar cells, it "uncovers" recessive endosperm markers for the deficient region. If the zero B^A sperm fertilizes the egg, it "uncovers" recessive plant markers.

Beckett (1982) described another B chromosome accumulation mechanism. He found that pollen that contains B-A translocations outcompetes normal pollen for fertilization of the egg. Therefore, genes that are linked to B-A translocation chromosomes are transmitted in excess through the male parent.

The second category of biased transmission is produced by lethal or semilethal deficiencies. When a deficiency involves many genes, the chromosome carrying it is lethal in both the male and female gametophyte. The deficient chromosome cannot be propagated, making its study difficult. As a result, the only commonly observed case of large deficiencies occurs in work with B-A translocations. In crosses of the B-A translocations, deficient A^B sperm are regularly produced by B^A nondisjunction. When the A^B sperm is combined with a normal egg, an A A^B individual is produced. This plant type gives two meiotic products: A and A^B. The A^B spore class is both male and female lethal. It lacks genes that are required for mega- and microgametophyte viability. Alleles linked to the A^B chromosome are transmitted at reduced frequency through both male and female parents.

Another type of deficiency is female-viable, although lethal in the male. This type of deficiency usually affects a small number of genes. Deficiencies that are female-viable can be maintained as heterozygotes. Genes that are linked to such deficiencies show reduced or no transmission through the pollen.

The third method for modifying transmission of genes and chromosomes depends on a specific class of translocation. This type of translocation produces female-viable, male-lethal deficiencies similar to those discussed above. The deficiencies arise by adjacent segregation from the translocation heterozygote. A review of this translocation group was given by Phillips et al. (1971). Typically, one breakpoint of the translocation is distal, so one arm of the translocation "cross" is very short. An example is T9–10a. It has breakpoints of 9L .14 and 10L .92. The viable gametes are the alternate products (9, 10 and 9^{10}, 10^9) plus one of the adjacent products (9, 10^9). The last gamete class is viable in the female because the deficiency of genes on chromosome 10 is not vital to megagametophyte development. Crosses with the translocation heterozygote as the female parent can produce off-ratios in two different ways: 1) increased transmission of genes that are linked to the viable adjacent 1 gamete, or 2) "uncovering" of recessive phenotypes by the deficient region of the adjacent 1 gamete. (See Below.)

The fourth category of biased transmission is produced by a collection of single genes that affect zygotic or gametophytic survival. Any vital gene can produce a recessive lethal allele (zygotic lethal) by mutation to loss-of-function. Linkage of a gene to such a lethal recessive will affect genetic ratios. More interesting are genes that affect viability of the gametophyte. They control gametophytic development and/or functioning. Gametophytic genes modify the transmission of genes linked to them (Coe et al. 1988, p 196).

CONSTRUCTING EXAMPLES OF BIASED TRANSMISSION

Abnormal chromosome 10 provides a simple system for demonstrating deviations from Mendelian ratios. For example, the knob on abnormal 10 is closely linked to the endosperm color factor *R*. A homozygous abnormal 10 stock marked by *R R* can be crossed to a normal *r r* stock. Next, the abnormal 10/normal 10 heterozygote is backcrossed as female to the *r r* stock. The result will be an ear with approximately 70% colored (*R*) kernels. Because the *R* locus is tightly linked to the knob of abnormal 10, the rate of *R* transmission is equivalent to the rate of knob transmission. Genes that are less closely linked to the knob show a less extreme deviation from a 1:1 ratio. Similar crosses can be constructed for knobs on other chromosomes. For example, the *ligueless-2* gene *(lg2)* is linked to a knob on the long arm of chromosome 3. The *yellow-green-2* gene (*yg2*) is linked to the terminal knob on 9S (Rhoades and Dempsey 1966). These genes will show deviations from a 1:1 ratio in the female testcross, provided that abnormal 10 is present.

A second system that can be readily used to produce off-ratios is the set of translocations that give female-viable gametes from adjacent 1 segregation. An example of this type of translocation is shown in Figure 33.2. The translocation, T8–9(4453), produces the deficient-viable gamete, $8 + 9^8$. A heterozygote of the type shown can be produced by crossing a homozygous translocation stock to a *yg2 yg2 wx wx* tester. Next, reciprocal backcrosses to the *yg2 yg2 wx wx* tester are made. When the heterozygote is crossed as male parent to the tester, a 1:1 ratio will be found for both genes. However, when the translocation stock is crossed as female, both genes will show a deviation from the Mendelian ratio. The recessive *yg2* and the dominant *Wx* phenotypes will be expressed more than 50% of the time. These are both effects of survival of the $8 + 9^8$ gamete. This

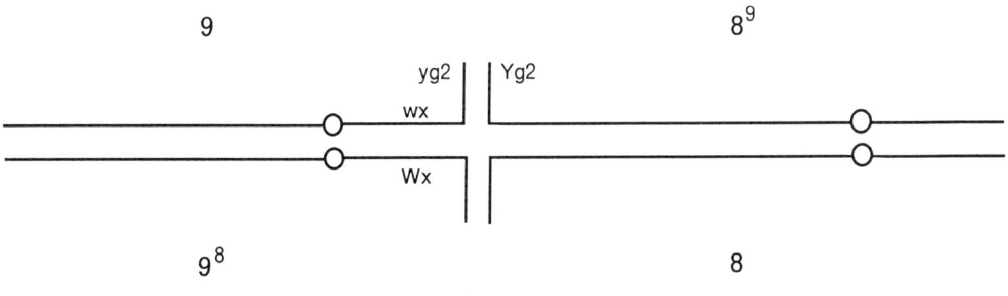

FIGURE 33.2 Translocation "cross" in pachytene of a T8–9(4453) heterozygote. Dominant alleles of *Wx* and *Yg2* are shown on the translocation chromosomes and recessives on chromosome 9.

gamete is deficient for all genes distal to the breakpoint on 9S, including the *Yg2* gene. It therefore behaves as a recessive in a testcross and increases the *yg2* phenotype frequency above 50%. The *Wx* gene shows an off-ratio for a different reason. The dominant allele is linked to the viable adjacent 1 gamete ($8 + 9^8$), whereas the *wx* allele is linked to the inviable reciprocal gamete ($8^9 + 9$). Therefore, *Wx* frequency is greater than 50%.

Interpreting Abnormal Ratios that Arise in Crosses

When an unusual pattern of inheritance is found, the number of hypothetical causes can be narrowed down with testcrosses. The affected stock is crossed to an appropriate tester. Next, reciprocal backcrosses to the same tester are made. The reciprocal crosses demonstrate whether pollen transmission, egg transmission, or both are affected.

Another type of cross is used to detect the "uncovering" of recessives by non-Mendelian systems. In this cross, the parent being tested is homozygous dominant for a specific gene and the other parent is homozygous recessive. If recessive phenotypes are produced in the cross, they result from transmission of deficient gametes by the dominant parent. Recessive genes are "uncovered" in the female by translocation heterozygotes that transmit viable adjacent 1 gametes. Recessive genes are "uncovered" in the pollen by B-A translocations.

A third approach to understanding unusual patterns of inheritance is a cytological study. Examination of meiosis can quickly identify abnormal chromosome 10 or a B-type chromosome or a translocation. Mitotic preparations are useful mainly for identifying B-type chromosomes.

ACKNOWLEDGMENTS

This material is based upon work supported by the Cooperative State Research Service, U.S. Dept. of Agriculture, under agreements # 89–37140–4464 and 92–37301–7679.

REFERENCES

Beckett JB (1982) An additional mechanism by which B chromosomes are maintained in maize. J Hered 73: 29–34

Beckett JB (1991) Cytogenetic, genetic and plant breeding applications of B-A translocations in maize. In Gupta PK, Tsuchiya T (eds) Chromosome Engineering in Plants: Genetics, Breeding, Evolution. Part A, Elsevier, Amsterdam, pp 493–529

Carlson WR (1986) The B chromosome of maize. Crit Rev Plant Sci 3: 201–226

Carlson WR (1988) The cytogenetics of corn. In Sprague GF, Dudley JW (eds) Corn and Corn Improvement, Third Edition, Am Soc Agronomy, Madison, WI, pp 259–343

Coe EH, Jr, Neuffer MG, Hoisington DA (1988) The genetics of corn. In Sprague GF, Dudley JW (eds) Corn and Corn Improvement, Third edition, Am Soc Agronomy, Madison, WI, pp 81–258

Phillips RL, Burnham CR, Patterson EB (1971) Advantages of chromosomal interchanges that generate haplo-viable deficiency-duplications. Crop Science 11: 525–528

Rhoades MM, Dempsey E (1966) The effect of abnormal chromosome 10 on preferential segregation and crossing over in maize. Genetics 53: 989–1020

Rhoades MM, Dempsey E (1985) Structural heterogeneity of chromosome 10 in races of maize and teosinte. In Freeling M (ed) Plant Genetics, Alan R Liss, New York, pp 1–18

34

Anthocyanin Genetics

E. H. Coe

Genetic control of anthocyanin biosynthesis has been the subject of more intensive research than has any other gene system, with the consequence that so much information is available that it may appear too complex to learn easily, despite underlying simplicity. Some tips and clues relevant to genotype and phenotype follow. For some additional details, including pigment distribution, tissue by tissue and genotype by genotype, see Coe (1985) and Coe et al. (1988).

The sequence of biosynthetic steps and the action of enzymes encoded by structural genes is evidently in the order (*C2* or *Whp1*) *A1 A2 Bz1 Bz2* in a linear pathway that requires each of these functions for pigment to be formed; thus, if one or more of the first three functions fails (i.e., if the genotype is homozygous recessive), anthocyanin is not produced. If *Bz1* or *Bz2* is recessive, only reddish brown (bronze) pigment is produced.

The functioning of the structural genes is regulated (i.e., is turned on):

In the aleurone tissue by *C1* in combination with *R1*
In sheaths, husks, and culm by *Pl1* with *B1* or *Lc1*

The Maize Handbook—M. Freeling, V. Walbot, eds.

In seedling coleoptiles, leaf sheaths, leaf tips, and leaf margins by *Pl1* with *B1* (or with *R1-r* or *r1-r*)

In the anther wall by *Pl1* with *R1-r* or *r1-r*

If one or more of these is recessive, anthocyanin is not produced in the specified tissue. Light exposure can substitute for *Pl1* in vegetative tissues, or for *C1* in the aleurone tissue (but only with a specific allele, *c1-p*). Dominant inhibition of pigment biosynthesis in the aleurone tissue is characteristic of *C1-I, C2-Idf,* and *In1-D.*

Anthocyanins and their intermediates are phenolic compounds. These compounds will oxidize and polymerize into brown products (phlobaphenes and tannins) unless they are glycosyated and are stabilized by complexing with other cell constituents. Anthocyanin color is affected by differences in hydroxylation, methylation, glycosylation, deposition, and complexing. The lacquer-red and brick-red pigments of cobs and kernel pericarps are deoxyanthocyanins, a class of anthocyanins with one fewer specific hydroxylation than the rose-to-bluish anthocyanins. Other stable phenolics, relatives of the anthocyanins, include colorless, ivory, and yellow compounds.

Plants that are *A1 A2 Bz1 Bz2 C2 B1 Pl1* have dark blackish-purple sheaths, husks, and culm. Strongly pigmented strains must be maintained by selection for strong expression.

Plants that are *A1 A2 Bz1 Bz2 C2 B1* *pl1* have reddish-purple sheaths, husks, and culm only where those tissues have been exposed directly to light ("sun-red" phenotype).

Plants that are *a1* *A2 Bz1 Bz2 C2 B1 Pl1* have light brown sheaths, husks, and culm. The brown pigment is sun-brown with *pl1.*

Plants that are *A1* *a2* Bz1 Bz2 C2 B1 Pl1 have brown-necrotic sheaths, husks, and culm. Homozygous strains are very sensitive to decay and have poor vigor. The brown pigment is sun-brown with *pl1.*

Plants that are *A1 A2* *bz1* Bz2 C2 B1 Pl1 have dark brown-necrotic sheaths, husks, and culm. Homozygous strains have brittle tissues and poor vigor. The brown pigment is sun-brown with *pl1.*

Plants that are *A1 A2 Bz1* *bz2* C2 B1 Pl1 have dark brown sheaths, husks, and culm. Strongly pigmented strains have poor vigor. The brown pigment is sun-brown with *pl1.*

Plants that are *A1 A2 Bz1 Bz2* *c2* *B1 Pl1* have light purple to brown sheaths, husks, and culm, with a distinctive peripheral-pigment pattern. Strongly pigmented strains must be maintained by selection for strong expression. The pigment is sun-brown with *pl1.*

Plants that are *A1 A2 Bz1 Bz2 C2* *b1* Pl1 have no anthocyanin in sheaths, husks, and culm (although with *R1-r* or *r1-r*, some moderate pigmentation occurs in lower sheaths and in leaf margins).

Aleurone tissues that are *A1 A2 Bz1 Bz2 C1 C2 R1 Pr1* have blackish-purple to blue pigmentation.

Aleurone tissues that are *A1 A2 Bz1 Bz2 C1 C2 R1* *pr1* have red to maroon pigmentation.

Aleurone tissues that are *A1 A2* *bz1* Bz2 C1 C2 R1 Pr1 have purplish-brown (bronze) pigmentation. Likewise if *bz2*.

Aleurone tissues that are *A1 A2* *bz1* Bz2 C1 C2 R1 *pr1* have reddish-brown pigmentation. Likewise if *bz2*.

Aleurone tissues that are pigmented have considerably more pigment when they are recessive for *in1*, to such extent that they may appear black rather than purple or red.

Seedlings that are *A1 A2 Bz1 Bz2 C2 Pl1* and *R1-r* or *r1-r* have reddish to purplish coleoptiles, leaf sheaths, leaf tips, and roots.

Seedlings that are *A1 A2 Bz1 Bz2 C2 pl1* and *R1-r* or *r1-r* have reddish to purplish coleoptiles, leaf sheaths, and leaf tips, but no root color in the absence of light.

Anthers that are *A1 A2 Bz1 Bz2 C1 C2 R1-r* or *r1-r Pl1 Pr1* have blackish-purple pigmentation.

Anthers that are *A1 A2 Bz1 Bz2 C1 C2 R1-r* or *r1-r Pl1 pr1* have maroon pigmentation.

Anthers that are *A1 A2 Bz1 Bz2 C1 C2 R1-r* or *r1-r pl1* Pr1 have sun-dependent light red pigmentation.

Silks that are *A1 A2 Bz1 Bz2 C2* and *R1-r* or *r1-r* have pink silk hairs.

Silks that are *P1-rr* or *P1-wr* and *Sm1* are light brick-red.

Silks that are *P1-rr* or *P1-wr* and *sm1* are salmon-colored.

Silks that are *P1-ww* and *sm1* are tan to brown.

Pericarps and cobs that are *P1-ww* are clear and colorless.

Pericarps that are *A1 P1-wr* can be clear or tinged with lacquer-red or brick-red; cobs, brick-red.

Pericarps that are *A1 P1-rr* are lacquer-red; cobs, brick-red.

Pericarps and cobs that are *A1 P1-vv* are striped ("calico corn").

Pericarps and cobs that are *a1* or *A1-b* and *P1-rr* are brown or tan.

Pericarps and cobs that are *Ch1* are brown or tan, clearest if *P1-ww*.

Pericarps that are *A1 A2 Bz1 Bz2 C2 Lc1 Pl1* are oxblood to blackish-purple.

Embryo plumules that are *A1 A2 Bz1 Bz2 C1 C2* with *R1-nj* or *B1-peru* or certain other *r1-* alleles are purple or red.

Scutellums that are *A2 Bz1 Bz2 C1 C2* with appropriate *R1* alleles and other factor(s) are purple or red.

Pollen does not develop anthocyanin naturally, but under sustained humid conditions it is sometimes stained by anthocyanins bleeding from the anther wall of intensely pigmented genotypes.

REFERENCES

Coe EH, Jr (1985) Phenotypes in corn: control of pathways by alleles, time and place. In Freeling M (ed) Plant Genetics, Alan R Liss, New York, pp 509–521

Coe EH, Jr, Neuffer MG, Hoisington DA (1988) The genetics of corn. In Sprague GF, Dudley JW (eds) Corn and Corn Improvement, Am Soc Agron, Madison, WI, pp 81–258

35

Cloned Anthocyanin Genes and Their Regulation

KAREN CONE

Anthocyanin synthesis in the kernel and in the plant body requires the action of at least five structural genes—*A1, A2, Bz1, Bz2, C2*—that encode the biosynthetic enzymes of the pathway. Regulation of the coordinate expression of the structural genes occurs mainly at the level of transcriptional activation. The anthocyanin regulatory genes fall into two gene families, both of which encode proteins with homology to other eukaryotic transcription factors. (See Table 35.1 footnote.) Anthocyanin synthesis requires a member of the *C1/Pl* family *and* a member of the *R/B* family. This dual requirement does not simply reflect the interdependence of the regulatory genes for their own expression, because colorless kernels carrying recessive *c1* still express *R-S* mRNA and colorless *r* kernels still express *C1* mRNA. Rather, the requirement for regulatory genes from both families supports the idea that the *C1/Pl* proteins and *R/B* proteins interact with each other to effect transcriptional activation of the structural genes.

The tissue and cell specificity of anthocyanin synthesis appears to reflect the patterns of expression of the diverse members of the *regulatory* gene families. Although members of a gene family share extensive homology within their protein coding regions, the sequences flanking those coding regions are very diverse. These flanking sequences presumably contain the regulatory signals that control the tissue-specific expression of the different family members. Thus far, molecular regulation of only one anthocyanin regulatory gene, *C1*, has been examined in detail. Anthocyanin expression in the kernel requires *C1* and *R-S*. Expression of *C1* is dependent on the *Vp1* gene; kernels carrying the recessive *vp1* allele do not express *C1* mRNA. *Vp1* encodes a putative transcription factor that is involved in the expression of genes during seed maturation. *Vp1* control of *C1* mRNA expression is modulated by a conserved nucleotide sequence upstream of *C1* that resembles sequences found upstream of a number of abscisic-acid-regulated genes (Hattori et al. 1992). The kernel specificity of *C1* expression results because *Vp1* expression is limited to the endosperm (aleurone) and embryo (McCarty et al. 1989).

The Maize Handbook—M. Freeling, V. Walbot, eds.

Table 35.1. Cloned anthocyanin genes

Gene	Gene product	References
C2	chalcone synthase	Wienand et al. 1986
Whp	chalcone synthase	Franken et al. 1991
A1	dihydroflavonol reductase	O'Reilly et al. 1985
		Schwarz-Sommer et al. 1987
A2	flavan-3-hydroxylase?[a]	Menssen et al. 1991
Bz1	UDP-glucose:flavonoid-3-O-glucosyl transferase	Fedoroff et al. 1984
		Ralston et al. 1988
		Furtek et al. 1988
Bz2	unknown	McLaughlin and Walbot 1987
		Theres et al. 1987
		Nash et al. 1990
C1[b]	transcriptional activator	Cone et al. 1986
		Paz-Arez et al. 1986
		Paz-Arez et al. 1987
Pl[b]	transcriptional activator	Cone and Burr 1989
R-S, R-P[c]	transcriptional activator	Dellaporta et al. 1988
		Perrot and Cone 1989
		Robbins et al. 1991
Lc[c]	transcriptional activator	Ludwig et al. 1989
Sn[c]	transcriptional activator	Tonelli et al. 1991
		Consonni et al. 1992
B[c]	transcriptional activator	Chandler et al. 1989
		Radicella et al. 1991
Vp1	transcriptional activator	McCarty et al. 1989
		McCarty et al. 1991

[a]The *A2* gene product shares homology with the flavan-3-hydroxylase product of the *Inc* gene from *Antirrhinum majus* (Menssen et al. 1990).

[b]*C1* and *Pl* comprise the *C1* family and encode functionally duplicate proteins that contain regions of homology to the DNA binding domain of *myb* oncoproteins and to the transcriptional activation domains of a number of eukaryotic transcription factors.

[c]*R-S, R-P, Lc, Sn,* and *B* comprise the *R/B* gene family and encode functionally similar proteins that contain a region of homology to the basic-helix-loop-helix DNA binding/dimerization domain of *myc* proteins.

REFERENCES

Chandler VL, Radicella JP, Robbins TP, Chen J, Turks D (1989) Two regulatory genes of the maize anthocyanin pathway are homologous: Isolation of *B* utilizing *R* genomic sequences. Plant Cell 1: 1175–1183

Cone KC, Burr FA, Burr B (1986) Molecular analysis of the maize anthocyanin regulatory locus *C1*. Proc Natl Acad Sci USA 83: 9631–9635

Cone KC, Burr B (1989) Molecular and genetic analyses of the light requirement for anthocyanin synthesis in maize. In Styles DE, Gavazzi GA, Racchi ML (eds) The Genetics of Flavonoids: Proceedings of a Post Congress Meeting of the XVI International Congress of Genetics, Edizoni Unicopli, Milano, Italy

Consonni G, Viotti A, Dellaporta SL, Tonelli C (1992) cDNA nucleotide sequence of *Sn*, a regulatory gene in maize. Nucl Acids Res 20: 373

Dellaporta, SL, Greenblatt IM, Kermicle JL, Hicks JB, Wessler SR (1988) Molecular cloning of the *R-nj* gene by transposon tagging with *Ac*. In Gustafson JP, Appels R (eds) Chromosome Structure and Function: Impact of New Concepts, Plenum Press, New York, pp 263–281

Fedoroff NV, Furtek DB, Nelson OE (1984) Cloning of the *bronze* locus in maize by a simple and generalizable procedure using the transposable controlling element *Activator* (*Ac*). Proc Natl Acad Sci USA 81: 3825–3829

Franken P, Niesbach-Klösgen U, Weydemann U, Maréchal-Drouard L, Saedler H, Wienand U (1991) The duplicated chalcone synthase genes *C2* and *Whp* (*white pollen*) of *Zea mays* are independently regulated: evidence for translational control of *Whp* expression by the anthocyanin intensifying gene *in*. EMBO J 10: 2605–2612

Furtek D, Schiefelbein JW, Johnston F, Nelson, OE Jr (1988) Sequence comparisons of three wild-type *Bronze-1* alleles from *Zea mays*. Plant Mol Biol 11: 473–482

Ludwig SR, Habera LF, Dellaporta SL, Wessler SR (1989) *Lc*, a member of the maize *R* gene family responsible for tissue-specific anthocyanin production, encodes a protein similar to transcriptional activators and contains the *myc*-homology region. Proc Natl Acad Sci USA 86: 7092–7096

McCarty DR, Carson CB, Stinard PS, Robertson DS (1989) Molecular analysis of *viviparous-1*: an abscisic acid-insensitive mutant of maize. Plant Cell 1: 523–532

McCarty DR, Hattori T, Carson, CB, Vasil V, Lazar M, Vasil IK (1991) The *Viviparous-1* developmental gene of maize encodes a novel transcriptional activator. Cell 66: 895–905

Hattori T, Vasil V, Rosenkrans L, Hannah C, McCarty DR (1992) A regulatory hierarchy in maize: The *Viviparous-1* gene and abscisic acid activate the *C1* regulatory gene for anthocyanin biosynthesis during seed maturation in maize. Genes and Dev. 6: 609–618

McLaughlin M, Walbot V (1987) Cloning of a mutable *bz2* allele of maize by transposon tagging and differential hybridization. Genetics 117: 771–776

Menssen, A, Höhmann S, Martin W, Schnable PS, Peterson PA, Saedler H, Gierl A (1990) The *En/Spm* transposable element of *Zea mays* contains splice sites at the termini generating a novel intron from an element in the *A2* gene. EMBO J 9: 3051–3057

Nash J, Luehrsen KR, Walbot V (1990) *Bronze-2* gene of maize: reconstruction of a wild-type allele and analysis of transcription and splicing. Plant Cell 2: 1039–1049

O'Reilly C, Shepherd N, Pereira A, Schwarz-Sommer Zs, Bertram I, Robertson DS, Peterson PA, Saedler H (1985) Molecular cloning of the *a1* locus of *Zea mays* using the transposable elements *En* and *Mu1*. EMBO J 4: 877–882

Paz-Ares J, Wienand U, Peterson PA, Saedler H (1986) Molecular cloning of the *c* locus of *Zea mays*: a locus regulating the anthocyanin pathway. EMBO J 5: 829–833

Paz-Ares J, Ghosal D, Wienand U, Peterson PA, Saedler H (1987) The regulatory *c1* locus of *Zea mays* encodes a protein with homology to *myb* proto-oncogene products and with structural similarities to transcriptional activators. EMBO J 6: 3353–3358

Perrot GH, Cone KC (1989) Nucleotide sequence of the maize *R-S* gene. Nucl Acids Res 17: 8003

Radicella JP, Turks D, Chandler VL (1991) Cloning and nucleotide sequence of a cDNA encoding *B-Peru*, a regulatory protein of the anthocyanin pathway in maize. Plant Mol Biol 17: 127–130

Ralston EJ, English JJ, Dooner HK (1988) Sequence of three *bronze* alleles of maize and correlation with the genetic fine structure. Genetics 119: 185–197

Robbins TP, Walker EL, Kermicle JL, Alleman M, Dellaporta SL (1991) Meiotic instability of the *R-r* complex arising from displaced intragenic exchange and intrachromosomal rearrangement. Genetics 129: 271–283

Schwarz-Sommer Z, Shepherd N, Tacke E, Gierl A, Rohde W, Leclercq L, Mattes M, Berndtgen R, Peterson PA, Saedler H (1987) Influence of transposable elements on the structure and function of the *A1* gene of *Zea mays*. EMBO J 6: 287–294

Theres N, Scheele T, Starlinger P (1987) Cloning of the *Bz2* locus of *Zea mays* using the transposable element *Ds* as a gene tag. Mol Gen Genet 209: 193–197

Tonelli C, Consonni G, Dolfini SF, Dellaporta SL, Viotti A, Gavazzi G (1991) Molecular studies of light-dependent and light-independent alleles of the regulatory *Sn* locus in maize. Mol Gen Genet 225: 401–410

Wienand U, Weydemann U, Niesbach-Klösgen U, Peterson PA, Saedler H (1986) Molecular cloning of the *c2* locus of *Zea mays*, the gene coding for chalcone synthase. Mol Gen Genet 203: 202–207

Maize and Puccinia sorghi: A System for the Study of the Genetic and Molecular Basis of Host-Pathogen Interactions

Tony Pryor

Several rust diseases occur on maize. The best studied is the common rust of maize caused by the basidiomycete *Puccinia sorghi* Schw. First described by Schweinitz on a seedling initially identified as sorghum, *P. sorghi* in fact does not grow on sorghum but produces uredia and telia on maize and closely related teosintes such as *Zea mexicana*, *Z. perennis*, and *Z. diploperennis*. Pycnial and aecial stages of the life cycle occur on several species of *Oxalis*, notably *O. corniculata* and *O. europaea*. Axenic culture of *P. sorghi* has not been successful (Maclean 1982).

Our present-day understanding of the genetics of resistance in maize against common rust depends largely on the work of Hooker and his students at Illinois and has been reviewed several times (Hooker 1985; Hooker and Saxena 1971). The gene-for-gene model for resistance that characterizes most major gene resistance in plants against their obligate fungal hosts was described by Flor in the flax-rust interaction (Flor 1956). The genetic basis for resistance of maize against rust can be understood in this framework.

GENETICS OF RESISTANCE TO *PUCCINIA SORGHI*

Many genes influence the development of rust disease in maize. The extent of the development of disease (measured as the number of uredia per unit leaf area) shows continuous variation in field plants. Resistance, sometimes termed adult plant resistance, shows high combining ability and heritability (Hooker 1985) and consequently is im-

The Maize Handbook—M. Freeling, V. Walbot, eds.

portant in agriculture. However, polygenic inheritance makes adult plant resistance difficult to study genetically.

Major genes specifying resistance against *Puccinia sorghi* (called *Rp*) are identified on the basis of the infection type produced on seedlings. The infection type is scored on a 0–4 scale, in which type 0 is immune and resistant and type 4 is fully susceptible (Hooker 1985). The infection type is characteristic of the particular *Rp* gene and can be used to follow different *Rp* genes in segregating populations.

Using this approach, dominant genes for rust resistance have been mapped to three chromosomal locations in maize. The *Rp3* locus with six identified alleles (*Rp3a* to *Rp3f*) maps to chromosome 3 (Wilkinson and Hooker 1968), the *Rp4* locus with two alleles (*Rp4a* and *Rp4b*) maps to chromosome 4 (Hooker 1985; Wilkinson and Hooker 1968), and finally there is a complex of resistance genes mapping to a three map unit segment at the tip of the short arm of chromosome 10 delineated by RFLP markers *BNL3:04* and *NPI422* (Figure 36.1). The *Rp5* and *Rp6* loci map 1.1 and 2.1 map units, respectively, from *Rp1* and 3.3 units from each other (Wilkinson and Hooker 1968). Some 14 distinguishable resistance specificities were mapped to the region of the *Rp1* gene and were initially considered as alleles at this locus and designated *Rp1-A* to *Rp1-N* (Hagan and Hooker 1965; Hooker 1985). However, subsequent work indicates that many of these specificities are probably not allelic and represent closely linked loci (Hagan and Hooker 1965; Hooker 1985; Hulbert and Bennetzen 1991).

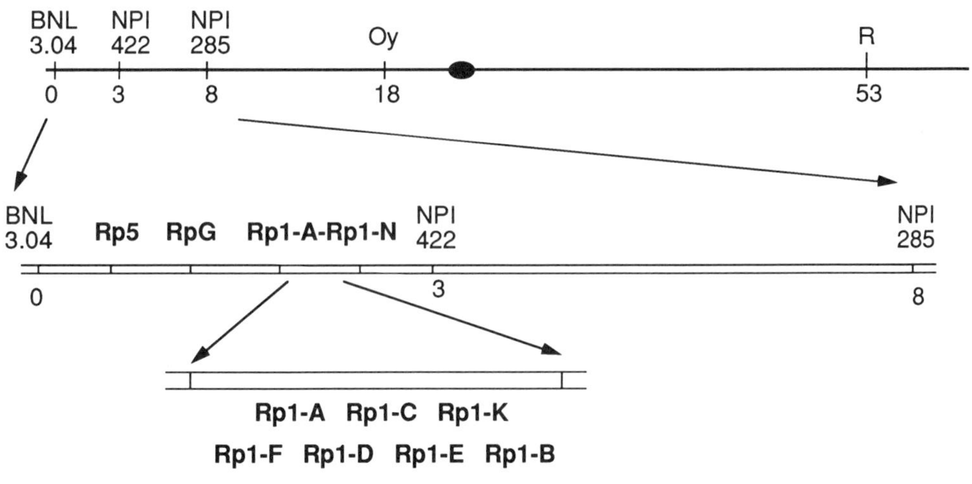

FIGURE 36.1 A genetic map of the rust resistance gene complex on chromosome 10. The map distances are approximate and vary in different genetic backgrounds. The gene symbols are *R*—colored aleurone; *Oy*—oil yellow seedling; *BNL3:04*, *NPI422*, and *NPI285* are RFLP markers; *Rp*—resistance to *Puccina sorghi*.

WORKING WITH *PUCCINIA SORGHI*

Identification of Rust Races

Natural populations of rusts are frequently mixtures of genotypes. Urediospores multiplied from a single pustule isolate can be characterized by their growth on seedlings of a set of differentials (maize lines that each carry a single *Rp* gene) developed by Hooker and co-workers at Illinois. These near-isogenic lines, generated by about 10 backcross generations of most of the *Rp* genes into the R168 inbred background, provide the basis for race identification in *P. sorghi*. The reaction type can be scored as simply resistant or susceptible but it is more useful to score the level of infection on the $0 \rightarrow 4$ rankings (Hooker 1985). Rust races that are distinguished by their virulence on plants containing a particular *Rp* gene can be maintained on this differential although better yields of urediospores are obtained from universally susceptible lines such as some of the Hopi Indian varieties. Contamination can be a serious problem on these lines. A successful strategy is the maintenance of the rust on a differential and single-generation amplifications on the Hopi lines for those instances in which larger amounts of spores are required.

The *P. sorghi* races used by Hooker to identify the various *Rp* genes are no longer available. Recently a number of new isolations have been made that are capable of differentiating most *Rp* genes (Hulbert et al. 1991). One exception is the *Rp6* gene, for which there is at present no rust race capable of recognizing this resistance.

Urediospore Storage

Under dry conditions in a desiccator *P. sorghi* urediospores retain viability for up to 3–4 months. For longer periods, desiccator-dried spores sealed in tin-foil packages or screw capped eppendorf tubes have been stored at −80°C for several years. Storage in liquid nitrogen and under vacuum in sealed glass vials enhances the long term survival. Cold dormant spores require a heat shock of 5 minutes at 40°C to give good germination rates (Melching et al. 1991).

Methods of Inoculation

Seedlings can be inoculated by dusting leaves with spores that can be diluted in talc powder. The leaves are then sprayed with fine water droplets and incubated overnight in a wet plastic chamber. To inoculate large numbers of seedlings, it is more convenient to suspend the spores in a 1/5,000 dilution of Tween 80 wetting agent. The suspension is applied directly as a fine spray to the leaves followed by overnight incubation. The temperature of inoculation is important and infection is greatly reduced at temperatures above 25°C (LeRoux and Dickson 1957). Pustules develop in 9–14 days depending on the growth temperature and urediospores can be collected every second day for 2 weeks.

Isolation of Nucleic Acids

Urediospores are difficult to break and although they can be disrupted by vortexing in the presence of glass beads (Dickinson and Pryor 1989) this treatment substantially shears the nuclear DNA, making it unsuitable for most analyses. High-molecular-weight DNA and polyA-containing RNA can be recovered from germinating urediospores by standard techniques. The available evidence suggests that germ tube growth of rust urediospores does not involve DNA replication and there is relatively little RNA synthesis. Maize rust urediospores germinate well on a surface monolayer on water. Care must be taken to break up spore clumps, because clumps will not germinate. This inhibition is presumably from the presence of density-dependent inhibitors of germination (Wolf 1982) that normally function to prevent the germination of spores within the pustule. From renaturation kinetics it has been estimated that the genome size of the haploid nucleus is similar to most other basidiomycetes at about 4.7×10^7 base pairs (Anderson et al. 1992).

Sexual Cycle

On old or senescing maize plants, frequently on the leaf sheaths or close to the midrib, the brown uredia are replaced by black telia containing two-celled teliospores (Figure 36.2). This spore type is an overwintering stage and methods for inducing germination have involved exposure to freezing and thawing during early spring (Pavgi 1975). More controlled germination has used various combinations of freezing and thawing together

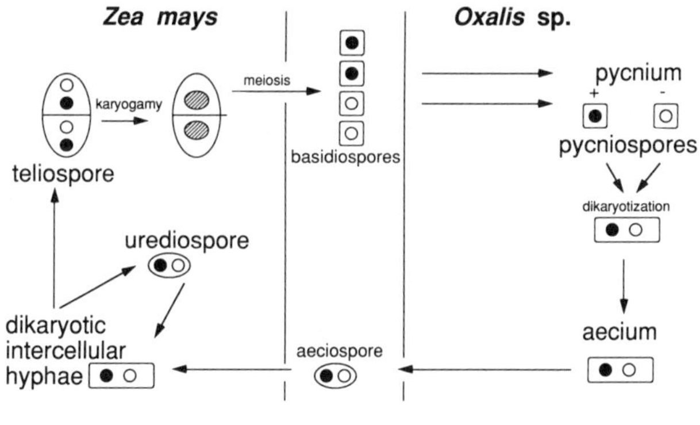

FIGURE 36.2 A diagrammatic representation of the life cycle of *Puccinia sorghi* showing nuclear behavior during the formation of urediopores and teliospores on *Zea mays* and the development of the pycnium and the aecium on *Oxalis*. After karyogamy, the teliospores germinate with the meiotic reduction division leading to the production of haploid basidiospores that segregate for positive and negative mating types. Dikaryotization between gametes from pycnia of opposite mating type reforms the dikaryon that subsequently develops into an aecium with aeciospores. Dikaryotic aeciospores reinfect the maize plant leading to either the asexual urediospore cycle or to the production of teliospores.

with extended periods of washing (Hooker and Yarwood 1966). Leaf sections containing telia with germinating teliospores are suspended over *Oxalis* plants for several (3–6) hours. The leaves on which the basidiospore have landed are sprayed with fine water droplets and incubated overnight in a wet plastic container. After 5–6 days pycnia develop. Exudate from different pycnia is mixed with a sterile toothpick, resynthesizing the dikaryon and leading to the production of the aecium and aeciospores after a further 5–6 days. Maize seedlings are inoculated with the aeciospores in much the same way as for urediospores. There is some evidence that the aeciospore germination has a lower and narrower temperature optimum around 12°C to 16°C compared to that of the urediospores around 2°C to 28°C.

REFERENCES

Anderson PA, Tyler BM, Pryor A (1992) Genome Complexity of the maize rust fugus, *Puccinia sorghi*. Exp. Mycol. 16: 302–307

Dickinson MJ, Pryor A (1989) Isometric virus-like particles encapsidate the double-stranded RNA found in *Puccinia striiformis*, *Puccinia recondita*, and *Puccinia sorghi*. Can J Bot 67: 3420–3425

Flor HH (1956) The complementary genic systems in flax and flax rust. Adv Genet 8: 29–54

Hagan WL, Hooker AL (1965) Genetics of reaction to *Puccinia sorghi* in eleven corn inbred lines from Central and South America. Phytopathology 55: 193–197

Hooker AL (1985) Corn and sorghum rusts. In Roelfs AP, Press WRBA (eds) The Cereal Rusts. Diseases, Distribution, Epidemiology and Control, Vol 2, Academic Press, San Diego, CA, pp 208–229

Hooker AL, Saxena KMS (1971) Genetics of disease resistance in plants. Annu Rev Genet 5: 407–424

Hooker AL, Yarwood CE (1966) Culture of *Puccinia sorghi* on detached leaves of corn and *Oxalis corniculata*. Phytopathology 56: 536–539

Hulbert SH, Lyons PC, Bennetzen JL (1991) Reactions of maize lines carrying *Rp* resistance genes to isolates of the common rust pathogen. Plant Dis 75: 1130–1133

LeRoux PM, Dickson JG (1957) Physiology, specialization, and genetics of *Puccinia sorghi* on corn and *Puccinia purpurea* on sorghum. Phytopathology 47: 101–107

Maclean DJ (1982) Axenic culture and metabolism of rust fungi. In Scott KJ, Chakravorty AK (eds) The Rust Fungi, Academic Press, London, pp 38–84

Melching JS, Bonde MR, Dowler WM (1991) The effect of free water on the potential germinability of cold-dormant uredospores of *Puccinia graminis* f. sp. *tritici*. Phyopathology 81: 734–738

Pavgi MS (1975) Teliospore germination and cytological aberrations in *Puccinia sorghi*. Schw Cytologia 40: 227–235

Wilkinson DR, Hooker AL (1968) Genetics of reaction to *Puccinia sorghi* in ten corn inbred lines from Africa and Europe. Phytopathology 58: 605–608

Wolf G (1982) Physiology and biochemistry of spore germination. In Scott KJ, Chakravorty AK (eds) The Rust Fungi, Academic Press, London, pp 152–178

37

Disease Lesion Mutants

M. G. NEUFFER

A frequently seen but very rare event, population-wise, is the single apparently diseased plant that appears in large populations in the farmer's corn field. These spontaneous variants, like similar frequent cases arising in small populations from treatment of pollen with the chemical mutagen ethyl methanesulfonate (EMS), are, in fact, mutants. The disease "symptoms" are caused by noninfectious Mendelian inherited disease lesion mimic mutations, most of which are dominant (Neuffer and Calvert 1975; Hoisington et al. 1982; Walbot et al. 1983). These mutants are called *les*. The dominant mutants in our laboratory now number 23. Seventeen of these have been located to nine chromosome arms, but only one likely pair of alleles was found among these 17 located mutants. Utilizing the birthday paradox statistic, the presence of one pair of alleles among the 17 mutants tested indicates that there are likely to be as many as 200 possible loci conferring the *Les* phenotype(s). [The birthday paradox is a calculation of the population size required to explain the number of "matches" in a sample from the population; with 365 days in the year, in a group of 20 people, there is likely to be one pair with the same birthday: (1 + 2 + 3 . . . 19)/365.]

These mutants typically are expressed as necrotic or chlorotic spots (lesions) on plant leaves and some other tissues. They vary in a number of phenotypic aspects, such as kind of lesion, whether chlorotic or necrotic or transition from one to the other; shape of lesion, whether circular, elliptical, or angular; size of lesion, from tiny, few cells, to large cutting across most of a leaf blade; frequency, from as few as 2–3 lesions per leaf to as many as 10 per cm^2 of leaf surface; timing, from initiation on the first seedling leaf to initiation just before flowering; distribution, from fairly uniform patterns of similar lesions to clusters and rings of various sizes and shapes to diurnal crossbands of variations in frequency and type of lesions; response to genetic modifications, some enhanced by certain inbred backgrounds and inhibited by others; and responses to environmental differences, some inhibited by high temperatures and enhanced by low temperature, etc. Another general feature in the expression of the dominant mutants is that the homozygote (*Les Les*) condition is almost always more extreme in expression and broader in response to conditions for expression than is the heterozygote (*Les*/+). Categorization of the known

The Maize Handbook—M. Freeling, V. Walbot, eds.

Table 37.1. Variations in phenotypic expression of dominant and recessive disease lesion mutants[a]

Mutant	Characteristics							
	Source	Location	Type	Size	Time	Frequency	Place	Other
(1)	(2)	(3)	(4)	(5)	(6)	(7)	(8)	(9)
Dominants								
Les1-843	E	2S	n	M	e	6	b	rings and satellites
Les2-845	E	1S	nw	S	e	2	b	white spots
Les3-Ullstrup	S	10	n	ML	1	1,3*	b	rings in homozygotes
Les4-1375	E	2L	n	SML	l	5–8	bs	clusters, rings, 2 sizes
Les5-1449	E	1S	nw	S	m	3	b	
Les6-1451	E	10S	cn	M	m	8	bs	angular shape
Les7-1461	E	1L	c	S	ml	8	b	
Les8-2005	E	9S	c-n	S	m	9	bs	
Les9-2008	E	7	c-n	S	m	10	b	pg plant, white spots, rings
Les10-Kermicle	S	2L	cn	SM	em	9	bs	
Les11-1438	E	2S	n	SM	em	5–6	b	
Les12-1453	E	10S	c-n	S-M	em	4	b	small pg plant, clusters
Les13-2003	E	6L	n	S-M	e	5–9	bsc	
Les14-2004	S	3L	c-n	S	e	3–5	bs	
Les15-2007	E	2S	c-n	M	m	9–10	b	tiny pg plants
Les16-2016	E	10S	c	S	l	—	b	pg plant
Les17-2345 (A762)	S	3L	cn	SM	l	9	bs	pg plant, normal sectors

Mutant (1)	Origin (2)	Location (3)	Type (4)	Size (5)	Time (6)	Freq (7)	Place (8)	(9)
*Les**-1378	E	2L	n	SML	l	8–10	bs	clusters, rings, 2 sizes
*Les**-1442	E	—	cn	S-M	m	8	b	fuzzy angular lesions
*Les**-2006	E	—	—	—	—	—	b	
*Les**-7145 (Beckett)	S	—	cn	V	m	8–10	b	necrotic streaks
Recessives								
lls1-Troyer	S	1S	n	M-L	m	6–8	b	
*les**-A467 Blanco	S	—	n	EL	l	1–5	b	clusters
*les**-721	S	—	n	EL	l	1–5	b	
lls1-501B	E	1S	n	L	m	—	b	
*les**-1395	E	—	c	S	l	—	b	
*les**-1521C	E	—	c	S	l	—	b	
*les**-2012	S	—	n	M	l	—	b	
*les**-2013	S	—	c	S	l	9	b	

[a](1) Mutant: designated gene symbol and/or lab number and source.
(2) Origin, causes: E = ethyl methanesulfonate; S = spontaneous.
(3) Location = to chromosome arm or linked to translocation breakpoint in arm indicated.
(4) Type = chlorotic (c); necrotic (n); both (cn); transition (c-n).
(5) Size = small (S); medium (M); large (L); extralarge (EL); combination (SM); transition (S-L); variable (V).
(6) Time = early seedling (e); medium juvenile (m); late, at flowering (l).
(7) Freq = scale of 1—few, scattered, to 10—crowded mass of lesions. * Heterozygote, homozygote expression.
(8) Place = leaf blade (b); sheath (s); culm (c); roots (r).
(9) pg = pale green.

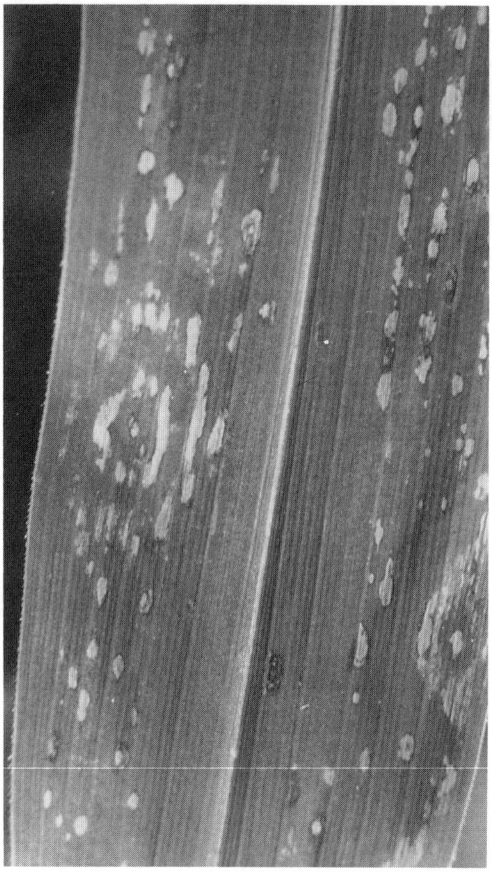

FIGURE 37.1 *Les1–* Leaf section of a *Les1/+* plant showing individual brownish necrotic lesions in various stages of development and also concentric rings of lesions.

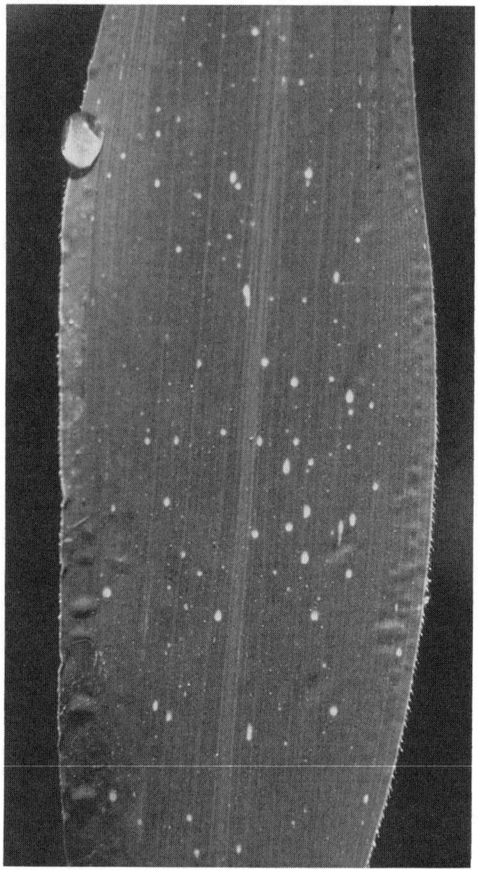

FIGURE 37.2 *Les2–* Leaf section of a *Les2/+* plant showing small round whitish necrotic lesions.

dominant mutants and some of the less-studied recessive lesion mutants with regard to these variables is shown in Table 37.1. Examples of four distinctive lesion mutants, *Les1-843*, *Les2-845*, *Les4-1375*, and *Les10-Kermicle*, are shown in Figures 37.1–4 respectively.

Intensive study of one mutant (*Les1*) has revealed several interesting features that may be true for the whole collection. *Initiation*: when *Les1* plants are grown in favorable (for lesion formation) conditions, a developmental window of opportunity for lesion initiation moves along the leaves as they mature—from leaf tip to base. Lesions usually form only at a specific developmental stage. If the stage is passed, generally no more lesions are formed. During that "sensitive" stage, at least three factors were shown to cause lesion initiation. (1) *Wounding* (Hoisington et al. 1982): lesions will form at a wound

FIGURE 37.3 *Les4–* Mature plant showing many medium to large tan necrotic lesions on both the leaf blade and sheath.

site if the wound occurs at the right developmental stage. Wounds in tissue outside the developmental window in mutant plants and in nonmutant plants do not develop lesions. (2) *Treatment with* n-*propanol* (Ray and Walbot 1984): surface application of *n*-propanol to a *Les1* leaf during the susceptible developmental stage stimulates profuse lesion formation in the treated area while only the normal number of lesions occur in the untreated mutant leaf and none occur in nonmutant genotypes. (3) *Sunlight* (Echt 1986): if mutant leaf tissue that has not yet arrived at the initiation stage is masked from sunlight with opaque paper or metal foil during the initiation stage, no lesion initiation occurs until the mask is removed. After sunlight exposure, profuse lesions arise in the now unmasked area within 24 hours.

A possible mechanism to explain the large number and dominance of *les* loci has been proposed but has not been subjected to rigorous testing (Walbot et al. 1983). In this

FIGURE 37.4 *Les10*– Leaf section of a *Les10*/+ plant showing many small chlorotic lesions so frequent and closely spaced that the leaf has an overall pale green appearance.

model the dominant allele of each *les* locus is hypothesized to lead to overproduction of a metabolite or metabolites involved in disease response. Transgenic dicot plants overproducing such compounds can develop lesion symptoms.

REFERENCES

Echt C (1986) Photo-induction of leaf lesions in *Les1* plants. Maize Genetics Coop News Letter 60: 49

Hoisington DA, Neuffer MG, Walbot V (1982) Disease lesion mimics in maize. I. Effect of genetic background, temperature, developmental age, and wounding on necrotic spot formation with *Les1*. Dev Biol 93: 381–388

Neuffer MG, Calvert OH (1975) Dominant disease lesion mimics in maize. J Hered 66: 265–270

Ray N, Walbot, V (1984) Marking pens can cause lesions in *Les* mutants. Maize Genetics Coop News Letter 58: 190

Walbot V, Hoisington DA, Neuffer MG (1983) Disease lesion mimic mutations. In Kosuge T, et al (eds) Genetic Engineering of Plants, Plenum Press, New York, pp 431–442

38

Classification of Pollen Abortion in the Field

R. L. Phillips

Chromosomal aberrations are useful in a wide variety of maize genetics experiments. Most aberrations in heterozygous condition can be followed simply by the classification of pollen. The degree of pollen abortion depends on the aberration but is usually quite distinctive. Plants heterozygous for certain aberrations may produce pollen grains completely or partially devoid of starch in addition to normal grains. Others may produce normal and distinctly smaller grains that are completely filled with starch.

The identification of plants heterozygous for a particular aberration is very simple with a "pocket microscope" (Miller 1960; Burnham 1961, 1982). With this device, pollen sterility and size can be assessed in 30 seconds per plant. No staining is required. A *freshly* extruded anther not previously dehisced is removed from the tassel. The anther tip is removed by holding the anther between the thumb and finger of one hand and pinching the anther between the fingernail of the thumb and finger of the other hand. The pollen can be forced out of the anther onto a glass slide by squeezing the anther. The slide is placed into the pocket microscope. The pollen is observed by holding the microscope up to the light; the pollen will adhere to the slide and starch-engorged pollen will be black. Slight tapping of the slide will generate a monolayer of pollen. The optics may make abortive grains appear like doughnuts having a dark ring with light passing through the center.

Several versions of magnifying devices exist. The pocket microscope called Midgard #118 made by Nippon Microscope Works performs extremely well. The 1992 price is $40.00 each. The address is Nippon Microscope Works, LTD., 2-4-16 Minami-Aoyama, Minato-ku, Tokyo 107, Japan. Magnification is about 40×.

The Maize Handbook—M. Freeling, V. Walbot, eds.

REFERENCES

Burnham CR (1961) Notes on the pocket microscope. Maize Genetics Cooperation News Letter 35: 88

Burnham CR (1982) Details of the smear technique for studying chromosomes in maize. In Sheridan WF (ed) Maize for Biological Research, Plant Molec Biol Assoc, University of North Dakota Press, Grand Forks, ND, pp 107–118

Miller OL, Jr (1960) A hand scope for pollen examination. Maize Genetics Cooperation News Letter 34: 89

39

Genetic Fine Structure as Revealed in Pollen Assays

OLIVER E. NELSON

The sine qua non of attempts to investigate genetic fine structure in any experimental organism is the ability to handle the very large populations necessary to reveal intragenic recombination. This is the reason that the first reports of intragenic recombination in eukaryotic organisms came from studies of the filamentous fungi, *Aspergillus* and *Neurospora*, using auxotrophic mutants in the laboratories of Pontecorvo and Giles, respectively. These reports raised the question of whether such recombinational events occurred in the higher organisms that had been so important in the development of genetic theory over the previous half-century. An attempt to test the occurrence of intragenic recombination in maize, the best characterized higher plant genetically, targeted the *waxy* locus as the most suitable at which to explore the question (Nelson 1957). Brink and MacGillvray (1924) and Demerec (1924) had reported that 50% of the pollen grains produced by a *Wx/wx* plant were *Wx* in phenotype (stained black with a I_2/KI stain) indicating that the phenotype of a pollen grain depended on its own genotype at the *wx* locus rather than that of the plant. Thus, each pollen grain can be a unit of genetic observation, and the more than 2×10^7 pollen grains produced by a vigorous plant can potentially be sampled. The F_1 hybrid between two *wx* mutants of independent origin where the mutational lesions affected different segments of the coding sequence should produce pollen with a few *Wx* (black-staining) pollen grains scattered among the numerous *wx* (tan-staining) pollen grains if recombination occurs between the mutant sites.

A standard technique was developed for collecting and assaying pollen samples. A section of the tassel should be collected just as the glumes are ready to open and be fixed in 70% ethanol. A curing period of several weeks in the ethanol solution facilitates the uptake of the stain by the pollen grains. Following this period, 24 anthers are removed from eight unopened florets to minimize the possibility of a stray wind-blown pollen grain being included in the sample. Maximum differentiation between *Wx* and *wx* pollen grains comes from selecting the less-mature floret in a glume that is about to open. The anthers are placed in the small stainless steel cup of a Virtis Microhomogenizer, cut apart with scissors, and then homogenized with 0.75 ml of stain for 2 minutes. The homogenate is then strained through two layers of cheesecloth onto an 80 × 100-mm slide and distributed as evenly as possible over the area that is to be covered by a 50 × 75-mm coverslip. After the stain has solidified, the edges of the coverslip are sealed with colorless fingernail polish. Maximum distinction between *Wx* and *wx* pollen is seen about 24 hours after the slide is made, but such preparations should keep for several weeks.

The stain formulation that has been most satisfactory is 250 mg KI and 45 mg I_2 in 25 ml water. After solution of these compounds, add one drop of Tween 80 and 0.5 g Baker's gelatin, and heat on a hot plate to solubilize the gelatin. Such a preparation remains liquid in a flask for a day at room temperatures but gels within an hour when spread thinly over the surface of a slide and topped by a coverslip. The coverslip should be lowered carefully from one side to exclude air bubbles.

The slides are best viewed at low magnification with both transmitted and reflected light such as is provided by an AO Cycloptic Binocular with a substage mirror and a lamp inserted through the illumination port behind the objective. Estimates of the number of pollen grains on a slide can be obtained by multiplying times a constant the sum of a number of counts (15 were used) through holes in a mask laid over the surface. The locations of *Wx* pollen grains can be marked with Kodak Opaque using a finely pointed paintbrush for easy counting and rechecking. More complete directions for the preparation of slides are given by Nelson (1968).

With the method described, it was possible to demonstrate with crosses between a set of *wx* mutations of independent origin that each heteroallelic cross had a characteristic frequency of *Wx* pollen grains ranging from ca. 1×10^{-5} (a value typical of the homoallelic stocks) to 100×10^{-5} (Nelson 1959). It was also possible to show that the distribution of numbers of pollen grains per anther accorded with a Poisson distribution as would be expected if the occurrences were unrelated and arose as meiotic events. A further observation was that backcrosses of the heteroallelic combination (90 × C) with the highest frequency of *Wx* pollen grains to either homoallelic parent yielded progenies in which about 50% of the plants had a *Wx* frequency comparable to that of the homoallelic parent while the remainder of the plants had *Wx* frequencies comparable to the heteroallelic parent but somewhat more variable. These data made it highly probable that intragenic recombination occurred between some pairs of *wx* mutations, and a tentative map of the mutations within the locus was made based on the observed frequencies of *Wx* pollen grains. The map was later shown to require revision (Nelson 1968).

If the observed frequency of *Wx* pollen grains observed in a cross between two *wx* mutants of independent origin were a valid measure of recombination between the mutant sites, it should be possible to observe the same frequency in seeds produced by pollinating a *wx/wx* tester stock by a heteroallelic male parent. At the same time, the use of stocks with flanking markers would allow ordering the mutant sites within the *wx* locus and answer the question of what proportion of the *Wx* gametes were recombinant for the flanking markers. Tests reported by Nelson (1962) and in 1968 with a better system for detecting the *wx/wx/Wx* seeds produced and excluding contamination by wind-blown *Wx* pollen indicated that the conventional genetic tests did validate the frequency of *Wx* pollen grains observed in the pollen assay for the heteroallelic combination, *wx-C/wx-90.* In the tests reported in 1968, the majority of the *Wx* recombinants (62%) had the distal flanking marker (*Bz1*) that entered the cross linked to *wx-90* and the proximal marker (*v1*) originally linked to *wx-C* thus indicating that the order of the mutant sites is (*bz1) wx-C wx-90* (*v1*). The next most numerous class of *Wx* recombinants (31%) was that of those that had the flanking markers that entered the cross linked to *wx-C*; the other combinations of flanking markers were each only 3.4%, or one incident each.

Nelson (1968) reported the location of 24 *wx* alleles within the *wx* locus. It was not possible to map the alleles using recombination frequencies as the basis because it was clear that recombination frequencies had only a tenuous relation to physical distance. Presumably this was true because the alleles were not in the same genetic background, having been collected from a number of different sources. It was possible, however, to place the mutant sites in an unambiguous order within the locus using the method of overlapping deletions developed by Benzer (1959) with bacteriophage T4 and the information from conventional genetic analyses that the order of mutant sites was (*bz1) wx -H21 wx-C wx-90* (*v1*). In the overlapping deletion method, the datum used is whether or not two alleles recombine, and the frequency of recombination is immaterial. It was possible to employ this method with these *wx* alleles because many of them behaved in these tests as though they possessed a physical size. Wessler and Varagona (1985) in an investigation of many of these alleles by molecular techniques corroborated the location of all but one allele within the locus as well as the physical size, showing that the majority of the mutations had resulted from either a deletion or an insertion. It might be noted that the original impetus for this particular investigation was to test whether the *wx* alleles that had resulted from the insertion of a transposable element were capable of recombining with each other and with the *wx* alleles in whose origin transposable element insertion had not been shown to be involved. The transposable element-induced alleles did recombine with other transposable element-induced alleles as well as with alleles not of that origin, so it was demonstrated that the transposable elements were affecting different regions of the locus.

In discussing the conventional tests of recombination between *wx-C* and *wx-90*, it has been noted that *Wx* gametes with the flanking markers from the *wx-C* parent constituted 31% of the total. This percentage was higher than might have been expected from two crossovers, one within the *wx* locus and the other between the locus and one of the flanking loci, suggesting that these *Wx* gametes could have arisen by gene conversion. It is not possible to make a direct test of gene conversion in maize by the examination of tetrads, but it is feasible to test whether the gametes in question could have arisen in most cases by double crossovers. This was done by examining recombination between *wx-c* and *wx-90* in plants that were heterozygous for one of three rearrangements involving chromosome 9. Heterozygosity for any of these three reduces recombination on the short arm of 9. These were *Tp9*, an insertional translocation in which a segment of chromosome 3 is inserted into the short arm of 9 between *wx* and *bz1*; *Inv9a*, which is a long pericentric inversion with the *wx* locus in the inverted segment; and *Rearr. 9*, a complex rearrangement of chromosome 9. Heterozygosity for any of these rearrangements reduces the frequency of *Wx* gametes. In heterozygotes for *Tp9*, the proportion of *Wx* gametes that had the flanking markers from the *wx-C* parent was enhanced. For the *Inv9a* heterozygote, the frequency of *Wx* gametes recombinant for flanking markers was markedly decreased, but the frequency of *Wx* gametes with the flanking markers from *wx-C* was unaffected. The frequency of *Wx* gametes was drastically reduced in the *Rearr. 9* heterozygote, and all had the *wx-C* flanking markers. The data from all three rearrangements agree in showing that the *Wx* gametes with the flanking markers from *wx-C* are unlikely in most cases to have resulted from double crossovers and are probably the result of gene conversion (Nelson, 1985).

It is apparent that a gene expressed postmeiotically in pollen and producing a visible and easily scored phenotype (with or without the use of a stain) offers the best opportunity to deal expeditiously with populations of sizes that make fine-structure investigations possible. Unfortunately, such genes are rare. In addition to *waxy*, only *alcohol dehydrogenase1* (*adh1*) has been shown to be as readily amenable to such investigations. Freeling (1976) has reported the methods for using null or low activity mutants of *Adh1* in such investigations. The parental *Adh1*- alleles from which his mutants were derived were apparently so divergent that recombination between mutants to yield *Adh1* gametes was much restricted if the mutants came from different parental alleles. Johns et al. (1983) documented the extensive sequence polymorphism found in the DNA flanking the coding regions of these *Adh1* alleles but not within the coding regions. The requirement that mutant alleles come from the same parental allele was not a constraint noted in crosses between *wx* alleles where nearly all the mutations occurred in different stocks (Nelson 1968), and it is not understood why the *adh1* mutants behave so differently.

It should also be noted that Moore and Creech (1972) were able to map five independently occurring *amylose extender* mutants from the results of pollen assays of heteroallelic plants that were also *wx*/*wx*. In their technique, the pollen samples were greatly overstained with a KI/I_2 solution and then processed through several rounds of destaining to reveal the red-staining *Ae*; *wx* pollen grains among the blacker *ae*; *wx* pollen.

REFERENCES

Benzer S (1959) On the topology of genetic fine structure. Proc Natl Acad Sci USA 45: 1607–1620

Brink RA, MacGillivray JH (1924) Segregation for the *waxy* character in maize pollen and differential development of the male gametophyte. Am J Bot 11: 465–469

Demerec M (1924) A case of pollen dimorphism in maize. Am J Bot 11: 461–464

Freeling M (1976) Intragenic recombination in maize: pollen analysis methods and the effect of parental *Adh1* alleles. Genetics 83: 701–717

Johns MA, JN Strommer, M Freeling (1983) Exceptionally high levels of restriction site polymorphism in DNA near the maize *Adh1* gene. Genetics 105: 733–743

Moore CW, Creech RG (1972) Genetic fine structure analysis of the *amylose extender* locus in *Zea mays* L. Genetics 70: 611–619

Nelson OE (1957) The feasibility of investigating "genetic fine structure" in higher plants. Am Nat XCI: 331–332

Nelson OE (1959) Intracistron recombination in the *Wx/wx* region in maize. Science 130: 794–795

Nelson OE (1962) The *waxy* locus in maize. I. Intralocus recombination frequency estimates by pollen and by conventional analyses. Genetics 47: 737–742

Nelson OE (1968) The *waxy* locus in maize. II. The location of the controlling element alleles. Genetics 60: 507–524

Nelson OE (1975) The *waxy* locus in maize. III. Effect of structural heterozygosity on intragenic recombination and flanking marker assortment. Genetics 79: 31–44

Wessler SR, Varagona M (1985) Molecular basis of mutations at the *waxy* locus of maize: correlation with the genetic fine structure map. Proc Natl Acad Sci USA 82: 4117–4122

Genetic Fine Structure from Testcross Progeny Analysis

HUGO K. DOONER

For decades, geneticists have analyzed the fine structure of genes by taking advantage of the cell's own DNA recombination machinery. Genetic recombinational analysis has led to the construction of gene maps that consist of linear arrays of mutations separated from each other by measured genetic intervals along the length of the gene. Today, the ability to dissect genes recombinationally is mostly of historic interest since, once a gene is isolated, much more precise information on gene structure can often be obtained from the analysis of restriction digests and, eventually, of course, from nucleotide sequencing. The combination of genetic and molecular analysis of genetic fine structure remains, however, a powerful approach that has enabled researchers to gain information that would have been difficult or impossible to gain purely from molecular studies. I will briefly discuss three such examples and will then outline the basic steps involved in carrying out a genetic fine structure study based on progeny analysis in maize.

CORRELATION BETWEEN PHYSICAL AND GENETIC DISTANCE

Intragenic recombination has been measured for several maize genes that have also been cloned. When the genetic and molecular maps of these genes are compared, it is clear that the frequency of recombination per kilobase within genes is as much as 100 times higher than the average for the entire genome: *bz* (Table 40.1) (Dooner et al. 1985; Dooner 1986); *wx* (Nelson 1962; Wessler and Varagona 1985); *Adh1* (Freeling 1978; Sachs et al. 1986); *a1* (Brown and Sundaressan 1991); and *R* (J.L. Kermicle, personal communication). These observations lend support to the hypothesis (Thurieaux 1977) that recombination in eukaryotic genomes is confined to genes.

The Maize Handbook—M. Freeling, V. Walbot, eds.

Table 40.1. *Bz* selections from *bz* heteroallelic plants pollinated with *sh bz-R wx*

Genotype of heteroallelic parent	No. of seeds screened	*Bz* seed selections				Inferred order sh [ ] wx	Estimated map distance
		sh wx	*Sh Wx*	*sh Wx*	*Sh wx*		
sh bz-m1 wx / *Sh bz-E2 Wx*	114,650	0	0	7	0	[*E2 m1*]	0.012
sh bz-m2(DI) wx / *Sh bz-E2 Wx*	217,820	0	0	0	6	[*m2(DI) E2*]	0.006

ORIENTATION OF THE PHYSICAL MAP OF THE GENE RELATIVE TO THE CENTROMERE

If the centromere proximal-distal orientation of a genetic fine structure map is known, it is possible to orient the gene's physical map relative to the centromere by placing two genetically mapped mutations in the physical map. Dooner et al. (1985) and Dooner (1986) placed several transposon insertion mutations in the *bz* genetic map. The genetic map was oriented within the short arm of chromosome *9* in the mapping experiment by using the flanking markers *sh* and *wx* to mark the location of the telomere and the centromere, respectively. The insertions were also placed in the physical (restriction and transcription) map and the orientation of the *bz* gene in *9S* was determined from a comparison between the order of the mutations in the genetic and physical maps. The *bz* promoter was found to lie at the proximal end of the gene, so the direction of transcription of *bz* in a normal chromosome is from centromere to telomere.

ANALYSIS OF COMPLEX LOCI

A complex locus consists of a series of structurally and functionally related components that are very closely linked. Molecularly, a complex locus corresponds to a clustered gene family. The functional analysis of the various members of a gene family is facilitated when prior genetic analysis has resulted in a structural dissection of the locus. In maize, *R* constitutes the prime example of a complex locus. Several of its components have been resolved genetically and separated from each other in work dealing with the genetic fine structure of the locus (e.g., Stadler and Neuffer 1953; Dooner and Kermicle 1971, 1976; Kermicle 1980). Today, the molecular analysis of the locus is aided by the availability of the many derivative alleles that arose in prior genetic studies. The current work has confirmed the existence of the components that were predicted genetically and has established the existence of new components, such as a cryptic (truncated) gene, that would have been difficult to predict from a pure genetic analysis (Robbins et al. 1991).

HOW TO CONSTRUCT A GENETIC FINE-STRUCTURE MAP

A fine-structure map gives the order of mutations within a gene and the genetic distances between them. Typically, heterozygous individuals carrying two different mutations of the same gene (heteroalleles) are synthesized, and their meiotic products are screened for wild-type recombinants in progeny produced from a testcross.

Testcross progenies are produced by crossing heteroallelic plants that are also heterozygous for flanking genetic markers to plants that carry a third mutation of the gene and are homozygous recessive for the flanking markers. Because intragenic recombination frequencies are low, usually ranging from 10^{-3} to 10^{-5}, large testcross populations must be generated in order to detect the rare intragenic recombinants. Large seed populations are produced most easily in isolated detasseling plots (Stadler 1942). In such a plot, the heteroallelic plants serve as female parents and the plants homozygous for a distinct third mutation serve as male parents. They are planted in either a 2:1 or a 3:1 female:male ratio. The plants in the female rows are hand detasselled by removing the tassels (together with the top one or two leaves) before they start to shed pollen. The field should be walked 2–3 times to ensure that no plants will shed in the female rows. (Tillers can be problematic and are best removed early.) The field is allowed to open-pollinate, but seed set can be improved by some hand pollination.

In commercial seed fields, the minimum distance for isolation is considered to be 201 m. Other factors, besides isolation, that help to reduce pollen contamination are: male border rows (2–3 on each side), natural barriers, such as trees, upwind placement of the isolation plot relative to other corn fields, and good "nicking" between female and male rows (Wych 1988). To ensure that pollen contaminants can be identified and eliminated from the analysis, it is essential that some distinct genetic marker, not present in the female rows, be used to identify the pollen parent. Such a marker could be a unique mutation of the gene being studied (with a distinguishable visible or molecular phenotype) or an easily classified endosperm marker (e.g., *y*, *wx*, *sh*).

The order of the mutations is established from an analysis of the distribution of flanking markers among the wild-type recombinants. Of the two possible recombinant arrangements of flanking markers, one will predominate. This constitutes the majority recombinant class and serves to establish the order and relative distances of the mutations in the gene. The example in Table 40.1 is adapted from Dooner et al. (1985). Map distances have been estimated by doubling the percentage of wild-type recombinants to account for the unrecovered reciprocal recombinant class (double mutant).

REFERENCES

Brown J, Sundaressan V (1991) A recombinational hotspot in the maize *A1* intragenic region. Theor Appl Genet 81: 185–188

Dooner HK (1986) Genetic fine structure of the *bronze* locus in maize. Genetics 113: 1021–1036

Dooner HK, Kermicle JL (1971) Structure of the *R-r* tandem duplication in maize. Genetics 67: 427–436

Dooner HK, Kermicle JL (1974) Reconstitution of the *R-r* compound allele in maize. Genetics 78: 691–701

Dooner HK, Kermicle JL (1976) Displaced and tandem duplications in the long arm of chromosome 10 in maize. Genetics 82: 309–322

Dooner HK, Weck E, Adams S, Ralston E, Favreau M, English J (1985) A molecular genetic analysis of insertions in the *bronze* locus in maize. Mol Gen Genet 200: 240–246

Freeling M (1978) Allelic variation at the level of intragenic recombination. Genetics 89: 211–224

Kermicle JL (1980) Probing the component structure of a maize gene with transposable elements. Science 208: 1457–1459

Nelson OE (1962) The *waxy* locus in maize. I. Intralocus recombination frequency estimates by pollen and by conventional analyses. Genetics 47: 737–742

Robbins T, Walker EL, Kermicle JL, Alleman M, Dellaporta SL (1991) Meiotic instability of the *R-r* complex arising from intragenic exchange and intrachromosomal rearrangement. Genetics 129: 271–283

Sachs M, Dennis E, Gerlach W, Peacock WJ (1986) Two alleles of maize *Alcohol dehydrogenase-1* have 3′structural and poly(A) addition polymorphisms. Genetics 113: 449–467

Stadler LJ (1942) Some observations on gene variability and spontaneous mutation. The Spragg Memorial Lectures (3rd Ser.), Michigan State College, East Lansing, MI

Stadler LJ, Nuffer MG (1953) Problems of gene structure II. Separation of the *R-r* elements (S) and (P) by unequal crossing over. Science 117: 471–472

Thurieaux P (1977) Is recombination confined to structural genes on the eukaryotic genome? Nature 268: 460–462

Wessler SR, Varagona R (1985) Molecular basis of mutations at the *waxy* locus of maize: correlation with the fine structure genetic map. Proc Natl Acad Sci USA 82: 4177–4181

Wych RD (1988) Production of hybrid seed corn. In Sprague GF, Dudley JW (eds) Corn and Corn Improvement, Third Edition, Am Agronomy Soc, Madison, WI, pp 565–607

41

Trisomic Manipulation

JAMES A. BIRCHLER

Primary trisomics for each of the ten maize chromosomes have been isolated. Stocks are available from the stock center at the University of Illinois. McClintock first recovered trisomics from the progeny of a triploid individual (1929) and used them to assign linkage groups to the corresponding chromosome (McClintock and Hill 1931). Present day uses of trisomics include assignment of molecularly defined loci to chromosome, whole chromosome dosage studies, and a means to force recovery of homologues in the same megaspore.

The chromosome pairing configuration in primary trisomics is such that at any one point, two chromosomes are synapsed and the third homolog is free. The synapsed pair usually switches along the length of the chromosome. Under ideal conditions, one would predict that the progeny of a trisomic would be half trisomics and half normal. In practice, however, the univalent chromosome is often lost in meiosis and is not included in the spores produced at the conclusion of meiosis. The predicted segregation from a trisomic, say, of A/A/a genotype would be A; A; a; Aa; Aa; AA. If such a plant were used as a female parent in a cross by a recessive tester, a ratio of 5A:1a would be produced. If the same plant were used as a male onto the same tester, then a ratio of 2A:1a would result. The reason the latter ratio differs from the former is that pollen tubes that have a genetic duplication grow more slowly and are unsuccessful in achieving fertilization in competition with the normal types. These ratios are hypothetical and are affected by the particular trisomic under consideration and the genetic map distance of the marker being followed from the respective centromere (Einset 1943). The disparity of female and male ratios, however, is characteristic of primary trisomics and is the standard genetic means of identification. Morphologically, trisomics 1, 5, and 7 can be easily distinguished from segregating siblings, although more intimate scutiny will allow the distinction of all of the trisomics. Some examples of the uses of trisomics include localization work (Tsaftaris et al. 1981) and analysis of preferential pairing in meiosis (Doyle 1982).

The Maize Handbook—M. Freeling, V. Walbot, eds.

REFERENCES

Doyle G (1982) The allotetraploidization of maize. Part 3. Gene segregation in trisomic heterozygotes. Theor Appl Genet 61: 81–89

Einset J (1943) Chromosome length in relation to transmission frequency of maize trisomes. Genetics 28: 349–364

McClintock B (1929) A cytological and genetical study of triploid maize. Genetics 14: 180–222

McClintock B, Hill HE (1931) The cytological identification of the chromosome associated with the R-G linkage group in *Zea mays*. Genetics 16: 175–190

Tsaftaris AS, Scandalios JG, McMillin DE (1981) Gene dosage effects on catalase expression in maize. J Hered 72: 11–14

42

B-A Translocation Manipulation*

WAYNE R. CARLSON

B-A translocations have unique properties that distinguish them from standard (A-A) translocations. One of these is the capacity for nondisjunction. Standard B chromosomes undergo nondisjunction very frequently at the second pollen mitosis. In B-A translocations, the chromosome with the B centromere acquires this property. Therefore, B^A chromosomes show dosage variation when transmitted through the pollen. B-A translocated chromosomes are usually designated in the form B^A or A^B. The superscripts denote the arm without its centromere.

A second unusual characteristic of B-A translocations is the nonessential nature of one member of the translocation. The B chromosome is completely dispensable to the organism. As a result, B-A translocation heterozygotes are often propagated in the absence of the normal B chromosome. This means that meiotic pairing in the heterozygote produces

*The classical terminology for B-A translocations is used in this chapter. The chromosome listed first carries the centromere; the second chromosome (an arm donor) is written as a full-sized superscript character, i.e., B^A or A^B. New terminology has been proposed that will probably render superscripts obsolete (B^A becomes B-A). When a final decision is reached, the new "rules" will be published in the *Maize Genetics Cooperation News Letter*. The Editors

The Maize Handbook—M. Freeling, V. Walbot, eds.

a trivalent (A A^B B^A) rather than a quadrivalent. The dispensable nature of B chromosomes also affects the type of gametes produced by translocation stocks. It allows B-A translocation heterozygotes to produce a wider variety of viable gametes than do standard translocations. One class of adjacent 1 segregation, the A B^A chromosome type, is viable because its only deficiency is for a segment of B chromatin. From adjacent 2 segregation, the A A^B spore is viable for the same reason.

CHROMOSOME COMBINATIONS OF B-A TRANSLOCATIONS

Because of these factors, B-A translocation chromosomes are found in a larger number of chromosome combinations than are standard translocations. Two major chromosomal groups can be identified according to their homozygosity or heterozygosity for the A^B chromosome. Homozygotes for A^B include (1) A^B A^B B^A, (2) A^B A^B B^A B^A, and (3) A^B A^B B^A B^A B^A. Heterozygotes for A and A^B are (1) A A^B, (2) A A^B B^A, and (3) A A^B B^A B^A. One additional class is the tertiary trisomic, which consists of the A A B^A and A A A^B chromosome types. The only genetically balanced constitutions listed above are the A^B A^B B^A B^A homozygote and the A A^B B^A heterozygote. All others have either three doses of an A segment (hyperploid) or one dose (hypoploid).

IDENTIFYING CHROMOSOMAL TYPES

As shown above, a great variety of chromosomal types can be produced by B-A translocation crosses. Identifying the exact constitution of a plant may require the use of several techniques. The following methods are useful:

1. *Utilize appropriate marker genes.* Gene linkage can be used to follow the inheritance of A^B chromosomes. Linkage is strongest when the marker locus is near the translocation breakpoint. For example, the *Wx* gene is very near the breakpoint on chromosome 9 for TB-9Sb (Robertson 1967). When 9^B carries the *Wx* allele and 9 carries *wx*, selection of the dominant phenotype in testcrosses is virtually error-proof in selecting for 9^B.

   Marker linkage can also be used for tracing inheritance of B^A chromosomes, but primarily when transmission is through the female parent. In pollen parent crosses, gene linkage to the B^A can be obscured by nondisjunction. In the male, therefore, B^A's are usually identified with a test of nondisjunction. The transmission of B^A's is revealed when a recessive phenotype is "uncovered." In this test, the male parent carries only dominant alleles. For example, the *Su* gene can be used in a test of nondisjunction by TB-4Sa. For a TB-4Sa heterozygote, an appropriate cross is *su su* × 4(*Su*) 4^B B^4(*Su*). In the progeny, kernels with the recessive *sugary* endosperm phenotype represent one class of nondisjunction. The endosperm lacks the B^4 chromosome, giving a recessive endosperm phenotype. The embryo, therefore, has two B^4's and its chromosome constitution is 4 4^B B^4 B^4. Kernels with the nonsugary endosperm phenotype arise from several gamete types. The addition of a seedling phenotype gene to the cross would allow identification of those nonsugary kernels that arise by nondisjunction. The latter kernels have two B^4's in the endosperm and none in the embryo. They produce plants with the recessive seedling phenotype and the 4 4^B chromosome constitution.

2. *Classify rate of pollen abortion.* Meiosis in some B-A translocation chromosome types produces significant numbers of inviable meiotic products. Inviability is usually the result of segregating the A^B chromosome to one pole without an accompanying B^A. Less frequently, an inviable spore that contains the B^A but not the A^B chromosome is produced. The latter spore is only produced by the A A^B B^A heterozygote and only with certain translocations (Kindiger et al. 1991). Chromosome types that produce large numbers of inviable spores, along with approximate rates of pollen abortion, are as follows:

   a. A^B A^B B^A 50% abortion
   b. A A^B 50% abortion
   c. A A^B B^A 25% abortion

3. *Analyze mitotic metaphase.* The number of B^A's in a plant can be determined directly in root-tip metaphase preparations. B^A chromosomes appear telocentric in metaphase, even though they probably have a minute short arm. The number of telocentrics equals the B^A number, because none of the standard chromosomes are telocentric. Chromosome count can also be used to identify B^A's. Any number in excess of 20 is equal to the number of B^A chromosomes. It is notable that some maize genetic stocks carry standard B chromosomes, and these are difficult to distinguish from B^A chromosomes. Therefore, stocks used in B-A translocation crosses should lack the standard B.

VIABILITY OF GENETICALLY UNBALANCED MICROSPORES, MEGASPORES, AND GAMETES

A distinction must be made between genetically unbalanced spores (meiotic segregants) that give rise to unbalanced gametes and genetically balanced spores that produce unbalanced gametes. Meiotic segregation in the various B-A translocation constructs produces a variety of unbalanced spore types. The most common of these are A B^A and A^B. The A B^A chromosome type is duplicate for a segment of the A chromosome and deficient for a B segment. As the B chromosome does not carry any vital genes, this meiotic segregant is both male and female viable. However, the male parent transmits the A B^A microspore class at low frequency. Pollen transmission is low as a result of the negative effects of gene duplication. Normal pollen tends to outcompete duplicate pollen for fertilization of the egg. Concerning the A^B meiotic segregant, it is inviable in both male and female parents.

The genetically balanced microspore class, A^B B^A, produces unbalanced gametes by nondisjunction. The unbalanced gametes are fully transmissible through the male parent, unlike those that arise by meiotic segregation. The viability and transmissibility of these gametes depends on their origin very late in development. Nondisjunction occurs at the second division in the pollen. Consequently, the balanced chromosome combination, A^B B^A, is present in the tube cell (vegetative nucleus) while A^B B^A B^A and A^B are present in the two sperm cells. Because the tube cell is responsible for metabolic activity of the pollen, and the sperm cells are genetically inert, the unbalanced sperm cells are easily transmitted. (Note: nondisjunction does not occur in A B^A microspores, because of the absence of a region on the A^B that controls B^A nondisjunction; see next chapter by Beckett also.)

MAINTENANCE OF B-A TRANSLOCATION STOCKS

Three methods are commonly used in the maintenance of B-A translocations. Each has its own advantages and disadvantages. (See also Beckett 1978, 1991.)

1. *Inbreed homozygous stocks.* Normally, the sib or self-pollination of homozygous stocks is a very simple procedure. However, the nondisjunction of B^A chromosomes complicates the process for B-A translocations. Self-pollination of the A^B A^B B^A B^A balanced homozygote produces three types of progeny: A^B A^B B^A, A^B A^B B^A B^A, and A^B A^B B^A B^A B^A. Consequently, a truly homozygous stock cannot be established. However, a functionally homozygous stock can be maintained by inbreeding, if the number of B^A's present in each plant is not critical. Each of the individual chromosome types shown above produces the balanced chromosome combination (A^B B^A) by meiotic segregation. In addition, plants with only one B^A produce substantial numbers of A^B spores. The latter spores are both male and female lethal and will not be transmitted. Plants with more than two B^A chromosomes produce a duplicate spore type containing A^B B^A B^A. The latter chromosome combination is female transmissible, but is seldom transmitted through the pollen. Thus, both duplicate and deficient spores are produced by an inbred "homozygous" stock, but selection eliminates many of them. In the female, the main gametes transmitted by a homozygous stock are A^B B^A and A^B B^A B^A. In the male, the A^B B^A combination is the main viable microspore. It produces the A^B B^A gamete plus two nondisjunctional gametes: A^B and A^B B^A B^A. Random combinations of male and female gametes give a "homozygous" stock with a variable number of B^A's. However, the range of B^A number is limited by selection.

   When a "homozygous" stock is used for experimental crosses, it is preferable to eliminate A^B A^B B^A B^A B^A (and higher ploidy) plants that produce duplicate spores. This can be accomplished by selecting plants in the "homozygous" population that show semisterile pollen. Semisterile plants have the A^B A^B B^A constitution. They produce no duplication-type spores and are equivalent to the homozygote in terms of gamete production.

Maintenance of B-A stocks by inbreeding has two advantages. First, the maintenance cross is very simple. Second, crosses with the stock transmit the translocation to progeny 100% of the time. The main disadvantages deal with the need to select A^B A^B B^A plants for outcrosses. Pollen classification adds to fieldwork. In addition, semisterile plants give poor seed set when crossed as females.

2. *Maintain heterozygous A A^B B^A plants.* When a heterozygote is crossed as female to a chromosomally normal stock, three major classes of progeny result. They are (a) the heterozygote, (b) the A A normal chromosome type, and (c) the A A B^A tertiary trisomic (Robertson 1967). A fourth class, the A A A^B tertiary trisomic, may also be produced in significant numbers (Kindiger et al. 1991). Heterozygotes can be selected from this population because they show pollen semisterility, unlike other classes. Meiotic segregation in the heterozygote produces approximately 25% lethal-deficient spores from adjacent segregation. Maintenance, therefore, requires crossing A A^B B^A heterozygotes as female to normal plants and selecting semisterile progeny.

   The advantages of maintaining B-A translocations as heterozygotes include (a) absence of B^A-dosage variation in crosses through the female parent (no nondisjunction), (b) lack of a requirement for genetic markers to classify the translocation, and (c) propagation by outcrossing, which tends to add vigor. In addition, this is the best method for transferring a translocation into a particular inbred line, as noted by Beckett (1978). Disadvantages of this method are (a) the time-consuming nature of classifying pollen sterility and (b) complications when using heterozygous plants experimentally, as a result of transmission of many different gametic types.

3. *Maintain stocks as the A A^B B^A B^A hyperploid.* In this case, the stock is maintained by pollen parent transmission of the B-A translocation and selection for cases of nondisjunction. Chromosomally normal plants are crossed as female parents to a hyperploid male. Meiosis in the hyperploid produces two major segregants: A^B B^A and A B^A. The A B^A chromosome combination is duplicate and is seldom transmitted through the pollen. The A^B B^A microspore can produce pollen with one B^A in both sperm (absence of nondisjunction) or zero B^A's and two B^A's in the sperm (nondisjunction). Selection of hyperploids from the maintenance cross depends on the ability to recognize cases of nondisjunction in which the two-B^A sperm fertilizes the embryo and the zero-B^A sperm fertilizes the polar cells. This is accomplished by using an endosperm-phenotype gene. For example, the *C* locus on chromosome 9 controls endosperm color. It is present on the B^9 chromosome in TB-9Sb. A maintenance cross for TB-9Sb is 9(*c*) 9(*c*) × 9(*C*) 9^B B^9(*C*) B^9(*C*). In the progeny, white (recessive) kernels are selected. They have zero B^A's in the endosperm, allowing the recessive phenotype to be "uncovered." The embryo received the 9^B B^9 B^9 sperm and is, therefore, hyperploid.

   The hyperploid method of maintenance has several advantages. These include (a) high-frequency transmission of the translocation through the male parent due to selection against duplicate A B^A pollen, (b) simplicity of selecting hyperploids with an endosperm marker, and (c) propagation by outcrossing, which tends to give vigorous stocks. Disadvantages of the technique are (a) the requirement for an appropriate endosperm marker on the B^A and (b) errors that occur in the selection of hyperploids in a maintenance cross.

Errors that arise in the propagation of hyperploid stocks result from the use of a recessive endosperm marker for classification. In the example above of a TB-9Sb cross, errors in selection should be minimal. White kernels should all contain hyperploid embryos, except for cases of self-contamination. However, after one generation of propagation the hyperploid TB-9Sb stock will carry the recessive allele on chromosome 9: 9(*c*) 9^B B^9(*C*) B^9(*C*). Future crosses of recessive *c c* × hyperploid stock could give white kernels resulting from transmission of 9(*c*) from the male parent rather than nondisjunction. However, meiosis in the hyperploid produces two major microspore classes: 9^B B^9(*C*) and 9(*c*) B^9(*C*). The 9(*c*) chromosome is covered by the B^9(*C*). Nevertheless, in a small percent (1–2%) of cases, meiosis produces a 9(*c*) spore. The errors that occur in this way are few in number. In addition, incorrect selections are easy to recognize in crosses, because they are homozygous recessive: 9(*c*) 9(*c*). If selection needs to be error-free, recessive testers can be alternated each generation (Beckett, 1978). However, this requires the availability of two endosperm markers for the B^A chromosome.

CALCULATING NONDISJUNCTION AND PREFERENTIAL FERTILIZATION

Nondisjunction of the B^A chromosome at the second pollen mitosis requires the presence of the A^B chromosome. Consequently, the A^B B^A microspore is capable of nondisjunction, whereas the A B^A class is not. The calculation of nondisjunctional rates is simplest when a homozygous B-A translocation stock is used. All the microspores produced in the cross are the A^B B^A type. The rate of nondisjunction equals the total number of kernels that receive the A^B and A^B B^A B^A nondisjunctional sperm, per total kernels. Cases of nondisjunction are identified by the ability of A^B sperm to "uncover" recessive phenotypes. For example, with TB-9Sb the 9S markers *Yg2* and *Bz* can be used to calculate nondisjunction. The *Yg2* allele produces a normal green plant whereas *yg2* gives a yellow-green phenotype. The *Bz* allele conditions a purple endosperm phenotype while the recessive allele gives a bronze phenotype. A test of nondisjunction is 9(*yg2 bz*) 9(*yg2 bz*) × 9^B 9^B B^9(*Yg2 Bz*) B^9(*Yg2 Bz*). Kernels that uncover the bronze endosperm phenotype represent one class of nondisjunction. They have green seedlings. Purple kernels that give yellow-green seedlings are the second class of nondisjunction. The rate of nondisjunction equals the number of *bz Yg* plus *Bz yg* phenotypes divided by total kernels.

The determination of nondisjunction rates is straightforward when a homozygous B-A translocation is used. However, for heterozygous or hyperploid stocks the calculation is much more difficult. Only for TB-9Sb, where an excellent marker (*Wx*) for the 9^B chromosome is available, have measurements of nondisjunction been made in nonhomozygous stocks. The *Wx* allele provides an accurate measure of gametes derived from the 9^B B^9 microspore (Carlson 1969).

Preferential fertilization refers to the competition between nondisjunctional sperm for fertilization of the egg. The A^B B^A B^A sperm fertilizes the egg more often than does the A^B sperm. The B^A chromosome confers an advantage in fertilization. The rate of preferential fertilization, using the example above for TB-9Sb, is equal to the number of *bz Yg* phenotypes per total cases of nondisjunction (*bz Yg* + *Bz yg*). The rate of preferential fertilization can be calculated for any type of B-translocation cross, unlike calculations of nondisjunction.

Further information about B-A translocations is available in review articles by Beckett (1978, 1991) and Carlson (1978, 1986).

ACKNOWLEDGMENT

This material is based upon work supported by the Cooperative State Research Service, U.S. Dept. of Agriculture, under agreements # 89-37140-4464 and # 92-37301-7679.

REFERENCES

Beckett JB (1978) B-A translocations in maize. I. Use in locating genes by chromosome arms. J Hered 69: 27–36

Beckett JB (1991) Cytogenetic, genetic and plant breeding applications of B-A translocations in maize. In Gupta PK, Tsuchiya T (eds) Chromosome Engineering in Plants: Genetics, Breeding, Evolution. Part A, Elsevier, Amsterdam, pp 493–529

Carlson WR (1969) A test of homology between the B chromosome of maize and abnormal chromosome 10, involving the control of nondisjunction in B's. Mol Gen Genet 104: 59–65

Carlson WR (1978) The B chromosome of corn. Annu Rev Genet 12: 5–23

Carlson WR (1986) The B chromosome of maize. Crit Rev Plant Sci 3: 201–226

Kindiger B, Curtis C, Beckett JB (1991) Adjacent II segregation products in B-A translocations of maize. Genome 34: 595–602

Robertson DS (1967) Crossing over and chromosomal segregation involving the B^9 element of the A-B translocation B-9b in maize. Genetics 55: 433–449

43

Locating Recessive Genes to Chromosome Arm with B-A Translocations

J. B. BECKETT

The B-A translocations of maize permit recessive genes to be located to the correct chromosome arm in the F_1. RFLPs, codominants, and dominant genes with extreme phenotypes when homozygous or hemizygous can also be located to the appropriate chromosome arms by this system. See previous chapter by Carlson for a description of B-A translocation behavior.

B-A translocations are reciprocal translocations between the basic, or A, set of chromosomes and the supernumerary B chromosome (Figure 43.1). At the second pollen division, the B chromosome nondisjoins (i.e., both products of mitosis remain together), so that one sperm cell of a pollen grain has two B chromosomes and the other has none. If part of an A chromosome arm has been translocated onto the centromere-bearing portion of the B chromosome, then nondisjunction of this B-A chromosome results in one sperm with two copies of the A arm and the other sperm with none (Figure 43.2). (Note that a complete B-A translocation consists of two parts, the B-A chromosome, bearing the B centromere, and the A-B chromosome with the A centromere. The distal portion of the B chromosome carries a factor for nondisjunction, so the A-B chromosome must be present before the B-A chromosome can nondisjoin.) If the deficient sperm fertilizes the egg, the resulting embryo (and subsequently the resulting plant) will be hemizygous (hypoploid) for part of the A chromosome arm. On the other hand, if the deficient sperm unites with the polar nuclei, the resulting endosperm will lack the B-A chromosome. If a dominant gene for color resides on the translocated A arm, and the female parent carries the recessive allele, then noncorrespondence between the genotypes of the embryo and the endosperm can be demonstrated (Figure 43.3). More complete discussions of B-A translocations (Beckett 1978, 1991; Birchler 1983) and factors affecting nondisjunction (Carlson 1978, 1986) are available.

This chapter is concerned primarily with providing a list of the B-A translocations

The Maize Handbook—M. Freeling, V. Walbot, eds.

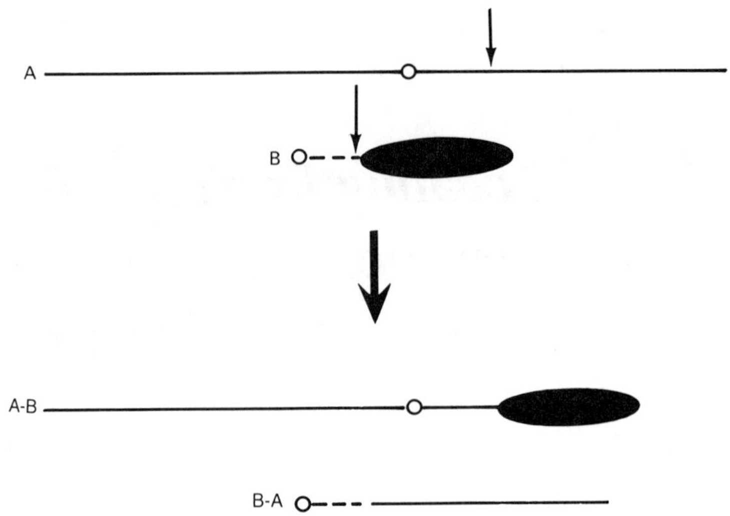

FIGURE 43.1 Production of a B-A translocation by breakage of a normal (A) chromosome and a B chromosome at the sites indicated by arrows, followed by rejoining of broken ends as illustrated, giving an A-B chromosome and a B-A chromosome. (From Beckett 1991.)

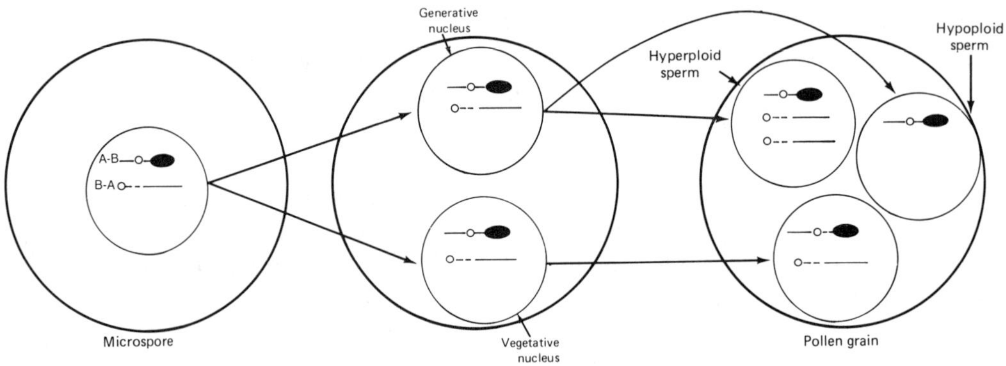

FIGURE 43.2 Development of a pollen grain with nondisjunction of the B-A chromosome at the second division of the microspore. (From Beckett 1991.)

and discussing their use in locating genes to chromosome arm. Please consult the reviews by Beckett, Birchler, and Carlson (see preceding paragraph) for references to original publications. Other papers are cited only if their contents are not covered in the reviews. Table 48.1 lists the B-A translocations available. The compound B-A translocations listed in the table were developed by crossing regular reciprocal translocations with existing

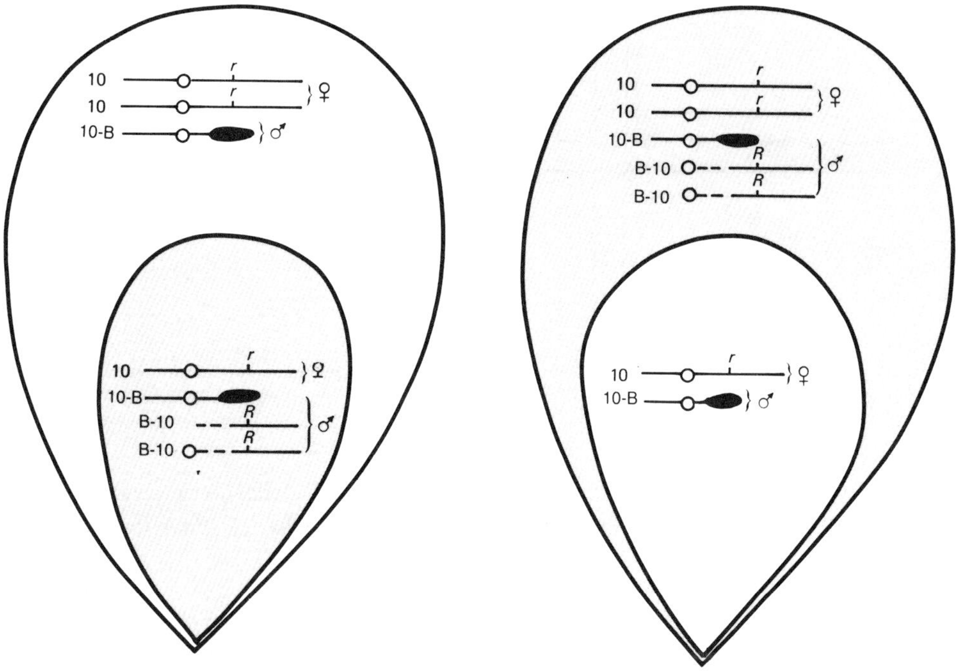

Egg fertilized by hyperploid (10-B B-10*R* B-10*R*) gamete

Egg fertilized by hypoploid (10-B) gamete

FIGURE **43.3** Alternative results of pollinating colorless (*rr*) with a pollen grain bearing gametes of the constitution 10-B B-10(*R*) B-10(*R*) and 10-B, respectively. Male (♂) and female (♀) contributions to embryo and endosperm are indicated. Presence of anthocyanin color is indicated by shading. (From Beckett 1991.)

B-A translocations and selecting crossover products with part of a new chromosome arm translocated onto the B-A chromosome (Figure 43.4).

LIST OF B-A TRANSLOCATIONS

A list of all B-A translocations is provided in Table 48.1. A basic set of B-A translocations, selected primarily to give maximum coverage, is designated in Table 48.1 and listed again below for convenience. All chromosome arms except the short arm of chromosome 8 (8S) are involved.

TB-1Sb	TB-9Sb-4L6222	TB-7Lb
TB-1La	*TB-4Lc	TB-8Lc
TB-3La-2S6270	TB-1La-5S8041	TB-9Sd

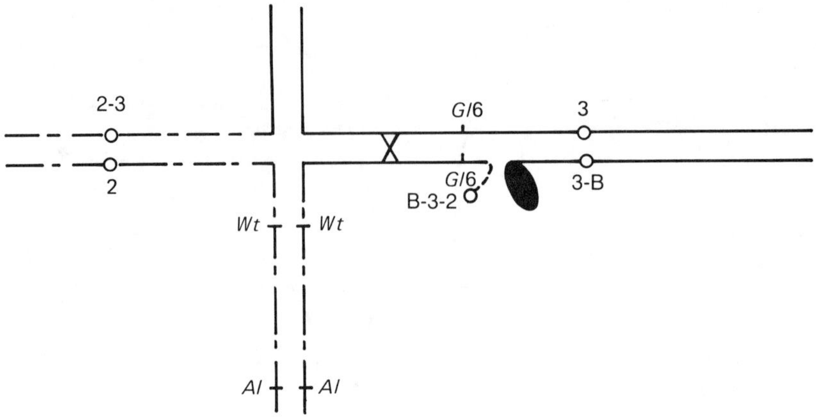

FIGURE 43.4 Heterozygous TB-3La-2S6270 at pachytene. Note that if a crossover occurs in the region indicated, TB-3La and reciprocal translocation 2–3(6270) will be reconstituted. (From Beckett 1991.)

*TB-2Sa	*TB-5Sc	TB-9Lc
TB-1Sb-2L4464	TB-5La	TB-10Sc
TB-3Sb	TB-6Sa	TB-10L19
TB-3La	TB-6Lc	
TB-4Sa	TB-7Sc	

*Alternative B-A translocations are designated for three chromosome arms; TB-2Sa may uncover nearly as much of 2S as the compound TB-3La-2S6270, but is not as dependable (see later discussion); TB-4Lc may uncover as much of 4L as TB-9Sb-4L6222, and it gives more vigorous hypoploids. (TB-4Ld, TB-4Le, and TB-4Lf also may be as good as TB-9Sb-4L6222.) The breakpoint of TB-5Sc is farther out on 5S than TB-1La-5S8041, so its hypoploids are more vigorous.

LOCATING RECESSIVE GENES TO CHROMOSOME ARM

In the simplest case, determining the location of a recessive or codominant factor (or an RFLP) to chromosome arm requires only that plants from a stock homozygous for that factor be pollinated by the basic set of B-A translocations. The cross that gives endosperms or seedlings that segregate for the factor in the F_1 identifies the chromosome arm involved. If the test fails to locate the factor, then the locus is proximal to one of the translocation breakpoints (i.e., between the breakpoint and the centromere) or is on 8S.

In practice, locating a gene may be more involved. Because homozygotes of most recessive genes lack vigor, heterozygous plants make better female parents and should generally be used. The frequency of recessives that segregate in the critical cross will be halved, but the extra seed produced will more than compensate.

A lethal or sublethal factor that is maintained by self-pollinating heterozygotes obviously will require several crosses to help insure that at least one heterozygote is tested for each chromosome arm.

Genes located to chromosome arm by a particular B-A translocation are said to be "uncovered" by that translocation.

Whenever translocation stocks are used to locate new mutants, each plant used should be testcrossed (as male) onto an appropriate tester stock (Table 43.1) to confirm that it is indeed producing hypoploid progeny.

TYPES OF TRANSLOCATION STOCKS SUPPLIED

B-A translocations are supplied in various forms. Among them are (1) homozygous stocks, (2) selected kernels that should produce hyperploid heterozygotes, (3) progenies segregating heterozygotes, and (4) progenies segregating heterozygotes and hyperploid heterozygotes.

1. Homozygous B-A translocation stocks are especially desirable because all plants carry the full translocation (i.e., both the A-B and the B-A chromosomes), but they are often weak because of inbreeding and aneuploidy. In homozygous stocks, A-B chromosomes are transmitted normally but B-A chromosomes vary in number because of nondisjunction; plants with one B-A are weaker than normal, those with three are often somewhat weaker, and those with four or more are usually dwarfed and sterile. If the nondisjunction rate approaches 100%, which is not uncommon, B-A chromosomes will tend to accumulate. In contrast, the homozygous stocks that are in an *A C R-nj* (or *R-sc*) W23 background tend to be reasonably vigorous because the rate of nondisjunction is probably closer to 50%. Homozygous translocation stocks, especially those involving TB-6Sa, may appear to lose the ability to nondisjoin because they produce very few hypoploid offspring when testcrossed. The 6S segment transferred to the B chromosome is very short, so apparently several can accumulate without affecting vigor. If two or three B-A's are present at the second mitosis of a pollen grain, most of the resulting sperm cells will carry one or more B-A chromosomes and thus be unable to produce hypoploid offspring.

   Confirmed and probable homozygous stocks of B-A translocations are listed in Table 43.2.

2. In general, selected kernels are supplied for use in uncovering mutants. The selection of small kernels works well for some translocations, especially the TB-1L's, the more proximal TB-10L's, and most of the compound TB's, but selected kernels involving endosperm markers are generally more dependable. Almost any classifiable recessive endosperm gene with a locus distal to a B-A translocation breakpoint can be used for this purpose; when plants homozygous or heterozygous for such a gene are pollinated by an appropriate B-A translocation, the recessive kernels that segregate on the ear yield the desired hyperploid heterozygotes. However, there is always the pos-

TABLE 43.1 Selected kernel and seedling factors for confirming the presence of the current basic set of B-A translocations: genes listed in approximate order of distance from centromere

Translocation	Kernel factors	Seedling factors
TB-1Sb	dek1[b] vp5	vp5
TB-1La[a]	bz2 lw1	lw1, also hypoploid phenotype[d]
TB-3La-2S6270	al1	wt1[c] gl11 d5 gl2 lg1 al1[c]
TB-2Sa	al1 b1[b]	d5 gl2 lg1 al1[c]
TB-1Sb-2L4464[a]	w3	v4[c] w3 spt1
TB-3Sb	cl1	cl1 d1 g2-m
TB-3La	vp1[b] y10 a1[b] sh2 et1	gl6 lg2 y10 et1[c]
Tb-4Sa	bt2 su1 dek7	dek7 spt2 hcf23[c]
TB-9Sb-4L6222	c2[b]	gl4 gl3 dp1[c]
TB-4Lc	c2[b]	gl4 gl3 dp1[c]
TB-1La-5S8041[a]	ps1 vp2 a2[b] anl1[b]	ps1 vp2 gl17
TB-5Sc	a2[b] anl1[b]	gl17
TB-5La	pr1 lw2 sh4	gl8 lw2 v2[c]
TB-6Sa	dek28	hcf26[c] oro1
TB-6Lc[a]	y1	w15 l12 py1
TB-7Sc	vp9 o2	vp9
TB-7Lb[a]	o5	gl1 o5 ij1[c]
TB-8Lc[a]	pro1	v16[c] v21[c]
TB-9Sd	wx1 bz1 sh1 c1[b]	d3 yg2[c] wd1
TB-9Lc	wc1[e]	v1[c] gl15 Bf1
TB-10Sc	y9	oy1[c] sr3[c]
TB-10L19[a]	r1[b]	bf2 g1

[a]Small grains on ears pollinated by this translocation will usually have hypoploid endosperms and hyperploid embryos.
[b]Gene capable of giving diagnostic colorless kernels with colored scutella.
[c]Expressed best when germinated in unheated sand bench.
[d]Hypoploid seedlings small with small primary leaves.
[e]Normal yellow (wc1) stocks make suitable testers for TB-9Lc stocks that carry the allele Wc1 on the B-9 chromosome.

sibility that some of these kernels are the result of self-pollination or other outcrossing. For this reason, it is safer to use pollen from more than one presumed translocation-bearing plant, or even to bulk pollen from two or more plants, to test new genes. If desired, the presumed translocation-bearing plants can be testcrossed by pollinating one or more plants of a tester carrying a recessive endosperm or seedling marker that is located beyond the translocation breakpoint. (If the presumed translocation-bearing plants carry a recessive allele of any of these marker genes, then that gene usually cannot be used to test for the presence of the translocation.)

For chromosome arms that carry color genes, noncorresponding kernels with colorless endosperms and colored scutella can be produced for distribution to others as translocation sources. To produce this seed, the dominant (+) allele of the color gene involved is required on the B-A chromosome of the male (translocation-bearing) parent, dominant alleles of all other color genes (except *B-Peru*) must be present in one or both parents, and a self-color allele of *r* (*R-sc* or *R-scm*) is normally required. TB-2Sa is a special case because *B-Peru*, the *b* allele present on the B-2S chromosome, acts as a duplicate factor of *R-sc*. Therefore, a recessive seed-color allele must be present at the *r* locus when using TB-2Sa to produce colorless kernels with colored scutella. Plants grown from noncorresponding kernels can be tested by crossing onto silks of plants carrying the same recessive color factor and observing for the segregation of additional colorless kernels with colored scutella. TB-10L18, with the breakpoint on the short arm of the B, is an exception. (See later discussion.) Rarely, even a plant grown from noncorresponding kernels may fail to carry the full B-A translocation because of an event such as heterofertilization.

Lethal endosperm traits, such as *vp5* and *dek28*, can be used repeatedly to testcross for the presence of B-A translocations. For example, white kernels from the cross *vp5* × (*vp5* × TB-1Sb) are of two types, those that are homozygous *vp5* and those with hypoploid *vp5* endosperms and hyperploid heterozygote embryos. The former class gives lethal seedlings and the latter gives viable plants that again carry the full translocation.

3. If a uniform genetic background is needed in B-A translocation stocks, several are available. Except for many of the TB-10L's, nearly all simple and many compound B-A translocations are available in L289 and *A C R-nj* (or *R-sc*) W23 backgrounds. Simple B-A translocations involving 18 chromosome arms are available in A619, A632, B73, and Mo17 backgrounds. Such stocks are usually maintained by crossing heterozygous female parents by the appropriate recurrent line. Although differences do occur, it is usually assumed that one-third of the progeny of such a cross are heterozygous and capable of supporting nondisjunction (Figure 43.5). Therefore, every plant used to determine the location of new genes must also be tested for the presence of the full translocation. Both endosperm and seedling markers are suitable; testers carrying both are especially good. Unless vigorous homozygous testers are available, heterozygous testers produced by outcrossing to unrelated stocks are generally best.

Table 43.2 List of homozygous B-A translocation stocks: inbred backgrounds are specified whenever possible (P = probably homozygous; x = homozygous)

	Background		
Translocation	nj W23[a]	L289	Undefined
TB-1Sb			x
TB-1Lc	x		
TB-1Sb-2L4464		P	x
TB-3Sb			x
TB-3La			x
TB-3Ld	x		
TB-3Lg			P
TB-3Ll		P?	
TB-4Sa		x	x
TB-4Lc			P
TB-4Lf			x
TB-5Lb	P		
TB-6Sa		x	x
TB-6Lb		x	
TB-6Lc	P		
TB-7Lb		x	x
TB-8La	x		
TB-8Lc			P
TB-9Sb			x
TB-9Sd			x
TB-9Lc (Wc)			x
TB-10La		P	
TB-10L20			P
TB-10L32			p

[a] *A C R-nj* (or *R-sc*) in W23 background.

Ears from tester stocks pollinated by putative translocation-bearing plants can usually be examined for segregation of endosperm traits without being shelled. Seedling tests are usually conducted in greenhouse sand benches. Because nondisjunction rates vary, and because tester stocks are usually heterozygous, 100 seedlings are desirable for screening. Hypoploid seedlings may be slow to emerge, so a single observation may not be sufficient. Occasional pollen contamination is inevitable, so the presence of one or two seedlings with the recessive phenotype in a sand-bench progeny is not enough to confirm that the pollen parent carried the B-A translocation. Haploids with the recessive phenotype also occur. Hypoploid seedlings can sometimes be differ-

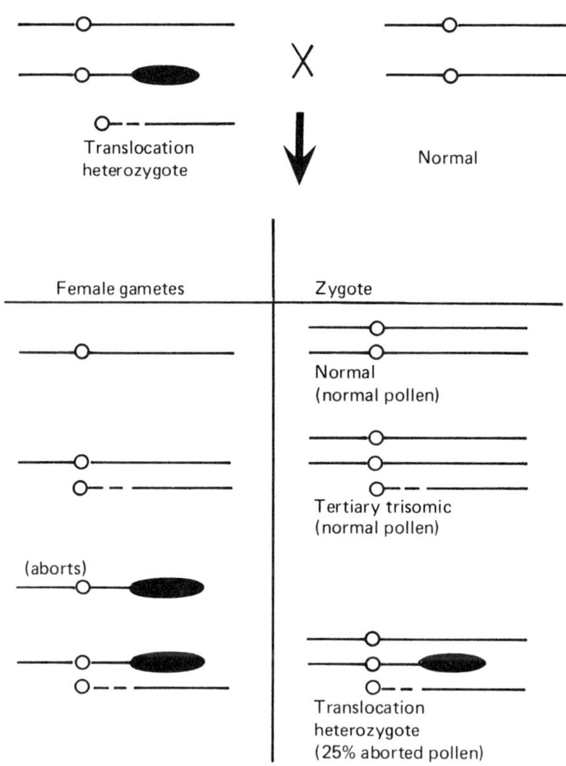

FIGURE 43.5 Classes of gametes and progeny expected from pollinating a B-A translocation heterozygote by a normal plant. (From Beckett 1978.)

entiated from euploid recessives by their delayed emergence, smaller size, slenderness, slightly altered leaf shape, etc., but the hypoploids for each arm (and even for different positions on the same arm) are unique, so a general description is not possible.

Many of the marker genes that are useful in testing translocation stocks are listed in Table 43.1. Markers that give yellowish or virescent seedlings are usually expressed more clearly in an unheated sand bench. Exceptions are *g1* and *g2*. Golden plants (*g1*) can be separated from normal by cutting the seedlings off above the coleoptilar node and examining the cut ends for yellow color; leaf striping makes *g2-mutable* the easiest *g2* allele to screen. Endosperm mutants such as *sugary-1* germinate better in a sand bench warmed with heating cables. Perhaps the best general procedure is to use heat until the first coleoptile tips can be seen, and then to discontinue heating if virescent hypoploids are to be screened.

When a translocation stock is segregating heterozygous plants, the earlier plants generally include most of the heterozygotes, presumably because linked chromatin from the nonrecurrent parent produces some degree of hybrid vigor. Heterozygotes, which have approximately 25% aborted pollen, can often be selected by placing fresh pollen (from fresh anthers that have not yet shed) onto a dark surface and observing with a hand microscope. Some translocations are more difficult to screen than others, and mistakes in classification are inevitable, so testcrossing is still necessary.

4. A B-A translocation may occasionally be supplied in a progeny produced by pollinating some stock by a translocation-bearing plant. If the male parent of such a progeny was homozygous for the translocation, only hypoploids, hyperploid heterozygotes, and sometimes heterozygotes are expected. If the hypoploids, which are semisterile and usually weaker and slower-growing, can be eliminated, most or all of the remaining plants will carry the full translocation.
   If the male parent of the progeny received was heterozygous for the translocation and had 100% nondisjunction, then a ratio of 4 "normals":1 hyperploid heterozygote:1 hypoploid could occur, although about 20:13:7 is more likely, especially if an excess of viable pollen was used to produce the progeny. This is because (a) half of the "normal" pollen grains carry an extra B-A element and usually compete poorly with normal pollen grains and (b) the egg is fertilized about 65% of the time by the sperm carrying two B-A chromosomes (instead of the sperm with no B-A's).
   Furthermore, pollen grains carrying the full B chromosome (i.e., the A-B plus the B-A) effect fertilization 56–59% of the time in competition with normal pollen, so an even higher frequency of hyperploid heterozygotes and of hypoploids can be expected (Beckett 1982).
   If the nondisjunction rate of the heterozygous pollen parent (of the translocation stock received) was less than 100%, heterozygotes will appear in the progeny and the frequency of hypoploids and hyperploid heterozygotes will decrease. This causes no real problem, however.
   If the pollen parent of the translocation stock received was a hyperploid heterozygote with a nondisjunction rate of 100%, the progeny could contain about 2 "normals":1 hyperploid heterozygote:1 hypoploid; the ratio should approach 0:13:7 if pollen competition and preferential fertilization had occurred. If the hypoploid plants can be recognized and removed or ignored, then most of the remaining plants should carry the full translocation.
   If the translocation stock (as received) was produced by pollinating a tester stock carrying a seedling or mature plant trait by a translocation-bearing plant, then any recessive plants should be hypoploids (or contaminants) and should be discarded, thus increasing the frequency of useful heterozygotes and hyperploid heterozygotes.

LOCATING MUTANT GENES MORE PRECISELY

After a gene has been assigned to a particular chromosome arm by use of B-A translocations, it is sometimes possible to determine its position more precisely by crossing it by other B-A translocations that are located farther out on the same arm. Procedures are identical, but the tester genes used for each translocation must be distal to the breakpoint.

PROCEDURE IF MUTANT PHENOTYPE FAILS TO SEGREGATE IN F_1

E. H. Coe, Jr. (personal communication 1992) estimates that 88% of mapped loci are uncovered by the basic set of B-A translocations. The remainder presumably lie on 8S, which lacks a B-A translocation, or on one of the other arms between the centromere and the translocation breakpoint. Therefore, if the mutant phenotype fails to appear in the F_1, it is necessary to self-pollinate the hypoploid plants in order to locate the recessive mutant. If the female parent of the hypoploids was homozygous for the recessive mutant allele, all hypoploid plants will be heterozygous for the recessive; if the recessive allele was heterozygous, half of the hypoploids will carry the recessive and half will not. If the gene is not on the same chromosome as the translocation, then self-pollination will yield the usual 3:1 ratio. If the locus of the mutant is just proximal to the translocation breakpoint, the F_2 progeny will be almost wholly recessive for the mutant because the A-B chromosome is rarely transmitted (Kindiger et al. 1991) (Figure 43.6). If the mutant locus is more proximal to the translocation breakpoint, or is even on the other arm, the ratios still will be (or may be) biased in favor of the recessive allele.

COMPOUND B-A TRANSLOCATIONS

Compound B-A translocations present special problems in use and maintenance, and so require special instructions. Breakpoint positions likely have great effect on pairing, crossing-over, and segregation, but the presence of two heterozygous translocations should usually give high sterility. In the simplest, possibly inaccurate case, the A-A translocation should give 50% sterility and the B-A translocation an additional 12.5%.

The maintenance of compound B-A translocations by pollinating translocation heterozygotes by the recurrent parent is unusually difficult because crossovers often reconstitute the two original translocations (Figure 43.4). For that reason, several compound

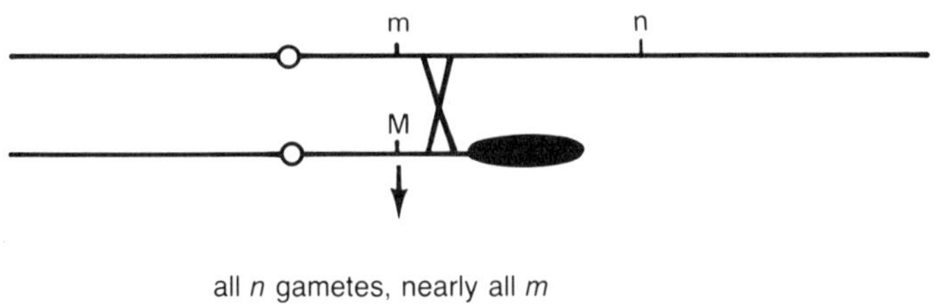

FIGURE 43.6 Diagram of the chromosomes of a hypoploid plant and the expected gametes. The mutant gene (*n*) is beyond the breakpoint, so its phenotype is immediately expressed ("uncovered") because the B-A chromosome, carrying the dominant allele, is absent. The mutant (*m*), which is proximal to the break, is not expressed in the F_1. All viable gametes carry *n* and usually *m*. *M* is transmitted if it is transferred to the A chromosome by crossing-over.

translocations have received few or no backcrosses to standard lines; instead, they are crossed each generation onto an appropriate color tester to produce colorless kernels with colored scutella.

Sand-bench and field tests involving compound translocations are sometimes difficult because the hypoploids lack portions of two chromosome arms and are especially weak. This can lead to a scarcity of viable hypoploid plants. Normally, if a recessive mutant cannot be assigned to chromosome arm in the F_1, self-pollination of hypoploids generally locates the mutant in the F_2. In the case of compound hypoploids, however, many are so weak that neither pollen nor seed can be obtained. Furthermore, it is often difficult to differentiate compound hypoploids from duplicate-deficient plants generated by the A-A reciprocal translocations.

Because compound translocations are B-A translocations with another chromosome arm replacing part of the original B-A chromosome, they are able to uncover genes on the original translocated arm up to the point of the second translocation. Therefore, any new locus uncovered by a compound translocation may actually be on the original chromosome arm. For example, translocation TB-3La-2S6270 uncovers genes on both 3L and 2S, so tests involving TB-3La alone must give negative results before the new mutant can be confidently assigned to 2S.

ADDITIONAL DETAILS

Appropriate tester stocks for the basic set of B-A translocations are listed in Table 43.1. If other translocations are used, it is important to select a tester stock that carries an endosperm or seedling trait beyond the breakpoint of the translocation involved. Table 48.1 gives the first distal gene known to be uncovered by each translocation, but it is not necessarily a good tester gene. Consult the most recent *Maize Genetics Cooperation News Letter* for chromosome maps giving relative positions of gene loci and translocation breakpoints.

The phenotype of a few of the genes used in seedling tests is expressed relatively late; among them are *spt1*, *spt2*, *gl15*, and *Bf1*. Because the seedlings need to grow longer than usual, kernels should be spaced in the sand bench so that the seedlings will receive enough light. *Bf1* is expressed most clearly on fully expanded leaves on a sunny day, but the plants must be screened under UV light in darkness. They may be examined in the evening as soon as darkness falls or under a "tent" during the day. They also may be excised and screened in a darkened enclosure or room. In contrast to *Bf1*, *bf2* should be scored soon after the coleoptiles emerge. Progenies involving *gl15* must be grown for a relatively long period; the glossy phenotype usually appears first on the lower half of the third leaf or on the fourth leaf. Because hypoploid seedlings tend to die early in the sand bench, watering at least once with a dilute fungicide will aid in the identification of progenies segregating for *gl15* hypoploids.

TB-10L18 requires special handling because its B chromosome was broken in the short arm instead of the long. As a result, its B-A chromosome possesses the terminal portion of the B and can therefore nondisjoin in the absence of the A-B chromosome. If care is not taken, the A-B chromosome may be lost even though testcrosses onto *r* produce colored kernels with colored scutella. Once the A-B chromosome has been lost, the B-A can persist only as a partial trisome (A A B-A) and so is useless for uncovering recessive genes. Two procedures can be employed to avoid loss of the A-B element. First, the use of tester stocks with seedling markers will determine whether hypoploids are still being produced. Second, *r1*-carrying testers clearly segregate small colorless kernels with colored scutella when the pollen parent carries the full translocation. Only these small kernels should be used to propagate the translocation.

Although TB-2Sa uncovers much of the short arm of chromosome 2 and is capable of giving colorless kernels with colored scutella, it is undependable. Even when colorless kernels with colored scutella are selected, many of the resulting plants behave as if the translocation were absent; at least the ability to nondisjoin is lost. Therefore, it is suggested that several plants be used in tests and that all be crossed onto a tester that involves a 2S seedling marker in order to confirm that hypoploids are being produced.

Birchler (this volume) is perfecting a set of compound translocations, derived from crosses of TB-10L18 with the basic set of B-A translocations, that has the special virtue of segregating colored kernels with colored scutella for nearly all chromosome arms in testcrosses that involve *r1*.

The author's stocks are in the process of being transferred to the Maize Genetics Cooperation Stock Center. Please contact Marty Sachs, S-108 Turner Hall, Agronomy Department, University of Illinois, 1102 S. Goodwin Avenue, Urbana, IL 61801, for seed of translocations and testers.

For seed of translocation stocks in A619, A632, B73, and Mo17 backgrounds, contact E.H. Coe, Jr., 210 Curtis Hall, University of Missouri, Columbia, MO 65211.

REFERENCES

Beckett JB (1978) B-A translocations in maize. I. Use in locating genes by chromosome arms. J Hered 69: 27–36

Beckett JB (1982) An additional mechanism by which B chromosomes are maintained in maize. J Hered 73: 29–34

Beckett JB (1991) Cytogenetic, genetic and plant breeding applications of B-A translocations in maize. In Gupta PK, Tsuchiya T (eds) Chromosome Engineering in Plants: Genetics, Breeding, Evolution, Vol 2A, Elsevier, Amsterdam, pp 493–529

Birchler JA (1983) Chromosomal manipulation in maize. In Swaminathan MS, Gupta PK, Sinha U (eds) Cytogenetics of Crop Plants, Macmillan India, Delhi, pp 379–403

Carlson WR (1978) The B chromosome of corn. Annu Rev Genet 16: 5–23

Carlson WR (1986) The B chromosome of maize. CRC Crit Rev Plant Sci 3: 201–226

Kindiger, B, Curtis, C, Beckett, JB (1991) Adjacent II segregation products in B-A translocations of maize. Genome 34: 595–602

44

Dosage Analysis Using B-A Translocations

JAMES A. BIRCHLER

Dosage studies are useful for localizing the structural genes for certain gene products, most notably enzymes; for the study of dosage-sensitive regulatory effects (Birchler and Newton 1981); for determination of mutants as null or leaky or gain of function (Poethig, 1988); or for introduction of more than one copy from a single parent as a test of parental imprinting (Kermicle 1970). The tools available in maize for dosage studies are the most extensive in any diploid plant species and for overall utility, rival those in any higher eukaryote. The basis for the dosage series depends on the nondisjunction property of the B chromosome centromere described in previous chapters on B-A translocations.

The extensive use of B-A translocations for gene localizations does not require any marker system beyond the pollen sterility characteristics of the B-A chromosome configuration used. Their use for dosage studies, however, requires a means to distinguish the one-, two-, and three-dose individuals that are all present in progeny that result when B-A translocations are used as male parents. Among the possible ways to determine the dosages include the use of molecular markers that give different patterns following electrophoretic separation. The most notable of these are allozyme markers (see Wendel 1989 for those available), although restriction-fragment-length polymorphisms might also be used. These types of classifications have some advantages, particularly when one wishes to compare different doses introduced through the female using hyperploid or euploid heterozygotes for which other types of markers are not useful. The use of anthocyanin color markers has recently rendered such molecular classifications cumbersome by comparison. The use of color markers, however, requires a specific tester stock that may not be available, together with another mutant under scrutiny. The available stocks are described in the chapter on marker systems for B-A translocations (next chapter).

A further development involves using a compound A-B-A translocation with TB-10L18 that is marked by *R-scm3*. Since most inbred stocks developed in breeding programs are recessive for *r* (and *c*), it is likely that a particular mutant or trait under study could

The Maize Handbook—M. Freeling, V. Walbot, eds.

be used with this color marker without first introducing the two mutants into the same stock.

Allozymes can be used if a different electrophoretic variant is present on the B-A translocation than in the female stock. When the translocation is crossed to the different variant, the one dose individuals will only have the maternal form. The two dose individuals will have the maternal and paternal in equal contribution. The three dose cases will have the maternal and paternal forms in unequal quantities, the latter at twice the former. For experiments that involve introducing extra copies through the female side, hyperploid heterozygotes with one variant on the normal chromosome and a second on the translocation can be crossed by a pollen source with a third marker. The class of gamete from the female side that is a balanced translocation will have the variant linked to the translocation and the paternal variant. The three-dose cases inherit the normal chromosome, a B-A, and also receive the paternal chromosome. These will consequently have all three variants. Further details can be found in Birchler (1979, 1980).

Allozyme markers can also be used to classify individuals with four doses of particular chromosome arms. If a situation exists in which enough variants are available such that two hyperploid heterozygotes with different variants on the normal chromosomes and the translocations (four variants total) can be produced, crossing the two together will give some progeny with four copies. These can be distinguished on the basis of the variants present and their band ratios (Birchler 1979, 1980).

REFERENCES

Birchler JA (1979) A study of enzyme activities in a dosage series of the long arm of chromosome one in maize. Genetics 92: 1211–1229

Birchler JA (1980) The genetic basis of dosage compensation of alcohol dehydrogenase-1 in maize. Genetics 97: 625–637

Birchler JA, Newton KJ (1981) Modulation of protein levels in chromosomal dosage series of maize: the biochemical basis of aneuploid syndromes. Genetics 99: 247–266

Kermicle JL (1970) Dependence of the R-mottled aleurone phenotype in maize on mode of sexual transmission. Genetics 66: 69–85

Poethig RS (1988) Heterochronic mutations affecting shoot development in maize. Genetics 119: 959–973

Wendel JF (1989) Chromosomal locations of isozyme loci in maize (*Zea mays* L.) Isozyme Bulletin 22: 33–35

45

Marker Systems for B-A Translocations

JAMES A. BIRCHLER

A number of marker systems have recently been developed that will allow the distinction of 1, 2, and 3 doses of the regions translocated to the B chromosome in the the standard set of translocations (Birchler and Alfenito, 1993). Lin (1978) produced a series of 10L translocations marked by *R-scm3*. Echt (1987) produced a 2S translocation marked by *B-peru*. The complete set of markers developed will allow the recognition of dosage series for 14 of the 18 chromosome arms for which there exists a simple B-A translocation. By use of transposable element systems, the available testers should eventually permit the remaining four arms to be marked similarly in the future. At the present, this system represents the most comprehensive set of genetic markers for determination of chromosomal dosage in any plant species.

The basis on which this collection works is as follows. The use of a female tester that has an allele of the *R* locus that conditions pigmentation in the scutellum of the embryo as well as in the aleurone layer of endosperm is present throughout, with the exception of those that use *R-scm* or *B-peru* alleles directly. The other testers are all recessive for some gene that is required for full pigmentation. The B-A translocation supplies the dominant form of this missing gene. As described in more detail elsewhere in this volume, the B centromere nondisjoins at a higher rate at the second microspore division. This produces sperm with 0, 1, or 2 copies of the B centric chromosome, including the A portion translocated to it. Therefore, following fertilization and growth of the kernel, the respective doses can be discerned. The one dose individuals will exhibit the recessive phenotype in the embryo and the dominant in the endosperm. The two-dose individuals will have the dominant phenotype in both tissues. The three-dose individuals will have the dominant phenotype only in the embryo.

The marker systems are listed in Table 45.1

The Maize Handbook—M. Freeling, V. Walbot, eds.

TABLE 45.1 B-A translocation marker systems[a]

Chromosome arm	Genotype of tester	Marker locus
1S	*A A2 C C2 r-m3 Pr Bz Bz2 In*	*Ac(P-vv); r-m3*
	A A2 C C2 R-scm2 Pr Bz bz2-m In	*Ac(P-vv); bz2-m*
1L	*A A2 C C2 R-scm2 Pr Bz bz2-m In*	*Bz2*
2S	*A A2 C C2 r-g Pr Bz Bz2 In*	*B-peru*
3L	*a-m1 A2 C C2 R-scm2 Pr Bz Bz2 In*	*A*
	a-m1-5719 A2 C C2 R-scm2 Pr Bz Bz2 In	*(Spm)*
4S	*a-m1 A2 C C2 R-scm2 Pr Bz Bz2 In*	*Dt6*
4L	*A A2 C c2 R-scm2 Pr Bz Bz2 In*	*C2*
	A A2 C c2-m2 R-scm2 Pr Bz Bz2 In	*(Spm)*
5S	*A a2 C C2 R-scm2 Pr Bz Bz2 In*	*A2*
5L	*A A2 C C2 R-scm2 pr Bz Bz2 In*	*Pr*
6L	*a-m1 A2 C C2 R-scm2 Pr Bz Bz2 In*	*Dt2*
7S	*A A2 C C2 R-scm2 pr Bz Bz2 In-D*	*In*
7L	*a-m1 A2 C C2 R-scm2 Pr Bz Bz2 In*	*Dt3*
8L	*A A2 C C2 R-scm2 Pr Bz bz2-m In*	*Ac2*
9S	*A A2 c C2 R-scm2 Pr Bz Bz2 In*	*C*
	A A2 C C2 R-scm2 Pr bz Bz2 In	*Bz*
10L	*A A2 C C2 r-g Pr Bz Bz2 In*	*R-scm3*
	A A2 C C2 r-m3 Pr Bz Bz2 In	*r-nj:m1*

[a]The alleles *r-m3* and *bz2-m* both contain insertions of *Ds* that respond to *Ac*, or, in the latter case, also to *Ac2*. The allele *a-m1* responds to the individual *Dotted* loci. The alleles, *a-m1-5719* and *c2-m2* respond to *Spm*. The allele *r-nj:m1* contains an *Ac* insertion into *R-nj*; this allele conditions pigment in the embryo.

REFERENCES

Birchler JA, Alfenito MR (1993) Marker systems for B-A translocations in maize. Journal of Heredity, in press

Echt C (1987) A new B-A translocation: TB-2Sa. Maize Genetics Cooperation News Letter 61: 94

Lin B-Y (1978) Regional control of nondisjunction of the B chromosome in maize. Genetics 90: 613–627

46

Construction of Compound B-A Translocations

JAMES A. BIRCHLER

Compound translocations combine a simple B-A translocation and an A-A translocation so that portions of two different chromosome arms are appended to the B chromosome centromere (Rakha and Robertson 1970). They are produced by recombination in the region in common between a B-A translocation and the appropriate A-A translocation. They are useful in genetic studies for varying the dosage of two different regions of the genome simultaneously and for extending the regions of the genome that can be uncovered using B-A chromosome techniques (e.g., Rakha and Robertson 1970). They also can partition chromosome arms into very refined subdivisions for gene localizations and dosage studies (e.g., Birchler, 1980, 1981). Studies on chromosome pairing in meiosis have also utilized them (e.g., Maguire 1984, 1985, 1986).

To produce a compound, one would start with a hyperploid heterozygote of the appropriate simple B-A translocation. (See preceding chapters on B-A translocations for introductory comments.) These plants would be crossed as a female by the stock homozygous for the A-A translocation. The hyperploid heterozygotes produce about half balanced B-A translocation gametes and half hyperploid. The balanced euploid gamete when joined at fertilization with the A-A translocation will form a zygote that is

The Maize Handbook—M. Freeling, V. Walbot, eds.

heterozygous for both the B-A translocation and the A-A translocation. If a crossover occurs between the B-A breakpoint and the A-A breakpoint, one product will be a B-A-A chromosome. If this chromosome segregates with the A-B chromosome and nondisjoins at the second microspore division, the potential then exists for a fertilization to occur that will uncover a genetic marker in the endosperm in the region of the genome distal to the A-A breakpoint in the chromosome other than that involved in the B-A chromosome. Because the uncovering in the endosperm is indicative of a fertilization in which the two B chromosome centromeres fertilized the egg nucleus, the newly formed chromosome can be selected and recovered.

In practice then the heterozygotes are crossed as males onto a tester stock that has an endosperm mutant distal to the A-A breakpoint to be recombined onto the B-A. The uncovering of the marker will select for the recombinant hyperploid heterozygotes. These will be rare, but the selection is strong. A caveat, as always, is that such candidates might be self-contaminants. If, however, one uses anthocyanin testers that carry an *R-scm2* allele, then the embryo of the kernel will be colored and will be distinguishable from self-contaminants, which will be completely colorless.

Compounds uncover the segments from the B-A breakpoint to the A-A breakpoint on one chromosome and from the A-A breakpoint to the tip of the chromosome arm on the other chromosome. Therefore, if one uses a compound for a localization study, it is necessary to include in the set of crosses the simple translocation from which the compound was made in order to be confident of the correct location.

Compounds are usually maintained as hyperploid heterozygotes. The reason for this is that recombination between the B-A-A and the normal A chromosome will regenerate the starting simple B-A and A-A translocations. Hyperploid heterozygotes, as opposed to euploid heterozygotes, are less likely to have such recombination. Of course, if a compound is maintained as a "homozygote," there is no problem. Homozygous lines for B-A translocations have in the past been denigrated, presumbly because many aneuploid types are present and only some of the plants in a row are sufficiently vigorous and fertile. However, all plants that do produce pollen are equally useful and produce a very high frequency of marker uncovering in the progeny.

The degree of chromosome division that can be achieved with compounds is extensive given that there are simple B-A translocations for 18 of the 20 chromosome arms and nearly 900 A-A translocations.

REFERENCES

Birchler JA (1980) The cytogenetic localization of the alcohol dehydrogenase-1 locus in maize. Genetics 94: 687–700

Birchler JA (1981) The genetic basis of dosage compensation of alcohol dehydrogenase-1 in maize. Genetics 97: 625–637

Maguire MP (1984) The pattern of pairing that is effective for crossing over in complex B-A chromosome rearrangements in maize. Chromosoma 89: 18–23

Maguire MP (1985) The pattern of pairing that is effective for crossing over in complex B-A chromosome rearrangements in maize. II Chromosoma 91: 101–107

Maguire MP (1986) The pattern of pairing that is effective for crossing over in complex B-A chromosome rearrangements in maize. III Chromosoma 94: 71–85

Rakha FA, Robertson DS (1970) A new technique for the production of A-B translocations and their use in genetic analysis. Genetics 65: 223–240

47

A-B-A Compound Chromosomes

JAMES A. BIRCHLER

The marker systems presently available for B-A translocations permit their use for genetic tracking for most of the genome. These systems, however, require individual testers for individual chromosome arms. For some applications, this situation would be desired. However, recently a system has been initiated to mark all of the standard set of B-A translocations with a uniform marker that will allow discrimination of all doses produced at the mature kernel stage. Such chromosomes could also be synthesized using other B-A translocations for specific applications. The complete set of such translocations would greatly facilitate the handling and use of the B-A translocation set for gene localizations and dosage studies. The complete set would allow extensive coverage of the genome with selection of 1, 2, and 3 doses of the respective regions in each case.

This uniform marking system utilizes TB-10L18, a unique translocation that is broken in the minute short arm of the B chromosome (Lin 1979). All of the remaining B-A translocations are broken in the major long arm. Consequently, there is a region in common within the B chromosome sequences between TB-10L18 and any other TB-A; these shared sequences can recombine in a heterozygote to give a metacentric compound chromosome. This metacentric combines the long arm of chromosome 10, marked by *R-scm3*, with the other chromosome arm with both attached to the B chromosome centromere. The *R-scm3* marker conditions pigment in both the embryo and endosperm and can be used onto a *r-g* tester for any translocation so marked.

An example involving TB-1La and TB-10L18 was produced (Birchler et al. 1990).

The Maize Handbook—M. Freeling, V. Walbot, eds.

Others are in various stages of completion and confirmation. With the original, the two translocations were crossed together. This brings together the regions of the B in common. In the heterozygote, recombination in the B chromosome will generate a chromosome that has the long arm of chromosome 1 linked to the long arm of chromosome 10 through the B centromere. Also present are the two A-B chromosomes that are required for nondisjunction of the B centromere. When the heterozygote was crossed to a tester, individuals in the progeny were selected that had pigmented embryos and nonpigmented endosperm. These were crossed again to a tester for alcohol dehydrogenase; this enzyme was used as a marker for 1L in the construction. An example was found that linked *R-scm3* to *Adh*, and these two loci behaved as they were both linked to the B centromere. Several cytological tests confirmed the configuration.

Subsequent extension to other chromosome arms has relied on the simultaneous uncovering of an endosperm marker on one chromosome arm and the recessive 10L marker *r*; most colorless stocks carry *r*. The heterozygotes of the two translocations are crossed as a male onto the respective tester and individual kernels are selected that uncover the appropriate endosperm marker for that arm and that have colored embryos. This is the expected phenotype of the simultaneous uncovering of both arms. This, of course, could occur with or without recombination, so these individuals are tested for linkage of the two traits as the first test for the formation of a new chromosome.

Probably the best means to maintain the set of A-B-A chromosomes is by self- or sib pollination of homozygous stocks. This is in contrast to hyperploid heterozygotes. In the latter, recombination between the A-B-A chromosomes and the normal chromosome 10 carrying the recessive marker, *r*, will remove *R-scm3*, thus eliminating this useful marker. In homozygous stocks, this is not a problem. However, one should be aware that in such stocks there are a variety of chromosomal constitutions, including doubly segmental monosomics and doubly segmental tetrasomics, that can generate severely depauperate plants; these plants are often sterile. Nevertheless, euploids are present that are fully fertile and give high levels of uncovering when outcrossed to a tester. While not every plant in a homozygous line is useful, any that are fertile will have none of the complications associated with hyperploid heterozygotes.

REFERENCES

Birchler JA, Chalfoun DJ, Levin DM (1990) Recombination in the B chromosome of maize to produce A-B-A chromosomes. Genetics 126: 723–733

Lin B-Y (1979) Two new translocations involved in the control of nondisjunction of the B chromosome in maize. Genetics 92: 931–945

Comprehensive List of B-A Translocations in Maize

J. B. BECKETT

Much of the information in the accompanying compilation (Table 48.1) was taken from numerous unpublished notes in the Maize Genetics Cooperation News Letter. Further information on translocation and gene location is summarized in recent News Letters. Additional breakpoint data (and extensive references) may be found in Beckett (1991). Chapter 43, this volume, discusses the use of B-A translocations in locating recessive genes. Additional uses are described in Beckett (1991).

The Maize Handbook—M. Freeling, V. Walbot, eds.

Table 48.1 List of B-A translocations in maize[a]

Designation	Arm(s) uncovered by translocation[b]	Closest known gene(s) proximally (not uncovered)[c]	Closest-known gene(s) distally (uncovered)[c]
*TB-1Sb	1S.05 to end	*as1, rs2*	*nec2*
*TB-1La	1L.2 to end	*nec2*	*hm1*
TB-1Lc	1L, perhaps distal to TB-1La	*nec2*	*br1*
*TB-3La-2S6270	3L.1-.60, 2S.46 to end	*gs2?*[d], *v4*	*wt1*
*TB-2Sa	Part of 2S	*wt1*	*b1*
TB-2Sb	Part of 2S, may be distal to TB-2Sa	*sk1?*[d]	*d5*
TB-3La-2L7285	3L.1-.39, 2L.26 to end	*sk1*	*v4*
*TB-1Sb-2L4464	1S.05-.53, 2L.28 to end	*sk1*	*v4*
TB-1Sb-2Lc	1S.05-.77, 2L.33 to end	*v4*	*w3*
*TB-3Sb	3S.5 to end	*Tpi4*	*cl1*
*TB-3La	3L.1 to end	*Tpi4*	*gl6, ys3*
TB-3Lg	3L, like TB-3La	*Tpi4*	*gl6*
TB-3Lf	3L, prob distal to TB-3La	*Tpi4*	*gl6*
TB-3Ld	3L, distal to TB-3Lf	*ts4, pm1*	*vp1*
TB-3Lh	3L, like TB-3Ld	*gl6*	*vp1*
TB-3Li	3L, like TB-3Ld	*vp1, pm1*	*lg2*
TB-3Lj	3L, like TB-3Ld	*vp1, pm1*	*lg2*
TB-5La-3L5521	5L.1-.48, 3L.17 to end		*a1*
TB-1La-3L4759–3	1L.2-.39, 3L.20 to end	*gl6*	*lg2*
TB-1La-3Le	1L.2-.58, 3L.45 to end	*gl6*	*lg2*
TB-3Lc	3L.55-.65 to end; distal to TB-3Ld	*lg2*[e]	*ba1*
TB-3Lk	3L, like TB-3Lc	*lg2*	*a3*
TB-3Lm	3L, distal to TB-3Lc	*lg2*	*a1*
TB-3Ll	3L, distal to TB-3Lc	*na1*	*a3*
TB-5La-3Lb	5L.1-.57, 3L.61 to end		*a1*
TB-5La-3L7043	5L.1-.61, 3L.63 to end		*a1*
TB-1La-3L5242	1L.2-.90, 3L.65 to end	*gl6*	*lg2*
TB-1La-3L5267	1L.2-.72, 3L.73 to end		*Mdh3, a1*
*TB-4Sa	4S.25 to end	*bm3*	*bt2*

TABLE 48.1 (Continued)

Designation	Arm(s) uncovered by translocation[b]	Closest known gene(s) proximally (not uncovered)[c]	Closest-known gene(s) distally (uncovered)[c]
TB-4Sg	Like TB-4Sa		*bt2*
*TB-9Sb-4L6222	9S.4-.68, 4L.03 to end	*su1*	*gl4*
TB-9Sb-4L6504	9S.4-.83, 4L.09 to end	*bm3*	*gl4*
TB-1Sb-2L4464–4Lf	1S.05-.53, 2L.28-.75, 4L.12 to end		*gl3*
TB-1La-4L4692	1L.2-.46, 4L.15 to end	*bm3*	*gl4*
*TB-4Lc	4L.15-.20 to end		*gl4*
TB-4Ld	4L, like TB-4Lc	*bm3*	*o1*
TB-4Le	4L, like TB-4Lc	*bm3*	*o1*
TB-4Lf	4L.15-.20 to end	*bm3*	*gl4*
TB-4Lb	4L, prob distal to TB-4Lc & TB-4Lf	*bm3*	*gl4*
TB-4Lh	Part of 4L		*gl3*
TB-4Li	Part of 4L, may be a compound translocation[f]		*gl3*
TB-7Lb-4L4698	7L.3-.74, 4L.08 to end	*gl3*[g]	*c2, gl3*[g]
*TB-1La-5S8041	1L.2-.80, 5S.10 to end		*bm1, nec3*
*TB-5Sc	5S.3 to end		*a2*
*TB-5La	5L.1	*bt1, v3?*[d]	*bv1*
TB-5Lb	5L, like TB-5La	*bt1, v3?*[d]	*bv1*
TB-5Ld	5L, like TB-5La	*bt1, td1*	*bv1*
*TB-6Sa	6S.5 (middle of nucleolus organizer) to end	*rgd1, Pgd1, mn3*	*po1, dek28, hcf26*
TB-6Ld	6L, near centromere[h]		*w15*
*TB-6Lc	6L.11 to end	*rgd1, Pgd1, mn3*	*w15*
TB-6Lb	6L.65 to end; distal to TB-6Lc	*sm1*	*py1*
*TB-7Sc	Part of 7S	*ra1*	*vp9*
*TB-7Lb	7L.3 to end; 7L.34 to end	*vp9*	*ra1, ms7, o5*
*TB-8Lc	8L.24 ± .05 to end	*Mdh1*	*pro1, Idh1*
TB-8La	8L.7 to end	*pro1, nec1*	*v16*
TB-8Lb	8L, like TB-8La	*pro1, nec1*	*v16*
*TB-9Sd	9S, centric heterochromatin to end; 9S.08 ± 0.24 to end	*v1*	*d3*

TABLE 48.1 (Continued)

Designation	Arm(s) uncovered by translocation[b]	Closest known gene(s) proximally (not uncovered)[c]	Closest-known gene(s) distally (uncovered)[c]
TB-9Sb	9S.4 to end; 9S.49 to end	*wx1, w11*	*lo2, baf1*
TB-9S(Saraiva)[i]	9S, distal 1–2 chromomeres to end	*bz1*	*(yg2)*
*TB-9Lc	9L.1 to end	*d3*	*ar1*
TB-9La	9L.4 to end; 9L.47 to end; or 9L.49	*gl15*	*bk2*
TB-10Sc	10S.3 (or closer) to end	*zn1*	*y9*
TB-10L18	10L, centromere to end	*(y9)*	*zn1*
*TB-10L19	10L, centromere to end	*(y9)*	*zn1*
TB-10L26	10L, distal to TB-10L19	*zn1*	*du1*
TB-10Lb	10L.34 ± 0.17 to end[j]; distal to TB-10L26	*du1*	*bf2, li1*
TB-10L22	10L, distal to TB-10L26	*du1*	*bf2, li1*
TB-10L36	10L, distal to TB-10L22	*bf2, li1*	*ms10*
TB-10L20	10L, distal to TB-10L36	*bf2, li1*	*ms10*
TB-10L1	10L, distal to TB-10L20	*bf2, li1*	*ms10*
TB-10L3	10L, like TB-10L1	*bf2, li1*	*ms10*
TB-10L4	10L, like TB-10L1	*bf2, li1*	*ms10*
TB-10L5	10L, like TB-10L1	*bf2, li1*	*g1*
TB-10L7	10L, like TB-10L1	*bf2, li1*	*ms10*
TB-10L9	10L, like TB-10L1	*bf2, li1*	*g1*
TB-10L10	10L, like TB-10L1	*bf2, li1*	*ms10*
TB-10L25	10L, like TB-10L1	*bf2, li1*	*ms10*
TB-10L28	10L, like TB-10L1	*bf2, li1*	*ms10*
TB-10L31	10L, like TB-10L1	*bf2, li1*	*ms10*
TB-10L37	10L, like TB-10L1	*bf2, li1*	*g1*
TB-10L6	10L, distal to TB-10L1	*bf2, li1*	*g1*
TB-10L8	10L, like TB-10L6	*bf2, li1*	*ms10*
TB-10L11	10L, like TB-10L6	*bf2, li1*	*g1*
TB-10L12	10L, like TB-10L6	*bf2, li1*	*g1*
TB-10L14	10L, like TB-10L6	*bf2, li1*	*g1*
TB-10L16	10L, like TB-10L6	*bf2, li1*	*g1*
TB-10L17	10L, like TB-10L6	*bf2, li1*	*g1*

TABLE 48.1 List of B-A translocations in maize[a]

Designation	Arm(s) uncovered by translocation[b]	Closest known gene(s) proximally (not uncovered)[c]	Closest-known gene(s) distally (uncovered)[c]
TB-10L24	10L, like TB-10L6	*bf2, li1*	*g1*
TB-10L27	10L, like TB-10L6	*bf2, li1*	*g1*
TB-10L29	10L, like TB-10L6	*bf2, li1*	*g1*
TB-10L30	10L.13 to end; like TB-10L6[j,k]	*bf2, li1*	*g1*
TB-10L34	10L, like TB-10L6	*bf2, li1*	*g1*
TB-10L35	10L, like TB-10L6	*bf2, li1*	*g1*
TB-10L38	10L, like TB-10L6	*bf2, li1*	*g1*
TB-10L2	10L, distal to TB-10L6	*bf2, li1*	*g1*
TB-10L21	10L, like TB-10L2	*bf2, li1*	*g1*
TB-10L23	10L, like TB-10L2	*bf2, li1*	*g1*
TB-10La	10L.35 to end; distal to TB-10L20, prox. to TB-10L32	*ms10*	*g1*
TB-10L13	10L, dist to TB-10L20, prox to TB-10L32	*bf2, li1*	*g1*
TB-10L15	10L, like TB-10L13	*bf2, li1*	*g1*
TB-10L33	10L, like TB-10L13	*bf2, li1*	*g1*
TB-10Ld	10L, proximal to TB-10L32	*bf2, v18*	*g1*
TB-10L32	10L.74 to end; distal to TB-10L2	*g1*	*r1*

*Basic B-A translocation set for locating mutant genes. Alternative translocations are designated for three chromosome arms.

[a]Breakpoints on A chromosomes are given whenever possible. When two or more translocations are available on a particular chromosome arm, they are listed in order of breakpoint, beginning at the centromere. In many cases, order is not established, so they are listed in groups to the extent possible.

[b]Each breakpoint is given as a decimal fraction of the distance from the centromere to the end of the chromosome arm; the portion of the arm uncovered by the translocation extends from that point to the end of the arm. The compound translocations uncover portions of two chromosome arms: for the first arm listed, the portion between the two breakpoints is uncovered; for the second arm, the portion beyond the breakpoint is uncovered. TB-1Sb-2L4464–4Lf uncovers portions of three chromosome arms.

Table 48.1 (Continued)

[c]Two or more genes are listed if gene order is not established.
[d]"?" indicates preliminary data.
[e]Original tests failed to uncover *lg2*, but recent tests do. An error in pedigree may have occurred, so all current stocks should be tested.
[f]May involve 6S also (Beckett unpublished 1991, *Maize Genetics Cooperation News Letter* 65: 57), but further genetic tests (Beckett JB unpublished 1991) failed to uncover *l11, rgd1, oro1, hcf26,* or *dek28.*
[g]4L breakpoint or genetic data may be incorrect; genetic data originally placed the breakpoint between *gl3* and *c2* (Rakha and Robertson 1970); recent data (Beckett 1990, unpublished) indicate that *gl3* is uncovered, so further testing is needed.
[h]Probably not a simple reciprocal B-A translocation.
[i]A half-translocation; only the 9-B chromosome was recovered.
[j]Breakpoint and genetic data for TB-10Lb and TB-10L30 conflict.
[k]The 10-B chromosome of TB-10L30 carries a duplication of part of the B chromosome, including the centromere, but it lacks the distal portion. Therefore it cannot undergo nondisjunction unless another source, such as an intact B chromosome, is present.

REFERENCES

Beckett JB (1991) Cytogenetic, genetic and plant breeding applications of B-A translocations in maize. In Gupta PK, Tsuchiya T (eds) Chromosome Engineering in Plants: Genetics, Breeding, Evolution, Vol 2A, Elsevier, Amsterdam, pp 493–529

Rakha FA, Robertson DS (1970) A new technique for the production of A-B translocations and their use in genetic analysis. Genetics 65: 223–240.

49

Chromosomal Translocations Involving the Nucleolus Organizer Region or Satellite of Chromosome 6

R. L. PHILLIPS

The 18S and 26S ribosomal RNA genes of maize are located at the nucleolus organizer region (NOR) of chromosome 6 (Phillips 1978; Phillips et al. 1971, 1974, 1979, 1983; Givens and Phillips 1976; Phillips and Wang 1982; Ramirez and Sinclair 1975). Deficient-duplicate chromosomes representing deficiencies for limited terminal segments of the short arm of chromosome 6 can be transmitted through the ovule (Phillips et al. 1971, 1984; Phillips and Thompson 1979, 1980). An extensive set of translocations with breaks in either the chromosome 6 NOR or satellite is available (Table 49.1). Heterozygotes for most of these translocations will generate female-transmissible deficiency-duplications (Phillips 1976; Phillips et al. 1977). The regions duplicated depend on the translocation; all the other chromosomes except chromosome 8 are involved in translocations with chromosome 6 in this set. Pollen sterility of the heterozygotes ranges from 25% to nearly 50% depending on the translocation (Phillips 1976). The deficiency-duplication heterozygotes can often be identified on the ear of a translocation heterozygote by somewhat smaller seed. Planting these smaller seeds (approximately 75% the normal size) will increase the frequency of plants that are deficiency-duplication heterozygotes. Depending on the duplication, such deficient-duplicate heterozygous plants may have a distinctive morphology. Pollen of the deficiency-duplication heterozygotes may be 50% smaller and filled with starch and 50% normal. In many cases abortive grains may also be present.

Translocations with breaks near the ends of chromosomes and/or for dispensable terminal segments are useful for genetic linkage studies. Linkage of a mutation to one of the two breakpoints of the translocation can be detected simply by a distorted ratio of the respective alleles when the F_1 (translocation × mutant) is used as the female but not as the male in backcrosses (Phillips et al. 1971). No classification for the translocation is required. Such reciprocal backcross tests also may be useful in mapping major loci for quantitative traits. Finally, heterozygous translocations that generate viable deficiency-

The Maize Handbook—M. Freeling, V. Walbot, eds.

TABLE 49.1 NOR and satellite translocations and breakpoints

Translocation	Breakpoints	
	6	Other
Satellite-interchanges		
2–6(001–15)	sat.[a]	2S.72[b]
6–7(7036)	sat.	7L.63
3–6b	sat.	3S.73
4–6c	sat.	4S.33
6–10f	sat.	10S.28
4–6 (5227)	sat.	4S.46
4–6 (003–16)	sat.	4L.50
4–6 (7328)	sat.	4S.53
6–9 (017–14)	sat.	9L.50
5–6b	sat.	5S.10
5–6d	sat.	5S.64
1–6b	sat.	1L.25
5–6 (8219)	sat.	5L.69
NOR-interchanges		
3–6 (032–3)	S.C.-midway	3S.34
5–6f	S.C.-midway	5S.23
5–6 (8696)	S.C.-midway	5L.79
2–6 (5419)	S.C.-.25	2L.82
3–6 (030–8)	S.C.-.25	3S.05
6–7 (035–3)	S.C.-.25	7L.59
1–6Li	S.C.-prox.	1L.81
1–6 (4986)	S.C.-prox.	1S.11
1–6 (8415)	S.C.-prox.	1L.31
2–6 (8441)	S.C.-prox.	2L.95
2–6 (027–4)	S.C.-prox.	2L.04
6–10 (5519)	S.C.-prox.	10L.10
6–9 (4778)	Het .95	9L.30
4–6 (7037)	Het .90	4L.61
2–6 (8786)	Het .88	2S.97
6–7 (5181)	Het .71	7L.85
6–9a	Het .67	9L.32
4–6 (4341)	Het .50	4S.36
6–9d	Het .46	9L.84
6–7 (4964)	Het .32	7L.67
6–10 (5253)	Het .30	10L.41
1–6 (6189)	Het .10	1S.50

TABLE 49.1 (Continued)

[a]sat. = chromosome 6 satellite; S.C.-midway, S.C.-prox., and S.C.-.25 = breakpoint midway, proximal portion, or between midway and proximal portions of the NOR secondary constriction, respectively; Het .95 = breakpoint in NOR-heterochromatin 95% of the distance from proximal to distal end of the heterochromatic segment. See Phillips and Wang (1977) for tentative satellite breakpoint positions.
[b]Breakpoints (determined by Longley 1961) represent the relative distance of the breakpoint from the centromere to the end of the chromosome arm.

duplications can be used to uncover recessive mutations and, thereby, place the mutation in the deficient segment (Phillips and Thompson 1979, 1980). Homozygous translocations with a break in the NOR have been used for locating ribosomal RNA genes to the NOR-heterochromatin and the NOR secondary constriction by ín situ hybridization (Phillips et al. 1979). The translocations with a break in the satellite (Phillips and Wang 1977) have been used to locate *polymitotic* (*po*), probably to the first chromomere of the satellite (Phillips et al. 1977). One of the translocations with a break in the satellite (T5–6b) is used to produce lines with nearly all male-sterile (*polymitotic*) plants (Phillips 1978).

Most of the translocations listed in Table 49.1 have been backcrossed six times to A188, W23, A619, and A632 (Phillips and Wang 1983). Seed of self-pollinations of sixth backcross heterozygous plants is available from the University of Minnesota.

REFERENCES

Givens JF, Phillips RL (1976) The nucleolus organizer region of maize (*Zea mays* L.): Ribosomal RNA gene distribution and nucleolar interactions. Chromosoma (Berl) 57: 103–117

Longley AE (1961) Breakage points for four corn translocation series and other corn chromosome aberrations. US Dept Agric, Agric Res Serv ARS-34–16: 1–40

Phillips RL, Burnham CR, Patterson EB (1971a) Advantages of chromosomal interchanges that generate haplo-viable deficiency-duplications. Crop Sci 11: 525–528

Phillips RL, Kleese RA, Wang SS (1971b) The nucleolus organizer region of maize (*Zea mays* L.): Chromosomal site of DNA complementary to ribosomal RNA. Chromosoma (Berl) 36: 79–88

Phillips RL, Weber DF, Kleese RA, Wang SS (1974) The nucleolus organizer region of maize (*Zea mays* L.): Tests for ribosomal gene compensation or magnification. Genetics 77: 285–297

Phillips RL (1976) Transmission of nucleolus organizer region deficiences. Maize Genetics Cooperation News Letter 50: 79–82

Phillips RL, Wang AS (1977) Interchange breakpoints in the chromosome 6 satellite. Maize Genetics Cooperation News Letter 51: 52

Phillips RL, Patterson EB, Buescher PJ (1977) Cytogenetic mapping of genes in the nucleolus organizer-satellite region of chromosome 6. Maize Genetics Cooperation News Letter 51: 49–52

Phillips RL (1978) Molecular cytogenetics of the nucleolus organizer region. In Maize Breeding and Genetics, Chap 43, John Wiley & Sons, New York, pp. 711–741

Phillips RL (1978) Development of a nuclear male-sterility system for hybrid corn production.

Maize Genetics Cooperation News Letter 52: 67–70

Phillips RL, Thompson SA (1979) Cytogenetic localization of a high chlorophyll fluorescence mutation (hcf-26) within the chromosome 6 satellite. Maize Genetics Cooperation News Letter 53: 115–116

Phillips RL, Wang AW, Rubenstein I, Park WD (1979) Hybridization of ribosomal RNA to maize chromosomes. Maydica 24: 7–21

Phillips RL, Thompson SA (1980) Chromosome 6 satellite location of a high chlorophyll fluorescence mutation (hcf-26). Maize Genetics Cooperation News Letter 54: 110

Phillips RL, Wang AS (1982) *In situ* hybridization of maize meiotic cells. In Sheridan WF (ed) Maize for Biological Research, Plant Molec Biol Assoc, University of North Dakota Press, Grand Forks, ND, pp 121–122

Phillips RL, Wang AS (1983) Near-isogenic lines of various genetic markers and interchanges. Maize Genetics Cooperation News Letter 57: 132–133

Phillips RL, Wang AS, Kowles RV (1983) Molecular and developmental cytogenetics of gene multiplicity in maize. Stadler Symp 15: 105–118

Phillips RL, Wang AS, Bullock WP (1984) Transmission of a deficiency for nearly the entire nucleolus organizer region. Maize Genetics Cooperation News Letter 58: 181

Ramirez SA, Sinclair JH (1975) Ribosomal gene localization and distribution (arrangement) within the nucleolar organizer region of *zea maya*. Genetics 80: 505–518

50

Inversions and List of Inversions Available

G. G. Doyle

Inversions are very important and useful chromosome aberrations. An inverted chromosome has a segment that is rearranged in reverse order. If this segment includes the centromere, it is a pericentric inversion. If the event does not include the centromere, it is a paracentric inversion. The cytogenetics of inversions are very complicated. Rather than discuss the behavior of inversions here and distort the subject by oversimplifications and omissions imposed by space limitations the reader is directed to cytogenetic texts by Burnham (1962) and others and to the original papers by McClintock (1931, 1938), Rhoades and McClintock (1935), Russell and Burnham (1950), Morgan (1950), Rhoades and Dempsey (1953), and others.

The handling of inversions requires some special techniques. Inversions may be followed in experiments by genetic markers. This is not reliable because there may be an exchange of genetic material in the inverted segment and the corresponding segment in the normal chromosome by double crossovers that result from two- and three-strand double exchanges. While this does not happen frequently, particularly in small inversions, it must be considered. It is better and easier to follow the inversion by pollen abortion examinations. These can be done in the field using a pocket microscope. One fresh anther is taken from the tassel. The end is snipped off with fingernails and the pollen is sprinkled on a black or dark surface. A bottlecap works well for this purpose. Some people paint their thumbnails black. See the chapter by Phillips on the pocket microscope for additional hints.

Aborted pollen will be shrunken, only partially filled with starch, large and clear, or abnormal in some respect. There are several types of aborted pollen that probably indicate blockage in starch, protein, or other syntheses that result from losses of genetic material. Normal pollen looks like pearls. When evaluating pollen, use just one anther at a time. If two are combined, one may be full of dead pollen as a result of it being old or damaged by heat. Mixing such a bad anther with a good one would give the appearance of 50% pollen abortion. The rate of pollen abortion will vary from 5 to 50% depending

The Maize Handbook—M. Freeling, V. Walbot, eds.

on the frequency of chiasmata formation in the inverted region. Longer inversions or inversions in regions of high crossover potential will have more aborted pollen than short inversion or inversions in low-crossover areas. Because there is a variable rate of pollen abortion (about 2–5%) in normal plants, it is difficult to follow inversions with low pollen abortion rates.

Before working with any inversion it should be cytologically verified. Don't trust anybody. Verification requires examination at pachynema. The beautiful inversion loops that are pictured in textbooks are rarely seen. Frequently there is asynapsis in the inverted segment or in distal or proximal segments. Very often there is nonhomologous pairing—this appears as intimate as homologous pairing except that knobs and chromomeres do not match. This has been studied by McClintock (1932). The shorter the chromosome the more likely it is to pair nonhomologously. There may be a great number of very small inversions in the maize genome that remain undetected.

If the inversion is a pericentric that greatly changes the arm ratio, it is possible that it could be recognized at mitosis as found in root tips.

It is very important to be aware of the problem of deficient-duplicate chromosome production by single crossovers between normal and inverted chromosomes. These Df-Dp chromosomes are often viable in the female gametophyte generation but are not viable in male gametophyte generation. To keep the inversion intact, the inversion chromosome should be kept as a homozygote. If it is necessary to use it in heterozygous conditions employ it as the male parent only.

USES OF INVERSIONS

There are many uses for inversions. They may be used for the placement of qualitative genes. Because reciprocal translocations and B-A translocations are easier to use, inversions are rarely used for this purpose in corn.

Because an inversion keeps a block of genes together they may be used to place qualitative genes as suggested by Dobzhansky and Rhoades (1938). However, because inversion/normal heterozygotes have some ovule abortion, this method is not very useful in detecting genes that affect yield.

If a paracentric inversion is put into an isochromosome, crossing-over will produce ring chromosomes. This has been done in *Drosophila* (Sturtevant and Beadle 1936), but not in corn. Ring chromosomes are stochastically lost during mitosis, and hence can be useful in creating chimeras (i.e., if the ring carries a wild-type gene, its loss can uncover a recessive on the nonring chromosome).

Duplications can be produced by crossing-over in the common segment of overlapping inversions.

Inversions may be used to keep gene combinations together during backcrossing programs.

Table 50.1 gives a list of inversions that I have maintained. Some of these are available through the Maize Co-op. I will send all my inversion stocks to the Co-op and retain

TABLE 50.1 List of inversions[a]

Inversion	Breakpoints	Source	Inversion	Breakpoints	Source
Inv 1a SL	S.30-L.50	F.C.	Inv 4e L	L.16-L.81	R.M.
Inv 1c SL	S.70-L.01	R.M.	Inv 4f L	L.17-L.63	R.M.
Inv 1d L	L.55-L.92	R.M.	Inv 4g SL	S.89-L.72	4791-5
Inv 1f SL	S.85-L.56	4345	Inv 4h L	L.16-L.56	5887-2
Inv 1g SL	S.41-L.35	4932-5	Inv 4i L	L.19-L.66	5965-8
Inv 1h L	L.70-L.87	5083	Inv 4j L	L.24-L.66	5876
Inv 1j SL	S.37-L.05	5517	Inv 4k SL	S.23-L.39	6323-4
Inv 1k L	L.46-L.82	5131	Inv 5a SL	S.05-L.72	D.M.
Inv 1l SL	S.82-L.46	5828	Inv 5b SL	S.80-L.91	5431
Inv 1m SL	S.81-L.10	061-6	Inv 5c SL	S.78-L.63	5787
Inv 2a SL	S.70-L.80	B.M.	Inv 5d SL	S.42-L.63	7088
Inv 2b SL	S.06-L.15	8865	Inv 5e SL	S.21-L.75	5068
Inv 2d SL	S.93-L.65	R.M.	Inv 5f SL	S.67-L.69	8611
Inv 2e S	S.44-S.84	R.M.	Inv 5g L	L.65-L.87	8463
Inv 2f SL	S.35-L.22	R.M.	Inv 6a SL	S.76-L.63	R.M.
Inv 2g SL	S.88-L.50	4333	Inv 6b SL	S.38-L.92	6669-6
Inv 2h L	L.13-L.51	5393	Inv 6c SL	S.80-L.20	014-11
Inv 2i SL	S.91-L.50	5407	Inv 6d SL	S.70-L.33	8452
Inv 2l SL	S.38-L.44	003-20	Inv 6e SL	S.80-L.32	8604
Inv 2m SL	S.69-L.58	005-19	Inv 7a L	L.05-L.95	5262
Inv 2o SL	S.73-L.70	48-51-4	Inv 7b SL	S.32-L.30	R.M.
Inv 2p SL	S.76-L.86	5343-6	Inv 7c L	L.34-L.52	R.M.
Inv 3a L	L.38-L.95	M.R.	Inv 7d SL	S.51-L.72	5250
Inv 3b L	L.21-L.70	6784	Inv 7e SL	S.89-L.93	020-15
Inv 3c L	L.05-L.95	R.M.	Inv 7f L	L.17-L.61	5803
Inv 3d SL	S.72-L.42	6837-5	Inv 8a SL	S.30-L.15	R.M.
Inv 3e SL	S.46-L.80	025-6	Inv 8b L	L.10-L.42	5949
Inv 3h L	L.19-L.72	5272	Inv 8c S	S.51-S.77	002-16
Inv 4a L	L.40-L.96	D.M.	Inv 9a SL	S.70-L.90	C.L.
Inv 4b SL	S.10-L.12	R.M.	Inv 9b SL	S.05-L.87	R.M.
Inv 4c SL	S.89-L.62	R.M.	Inv 9c SL	S.10-L.67	R.M.
Inv 4d L	L.40-L.96	R.M.	Inv 10a SL	S.57-L.86	5570

[a]Sources: numbers = Longley; F.C. = Frances Clark; R.M. = Rosalind Morris; B.M. = Barbara McClintock; M.R. = Marcus Rhoades; D.M. = Delbert Morgan; and C.L. = Li.

a backup supply. I have given letter symbols to some of the Co-op stocks that are identified by Longley's culture numbers.

REFERENCES

Burnham CR (1962) Discussion in Cytogenetics, Burgess Publishing, Minneapolis, MN, 375 pp

Dobzhansky T, Rhoades MM (1938) A possible method for locating favorable genes in maize. J Am Soc Agron 30: 668–675

McClintock B (1931) Cytological observations of deficiencies involving known genes, translocations, and an inversion in *Zea mays*. Missouri Agr Exp Res Bull 163: 1–30

McClintock B (1932) Cytological observations in Zea on the intimate association of non-homologous parts of chromosomes in the mid-prophase of meiosis and its relationship to diakinesis configurations. Proc VI Internat Cong Genetics 2: 126–128

McClintock B (1938) The fusion of broken ends of sister half-chromatids following chromatid breakage at meiotic anaphase. Missouri Agr Exp Sta Bull 290: 48–60

Morgan DT (1950) A cytogenetic study of inversions in *Zea mays*. Genetics 35: 153–174

Rhoades MM, McClintock B (1935) The cytogenetics of maize. Bot Rev 1: 292–325

Rhoades MM, Dempsey E (1953) Cytogenetic studies of deficient-duplicate chromosomes derived from inversion heterozygotes in maize. Am J Bot 40: 405–424

Russell WA, Burnham CR (1950) Cytogenetic studies of an inversion in maize. Scient Agr 30: 93–111

Sturtevant AH, Beadle GW (1936) The relation of inversions in the X chromosome of *Drosophila melanogaster* to crossing over and disjunction. Genetics 21: 554–604

51

Use of Maize Monosomics for Gene Localization and Dosage Studies

David F. Weber

Monosomics are individuals in which one chromosome of a homologous pair is missing. A monosomic maize plant contains one copy of one chromosome and two copies of each of the other nine chromosomes ($2n - 1 = 2x - 1 = 19$ chromosomes).

Monosomics are perhaps the most interesting aneuploid type because a chromosome is unpaired in each meiotic cell and because genes on an entire chromosome are present singly in somatic cells. Monosomics in diploids have seldom been used because they are rarely produced, and once produced, the monosomic condition is not transmitted to progeny of monosomics.

Luckily there is a genetic system in maize, the *r-X1* system (Satyanarayana unpublished), that generates a high frequency of monosomics (reviewed in Weber 1983, 1986, 1991, 1992). We have been working with the *r-X1* system for over two decades in our lab and have produced several thousand maize monosomics that include monosomics for each of the 10 chromosomes. This is the only series of its type that has been produced in any higher diploid form. The monosomics are an extremely powerful tool for many studies.

PRODUCING MONOSOMICS USING THE *r-X1* SYSTEM

The *r-X1* Deficiency

The *r-X1* deficiency, a deficiency that includes the *R* locus on chromosome 10 in maize, was X-ray induced by L. J. Stadler (unpublished). The dominant allele of the *R* locus is necessary for anthocyanin production. If *R*/*r-X1* plants (that are heterozygous for the *r-X1* deficiency) are testcrossed as female parents, about 55–60% of the kernels produced are colored (with *R*/*R*/*r* endosperm and *R*/*r* embryos), and the remaining 40–45% are

The Maize Handbook—M. Freeling, V. Walbot, eds.

colorless (with *r/r/r-X1* endosperm and *r/r-X1* embryos). Thus, the deficiency is transmitted through the female parent with a high efficiency; however, it is not transmitted through the pollen. Plants germinated from the colored kernels (that lack the *r-X1* deficiency) are diploid while those from colorless kernels (that contain the *r-X1* deficiency) include 10–18% monosomics and 10–18% trisomics (Weber 1983). Thus, a high rate of nondisjunction is taking place during the embryo sac divisions in gametophytes with the *r-X1* deficiency but not in gametophytes lacking the deficiency. Clearly, a factor on chromosome 10 that is necessary for normal chromosome disjunction is deleted in the *r-X1* deficiency. Nearly all of the remaining plants from colorless kernels are diploid. However, a low frequency of multiply aneuploid individuals (Weber 1973) and individuals with deficiencies (Weber 1983) is also recovered.

Nearly all of the monosomics produced by the *r-X1* system have a chromosome missing from the entire plant. Each of the monosomic types is distinctively smaller than its diploid siblings. However, they are remarkably vigorous; most monosomic types are between 1 and 2 m tall at maturity in a genetic background where their diploids siblings are about 2.7 m tall. Excellent meiotic samples can be collected from these plants, and most monosomic types can be crossed. Thus, genetic and cytological analyses are possible with these plants. Each of the monosomic types has at least 50% aborted pollen because the univalent chromosome in a monosomic can be contributed to no more than half of its meiotic products.

Propagating the r-X1 Deficiency

Because the *r-X1* deficiency is not male transmissible, it must be maintained in the heterozygous condition. The *r-X1* deficiency can be propagated by crossing *R/r-X1* plants as female parents by *R/R* male parents. The resulting kernels that contain the *r-X1* deficiency (with *R/r-X1/r-X1* endosperm and *R/r-X1* embryos) can be distinguished from sibling kernels lacking the deficiency (with *R/R/R* endosperm and *R/R* embryos) because the former are purple with small colorless sectors and the latter are uniformly purple colored (when other factors necessary for anthocyanin pigmentation in the endosperm are present). Moreover, a self of an *R/r-X1* plant produces the same types and frequencies of progeny as the previous cross because it only produces functional *R* male gametes (because the *r-X1* deficiency is not male transmissible).

Some Characteristics of the *r-X1* System

Some of the characteristics of the *r-X1* deficiency are as follows: (1) It is submicroscopic (too small to be detected using the light microscope). (2) It is transmitted with a high efficiency through female gametes; however, pollen grains containing it do not function even though they appear to be morphologically normal. (3) It causes nondisjunction postmeiotically at the second of the three embryo sac divisions (Lin and Coe 1986; Simcox et al. 1987) and at the first male gametophyte division (Zhao and Weber 1988). (4) The genetic background influences the activity of the *r-X1* system. The *r-X1* deficiency produces monosomics at a 10–18% frequency in a W22 inbred background; however, it produces much lower frequencies of monosomics (often approaching zero) in some of the other genetic backgrounds that we have tested. (5) Deficiencies are also produced by the *r-X1* system (Weber 1983; Lin 1987; Weber et al. 1993) that involve most, probably all, of the chromosome arms with breakpoints at different positions within the same chromosome arm.

Identifying Monosomics

To select for specific monosomic types, one can cross a male parent with a recessive plant-expressed mutation onto a *R/r-X1* female parent. Endosperm-expressed mutations cannot be used for this purpose because kernels with monosomic embryos do not have endosperms that are monosomic for the same chromosome. F_1 progeny expressing the recessive phenotype of this mutation are identified as putative monosomics for the chromosome that carries the marker mutation. Most of these are monosomics; however, a few plants with deficiencies in the chromosome arm with the marker mutation are also recovered. Ideally, one would select several different plants each expressing a specific marker mutation from this type of cross and only use plants that are similar in morphology and stature in studies. The plants with deficiencies including the marker locus can usually be distinguished from monosomics for this chromosome because they usually flower earlier and the plants are larger than the monosomics. Cytological analysis of chromosomes or RFLP analysis using markers on both arms of the putative monosomic chromosome can be carried out on individuals expressing the mutant phenotype to determine which putative monosomics are, in fact, monosomic.

To select for monosomics for each of the 10 chromosomes, one could use a series of tester stocks with plant-expressed mutations on each chromosome. If all of the testers were in the same genetic background, monosomics for all chromosomes recovered would be similar genetically except for monosomy.

We typically select each of the monosomic types from progeny of a cross between two inbred lines, Mangelsdorf's tester male parents and *R/r-X1* female parents in the inbred W22. Mangelsdorf's tester has a recessive allele on each of its chromosomes and is *r/r*, and the *R/r-X1* female parent is homozygous for the corresponding dominant alleles. This cross is as in Table 51.1.

The colorless *r-X1* deficiency-carrying kernels from this cross are selected and planted directly into a research nursery. Monosomic frequencies from these plantings are equivalent to those from sand-beach-planted kernels; therefore, the monosomics do not appear to be selected against under field conditions. We are usually able to recover more than one whole plant monosomic for each of the maize chromosomes from a field planting of 2,000–3,000 colorless kernels from this cross. The frequencies given in the Table 51.1 are of plants expressing marker mutants, distinctive monosomic morphologies, and/or confirmed from testcrosses. We realize that a few plants with deficiencies are included; however, the proportion of plants with deficiencies is small. We are currently exploring

TABLE 51.1. Genetic markers and frequencies of monosomic types produced by the *R/r-X1* × Mangelsdorf's tester cross

Chromosome	Female parent	Male parent	Marker gene name	Frequency recovered (%)
	(*R/r-X1*)	(*r/r*)		
1	*Bm2*	*bm2*	*brown midrib-2*[a]	0.03[b]
2	*Lg*	*lg*	*liguleless*[a]	1.24[b]
3	*A*	*a*	*anthocyaninless*[c]	0.21[d]
4	*Su*	*su*	*sugary endosperm*[c]	0.29[b]
5	*Pr*	*pr*	*red aleurone*[c]	0.06[d]
6	*Y*	*y*	*yellow endosperm*[c]	1.89[b]
7	*Gl*	*gl*	*glossy seedling*[a]	1.01[b]
8	*J*	*j*	*japonica*[a]	3.46[b]
9	*Wx*	*wx*	*waxy endosperm*[c]	0.49[b]
10	*G*	*g*	*golden stalk*[a]	1.57[b]
			Total	10.25

[a]Mutant phenotype expressed in the plant.
[b]Frequencies for these types are from Weber (1983).
[c]Mutant phenotype expressed in the endosperm.
[d]Frequencies for these types are from a 1990 planting.

progeny of this cross with RFLP markers on both chromosome arms to determine the proportion of plants that contain deficiencies.

We first identify plants expressing the *bm2*; *lg*; *gl*; *j*; or *g* recessive phenotypes as presumptive monosomics for chromosomes 1, 2, 7, 8, or 10, respectively. Most of these plants are monosomic. The other five recessive alleles in Mangelsdorf's tester are expressed in the endosperm of kernels. Kernels with embryos monosomic for chromosomes with these genes do not express these mutant phenotypes because loss of a chromosome in the embryo of a kernel is not accompanied by loss of the same chromosome from the endosperm. We identify monosomics for these chromosomes in the following way. Semisterile (50% or greater pollen abortion) plants of subnormal stature are identified as possible monosomic plants. These are testcrossed with a line that is *a*; *su*; *pr*; *y*; *wx* (on chromosomes 3, 4, 5, 6, and 9, respectively); and *R*. All kernels produced by a monosomic-3, -4, -5, -6, or -9 plant will express one of these recessive phenotypes, corresponding to the marker gene on the monosomic chromosome. Diploids and all other monosomic types will give a 1:1 ratio for that gene. For example, monosomic-6 plants will produce all kernels with white (*y/y/y*) endosperms when testcrossed while all other monosomic types and diploids produce a 1:1 ratio of white (*y/y/y*) to yellow (*Y/Y/y*) kernels. In addition, monosomic-3, -4, -5, -6, and -9 plants each have distinctive plant morphologies that we have learned to recognize, as described in Table 51.2.

USES OF MONOSOMICS

Monosomics are extremely powerful experimental tools that can be utilized in numerous ways. Some of these are given below.

Analysis of Univalent Chromosome Behavior

The behavior of univalent chromosomes is poorly understood. Monosomics in diploid species provide the ideal opportunity to analyze univalent behavior because each meiotic cell in a monosomic contains a univalent. Studies of the behavior of univalents in maize monosomics have been summarized previously (Weber 1983). In addition, maize plants monosomic for two different chromosomes have been recovered, and tests for the possible interactions between nonhomologous univalent chromosomes have been carried out with these plants (Weber 1970, 1973).

Mapping Unplaced Genes to Chromosomes

Monosomics can be used in several different ways to assign unplaced genes to specific chromosomes.

Morphological Loci

Maize monosomics produced by the *r-X1* system can be used to identify the chromosomes that carry plant-expressed morphological loci. *R/r-X1* plants that are homozygous for the dominant allele of a gene can be crossed as female parents by plants with the recessive allele of an unplaced gene. Almost all of the F_1 progeny expressing the recessive phenotype (pseudodominants) are monosomic for the chromosome that carries this gene. The monosomic chromosome can be determined by cytological or RFLP analysis. Simcox and Weber (1985) used this approach to assign the *bx* locus to chromosome 4 in maize.

Loci With Different Electrophoretic Mobilities

Monosomics generated by the *r-X1* system can be used to determine the chromosomal locations of genetic loci whose gene products have different electrophoretic mobilities. Maize plants with the *r-X1* deficiency, an allele of an unplaced gene whose product has one electrophoretic mobility and dominant morphological tester alleles can be crossed as female parents with plants with an allele for a different electrophoretic mobility and recessive tester alleles. Monosomics are identified using the morphological markers. The monosomic type displaying only the electrophoretic allele from the male parent is monosomic for the chromosome that carries the unplaced locus (because the monosomic chromosome is contributed by the male parent). Stout and Phillips (1973) and Weber and

Table 51.2. Characteristics of maize monosomics produced by the *R/r-X1* × Mangelsdorf's tester cross

1	Expresses the brown-midrib phenotype, less than 1 m tall, typically too small to be crossed
2	Expresses the liguleless phenotype, highly variable height but up to 2 m tall, male-sterile, sets seed as female parent
3	Has thick, leathery, dark green leaves that are slightly narrower, up to 1.5 m tall, sets seed as female parent—some produce a little pollen
4	Seedlings have blue-green leaves; the upper leaves of mature plants have wide midveins that are flat (the midvein is not recessed), up to 1.7 m tall; tassel is partially retained within the leaf whorl; anthers are extruded irregularly; some viable pollen is produced; sets seed as female parent; ear is quite large
5	Extremely narrow leaves, about 1 m tall, too small to cross as female parent; a small amount of viable pollen may be produced
6	Leaves are more upright; internodes are shorter; sheds abundant pollen and sets seeds as female parent; up to 1.5 m tall, the slowest-maturing monosomic type
7	Expresses the glossy phenotype; leaves are thin and wrinkled in mature plants; up to 2 m tall; sheds some pollen and sets a few seeds as female parent
8	Expresses the japonica phenotype, stalks are thinner, up to 1.8 m tall, sheds pollen and sets seed as female parent, often has several tillers
9	Leaves are somewhat thinner and stiff, up to 1.8 m tall, sheds no pollen, occasionally sets a few seed as female parent; most rapidly maturing monosomic type
10	Expresses the golden phenotype, up to 1.7 m tall, sheds abundant pollen and sets seed as female parent

Brewbaker (1983) have used this strategy to assign the *H1a* and *px3* loci to chromosomes 1 and 7, respectively, in maize. This procedure does not require the cytological identification of the monosomic chromosome. Another genetic approach using maize monosomics has also been proposed (Weber 1974).

Maize monosomics generated with the *r-X1* system have been especially useful for assigning RFLP loci to specific maize chromosomes. We (Helentjaris et al. 1986, 1988)

examined DNAs of *R/r-X1* W22 plants and Mangelsdorf's tester using different RFLP probes and restriction enzymes to identify probe-restriction enzyme combinations that revealed a polymorphism between the two parents. Monosomics for the various maize chromosomes were then selected from a cross between these two inbred lines as previously described, and the monosomic type that displayed only the RFLP allele from the male parent is monosomic for the chromosome that carries this locus. This work identified the linkage groups for most of maize chromosomes on the RFLP map and facilitated the development of the first comprehensive RFLP map for maize. Recently, we (Chaubet et al. 1992) have used monosomics to map histone H3 and H4 loci in maize. H3 and H4 genes were found to be located on most, possibly all, of the chromosomes in the maize genome.

Altering the Number of Copies of Known Genetic Loci

Monosomics can be used to alter the number of copies of previously mapped genetic loci.

One vs. Two Copies

Monosomics can be compared with diploids to determine the effects of one vs. two copies of previously mapped genetic loci. For example, monosomics have been used to compare one vs. two copies of the (NOR) nucleolar organizing region. The genes for the 18S and 28S rRNAs are located at the NOR in maize and in other organisms where they have been studied. In *Drosophila melanogaster,* the number of 18S and 28S templates per NOR is greater in flies with one NOR than with two NORs. This phenomenon is termed "gene compensation" (Tartof 1971). The NOR is on chromosome 6 in maize. We (Phillips et al. 1974) compared DNAs of monosomic-6 (with one NOR) and diploid (with two NORs). We found that the multiplicity of 18S and 28S rRNA templates per NOR was essentially the same in DNAs of both plant types; therefore, the phenomenon of gene compensation was not found to occur in maize.

Zero vs. One Copy

A monosomic plant produces haploid (x) spores and spores that are nullosomic (x-1) for the monosomic chromosome. By comparing these daughter haploid and nullosomic cells, one compares one vs. zero copies of all loci on a specific chromosome. This approach was used to screen the maize genome for factors necessary for formation of the nucleolus (Weber 1978a), the site of rRNA synthesis. Quartets of microspores from plants monosomic for chromosomes 1, 2, 4, 5, 6, 7, 8, 9, and 10 were examined. As expected, only 2 of the 4 spores in quartets from monosomic-6 plants contained nucleoli, because only 2 of the 4 members have a chromosome 6, the site of the NOR. The multiple templates for 5S rRNA (a component of ribosomes) are located on the long arm of chromosome 2 in maize (Wimber et al. 1974). Quartets of monosomic-2 plants had nucleoli in all 4 spores even though 2 of these 4 were nullosomic for chromosome 2 (and lacked 5S templates); therefore, the 5S rRNA templates are not necessary for formation of the nucleolus.

Quartets of all other monosomic types analyzed contained nucleoli in all four spores; therefore, factors located on these chromosomes were not required for nucleolar formation at the quartet stage.

Exploring the Genome for Gene Dosage Effects

By comparing a plant monosomic for a specific chromosome with its diploid siblings, one compares the effects of one vs. two copies of all genes on a specific chromosome. If a gene is present on the monosomic chromosome that expresses dosage effects, a difference will be found between these two plant types for the trait specified by this gene. In this way, one analyzes all genetic loci on an entire chromosome without the use of mutant alleles. This powerful methodology has been used to identify genetic determinants for several types of traits in maize.

Plewa and Weber (1973) used this approach to screen the maize genome for dosage-sensitive loci that affect lipid quantity. We found that embryos monosomic for chromosomes 2, 6, or 10 contained significantly less lipid than in diploid control embryos. Thus, dosage-sensitive loci (either structural or regulatory) controlling the level of lipids are present on these chromosomes. We (Plewa and Weber 1975) also determined the effects of monosomy on the relative amounts of different fatty acids in maize embryos. We found that monosomic-2 embryos had significantly more oleic acid and significantly less linoleic acid than diploid control embryos. Because the conversion from oleic acid to linoleic acid is a single-enzyme-mediated reaction, the data suggest that a dosage-sensitive gene involved in this conversion is located on chromosome 2. Monosomic-diploid comparisons have also been used to study the genetic control of acid-extractable amino acids (the free amino acid pool) (Cook and Weber 1976; Cook 1977), intergenic recombination (Weber 1971, 1976, 1979), and intragenic recombination (Weber 1978b).

In summary, monosomics for each of the maize chromosomes can be efficiently generated utilizing the *r-X1* system. Monosomics are exceedingly powerful experimental tools that can and have been used in diverse ways to explore the maize genome.

REFERENCES

Chaubet N, Philipps G, Gigot C, Guitton C, Bouvet N, Freyssinet G, Schneerman M, Weber D (1992) Subfamilies of histone *H3* and *H4* genes are located on most, possibly all of the chromosomes in maize. Theor Appl Genet, 84: 555–559

Cook JW (1977) Effects of monosomy upon free amino acid profiles in *Zea mays* leaves. MS Thesis, Illinois State Univ, Normal, IL

Cook JW, Weber DF (1976) Monosomic analysis of the acid extractable amino acids (free amino acid pool) in maize leaves. Maize Genetics Cooperation News Letter 50: 40–42

Helentjaris T, Weber DF, Wright S (1986) Use of monosomics to map cloned DNA fragments in maize. Proc Natl Acad Sci USA 83: 6035–6039

Helentjaris T, Weber DF, Wright S (1988) Identification of the genomic locations of duplicate nucleotide sequences in maize by analysis of restriction fragment length polymorphisms. Genetics 118: 353–363

Lin B-Y (1987) Cytological evidence of terminal deficiencies produced by the *r-X1* deficiency in maize. Genome 29: 718–721

Lin B-Y, Coe EH (1986) Monosomy and trisomy induced by the *r-X1* deletion in maize, and associated effects on endosperm development. Can J Genet Cytol 28: 831–834

Phillips RL, Weber DF, Kleese RA, Wang SS (1974) The nucleolus organizer region of maize (*Zea mays* L.): tests for ribosomal gene compensation or magnification. Genetics 77: 285–297

Plewa MJ, Weber DF (1973) The use of monosomics to detect genes conditioning lipid content in *Zea mays* L. embryos. Can J Genet Cytol 15: 313–320

Plewa MJ, Weber DF (1975) Monosomic analysis of fatty acid composition in embryo lipids of *Zea mays* L. Genetics 81: 277–286

Simcox KD, Shadley JD, Weber DF (1987) Detection of the time of occurrence of non-disjunction induced by the *r-X1* deficiency in *Zea mays* L. Genome 29: 782–785

Simcox KD, Weber DF (1985) Location of the *benzyxanzinless* (*bx*) locus in maize by monosomic and B-A translocational analyses. Crop Sci 25: 827–830

Stout JT, Phillips RL (1973) Two independently inherited electrophoretic variants of the lysine-rich histones of maize (*Zea mays*). Proc Natl Acad Sci USA 70: 3043–3047

Tartof KD (1971) Increasing the multiplicity of ribosomal RNA genes in *Drosophila melanogaster.* Science 171: 294–297

Weber DF (1970) An attraction between nonhomologous univalent chromosomes and further tests of distributive pairing in *Zea mays.* Genetics 64: s65

Weber DF (1971) The use of monosomy to detect genes altering recombination in *Zea mays.* Maize Genetics Cooperation News Letter 45: 32–35

Weber DF (1973) A test of distributive pairing in *Zea mays* utilizing doubly monosomic plants. Theor Appl Genet 43: 167–173

Weber DF (1974) A monosomic mapping method. Maize Genetics Cooperation News Letter 48: 49–52

Weber DF (1976) Effect of monosomy on recombination of chromosome 9 of *Zea mays.* Genetics 83: s81

Weber DF (1978a) Nullosomic analysis of nucleolar formation in *Zea mays.* Can J Genet Cytol 29: 97–100

Weber DF (1978b) Monosomic analysis of intragenic recombination at the waxy locus in maize. Genetics 88: s109–s110

Weber DF (1979) Monosomic regulation of intergenic recombination in chromosomes 2 and 9 in *Zea mays.* Genetics 91: s135

Weber DF (1983) Monosomic analysis in diploid crop plants. In Swaminathan MS, Gupta PK, Sinha U (eds) Cytogenetics of Crop Plants, MacMillan India, New Delhi, pp 351–378

Weber DF (1986) The production and utilization of monosomic *Zea mays* in cytogenetic studies. In Reddy GM, Coe EH (eds) Gene Structure and Function in Higher Plants, Oxford and IBH, New Delhi, India, pp 191–204

Weber DF (1991) Monosomic analysis in maize and other diploid crop plants. In Gupta PK, Tsuchiya T (eds) Chromosome Engineering in Plants: Genetics, Breeding, and Evolution Part A, Elsevier, Amsterdam, pp 181–209

Weber DF, Helentjaris T, Zhao Z (1993) RFLP analysis of deletions produced by maize plants with the *r-X1* deficiency. Genetics, *in press.*

52

Marker Systems for r-x1

James A. Birchler and E. H. Coe

Previous systems for recognition of monosomic individuals generated by the *r-x1* system used the Mangelsdorf tester. This stock has a marker on each of the ten chromosomes. Some of these are seedling/plant characters while others affect the endosperm. For the cases of seedling/plant characters, the uncovering of one of the markers in the F_1 is an indication that that individual is a monosomic for the respective chromosome. To distinguish the other monosomics, characteristic morphological phenotypes are used. Self pollination of the latter would give ears with 50% ovule abortion that would be homozygous for the recessive endosperm marker. Such a system is effective but requires large populations to be grown in order to recover sufficient monosomics. As monosomics are invaluable in gene mapping studies using RFLP technology (see chapter by Weber), a more economical method, allowing selection of kernels likely to be monosomic, is highly desirable.

Our labs are attempting to develop a system for recognition of individual monosomics at the mature kernel stage. This is based on the use of the *R-scm2* or *R-nj* allele of the *R* locus that conditions anthocyanin pigmentation in the embryo and the endosperm, in combination with individual anthocyanin genes on each of the chromosomes. The various *r-x1* stocks when crossed as females by the respective tester would produce progeny that would have colored scutella except for those individuals that have lost the dominant form of the gene from the maternal parent. These would have no pigmentation in the scutellum and would be a candidate for the desired monosomic. The stocks for each chromosome are listed in Table 52.1.

The principle of the above stocks is that, whenever the *r-x1* deficiency induces the loss of a particular chromosome, the dominant marker will be lost. When crossed by the respective male parent the individual monosomics can be distinguished by the loss of the marker in the embryo but not the endosperm. The female stocks can be maintained by crossing with the respective maintenance stocks. For chromosomes 1 and 3–9, they work on the principle that endosperms with only a copy of *R* from the male parent exhibit a mottled phenotype. This will select for the *r-x1/R* plants. The *Dotted2* and *3* and *Activator 2* material still allows this type of selection. For chromosome 2, the *B-peru* marker will not give this type of selection; therefore, it will be necessary to select the *r-x1* bearing

The Maize Handbook—M. Freeling, V. Walbot, eds.

TABLE 52.1 Marker System for monosomic selection

Chromosome	Female stock	Maintenance stock	Selection stock
1	*Bz2; r-x1/R-r*	*R-r*	*bz2; R-scm2*
2	*B-peru; r-x1/r-g*	*B-peru; r-g*	*b; r-g*
3	*A; r-x1/R-r*	*R-r*	*a-m1; R-scm2*
4	*C2; r-x1/R-r*	*R-r*	*c2; R-scm2*
5	*A2; r-x1/R-r*	*R-r*	*a2; R-scm2*
6	*a-m1; Dt2; r-x1/R-r*	*a-m1; Dt2; R-r*	*a-m1; R-scm2*
7	*a-m1; Dt3; r-x1/R-r*	*a-m1; Dt3; R-r*	*a-m1; R-scm2*
8	*bz2-m; Ac2; r-x1/R-r*	*bz2-m; Ac2; R-r*	*bz2-m; R-scm2*
9	*Bz; r-x1/R-r*	*R-r*	*bz1; R-scm2*
10	*Ac—r-x1/r-m3*	*r-m3*	*r-m3*

ears on the criterion of numerous defective kernels. The marker system for chromosome 10 is not yet available, but will involve the tight linkage of an *Ac* element near *r-x1*. By crossing by a *Ds* insertion into *R-sc*, namely *r-m3*, the stock can be maintained as well as used for selection of the loss of chromosome 10 from the maternal parent.

The above stocks can be used for selection of the respective monosomics at the mature kernel stage. In addition, for chromosomes 1 and 3–9, they can also be used to select monosomics at the mature plant stage if instead of the *R-scm2* tester stock an *R-nj* tester is used. This works because the *R-nj* allele conditions not only pigment in the embryo and aleurone but also in the silks and anthers. Consequently, loss of the maternal dominant will result in anthers with the recessive phenotype. The stock listed in Table 52.1 for chromosome 2 could be used for mature plant selection, because *B-peru* will pigment the brace roots, leaf sheath, and glumes. The chromosome 10 tester might be used at the seedling stage because *R-sc* alleles pigment the coleoptile.

While the above stocks have the advantage that they will allow the selection of individual monosomics at the mature kernel or mature plant stage, it should be noted that doubly monosomic plants will not be genetically identified in this system. Fortunately, such types are rare and when they do occur, they exhibit an extreme phenotype that might serve to identify them if the individuals are grown into plants. Another caveat of the *r-x1* system is that there are also trisomics produced that will not be genetically distinguished in these systems. Still another caveat is that chromosome breakages are common with *r-x1*, representing perhaps 50% of the gene loss events. If such considerations are critical to individual investigations, the proper steps should be taken to further identify the singly monosomic kernels from among those selected with the above system.

53

Translocations as Genetic Markers

E. B. PATTERSON

If two nonhomologous chromosomes interchange ends following chromosome breakage, the products of this event are termed a reciprocal interchange or reciprocal translocation. The two rearranged chromosomes that result are fully stable and, when transmitted together, contain all the chromosomal material that was present in the two normally arranged chromosomes from which they arose.

In maize these physical points of interchange may be determined by cytological observation of the chromosomes at pachynema of microsporogenesis in plants heterozygous for a translocation. Pairing of the two rearranged chromosomes with homologous portions of their two normally arranged counterparts leads typically to formation of a cross configuration in which the center of the cross marks the points of interchange in the original standard chromosomes.

In plants heterozygous for a reciprocal translocation, typically about half the products of meiosis carry chromosome imbalance that is expressed as spore abortion (Figure 53.1). The semisterile phenotype resulting from pollen or ovule abortion is the usual basis for identifying heterozygotes in segregating progenies. Chromosomally balanced spores produced by translocation heterozygotes contain either both interchanged or both normally arranged homologs. In crosses in which unbalanced chromosome complements are not transmitted, a complete artificial linkage is established between the two interchange points as well as between the corresponding points on their normal homologs. This complete linkage between points on nonhomologous chromosomes is not the result of linkage during meiotic assortment but rather is a direct consequence of nontransmission to progeny of other combinations of these points. Because of their complete association in transmission, the two interchange points may be considered as one in inheritance. The symbol T is used to designate the combination of rearranged chromosomes or the interchange points individually. The symbol + is commonly used to designate the alternative combination of normal chromosomes or their standard structures.

In crosses of standard maize by plants heterozygous for a reciprocal translocation, the translocation may be followed as if it were a dominant gene for semisterility. If plants heterozygous for a translocation are self-pollinated, the progeny are of three types with

The Maize Handbook—M. Freeling, V. Walbot, eds.

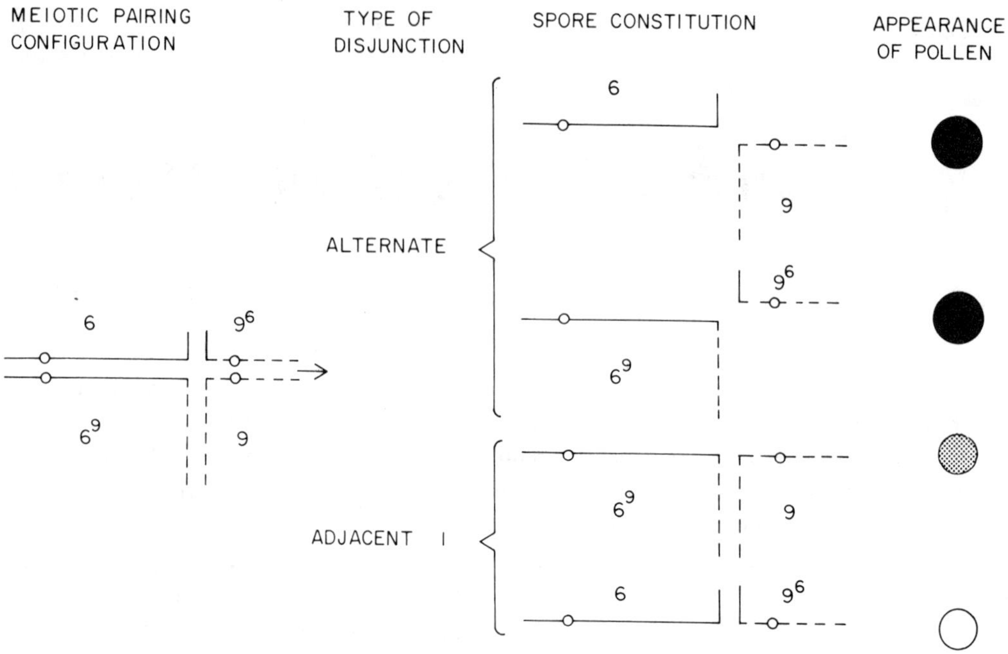

FIGURE 53.1 Chromosome pairing configuration and products of meiosis in a maize plant heterozygous for a reciprocal translocation. Phenotypes of pollen grains with unbalanced chromosome complements reflect specific imbalances in different reciprocal translocations.

regard to chromosome constitution, provided there is no transmission of unbalanced complements. These occur in the proportions one-half heterozygous translocation: one-fourth homozygous translocation: one-fourth homozygous normal. The latter two types are fully fertile and may be distinguished by testcrosses to plants carrying standard chromosomes. Testcrosses of homozygous translocation plants yield all semisterile plants; progeny from testcrossing plants with normal chromosomes are all fully fertile.

The map position of a reciprocal translocation may be determined with respect to loci in either of the two parental standard chromosomes independently. If transmission of the translocation is followed simultaneously with respect to markers on both chromosomes, the resulting linkage relationships can be represented as a two-dimensional, cross-shaped linkage map in which both interchange points occur at the intersection of the cross configuration. Because reciprocal translocations alter normal patterns of pairing and crossing over, values on crossing over frequencies obtained through the use of reciprocal translocations may not be incorporated directly into standard linkage maps. However, information on the sequence of interchange points relative to gene loci is reliable. If two loci linked in standard strains are separated as a result of an interchange between them, they will assort independently in plants homozygous for the translocation.

Reciprocal translocations have been used extensively to locate genes for both qualitative and quantitative traits in maize. Examples of applications and proposed uses were presented by Anderson (1956). Translocations have also been used to control or monitor gene transmission by actual or artificial linkage (Patterson 1978).

Because the interchange points of reciprocal translocations in the physical chromosomes can be correlated with positions in the linkage maps, it is possible to make inferences regarding the positions of gene loci in the physical chromosomes. Currently, translocations are also being used as tools to aid in ordering RFLP markers relative to the chromosome structure.

Longley (1961) published a listing of the cytological positions of 1,003 reciprocal translocations. About 875 have been perpetuated and seed samples are available upon request. Table 54.1 in the next chapter lists the stocks available in 1992. A listing of the current collection may be obtained from the following address:

Director, Maize Genetics Stock Center
S-123 Turner Hall, Agronomy Department
University of Illinois
1102 S. Goodwin Avenue
Urbana, IL 61801
Phone: 217-333-6631 FAX: 217-333-9817

The reciprocal translocations in the current collection permit complete artificial linkage of some 875 pairs of points in the maize genome. Alternatively, they represent about 1,750 genetic loci, all classified by the same trait, semisterility.

REFERENCES

Anderson EG (1956) The application of chromosomal techniques to maize improvement. Brookhaven Symp Biol No 9, Genetics in plant breeding, pp 23–36

Longley AE (1961) Breakage points for four corn translocation series and other corn chromosome aberrations. USDA Agr Res Serv Crops Res Bull No. 34–16, 40 pp

Patterson EB (1978) Properties and uses of duplicate-deficient chromosome complements in maize. In Walden DB (ed) Maize Breeding and Genetics, John Wiley and Sons, New York, pp 693–710

54

A-A Translocations: Breakpoints and Stocks

E. H. COE

A remarkable resource of cytogenetic modifications, far beyond those available in most other experimental species, is available in the reciprocal translocations between A chromosomes of maize. Translocations were originally found in the 1920s (Brink 1927; Burnham 1930) as spontaneously arising, segregating "semisteriles" (i.e., plants with 50% aborted pollen and eggs). Physical exchange of chromosome segments was revealed by cross-shaped configurations of four elements at pachytene (McClintock 1930). Numbers of translocations were isolated following X-irradiation (Anderson 1935; 64 additional translocations). The resource was greatly expanded with interchanges induced by exposure to nuclear explosions; these and the other collections were all characterized cytologically by Longley (1961). See the previous chapter by Patterson for a description of the transmission genetics of translocations.

Table 54.1 lists the translocations: Symbol for chromosomes interchanged; the Symbol identifier (ID) letter or number for the aberration; arms and cytological positions at which the breakpoints were identified (e.g., 0.9 indicates that the breakpoint was identified at a position 90% of the distance from the centromere to the end of that chromosome arm); stock No. in the Maize Genetics Cooperation Stock Center; synonyms; and remarks relating to breakpoint determinations. Symbol IDs marked with ** indicate changes in information. The accompanying figure displays the distribution of breakpoints of translocations.

The Maize Handbook—M. Freeling, V. Walbot, eds.

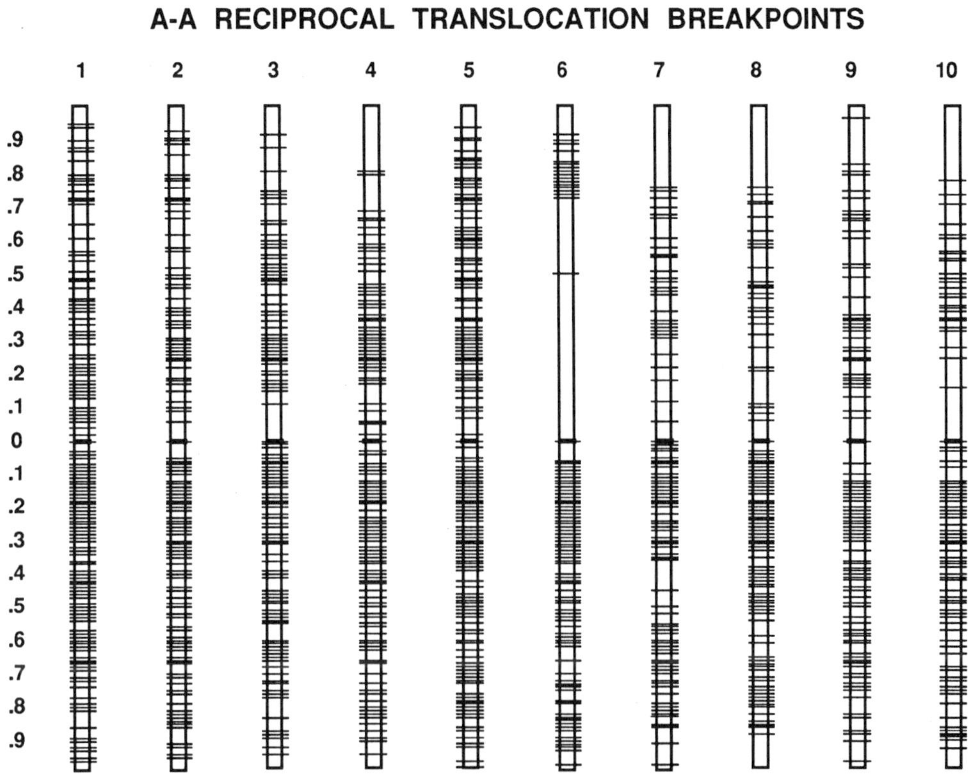

FIGURE 54.1 A-A reciprocal translocation breakpoints

TABLE 54.1. A-A translocations: breakpoints and stocks

Symbol	Symbol ID	Arm	Pos.	Arm	Pos.	Stock No.	Synonyms	Remarks
T1-2	7039	1S	0.9	2S	0.89	1041A	T1-2(7309)	Longley breakpoints (1961)
T1-2	d	1S	0.78	2L	0.56	1041D	T1-2(17)	Longley breakpoints (1961)
T1-2	c	1S	0.77	2L	0.33	1041B		Longley breakpoints (1961)
T1-2	e	1S	0.61	2L	0.47	1041E	T1-2(B-75)	Longley breakpoints (1961)
T1-2	4464	1S	0.53	2L	0.28	1041C		Longley breakpoints (1961)
T1-2	b	1S	0.43	2S	0.36	1041F		Longley breakpoints (1961)
T1-2	036-7	1S	0.37	2L	0.33	1042A		Longley breakpoints (1961)
T1-2	5255	1S	0.25	2S	0.31	1042B		Longley breakpoints (1961)
T1-2	5896	1S	0.22	2S	0.3	1042C		Longley breakpoints (1961)
T1-2	6891	1S	0.14	2S	0.83			Longley breakpoints (1961)
T1-2	004-11	1S	0.13	2S	0.36	1042D		Longley breakpoints (1961)
T1-2	8628	1	0	2L	0.49	1042E		Longley breakpoints (1961)
T1-2	5946	1	0	2	0	1042F		Longley breakpoints (1961)
T1-2	028-17	1	0	2	0	1043A		Longley breakpoints (1961)
T1-2	4937	1L	0.1	2S	0.15	1043B		Longley breakpoints (1961)
T1-2	5453	1L	0.11	2S	0.58	1043C		Longley breakpoints (1961)
T1-2	051-1	1L	0.16	2S	0.3	1043D		Longley breakpoints (1961)
T1-2	018-18	1L	0.16	2L	0.44	1043E		Longley breakpoints (1961)
T1-2	(9-10a)	1L	0.21	2L	0.37			Longley breakpoints (1961)
T1-2	5539	1L	0.21	2L	0.61	1043F		Longley breakpoints (1961)
T1-2	015-18	1L	0.26	2L	0.26			Longley breakpoints (1961)
T1-2	5523	1L	0.27	2S	0.62	1044A		Longley breakpoints (1961)
T1-2	041-9	1L	0.27	2S	0.57	1044B		Longley breakpoints (1961)
T1-2	6892	1L	0.3	2L	0.35	1044C		Longley breakpoints (1961)
T1-2	6427	1L	0.43	2L	0.5	1044D		Longley breakpoints (1961)
T1-2	a	1L	0.5	2L	0.4			Longley breakpoints (1961)
T1-2	7211	1L	0.57	2L	0.79	1044E		Longley breakpoints (1961)
T1-2	6883	1L	0.63	2L	0.52	1044F		Longley breakpoints (1961)
T1-2	017-3	1L	0.67	2S	0.58	1045A		Longley breakpoints (1961)
T1-2	5376	1L	0.77	2L	0.08	1045B		Longley breakpoints (1961)
T1-2	8347**	1		2		1045C	T1-5(8347)	Burnham breakpoints (1981)
T1-2	018-5**	1		2		1045D	T1-5(018-5)	Burnham breakpoints (1981)
T1-2	024-5**	1		2		1045E	T1-5(024-5)	Burnham breakpoints (1981)
T1-2	6178**	1		2		1045F	T1-5(6178)	Burnham breakpoints (1981)
T1-2	8388**	1		2		1045G	T1-5(8388)	Burnham breakpoints (1981)
T1-3	5883	1S	0.88	3S	0.6	1045H		Longley breakpoints (1961)
T1-3	5597	1S	0.77	3L	0.48	1045I		Longley breakpoints (1961)
T1-3	5982	1S	0.77	3L	0.66	1045J		Longley breakpoints (1961)
T1-3	021-7	1S	0.72	3L	0.41			Longley breakpoints (1961)
T1-3	8995	1S	0.49	3L	0.06	1045K		Longley breakpoints (1961)
T1-3	k	1S	0.17	3L	0.34	1046A	T1-3(G-3)	Longley breakpoints (1961)
T1-3	a	1S	0.15	3L	0.17	1046B		Longley breakpoints (1961)
T1-3	c	1S	0.14	3L	0.14	1046C		Longley breakpoints (1961)
T1-3	h	1S	0.06	3L	0.04	1046D	T1-3(C-15)	Longley breakpoints (1961)
T1-3	013-9	1	0	3	0	1046E		Longley breakpoints (1961)
T1-3	6861	1L	0.04	3L	0.65	1046F		Longley breakpoints (1961)
T1-3	8048	1L	0.11	3S	0.18	1047A		Longley breakpoints (1961)
T1-3	j	1L	0.11	3L	0.13	1047B	T1-3(F-10)	Longley breakpoints (1961)
T1-3	f	1L	0.17	3L	0.11			Longley breakpoints (1961)
T1-3	g	1L	0.17	3L	0.13			Longley breakpoints (1961)
T1-3	6884	1L	0.17	3L	0.19	1047C		Longley breakpoints (1961)
T1-3	024-14	1L	0.27	3S	0.49	1047D		Longley breakpoints (1961)
T1-3	8637	1L	0.37	3S	0.5	1047E		Longley breakpoints (1961)
T1-3	4759	1L	0.39	3L	0.2	1047F		Longley breakpoints (1961)
T1-3	e	1L	0.58	3L	0.45	1048A	T1-3(A-33)	Longley breakpoints (1961)
T1-3	8405	1L	0.6	3L	0.31	1048B		Longley breakpoints (1961)
T1-3	d	1L	0.61	3S	0.75	1048C		Longley breakpoints (1961)
T1-3	5476	1L	0.66	3L	0.87	1048D		Longley breakpoints (1961)
T1-3	i	1L	0.68	3S	0.3	1048E	T1-3(C-43)	Longley breakpoints (1961)
T1-3	5267	1L	0.72	3L	0.73	1048F		Longley breakpoints (1961)
T1-3	4314	1L	0.81	3L	0.89	1049A		Longley breakpoints (1961)
T1-3	5242	1L	0.9	3L	0.65	1049B		Longley breakpoints (1961)
T1-4	h	1S	0.94	4L	0.52	1049C	T1-4(X-22-61)	Longley breakpoints (1961)
T1-4	002-19	1S	0.87	4L	0.42	1049D		Longley breakpoints (1961)
T1-4	5688	1S	0.87	4L	0.45			Longley breakpoints (1961)
T1-4	002-19	1S	0.87	4L	0.42			Longley breakpoints (1961)
T1-4	4308	1S	0.65	4L	0.58	1049E		Longley breakpoints (1961)
T1-4	b	1S	0.55	4L	0.83	1049F	T1-4(Conn. R-29)	Longley breakpoints (1961)
T1-4	8598	1S	0.41	4L	0.81			Longley breakpoints (1961)
T1-4	8602	1S	0.41	4L	0.81	1050A		Longley breakpoints (1961)
T1-4	064-20	1S	0.23	4L	0.19	1050B		Longley breakpoints (1961)
T1-4	5566	1S	0.21	4L	0.26	1050C		Longley breakpoints (1961)
T1-4	8368	1S	0.14	4S	0.3	1050D		Longley breakpoints (1961)
T1-4	i	1S	0.13	4S	0.42	1050F		Longley breakpoints (1961)
T1-4	8663	1S	0.09	4S	0.36	1050G		Longley breakpoints (1961)
T1-4	8664	1S	0.09	4S	0.36			Longley breakpoints (1961)
T1-4	5910	1S	0.03	4S	0.04			Longley breakpoints (1961)
T1-4	007-1	1L	0.04	4L	0.06			Longley breakpoints (1961)
T1-4	5629	1L	0.1	4L	0.1	1051A		Longley breakpoints (1961)
T1-4	039-15	1L	0.14	4S	0.26	1051B		Longley breakpoints (1961)
T1-4	6422	1L	0.16	4S	0.11	1051C		Longley breakpoints (1961)
T1-4	5373	1L	0.17	4S	0.29	1051D		Longley breakpoints (1961)
T1-4	f	1L	0.25	4L	0.16	1051E	T1-4(C-46)	Longley breakpoints (1961)
T1-4	8249	1L	0.26	4L	0.63	1051F		Longley breakpoints (1961)
T1-4	d	1L	0.27	4L	0.3	1052A		Longley breakpoints (1961)
T1-4	c	1L	0.33	4S	0.23	1052B	T1-4(A-57)	Longley breakpoints (1961)
T1-4	8563	1L	0.39	4S	0.21	1052C		Longley breakpoints (1961)
T1-4	8592	1L	0.43	4L	0.69			Longley breakpoints (1961)
T1-4	e	1L	0.45	4S	0.27			Longley breakpoints (1961)
T1-4	4692	1L	0.46	4L	0.15	1052D		Longley breakpoints (1961)
T1-4	015-15	1L	0.49	4S	0.4			Longley breakpoints (1961)
T1-4	a	1L	0.51	4S	0.69	1052E		Longley breakpoints (1961)
T1-4	5756	1L	0.89	4L	0.81			Longley breakpoints (1961)
T1-4	5438	1L	0.93	4L	0.81	1052F		Longley breakpoints (1961)
T1-4	g	1L	0.95	4L	0.35	1053A	T1-4(C-49)	Longley breakpoints (1961)
T1-5	5045	1S	0.94	5L	0.5	1053B		Longley breakpoints (1961)
T1-5	058-2	1S	0.88	5S	0.62	1053C		Longley breakpoints (1961)
T1-5	8347	1S	0.84	5L	0.51		T1-2(8347**)	Longley breakpoints (1961); T1-2 per Burnham
T1-5	4613	1S	0.78	5L	0.22	1053E		Longley breakpoints (1961)
T1-5	5525	1S	0.75	5S	0.53	1053F		Longley breakpoints (1961)
T1-5	004-14	1S	0.75	5L	0.33			Longley breakpoints (1961)
T1-5	i	1S	0.71	5S	0.74	1054A	T1-5(X-23-2)	Longley breakpoints (1961)

Table 54.1 (Continued)

T1-5	8972	1S	0.56	5S	0.29	1054B		Longley breakpoints (1961); 1 not correct per Burnham
T1-5	018-5	1S	0.53	5L	0.52		T1-2(018-5**)	Longley breakpoints (1961); T1-2 per Burnham
T1-5	6899	1S	0.32	5S	0.2		T1-5(6899**)	Longley breakpoints (1961)
T1-5	6899**	1S	0.4	5L	0.1	1054C	T1-5(6899)	Burnham breakpoints (1981)
T1-5	055-4	1S	0.32	5L	0.31			Longley breakpoints (1961); 1 or 5 not correct per Burnham
T1-5	4832	1S	0.2	5L	0.12			Longley breakpoints (1961); 1 or 5 not correct per Burnham
T1-5	b	1S	0.17	5L	0.1	1054E		Longley breakpoints (1961)
T1-5	040-3	1S	0.17	5L	0.61	1055A		Longley breakpoints (1961); 1 or 5 not correct per Burnham
T1-5	5537	1S	0.11	5S	0.15			Longley breakpoints (1961); 1 or 5 not correct per Burnham
T1-5	043-15	1S	0.1	5L	0.63	1055B		Longley breakpoints (1961)
T1-5	024-5	1S	0.09	5L	0.98		T1-2(024-5**)	Longley breakpoints (1961); T1-2 per Burnham
T1-5	5512	1S	0.08	5L	0.7	1055D		Longley breakpoints (1961)
T1-5	6197	1S	0.02	5L	0.02	1055E		Longley breakpoints (1961)
T1-5	8782	1	0	5	0	1055F		Longley breakpoints (1961); 1L-5S or 1S-5L per Burnham
T1-5	e	1L	0.03	5L	0.09	1055B	T1-5(A-90)	Longley breakpoints (1961)
T1-5	4331	1L	0.03	5S	0.02		T7-10(4331**)	Longley breakpoints (1961); T7-10 per Burnham
T1-5	6178	1L	0.04	5L	0.05		T1-2(6178**)	Longley breakpoints (1961); T1-2 per Burnham
T1-5	e**	1S	0.13	5S	0.24		T1-5e	Burnham breakpoints (1981)
T1-5	044-10	1L	0.05	5S	0.83	1056D		Longley breakpoints (1961)
T1-5	f	1L	0.07	5L	0.09	1056E	T1-5(D-5)	Longley breakpoints (1961)
T1-5	6401	1L	0.14	5S	0.2	1056F		Longley breakpoints (1961)
T1-5	7219	1L	0.15	5S	0.19		T1-5(7219**)	Longley breakpoints (1961)
T1-5	7219**	1S	0.18	5L	0.39	1057A	T1-5(7219)	Burnham breakpoints (1981)
T1-5	h	1L	0.18	5L	0.53	1057B	T1-5(X-1-37)	Longley breakpoints (1961)
T1-5	48-34-2	1L	0.19	5L	0.76	1057C		Longley breakpoints (1961)
T1-5	5813	1L	0.21	5S	0.96			Longley breakpoints (1961)
T1-5	022-11	1L	0.27	5L	0.56			Longley breakpoints (1961)
T1-5	8388	1L	0.3	5S	0.25		T1-2(8388**)	Longley breakpoints (1961); T1-2 per Burnham
T1-5	c	1L	0.34	5L	0.29	1057E		Longley breakpoints (1961)
T1-5	070-12	1L	0.39	5S	0.71	1057F		Longley breakpoints (1961)
T1-5	7212	1L	0.44	5S	0.28	1058A		Longley breakpoints (1961)
T1-5	4597	1L	0.52	5S	0.43	1058B		Longley breakpoints (1961)
T1-5	a	1L	0.52	5S	0.42	1058C	T1-5a**	Longley breakpoints (1961)
T1-5	a**	1L	0.62	5L	0.44		T1-5a	Burnham breakpoints (1981)
T1-5	g	1L	0.58	5S	0.85	1058D	T1-5(I-24)	Longley breakpoints (1961)
T1-5	8041	1L	0.8	5L	0.15	1058E	T1-5(8041**)	Longley breakpoints (1961)
T1-5	8041**	1L	0.8	5S	0.1		T1-5(8041)	Burnham breakpoints (1981)
T1-5	7267	1L	0.92	5L	0.82	1058F		Longley breakpoints (1961)
T1-6	8452	1S	0.8	6L	0.52	1059A		Longley breakpoints (1961)
T1-6	8605	1S	0.79	6L	0.59			Longley breakpoints (1961)
T1-6	8609	1S	0.79	6L	0.59	1059B		Longley breakpoints (1961)
T1-6	028-13	1S	0.56	6L	0.54	1059C		Longley breakpoints (1961)
T1-6	7097	1S	0.46	6L	0.62	1059D		Longley breakpoints (1961)
T1-6	7352	1S	0.4	6L	0.6	1059E		Longley breakpoints (1961)
T1-6	e	1S	0.37	6L	0.21	1059F	T1-6e**; T1-6(A-80)	Longley breakpoints (1961)
T1-6	e**	1		6S			T1-6e	Burnham breakpoints (1981)
T1-6	5537	1S	0.31	6L	0.22			
T1-6	055-10	1S	0.29	6L	0.48	1060A		Longley breakpoints (1961)
T1-6	5013	1S	0.26	6L	0.28	1060B		Longley breakpoints (1961)
T1-6	5495	1S	0.25	6S	0.8	1060C		Longley breakpoints (1961)
T1-6	c	1S	0.25	6L	0.27	1060D		Longley breakpoints (1961)
T1-6	6189	1S	0.23	6L	0.17	1060E	T1-6(6189**)	Longley breakpoints (1961)
T1-6	6189**	1S	0.5	6S	0.52	1060E	T1-6(6189)	At .10 in NOR-heterochromatin per Phillips et al. 1977
T1-6	4986	1S	0.21	6S	0.78	1060F		Longley breakpoints (1961); 1S.11, 6S in proximal part of secondary constriction per Phillips et al. 1977
T1-6	5077	1S	0.2	6L	0.6	1061A		Longley breakpoints (1961)
T1-6	5072	1S	0.17	6L	0.45			Longley breakpoints (1961)
T1-6	h	1L	0.03	6L	0.17	1061B	T1-6(X-41-13)	Longley breakpoints (1961)
T1-6	d	1L	0.13	6S	0.74	1061C	T1-6(Conn. R-28)	Longley breakpoints (1961)
T1-6	g	1L	0.16	6L	0.84	1061D	T1-6(F-30)	Longley breakpoints (1961)
T1-6	a	1L	0.2	6L	0.54	1061E		Longley breakpoints (1961)
T1-6	b	1L	0.25	6S	0.8			In satellite; breakpoints per Phillips et al. 1977
T1-6	8415	1L	0.29	6S	0.82	1061F		Longley breakpoints (1961); 1L.31, 6S in proximal part of secondary constriction per Phillips et al. 1977
T1-6	f	1L	0.32	6L	0.42	1062A	T1-6(B-92); T1-6(84-2)	Longley breakpoints (1961)
T1-6	070-1	1L	0.4	6L	0.58	1062B		Longley breakpoints (1961)
T1-6	5225	1L	0.61	6L	0.72	1062C		Longley breakpoints (1961)
T1-6	4456	1L	0.71	6L	0.3	1062D		Longley breakpoints (1961)
T1-6	8650	1L	0.79	6L	0.91			Longley breakpoints (1961)
T1-6	8658	1L	0.79	6L	0.91	1062E		Longley breakpoints (1961)
T1-6	Li	1L	0.81	6S	0.7			Breakpoints per Phillips et al. 1977; 6S in proximal part of secondary constriction
T1-6	5648**	1		6		1062F	T2-6(5648)	Burnham breakpoints (1981)
T1-6	9002**	1		6		1062G	T2-6(9002)	Burnham breakpoints (1981)
T1-6	f**	1		6			T2-6f	Burnham breakpoints (1981)
T1-7	4742	1S	0.95	7L	0.03	1063A		Longley breakpoints (1961)
T1-7	g	1S	0.79	7S	0.22	1063B	T1-7(B-49)	Longley breakpoints (1961)
T1-7	4837	1S	0.73	7L	0.55	1063C		Longley breakpoints (1961)
T1-7	f	1S	0.72	7L	0.8	1063D	T1-7(A-69)	Longley breakpoints (1961)
T1-7	4444	1S	0.65	7L	0.5	1063E		Longley breakpoints (1961)
T1-7	4405	1S	0.43	7S	0.46	1063F		Longley breakpoints (1961)
T1-7	6796	1S	0.4	7S	0.39	1064A		Longley breakpoints (1961)
T1-7	010-12	1S	0.35	7L	0.57	1064B		Longley breakpoints (1961)
T1-7	i	1S	0.31	7L	0.26	1064C	T1-7(I-17)	Longley breakpoints (1961)
T1-7	021-7	1S	0.22	7S	0.56			Longley breakpoints (1961)
T1-7	4302-31	1S	0.15	7L	0.12	1064D		Longley breakpoints (1961)
T1-7	k	1L	0.1	7L	0.56	1064E		Longley breakpoints (1961)
T1-7	f	1S	0.1	7S	0.5			Longley breakpoints (1961)
T1-7	A-37	1L	0.1	7L	0.56			Longley breakpoints (1961)
T1-7	5871	1L	0.12	7L	0.24	1064F		Longley breakpoints (1961)
T1-7	48-83-15	1L	0.12	7L	0.25			Longley breakpoints (1961)
T1-7	4891	1L	0.12	7L	0.69	1065A		Longley breakpoints (1961)
T1-7	j	1L	0.2	7L	0.61	1065B	T1-7(X-55-16)	Longley breakpoints (1961)
T1-7	5339	1L	0.24	7L	0.14	1065C		Longley breakpoints (1961)
T1-7	a	1L	0.28	7L	0.13	1065D		Longley breakpoints (1961)
T1-7	e	1L	0.39	7L	0.11	1065E	T1-7(42)	Longley breakpoints (1961)
T1-7	c	1L	0.39	7L	0.14	1065F		Longley breakpoints (1961)
T1-7	h	1L	0.46	7L	0.19	1066A	T1-7(B-94)	Longley breakpoints (1961)
T1-7	4420	1L	0.47	7L	0.9	1066B		Longley breakpoints (1961)
T1-7	b	1L	0.53	7S	0.12	1066C	T1-7b**	Longley breakpoints (1961)
T1-7	b**	1		7L			T1-7b	Burnham breakpoints (1981)
T1-7	d	1L	0.81	7S	0.44	1066D		Longley breakpoints (1961)
T1-7	017-10	1L	0.82	7L	0.12			Longley breakpoints (1961)
T1-7	012-10	1L	0.91	7L	0.9			Longley breakpoints (1961)
T1-7	5693	1L	0.92	7L	0.18	1066E		Longley breakpoints (1961)
T1-8	8919	1S	0.53	8L	0.44	1067A		Longley breakpoints (1961)

Table 54.1 (Continued)

T1-8	4307-4	1S	0.42	8L	0.61	1067B		Longley breakpoints (1961)	
T1-8	001-13	1S	0.39	8L	0.67	1067C		Longley breakpoints (1961)	
T1-8	4685	1S	0.2	8L	0.21	1067D		Longley breakpoints (1961)	
T1-8	6591	1S	0.18	8S	0.43	1067E		Longley breakpoints (1961)	
T1-8	008-17	1S	0.16	8L	0.2	1067F		Longley breakpoints (1961)	
T1-8	5588	1S	0.1	8S	0.32	1068A		Longley breakpoints (1961)	
T1-8	055-23	1	0	8	0	1068B		Longley breakpoints (1961)	
T1-8	064-13	1	0	8	0	1068C		Longley breakpoints (1961)	
T1-8	4676	1L	0.04	8S	0.06	1068D		Longley breakpoints (1961)	
T1-8	5619	1L	0.07	8L	0.16	1068E		Longley breakpoints (1961)	
T1-8	5634	1L	0.08	8S	0.28	1068F		Longley breakpoints (1961)	
T1-8	5384	1L	0.1	8L	0.59	1069A		Longley breakpoints (1961)	
T1-8	8683	1L	0.11	8	0	1069B		Longley breakpoints (1961)	
T1-8	8640	1L	0.11	8L	0.16	1069C		Longley breakpoints (1961)	
T1-8	020-19	1L	0.11	8L	0.38	1069D		Longley breakpoints (1961)	
T1-8	4748	1L	0.12	8L	0.15	1069E		Longley breakpoints (1961)	
T1-8	036-40	1L	0.18	8L	0.59	1069F		Longley breakpoints (1961)	
T1-8	005-7	1L	0.22	8L	0.78	1070A		Longley breakpoints (1961)	
T1-8	7509	1L	0.28	8L	0.21	1070B		Longley breakpoints (1961)	
T1-8	5752	1L	0.36	8L	0.25	1070C		Longley breakpoints (1961)	
T1-8	a	1L	0.41	8S	0.52	1070D	T1-8(Conn. R-20)	Longley breakpoints (1961)	
T1-8	026-2	1L	0.49	8L	0.8	1070E		Longley breakpoints (1961)	
T1-8	6766	1L	0.54	8L	0.77	1070F		Longley breakpoints (1961)	
T1-8	b	1L	0.59	8L	0.82	1071A	T1-8(B-42)	Longley breakpoints (1961)	
T1-8	5821	1L	0.65	8L	0.31	1071B		Longley breakpoints (1961)	
T1-8	6697	1L	0.89	8L	0.52	1071C		Longley breakpoints (1961)	
T1-8	5910	1L	0.93	8L	0.67	1071D		Longley breakpoints (1961)	
T1-8	c	1L	0.94	8L	0.89		T1-8(B-49)	Longley breakpoints (1961)	
T1-8	5704	1L	0.96	8S	0.67	1071E		Longley breakpoints (1961)	
T1-8	055-4	1		8		1071F		Longley breakpoints (1961)	
T1-9	024-7	1S	0.71	9L	0.13	1072A		Longley breakpoints (1961)	
T1-9	8265	1S	0.58	9L	0.27			Longley breakpoints (1961)	
T1-9	8302	1S	0.55	9L	0.29	1072B		Longley breakpoints (1961)	
T1-9	8129	1S	0.53	9L	0.27			Longley breakpoints (1961)	
T1-9	48-34-2	1S	0.51	9L	0.24			Longley breakpoints (1961)	
T1-9	8001	1S	0.51	9L	0.24	1072C		Longley breakpoints (1961)	
T1-9	c	1S	0.48	9L	0.22			Longley breakpoints (1961)	
T1-9	7535	1S	0.33	9S	0.27	1072D		Longley breakpoints (1961)	
T1-9	8918	1S	0.21	9L	0.2	1072E		Longley breakpoints (1961)	
T1-9	6762	1S	0.16	9L	0.53	1072F		Longley breakpoints (1961)	
T1-9	a	1S	0.13	9L	0.15	1073A		Longley breakpoints (1961)	
T1-9	8460	1S	0.13	9L	0.24	1073B		Longley breakpoints (1961)	
T1-9	8824	1S	0.06	9L	0.83			Longley breakpoints (1961)	
T1-9	5622	1L	0.1	9L	0.12	1073C		Longley breakpoints (1961)	
T1-9	4995	1L	0.19	9S	0.2			Longley breakpoints (1961)	
T1-9	9021	1L	0.26	9L	0.83			Longley breakpoints (1961)	
T1-9	8886	1L	0.33	9L	0.23	1073D		Longley breakpoints (1961)	
T1-9	8889	1L	0.33	9L	0.23			Longley breakpoints (1961)	
T1-9	4997	1L	0.37	9S	0.28	1073E		Longley breakpoints (1961)	
T1-9	d	1L	0.42	9L	0.25	1073F	T1-9(I-9)	Longley breakpoints (1961)	
T1-9	b	1L	0.5	9L	0.6	1074A		Longley breakpoints (1961)	
T1-9	4398	1L	0.51	9S	0.19	1074B		Longley breakpoints (1961)	
T1-9	8389	1L	0.74	9L	0.13			Longley breakpoints (1961)	
T1-9	6328	1L	0.79	9L	0.4			Longley breakpoints (1961)	
T1-9	035-10	1L	0.89	9S	0.67	1074C		Longley breakpoints (1961)	
T1-10	g	1S	0.8	10L	0.21	1074D	T1-10(C-47)	Longley breakpoints (1961)	
T1-10	007-19	1S	0.07	10L	0.08	1074E		Longley breakpoints (1961)	
T1-10	f	1S	0.04	10L	0.3	1075A	T1-10(C-36)	Longley breakpoints (1961)	
T1-10	4885	1	0	10	0	1075B		Longley breakpoints (1961)	
T1-10	8770	1L	0.09	10L	0.38	1075C		Longley breakpoints (1961)	
T1-10	e	1L	0.16	10L	0.31	1075D	T1-10(B-98)	Longley breakpoints (1961)	
T1-10	068-14	1L	0.16	10L	0.79	1075E		Longley breakpoints (1961)	
T1-10	5273	1L	0.17	10L	0.69	1075F		Longley breakpoints (1961)	
T1-10	b	1L	0.19	10S	0.39	1076A	T1-10(Conn. R-41)	Longley breakpoints (1961)	
T1-10	a	1L	0.29	10L	0.33	1076B		Longley breakpoints (1961)	
T1-10	c	1L	0.43	10L	0.74	1076C	T1-10(A-50)	Longley breakpoints (1961)	
T1-10	8465	1L	0.45	10L	0.76			Longley breakpoints (1961)	
T1-10	8491	1L	0.45	10L	0.76	1076D		Longley breakpoints (1961)	
T1-10	d	1L	0.5	10L	0.68	1076E	T1-10(A-84)	Longley breakpoints (1961)	
T1-10	015-9	1L	0.67	10S	0.46	1076F		Longley breakpoints (1961)	
T1-10	8375	1L	0.69	10L	0.64	1077A		Longley breakpoints (1961)	
T1-10	001-3	1L	0.86	10L	0.48	1077B		Longley breakpoints (1961)	
T1-10	13-29	1L	0.93	10L	0.26			Longley breakpoints (1961)	
T2-3	a	2S	0.9	3L	0.6				
T2-3	023-5	2S	0.8	3L	0.7	1077C		Longley breakpoints (1961)	
T2-3	8662	2S	0.78	3L	0.83	1077D		Longley breakpoints (1961)	
T2-3	e	2S	0.76	3L	0.48	1077E		Longley breakpoints (1961)	
T2-3	5800	2S	0.73	3S	0.81	1077F		Longley breakpoints (1961)	
T2-3	5304	2S	0.62	3L	0.29	1078A		Longley breakpoints (1961)	
T2-3	064-2	2		3		1078B		Longley breakpoints (1961)	
T2-3	c	2S	0.46	3S	0.52	1078C		Longley breakpoints (1961)	
T2-3	6270	2S	0.46	3L	0.6	1078D		Longley breakpoints (1961)	
T2-3	014-12	2S	0.43	3L	0.51	1078E		Longley breakpoints (1961)	
T2-3	6862	2S	0.39	3L	0.2	1078F		Longley breakpoints (1961)	
T2-3	4369	2S	0.19	3S	0.26	1079A		Longley breakpoints (1961)	
T2-3	010-10	2S	0.17	3L	0.13	1079B		Longley breakpoints (1961)	
T2-3	8857	2S	0.08	3L	0.24			Longley breakpoints (1961)	
T2-3	4301-111	2	0	3	0	1079C		Longley breakpoints (1961)	
T2-3	023-2	2	0	3	0	1079D		Longley breakpoints (1961)	
T2-3	055-7	2L	0.1	3S	0.31	1079E		Longley breakpoints (1961)	
T2-3	005-14	2L	0.12	3S	0.29	1079F		Longley breakpoints (1961)	
T2-3	h	2L	0.14	3L	0.07	1080A	T2-3(K-7)	Longley breakpoints (1961)	
T2-3	8483	2L	0.14	3L	0.12	1080B		Longley breakpoints (1961)	
T2-3	i	2L	0.19	3S	0.51	1080C		Longley breakpoints (1961)	
T2-3	g	2L	0.21	3S	0.21	1080D	T2-3(F-35)	Longley breakpoints (1961)	
T2-3	7285	2L	0.26	3L	0.39	1080E		Longley breakpoints (1961)	
T2-3	8553	2L	0.26	3L	0.23			Longley breakpoints (1961)	
T2-3	8556	2L	0.26	3L	0.23			Longley breakpoints (1961)	
T2-3	033-4	2L	0.27	3L	0.23	1080F		Longley breakpoints (1961)	
T2-3	f	2L	0.35	3S	0.6	1081A	T2-3(A-61)	Longley breakpoints (1961)	
T2-3	b	2L	0.45	3L	0.08	1081B		Longley breakpoints (1961)	
T2-3	8628	2L	0.54	3L	0.44			Longley breakpoints (1961)	
T2-3	d	2L	0.67	3L	0.48	1081C		Longley breakpoints (1961)	

Table 54.1 (Continued)

T2-3	4303-74	2L	0.73	3L	0.68	1081D		Longley breakpoints (1961)
T2-3	6750	2L	0.76	3S	0.53	1081E		Longley breakpoints (1961)
T2-3	011-6	2L	0.77	3L	0.48			Longley breakpoints (1961)
T2-3	(3-6a)	2L	0.79	3S	0.9			Longley breakpoints (1961)
T2-3	6284	2L	0.81	3L	0.75	1081F		Longley breakpoints (1961)
T2-4	5157	2S	0.86	4L	0.07	1082A		Longley breakpoints (1961)
T2-4	4413	2S	0.84	4S	0.32			Longley breakpoints (1961)
T2-4	8682	2S	0.7	4S	0.25			Longley breakpoints (1961)
T2-4	6994	2S	0.59	4L	0.95			Longley breakpoints (1961)
T2-4	8865	2S	0.52	4L	0.27	1082B		Longley breakpoints (1961)
T2-4	060-8	2S	0.5	4L	0.37	1082C		Longley breakpoints (1961)
T2-4	018-3	2S	0.38	4L	0.47	1082D		Longley breakpoints (1961)
T2-4	057-21	2S	0.28	4L	0.83			Longley breakpoints (1961)
T2-4	5495	2S	0.27	4L	0.1	1082E		Longley breakpoints (1961)
T2-4	j	2	0	4	0	1082F	T2-4(K-10)	Longley breakpoints (1961)
T2-4	8407	2	0	4	0	1083A		Longley breakpoints (1961)
T2-4	052-13	2	0	4	0			Longley breakpoints (1961)
T2-4	g	2L	0.13	4S	0.31	1083B	T2-4(C-31)	Longley breakpoints (1961)
T2-4	010-4	2L	0.13	4S	0.31	1083C		Longley breakpoints (1961)
T2-4	k	2L	0.13	4L	0.04	1083D	T2-4(X-1-1)	Longley breakpoints (1961)
T2-4	004-13	2L	0.14	4S	0.51	1083E		Longley breakpoints (1961)
T2-4	4374	2L	0.15	4L	0.23	1083F		Longley breakpoints (1961)
T2-4	8027	2L	0.15	4L	0.43	1084A		Longley breakpoints (1961)
T2-4	d	2L	0.17	4L	0.45	1084B		Longley breakpoints (1961)
T2-4	5951	2L	0.18	4S	0.26	1084C		Longley breakpoints (1961)
T2-4	a	2L	0.3	4L	0.21	1084D		Longley breakpoints (1961)
T2-4	e	2L	0.31	4S	0.47	1084E	T2-4(Conn. R-42)	Longley breakpoints (1961)
T2-4	m	2L	0.34	4S	0.47	1084F	T2-4(X-47-41)	Longley breakpoints (1961)
T2-4	057-19	2L	0.35	4S	0.51	1085A		Longley breakpoints (1961)
T2-4	017-18	2L	0.39	4L	0.19	1085B		Longley breakpoints (1961)
T2-4	6266	2L	0.4	4L	0.27	1085C		Longley breakpoints (1961)
T2-4	011-7	2L	0.53	4L	0.76	1085D		Longley breakpoints (1961)
T2-4	l	2L	0.59	4S	0.4	1085E	T2-4(X-2-64)	Longley breakpoints (1961)
T2-4	f	2L	0.75	4L	0.12	1085F	T2-4(A-29)	Longley breakpoints (1961)
T2-4	052-15	2L	0.75	4L	0.66	1086A		Longley breakpoints (1961)
T2-4	016-19	2L	0.81	4S	0.79			Longley breakpoints (1961)
T2-4	014-11	2L	0.81	4S	0.16			Longley breakpoints (1961)
T2-4	c	2L	0.81	4S	0.09	1086B		Longley breakpoints (1961)
T2-4	b	2L	0.81	4L	0.53	1086C		Longley breakpoints (1961)
T2-5	g	2S	0.79	5S	0.24	1086D	T2-5(X-14-122)	Longley breakpoints (1961)
T2-5	059-17	2S	0.73	5S	0.61	1086E		Longley breakpoints (1961)
T2-5	019-1	2S	0.67	5S	0.51	1087A		Longley breakpoints (1961)
T2-5	062-16	2S	0.56	5L	0.04			Longley breakpoints (1961)
T2-5	4741	2S	0.47	5L	0.47	1087B		Longley breakpoints (1961)
T2-5	009-19	2S	0.29	5L	0.11	1087C		Longley breakpoints (1961)
T2-5	025-4	2S	0.26	5L	0.78	1087D		Longley breakpoints (1961)
T2-5	e	2S	0.19	5S	0.28	1087E	T2-5(B-69)	Longley breakpoints (1961)
T2-5	010-17	2S	0.19	5L	0.21			Longley breakpoints (1961)
T2-5	059-1	2	0	5S	0.84	1087F		Longley breakpoints (1961)
T2-5	b	2L	0.06	5S	0.09	1088A		Longley breakpoints (1961)
T2-5	6580	2L	0.09	5S	0.09	1088B		Longley breakpoints (1961)
T2-5	5098	2L	0.13	5S	0.23	1088C		Longley breakpoints (1961)
T2-5	a	2L	0.14	5L	0.15	1088D		Longley breakpoints (1961)
T2-5	c	2L	0.16	5S	0.48	1088E	T2-5(Conn. R-50)	Longley breakpoints (1961)
T2-5	015-3	2L	0.16	5S	0.69	1088F		Longley breakpoints (1961)
T2-5	002-16	2L	0.25	5L	0.35	1089A		Longley breakpoints (1961)
T2-5	023-15	2L	0.28	5S	0.3	1089B		Longley breakpoints (1961)
T2-5	A-16	2L	0.34	5S	0.2			Longley breakpoints (1961)
T2-5	032-9	2L	0.4	5S	0.31	1089C		Longley breakpoints (1961)
T2-5	062-3	2L	0.45	5S	0.34	1089D		Longley breakpoints (1961)
T2-5	5876	2L	0.47	5L	0.46	1089E		Longley breakpoints (1961)
T2-5	5645	2L	0.6	5S	0.85	1089F		Longley breakpoints (1961)
T2-5	6885	2L	0.63	5S	0.79	1090A		Longley breakpoints (1961)
T2-5	5602	2L	0.73	5L	0.77	1090B		Longley breakpoints (1961)
T2-5	8321	2L	0.86	5L	0.11	1090C		Longley breakpoints (1961)
T2-5	f	2L	0.91	5L	0.1	1090D	T2-5(K-3)	Longley breakpoints (1961)
T2-5	d	2L	0.91	5L	0.86	1090E	T2-5(A-74)	Longley breakpoints (1961)
T2-5	4578	2L	0.92	5S	0.71	1090F		Longley breakpoints (1961)
T2-6	4394	2S	0.91	6L	0.12		T2-6(4394**)	Longley breakpoints (1961); T4-6 per Burnham
T2-6	8786	2S	0.9	6S	0.77	1091B		Longley breakpoints (1961); 2S.97, 6S at .88 in NOR-heterochromatin per Phillips et al. 1977
T2-6	001-5	2S	0.72	6S	0.87	1091C		Longley breakpoints (1961); in satellite per Phillips et al. 1977
T2-6	001-5	2S	0.72	6S	0.87	1091D		Longley breakpoints (1961); in satellite per Phillips et al. 1977
T2-6	b	2S	0.69	6L	0.49			Longley breakpoints (1961)
T2-6	Burnham6052	2S	0.6	6L	0.6			
T2-6	008-5	2S	0.4	6L	0.26			Longley breakpoints (1961)
T2-6	4372	2S	0.37	6S	0.81			Longley breakpoints (1961)
T2-6	060-5	2S	0.36	6L	0.18	1091E		Longley breakpoints (1961)
T2-6	027-4	2S	0.34	6L	0.11	1091F	T2-6(027-4**)	Longley breakpoints (1961)
T2-6	027-4**	2L	0.1	6S	0.7	1092A	T2-6(027-4)	Breakpoints per Phillips et al. 1977; 6S in proximal part of secondary constriction
T2-6	5472	2S	0.25	6L	0.15			Longley breakpoints (1961)
T2-6	5473	2S	0.25	6L	0.15			Longley breakpoints (1961)
T2-6	6671	2S	0.22	6L	0.22		T5-6(6671**)	Longley breakpoints (1961); T5-6 per Burnham
T2-6	Burnham6049	2S	0.15	6L	0.06			
T2-6	e	2L	0.18	6L	0.2	1092B	T2-6e**	Longley breakpoints (1961)
T2-6	e**	2		6S			T2-6e	Burnham breakpoints (1981)
T2-6	6931	2L	0.24	6L	0.23	1092C		Longley breakpoints (1961)
T2-6	g	2L	0.24	6L	0.29	1092D		Longley breakpoints (1961)
T2-6	"78"	2L	0.24	6L	0.29			Longley breakpoints (1961)
T2-6	5648	2L	0.25	6L	0.19		T1-6(5648**)	Longley breakpoints (1961); T1-6 per Burnham
T2-6	a	2L	0.28	6L	0.2	1093A	T2-6a**	Longley breakpoints (1961)
T2-6	a**	2S	0.4	6S	0.5		T2-6a	Burnham breakpoints (1981)
T2-6	c	2L	0.37	6L	0.25	1093B	T2-6c**	Longley breakpoints (1961)
T2-6	c**	2S		6			T2-6c	Burnham breakpoints (1981)
T2-6	d	2L	0.41	6L	0.45	1093C		Longley breakpoints (1961)
T2-6	4717	2L	0.27	6L	0.27	1093D		Longley breakpoints (1961)
T2-6	9002	2L	0.57	6L	0.5		T1-6(9002**)	Longley breakpoints (1961); T1-6 per Burnham
T2-6	9009	2L	0.57	6L	0.5			Longley breakpoints (1961)
T2-6	5419	2L	0.82	6S	0.79	1094A		Longley breakpoints (1961); 6S at .25 in secondary constriction per Phillips et al. 1977
T2-6	f	2L	0.79	6L	0.87		T1-6(f**); T2-6(84-2)	Longley breakpoints (1961); T1-6 per Burnham
T2-6	014-11	2L	0.81	6L	0.2			Longley breakpoints (1961); T1-6 per Burnham
T2-6	8441	2L	0.94	6S	0.79	1094B	T2-6(8414)	Longley breakpoints (1961); 2L.95, 6S in proximal part of secondary constriction per Phillips et al. 1977
T2-7	5279	2S	0.93	7L	0.25	1094C		Longley breakpoints (1961)
T2-7	5144	2S	0.35	7L	0.08	1094D		Longley breakpoints (1961)

Table 54.1 (Continued)

T2-7	022-4	2S	0.3	7L	0.24	1094E		Longley breakpoints (1961)
T2-7	8045	2S	0.12	7L	0.06	1094F		Longley breakpoints (1961)
T2-7	d	2L	0.16	7L	0.18	1095A	T2-7(B-108)	Longley breakpoints (1961)
T2-7	48-74-1	2L	0.22	7S	0.47			Longley breakpoints (1961)
T2-7	4400	2L	0.24	7L	0.32	1095B		Longley breakpoints (1961)
T2-7	8465	2L	0.27	7S	0.56			Longley breakpoints (1961)
T2-7	f	2L	0.3	7L	0.68	1095C	T2-7(F-29)	Longley breakpoints (1961)
T2-7	b	2L	0.37	7L	0.12	1095D		Longley breakpoints (1961)
T2-7	c	2L	0.47	7S	0.34	1095E		Longley breakpoints (1961)
T2-7	4519	2L	0.65	7L	0.66	1095F		Longley breakpoints (1961)
T2-7	5783	2L	0.66	7L	0.1	1096A		Longley breakpoints (1961)
T2-7	038-12	2L	0.75	7S	0.68	1096B		Longley breakpoints (1961)
T2-7	8322	2L	0.76	7L	0.74	1096C		Longley breakpoints (1961)
T2-7	e	2L	0.82	7L	0.63	1096E	T2-7(C-44)	Longley breakpoints (1961)
T2-7	3692-1	2S	0.1	7L	0.02	1097A		Longley breakpoints (1961)
T2-8	013-17	2S	0.89	8L	0.61	1097B		Longley breakpoints (1961)
T2-8	4711	2S	0.86	8L	0.67	1097C		Longley breakpoints (1961)
T2-8	011-20	2S	0.58	8L	0.28	1097D		Longley breakpoints (1961)
T2-8	8458	2S	0.22	8L	0.32			Longley breakpoints (1961)
T2-8	c	2S	0.15	8S	0.11	1097E	T2-8(A-36)	Longley breakpoints (1961)
T2-8	f	2	0	8	0	1097F	T2-8(C-57)	Longley breakpoints (1961)
T2-8	006-10	2	0	8	0	1097G		Longley breakpoints (1961)
T2-8	d	2L	0.05	8L	0.1		T2-8(C-24)	Longley breakpoints (1961)
T2-8	e	2L	0.07	8L	0.1		T2-8(C-40)	Longley breakpoints (1961)
T2-8	4414	2L	0.12	8L	0.14			Longley breakpoints (1961)
T2-8	7069	2L	0.13	8L	0.14			Longley breakpoints (1961)
T2-8	8428	2L	0.16	8L	0.1			Longley breakpoints (1961)
T2-8	003-5	2L	0.19	8S	0.72			Longley breakpoints (1961)
T2-8	b	2L	0.2	8L	0.18		T2-8(A-1)	Longley breakpoints (1961)
T2-8	5454	2L	0.21	8S	0.39	1099A		Longley breakpoints (1961)
T2-8	h	2L	0.23	8L	0.22		T2-8(X-42-32)	Longley breakpoints (1961)
T2-8	5484	2L	0.24	8S	0.58			Longley breakpoints (1961)
T2-8	031-7	2L	0.3	8S	0.44	1099D		Longley breakpoints (1961)
T2-8	i	2L	0.32	8L	0.3	1099E	T2-8(84)	Longley breakpoints (1961)
T2-8	051-15	2L	0.62	8L	0.48	1099F		Longley breakpoints (1961)
T2-8	062-15	2L	0.7	8L	0.26			Longley breakpoints (1961)
T2-8	g	2L	0.71	8S	0.71	2000B	T2-8(G-2)	Longley breakpoints (1961)
T2-8	051-7	2L	0.83	8L	0.74	2000C		Longley breakpoints (1961)
T2-8	48-45-6	2L	0.84	8L	0.68			Longley breakpoints (1961)
T2-8	8376	2L	0.95	8L	0.03			Longley breakpoints (1961)
T2-8	037-5	2L	0.95	8L	0.54	2000F		Longley breakpoints (1961)
T2-9	7096	2S	0.57	9L	0.66	2001A		Longley breakpoints (1961)
T2-9	c	2S	0.49	9S	0.33		T2-9(C-61)	Longley breakpoints (1961)
T2-9	a	2S	0.36	9L	0.58	2001B		Longley breakpoints (1961)
T2-9	055-14	2S	0.28	9L	0.27	2001C		Longley breakpoints (1961)
T2-9	5711	2S	0.24	9L	0.23	2001D		Longley breakpoints (1961)
T2-9	b	2S	0.18	9L	0.22			Longley breakpoints (1961)
T2-9	062-11	2L	0.21	9S	0.53	2001E		Longley breakpoints (1961)
T2-9	5257	2L	0.28	9L	0.2	2001F		Longley breakpoints (1961)
T2-9	6656	2L	0.32	9S	0.31	2002A		Longley breakpoints (1961)
T2-9	5208	2L	0.76	9L	0.68			Longley breakpoints (1961)
T2-9	d	2L	0.83	9L	0.27		T2-9(H-7)	Longley breakpoints (1961)
T2-10	043-10	2S	0.89	10L	0.4			Longley breakpoints (1961)
T2-10	5651	2S	0.71	10L	0.62			Longley breakpoints (1961)
T2-10	b	2S	0.5	10L	0.75			Longley breakpoints (1961)
T2-10	8864	2S	0.1	10L	0.76	2003		Longley breakpoints (1961)
T2-10	4484	2S	0.09	10L	0.14			Longley breakpoints (1961)
T2-10	5830	2L	0.12	10L	0.12			Longley breakpoints (1961)
T2-10	a	2L	0.16	10L	0.55			Longley breakpoints (1961)
T2-10	c	2L	0.3	10S	0.4	2004B		Longley breakpoints (1961)
T2-10	5561	2L	0.35	10S	0.16			Longley breakpoints (1961)
T2-10	011-9	2L	0.39	10L	0.76			
T2-10	8219	2L	0.5	10L	0.35			Longley breakpoints (1961)
T2-10	6853	2L	0.79	10L	0.86			Longley breakpoints (1961)
T2-10	035-2	2L	0.85	10L	0.49			Longley breakpoints (1961)
T2-10	6061**	2		10			T5-10(6061)	Burnham breakpoints (1981)
T3-4	8969	3S	0.75	4L	0.75	2005A		Longley breakpoints (1961)
T3-4	8397	3S	0.74	4S	0.55	2005B		Longley breakpoints (1961)
T3-4	8634	3S	0.71	4L	0.75			Longley breakpoints (1961)
T3-4	5156	3S	0.47	4L	0.67	2005D		Longley breakpoints (1961)
T3-4	5920	3S	0.28	4L	0.73			Longley breakpoints (1961)
T3-4	012-16	3S	0.27	4S	0.3			Longley breakpoints (1961)
T3-4	4662	3S	0.24	4S	0.67			Longley breakpoints (1961)
T3-4	4726	3S	0.16	4L	0.15	2006B		Longley breakpoints (1961)
T3-4	5891	3	0	4	0			Longley breakpoints (1961)
T3-4	5074-6	3	0	4	0	2007C		Longley breakpoints (1961)
T3-4	a	3L	0.07	4L	0.85			Longley breakpoints (1961)
T3-4	006-17	3L	0.1	4S	0.45	2006E		Longley breakpoints (1961)
T3-4	037-9	3L	0.1	4L	0.14			Longley breakpoints (1961)
T3-4	8443	3L	0.12	4L	0.13			Longley breakpoints (1961)
T3-4	4713	3L	0.21	4S	0.59			Longley breakpoints (1961)
T3-4	6534	3L	0.48	4L	0.89	2007B		Longley breakpoints (1961)
T3-5	4635	3S	0.44	5S	0.48			Longley breakpoints (1961)
T3-5	e	3S	0.34	5S	0.16	2010D	T3-5(A-101)	Longley breakpoints (1961)
T3-5	6473	3S	0.32	5S	0.26	2008A		Longley breakpoints (1961)
T3-5	6462	3S	0.31	5L	0.47	2008B		Longley breakpoints (1961)
T3-5	4873	3S	0.24	5S	0.16			Longley breakpoints (1961)
T3-5	4880	3	0	5	0	2008C		Longley breakpoints (1961)
T3-5	4898	3	0	5	0			Longley breakpoints (1961)
T3-5	6695	3	0	5	0	2008E		Longley breakpoints (1961)
T3-5	g	3L	0.01	5S	0.73	2008F	T3-5(X-4-108)	Longley breakpoints (1961)
T3-5	8104	3L	0.05	5L	0.08	2009A		Longley breakpoints (1961)
T3-5	8528	3L	0.06	5L	0.72			Longley breakpoints (1961)
T3-5	039-13	3L	0.13	5L	0.14			Longley breakpoints (1961)
T3-5	5874	3L	0.16	5L	0.21			Longley breakpoints (1961)
T3-5	i	3L	0.23	5L	0.2		T3-5(B-104)	Longley breakpoints (1961)
T3-5	5521	3L	0.17	5L	0.48			Longley breakpoints (1961)
T3-5	015-18	3L	0.18	5S	0.21			Longley breakpoints (1961)
T3-5	B-104	3L	0.23	5L	0.2			Longley breakpoints (1961)
T3-5	a	3L	0.28	5L	0.6			Longley breakpoints (1961)
T3-5	h	3L	0.55	5L	0.22	2010B	T3-5(X-7-38)	Longley breakpoints (1961)
T3-5	b	3L	0.61	5L	0.57	2010C		Longley breakpoints (1961)
T3-5	c	3L	0.62	5L	0.27			Longley breakpoints (1961)

Table 54.1 (Continued)

T3-5	7043	3L	0.63	5L	0.61	2010E		Longley breakpoints (1961)
T3-5	8351	3L	0.75	5L	0.68			Longley breakpoints (1961)
T3-5	6346	3L	0.94	5L	0.83	2011A		Longley breakpoints (1961)
T3-6	b	3S	0.73	6S	0.82	2011B		Longley breakpoints (1961); in satellite per Phillips et al. 1977
T3-6	060-4	3S	0.62	6L	0.08	2011C		Longley breakpoints (1961)
T3-6	4349	3S	0.58	6L	0.7	2011D		Longley breakpoints (1961)
T3-6	c	3S	0.56	6L	0.54		T3-6(Conn. R-34)	Longley breakpoints (1961)
T3-6	016-17	3S	0.48	6L	0.3	2012A		Longley breakpoints (1961)
T3-6	032-3	3S	0.41	6S	0.78	2012B		Longley breakpoints (1961); 3S.34, 6S midway in secondary constriction per Phillips et al. 1977
T3-6	030-8	3S	0.27	6S	0.81	2012C		Longley breakpoints (1961); 3S.05, 6S at .25 in secondary constriction per Phillips et al. 1977
T3-6	8963	3S	0.23	6L	0.14	2012D		Longley breakpoints (1961)
T3-6	a	3L	0.06	6L	0.3	2012E		Longley breakpoints (1961)
T3-6	7067	3L	0.07	6L	0.75	2012F		Longley breakpoints (1961)
T3-6	6349	3L	0.1	6L	0.15	2013A		Longley breakpoints (1961)
T3-6	055-5	3L	0.16	6L	0.32			Longley breakpoints (1961)
T3-6	8145	3L	0.17	6L	0.26	2013C		Longley breakpoints (1961)
T3-6	5368	3L	0.22	6L	0.2			Longley breakpoints (1961)
T3-6	d	3L	0.23	6L	0.82	2013E	T3-6(A-53)	Longley breakpoints (1961)
T3-6	5201	3L	0.26	6L	0.21			Longley breakpoints (1961)
T3-6	003-2	3L	0.35	6L	0.15			Longley breakpoints (1961)
T3-6	6566	3L	0.41	6L	0.35			Longley breakpoints (1961)
T3-6	8672	3L	0.47	6L	0.87	2014A		Longley breakpoints (1961)
T3-6	7162	3L	0.52	6L	0.53			Longley breakpoints (1961)
T3-6	054-12	3L	0.72	6L	0.75	2014C		Longley breakpoints (1961)
T3-6	6266	3L	0.85	6S	0.8			Longley breakpoints (1961)
T3-7	b	3S	0.92	7L	0.03	2015A		Longley breakpoints (1961)
T3-7	001-15	3S	0.38	7L	0.3	2015B		Longley breakpoints (1961)
T3-7	004-7	3S	0.38	7L	0.26	2015C		Longley breakpoints (1961)
T3-7	a	3S	0.25	7L	0.18	2015D		Longley breakpoints (1961)
T3-7	4670	3S	0.2	7L	0.76	2015E		Longley breakpoints (1961)
T3-7	4773	3S	0.11	7L	0.07	2015F		Longley breakpoints (1961)
T3-7	5724	3	0	7	0	2016A		Longley breakpoints (1961)
T3-7	8541	3	0	7	0			Longley breakpoints (1961)
T3-7	6557	3L	0.06	7S	0.59			Longley breakpoints (1961)
T3-7	5955	3L	0.1	7L	0.58	2016B		Longley breakpoints (1961)
T3-7	029-3	3L	0.11	7L	0.13	2016C		Longley breakpoints (1961)
T3-7	5378	3L	0.13	7L	0.73	2016D		Longley breakpoints (1961)
T3-7	e	3L	0.25	7S	0.56	2016E	T3-7(F-25)	Longley breakpoints (1961)
T3-7	6466	3L	0.36	7L	0.14	2016F		Longley breakpoints (1961)
T3-7	c	3L	0.46	7L	0.45	2017A		Longley breakpoints (1961)
T3-7	5471	3L	0.64	7L	0.58	2017B		Longley breakpoints (1961)
T3-7	d	3L	0.64	7L	0.81	2017C	T3-7(C-75)	Longley breakpoints (1961)
T3-7	8006	3L	0.88	7L	0.9	2017D		Longley breakpoints (1961)
T3-8	024-11	3S	0.65	8L	0.49	2018A		Longley breakpoints (1961)
T3-8	6373	3S	0.53	8L	0.68	2018B		Longley breakpoints (1961)
T3-8	e	3S	0.36	8L	0.21	2018C	T3-8(A-22)	Longley breakpoints (1961)
T3-8	4626	3S	0.3	8L	0.31	2018D		Longley breakpoints (1961)
T3-8	6439	3S	0.3	8L	0.15	2018E		Longley breakpoints (1961)
T3-8	8666	3S	0.3	8L	0.14	2018F		Longley breakpoints (1961)
T3-8	8667	3S	0.3	8L	0.14	2019A		Longley breakpoints (1961)
T3-8	8670	3S	0.3	8L	0.14	2019B		Longley breakpoints (1961)
T3-8	8367	3S	0.28	8S	0.52	2019C		Longley breakpoints (1961)
T3-8	5558	3S	0.26	8S	0.74	2019D		Longley breakpoints (1961)
T3-8	c	3S	0.23	8L	0.85	2019E	T3-8(Burnham)	Longley breakpoints (1961)
T3-8	015-17	3S	0.12	8S	0.13			Longley breakpoints (1961)
T3-8	5830	3S	0.06	8L	0.13			Longley breakpoints (1961)
T3-8	4303-12	3	0	8	0	2019F		Longley breakpoints (1961)
T3-8	4872	3	0	8	0	2020A		Longley breakpoints (1961)
T3-8	043-14	3L	0.02	8S	0.4	2020B		Longley breakpoints (1961)
T3-8	5583	3L	0.06	8L	0.05			Longley breakpoints (1961)
T3-8	7362	3L	0.07	8L	0.69	2020C		Longley breakpoints (1961)
T3-8	f	3L	0.08	8L	0.1	2020D	T3-8(A-104)	Longley breakpoints (1961)
T3-8	g	3L	0.12	8L	0.19	2020E	T3-8(B-37)	Longley breakpoints (1961)
T3-8	007-8	3L	0.14	8L	0.11			Longley breakpoints (1961)
T3-8	b	3L	0.16	8L	0.23	2020F		Longley breakpoints (1961)
T3-8	8023	3L	0.18	8L	0.16	2021A		Longley breakpoints (1961)
T3-8	4874	3L	0.28	8L	0.32	2021B		Longley breakpoints (1961)
T3-8	012-10	3L	0.39	8L	0.23			Longley breakpoints (1961)
T3-8	a	3L	0.41	8L	0.61	2021C		Longley breakpoints (1961)
T3-8	6261	3L	0.49	8L	0.4	2021D		Longley breakpoints (1961)
T3-8	h	3L	0.53	8S	0.46	2021E	T3-8(X-23-26)	Longley breakpoints (1961)
T3-8	B-31	3L	0.63	8S	0.06			Longley breakpoints (1961)
T3-8	8350	3L	0.75	8S	0.6	2021F		Longley breakpoints (1961)
T3-8	4340	3L	0.88	8L	0.72	2022A		Longley breakpoints (1961)
T3-8	4301-39	3L	0.92	8L	0.82	2022B		Longley breakpoints (1961)
T3-8	3687	3L	0.25	8L	0.88	2021G		Longley breakpoints (1961)
T3-9	054-18	3S	0.88	9L	0.82	2022C		Longley breakpoints (1961)
T3-9	6722	3S	0.66	9S	0.66	2022D		Longley breakpoints (1961)
T3-9	7041	3S	0.59	9L	0.7	2022E		Longley breakpoints (1961)
T3-9	5643	3S	0.55	9L	0.64	2022F		Longley breakpoints (1961)
T3-9	8447	3S	0.44	9L	0.14			Longley breakpoints (1961)
T3-9	030-2	3S	0.39	9L	0.3	2023A		Longley breakpoints (1961)
T3-9	8465	3S	0.27	9L	0.41	2023B		Longley breakpoints (1961)
T3-9	8032	3S	0.26	9L	0.96	2023C		Longley breakpoints (1961)
T3-9	020-5	3	0	9	0	2023D		Longley breakpoints (1961)
T3-9	e	3L	0.02	9L	0.29	2023E	T3-9(A-94)	Longley breakpoints (1961)
T3-9	5775	3L	0.09	9S	0.24	2023F		Longley breakpoints (1961)
T3-9	c	3L	0.09	9L	0.12		T3-9c**	Longley breakpoints (1961)
T3-9	c**	3S	0.15	9S	0.2		T3-9c	Burnham breakpoints (1981)
T3-9	h	3L	0.09	9L	0.33	2024A	T3-9(X-23-158)	Longley breakpoints (1961)
T3-9	a	3L	0.11	9L	0.16	2024B		Longley breakpoints (1961)
T3-9	d	3L	0.13	9L	0.26	2024C	T3-9(A-41)	Longley breakpoints (1961)
T3-9	064-11	3L	0.46	9S	0.36	2024E		Longley breakpoints (1961)
T3-9	g	3L	0.4	9L	0.14	2024D	T3-9(F-24)	Longley breakpoints (1961)
T3-9	034-11	3L	0.46	9S	0.36			Longley breakpoints (1961)
T3-9	b	3L	0.48	9L	0.53	2024F		Longley breakpoints (1961)
T3-9	5285	3L	0.51	9L	0.49	2025A		Longley breakpoints (1961)
T3-9	4727	3L	0.54	9L	0.42	2025B		Longley breakpoints (1961)
T3-9	f	3L	0.63	9S	0.69	2025C	T3-9(B-103)	Longley breakpoints (1961)
T3-9	8562	3L	0.65	9L	0.22			Longley breakpoints (1961)
T3-9	8568	3L	0.65	9L	0.22			Longley breakpoints (1961)
T3-9	8572	3L	0.65	9L	0.22			Longley breakpoints (1961)
T3-9	8574	3L	0.65	9L	0.22			Longley breakpoints (1961)

Table 54.1 (Continued)

T3-9	4963	3L	0.76	9L	0.57	2025D		Longley breakpoints (1961)
T3-10	7464	3S	0.49	10L	0.6			Longley breakpoints (1961)
T3-10	8412	3S	0.39	10S	0.36	2026B		Longley breakpoints (1961)
T3-10	8349	3S	0.38	10	0			Longley breakpoints (1961)
T3-10	4382	3S	0.38	10L	0.29			Longley breakpoints (1961)
T3-10	063-1	3S	0.25	10L	0.71			Longley breakpoints (1961)
T3-10	4383	3S	0.23	10L	0.3			Longley breakpoints (1961)
T3-10	5892	3S	0.17	10L	0.25	2026E		Longley breakpoints (1961)
T3-10	a	3L	0.16	10L	0.22	2026F		Longley breakpoints (1961)
T3-10	b	3L	0.19	10L	0.27			Longley breakpoints (1961)
T3-10	c	3L	0.22	10L	0.3	2027B		Longley breakpoints (1961)
T3-10	6691	3L	0.3	10L	0.87			Longley breakpoints (1961)
T3-10	036-15	3L	0.48	10L	0.64			Longley breakpoints (1961)
T3-10	044-10	3L	0.77	10L	0.72			Longley breakpoints (1961)
T4-5	e	4S	0.41	5L	0.32	2028A	T4-5(Conn. R-18)	Longley breakpoints (1961)
T4-5	g	4S	0.38	5L	0.3	2028B	T4-5(Conn. R-32)	Longley breakpoints (1961)
T4-5	8108	4S	0.37	5S	0.72	2028C		Longley breakpoints (1961)
T4-5	8006	4S	0.37	5S	0.18			Longley breakpoints (1961)
T4-5	5529	4S	0.37	5L	0.46	2028D		Longley breakpoints (1961)
T4-5	8069	4S	0.34	5S	0.71	2028E		Longley breakpoints (1961)
T4-5	c	4S	0.32	5L	0.27	2028F		Longley breakpoints (1961)
T4-5	6831	4S	0.32	5S	0.59			Longley breakpoints (1961)
T4-5	6560	4S	0.29	5S	0.21	2029B		Longley breakpoints (1961)
T4-5	002-12	4S	0.27	5S	0.36	2029C		Longley breakpoints (1961)
T4-5	4305-8	4S	0.25	5L	0.28	2029D		Longley breakpoints (1961)
T4-5	4472	4S	0.19	5S	0.19	2029E		Longley breakpoints (1961)
T4-5	d	4S	0.06	5L	0.22	2029F		Longley breakpoints (1961)
T4-5	k	4S	0.05	5L	0.13	2030A	T4-5(X-19-5)	Longley breakpoints (1961)
T4-5	8891	4	0	5	0			Longley breakpoints (1961)
T4-5	7078	4L	0.1	5L	0.1	2030B		Longley breakpoints (1961)
T4-5	i	4L	0.15	5S	0.15	2030C	T4-5(B-74)	Longley breakpoints (1961)
T4-5	h	4L	0.08	5L	0.08	2030D	T4-5(B-2)	Longley breakpoints (1961)
T4-5	8955	4L	0.17	5S	0.32			Longley breakpoints (1961)
T4-5	a	4L	0.29	5S	0.29	2030E		Longley breakpoints (1961)
T4-5	j	4L	0.36	5L	0.36	2030F	T4-5(X-6-77)	Longley breakpoints (1961)
T4-5	007-18	4L	0.22	5L	0.39			Longley breakpoints (1961)
T4-5	g	4L	0.27	5S	0.7			Longley breakpoints (1961)
T4-5	8622	4L	0.52	5L	0.52	2031A		Longley breakpoints (1961)
T4-5	044-8	4L	0.38	5L	0.33			Longley breakpoints (1961)
T4-5	006-7	4L	0.25	5S	0.25	2031B		Longley breakpoints (1961)
T4-5	7136	4L	0.45	5L	0.33	2031C		Longley breakpoints (1961)
T4-5	f	4L	0.5	5L	0.8	2031D		Longley breakpoints (1961)
T4-5	6743	4L	0.56	5S	0.59	2031E		Longley breakpoints (1961)
T4-5	018-4	4L	0.61	5L	0.67	2031F		Longley breakpoints (1961)
T4-5	027-10	4L	0.61	5L	0.79	2032A		Longley breakpoints (1961)
T4-5	021-3	4L	0.62	5S	0.71	2032B		Longley breakpoints (1961)
T4-5	8395	4L	0.63	5S	0.82	2032C		Longley breakpoints (1961)
T4-5	b	4L	0.76	5L	0.68	2032D		Longley breakpoints (1961)
T4-6	4461	4S	0.86	6L	0.17			Longley breakpoints (1961)
T4-6	b	4S	0.8	6L	0.16	2032F		Longley breakpoints (1961)
T4-6	e	4S	0.62	6L	0.56	2033A	T4-6(X-57-31)	Longley breakpoints (1961)
T4-6	7328	4S	0.53	6S	0.89	2033B		Longley breakpoints (1961); in satellite per Phillips et al. 1977
T4-6	8380	4S	0.47	6L	0.18	2033C		Longley breakpoints (1961)
T4-6	5227	4S	0.46	6S	0.84	2033D		Longley breakpoints (1961); in satellite per Phillips et al. 1977
T4-6	025-12	4S	0.44	6L	0.34	2033E	T4-6(025-12**)	Longley breakpoints (1961)
T4-6	025-12**	4		6S			T4-6(025-12)	Burnham breakpoints (1981)
T4-6	4341	4S	0.37	6S	0.81	2033F		Longley breakpoints (1961); 4S.36, 6S at .50 in NOR-heterochromatin per Phillips et al. 1977
T4-6	c	4S	0.33	6S	0.83	2034A		Longley breakpoints (1961); in satellite per Phillips et al. 1977
T4-6	011-16	4S	0.31	6L	0.33		T4-6(011-16**)	Longley breakpoints (1961)
T4-6	011-16**	4		6S			T4-6(011-16)	Burnham breakpoints (1981)
T4-6	4447	4S	0.28	6L	0.14	2034C		Longley breakpoints (1961)
T4-6	6623	4L	0.18	6L	0.31	2034E		Longley breakpoints (1961)
T4-6	8591	4L	0.17	6L	0.24	2034D		Longley breakpoints (1961)
T4-6	8591**	4		6S			T4-6(8591)	Burnham breakpoints (1981)
T4-6	8593	4L	0.17	6L	0.24	2034D		Longley breakpoints (1961)
T4-6	055-8	4L	0.29	6L	0.25	2034F		Longley breakpoints (1961)
T4-6	8428	4L	0.32	6L	0.28	2035A		Longley breakpoints (1961)
T4-6	8764	4L	0.32	6L	0.9	2035B		Longley breakpoints (1961)
T4-6	a	4L	0.37	6L	0.43	2035C		Longley breakpoints (1961)
T4-6	d	4L	0.49	6L	0.53	2035D	T4-6(Conn. R-43)	Longley breakpoints (1961)
T4-6	003-16	4L	0.5	6S	0.9	2035E	T4-6(033-16)	Longley breakpoints (1961); in satellite per Phillips et al. 1977
T4-6	7037	4L	0.61	6S	0.77	2035F		Longley breakpoints (1961); 6S at .90 in NOR-heterochromatin per Phillips et al. 1977
T4-6	8927	4L	0.7	6L	0.18	2036A		Longley breakpoints (1961)
T4-6	038-11	4L	0.78	6L	0.29	2036B		Longley breakpoints (1961)
T4-6	8339	4L	0.87	6L	0.79	2036C		Longley breakpoints (1961)
T4-6	4394**	4		6			T2-6(4394)	Burnham breakpoints (1981)
T4-7	8103	4S	0.81	7L	0.76	2036D		Longley breakpoints (1961)
T4-7	3686	4S	0.81	7S	0.06	2036E		Longley breakpoints (1961)
T4-7	6575	4S	0.38	7S	0.32	2036F		Longley breakpoints (1961)
T4-7	a	4S	0.32	7L	0.06	2037A		Longley breakpoints (1961)
T4-7	48-40-8	4S	0.32	7L	0.64	2037B		Longley breakpoints (1961)
T4-7	7347	4S	0.31	7L	0.66	2037C		Longley breakpoints (1961)
T4-7	7108	4S	0.17	7S	0.45	2037D		Longley breakpoints (1961)
T4-7	4698	4L	0.08	7L	0.74	2037E		Longley breakpoints (1961)
T4-7	7067	4L	0.17	7S	0.6	2037F		Longley breakpoints (1961)
T4-7	027-17	4L	0.17	7L	0.31	2038A		Longley breakpoints (1961)
T4-7	8374	4L	0.24	7L	0.55	2038B		Longley breakpoints (1961)
T4-7	052-13	4L	0.27	7L	0.25			Longley breakpoints (1961)
T4-7	4483	4L	0.39	7L	0.61	2038C		Longley breakpoints (1961)
T4-7	008-16	4L	0.86	7L	0.17			Longley breakpoints (1961)
T4-8	036-16	4S	0.66	8L	0.69	2038D		Longley breakpoints (1961)
T4-8	a	4S	0.59	8L	0.19	2038E		Longley breakpoints (1961)
T4-8	8987	4S	0.58	8L	0.76	2039A		Longley breakpoints (1961)
T4-8	8603	4S	0.42	8L	0.35			Longley breakpoints (1961)
T4-8	8606	4S	0.42	8L	0.35			Longley breakpoints (1961)
T4-8	8607	4S	0.42	8L	0.35	2039B		Longley breakpoints (1961)
T4-8	8608	4S	0.42	8L	0.35			Longley breakpoints (1961)
T4-8	4356	4S	0.4	8L	0.14			Longley breakpoints (1961)
T4-8	8004	4S	0.27	8L	0.84	2039C		Longley breakpoints (1961)
T4-8	8677	4S	0.24	8L	0.79			Longley breakpoints (1961)
T4-8	5339	4S	0.22	8L	0.71	2039D		Longley breakpoints (1961)
T4-8	8456	4S	0.22	8L	0.75	2039E		Longley breakpoints (1961)
T4-8	b	4S	0.18	8L	0.16	2039F	T4-8(X-17-108)	Longley breakpoints (1961)

Table 54.1 (Continued)

T4-8	8163	4S	0.17	8L	0.86			Longley breakpoints (1961)
T4-8	6063	4S	0.02	8L	0.05	2040A		Longley breakpoints (1961)
T4-8	004-12	4L	0.1	8L	0.1			Longley breakpoints (1961)
T4-8	016-6	4L	0.42	8S	0.45			Longley breakpoints (1961)
T4-8	5412	4L	0.59	8L	0.95			Longley breakpoints (1961)
T4-8	6926	4L	0.6	8L	0.71	2040B		Longley breakpoints (1961)
T4-8	6363	4L	0.76	8L	0.3	2040C		Longley breakpoints (1961)
T4-8	057-21	4L	0.83	8S	0.49			Longley breakpoints (1961)
T4-9	e	4S	0.53	9L	0.26		T4-9(A-52)	Longley breakpoints (1961)
T4-9	4307-12	4S	0.53	9L	0.55	2040D		Longley breakpoints (1961)
T4-9	g	4S	0.27	9L	0.27		T4-9(F-22)	Longley breakpoints (1961)
T4-9	5918	4S	0.24	9L	0.18	2040E		Longley breakpoints (1961)
T4-9	4304-82	4S	0.22	9L	0.37	2040F		Longley breakpoints (1961)
T4-9	6222	4L	0.03	9S	0.68	2041A		Longley breakpoints (1961)
T4-9	6504	4L	0.09	9S	0.83	2041B		Longley breakpoints (1961)
T4-9	d	4L	0.12	9L	0.17	2041C	T4-9(A-26)	Longley breakpoints (1961)
T4-9	a	4L	0.16	9L	0.58	2041D		Longley breakpoints (1961)
T4-9	004-7	4L	0.28	9L	0.26	2041E		Longley breakpoints (1961)
T4-9	4373	4L	0.29	9L	0.39	2041F		Longley breakpoints (1961)
T4-9	5657	4L	0.33	9S	0.25			Longley breakpoints (1961)
T4-9	5884	4L	0.4	9L	0.49	2042A		Longley breakpoints (1961)
T4-9	4333	4L	0.41	9L	0.93			Longley breakpoints (1961)
T4-9	f	4L	0.55	9L	0.18	2042B	T4-9(D-25)	Longley breakpoints (1961)
T4-9	5828	4L	0.61	9L	0.27			Longley breakpoints (1961)
T4-9	5788	4L	0.72	9L	0.82			Longley breakpoints (1961)
T4-9	5574	4L	0.8	9L	0.87	2042C		Longley breakpoints (1961)
T4-9	c	4L	0.82	9L	0.29	2042D	T4-9(bp)	Longley breakpoints (1961)
T4-9	b	4L	0.9	9L	0.29			Longley breakpoints (1961)
T4-9	8636	4L	0.94	9S	0.09	2042E		Longley breakpoints (1961)
T4-9	8649	4L	0.94	9S	0.09	2042F		Longley breakpoints (1961)
T4-10	c	4S	0.64	10L	0.18	2043A	T4-10(B-45)	Longley breakpoints (1961)
T4-10	9028	4S	0.57	10L	0.89			Longley breakpoints (1961)
T4-10	9029	4S	0.57	10L	0.89			Longley breakpoints (1961)
T4-10	8541	4S	0.45	10	0	2043C		Longley breakpoints (1961)
T4-10	d	4S	0.36	10L	0.36	2043D	T4-10(G-1)	Longley breakpoints (1961)
T4-10	6662	4L	0.04	10L	0.03	2043E		Longley breakpoints (1961)
T4-10	e	4L	0.14	10L	0.14	2043F	T4-10(K-17)	Longley breakpoints (1961)
T4-10	b	4L	0.15	10L	0.6	2044A		Longley breakpoints (1961)
T4-10	021-5	4L	0.34	10L	0.33			Longley breakpoints (1961)
T4-10	073-8	4L	0.41	10S	0.74	2044C		Longley breakpoints (1961)
T4-10	6587	4L	0.55	10L	0.51			Longley breakpoints (1961)
T4-10	057-14	4L	0.56	10S	0.48	2044E		Longley breakpoints (1961)
T4-10	024-16	4L	0.75	10L	0.18			Longley breakpoints (1961)
T4-10	f	4L	0.94	10L	0.14		T4-10(X-12-57)	Longley breakpoints (1961)
T5-6	5622	5S	0.94	6L	0.92	2045B		Longley breakpoints (1961); 5S.87, 6L.47 per Burnham
T5-6	8818	5S	0.91	6L	0.93	2045C	T5-6(8818**)	Longley breakpoints (1961)
T5-6	8818**	5L	0.91	6L	0.93		T5-6(8818)	Burnham breakpoints (1981)
T5-6	6522	5S	0.87	6L	0.7	2045D		Longley breakpoints (1961)
T5-6	6559	5S	0.72	6L	0.09	2045E		Longley breakpoints (1961)
T5-6	d	5S	0.64	6S	0.89	2046A	T5-6(A-75)	Longley breakpoints (1961); 5S.58, 6S in satellite per Phillips et al. 1977
T5-6	040-1	5S	0.48	6S	0.82	2046B		Longley breakpoints (1961)
T5-6	6671**	5S	0.49	6L	0.35	2049G	T2-6(6671)	Burnham breakpoints (1981)
T5-6	f	5S	0.37	6S	0.76	2046C	T5-6(X-23-41)	Longley breakpoints (1961); 5S.23, 6S midway in secondary constriction per Phillips et al. 1977
T5-6	8590	5S	0.29	6L	0.25	2046D	T5-6(8593)	Longley breakpoints (1961); 5S.25, 6L.61 per Burnham
T5-6	4933	5S	0.23	6L	0.89	2046E		Longley breakpoints (1961)
T5-6	6482	5S	0.22	6S	0.77			Longley breakpoints (1961)
T5-6	5765	5S	0.19	6L	0.32	2046F		Longley breakpoints (1961)
T5-6	5906	5S	0.15	6L	0.13	2047A		Longley breakpoints (1961)
T5-6	e	5L	0.11	6L	0.6	2047B	T5-6(A-77)	Longley breakpoints (1961)
T5-6	4669	5L	0.13	6L	0.4	2047C		Longley breakpoints (1961)
T5-6	6062	5L	0.2	6L	0.78	2047D		Longley breakpoints (1961)
T5-6	5685	5L	0.27	6L	0.2	2047E	T5-6(5685**)	Longley breakpoints (1961)
T5-6	5685**	5S	0.24	6L	0.23	2047E	T5-6(5685)	Burnham breakpoints (1981)
T5-6	4934	5L	0.34	6L	0.89	2047F		Longley breakpoints (1961)
T5-6	a	5L	0.35	6L	0.43	2048A		Longley breakpoints (1961)
T5-6	4666	5L	0.35	6L	0.86	2048B		Longley breakpoints (1961)
T5-6	8665	5L	0.58	6L	0.25	2048C		Longley breakpoints (1961); independent of chrom. 5 genes per Burnham
T5-6	004-17	5L	0.6	6L	0.24	2048D		Longley breakpoints (1961)
T5-6	b	5L	0.72	6L	0.21	2048E	T5-6(McCl)	Longley breakpoints (1961)
T5-6	8219	5L	0.76	6S	0.84	2048F		Longley breakpoints (1961); 5L.69, 6S in satellite per Phillips et al. 1977
T5-6	c	5L	0.81	6L	0.08	2049A	T5-6c**	Longley breakpoints (1961)
T5-6	c**	5L	0.89	6S	0		T5-6c	Burnham breakpoints (1981)
T5-6	8696	5L	0.89	6S	0.8	2049B		Longley breakpoints (1961); 5L.79, 6S midway in secondary constriction per Phillips et al. 1977
T5-6	8379	5L	0.89	6L	0.67			Longley breakpoints (1961); 5L.79, 6S midway in secondary constriction per Phillips et al. 1977
T5-7	d	5S	0.63	7S	0.33	2049C		Longley breakpoints (1961)
T5-7	064-18	5S	0.61	7S	0.49	2049D		Longley breakpoints (1961)
T5-7	061-4	5S	0.54	7L	0.3	2049E		Longley breakpoints (1961)
T5-7	5143	5S	0.51	7L	0.1	2049F		Longley breakpoints (1961)
T5-7	3699	5S	0.46	7L	0.07	2049H		Longley breakpoints (1961)
T5-7	e	5S	0.4	7S	0.18	2050A	T5-7(B-21)	Longley breakpoints (1961)
T5-7	013-3	5S	0.36	7S	0.35	2050B		Longley breakpoints (1961)
T5-7	4306-4	5S	0.32	7L	0.35	2050C		Longley breakpoints (1961)
T5-7	6372	5S	0.2	7L	0.14			Longley breakpoints (1961)
T5-7	3695	5S	0.27	7S	0.49	2050G		Longley breakpoints (1961)
T5-7	8679	5S	0.09	7S	0.26	2050D		Longley breakpoints (1961)
T5-7	8491	5	0	7	0			Longley breakpoints (1961)
T5-7	062-18	5	0	7	0	2050E		Longley breakpoints (1961)
T5-7	023-13	5L	0.12	7L	0.15	2050F		Longley breakpoints (1961)
T5-7	3690	5L	0.17	7L	0.61			Longley breakpoints (1961)
T5-7	b	5L	0.18	7S	0.36	2051A		Longley breakpoints (1961)
T5-7	6293	5L	0.26	7L	0.63	2051B		Longley breakpoints (1961)
T5-7	8630	5L	0.38	7L	0.24	2051C		Longley breakpoints (1961)
T5-7	c	5L	0.42	7L	0.72	2051D		Longley breakpoints (1961)
T5-7	5179	5L	0.55	7L	0.73	2051E		Longley breakpoints (1961)
T5-7	a	5L	0.78	7L	0.72	2051F		Longley breakpoints (1961)
T5-7	f	5L	0.8	7L	0.85	2052A	T5-7(X-27-44)	Longley breakpoints (1961)
T5-7	8671	5L	0.96	7L	0.67	2052B		Longley breakpoints (1961)
T5-8	8420	5S	0.9	8L	0.33	2052C		Longley breakpoints (1961)
T5-8	8746	5S	0.84	8L	0.25	2052D		Longley breakpoints (1961)
T5-8	8458	5S	0.68	8L	0.32			Longley breakpoints (1961)
T5-8	5013	5S	0.67	8L	0.59			Longley breakpoints (1961)
T5-8	6612	5S	0.59	8L	0.66	2053B		Longley breakpoints (1961)
T5-8	013-11	5S	0.59	8S	0.63	2053C		Longley breakpoints (1961)

Table 54.1 (Continued)

T5-8	d	5S	0.55	8L	0.12	2053D	T5-8(B-18)	Longley breakpoints (1961)
T5-8	5570	5S	0.47	8S	0.35	2053E		Longley breakpoints (1961)
T5-8	8513	5S	0.34	8L	0.24			Longley breakpoints (1961)
T5-8	c	5S	0.24	8L	0.2	2053F	T5-8(B-10)	Longley breakpoints (1961)
T5-8	b	5S	0.23	8L	0.23	2054A	T5-8(84)	Longley breakpoints (1961)
T5-8	5575	5S	0.21	8S	0.22	2054B		Longley breakpoints (1961)
T5-8	7068	5S	0.18	8L	0.18	2054C		Longley breakpoints (1961)
T5-8	5777	5S	0.13	8L	0.19	2054D		Longley breakpoints (1961)
T5-8	017-8	5S	0.07	8S	0.25			Longley breakpoints (1961)
T5-8	6402	5S	0.07	8L	0.07	2054E		Longley breakpoints (1961)
T5-8	A-50	5S	0.07	8L	0.11	2054F		Longley breakpoints (1961)
T5-8	6406	5	0	8	0	2055A		Longley breakpoints (1961)
T5-8	f	5L	0.02	8S	0.08	2055B	T5-8(C-52)	Longley breakpoints (1961)
T5-8	4302-116	5L	0.06	8S	0.11			Longley breakpoints (1961)
T5-8	6289	5L	0.06	8L	0.54	2055C		Longley breakpoints (1961)
T5-8	045-6	5L	0.08	8L	0.13	2055D		Longley breakpoints (1961)
T5-8	002-17	5L	0.11	8L	0.28	2055E		Longley breakpoints (1961)
T5-8	8997	5L	0.16	8L	0.08	2055F		Longley breakpoints (1961)
T5-8	8955	5L	0.17	8L	0.83			Longley breakpoints (1961)
T5-8	014-5	5L	0.19	8L	0.18	2056A		Longley breakpoints (1961)
T5-8	053-4	5L	0.21	8S	0.48	2056B		Longley breakpoints (1961)
T5-8	4636	5L	0.23	8L	0.79	2056C		Longley breakpoints (1961)
T5-8	g	5L	0.28	8S	0.44	2056D	T5-8(X-27-87)	Longley breakpoints (1961)
T5-8	007-17	5L	0.32	8S	0.47	2056E		Longley breakpoints (1961)
T5-8	5866	5L	0.32	8L	0.77	2056F		Longley breakpoints (1961)
T5-8	7102	5L	0.48	8S	0.1	2057A		Longley breakpoints (1961)
T5-8	030-1	5L	0.48	8L	0.78	2057B		Longley breakpoints (1961)
T5-8	a	5L	0.49	8S	0.58	2057C		Longley breakpoints (1961)
T5-8	8806	5L	0.72	8S	0.59	2057D		Longley breakpoints (1961)
T5-8	8796	5L	0.76	8L	0.11	2057E		Longley breakpoints (1961)
T5-8	055-20	5L	0.81	8L	0.67	2057F		Longley breakpoints (1961)
T5-8	068-12	5S	0.71	9L	0.86			Longley breakpoints (1961)
T5-9	8854	5S	0.33	9S	0.36	2058A		Longley breakpoints (1961)
T5-9	022-11	5S	0.3	9L	0.27	2058B		Longley breakpoints (1961)
T5-9	B-91	5S	0.23	9L	0.21	2058C		Longley breakpoints (1961)
T5-9	B-94	5S	0.17	9L	0.27			Longley breakpoints (1961)
T5-9	6057	5S	0.15	9S	0.52	2058D		Longley breakpoints (1961)
T5-9	8591	5S	0.09	9L	0.25	2058E		Longley breakpoints (1961)
T5-9	c	5S	0.07	9L	0.1		T5-9(X-10-6)	Longley breakpoints (1961)
T5-9	020-7	5	0	9	0	2058F		Longley breakpoints (1961)
T5-9	4817	5L	0.06	9S	0.07			Longley breakpoints (1961)
T5-9	5614	5L	0.09	9L	0.06	2059A		Longley breakpoints (1961)
T5-9	d	5L	0.14	9L	0.1		T5-9(X-11-73)	Longley breakpoints (1961)
T5-9	032-8	5L	0.19	9L	0.7	2059B		Longley breakpoints (1961)
T5-9	7205	5L	0.21	9L	0.9			Longley breakpoints (1961)
T5-9	008-18	5L	0.29	9L	0.26	2059C		Longley breakpoints (1961)
T5-9	044-8	5L	0.33	9L	0.55			Longley breakpoints (1961)
T5-9	4790	5L	0.34	9L	0.45	2059D		Longley breakpoints (1961)
T5-9	8704	5L	0.35	9L	0.85			Longley breakpoints (1961)
T5-9	8895	5L	0.37	9L	0.11	2059E		Longley breakpoints (1961)
T5-9	4305-12	5L	0.41	9L	0.64			Longley breakpoints (1961)
T5-9	4305-22	5L	0.42	9L	0.15	2059F		Longley breakpoints (1961)
T5-9	8936	5L	0.43	9L	0.8			Longley breakpoints (1961)
T5-9	e	5L	0.46	9L	0.74	2060A	T5-9(X-14-111)	Longley breakpoints (1961)
T5-9	4352	5L	0.48	9L	0.61	2060B		Longley breakpoints (1961)
T5-9	015-10	5L	0.5	9L	0.2	2060C		Longley breakpoints (1961)
T5-9	013-9	5L	0.51	9L	0.82			Longley breakpoints (1961)
T5-9	b	5L	0.68	9L	0.44	2060E	T5-9(X-7-39)	Longley breakpoints (1961)
T5-9	a	5L	0.69	9S	0.17			Longley breakpoints (1961)
T5-9	4871	5L	0.71	9S	0.38	2060F		Longley breakpoints (1961)
T5-9	8457	5L	0.78	9S	0.83	2061A		Longley breakpoints (1961)
T5-9	6200	5L	0.81	9L	0.71	2061B		Longley breakpoints (1961)
T5-9	8386	5L	0.87	9S	0.13			Longley breakpoints (1961)
T5-9	5368	5L	0.87	9L	0.95			Longley breakpoints (1961)
T5-10	6760	5S	0.78	10S	0.4	2061C		Longley breakpoints (1961)
T5-10	5355	5S	0.77	10L	0.45			Longley breakpoints (1961)
T5-10	5653	5S	0.76	10L	0.71	2062A		Longley breakpoints (1961)
T5-10	6061	5S	0.6	10L	0.57		T2-10(6061**)	Longley breakpoints (1961); T2-10 per Burnham
T5-10	031-18	5S	0.58	10S	0.55	2062B		Longley breakpoints (1961)
T5-10	c	5S	0.42	10L	0.42		T5-10(X-57-16)	Longley breakpoints (1961)
T5-10	5679	5S	0.16	10L	0.15	2062D		Longley breakpoints (1961)
T5-10	6830	5	0	10	0			Longley breakpoints (1961)
T5-10	b	5L	0.09	10S	0.25	2062F	T5-10(B-70)	Longley breakpoints (1961)
T5-10	5358	5L	0.1	10L	0.76			Longley breakpoints (1961)
T5-10	073-6	5L	0.13	10S	0.41	2063B		Longley breakpoints (1961)
T5-10	4384	5L	0.13	10L	0.79			Longley breakpoints (1961)
T5-10	a	5L	0.14	10S	0.54	2063D	T5-10(A-49)	Longley breakpoints (1961)
T5-10	5188	5L	0.37	10S	0.65	2063E		Longley breakpoints (1961)
T5-10	006-11	5L	0.49	10L	0.52			Longley breakpoints (1961)
T5-10	022-20	5L	0.65	10S	0.62	2064A		Longley breakpoints (1961)
T5-10	3693	5L	0.67	10L	0.51			Longley breakpoints (1961)
T5-10	7142	5L	0.73	10L	0.17			Longley breakpoints (1961)
T5-10	5290	5L	0.78	10S	0.49	2064C		Longley breakpoints (1961)
T5-10	5688	5L	0.78	10L	0.53			Longley breakpoints (1961)
T5-10	8345	5L	0.87	10S	0.61	2064E		Longley breakpoints (1961)
T5-10	4801	5L	0.91	10L	0.23			Longley breakpoints (1961)
T5-10	5557	5L	0.92	10S	0.39	2065A		Longley breakpoints (1961)
T6-7	7036	6S	0.9	7L	0.63	2065B		Longley breakpoints (1961); in satellite per Phillips et al. 1977
T6-7	035-3	6S	0.8	7L	0.2	2065C		Longley breakpoints (1961); 7L.59, 6S at .25 in secondary constriction per Phillips et al. 1977
T6-7	5181	6S	0.79	7L	0.86	2065D		Longley breakpoints (1961); 7L.85, 6S at .71 in NOR-heterochromatin per Phillips et al. 1977
T6-7	4964	6S	0.76	7L	0.72	2065E		Longley breakpoints (1961); 7L.67, 6S at .32 in NOR-heterochromatin per Phillips et al. 1977
T6-7	054-6	6L	0.1	7L	0.6	2066A		Longley breakpoints (1961)
T6-7	6498	6L	0.16	7S	0.48	2066B	T6-7(6498**)	Longley breakpoints (1961)
T6-7	6498**	6L	0.23	7S	0		T6-7(6498)	Burnham breakpoints (1981)
T6-7	4573	6L	0.22	7L	0.27	2066C		Longley breakpoints (1961)
T6-7	4545	6L	0.25	7S	0.73	2066D	T6-7(4545**)	Burnham breakpoints (1981)
T6-7	4545**	6L	0.07	7S	0		T6-7(4545)	Longley breakpoints (1961)
T6-7	7380	6L	0.29	7L	0.45	2066E		Longley breakpoints (1961)
T6-7	011-11	6L	0.29	7L	0.29	2066F		Longley breakpoints (1961)
T6-7	013-8	6L	0.31	7L	0.22	2067A	T6-7(013-8**)	Longley breakpoints (1961)
T6-7	013-8**	6L	0.27	7L	0.63		T6-7(013-8)	Burnham breakpoints (1981)
T6-7	6885	6L	0.33	7S	0.58	2067B		Longley breakpoints (1961)
T6-7	8143	6L	0.35	7L	0.36	2067C	T6-7(8143**)	Longley breakpoints (1961)

Table 54.1 (Continued)

T6-7	8143**	6L	0.18	7L	0.16		T6-7(8143)	Burnham breakpoints (1981)
T6-7	4337	6L	0.37	7L	0.13	2067D		Longley breakpoints (1961)
T6-7	6598	6L	0.43	7L	0.61	2067E		Longley breakpoints (1961)
T6-7	4594	6L	0.52	7S	0.67	2067F		Longley breakpoints (1961)
T6-7	027-6	6L	0.66	7L	0.97	2068A		Longley breakpoints (1961)
T6-7	a	6L	0.73	7L	0.68	2068B		Longley breakpoints (1961)
T6-7	7402	6L	0.97	7L	0.14	2068C		Longley breakpoints (1961)
T6-8	017-14	6S	0.92	8L	0.83			Longley breakpoints (1961)
T6-8	5386	6S	0.78	8L	0.83			Longley breakpoints (1961)
T6-8	058-1	6	0	8L	0.46	2068D		Longley breakpoints (1961)
T6-8	6187	6L	0.19	8L	0.51	2068E		Longley breakpoints (1961)
T6-8	6873	6L	0.21	8L	0.29	2069A		Longley breakpoints (1961)
T6-8	5028	6L	0.21	8L	0.31	2069B		Longley breakpoints (1961)
T6-8	c	6L	0.27	8L	0.5	2069C	T6-8(C-59)	Longley breakpoints (1961)
T6-8	5605	6L	0.36	8L	0.22	2069D		Longley breakpoints (1961)
T6-8	a	6L	0.41	8L	0.8	2069E		Longley breakpoints (1961)
T6-8	024-1	6L	0.42	8L	0.74	2069F		Longley breakpoints (1961)
T6-8	d	6L	0.51	8L	0.77	2070A	T6-8(D-1)	Longley breakpoints (1961)
T6-8	b	6L	0.79	8S	0.76	2070B	T6-8(B-83)	Longley breakpoints (1961)
T6-9	017-14	6S	0.8	9L	0.5	2070C		Longley breakpoints (1961); in satellite per Phillips et al. 1977
T6-9	4778	6S	0.8	9L	0.3			Longley breakpoints (1961); 6S at .95 in NOR-heterochromatin per Phillips et al. 1977
T6-9	a	6S	0.79	9L	0.4			Longley breakpoints (1961); 9L.32, 6S at .67 in NOR-heterochromatin per Phillips et al. 1977
T6-9	d	6S	0.73	9L	0.82	2070F	T6-9(C-23)	Longley breakpoints (1961); 9L.84, 6S at .46 in NOR-heterochromatin per Phillips et al. 1977
T6-9	067-6	6S	0.79	9L	0.47	2071A		Longley breakpoints (1961)
T6-9	5454	6	0	9S	0.75	2071B		Longley breakpoints (1961)
T6-9	6566	6	0	9L	0.9	2071C		Longley breakpoints (1961)
T6-9	84-39	6L	0.06	9S	0.73	2071D		Longley breakpoints (1961)
T6-9	b	6L	0.1	9S	0.37			Longley breakpoints (1961)
T6-9	4505	6L	0.13	9	0			Longley breakpoints (1961)
T6-9	c	6L	0.15	9L	0.29	2071E	T6-9(A-66)	Longley breakpoints (1961)
T6-9	8536	6L	0.18	9S	0.81	2071F		Longley breakpoints (1961)
T6-9	e	6L	0.18	9L	0.24	2072A	T6-9(X-25-78)	Longley breakpoints (1961)
T6-9	8592	6L	0.19	9S	0.41			Longley breakpoints (1961)
T6-9	6270	6L	0.19	9L	0.28	2072B		Longley breakpoints (1961)
T6-9	018-12	6L	0.21	9L	0.25			Longley breakpoints (1961)
T6-9	6019	6L	0.27	9L	0.26	2072C		Longley breakpoints (1961)
T6-9	5831	6L	0.27	9L	0.3			Longley breakpoints (1961)
T6-9	8906	6L	0.27	9L	0.59	2072D		Longley breakpoints (1961)
T6-9	043-1	6L	0.36	9L	0.36	2072E		Longley breakpoints (1961)
T6-9	5964	6L	0.47	9L	0.83	2072F		Longley breakpoints (1961)
T6-9	8768	6L	0.89	9S	0.61			Longley breakpoints (1961)
T6-10	f	6S	0.92	10S	0.28	2073B	T6-10(I-22)	Longley breakpoints (1961); in satellite per Phillips et al. 1977
T6-10	5253	6S	0.8	10L	0.41			Longley breakpoints (1961); 6S at .30 in NOR-heterochromatin per Phillips et al. 1977
T6-10	5519	6S	0.75	10L	0.17	2073D		Longley breakpoints (1961); 10L.10, 6S in proximal part of secondary constriction per Phillips et al. 1977
T6-10	McClintock	6S	0.5	10L	0.58			
T6-10	b	6L	0.12	10L	0.29	2073E		Longley breakpoints (1961)
T6-10	e	6L	0.14	10S	0.43	2073F	T6-10(D-13)	Longley breakpoints (1961)
T6-10	d	6L	0.16	10L	0.29		T6-10(C-27)	Longley breakpoints (1961)
T6-10	8645	6L	0.21	10L	0.28			Longley breakpoints (1961)
T6-10	8651	6L	0.27	10L	0.48			Longley breakpoints (1961)
T6-10	h	6L	0.47	10L	0.87	2074C	T6-10(X-46-13)	Longley breakpoints (1961)
T6-10	044-8	6L	0.48	10L	0.51			Longley breakpoints (1961)
T6-10	c	6L	0.51	10S	0.36	2074E	T6-10(A-23)	Longley breakpoints (1961)
T6-10	8904	6L	0.51	10L	0.83			Longley breakpoints (1961)
T6-10	4347	6L	0.65	10S	0.81			Longley breakpoints (1961)
T6-10	4307-12	6L	0.74	10S	0.71	2075A		Longley breakpoints (1961)
T6-10	a	6L	0.75	10L	0.15			Longley breakpoints (1961)
T6-10	4833	6L	0.83	10S	0.78	2075C		Longley breakpoints (1961)
T6-10	g	6L	0.85	10L	0.2		T6-10(X-17-15)	Longley breakpoints (1961)
T6-10	5780	6L	0.93	10L	0.13			Longley breakpoints (1961)
T7-8	8346	7S	0.49	8S	0.3			Longley breakpoints (1961)
T7-8	5828	7S	0.31	8L	0.1	2076A		Longley breakpoints (1961)
T7-8	6531	7	0	8	0	2076B		Longley breakpoints (1961)
T7-8	6981	7	0	8	0	2076C		Longley breakpoints (1961)
T7-8	8580	7	0	8	0	2076D		Longley breakpoints (1961)
T7-8	8583	7	0	8	0			Longley breakpoints (1961)
T7-8	004-3	7	0	8	0	2076E		Longley breakpoints (1961)
T7-8	016-15	7	0	8	0	2076F		Longley breakpoints (1961)
T7-8	034-17	7L	0.05	8S	0.59	2077A		Longley breakpoints (1961)
T7-8	5499	7L	0.05	8L	0.08	2077B		Longley breakpoints (1961)
T7-8	5413	7L	0.09	8L	0.77			Longley breakpoints (1961)
T7-8	062-16	7L	0.15	8L	0.17	2077C		Longley breakpoints (1961)
T7-8	014-17	7L	0.18	8L	0.3			Longley breakpoints (1961)
T7-8	4536	7L	0.34	8L	0.47	2077E		Longley breakpoints (1961)
T7-8	038-8	7L	0.52	8L	0.46	2077F		Longley breakpoints (1961)
T7-8	7149	7L	0.56	8L	0.65	2078A		Longley breakpoints (1961)
T7-8	5479	7L	0.7	8S	0.21	2078B		Longley breakpoints (1961)
T7-8	021-1	7L	0.72	8L	0.49	2078C		Longley breakpoints (1961)
T7-8	4824	7L	0.83	8L	0.25	2078D		Longley breakpoints (1961)
T7-8	6427	7L	0.96	8L	0.28			Longley breakpoints (1961)
T7-9	b	7S	0.76	9S	0.19	2079A	T7-9(F-11)	Longley breakpoints (1961)
T7-9	071-1	7S	0.7	9L	0.07	2079B		Longley breakpoints (1961)
T7-9	8659	7S	0.55	9S	0.35	2079C		Longley breakpoints (1961)
T7-9	053-8	7S	0.51	9L	0.77	2079D		Longley breakpoints (1961)
T7-9	5074	7S	0.48	9L	0.53	2079E		Longley breakpoints (1961)
T7-9	8558	7S	0.22	9L	0.16	2079F		Longley breakpoints (1961)
T7-9	4363	7	0	9	0			Longley breakpoints (1961)
T7-9	6225	7	0	9	0	2080B		Longley breakpoints (1961)
T7-9	8383	7	0	9	0	2080C		Longley breakpoints (1961)
T7-9	6482	7L	0.01	9S	0.97	2080D		Longley breakpoints (1961)
T7-9	7074	7L	0.03	9S	0.8	2080E		Longley breakpoints (1961)
T7-9	c	7L	0.14	9L	0.22	2080F	T7-9(X-56-86)	Longley breakpoints (1961)
T7-9	8611	7L	0.23	9L	0.18			Longley breakpoints (1961)
T7-9	008-16	7L	0.51	9L	0.62			Longley breakpoints (1961)
T7-9	4713	7L	0.6	9	0	2081A		Longley breakpoints (1961)
T7-9	027-9	7L	0.61	9S	0.18	2081B		Longley breakpoints (1961)
T7-9	6978	7L	0.62	9S	0.83	2081C		Longley breakpoints (1961)
T7-9	a	7L	0.63	9S	0.07	Wx38A	T7-9(A-76)	Longley breakpoints (1961)
T7-9	008-3	7L	0.8	9L	0.85			Longley breakpoints (1961)
T7-9	032-13	7L	0.82	9L	0.88	2081D		Longley breakpoints (1961)
T7-9	5381	7L	0.85	9L	0.78			Longley breakpoints (1961)
T7-10	7356	7S	0.75	10L	0.88	2082A		Longley breakpoints (1961)
T7-10	022-15	7	0	10	0			Longley breakpoints (1961)

TABLE 54.1 (Continued)

T7-10	015-12	7	0	10	0			Longley breakpoints (1961)
T7-10	019-3	7L	0.17	10L	0.47	2082C		Longley breakpoints (1961)
T7-10	a	7L	0.23	10L	0.06		T7-10(D-36)	Longley breakpoints (1961)
T7-10	4422	7L	0.79	10 0.0				Longley breakpoints (1961)
T7-10	4356	7L	0.7	10S	0.37	2082D		Longley breakpoints (1961)
T7-10	4331**	7		10			T1-5(4331)	Burnham breakpoints (1981)
T8-9	b	8S	0.67	9L	0.75	2083B		Longley breakpoints (1961)
T8-9	8661	8S	0.66	9L	0.92			Longley breakpoints (1961)
T8-9	4643	8S	0.37	9L	0.11	2083C	T8-9(4346)	Longley breakpoints (1961)
T8-9	c	8	0	9	0	2083D	T8-9(C-12)	Longley breakpoints (1961)
T8-9	8525	8L	0.06	9S	0.63	2083E		Longley breakpoints (1961)
T8-9	5391	8L	0.07	9S	0.33	2083F		Longley breakpoints (1961)
T8-9	d	8L	0.09	9S	0.16		T8-9(X-22-92)	Longley breakpoints (1961)
T8-9	a	8L	0.13	9L	0.38	2083G		Longley breakpoints (1961)
T8-9	8951	8L	0.13	9L	0.77	2084A		Longley breakpoints (1961)
T8-9	034-11	8L	0.14	9L	0.18			Longley breakpoints (1961)
T8-9	043-6	8L	0.17	9S	0.34	2084B		Longley breakpoints (1961)
T8-9	e	8L	0.32	9L	0.25	2084C	T8-9(X-26-8)	Longley breakpoints (1961)
T8-9	6673	8L	0.35	9S	0.31			Longley breakpoints (1961)
T8-9	4775	8L	0.42	9L	0.68	2084D		Longley breakpoints (1961)
T8-9	018-12	8L	0.52	9S	0.49			Longley breakpoints (1961)
T8-9	4593	8L	0.69	9L	0.65	2084E		Longley breakpoints (1961)
T8-9	6921	8L	0.85	9L	0.15	2084F		Longley breakpoints (1961)
T8-9	5300	8L	0.85	9S	0.43	2085A		Longley breakpoints (1961)
T8-9	4453	8L	0.86	9S	0.68	2085B		Longley breakpoints (1961)
T8-10	023-15	8S	0.42	10S	0.31			Longley breakpoints (1961)
T8-10	b	8	0	10	0	2085C		Longley breakpoints (1961)
T8-10	5585	8	0	10	0			Longley breakpoints (1961)
T8-10	6653	8L	0.04	10L	0.06	2086A		Longley breakpoints (1961)
T8-10	3697	8L	0.1	10L	0.18			Longley breakpoints (1961)
T8-10	K-7	8L	0.07	10S	0.22			Longley breakpoints (1961)
T8-10	9020	8L	0.13	10S	0.5	2086B		Longley breakpoints (1961)
T8-10	6488	8L	0.14	10S	0.34	2086C		Longley breakpoints (1961)
T8-10	5287	8L	0.17	10S	0.33	2086D		Longley breakpoints (1961)
T8-10	034-19	8L	0.24	10L	0.28	2086E		Longley breakpoints (1961)
T8-10	001-5	8L	0.3	10S	0.57	2086F		Longley breakpoints (1961)
T8-10	d	8L	0.39	10L	0.16	2087A	T8-10(F-1)	Longley breakpoints (1961)
T8-10	c	8L	0.41	10S	0.56	2087B		Longley breakpoints (1961)
T8-10	6128	8L	0.43	10S	0.49	2087C		Longley breakpoints (1961)
T8-10	a	8L	0.48	10S	0.48	2087D		Longley breakpoints (1961)
T8-10	5944	8L	0.75	10L	0.4			Longley breakpoints (1961)
T8-10	e	8L	0.84	10S	0.37	2087F	T8-10(F-33)	Longley breakpoints (1961)
T9-10	059-10	9S	0.31	10L	0.53			Longley breakpoints (1961)
T9-10	3688	9S	0.49	10L	0.02			Longley breakpoints (1961)
T9-10	8630	9S	0.28	10L	0.37			Longley breakpoints (1961)
T9-10	B-49	9S	0.21	10L	0.6			Longley breakpoints (1961)
T9-10	b	9S	0.13	10S	0.4			Longley breakpoints (1961)
T9-10	a	9L	0.14	10L	0.92			Longley breakpoints (1961)
T9-10	A-75	9L	0.24	10L	0.3			Longley breakpoints (1961)
T9-10	4303-9	9L	0.26	10S	0.44	2088C		Longley breakpoints (1961)
T9-10	5488	9L	0.57	10L	0.89	2089A		Longley breakpoints (1961)
T9-10	041-4	9L	0.67	10L	0.92			Longley breakpoints (1961)
T9-10	041-6	9L	0.7	10L	0.9			Longley breakpoints (1961)
T9-10	7103	9L	0.73	10L	0.88			Longley breakpoints (1961)

REFERENCES

Anderson EG (1935) Chromosomal interchanges in maize. Genetics 20: 70–83

Brink RA (1927) The occurrence of semi-sterility in maize. J Hered 18: 266–270

Burnham CR (1930) Genetical and cytological studies of semisterility and related phenomena in maize. Proc Natl Acad Sci USA 16: 269–277

Longley AE (1961) Breakage points for four corn translocation series and other corn chromosome aberrations. USDA-ARS Crops Research Bulletin No. 34–16, 40 pp

McClintock B (1930) A cytological demonstration of the location of an interchange between two non-homologous chromosomes of *Zea mays*. Proc Natl Acad Sci USA 16: 791–796

55

Segmental Aneuploid Analysis

JAMES A. BIRCHLER

Maize has an extensive set of tools available for segmental analysis. These include monosomy, trisomy, tetrasomy, centromeric trisomics, and terminal tetrasomics.

SEGMENTAL TRISOMY

Gopinath and Burnham (1956) described the conditions for producing interstitial deletions and duplications in gametes by the use of overlapping A-A translocations. Most combinations will produce some measure of nullisomy during the gametophyte generation and therefore are lethal and cannot be recovered. The exception is the combination in which the breakpoint in one chromosome is proximal and in the other chromosome is relatively distal in one of the two translocations. In the second translocation of the pair, the first break is relatively distal and in the other chromosome it is relatively proximal. When such a combination of translocations is crossed together and then backcrossed to normal, one class of gamete is produced that is duplicated for both regions between the breakpoints. (See Figure 55.1D.) Because there are no deficiencies produced, these gametes can survive. There are also the two different balanced translocations recovered (Figure 55.1A,B). The last product of meiosis is one in which the regions between the breakpoints are deleted (Figure 55.1C). This class will almost always abort in the gametophyte. The duplicate-duplicate class is the one with utility for segmental analysis. If the region between the breakpoints is marked by an allozyme or RFLP marker, the trisomic class can be distinguished. The two balanced translocation stocks serve as the diploid control.

The potential for segmental analysis in maize is great given the extensive number of reciprocal translocations that are available. (See list of available translocations—Table 54.1—in the previous chapter.) Further details of the production and behavior of the segmental trisomics can be found in Birchler (1980, 1991).

The Maize Handbook—M. Freeling, V. Walbot, eds.

SEGMENTAL TETRASOMY

For any pair of translocations for which a segmental trisomic can be produced, it is possible to extend the analysis to tetrasomy. An example has been described by Birchler et al. (1981). A segmental trisomic was first recovered as described above. These plants were self-pollinated. The duplicated gametes and the normal are produced in equal frequency through the female, but the duplicated ones are reduced in fertilization competition.

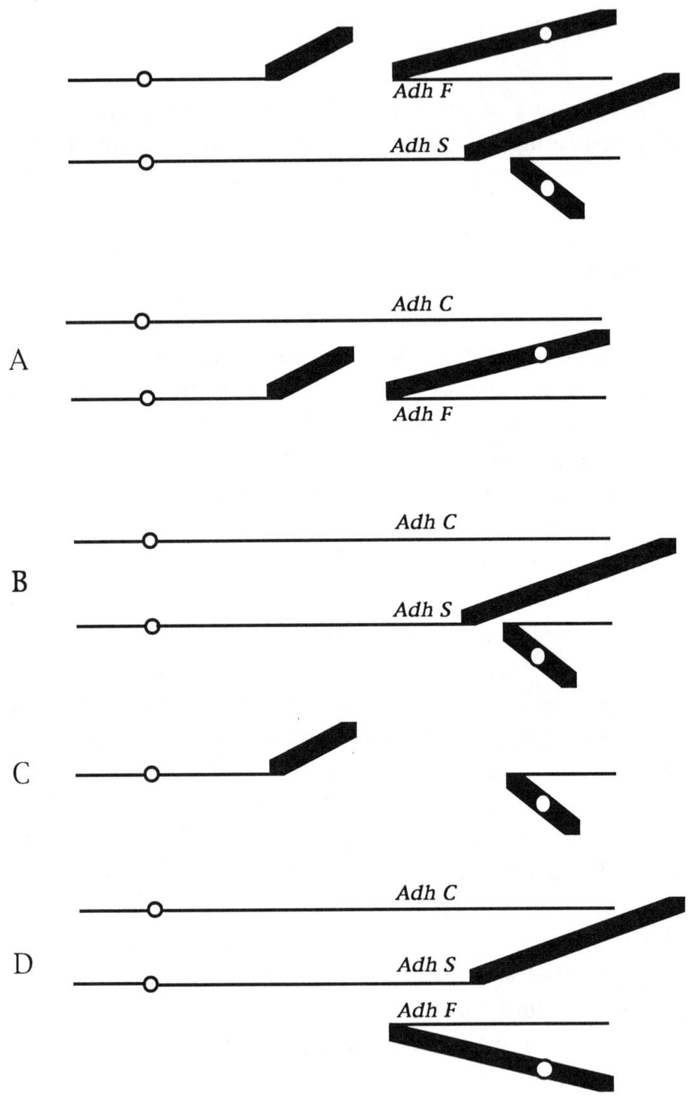

FIGURE 55.1

Nevertheless, it was possible to recover individuals that were homozygous for the duplication. The alcohol dehydrogenase allozymes were used to identify the appropriate cases (Figure 55.2). For regions with no allozyme markers, restriction fragment length polymorphism could potentially be used.

The tetrasomic stock was tested for the occurrence of recombination in the duplicated regions, which would produce normal gametes as one class. This particular case was very stable, suggesting that the two different homologous pairs of chromosomes pair with each other but not with the other pair despite a stretch of homology. Consequently, it appears that tetrasomics could be maintained by self- or sib pollination without too much concern for their spontaneous unravelling.

The advantages of tetrasomics over trisomics include the following: (1) a greater dosage difference from the diploid would make a dosage effect easier to recognize, and (2) they can be maintained by self-pollination without the need for selection. The disadvantages include the facts that the tetrasomic and diploid comparisons are not present on the same ear and might be subjected to slightly different environmental influences, tetrasomics take longer to produce if they are not already available, and the tetrasomics have greater effects on plant vigor than do trisomics. The latter effect might result in some indirect aneuploid effects.

STABLE TERMINAL TETRASOMICS

The above discussion of segmental trisomics and tetrasomics focused on interstitial segmental analysis. The procedure for producing the desired duplicate chromosomes used

FIGURE 55.1 Production of duplicated gametes from overlapping A-A translocations. At the top of the diagram is a schematic of the two types of translocations needed to produce viable duplicated gametes. The thin line represents one chromosome (chromosome 1 in this example) and the thick line a second. Circles denote centromeres. In the top translocation, the chromosome 1 breakpoint is proximally located in the arm while distally located in the other chromosome. In the second translocation, the relative position of the breakpoints is reversed so that the chromosome 1 breakpoint is now more distal and the breakpoint in the other chromosome is more proximal. Alcohol dehydrogenase (*Adh*) allozymes *F* (Fast) and *S* (Slow) are marked as an illustration of a marker system. In a heterozygote of the two translocations, which would have the genetic configuration at the top, four types of meiotic products are formed. The zygotes formed following crosses with normal(marked by *Adh1-C* in this example) are shown in A–D. A and B illustrate zygotes in which the two different translocations are respectively recovered. These are marked by *Adh1-F* and *-S* in this example. C shows the segregation of different elements of the two translocations that produce a deficiency for the regions between the two breakpoints. For most combinations of translocations, this class will abort because it is deleted for genes vital to the gametophyte generation. For this reason, plants heterozygous for two such translocations usually exhibit 25% ovule and pollen abortion. D illustrates the segregation of the elements of the two translocations that result in a duplication for all regions between the breakpoints. When the duplicated gamete is fertilized by a normal one a segmentally trisomic zygote is formed. With different molecular markers of the two translocations and a third in the normal parent, the different classes of progeny can be distinguished.

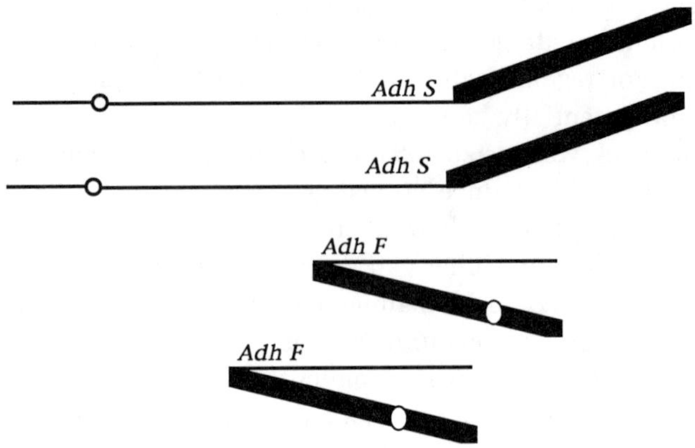

FIGURE 55.2 Configuration of the relevant chromosomes involved in a segmentally tetrasomic line. See Figure 1 for symbol definitions.

overlapping A-A translocations and consequently duplications are restricted to interstitial regions. Carlson and Roseman (1991) demonstrated a method for the production of terminal tetrasomics by using a combination of A-A and B-A translocations. Basically, it involves the use of the B-A element of a B-A translocation to recover the duplicate-deficient class of gamete from adjacent-1 segregation of an A-A translocation. The B-A chromosome covers the deficiency and allows it to pass through the gametophytes. Upon self-pollination the duplicated segment would now be present in quadruplicate and the B-A element would be homozygous and compensate for the terminal missing regions from the A-A translocation. The final construction is illustrated in Figure 55.3. Such constructions extend the ability to perform segmental analysis to the terminal regions of the chromosomes.

CENTROMERIC OVERLAPS

The other regions of the genome that cannot be analyzed using overlapping A-A translocations are the regions immediately surrounding the centromeres. Fortunately these regions can be varied in dosage by other means. Carlson and Curtis (1986) described the construction of centromeric overlaps for chromosomes 3 and 9. To produce these, B-A translocations from opposite arms were crossed together. Individuals were cytologically classified to identify those that were balanced for both of the two translocations. The doubly balanced plants were crossed by normal pollen. At a reasonable frequency, there was nondisjunction of the two A-B chromosomes. This produced a segmental trisomic for the centromeric region. This configuration is illustrated in Figure 55.4. Upon self-pollination, plants were selected that were homozygous for the overlap. Further analysis

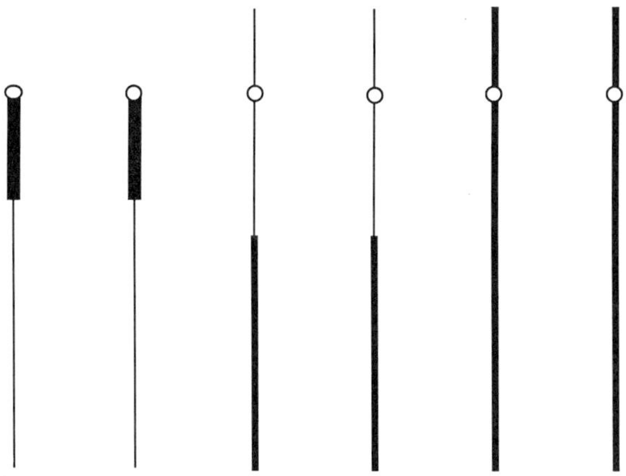

FIGURE 55.3 Stable terminal tetrasomic. At the left is a pair of B-A chromosomes involving the chromosome arm (thin line) that is deficient in the aneuploid segregant from a translocation heterozygote described in the text. The aneuploid segregant involves a normal chromosome (thick line) and one part of the A-A translocation (part thick–part thin). The B-A chromosome allows the passage of the deficiency for the distal part of the thin line chromosome to pass through the gametophyte generation. The terminal part of the thick-line chromosome is duplicated, so in the homozygous state (depicted here), a segmental terminal tetrasomic is formed. Circles denote the centromeres.

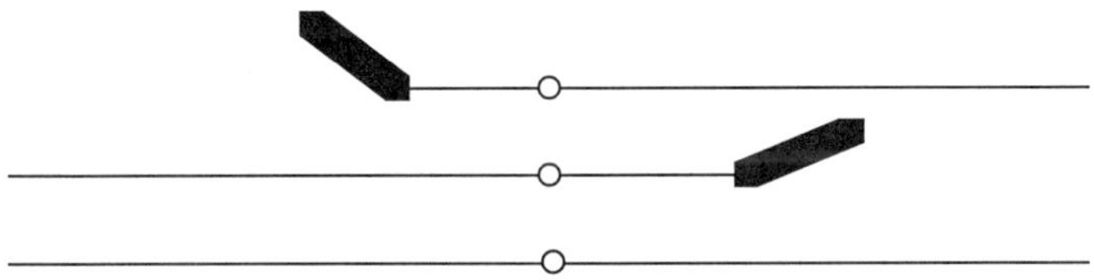

FIGURE 55.4 Centromeric overlap. Depicted here is a segmental trisomic for a region across the centromere. Such individuals can be produced by combining the A-B elements of opposite arm B-A translocations for the respective chromosome. The thick line represents the terminal portions of the B chromosome and the thin line the A chromosome. The opposite armed A-B elements are shown at the top and a normal A chromosome to complete the trisomic configuration is shown at the bottom.

of these constructs indicated that they were highly stable and apparently did not recombine to form normal chromosomes.

Carlson (1983) took one of these constructs a step further to produce a minichromosome for the region surrounding the centromere of chromosome 9. This was accomplished by selecting for recombinants in the trisomic overlap. Such were found and extend from the two different B-A translocation breakpoints in opposite arms through the centromere. Each end is capped by the respective terminal segment of the B chromosome. This chromosome was also made homozygous. This small chromosome is lost more often than normal chromosomes but is reasonably stable.

INTERSTITIAL DEFICIENCIES

Transmission of deficiencies from one generation to the next in diploid plants is rare. Only those cases that do not remove a gene vital to the haploid gametophytes can survive.

Consequently, the use of deficiencies is somewhat limited with the major exceptions of monosomics produced by the *r-x1* system and whole arm deficiencies generated by B-A translocations, both of which are described in detail elsewhere in this volume. A procedure also described in this volume has recently been developed for the construction of directed transpositions that would have the ability to cover deficiencies (Birchler and Levin 1991).

This section is included to note that it is technically possible to generate interstitial deficiencies using overlapping simple or compound B-A translocations. If a more distally broken hyperploid heterozygote is used as a female, there will be approximately half balanced translocation to half A B-A gametes produced. If this hyperploid heterozygote were crossed by a more proximally broken translocation and the A B-A female gamete joined with the deficient male gamete, an interstitial deficient zygote would be formed that would encompass the region between the breakpoints. This is illustrated in Figure 55.5. If allozymes or other molecular markers are available, the various classes can be distinguished.

SUMMARY

The ability to perform segmental analysis in maize is unparalleled in the plant kingdom. The combination of centromeric overlaps of B-A translocations, interstitial segmental tetrasomics, and terminal tetrasomics allows almost all of the regions of the maize chromosomes (with the exception of the short arm and centromeric region of chromosome 8 and the centromeric region of chromosome 2) to be present in four copies. There are over 900 A-A translocations available in maize, and these permit a very fine scale subdivision of the genome with these methods.

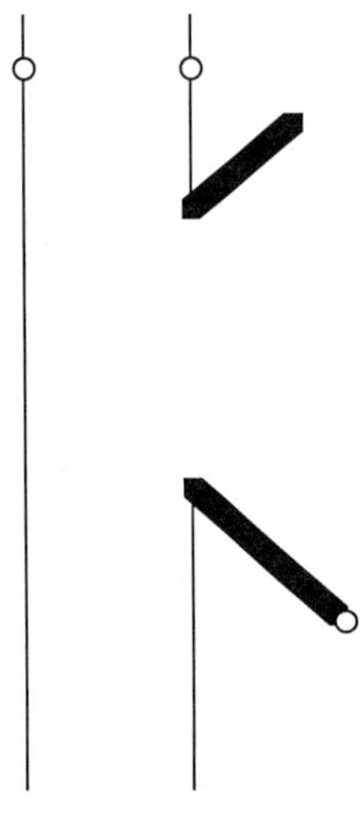

FIGURE 55.5 Interstitial deficiency. An interstitial deficiency individual can be formed by the union of an A B-A gamete from a distally broken TB-A with an A-B gamete from a more proximally broken TB-A. Thin lines represent the A chromosome, thick lines represent the B chromosome, and circles denote the centromeres. Note that the region between the breakpoints of the two B-A translocations is only present in one copy while the remainder of the A chromosome is present in two.

REFERENCES

Birchler JA (1980) The cytogenetic localization of the alcohol dehydrogenase-1 locus in maize. Genetics 94: 687–700

Birchler JA, Alleman M, Freeling M (1981) The construction of a segmental tetrasomic line of maize. Maydica 26: 3–9

Birchler JA (1991) Chromosome manipulation in maize. In Gupta PK, Tsuchiya T (eds) Chromosome Engineering in Plants: Genetics, Breeding and Evolution, Part A, Elsevier, New York, pp 531–559

Birchler JA, Levin DM (1991) Directed synthesis of a segmental chromosomal transposition: An approach to the study of chromosomes lethal to the gametophyte generation of maize. Genetics 127: 609–618

Carlson WR, Roseman RR (1991) Segmental duplication of chromosomal regions in maize. Genome 34: 537–542

Carlson WR (1983) Duplication of non-terminal A chromosome segments using B-A translocations. Maydica 28: 317–32

Carlson WR, Curtis C (1986) A new method for producing homozygous duplications in maize. Can J Genet Cytol 28: 1034–1040

Gopinath DM, Burnham CR (1956) A cytogenetic study in maize of deficiency-duplication produced by crossing interchanges involving the same chromosome. Genetics 41: 382–395

56

Directed Synthesis of Segmental Transpositions

James A. Birchler

Because of the haploid nature of the gametophyte generation, there are many mutations that can be induced but that cannot be recovered. The available evidence suggests that there are many genes that when mutant would be lethal to one or both gametophytes, but there is no way to know with certainty. Studies of messenger RNA complexities in shoots and pollen (Willing et al. 1988) indicate significant overlap between the sporophyte

The Maize Handbook—M. Freeling, V. Walbot, eds.

and gametophyte. The complexity studies indicate that there are well over 30,000 genes expressed in any one tissue, but genetic studies have only described about 3,000 mutants in mutagenized populations that contain one or more alleles for any known locus (G. Neuffer, personal communication). This is probably an underestimate of the number of mutants that can be recovered in maize because many subtle phenotypes undoubtedly have not been identified and many loci may not produce a phenotype when mutant. Also, the presence of gene families might reduce the number of mutants that can be recognized. At any rate, there is a disparity between the number of visible mutants that can be recovered and the estimated number of genes present. Some of this disparity may result from haploid lethality; a substantial fraction of mutants simply cannot be recovered with present day techniques.

In order to address this situation, a method was developed to create duplications that would cover mutations in the region of origin (Birchler and Levin 1991). The development of other segmental transpositions should permit deficiency analysis for the first time in plants. The details of the procedure will be given here only in protocol form.

1. The region to be duplicated is selected and from the list of translocations, two are selected that flank the desired segment. The orientation of the breakpoints must be such that in one translocation the breakpoint in the chromosome of interest is proximal and in the chromosome into which the insertion will be made the breakpoint is relatively distal. In the second translocation, the relative orientation of breakpoints is reversed.
2. These two translocations are crossed together.
3. The heterozygote between the two translocations is crossed by the respective B-A translocation for the region of the genome that is to be duplicated. This is done to eliminate any pairing competitors for the region between the breakpoints of the two translocations. Such recombination is the desired event to generate the transposition, so no pairing competition is wanted.
4. By using an appropriate marker system (phenotypic, allozyme, or RFLP), the segmental monosomic individuals are selected. These individuals will not have inherited the chromosome arm from the male parent. Among these there will be each of the two balanced translocations as well as the overlap of the two translocations. The latter are selected and used in crosses to chromosomally normal individuals. If used as a male, few would need to be selected in order to obtain sufficient numbers of progeny to recover recombinants.
5. The outcrossed individuals are again crossed by normal or self-pollinated. The nonrecombinants will exhibit 50% ovule and pollen abortion, while the recombinants will be fully fertile. Among the fully fertile plants will be chromosomally normal individuals as well as those that now have a transposition as a result of recombination in the region between the translocation breakpoints. The recovery of normals to transposition types is 2:1.
6. The transposition may now be used to recover deficiencies and gametophyte lethal

mutations in the region of origin. The transposition might be followed if it is maintained as a heterozygote with an inversion that is marked by a recessive marker. For example, Inversion 3c exists with the markers *lg2 a1*. If a transposition is inserted into 3L, then the inversion will prevent recombination along the arm and by repeated backcrosses to the inversion stock, sibling progeny with and without the transposition can be distinguished on the basis of the *A* vs. *a* phenotype. Lethal chromosomes could be induced by treating pollen from the homozygous inversion stock and crossing to the transposition/inversion heterozygotes. In the progeny, the *A* (transposition) kernels are selected and backcrossed to the inversion stock. If the ear segregates 2*A*:1*a* instead of the usual 1*A*: 1*a*, this is an indication that a lethal is present in the region that can be covered by the duplication.

This protocol could be used to induce cell lethal mutations that are lesions in metabolic or cell cycle/maintenance functions. Cell cycle defects could be directly observed in the first few mitotic divisions of the male gametophyte after meiotic segregation of the covering duplication. Excellent cytological techniques for chromosome analysis in the microspore divisions exist, as described elsewhere in this volume.

REFERENCES

Birchler JA, Levin DM (1991) Directed synthesis of a segmental chromosomal transposition: An approach to the study of chromosomes lethal to the gametophyte generation of maize. Genetics 127: 609–618

Willing RP, Bashe D, Mascarenhas JP (1988) An analysis of the quantity and diversity of messenger RNA's from pollen and shoots of *Zea mays*. Theor Appl Genet 75: 751–753

Practical Aspects of Haploid Production

JAMES A. BIRCHLER

There are currently two major systems for the production of haploids in maize. One involves the *indeterminate gametophyte* (*ig*) gene, which allows recovery of paternal haploids, and the other is *stock 6,* which allows recovery of maternal haploids. Haploids have been induced for innumerable types of investigations. Typically these plants are smaller in stature than their progenitor diploids and often possess striped leaves. They are highly sterile but produce kernels in a frequency much higher than predicted from the random segregation of the ten chromosomes. One possible reason for this is thought to be spontaneous doubling to produce diploid sectors. Another potential explanation might be failure of meiosis to produce unreduced gametes. It is also the case that there are occasionally fertile sectors in the tassel. As a result, it is possible to achieve self pollination at a low but workable frequency. The frequency of spontaneous haploid production in most lines is on the order of one in a thousand kernels. These are in the vast majority maternal in origin. The reader is referred to the treatise by Chase (1969) for an extensive discussion of haploids.

INDETERMINATE GAMETOPHYTE

Kermicle (1969) recognized that the *indeterminate gametophyte* (*ig*) mutation promoted the production of paternally derived haploids at a frequency of nearly 3%. These are typically recognized by the use of *ig* in a background of *R-nj* or *R-scm*, both of which yield anthocyanin in the embryo as well as the endopserm. These stocks are available from the Stock Center. To produce the haploids, the *ig* females are used in a cross with a stock that does not have factors that pigment the embryo. This will be the case with almost all common stocks. From the progeny, kernels are selected that exhibit the color marker in the endosperm but have colorless embryos. These are the putative paternal haploids. These haploids, which develop from the sperm nucleus, have the cytoplasm of the maternal parent.

The Maize Handbook—M. Freeling, V. Walbot, eds.

Normally, *ig/ig* individuals are male sterile, whereas *ig/+* are fertile. Therefore, it is necessary to maintain the stock by crossing *ig/ig* females by *ig/+* males. Alternatively, one could self-pollinate *ig/+* individuals, but this reduces the efficiency of the system because only half of the gametophytes are *ig* for haploid production.

Kindiger (1991) developed a method to maintain *ig* as a homozygous stock. The new stock has the B-A chromosome derived from TB-3Ld, which covers the *ig/ig* condition. This material is also homozygous for *R-nj*. The breakpoint of this translocation is very close to the *ig* locus and therefore there is very little recombination between the mutant and the extra chromosome that covers it. Transmission of the extra chromosome is about 50% through the female side and about 2% through the male. Consequently, by using the tertiary trisomic stock, there is approximately 98% transmission of *ig*, when used as a pollen parent. When crossed onto an *ig/ig* female stock, virtually all of the progeny would be homozygous for *ig*. This propagation method can greatly increase the utility of this system. These materials are available from the Stock Center.

Kindiger has used this system to develop a series of male sterile cytoplasms in lines that are homozygous for *ig* (Kindiger 1992). The cytoplasms include *cms-C*, *S*, *SD*, *CA*, *ME*, *Vg*, *Q*, *L*, and *MY*. The frequency of andogenetic haploids with this material can be as high as 9%, depending upon the male parent. Such a system can be used to transfer different inbred lines into other cytoplasms without extensive backcrosses. These materials are also available from the Stock Center.

ig has also been used to derive diploids from tetraploids.

STOCK 6

The second major means of producing haploids involves the line referred to as stock 6 (Coe 1959; Sarkar and Coe 1966). This stock, when self-pollinated, will produce maternal haploids at a frequency of approximately 3%. It appears to be the case that crossing this stock with others either as a male or female will foster the production of haploids.

Coe and Sarkar (1964) developed a system for the recognition of haploids in stock 6. This involves two lines, one of which has all of the necessary factors for scutellum anthocyanin pigment. The other has the dominant inhibitor of color, *C-I*. When the colored stock is used as a female and crossed by the colorless one, the maternal haploids can be recognized as kernels that have colorless endosperm and colored embryos (loss of *C-I*). These stocks are available from the Stock Center.

Chang (1992a,b) has developed a stock 6 that has incorporated the *R-nj* marker. In this case, a colorless line is used as a female stock and is crossed by the stock 6 *R-nj* males. The individuals that have a *R-nj* pattern in the endosperm but not in the embryo are candidates for haploids. A certain fraction of these appear to be doubled haploids as evidenced by their genetic constitution and fertility. The majority, however, are maternally derived haploids.

REFERENCES

Chang M-T (1992a) Stock 6 induced double haploidy is random. Maize Genetics Cooperation News Letter 66: 98

Chang M-T (1992b) Preferential fertilization induced from Stock 6. Maize Genetics Cooperation News Letter 66: 99

Chase SS (1969) Monoploids and monoploid derivatives of maize (Zea mays L.). Bot Rev 35: 117–167

Coe EH (1959) A line of maize with high haploid frequency. Am Nat 93: 381–382

Coe EH, Sarkar KR (1964) The detection of haploids in maize. J Hered 55: 231–233

Kermicle JL (1969) Androgenesis conditioned by a mutation in maize. Science 166: 1422–1424

Kindiger B (1991) Development of a tertiary trisomic (A A B-A) stock carrying indeterminate gametophyte (*ig*). Maize Genetics Cooperation News Letter 65: 64

Kindiger B (1992) Development of cytoplasmic male sterile lines of maize which are homozygous for indeterminate gametophyte (*ig*). Maize Genetics Cooperation News Letter 66: 47

Sarkar KR, Coe EH (1966) A genetic analysis of the origin of maternal haploids in maize. Genetics 54: 453–464

58

Indeterminate Gametophyte (ig): Biology and Use

JERRY L. KERMICLE

The pattern of angiosperm embryo sac development is strikingly determinate. As in a majority of species, the functional megaspore undergoes three complete cycles of mitotic division. Subsequent wall formation and cell differentiation produces the characteristic 7-celled, 8-nucleate mature female gametophyte comprising an egg with 2 associated synergids, the binucleate central cell, and 3 antipodals. In maize, the antipodals proliferate during embryo sac development and in some strains the synergids regress. This highly ordered sequence is perturbed in embryo sacs carrying a mutation at map position 90 on

The Maize Handbook—M. Freeling, V. Walbot, eds.

chromosome 3, leading to multiple embryological abnormalities, unusual fertilization events, and consequent ploidy variation in the resulting embryo and endosperm. The mutation was named *indeterminate gametophyte* because the number of eggs and of polar nuclei in the central cell are not fixed respectively at one and two but are indeterminate, and because action of the mutant correlated with the genotype of the embryo sac rather than the zygote or pollen (Kermicle 1971). The variant is designated as recessive because homozygous *ig* plants are male sterile in inbred W23, the background in which the mutation arose. *Ig/ig* plants segregate 1:1 for indeterminate female gametophytes.

EMBRYOLOGICAL OBSERVATIONS

Multiple embryological abnormalities have been described both in mature (Lin 1978) and in developing (Lin 1981) *ig* embryo sacs. Megasporogenesis and embryo sac development through the four nuclear stages did not depart detectably from wild type whereas subsequent development deviated in numerous ways. In some sacs not all nuclei divided a third time. In others, eight nuclei were present preceding cellularization but were not distributed with four at each pole as normally. Still others underwent a fourth cycle of mitosis. Some of these manifestations are interpretable as changes in control of the cell cycle. However, the altered pattern of nuclear migration and eventual cell differentiation suggest more general effects of the gene. The primary lesion has not been identified.

Lin (1981) analyzed 62 mature *ig* female gametophytes of inbred W23. Only two conformed to the cardinal features of wild type: a single micropylar cell (the synergids having regressed in inbred W23, leaving only an egg) and a binucleate central cell with the nuclei lying adjacent to the egg. The number of cells in the micropylar region in the *ig* sample ranged from zero to five, averaging 1.96. The number of polar nuclei averaged 3.42, with 18% of the sacs having one or more accessory central cells. Fertilization involving accessory central cells was not observed in the 10 cases analyzed. Correspondingly, endosperm doubleness has not been observed in mature seed. Lin also counted the chromosomes in the endosperm of kernels developing on *ig ig* plants. The number of maternally contributed chromosome sets corresponded to the number of polar nuclei appressed to the egg rather than the total number present in the primary central cell.

KERNEL ABNORMALITIES

Only about half of the kernels born on *ig ig* plants are full-sized (Kermicle 1971). Another discrete class, constituting approximately one-fourth of the total, are viable but miniature. These proportions did not differ appreciably between wind-pollinated and hand crossed samples involving W23 *Ig ig*, W23 *Ig Ig*, or three unrelated inbreds lines used as males. Nearly one-fourth of all kernels are abortive. By comparing frequencies of seed classes with endosperm ploidy levels Lin (1984) associated full-size kernels with the 3× endosperm class, miniature with 4×, and seed failure with 2×, 5×, and higher ploidy levels. The pollen parent regularly contributed a single chromosome set.

Special insight into the relation of seed failure to ploidy level was gained from analyzing the full-sized kernels (about 10%) resulting from pollinating *ig ig* plants with nonmutant tetraploid strains. Control *Ig Ig* ear parents uniformly produce abortive kernels in parallel crosses. By following the fate of different endosperm ploidy classes during development, Lin found the endosperm of the full-sized kernel class to be hexaploid. The embryos were triploid. Significantly, the ratio between ploidy levels of endosperm (6×), embryo (3×), and maternal tissue (2×) in these kernels departs in each case from the standard 3×:2×:2× genomic proportions. It was proposed that a ratio within the endosperm of two chromosome sets of maternal origin to one of paternal origin is required for normal development. That is, chromosome sets are "imprinted" according to their respective mode of parental transmission, and deviations from the normal ratio of 2:1 result in impaired development.

The most conspicuous effect of *ig* involving embryogenesis is polyembryony. Per 1,000 kernels screened, 67 extra embryos were found among the full-sized class and 104 among the miniature class. Members of multiple embryo sets were identical for color marker genes heterozygous in the female parent, indicating descent from a common embryo sac. The endosperm phenotype of such polyembryonic seeds was concordant with the embryo phenotype. When the pollen parent was heterozygous for the marker, by contrast, embryos frequently were of different phenotype, or, if similar, did not correspond to the endosperm phenotype. From studies using a male parent heterozygous at eight loci it was concluded that twins derive from separate fertilization events but sometimes are identical genetically because the two sperm from one pollen grain fertilize two eggs of one embryo sac. In that case a sperm from a second pollen grain unites with the central cell (Kermicle 1971, 1974). That normal constraints on fertilization are somewhat relaxed in *ig* embryo sacs is also suggested by an increase in heterofertilization from 2% to 7%. Nevertheless, many of the cells in the micropylar region of *ig* sacs either remain unfertilized or the resulting zygotes fail, based on the much higher frequency of multiple micropylar cells than multiple embryos.

ANDROGENESIS AND CYTOPLASMIC SUBSTITUTION

Crosses incorporating the *R-Navajo* (*R-nj*) embryo anthocyanin marker in the pollen to detect parthenogenesis revealed an increase in maternal haploids from 1.7 to 6.5×10^{-3} in *Ig Ig* compared with *ig ig* ear parents. It is perhaps not unexpected that a mutation that permits proliferation of nuclei and cells in the embryo sac also fosters embryogenesis without fertilization.

More exceptional was the recovery of patroclinous offspring, detected by reversing parentage of the color markers (Kermicle 1969). No instance of androgenesis was detected using *Ig Ig* females, hardly surprising in the small population tested since the frequency of parental haploids in wild type stocks has been estimated at 1/80,000 (Chase 1963).

This compares with 6.6 and 23.5 × 10^{-3} obtained from *Ig ig* and *ig ig* females. Frequencies did not differ between *Ig ig* and *Ig Ig* males of inbred W23. Many fewer patriclinous offspring were obtained when W22 *Ig Ig* was pollen parent, indicating the general genetic composition but not the specific *ig* constitution of the male gametophyte influenced the incidence of androgenesis. To induce androgenesis *ig* need be carried only by the embryo sac; thus a single stock, marked by an appropriate detection system such as *R-nj* embryo color, can be used to obtain haploids from various pollen sources.

Androgenesis occurred independently of twinning when one of the two embryos was biparental. Remarkably, a number of twins having one matroclinous and one patroclinous member have been observed. Although rare, this combination has occurred much more frequently than expected based on parthenogenesis and androgenesis as independent events. Cells of both gametophytes evidently can respond to embryogenic conditions present in certain *ig* embryo sacs.

To identify which nucleus in the male gametophyte undergoes androgenesis, plants homozygous for translocation B-10La were utilized as males. In the genetic background used, nondisjunction of the B-10 chromosome is virtually 100% at the division of the generative nucleus. Thus, the tube nucleus has 1 B-10 chromosome, 1 sperm has 2, and 1 sperm has none. The latter sperm class is, of course, incapable of giving androgenetic derivatives due to deficiency of chromosome arm 10L. Each of three androgenetic haploids had 12 chromosomes, including two B-10 chromosomes. The one androgenetic diploid in this sample had 24 chromosomes with four B-10s, likewise indicating involvement of a sperm but followed in this case by chromosome doubling (Kermicle 1974).

A subsequent experiment utilized contrasting cytoplasmic types (N vs. cms T) to trace the source of the cytoplasm. The frequency of androgenesis ranged from 13.5 to 20.8 × 10^{-3} for three inbred pollinators, with 15 diploid cases among 151 total (Table 58.1). Six of the 15 diploids were derived from *Rf1*/*rf1* pollen parents and could be tested for *Rf*:*rf* segregation through crosses to cms T, *rf1 rf1*. (*Rf1* is a nuclear gene required in

Table 58.1 Frequencies of androgenesis in crosses of three inbred lines as male to W23 *ig ig* females

Pollen parent	Kernel population	Androgenetic derivatives		
		Haploid	Diploid	Frequency (×10^{-3})
WA374	3,369	59	11	20.8
W23R	2,304	28	3	13.5
A632	3,583	49	1[a]	14.0
	9,256	136[b]	15	16.3

[a]Androgenetic diploid twins.

[b]Includes eight that occurred as one member of twins. In five the co-twin was a W23 hybrid; in three, the co-twin was a material haploid.

combination with *Rf2* for pollen fertility in *cms T* plants; R = restorer, f = fertility.) In no case was the androgenetic diploid heterozygous for the restorer as in the parent, again implying postmeiotic chromosome doubling. Attempts to establish diploid lines by pollinating the haploid plants with pollen from the maintainer line (*rf1* maintains sterility) of the respective inbred were successful in 52% (inbred WA374), 47% (W23R), and 82% (A632) of cases.

Plants in these lines as well as counterpart lines derived from androgenetic diploids were pollinated with a maintainer line of the other two inbreds, providing single cross material for evaluating male fertility and sensitivity to a crude toxin preparation from leaves infected with *Bipolaris maydis* race T, to which mitochondria of cms T cytoplasm are uniquely sensitive. All the hybrid progeny descended from 41 androgenetic individuals isolated from cms T, *ig ig* female × N, *Ig Ig* male crosses proved sensitive to toxin and all were male sterile. Hybrid progeny descended from seven cases of androgenesis in N, *ig ig* female × T, *Rf1 rf1 Ig Ig* male crosses were fertile and resistant to toxin. Similarly, the hybrid progeny of six control androgenetic lines developed from N, *ig ig* × N, *Ig Ig* crosses proved resistant to toxin and were fertile. In no case was evidence of paternal transmission of cytoplasm obtained. Thus *ig*-induced androgenesis appears to be an effective means of substituting the cytoplasm of an established line.

Albertsen and Trimnell (1990) report conversion of a number of lines to cms C and cms S by *ig*-induced androgenesis as well as through conventional backcrossing. Although plants in the cms C lines converted via androgenesis tended to be shorter than their fertile counterparts, comparison of the derived lines in hybrid combinations indicated few if any detrimental effects of *ig* haploidy.

MALE STERILITY

Although haploid *ig* and diploid *ig ig* plants undergo normal vegetative development, certain reproductive abnormalities are attributable to the sporophyte rather than to the embryo sac itself. On the female side, the incidence of ear tip fasciation is increased. On the male side, homozygous *ig* plants in inbred W23 background produce little if any functional pollen. Sterility is recessive in that *Ig ig* heterozygotes are fertile and the *ig* allele is transmitted by pollen in Mendelian proportion. The cytological basis for the sterility has not been reported. It would be interesting to learn how the cause of male sterility may be related to the various abnormalities of *ig* embryo sacs, including a low level (about 10%) of ovule abortion on *ig ig* plants (Lin 1981).

Two means of ameliorating the male sterility have been explored. Kindiger (1991) utilized the B-A translocation B-3Ld to construct *ig ig* B-3(Ig) partial trisomics. Because *Ig* lies near the translocation breakpoint, and since the B-3 chromosome rarely is transmitted by pollen, most pollen grains of the partial trisomic transmit *ig*. After self-pollination, 40–50% of the offspring are expected to be *ig ig*; crosses of the partial trisomic plants to *ig ig* females should give mostly homozygous offspring.

Another means of offsetting the male sterility surfaced while transferring *ig* and the

R-nj marker system to other inbred backgrounds by conventional backcrossing. Incorporation into inbred W64A was accompanied by little change in frequency of defective kernels, with complete male sterility of the homozygous derivative. In contrast, attempts to incorporate *ig* into inbred W22 were unsuccessful due to a decreased frequency of the aberrant kernel phenotypes and failure to obtain male sterile derivatives. During incorporation into A158 and M14 a high incidence of kernel defectiveness was retained but the derived homozygotes were male fertile, albeit somewhat reduced relative to the standard lines, permitting homozygous lines to be established. Small-scale tests using hybrids between *ig*-carrying lines of W23, W64A, A158, and M14 have produced androgenetic offspring in frequencies similar to *ig ig* plants of W23. Derivatives of these hybrids can be maintained with *ig* homozygous and should be adaptable to a much broader range of environments than the original inbred W23 *ig* stock.

Paper No. 3368 from the Laboratory of Genetics, University of Wisconsin.

REFERENCES

Albertsen MC, Trimnell MR (1990) Agronomic comparisons among *ig*-derived and backcross-derived CMS maize lines. Agron Abst, Am Soc Agron, Madison, WI, p 78

Chase SS (1963) Androgenesis—its use for transfer of maize cytoplasm. J Hered 54: 152–158

Kermicle JL (1969) Androgenesis conditioned by a mutation in maize. Science 116: 1422–1424

Kermicle JL (1971) Pleiotropic effects on seed development of the *indeterminate gametophyte* gene in maize. Am J Bot 58: 1–7

Kermicle JL (1974a) Identical twins of dizygotic origin. Maize Genetics Cooperation News Letter 48: 180–181

Kermicle JL (1974b) Origin of androgenetic haploids and diploids induced by the *indeterminate gametophyte* (*ig*) mutation in maize. In Kasha KJ (ed) Haploids in Higher Plants, The University of Guelph Press, Guelph, Canada, p 137

Kindiger B (1991) Development of a tertiary trisomic (A A B-A) stock carrying *indeterminate gametophyte* (*ig*). Maize Genetics Cooperation News Letter 65: 64

Lin B-Y (1978) Structural modifications of the female gametophyte associated with the *indeterminate gametophyte* (*ig*) mutant in maize. Can J Genet Cytol 20: 249–257

Lin B-Y (1981) Megagametogenetic alterations associated with the *indeterminate gametophyte* (*ig*) mutation in maize. Rev Brasil Biol 41: 557–563

Lin B-Y (1984) Ploidy barrier to endosperm development in maize. Genetics 107: 103–115

59

Production of a Ploidy Series

James A. Birchler

Maize is unique among plant species in having techniques available in which conversion from diploidy to tetraploidy can be accomplished by the use of two meiotic mutants that produce diploid unreduced gametes—namely, *asynaptic* (*as*) (e.g., Miller 1963, Alexander 1957) and *elongate* (*el*) (Rhoades and Dempsey 1966). Although theoretically these two mutants could be used to generate an extensive ploidy series, this has been accomplished only with the *elongate* mutant thus far.

Conversion of selected markers from diploid to tetraploid stocks can be accomplished by use of either of these two mutants. For example, to convert a certain allele of a gene of interest, it is first crossed by homozygous *elongate*. The next generation is self-pollinated. Among the progeny will be one-quarter that are homozygous for *el*. If these plants are used as females for the desired tetraploid line, the unreduced gametes from the *elongate* homozygotes will produce fully developed kernels. The reduced gametes also on the ear will be defective because of the ploidy hybridization barrier (Randolph 1935; Lin 1982). By repeated self-pollination, the mutant can be made homozygous in the tetraploid. The reader is referred to papers by Doyle (1973, 1979a,b, 1982, 1986) for discussions of the genetics of tetraploids.

Rhoades and Dempsey (1966) described the construction of a 2–7x series using *elongate*. Diploid stocks of *elongate* produce on the female side normal haploid eggs and to some degree diploid eggs. When these stocks are crossed by normal males, zygotes are produced that are diploid and triploid. The latter are associated with defective pentaploid endosperms, which cause a certain degree of problems with germination, but these problems are not insurmountable. If the same stock is crossed by a tetraploid male, then triploid and tetraploid individuals are produced. The triploids are associated with tetraploid endosperms that are somewhat defective, whereas the tetraploids have hexaploid endosperms that are normal. By self-pollinating the tetraploids, Rhoades and Dempsey recovered a stock that was a tetraploid homozygous for *elongate*. When this stock is used as a female for pollen from a diploid plant, triploid and pentaploid individuals are produced. If the same stock is crossed by pollen from a tetraploid male, tetraploid and hexaploid individuals were recovered. If the hexaploids are crossed as males onto the

The Maize Handbook—M. Freeling, V. Walbot, eds.

tetraploid *elongate* stock, then pentaploid and heptaploid individuals are produced. It is not possible to maintain a hexaploid *elongate* stock, because they are female sterile, but they do produce pollen that can be crossed onto the tetraploid. According to Rhoades and Dempsey, the vigor of the ploidy series is good until the hexaploid and heptaploid level, at which the plants are smaller and female-sterile.

If one wishes to produce a 1–7x series in the same background, haploids could be produced by crossing the diploid stock to females of *indeterminate gametophyte* (*ig*), which fosters androgenetic haploid production. The use of *ig* is described in more detail elsewhere in this volume.

REFERENCES

Alexander DE (1957) The genetic induction of autotetraploidy: A proposal for its use in corn breeding. Agron J 49: 40–43

Doyle GG (1973) Autotetraploid gene segregation. Theor Appl Genet 43: 139–146

Doyle GG (1979a) The allotetraploidization of maize. Part 1. The physical basis—differential pairing affinity. Theor Appl Genet 54: 103–112

Doyle GG (1979b) The allotetraploidization of maize. Part 2. The theoretical basis—the cytogenetics of segmental allotetraploids. Theor Appl Genet 54: 161–168

Doyle GG (1982) The allotetraploidization of maize. Part 3. Gene segregation in trisomic heterozygotes. Theor Appl Genet 61: 81–89

Doyle GG (1986) The allotetraploidization of maize. Part 4. Cytological and genetic evidence indicative of substantial progress. Theor Appl Genet 71: 585–594

Lin B-Y (1982) Association of endosperm reduction with parental imprinting in maize. Genetics 100: 475–486

Miller OL (1963) Cytological studies in *asynaptic* maize. Genetics 48: 1445–1466

Randolph LF (1935) Cytogenetics of tetraploid maize. J Agric Res 50: 591–605

Rhoades MM and Dempsey E (1966) Induction of chromosome doubling at meiosis by the elongate gene in maize. Genetics 54: 505–522

60

Absorption Cytophotometry of Nuclear DNA

RICHARD V. KOWLES, GEORGIA L. YERK, AND RONALD L. PHILLIPS

FEULGEN CYTOPHOTOMETRY

Cytophotometry allows quantitative analysis of cells and organelles. Measurements of DNA, RNA, and proteins can be made using techniques based upon either absorption or fluorescence. The following procedure outlines the preparation of tissue for the Feulgen reaction and absorption cytophotometry using the two-wavelength method (Mendelsohn 1958a; Berlyn and Cecich 1976). The technique can reveal the ploidy level of cells, the effects of chemical and physical agents upon ploidy levels, and other cellular kinetics related to DNA content. The technique is especially useful when it is important to maintain the position of cells in a tissue. Consult a microscope manufacturer about the availability of cytophotometry units appropriate for your microscope.

MATERIALS

Bleach

1 part 1 M HCl, 1 part 10% potassium metabisulfite (w/v), 18 parts distilled water. Make fresh.

Schiff's Reagent

1. Dissolve 1 g basic fuchsin into 200 ml boiling distilled water; do this slowly and carefully.

The Maize Handbook—M. Freeling, V. Walbot, eds.

2. Shake well.
3. Cool to approximately 50°C.
4. Filter and add 30 ml 1 M HCl to the filtrate.
5. Add 3 g potassium metabisulfite. Do this step in a fumehood.
6. Allow to bleach overnight in a tightly stoppered dark bottle placed in the dark at room temperature.
7. Add 0.5 g Norit carbon (vegetable carbon).
8. Filter the reagent.
9. Store in a dark bottle in the refrigerator. Under refrigeration, the Schiff's reagent will last for about 6 months, or even longer.

METHOD

1. Fix small pieces of the tissue overnight in 95% ethanol/glacial acetic acid, 3:1.
2. Squash in a small drop of 45% acetic acid on a slide and add a cover glass. Slides used in these preparations should be painstakingly cleaned. See the cell biology section of this volume for suggested methods of cleaning slides.
3. Place the slide on dry ice for about 5 minutes; then remove the cover glass by flipping it off with a razor or scalpel.
4. Pass the slides through the following Feulgen reaction schedule:

a. 70% ethanol	5 minutes
b. 50% ethanol	5 minutes
c. 30% ethanol	5 minutes
d. Distilled water	5 minutes
e. 5M HCl	Time based upon results of hydrolysis tests*
f. Distilled water	5 minutes
g. Schiff's reagent	60 minutes
h. Bleach	10 minutes
i. Distilled water	5 minutes
j. 30% ethanol	5 minutes
k. 50% ethanol	5 minutes
l. 70% ethanol	5 minutes
m. 95% ethanol	5 minutes
n. Absolute ethanol	5 minutes
o. Air dry	
p. Mount in Euparal or comparable medium	

*Hydrolysis tests: the hydrolysis time that will effect the maximum DNA absorption needs to be determined. (See subsequent section.)

HYDROLYSIS TESTS

1. Prepare a set of slides using the Feulgen technique as previously described. The series of slides should be hydrolyzed in 5 M HCl at room temperature for different time periods, beginning at about 6 minutes and sequentially increasing each preparation 2 additional minutes to about 30 minutes.
2. Measure the absorbance at 560 nm of at least 10 nuclei of the same size from each of the hydrolysis trials.
3. Select the optimum hydrolysis as the shortest hydrolysis time required to reach maximum absorbance.

SPECTRAL CURVE

The two-wavelength method of cytophotometry also requires the construction of a spectral curve to determine which two wavelengths to use. The two wavelength method reduces error from the non-uniform distribution of DNA within the nucleus—that is, distributional error.

1. Prepare slides as previously outlined using the appropriate hydrolysis time as determined from the hydrolysis tests.
2. Measure the absorbance of 10 nuclei over a range of 470 to 600 nm at 10-nm intervals.
3. Calculate the mean absorbance of the ten nuclei at each wavelength.
4. Plot the mean absorbances vs. wavelengths.
5. Select the two wavelengths as follows: (a) The high wavelength should be slightly off the peak absorbance. (b) The low wavelength is the point of half the maximum absorbance on the ascending part of the spectral curve. Figure 60.1 illustrates the spectral curve obtained for nuclei from maize endosperm tissue.
6. Use these selected wavelengths for all subsequent measurements.

DNA MEASUREMENTS

1. The transmittance of each nucleus is determined for the high wavelength. Use a clear area adjacent to the nucleus being measured to give 100% transmittance on the instrument.
2. Repeat step 1 for the low wavelength.

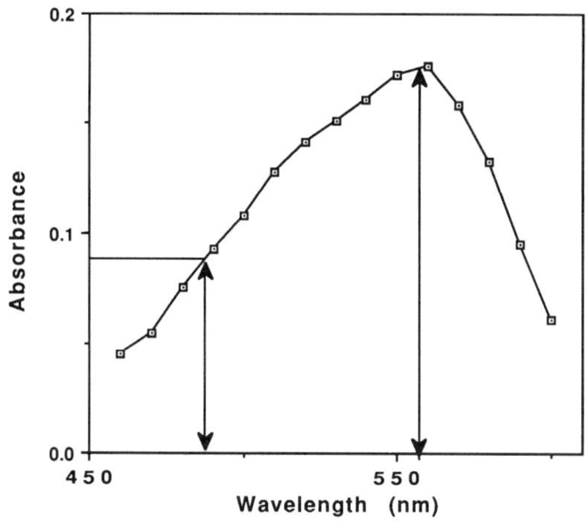

Figure 60.1. Spectral curve of nuclei from maize endosperm tissue. The two wavelengths selected were 560 and 488 nm.

3. The transmittance of the high and low wavelengths is used to obtain a value from tables found in Mendelsohn (1958b) or Berlyn and Miksche (1976).*
4. Multiply the table value by the aperature size used on the cytophotometer to gain DNA content in arbitrary units (A.U.).
5. Calculate the actual picograms of DNA per nucleus by measuring nuclei of chicken erythrocytes as a reference material. These preparations must be processed in exactly the same manner as the tissue being investigated. Then solve the following equation: 2.6 pg/A.U. for chicken erythrocytes = pg per nucleus/A.U. for nuclei being measured.

Notes

1. Many investigators use 5 M HCl at room temperature rather than 1 M HCl at 60° C for hydrolysis. The 5 M HCl step needs to be conducted in a glass vessel. Plastics, stainless steel, etc., will react with the acid.

*The equation used to generate Mendelsohn's table for use in two-wavelength cytophotometry can be programmed into a computer. Under these conditions, one only needs to enter the low- and high-wavelength transmittances to quickly obtain the arbitrary units of DNA.

$Mt = La\ C$	where Mt is the total amount of chromatophore
$La = 1 - T$	where T is the transmission of the low wavelength
$C = 1/(2 - Q) \times (\ln 1/Q - 1)$	
and	
$Q = 1 - T_b/1 - T_a$	where Tb is the transmission of the high wavelength and T_a is the transmission of the low wavelength

Mendelsohn's table gives La C directly.

2. Different values for the DNA content of chicken erythrocyte nuclei have been reported. Most of the values, however, range between 2.5 and 2.6 pg.
3. Some investigators place the chicken erythrocytes on the same slide with the tissue to be measured. This procedure eliminates slide-to-slide variation.

REFERENCES

Berlyn GP, Cecich RA (1976) In Miksche JP (ed) Modern Methods in Forest Genetics, Iowa State University Press, Ames, pp 1–8

Berlyn GP, Miksche JP (1976) Botanical Microtechnique and Cytochemistry, Iowa State University Press, Ames

Mendelsohn ML (1958a) The two-wavelength method of microspectrophotometry 1. A microspectrophotometer and tests on model systems. J Biophys Biochem Cytol 4: 407–414

Mendelsohn ML (1958b) The two-wavelength method of microspectrophotometry ll. A set of tables to facilitate calculations. J Biophys Biochem Cytol 4: 415–424.

61

Flow Cytometry for Endosperm Nuclear DNA

RICHARD V. KOWLES, GEORGIA L. YERK, LIANG SCHWEITZER, FREIDRICH SRIENC AND RONALD L. PHILLIPS

The following procedure may be used to stain endosperm nuclei for flow cytometry. Although many fluorochromes may be used, we prefer mithramycin A to obtain measurements of DNA content and cell numbers. It gives a good stable signal, does not stain RNA, is relatively inexpensive, and is easy to prepare. Mithramycin A is a GC-specific stain. When measurements using an AT-specific stain are desired, Hoechst 33258 can be used. Techniques are presented for both stains. The basic method is to macerate endosperm from individual kernels through a fine mesh to obtain relatively clean nuclear preparations. These preparations are then stained with a fluorochrome. Subsequently, fluorescence of individual nuclei is measured following excitation by a laser beam in a

The Maize Handbook—M. Freeling, V. Walbot, eds.

flow cytometer. These measurements can be utilized to calculate DNA content per nucleus, to measure DNA content in subpopulations of a sample, to determine cellular kinetics such as the proportion of cells in S vs. G phase, or to obtain information regarding changes in nuclear populations over development. Additional information may be obtained by including an internal standard of known concentration and DNA content, such as chicken red blood cells, in each sample. This allows for calculation of total cell number per endosperm and facilitates comparison among different samples.

METHODOLOGY

Pollinations

Samples to be compared should be taken from ears pollinated on the same date. Comparisons of nuclear DNA content of endosperm obtained from ears pollinated on different dates may show significant variation within an inbred line (Kowles et al. 1992).

Tissue Collection

Two methods may be utilized to obtain kernel samples for analysis. The choice of methods depends on the number of kernels needed for each sampling period, the number of sampling periods, the amount of remnant seed desired, and the number of pollinations that are feasible.

Method 1

This method of sampling is less time consuming but requires more pollinations if multiple samples are desired or if seed is needed for future growing seasons. Remove the ear from the plant you wish to sample. Using a scalpel, remove the desired number of kernels from the middle one-third of the ear, being careful to take the entire kernel. Kernels near the tip and base of the ear may not have the same developmental profile as the rest of the ear. Place kernels in fixative immediately.

Method 2

This method allows for multiple sampling from the same ear. It also decreases the number of pollinations that are necessary. Mature seed not taken for kernel samples may be stored for future planting. Slit the husk beginning at the silks to about 1 inch from the bottom of the ear. Carefully peel the husk back. Using a scalpel, remove the desired number of kernels from the middle one-third of the ear taking care to remove the entire kernel. Immediately, place kernels into fixative. Smooth the husk back into place. To prevent contamination, place masking tape around the ear to cover the incision. Cover the ear with the pollination bag again. In our experience, the time involved to apply fungicide or lanolin paste to each incision to reduce contamination is not justified. Two other factors that will reduce contamination are sterilizing the scalpel after cutting any infected ear and not sampling during rain or while the bags are wet.

FIXATIVE

Fixative should be freshly made just prior to sampling. Mix 3 parts of 95% ethanol to 1 part of propionic acid. Samples should be fixed in ample amounts overnight and then changed to 70% ethanol. Once in 70% ethanol, samples can be stored at −20°C until analysis.

SOLUTIONS FOR NUCLEI PREPARATIONS

Grinding Buffer

100 mM glycine
1.0% hexylene glycol (v/v)
0.1% Triton X-100 (v/v)
0.2% phenylmethylsulfonyl fluoride (w/v)

Mithramycin Buffer

45 mM $MgCl_2$
30 mM sodium citrate
20 mM morpholino propane sulfonic acid (MOPS)
0.1% Triton X-100 (v/v)
(pH to 7.0 with 1.0 M NaOH)

5× Mithramycin A Stock Solution

Dissolve 10 mg mithramycin A in 20 ml sterile mithramycin buffer. Store at −4°C in 1 ml aliquots.

Note: Mithramycin A may be carcinogenic so wear gloves at all times when working with it. Also, do not inhale stain powder.

Mithramycin A Working Solution

Dilute 5× mithramycin A stock solution, with mithramycin buffer, 1:4. Dilute enough stain for all samples in one quantity. This is important as it will reduce problems that could arise from differences among working stain solutions.

NUCLEAR PREPARATION

1. Dissect out the entire endosperm. Usually this can be easily accomplished by first cutting a thin slice from the top of the kernel. Then apply pressure at the base of the kernel with a curved dental hook and move toward the top of the kernel while keeping the pressure constant. The endosperm should slip out the cut opening. Then scrape the piece of endosperm from the slice initially cut from the top. Remove the embryo and any pericarp debris from the endosperm. Examine the inside of the kernel to be sure no endosperm has been left behind.
2. Using a flattened probe, macerate the endosperm through a 94-μm mesh screen (Bellco Biotechnology, Vineland, NJ) set on a small funnel in a 1.5-ml microfuge tube. Wash the mesh three times with 500 μl each of grinding buffer to free nuclei trapped in the mesh. Some debris will remain on the screen.
3. Screens and funnels should be soaked and shaken vigorously in undiluted bleach after each sample. They are then rinsed in distilled water several times and air dried.
4. Centrifuge the samples for 1 minute at 180*g*.
5. Pour off supernatant and resuspend pellet in 400 μl of mithramycin buffer. Do not vortex the samples to resuspend. Use gentle tapping of the tube.
6. Allow the samples to equilibrate for 30 minutes.
7. Centrifuge the samples for 1 minute at 180*g*.
8. Discard the supernatant. Resuspend the pellet in 200 μl of working stain solution. Cover the samples with foil and place in the refrigerator. The samples must stain a minimum of 5 hours. Staining overnight also works fine.

USE OF AN INTERNAL STANDARD

An internal standard may be included in the samples when measurement of DNA content in picograms is desired or when counts of total nuclei per endosperm are needed. Any fluorescent particle of known concentration may be utilized so long as it can be visualized as an entirely separate peak on the output from the flow cytometer. For calculations of DNA content in picograms, it is important to consider the base-pair specificity of the stain and the ratio of the particular bases in the standard vs. the nuclei being measured. We have used chicken red blood cells (CRBCs) to determine both DNA content in picograms and cell number in a given endosperm sample. The stained CRCBs should be added to the diluted stained endosperm nuclei preparations just prior to analysis on the flow cytometer. The volume of CRBCs to be added to the sample should be enough so that the CRBCs represent 25–30% of the total number of particles measured. Too many or too few cells will lead to biased estimates of the total number of nuclei in the endosperm preparation. Estimates of DNA content may also be affected if the proportion of CRBCs is too low.

Solutions for Preparing CRBCs

Phosphate-Buffered Saline (PBS)

Add the following to 800 ml distilled H_2O:

8 g NaCl
0.2 g KCl
1.44 g Na_2PO_4
0.24 g KH_2PO_4
pH to 7.4 with HCl

Bring the total volume to 1 liter with distilled H_2O. Autoclave for 20 minutes at 15 psi.

Mithramycin Buffer

See previous section.

1× Mithramycin Stain Solution

See previous section.

95% Ethanol

Fixing and Staining CRBCs

1. Dilute *fresh* CRBCs with 1 × PBS until they may be counted with a hemocytometer to obtain the concentration.
2. Add ethanol slowly to a final concentration of 25%. CRBCs may be stored this way for 1–2 weeks.
3. Count cells using a hemocytometer and determine their concentration.
4. Centrifuge a 200-μl aliquot of CRBCs at low speed for 5 minutes. Discard supernatant.
5. Resuspend the pellet in 200 μl of mithramycin buffer. Hold the CRBCs in this solution until all of the endosperm nuclei preparations are ready to be stained.
6. Centrifuge for 5 minutes at low speed. Discard the supernatant.
7. Add 200 μl of 1 × mithramycin to the CRBCs. This step must be conducted at the same time as the stain is added to the maize endosperm nuclei.
8. Just prior to analysis on the flow cytometer, add 800 μl of mithramycin buffer to the CRBCs.

STAINING WITH HOECHST 33258

Solutions

Grinding Buffer

See above.

Hoechst Stain

Dissolve 5 μg Hoechst 33258 per ml of Mcllvaine's buffer.

Mcllvaine's Buffer

Mix 17.6 ml of 0.1 M citric acid to 82.4 ml of 0.2 M disodium hydrogen phosphate. pH to 7.0 if necessary.

1. Dissect out endosperm as in the above procedure.
2. Macerate through a 94-μm mesh and rinse three times with 500 μl each of grinding buffer.
3. Centrifuge for 1 minute at 180*g*.
4. Discard supernatant. Add 400 μl of Mcllvaine's buffer and resuspend by gentle tapping.
5. Allow to equilibrate for 30 minutes.
6. Centrifuge for 1 minute at 180*g*. Discard supernatant. Resuspend in 400 μl of Hoechst stain. Allow to set for 1 hour at room temperature in light-tight container.
7. Centrifuge for 1 minute at 180*g*. Discard supernatant. Resuspend in 500 μl of Mcllvaine's buffer.
9. Centrifuge for 1 minute at 180*g*. Discard supernatant. Resuspend in 200 μl of Mcllvaine's buffer.

SETTINGS FOR THE FLOW CYTOMETER

The instrument we use is a Cytoflourograph IIs (Ortho Diagnostics, Westwood, MA) operating with an analytical quartz flow cell with a 200 μm square flow channel and an argon ion laser. The argon ion laser is tuned to 457 nm, and it is operated at a light intensity of 50–100 mW. To obtain DNA content and cell number measurements, the data are normally collected in logarithmic mode. The FALS (forward-angle light scatter) signal is detected on PMT3 (photomultiplier tube) using the standard configuration. The fluorescence signal is split using a green-transmitting, red-reflecting dichroic mirror (Ortho part #300.0271.002). Reflected fluorescence is detected by PMT2 after passing through a 570-nm cutoff filter, and it is processed in peak mode. This signal is used in combination with the FALS signal to trigger data acquisition and to exclude noise. Transmitted fluorescence is detected by PMT1 after passing through a 515-nm cutoff filter (Ortho part # 300.0281.003) and processed in area mode. This signal is used to quantify DNA content.

RUNNING SAMPLES

Dilute each sample at least 1:8 (sample: buffer) just prior to placing in the flow cytometer. For endosperm older than 18 days after pollination, dilute 1:16 or 1:24. Internal standards should be added at this time and mixed well. As mentioned earlier, the amount of standard needed depends upon the sample; hence, it will be necessary to run more than one aliquot to obtain optimum results. The standard should give a distinct peak on the histogram but should not be more than 30% of the total particles measured. Too few standard particles will lead to inaccurate estimates, and too many will make it difficult to obtain good histograms of the endosperm nuclei.

Some debris will be present in the nuclei preparations. Debris may be separated from the nuclei by gating in the cytogram of forward-angle light scatter vs. 90° light scatter. Flow rate should be maintained between 30 and 100 nuclei per second.

REFERENCE CITED

Kowles RV, Yerk GL, Srienc F, Phillips RL (1992) Maize endosperm tissue as an endoreduplication system. In Setlow M (ed) Genetic Engineering, Principles and Methods, Vol 14, Plenum Press, New York, pp 65–88.

ADDITIONAL REFERENCES

Kowles RV, Srienc F, Phillips RL (1990) Endoreduplication of nuclear DNA in the developing maize endosperm. Dev Genet 11: 125–132

Kowles RV, McMullen MD, Yerk G, Phillips RL, Kraemer S, Srienc F (1992) Endosperm mitotic activity and endoreduplication in maize affected by defective kernel mutations. Genome, 35: 68–77.

62

Allelism Testing of Lethal Mutations

JANICE K. CLARK AND WILLIAM F. SHERIDAN

When studying lethal mutations it is necessary to propagate and work with them as heterozygotes. Usually this involves planting normal kernels from a self-pollinated ear that segregates for the lethal mutation. In the case of *defective kernel* (*dek*) and *embryo-specific* (*emb*) mutations, visual examination serves to identify the kernels with homozygous mutant embryos; these can be avoided when selecting kernels for planting the next generation. When working with seedling or mature plant lethals including mutations resulting in barrenness or sterility, the mutant homozygotes become evident only after germination.

In the field the only useful plants will be those that are heterozygous for the mutation under study while the homozygous normal plants are usually of no value. Ideally, for many purposes it would be most convenient if every plant in a family of plants grown from a self-pollinated ear would be heterozygous, but this is not ordinarily possible. Usually a self-pollinated ear on a plant heterozygous for a mutation will, barring reduced sexual transmission of the mutant or normal allele, produce kernels with embryos that are $+/+$, $+/m$, and m/m in the ratio of 1:2:1. In the case of lethal mutations one-third of the plants available for pollinating will be of $+/+$ constitution and two-thirds will be of $+/m$ constitution, resulting in a probability of two-thirds that any particular plant is a heterozygote ($+/m$). In order to be certain of genotype, each plant must be self-pollinated to produce a scorable ear.

When examining a collection of mutations that produce similar phenotypes an early step in genetic analysis is a test for allelism. If the size of the mutant collection is more than a few, the number of crosses for an exhaustive allelism test can be greatly reduced by first determining the chromosome arm location for each of the mutations. This is evident from the following considerations: 3 crosses are required for an allelism test with three mutations; 15 crosses with 6 mutations; and thirty-six crosses with nine mutations.

Two sets of cytogenetic stocks have been developed to determine chromosome arm location. The T-*waxy* translocations use linkage between the gene of interest and the translocation breakpoint for each arm (which is linked in each case with the easily scored kernel trait *waxy*) to indicate arm locations. With the B-A translocations, a recessive allele

The Maize Handbook—M. Freeling, V. Walbot, eds.

of interest is uncovered by virtue of being hemizygous in the progeny of a cross by the arm-locating stock (termed a TB) for the arm on which the gene is located. Both the T-*waxy* and the B-A arm-locating sets are available from the Maize Stock Center. Instructions for propagation and use of T-*waxy* stocks are available in Anderson (1956) and Burnham (1982); for the B-A stocks consult Beckett (1978, 1991, and in this volume).

Because the entire basic set of arm-locating stocks must be grown in the experimental nursery regardless of the number of different mutations being tested, many mutations may be tested with almost the same efficiency as testing only one. Once chromosome arm locations have been determined, allelism tests are restricted to mutations on the same arm. During the time period needed to perform the chromosome arm analysis, the investigator can convert the mutant collection into a favorable genetic background for conducting the allelism tests. (See below.)

To conduct an unambiguous allelism test between two lethal mutations a known heterozygote of one mutation must be crossed with a known heterozygote of the other mutation. If the genetic stocks produce two ears suitable for pollination then the test is readily performed. Self-pollinate both plants, and cross one to the other. This is usually done on 2 successive days with the lower ear pollinated on the first day to assure its successful development. Stocks that regularly produce tillers could also be used. In those cases where examination of the progeny produced on the self-pollinated ears reveals that both parents were heterozygous for the mutations of interest, an examination of the cross-pollinated ear can provide an unambiguous test as to whether or not the mutations are at the same locus. If they are allelic they will ordinarily fail to complement, and approximately one-fourth of the kernels on the crossed ear will be obviously mutant (in the case of *dek* mutations), will have mutant embryos (in the case of *emb* mutations), or will show the lethal phenotype after germination.

ALLELISM TESTS WITH STOCKS BEARING ONE EAR

If the genetic stocks do not produce two good ears, double pollination of this single ear is required, i.e., self- and cross-pollination. One technique is described in detail in Sheridan and Clark (1987) with accompanying photographs and discussion of alternative procedures. Next we present the essential features of this procedure and its application, using as markers a self-pollination that produces colorless kernels and a cross-pollination that produces purple kernels.

DEVELOPING COLORED AND COLORLESS MUTANT STOCKS

Because both self-pollinated and cross-pollinated kernels are produced on the same ear it is important that they be readily distinguishable. This is a straightforward matter when the male parent carries all the factors for aleurone color while the female parent lacks one or more of these factors. The allelic combinations required for anthocyanin synthesis are reviewed in this volume and in Coe et al. (1988). Usually, stocks without anthocyanin are homozygous for the dominant alleles of the five structural genes required for anthocyanin synthesis (*A1*, *A2*, *Bz1*, *Bz2*, and *C2*) but homozygous for the recessive alleles at the regulatory loci (*c1* and *r1*). When a mutation arises in a colorless stock it is advantageous to promptly self-pollinate three plants in a family carrying the new mutation and cross each of these plants onto a plant of a homozygous colored stock. We recommend that the colored stock carry a self color allele of the *R1* locus such as *R1-scm2* or *R1-scm3*. When this *R* allele together with the other factors required for anthocyanin synthesis in the aleurone is present in the embryo, the scutellum of the embryo is deeply pigmented. This feature of the *R1-scm* alleles has proven to be useful in cytogenetic studies (Birchler et al. 1990) as well as in studies on *emb* lethal mutations (Clark and Sheridan 1991).

Because the probability that any one of the three pollen plants of the colorless stock does *not* carry the mutant allele is one-third, the probability that all three plants would fail to carry the allele is 1/3 × 1/3 × 1/3 or 1/27. Therefore the probability that *at least* one of the three plants is heterozygous for the mutant allele would be 26/27, which provides better than a 95% probability of successfully transmitting the mutant allele onto the colored standard stock. After harvesting and scoring the self-pollinated ears (or their progeny) an outcross ear (a colored ear crossed onto with pollen from a proven heterozygote for the new mutation) can be identified. This colored ear is used as a source of kernels that are planted in the next generation to propagate the new mutation. Commonly 8–10 kernels are planted and at least five plants are self-pollinated. Since the likelihood that an individual plant is heterozygous for the mutant allele is 1/2 (barring reduced sexual transmission) the probability that *none* of the five self pollinated plants carry the mutant allele is 1/2 × 1/2 × 1/2 × 1/2 × 1/2 or 1/32, which is about 3%. Consequently, if all five self-pollinated ears are recovered and bear kernels useful for scoring and propagation there is about a 97% probability of having successfully transferred the new mutation into the colored stock.

To obtain a pure-breeding colored stock containing the new mutation, one or more additional generations must be grown. Full colored normal kernels are selected from a self-pollinated ear segregating for the new mutation, avoiding kernels homozygous for the new mutation (in the case of *dek* or *emb* mutations), and these are planted. Assuming that the original colorless stock lacked anthocyanin because it was *c1/c1*, *r1/r1* the kernels produced on the outcross ear were all *C1/c1*, *R1/r1* (with half being +/+ and half being +/*m* with regard to the new mutation). Self-pollination of such a plant will result in an ear segregating in a 9:7 ratio for colored to colorless kernels. Among the 9 out of 16 normal kernels that are colored 1 will be *C1/C1*, *R1/R1* and therefore be pure-breeding colored in the next generation, and two-thirds of such kernels will be heterozygous (+/*m*) for the new mutation. (The frequency is two-thirds because only the normal kernels are being considered.) Of the remaining 8 colored kernels, 2 will be *C1/c1*, *R1/R1* and 2 will be *C1/C1*, *r1/R1*; these 4 kernels will produce plants that upon self-pollination will yield ears with a 3:1 ratio for colored to colorless kernels. Again, two-thirds of such ears will also segregate for the new mutation. The other 4 kernels will be *c1/C1*, *r1/R1* and these 4 kernels will produce plants that upon self-pollination bear ears segregating in a 9:7 ratio for colored to colorless kernels and two thirds of such ears will also segregate for the new mutation.

It is evident, therefore, that a large population of kernels from the original outcross would have to be grown and self-pollinated in order to recover an ear that bore all colored kernels and also segregated for the new mutation. But it is also evident that within one or two additional generations of self-pollinations such a stock could be readily recovered.

For allelism testing, kernels obtained from a self-pollinated ear segregating for the new mutation and segregating in a 3:1 (or even 9:7) ratio for colored vs. colorless kernels can be satisfactorily used to produce the pollen parent. Pollen from such a plant will produce one-half (or one-fourth) colored kernels on the cross-pollination side of the ear. This proportion of colored kernels is adequate to delineate the region of cross-pollination. Because only colored kernels can be scored without concern about accidental self-pollination, a pure-breeding colored pollen parent is an advantage. It is possible to use colored kernels from the original cross of the colored stock × mutant for allelism testing, but this is not advised because only one-half of such kernels carry the mutant allele instead of the two-thirds frequency obtained from a self-pollinated ear.

PREPARATION OF THE PARENT PLANTS FOR DOUBLE POLLINATION

On the day prior to performing the double-pollination procedure both the male parent plant and the female plant are visited to prepare their tassels and ears. When the male plant is visited its ear (which was previously covered with an ear shoot bag prior to emergence of silks in order to protect it from contaminating pollen grains) is examined and if the silks are fully emerged, the husks covering the ear tip and the silks are trimmed back by cutting through the husks just above the tip of the ear, and the ear shoot bag is replaced. Then a standard brown paper tassel bag is placed over the tassel and secured with a nonskid paper clip or a staple at its base. When the female plant is visited, its ear is examined and if the silks are fully emerged, the husks are trimmed back by cutting through them while taking care to be sure to cut on a level with or slightly below the tip of the ear. The knife is then used to make a vertical cut through the husks and ear tip to form a slit that bisects the cut silks and ear tip. A small paper square (cut from an index card or light-weight cardboard) is inserted into the slit. The size of the square is usually about 3/4 × 3/4 or 1 × 1 inch in size. It is inserted so that it is held firmly by the husks and ear tip and protrudes upward to form a divider that will completely and equally separate into two groups the silks that will subsequently emerge. The ear shoot bag is then replaced. During the same visit to the female plant or at a later time a standard brown tassel bag is placed over the tassel and secured. The cutting back of the ears on the male parent plant and the female parent plant should be done on the day prior to the planned pollinations to provide sufficient time for the growing silks to emerge far enough above the cut surfaces of the ear tips for adequate pollination. But the ear to be double pollinated should not be cut back so far ahead of the pollinating time that the silks will have grown above the top of the cardboard separator. During very warm weather, especially when the nights are warm, the ears on the female plants may best be cut back and prepared during the afternoon or even during the evening prior to the day of pollination. Consideration also should be given to the security of the pollen sources. If the weather is windy it may often be best to wait and set up the tassels by enclosing them in brown tassel bags (or one of their branches in a glassine pollinating bag) on the morning of the planned pollination as soon as the night's dew has evaporated from the tassels.

Pollinations are begun as soon as the pollen has commenced to shed. To perform a double pollination the pollen is first collected from the pollen parent plant and its ear is immediately self-pollinated with a portion of the pollen. The balance of the pollen is then carried in the bag to the female parent plant where a sufficient amount of pollen to accomplish full kernel set is applied to the silks on one side of the paper separator. Care must be taken in this procedure so that pollen is not inadvertently spilled across the separator onto the silks on the other side. But if this does happen it will be evident when the ear is examined following harvest because contaminating cross-pollinated kernels will be colored. In order to avoid contamination from wind-blown pollen and also to help prevent the wind from interfering with the proper application of the pollen only onto one-half of the silks, it is helpful to leave the ear shoot bag on the ear and to tear the top portion off of the shoot bag to form a chimney extending above the tip of the ear. Then the paper separator can be bent to one side to cover the silks to be self-pollinated, allowing specific delivery of pollen for cross-pollination. When more than one double pollination is to be performed with one pollen sample, the pollen should be transfered from the brown tassel bag to a glassine pollinating bag. Because this smaller bag allows for more precise application of the pollen we routinely use glassine bags for all double pollinations. (See chapter by Neuffer in this volume on the use and sources of glassine bags.)

Following application of the pollen for the cross, the open end of the torn ear shoot bag is folded over to close it, and the fold is secured with a paper clip. Once all of the cross-pollinations scheduled for that morning have been completed the self-pollinations are performed. The pollinator returns to each female plant and collects the pollen from the brown tassel bag. If the day is without wind and the pollinator is experienced, the self-pollination may be accomplished by careful application of pollen directly from the tassel bag. Otherwise it is best to transfer the pollen into a glassine bag prior to applying it to the silks. The paper clip is removed from the ear shoot bag and the paper separator is bent so as to cover the silks that were cross-pollinated. The pollen is then carefully applied to the remaining portion of silks to accomplish self-pollination. Upon completion of the double pollination, the brown tassel bag used to collect pollen from the tassel of the female plant is placed over the ear with the record of the cross- and self-pollinations written on the bag. The bag is secured on the plant by stapling it about the stem.

When the ears are mature they are harvested, dried, and tagged. The self-pollinated ears from pollen donors are matched up with the corresponding double-pollinated ears. The self-pollinated ears are scored for the presence of segregating kernels to determine if the pollen parent was heterozygous. When that proves to be the case the self-pollinated half of each corresponding double-pollinated ear is scored. In those cases where the female parent also proves to be heterozygous, the crossed side of each such ear is scored to obtain an unambiguous test for allelism. If the two mutations are allelic, approximately one-fourth of the kernels on the crossed side of the double-pollinated ear will be obviously mutant or will prove to be mutant when planted. But if the two mutations are nonallelic all of the kernels on the crossed side will be normal.

The above-described procedure can also be utilized with stocks that differ in carotenoid synthesis. A yellow kerneled stock (*Y/Y*) is used as the pollen parent and a white kerneled stock (*y/y*) is used as the female parent.

INCREASED EFFICIENCY BY USE OF BALANCED LETHALS

Ideally every plant in an allelism test would be heterozygous for the mutation of interest. Ordinarily only two-thirds of the plants from a self-pollinated heterozygote are heterozygous. A much higher frequency, approaching 100%, can be obtained with a balanced lethal stock. All of the 20 chromosome arms of maize (except *8S*) carry lethal mutations at *defective kernel* (*dek*) loci. In addition there are white kernel (i.e., *w3*, *y10*) lethal stocks, *viviparous* stocks, and lethal seedling stocks (i.e., *nec2*). The balanced lethal stock is produced by crossing a plant heterozygous for the lethal mutation under study on or by the second lethal stock. If one of the two lethal mutations is available in a colored stock (all 31 named *dek* mutations are available in a colored background from the Maize Stock Center) the double-pollination technique described above can be used. The goal is to produce kernels heterozygous for both mutations. The normal kernels produced by self-pollination of such double heterozygotes will be, barring crossing-over, heterozygous for both lethal mutations. Crossing-over will reduce the frequency of double heterozygotes depending on the map distance separating the two loci. But even if two *dek* loci are as much as 20 map units apart almost 84% of the plants grown from the nonmutant kernels would be heterozygous for the particular *dek* locus being studied. This is a significant improvement over the approximately 67% frequency expected amongst the progeny plants of nonmutant kernels from a self-pollinated simple heterozygote.

REFERENCES

Anderson EG (1956) The application of chromosomal techniques to maize improvement. Brookhaven Symp Biol 9: 23–36

Beckett JB (1978) B-A translocations in maize I. Use in locating genes by chromosome arms. J Hered 69: 27–36

Beckett JB (1991) Cytogenetic, genetic, and breeding application of B-A translocations in maize. In Gupta PK, Tsuchiya T (eds) Chromosome Engineering in Plants, Genetics, Breeding and Evolution, Part A, Elsevier, Amsterdam, pp 491–527

Birchler JA, Chalfoun DJ, Levin DM (1990) Recombination in the B-chromosome of maize to produce A-B-A chromosomes. Genetics 126: 723–733

Burnham CR (1982) The locating of genes to chromosomes by the use of chromosomal inter-

changes. In Sheridan WF (ed) Maize for Biological Research, Plant Mol Biol Assoc, Charlottesville, VA, pp 65–70
Clark JK, Sheridan WF (1991) Isolation and characterization of 51 *embryo-specific* mutations in maize. Plant Cell 3: 935–951
Coe EH Jr., Neuffer MG and Hoisington DA (1988) The genetics of corn. In Sprague GF, Dudley JW (eds) Corn and Corn Improvement, American Soc of Agronomy, Crop Sci Soc of America and Soil Sci Soc of America, Madison, WI, pp. 83–258.
Sheridan WF, Clark JK (1987) Allelism testing by double pollination of lethal maize *dek* mutants. J Hered 78: 49–50

63

Analysis of Cytoplasmically Inherited Mutants

KATHLEEN J. NEWTON

Although cytoplasmic mutants have been documented in higher plants, they have not been studied as well as in lower eukaryotes, such as yeast and *Chlamydomonas*. Very few potential chloroplast mutations have been described in maize, in contrast to several in *Chlamydomonas* and some in other higher plants. (See Börner and Sears 1986.) The only examples of cytoplasmically inherited, defective plastids in maize appear to result from the action of nuclear genes, such as *iojap*. (See Walbot and Coe 1979.) In plants homozygous for *iojap*, some plastids show a loss of chloroplast pigments and ribosomes. The ribosome-less plastids are then inherited maternally. Other nuclear gene mutations that affect photosynthetic function have been identified on the basis of a high chlorophyll fluorescence phenotype (*hcf*; Miles 1982). Over one hundred nuclear *hcf* mutations have now been identified in maize, and they are all inherited as typical, recessive Mendelian factors (Miles and Metz 1985).

In contrast to the scarcity of documented mutations in chloroplast DNA, several maize mitochondrial mutants have been reported. Some are spontaneous or tissue-culture-induced revertants of cytoplasmic male sterility traits. (See Laughnan and Gabay-Laughnan 1983.) These revertants usually result from rearrangements or small deletions in the CMS

The Maize Handbook—M. Freeling, V. Walbot, eds.

mitochondrial genome. Other mutants affect the plant more drastically and constitutively. One set of these abnormal growth mutants is known as nonchromosomal stripe (NCS; Shumway and Bauman 1967; Coe 1983; Newton and Coe 1986).

NCS plants have leaf stripes that vary from necrotic to yellow to pale green, depending on the specific lesion (Newton et al. 1989). Ears from NCS plants typically have sectors of aborted kernels. Somatic segregation of mutant from normal mitochondria within organs during development is the most likely explanation for such clonal sectoring. Although it has now been shown that NCS plants do carry a mixture of defective and normal mitochondria, this has not proved to be an obstacle to analyzing the mutations. MtDNA preparations greatly enriched for mutant genomes can be obtained from plants that have extreme phenotypes (heavy striping, very reduced stature).

NCS mutations are indeed within the mitochondrial DNA. There are also nuclear genes that appear to be involved in the generation or selection of mitochondrial DNA rearrangements. One line of maize, WF9, gives rise to NCS plants at a relatively high rate (approximately 1%, see Newton and Coe 1986). Any mitochondrial genotype is susceptible to NCS mutations: They have arisen in plants carrying cms-T, cms-S, fertile-revertant, and N mitochondrial genomes (Newton et al 1989).

RECOGNIZING A CYTOPLASMIC MUTATION

Two properties are characteristic of cytoplasmic mutations in maize. (1) They are inherited from the female parent only. In *Zea*, both mitochondrial and plastid DNA are transmitted only through the egg (Conde et al 1979; Pring and Levings 1978). (2) If the mutation is highly deleterious, or if a beneficial mutation has newly arisen, the plants exhibit some form of clonal sectoring or striping. Striping is the result of the "sorting out" of a heteroplasmic mixture of mutant and normal organelles during cell divisions. In the case of cytoplasmic reversions to fertility (a beneficial mutation to the plant, if not to the breeder), the sectors are detected as anthers or tassel branches that shed pollen. (See Laughnan and Gabay-Laughnan 1983.)

In maize, candidate cytoplasmic mutants are usually found as plants with altered phenotypes that are segregating in families in numbers that do not correspond to expected Mendelian ratios. The phenotypes one expects to see with putative cytoplasmic mutants include reduced growth and white, pale, or necrotic striping on the plant.

DETERMINING INHERITANCE PATTERNS

Each putative cytoplasmic mutant (e.g., a striped plant) is crossed reciprocally with a normal plant to see if only the maternal parent can transmit the mutant phenotype. This is a crucial experiment, and the results are usually clear, unless the mutant plant is highly defective. With a highly defective female plant, it is not unusual to find that the ear has just a few aborted kernels (while the same mutant plant used as a male in a reciprocal cross produces normal progeny). If there are less affected (e.g., slightly striped) siblings of the very defective plant, they should definitely be used in the reciprocal cross series, because they are more likely to transmit the mutation. It is not unusual to see great variation in plant height, vigor, and the amount of leaf striping among sibling mutant plants.

That ears from mutant plants show clusters (sectors) of aborted or defective kernels is more characteristic of mitochondrial than of chloroplast mutations. Most mitochondrially encoded functions are expected to be important during kernel development, whereas most of the plastid functions that are important during this stage appear to be nuclearly encoded.

MOLECULAR ANALYSES

Once a mutation has been shown to be maternally inherited, especially if clonal striping is easily observed in surviving plants, the determination of whether the mutation is in mitochondrial or chloroplast DNA is usually the next step.

The direct approach is a molecular one. One asks if there are any detectable alterations in the DNA that correlate with the mutant phenotype. Procedures for the isolation and analysis of organelle DNAs are included in other chapters of this volume. In some cases, the mutation is detectable as a restriction-fragment-length polymorphism (RFLP) within the DNA. For example, with NCS2 and NCS3, the relevant mitochondrial RFLPs were visible on the ethidium-bromide-stained agarose gels (Newton and Coe 1986). In the cases of the NCS5 and NCS6 mutations, the relevant RFLPs within the cytochrome oxidase subunit 2 gene were only detected following hybridizations with defined gene probes (Lauer et al. 1990; Newton et al. 1990). RFLP analyses are helpful in identifying mitochondrial mutations, because mitochondrial DNA tends to change by rearranging (Palmer and Herbon 1988). On the other hand, chloroplast DNAs tend to change by point mutations. Thus, the RFLP approach is less useful for identifying chloroplast mutations. Indeed, as stated earlier, there are no good examples of specific mutations in chloroplast DNA yet reported in maize, although some chloroplast mutations are well characterized in other higher plants.

Once a polymorphism in the DNA has been found and has been shown to correlate with the expression of the mutant phenotype, the next step is to identify the gene that is altered. Because the mutations are selected on the basis of (usually deleterious) phenotypes, it is expected that the mutations themselves identify important organellar genes. All available known gene clones are collected from various laboratories and are used as hybridization probes on DNA blots. In some cases this can help determine which gene is altered. (See Newton et al. 1990.) However, sometimes the altered RFLPs must be sequenced to identify a potential function for the mutant gene. (See Hunt and Newton 1991.)

The next step in the direct approach is to show that the alteration in the gene affects the expression of the gene (e.g., any transcripts from that gene are specifically eliminated or altered). Finally, it is usually possible to show that the function normally contributed by the product of the mutant gene (e.g., cytochrome oxidase) is altered.

The approach of finding an altered function in the mutant sector and working back to the gene can actually be more laborious and less successful than a direct approach. However, it has often been used to try to identify chloroplast mutations. Due to pleiotropy, a single disruption in chloroplast DNA sequence might affect multiple photosynthetic functions. A further difficulty is expected if pleiotropic effects extend to other organelles. A dysfunction in one organelle could affect the functioning of another type of organelle. For example, NCS2, which has been well characterized as a mutation in a mitochondrial Complex I gene, has pleiotropic effects on chloroplast ultrastructure and function; i.e., both electron transport after PSII and carbon fixation are interrupted in cells that are homoplasmic for the mitochondrial mutation (Roussell et al. 1991; J. Gu, D. Miles, and K. Newton, unpublished data). It is not known whether the converse occurs for any mutation affecting chloroplast function in maize. For example, it would be interesting to know if in *iojap* stripes, which lack chloroplast ribosomes, there is any effect on the respiratory properties of mitochondria.

NUCLEAR-CYTOPLASMIC INTERACTIONS

While inheritance from the female parent in a reciprocal maize cross implicates organelles, it is important to realize that nuclear genes can also be involved in affecting a specific organellar trait through nuclear-cytoplasmic interactions. (See Allen et al. 1989.) This possibility is especially strong if no sectoring for the defective phenotype is seen. For example, cytoplasmic male sterility is only observed if "permissive" nuclear genes are present. If the dominant alleles of the nuclear restorer genes (Rf) are present, CMS is not observed, despite the presence of the CMS mitochondrial genotype. This phenomenon is simply a reflection of the contributions that both nuclear and mitochondrial gene products make to organelle biogenesis and maintenance.

REFERENCES

Allen JO, Emenhiser GK, Kermicle JL (1989) Miniature kernel and plant: interaction between teosinte cytoplasmic genomes and maize nuclear genomes. Maydica 34: 277–290

Börner T, Sears BB (1986) Plastome mutants. Plant Molecular Biol Rep 4: 69–92

Coe EH (1983) Maternally inherited abnormal plant types in maize. Maydica 28: 151–157

Conde MF, Pring DR, Levings CS III (1979) Maternal inheritance of organelle DNA's in *Zea mays-Zea perennis* reciprocal crosses. J Hered 70: 2–4

Hunt MD, Newton KJ (1991) The NCS3 mutation: genetic evidence for the expression of ribosomal protein genes *rps3* and *rpl16* in *Zea mays* mitochondria. EMBO J 10: 1045–1052

Lauer M, Knudsen C, Newton KJ, Gabay-Laughnan S, Laughnan JR (1990) A partially deleted mitochondrial cytochrome oxidase 2 gene in the NCS6 abnormal growth mutant of maize. New Biolog 2: 179–186

Laughnan JR, Gabay-Laughnan S (1983) Cytoplasmic male sterility in maize. Ann Rev Genet 17: 27–48

Miles D (1982) The use of mutations to probe photosynthesis in higher plants. In Edelman M, Hallick RB, Chua N-H (eds) Methods in Chloroplast Molecular Biology, Elsevier Science, New York, pp 76–107

Miles D, Metz JG (1985) The role of nuclear genes of maize in chloroplast development. In Freeling M (ed) Plant Genetics, Alan R Liss, New York, pp 585–597

Newton KJ, Coe EH Jr (1986) Mitochondrial DNA changes in abnormal growth (nonchromosomal stripe) mutants of maize. Proc Natl Acad Sci USA 83: 7363–7366

Newton KJ, Coe EH Jr, Gabay-Laughnan S, Laughnan JR (1989) Abnormal growth phenotypes and mitochondrial mutations in maize. Maydica 34: 291–296

Newton KJ, Knudsen C, Gabay-Laughnan S, Laughnan JR (1990) An abnormal growth mutant in maize has a defective mitochondrial cytochrome oxidase gene. Plant Cell 2: 107–113

Palmer JD, Herbon LA (1988) Plant mitochondrial DNA evolves rapidly in structure but slowly in sequence. J Mol Evol 28: 87–97

Pring DR, Levings CS III (1978) Heterogeneity of maize cytoplasmic genomes among male-sterile cytoplasms. Genetics 89: 121–136

Roussell D, Thompson DL, Pallardy SG, Miles D, Newton KJ (1991) Chloroplast structure and function is altered in the NCS2 maize mitochondrial mutant. Plant Physiol 96: 232–238

Shumway LK, Bauman LF (1967) Nonchromosomal stripe of maize. Genetics 55: 33–38

Walbot V, Coe EH Jr (1979) Nuclear gene *iojap* conditions a programmed change to ribosome-less plastids in *Zea mays*. Proc Natl Acad Sci USA 76: 2760–2764

Male Sterility and Restorer Genes in Maize

SUSAN GABAY-LAUGHNAN AND JOHN R. LAUGHNAN

Male-sterile plants are those that fail to produce functional pollen grains. In maize, male sterility can result from either nuclear or mitochondrial gene mutation. The former is referred to as genic male sterility and the latter as cytoplasmic male sterility (CMS). Over 20 nuclear gene mutations that confer a male-sterile phenotype have been identified. (see Neuffer et al. 1993.) The majority of these are recessive and have been assigned the symbol *ms* (*Ms* for the dominant genes). Some mutations that produce aberrant cell or chromosome behavior, e.g., asynaptic (*as*) and ameiotic (*am*), may also result in male sterility. In contrast, only three major categories of CMS have been recognized—*cms-C* (Charrua), *cms-T* (Texas), and *cms-S* (USDA). (See reviews by Duvick 1965; Edwardson 1970; Laughnan and Gabay-Laughnan 1983; Newton 1988.) Genic male sterility is inherited according to Mendelian rules whereas CMS is transmitted maternally. Plants carrying one of the CMS-type cytoplasms are male-sterile unless they also carry a dominant nuclear restorer-of-fertility gene(s).

GENETIC NATURE OF RESTORER GENES

Restorer-of-fertility (*Rf*) genes override the male-sterile effect of the cytoplasm and are specific with regard to the CMS strains they restore (Table 64.1). In fact, the three types of CMS in maize were originally identified on the basis of their differing responses to the *Rf* genes.

The mode of restoration in *cms-C* and *cms-T* is sporophytic; it is the genotype of the plant (sporophyte) that determines whether normal pollen is produced. Thus, a *cms-T* plant that is heterozygous at both the *Rf1* and *Rf2* loci produces all normal pollen even though only one-fourth of those pollen grains carry both restoring alleles. The mode of restoration in *cms-S* is gametophytic; it is the genotype of the pollen grain (gametophyte) itself that determines whether the pollen is normal or aborted. Thus, a *cms-S* plant that

The Maize Handbook—M. Freeling, V. Walbot, eds.

TABLE 64.1. Characteristics of restorer-of-fertility genes in maize

Gene	Chromosome	CMS-type restored	Mode of restoration	Comments
Rf1	3S-near *Lg3*	T	sporophytic	Complementary to *Rf2*
Rf2	9-near *wx*	T	sporophytic	Complementary to *Rf1*
Rf3	2L-near *whp*	S	gametophytic	Spontaneous restorers of *cms-S* have been mapped to other chromosome sites
Rf4	2 or 8	C	sporophytic	There may be additional genes involved in *cms-C* restoration
Rf5		C	sporophytic	Both duplicate and complementary gene action may be involved

is heterozygous *Rf3 rf3* produces one-half normal and one-half aborted pollen grains. Such plants can be said to be semisterile although this term is usually reserved for the pollen (and ovule) abortion that results when plants are heterozygous for certain chromosome alterations, e.g., inversions, translocations.

RESTORER GENES IN INBRED LINES

Many inbred lines of maize have been analyzed as to which CMS strains they restore. The restoring characteristics of 42 lines commonly used in research are presented in Table 64.2.

Information on additional inbred lines is available (Beckett 1971; Gracen 1982). New inbred lines are constantly being developed, and the restorer gene constitution of a new line can be determined using the CMS versions of several inbred lines. The best analysis involves crossing sterile versions of each of the three CMS types as female parent by pollen from the inbred line being tested. The F_1 is then backcrossed by the same line. The progeny are crossed by that line as the recurrent male parent for a number of generations and then the fertility of the converted CMS is scored. A quick protocol involves the scoring of fertility in F_1 plants. Ears of *cms-C* WF9, *cms-T* WF9, and *cms-S* WF9 plants are crossed by pollen from the inbred line whose restorer constitution is unknown. If the F_1 progeny of the cross onto a particular CMS type are fertile, it can be said that the new line carries the restorer gene(s) for that CMS. If the new line fails to restore the

TABLE 64.2. Restoration patterns of 42 inbred lines of maize[a]

Inbred line	*CMS-type cytoplasm* C	T	S	Inbred line	*CMS-type cytoplasm* C	T	S
A619	+	−	±	M14	+	+	−
A632	−	−	±	Mo17	−	−	−
A634	−	−	+	N6	+	−	−
B14	−	−	±	N28	−	−	−
B14A	±	−	±	NY821 LERf	+	+	+
B37	−	−	−	Oh07	−		−
B73	−	−	−	Oh43	+	−	±
B77	−	−		Oh45	+	−	+
B84	+	−	−	Oh51A	−	−	−
C103	−	−		Oh545	−	+	−
C123	+	−	±	R138	−	+	−
CB59G	±	−	−	R177	+	+	+
CE1	+	−	+	R802A	−	−	−
CI21E	+	−	+	SD10	−	−	−
H95	+	−	+	SK2	−	−	−
I153	+	+	−	Tr	−	−	+
Ill A	−	−	−	Va26	+	−	−
K55	+	+	−	W23	+	−	−
Ky21	+	+	+	W64A	+	−	−
KYS	−		−	WF9	−	−	−
L317	+	−	−	38–11	+	−	−

[a]+ = fertile, − = sterile, ± = partially fertile.

C and S versions of WF9, i.e., the F_1 plants are sterile, it can be said that the line carries no S, and probably no C, restorer genes. If the new line fails to restore the T version of WF9, it can be said that the line does not carry both *Rf1* and *Rf2* but additional testing is necessary to determine its T restorer constitution because the genotype of WF9 is *rf1 rf1 rf2 rf2*. Inbred line nonrestorers of cms-T can have one of three possible genotypes—*rf1 rf1 rf2 rf2*, *Rf1 Rf1 rf2 rf2* or *rf1 rf1 Rf2 Rf2*. The majority of nonrestorers of *cms-T* that have been tested are *rf1 rf1 Rf2 Rf2* (Table 64.3).

Crosses of each of the *cms-T* diagnostic inbred lines as female parents with pollen from the new line will give different fertility patterns in the F_1 depending on the *Rf1-Rf2* constitution of the line being tested (Table 64.4).

Table 64.3. *Rf1-Rf2* constitutions of some non-T-restoring inbred lines

rf1 rf2	*rf1 Rf2*	*Rf1 rf2*
WF9	B37	R213
	Mo17	
	N6	
	W23	
	38–11	
	B73	

Table 64.4. Analysis of *cms-T* restorer gene constitution[a]

Diagnostic line (female parent)	Possible genotype of new line (male Parent)			
	rf1 rf2	*rf1 Rf2*	*Rf1 rf2*	*Rf1 Rf2*
cms-T WF9 *rf1 rf2*	−	−	−	+
cms-T B37 *rf1 Rf2*	−	−	+	+
cms-T R213 *Rf1 rf2*	−	+	−	+

[a] + = fertile, − = sterile.

GENETIC ANALYSIS OF NEWLY ARISEN MALE-STERILE TRAITS

Using the knowledge of the *Rf* constitution of various inbred lines, it is possible to determine the nature of a newly arisen male-sterile trait. Is the new trait genic or cytoplasmic in nature? If cytoplasmic, is it *cms-C*, *-T*, or *-S*? The protocol to answer these questions involves crossing ears of male-sterile plants by pollen from each of the diagnostic inbred lines (Table 64.5). The F_1 plants are scored for fertility vs sterility. If the new trait is genic, crosses by each of the diagnostic lines will result in male-fertile progeny. If the new trait is cytoplasmic, crosses by each of the diagnostic lines will give different fertility patterns depending on whether the CMS trait is *C*, *T*, or *S*.

MOLECULAR ANALYSIS OF CYTOPLASMIC MALE STERILITY

As mentioned previously in this chapter, the genetic defect in CMS strains of maize lies in the mitochondrial DNA (mtDNA). The mtDNA is the carrier of the genetic determiners

TABLE 64.5. Analysis of a newly arisen male-sterile trait[a]

New male sterile	Diagnostic line (male parent)			
	WF9	W23	K55	Tr
genic	+	+	+	+
cms-C	−	+	+	−
cms-T	−	−	+	−
cms-S	−	−	−	+

[a]+ = fertile, − = sterile.

of male fertility at the cytoplasmic level. The different maize cytoplasms, normal (fertile), *cms-C, cms-T*, and *cms-S*, can be distinguished by restriction enzyme digestion and agarose gel electrophoresis of the mtDNA (Pring and Levings 1978; Borck and Walbot 1982). Because restorer genes have no apparent effect on mtDNA organization, these procedures can be applied to fertile restored as well as male-sterile versions of the strain in question. Thus, there are available both genetic and molecular means of distinguishing CMS types.

CONCLUDING REMARKS

The inheritance of some types of male sterility in maize has been known to have an extranuclear basis for 60 years (Rhoades 1931). Only in the last 15 years, with the development of techniques for the study of mitochondrial DNA, has the molecular basis for cytoplasmic male sterility been amenable to analysis. Much of the research has been published, and it has not been the intent of this chapter to review those studies. Instead, emphasis has been placed on practical approaches to the study of CMS and *Rf* genes that are common knowledge to researchers in the field but have not been published previously.

REFERENCES

Beckett JB (1971) Classification of male-sterile cytoplasms in maize (*Zea mays* L.). Crop Sci 11: 724–727

Borck KS, Walbot V (1982) Comparison of the restriction endonuclease digestion patterns of mitochondrial DNA from normal and male sterile cytoplasms of *Zea mays* L. Genetics 102: 109–128

Duvick DN (1965) Cytoplasmic pollen sterility in corn. Adv Genet 13: 1–56

Edwardson JR (1970) Cytoplasmic male sterility. Bot Rev 36: 341–420

Gracen VE (1982) Type and availability of male sterile cytoplasms. In Sheridan WF (ed) Maize for Biological Research, Plant Molecular Biology Association, Charlottesville, VA, pp 221–224

Laughnan JR, Gabay-Laughnan S (1983) Cytoplasmic male sterility in maize. Ann Rev Genet 17: 27–48

Neuffer MG, Coe EH, Wessler S (eds) (1993) Mutants of Maize. Cold Spring Harbor, New York (in press)

Newton KJ (1988) Plant mitochondrial genomes: organization, expression and variation. Annu Rev Plant Physiol Plant Mol Biol 39: 503–532

Pring DR, Levings CS III (1978) Heterogeneity of maize cytoplasmic genomes among male sterility cytoplasms. Genetics 89: 121–136

Rhoades MM (1931) Cytoplasmic inheritance of male sterility in *Zea mays*. Science 73: 340–341

65

Inbred Lines of Maize and Their Molecular Markers

MICHAEL LEE

Inbred lines have been a rich resource for fundamental and applied investigations in maize. They constitute a sampling of the genetic diversity in *Zea mays* L. that has been captured and partitioned into an array of uniform, reproducible genotypes. Advances in genomic mapping with molecular markers, described in several chapters of this volume, have provided the means to characterize maize inbreds at the DNA level with unprecedented power and resolution. Such information has become increasingly valuable for planning and conducting research. The purpose of this chapter is to introduce and briefly summarize the relevant features of inbred lines, their development and origin, their description with molecular markers, and the utility of this information for basic and applied studies of maize.

DEVELOPMENT AND ORIGIN OF MAIZE INBRED LINES

Production of highly homogeneous, homozygous inbred lines has provided maize researchers with a large array of uniform, reproducible genotypes. Typically, the lines have been developed for use as parents in hybrid seed production or in genetic studies through successive generations of self-pollination with artificial selection for desired attributes. Modified versions of some inbreds have been developed through backcrossing and selection for the original inbred carrying a specific gene (e.g., *wx*, *y1*, *Ht1*) or chromosomal rearrangement. The number of generations, and thus the level of homozygosity, may vary considerably, although maize breeders in temperate climates usually release inbred lines after seven or more generations of self-pollination and selection. Subsequently, seed stocks may be maintained and increased by some combination of self- and sib-pollination, possibly with removal (roguing) for off-types each generation. Most inbred lines exhibit a high degree of genetic stability and reproducibility, although cases of unstable inbreds have been documented (Bogenschutz and Russell 1986).

Inbred lines have been derived from myriad source populations developed by different methods (Hallauer 1990). Many inbreds produced in the early decades of this century were selected directly from open-pollinated landraces. As breeding programs were organized, a preponderance of inbreds have been selected from populations created by crossing two or more parental inbreds according to some breeding scheme (Hallauer 1987). Information on pedigree and ancestry for many inbreds developed by North American breeding programs has been summarized in a forthcoming publication, *Compilation of North American Maize Breeding Germ Plasm* (Gerdes et al. 1993). Additional data may be located in the "Registration" section in the journal *Crop Science.* Such information can be particularly helpful in research that attempts to make a statement about an attribute or characteristic of maize germplasm; often, the validity and scope of a study's interpretation depend upon the germplasm included. Also, an informed sampling of inbreds could increase research efficiency by selecting groups of closely or distantly related lines as desired.

Collections of maize inbreds are maintained at several institutions. Small quantities of seed (50–100 kernels) are often available free or for a modest fee. Sources of inbreds adapted to temperate growing conditions may be found in the United States at state agricultural experiment stations, foundation seed companies, hybrid seed corn companies, the USDA North Central Plant Introduction Station (Ames, IA), the Maize Genetics Cooperative Stock Center, and the National Seed Storage Laboratory (Fort Collins, CO). Seed stocks of inbreds adapted to the tropics may be available through national programs, CIMMYT (International Maize and Wheat Improvement Center, El Batan, Texcoco, Mexico), IITA (International Institute for Tropical Agriculture, Ibadan, Nigeria), hybrid seed companies, and the University of Hawaii. In most cases, the collections have been systematically described for only a few attributes. Data on agronomic traits of some U.S. public inbreds of significance to maize breeding programs have been recorded in reports of the NCR-2 committee (North Central Committee on Quantitative Genetics in Maize Breeding) and by the Iowa State University–USDA cooperative corn breeding project. These reports have been maintained in collections at the Department of Agronomy, Iowa State University. Information may also be accessed through a germplasm database, GRIN (Germplasm Resources Information Network).

DESCRIBING MAIZE INBREDS WITH MOLECULAR MARKERS

Various methods have been used to survey molecular polymorphisms in the nuclear genome of maize inbreds including cytology (Chughtai and Steffensen 1989), isoelectric focusing of zeins (Nucca et al. 1978), high-performance liquid chromatography of zeins (Smith and Smith 1989), two-dimensional gel electrophoresis of proteins (Higginbotham et al. 1991), starch gel electrophoresis of enzymes (Stuber and Goodman 1983), and restriction-fragment-length polymorphisms (Melchinger et al. 1991); however, only the latter two have been widely used in genetic studies, so they will be the focus of the chapter. Development of more powerful methods based on the polymerase chain reaction (Martin et al. 1991; Welsh et al. 1991) appears imminent, but surveys of maize inbreds with these methods and reports of method repeatability have not been published.

Starch gel electrophoresis of isozymes has been used in a very wide array of basic and applied genetic studies of maize. The utility and power of these molecular markers may be attributed to several factors including their well-characterized genetic control and map location, known temporal and spatial expression patterns during plant development, codominant expression of alleles, relatively rapid and efficient detection, and generally neutral effect of allelic variation on plant growth and development. Also, most isozymes in routine use may be detected in a small quantity of coleoptilar tissue without destroying the plant. Information regarding sample preparation, equipment, protocols for gel staining and interpretation (with illustrated examples for commonly assayed systems and loci), allelic designations for standard inbred lines, notes on relative allelic frequencies, and other basics has been summarized in an excellent technical bulletin (Stuber et al. 1988).

Electrophoretic variants of enzymes have been placed to at least 77 loci (Coe et al. 1988); approximately 40 loci representing 21 enzymes may be routinely assayed with some diligent practice. Chromosome map positions have been established for most isozyme loci in routine use. The loci are distributed to the 10 maize chromosomes, albeit somewhat unevenly: 5 chromosomes (2, 4, 7, 9, and 10) are marked by only 1–3 loci and others (1 and 5) with more loci may have a very uneven distribution. Allelic constitution at most loci has been determined and summarized for hundreds of maize inbreds (Stuber and Goodman 1983). The number of alleles per locus ranges from one (*Me*) to at least 10 (*Glu1*) and allelic frequencies exhibit extreme variation. Considering all loci, the majority (>75%) of inbreds have unique allelic profiles although the percentage declines somewhat within groups of lines with common ancestry. In relatively few cases, lines with no known shared, recent ancestry may also have identical isozyme profiles at all loci. With such a database, isozymes have been used extensively and routinely to identify inbreds, monitor purity of inbred-line seed stocks, and assess quality control in hybrid seed production (Stuber et al. 1988); isozyme tests can reduce the time required for such tasks from a period of several months to days. Such information has also been valuable in planning and conducting experiments that use isozymes as genetic markers to locate genetic factors associated with expression of quantitative traits (Edwards et al. 1987) and as a means of assessing genetic diversity in samples of inbreds (Smith et al. 1985).

Studies with isozymes firmly established the utility of molecular markers and related databases of inbred lines for research with maize. However, they also identified some deficiencies of those markers for investigations benefiting from more thorough, discriminative, and possibly directed coverage of the genome. The limitations of isozymes regarding genome coverage have been described. In addition, many isozyme loci have few allelic variants among maize inbreds and some alleles have been detected at extremely low frequencies. These problems are exacerbated when widely used inbreds or their relatives are preferred for the experiment (Stuber and Goodman 1983; Smith et al. 1985). Therefore, molecular markers that detect variation directly at the DNA level, and at a higher frequency, have been developed with great anticipation.

The first surveys of maize inbreds and loci for restriction-fragment-length polymorphisms (RFLPs) revealed an exceptionally high degree of variation in comparison to other plant and animal species (Evola et al. 1986). Subsequently, maize RFLP linkage maps were rapidly constructed and implemented into many genetic investigations of basic and applied significance. Methods for RFLP map development, mapping, and laboratory protocols have been described in other chapters of this volume.

Published surveys for RFLPs among maize inbreds have supported the initial results (Lee et al. 1989; Godshalk et al. 1990; Melchinger et al. 1991; J. Boppenmaier et al. 1992). More comprehensive information should be forthcoming through surveys conducted by hybrid seed corn companies, private service laboratories, North Carolina State University, CIMMYT, and the USDA maize genome project (initial phases scheduled for completion during 1992). Generally, the studies have been conducted with mapped, single- to low-copy genomic and cDNA clones as probes against single digests (commonly, *Eco*RI, *Eco*RV, *Hin*dIII, *Bgl*II, *Bam*HI) of DNA prepared from leaf tissue of inbreds of current or historical importance to breeding programs. Conditions and protocols for electrophoresis, hybridization, and scoring autoradiograms vary slightly among laboratories but typically, the methods reliably resolve RFLPs among fragments differing by 100 to several thousand base pairs (approximately 5% of the fragment size) within a range of 2 to 20 kb. In autoradiograms, one to several bands exhibiting a range of exposure levels (reflecting the repetitive nature of the nuclear genome) may be observed for any given combination of probe and restriction enzyme for each inbred line. Combinations of public probes and restriction enzymes known to reliably detect relatively high levels of polymorphism among a selected group of inbreds have been proposed (Gardiner et al. 1991).

Among selected samples (ca. 20–40 lines per sample) of U.S. and European maize inbreds, the average number of RFLP variants per probe-enzyme combination has been estimated to be between three and five depending upon the number of lines and degree of common ancestry of lines included in the studies. This is nearly twice the level of isozyme polymorphism for comparable groups of inbreds (Godshalk et al. 1990; Smith et al. 1990; Melchinger et al. 1991; Boppenmaier et al. 1992). Monomorphic probe-enzyme combinations were detected only in rare instances in comparison with isozyme loci and in absolute terms (Messmer et al. 1991). This was somewhat surprising, and perhaps revealing of the frequency of DNA polymorphism in maize. Probes used in these surveys were selected from sets of mapped probes known to detect RFLPs between very few lines, typically those inbreds used to create the mapping populations. Restriction enzymes of the same class have detected variation at nearly equivalent frequencies (Lee et al. 1990a) indicating that much of the polymorphism seems attributable to insertion and deletion events reminiscent of transposable elements. At selected loci and alleles, the nucleotide diversity (number of nucleotides per thousand that distinguish two individuals taken at random) has been estimated at 0.082 and 0.048 for the *Adhl* and *Sh1* loci, respectively (Evola et al. 1986). Based on these observations and subsequent experience, most inbreds should be clearly distinguished from each other. Indeed, all lines included in published surveys have unique RFLP profiles; even very closely related, backcross-derived lines have been differentiated, although at a much lower frequency of polymorphism (Lee et al. 1990b).

RFLP surveys have been of immediate utility as a means of more direct assessment of the extent of diversity, genetic relation, and coancestry among inbreds. In lieu of molecular marker data, such inferences were made on the basis of pedigree records, morphological traits of unknown genetic control, quantitative genetic parameters estimated from mating designs, and studies of breeding behavior for combining ability and heterosis requiring years of field trials; all approaches represent rather lengthy and indirect, albeit pragmatic, evaluations of an inbred's genotype (Smith and Smith 1989; Smith et al. 1990). In at least some cases, RFLP data should be a helpful adjunct, while in others, they may be irreplaceable for reliable discrimination among inbreds. For example, pedigree relationships have been calculated on the basis of probability under the assumptions of proportional contribution of all parents (if they are known) and no selection during development of the inbreds (Smith et al. 1990; Melchinger et al. 1991). There has been excellent agreement (coefficient of determination of 0.81) between estimates of genetic relationship based on pedigree and RFLP data (Smith et al. 1990), suggesting that, in cases of ambiguous records or significant deviations from theoretical expectations, molecular markers could provide critical information. In a related application, inbreds were placed into groups using RFLP data in good agreement with germplasm groups (heterotic groups) previously developed and used by breeders (Lee et al. 1989; Godshalk et al. 1990; Melchinger et al. 1991). This result suggests RFLP profiles of inbreds at a sufficient number of loci (>80) could be used to assign new germplasm to appropriate groups in lieu of lengthy and often ambiguous field testing.

It is imperative to develop methods that facilitate objective, insightful, and efficient maintenance and utilization of maize germplasm. There may be at least 76,000 accessions in maize germplasm banks (Plucknett et al. 1983), but only an estimated 2% of the germplasm has been used in breeding programs (Hallauer and Miranda 1981). Furthermore, the diversity of on-farm germplasm may be in decline (Darrah and Zuber 1986). More informative assessments of current breeding germplasm (inbreds) could provide some much-needed direction for these tasks related to germplasm management.

Heterosis, or hybrid vigor expressed by F_1 generation progeny of crosses between pairs of parents (i.e., inbred lines), has been the basis for modern maize production worldwide and a profitable seed industry; yet it has remained one of the great mysteries of biology. Quantitative genetic theory and empirical study have indicated the degree of heterosis may be, in part, positively related to genetic diversity between the parents (Hallauer and Miranda 1981; Moll et al. 1965). Direct tests of this relationship with RFLP data have revealed excellent agreement with theory for crosses between lines related by pedigree (Smith et al. 1990). The coefficient of determination from regression of heterosis on genetic distance based on RFLPs was 0.76. However, in other studies weaker relations between heterosis and RFLP-based measures of diversity were observed for crosses strictly between unrelated inbred lines (Godshalk et al. 1990; Melchinger et al. 1991), perhaps reflecting limitations of theory, the utility of RFLP data, experimental design, or knowledge of other biological phenomena.

The exceptional level of molecular polymorphism in maize has been the basis of several recent achievements of fundamental significance. Initial RFLP map construction dramatically expanded the number of mapped loci (Helentjaris et al. 1986; Hoisington 1987; Burr et al. 1988) and improved knowledge of genome duplication (Helentjaris et al. 1988). Further integration of RFLP and previously established linkage maps consisting of translocation breakpoints and visible mutants (pigmentation, morphology, etc.) should enhance the current appreciation of physical and genetic distances (Meagher et al. 1988) and the basis of quantitative genetic variation (Beavis et al. 1991). Once in place, such knowledge combined with advances in cloning and mapping technology should provide opportunities to pursue the genetic components of any reliably measured phenotype in a directed manner.

Efforts to identify and clone genes have been facilitated through knowledge of molecular marker profiles of inbreds. Isolation of the *02* and *B* loci (Schmidt et al. 1987; Chandler et al. 1989, respectively) was confirmed, in part, on the basis of RFLPs between inbreds of appropriate allelic constitution for those genomic regions. Such an approach should contribute to future transposon-tagging efforts. Other qualitative factors with difficult-to-recognize-transposon-tagged phenotypes (e.g., resistance to maize dwarf mosaic virus) have been placed relative to RFLP loci (McMullen and Louie 1989); this provides the means for marker-based selection of desired alleles and perhaps the landmarks for map-based cloning. Factors with quantitative effects on traits of importance to maize evolution (Doebley et al. 1990) and breeding (Beavis et al. 1991; Reiter et al. 1991) have also been mapped with RFLP markers. The availability of maize inbreds with appropriate phenotypes and adequate frequency and placement of molecular polymorphism was critical to the design and interpretation of each study.

SOME NOTES OF CAUTION

Information provided by characterization of maize inbreds with molecular markers has been an important resource for a wide variety of investigations. However, as with any source of information or procedure, certain precautions may be needed as warranted by the objectives of the study. At this time, there are some open questions and unresolved issues regarding data collection and interpretation of RFLP profiles of maize inbreds.

Probably the first and most important step toward effective utilization of RFLP data of maize inbreds would be verification of identity and purity of seed stocks and clones. Given the large numbers of each, mixtures and mislabeling occur and may create much confusion. The USDA maize genome project has planned to establish a database containing RFLP profiles of selected inbreds and pertinent information on mapped probes. Ideally, the database would be available in pictorial form in the near future. This should minimize certain problems pertaining to identity and quality control of materials and data interpretation.

Collecting and analyzing RFLP data from maize inbreds has been a somewhat subjective endeavor. Protocols for electrophoresis, Southern transfer, hybridization, and related techniques have varied among laboratories and could produce novel, perhaps anomalous, observations. For example, many probes hybridize to more than one fragment in single digests of individual inbreds because of extensive duplications in the maize genome. Usually, the resulting bands in autoradiograms exhibit different levels of exposure. The uneven exposure could reflect differential homology, unequal transfer of fragments, and other factors. This situation creates some doubt about which bands were mapped in the original reports, and allelism among bands. As yet, universal solutions to these problems have not been devised, but the confusion may be ameliorated by including internal standards on the filters such as the inbreds used as parents in RFLP map construction and other indicators of hybridization conditions.

Standardization of many laboratory techniques could be an unreasonable goal; it may be simpler and more desirable to use common internal standards. This could be especially important for declaring matches of bands among lanes on autoradiograms within a study, describing maize RFLP profiles in terms of the molecular weights of bands, and comparing molecular weights of RFLP profiles among labs and experiments. Lack of appropriate molecular weight standards has been a serious issue for human DNA fingerprinting and diagnostics (Lander 1989) with important implications for use of this technology in other species. Initial surveys of maize inbreds for RFLPs included molecular weight standards in selected lanes in each gel (Lee et al. 1989; Smith et al. 1990; Melchinger et al. 1991); however, such an approach may not be adequate for larger surveys and public databases. Accurate comparisons between laboratories would be facilitated by including molecular weight standards in each lane (e.g., include known fragments of approximately 2 and 20 kb in every lane and possibly a ladder of standards in selected lanes). Such a system should be easy to implement in many laboratories and would greatly improve transfer and integration of information. The need for such a system should increase greatly as more discriminative DNA marker techniques are developed and implemented.

When asked why he robbed banks, the legendary Willie Sutton responded, "Because that's where the money is kept." The secret to his success (apparently he was wrong at least once) was knowing where and how to look for the prize. Information on the maize genome and a wide array of inbred lines with unique phenotypes has been accumulating for decades, recently at accelerated rates. The advent of DNA-based molecular markers has expanded the opportunities for describing and exploring the maize genome in great detail, possibly providing an important means of learning where to search for rather elusive targets. In many instances, the right combination of inbred lines and their DNA profiles may provide the key to a focused and successful expedition.

REFERENCES

Beavis WD, Grant D, Albertsen MC, Fincher R (1991) Quantitative trait loci for plant height in four maize populations and their associations with qualitative genetic loci. Theor Appl Genet 83: 141–145

Bogenschutz TG, Russell WA (1986) An evaluation for genetic variation within maize inbred lines maintained by sib-mating and self-pollination. Euphytica 35: 403–412

Boppenmaier J, Melchinger AE, Brunlaus-Jung E, Geiger HH, Herrmann RG (1992) Genetic diversity for RFLPs in European maize inbreds: relation to performance in flint x dent crosses for forage traits. Crop Sci, 32: 895–902

Burr B, Burr FA, Thompson KH, Albertsen MC, Stuber CS (1988) Gene mapping with recombinant inbreds in maize. Genetics 118: 519–526

Chandler VL, Radicella JP, Robbins TP, Chen J, Turks D (1989) Two regulatory genes of the maize anthocyanin pathway are homologous: Isolation of *B* utilizing *R* genomic sequences. Plant Cell 1: 1175–1183

Chughtai SR, Steffensen DM (1989) Knob constitution of inbred lines of maize (*Zea mays* L.) and its implications in maize breeding. SABRAO J 21: 21–26

Coe EH, Neuffer MG, Hoisington DA (1988) The genetics of corn. In Sprague GF, Dudley JW (eds) Corn and Corn Improvement, Third Edition, American Society of Agronomy, Madison, WI

Darrah LL, Zuber MS (1986) 1985 United States farm maize germplasm base and commercial breeding strategies. Crop Sci 26: 1109–1113

Doebley J, Stec A, Wendel J, Edwards M (1990) Genetic and morphological analysis of a maize-teosinte F2 population: implications for the origin of maize. Proc Natl Acad Sci 87: 9888–9892

Edwards MD, Stuber CW, Wendel JF (1987) Molecular-marker facilitated investigations of quantitative trait loci in maize: I. Numbers, genomic distribution and types of gene action. Genetics 116: 113–125

Evola SV, Burr FA, Burr B (1986) The suitability of restriction fragment length polymorphisms as genetic markers in maize. Theor Appl Genet 71: 765–771

Gardiner J, Coe EH, Melia-Hancock S, Hoisington DA, Chao S (1991) A set of core RFLP markers for maize. Maize Genetics Cooperation News Letter 65: 54–56

Gerdes JT, Behr CS, Coors JG, Tracy WF (1993) Compilation of North American Maize Breeding Germ Plasm, Crop Science Society of America, Madison, WI

Godshalk EB, Lee M, Lamkey KR (1990) Analysis of the relationship of restriction fragment length polymorphisms to maize single-cross hybrid performance. Theor Appl Genet 80: 273–280

Hallauer AR (1987) Maize. In Fehr WR (ed) Principles of Cultivar Development, Macmillan, New York, pp 249–294

Hallauer AR (1990) Methods used in developing maize inbreds. Maydica 35: 1–16

Hallauer AR, Miranda JB, Fo (1981) Quantitative Genetics in Maize Breeding. Iowa State University Press, Ames, IA

Helentjaris T, Slocum M, Wright S, Schaefer A, Nienhuis J (1986) Construction of genetic linkage maps in maize and tomato using restriction fragment length polymorphisms. Theor Appl Genet 72: 761–769

Helentjaris T, Weber DF, Wright S (1988) Duplicate sequences in maize and identification of their genomic locations through restriction fragment length polymorphisms. Genetics 118: 353–363

Higginbotham JG, Smith JSC, Smith OS (1991) Quantitative analysis of two-dimensional protein profiles of inbred lines of maize (*Zea mays* L.). Electrophoresis 12: 425–431

Hoisington DA (1987) Maize (*Zea mays* L.) RFLP clones and linkage map—A public set. Genetics 116: s27

Lander ES (1989) DNA fingerprinting on trial. Nature 339: 501–505

Lee EA, Lee M, Lamkey KR (1990a) Effectiveness of two restriction enzymes for detecting RFLPs in the Iowa Stiff Stalk Synthetic (BSSS) maize population. Maize Genetics Cooperation News Letter 64: 20

Lee EA, Lee M, Lamkey KR (1990b) RFLP analysis of isogenic lines B14 and B14A. Maize Genetics Cooperation News Letter 64: 20

Lee M, Godshalk EB, Lamkey KR, Woodman WW (1989) Association of restriction fragment length polymorphisms among maize inbreds with agronomic performance of their crosses. Crop Sci 29: 1067–1071

Martin GB, Williams JGK, Tanksley SD (1991) Rapid identification of markers linked to a *Pseudomonas* resistance gene in tomato by using random primers and near-isogenic lines. Proc Natl Acad Sci USA 88: 2336–2340

McMullen MD, Louie R (1989) The linkage of molecular markers to a gene controlling the symptom response in maize to maize dwarf mosaic virus. Mol Plant-Microbe Interactions 2: 309–314

Meagher RB, McLean MD, Arnold J (1988) Recombination within a subclass of restriction fragment length polymorphisms may help link classical and molecular genetics. Genetics 120: 809–818

Melchinger AE, Messmer MM, Lee M, Woodman WL, Lamkey KR (1991) Diversity and relationships among U.S. maize inbreds revealed by restriction fragment length polymorphisms. Crop Sci 31: 669–678

Messmer MM, Melchinger AE, Lee M, Woodman WL, Lee EA, Lamkey KR (1991) Genetic diversity among progenitors and elite lines from the Iowa Stiff Stalk Synthetic (BSSS) maize populations: comparison of allozyme and RFLP data. Theor Appl Genet 83: 97–107

Moll RH, Lonnquist JH, Veley Fortuna J, Johnson EC (1965) The relationship of heterosis and genetic divergence in maize. Genetics 52: 139–144

Nucca R, Soave C, Motto M, Salamini F (1978) Taxonomic significance of the zein isoelectric focusing pattern. Maydica 23: 239–249

Plucknett DL, Smith NJH, Williams JT, Anishetty NM (1983) Crop germ-plasm conservation and developing countries. Science 220: 163

Reiter RS, Coors JG, Sussman MR, Gabelman WH (1991) Genetic analysis of tolerance to low-phosphorus stress in maize using restriction fragment length polymorphisms. Theor Appl Genet 82: 561–568

Schmidt RJ, Burr FA, Burr B (1987) Transposon tagging and molecular analysis of the maize regulatory locus *opaque-2*. Science 238: 960–963

Smith JSC, Goodman MM, Stuber CW (1985) Genetic variability within U.S. maize germplasm. I. Historically important lines. Crop Sci 25: 550–555

Smith JSC, Smith OS (1989) The description and assessment of distances between inbred lines of maize: II. The utility of morphological, biochemical, and genetic descriptors and a scheme for the testing of distinctiveness between lines. Maydica 34: 151–161

Smith OS, Smith JSC, Bowen SL, Tenborg RA, Wall SA (1990) Similarities among a group of elite maize inbreds as measured by pedigree, F1 grain yield, heterosis, and RFLPs. Theor Appl Genet 80: 833–840

Stuber CW, Goodman MM (1983) Allozyme genotypes for popular and historically important inbred lines of corn. US Dept of Agriculture-Agric Res Serv, Southern Series, No. 16

Stuber CW, Wendel JF, Goodman MM, Smith JSC (1988) Techniques and scoring procedures for starch gel electrophoresis of enzymes from maize (*Zea mays* L.). Technical Bulletin 286, North Carolina Agricultural Research, Raleigh, NC

Welsh J, Honeycutt RJ, McClelland M, Sobral BWS (1991) Parentage determination in maize hybrids using the arbitrarily primed polymerase chain reaction (AP-PCR). Theor Appl Genet 82: 473–476

66

Traditional Analysis of Maize Pachytene Chromosomes

ELLEN DEMPSEY

ACETOCARMINE SQUASH TECHNIQUE

The acetocarmine squash technique of Belling (1926) and McClintock (1929a) has been widely used by maize workers in chromosome identification, determining chromosome numbers, analyses of chromosomal aberrations, studies of meiotic mutants, analysis of homologous and nonhomologous pairing, and in many other studies. Not only is carmine an excellent dye for distinguishing chromatin in a variety of states, but the squashing process aids in separation of chromosomes and the flattening of a three dimensional nucleus into a single plane is desirable for photomicrography. While the squash technique is extremely useful in analysis of chromosome structure and behavior, it must be recognized that distortions occur during the flattening and heating of cells. The nuclear membrane of prophase I meiotic stages is destroyed and the orientation of the chromosomes within the nucleus is modified. Associations of telomeres with the nuclear membrane, reported in other studies, cannot be verified by this technique.

Given below is a brief outline of the technique as applied to maize microsporogenesis. For a more detailed description, the reader is referred to the article by Burnham (1982).

The Maize Handbook—M. Freeling, V. Walbot, eds.

MATERIALS

1. Tassel sample. Sample is collected just before tassel emergence from the whorl, preserved in freshly mixed fixative containing 3 parts 95% ethanol: 1 part glacial acetic (or propionic) acid. After 1 day at room temperature, store at 4°C or −10°C, either in original fixative or in 70% ethanol.
2. 3 × 1-inch microscope slides and 22 mm square cover glasses, #1 thickness, cleaned in 70% ethanol and dried with cheesecloth.
3. 2 dissecting needles (replace standard needles with sewing needles bent at tip).
4. Petri dish.
5. Filter paper.
6. Alcohol lamp containing 95% ethanol.
7. Petri dish containing 1 part beeswax and 1 part paraffin mix.
8. Carmine (alum lake) certified for use in cytology. Gradually add 0.5 g to 100 ml of hot 45% acetic acid and boil gently for 2–10 minutes in a loosely covered flask. Cool and filter.
9. Microscope with 100× magnification

METHODS

1. Place tassel sample and fluid in a petri dish containing a round filter paper. A black surface below the dish aids in visualizing spikelet parts.
2. Isolate a single branch and progressively sample anthers from sequential spikelets to identify the desired stage.
3. Using needles, remove a single anther (usually from the larger set of three, see diagram), place it on a slide, and add a small drop of carmine (barely enough to reach the edges of the cover glass). See Figure 72.37 in chapter by Chang and Neuffer (this volume) for a staging system for anthers.

4. Cut the anther in half by crossing the two needles and gently press out the rows of pollen mother cells from the four cylindrical chambers of the anther. Separate the cells by vigorous stirring of the needles in the drop. Iron in non-stainless-steel needles darkens the stain and increases binding of the dye, so prolonged use of needles may lead to overstained preparations.

5. Examine cells without a coverslip under 100× magnification for identification of stage. Drops containing unwanted stages can be removed with cheesecloth and an older or younger anther can be selected for preparation on the same slide.

6. Remove fragments of anther wall: collect visible bits of tissue at far side of drop, tilt the slide so liquid portion separates from fragments, and scrape off and discard the unwanted tissue. Or use the two needles to lift anther debris from the drop.

7. Place a cover glass over the carmine drop and heat gradually over flame of alcohol lamp. Examine under 100× magnification to monitor flattening of cells and destaining of cytoplasm.

8. Invert slide on filter paper and press gently with thumbs on either side of cover glass. Do not shift cover glass. Add small amount of carmine to edge of cover glass to remove air bubbles.

9. For prolonged viewing, the preparation may be sealed with beeswax mixture applied with hot metal applicator (dissecting needle bent to exact dimensions of cover glass). After 2 or 3 days, the sealed preparation becomes overstained and carmine precipitates. Schedules are available for making permanent preparations (McClintock 1929a; Buck 1935).

STRUCTURE OF MAIZE CHROMOSOMES AT PACHYNEMA

The foundations of maize cytogenetics were laid in the late 1920s and early 1930s when the mitotic and meiotic chromosomes were first described (McClintock 1929b, 1933). Because the pachytene stage of microsporogenesis is the most appropriate stage for observing the structure of maize chromosomes, a review of the traditional lore as revealed by the squash technique may be useful to future maize researchers. The chromosomes at this stage are greatly extended and exhibit many features not recognizable in the short, condensed chromosomes of later stages. Moreover, the close pairing of homologues reduces the number of independent bodies from 20 to 10.

In some maize stocks, the pachytene chromosomes are not easily separated in squash preparations; when they become sufficiently shortened to allow individual pairs to be distinguished, details of structure may have been lost by progressive condensation. The inbred lines, L289, W22, and especially KYS, are recognized as superior in the spreading ability and quality of pachytene preparations. Crosses of unrelated stocks to KYS usually show immediate cytological improvement in the F_1. However, certain difficulties arise in converting experimental material to KYS background. KYS is a late maturing inbred, and it does not thrive in winter greenhouse environments in U.S. corn belt latitudes. Moreover, a backcross program with KYS as recurrent parent in some cases (depending on the genotype of the material to be converted) leads to a segregation of male-sterile individuals due to the *ms ms, s s* (suppressor of male sterility) genotype of KYS (Schwartz 1951; Dempsey 1957). A number of translocations and inversions are available in the KYS background. Some were introduced by repeated backcrossing; others resulted from irradiation of KYS pollen. Sporocyte samples from standard KYS and from plants containing these aberrations provide excellent material for classroom demonstration of chromosome mechanics in normal plants as well as in those carrying aberrations in heterozygous or homozygous condition.

The chromosomes of maize may be distinguished by their length, arm ratio (i.e., centromere position), possession of varying amounts of centric heterochromatin, and enlarged chromomeres or knobs in characteristic positions. Chromosome 6 contains the large nucleolus organizer region and is attached by this region to the nucleolus in prophase stages. All of the chromosomes are composed of a linear sequence of chromomeres (condensed or coiled chromatin); the largest and deepest staining of these are adjacent to the centromere and they become progressively smaller toward the termini. The distalmost chromomere (telomere) may be slightly enlarged. A few exceptions to the orderly gradation in chromomere size occur. (See prominent chromomeres in Figure 66.1.)

In acetocarmine preparations, the centromeres of pachytene chromosomes appear as clear areas a shade darker than the surrounding cytoplasm and delimited by large, deeply stained chromomeres (centric heterochromatin). Thus, centromeres occur in the darkest portion of the chromosome. The homologous centromeres are closely associated and usually cannot be distinguished as separate entities although in some preparations, paired centromeric chromomeres contributed by the two homologs have been described (Lima-de-Faria 1958). Centromere position probably does not vary from stock to stock, except in rare cases where small pericentric inversions may have occurred. Apparent deviations in arm ratio at pachynema in different stocks can be attributed to fluctuations in the coiling pattern during condensation or to the presence of heterochromatic knobs. Four of the chromosomes (2, 7, 9, and 10) have large, easily identified centromeres, often rectangular in shape. Smaller, circular centromeres occur on the other chromosomes. The latter type of centromere often protrudes slightly to one side of the linear axis of the paired homologues.

Most of the standard features of pachytene chromosomes can be seen in the micrograph of KYS chromosomes (Figure 66.2). An enumeration of identifying characteristics for each chromosome is given in Table 66.1.

In addition to these standard features showing little variation in maize races, the chromosomes may contain prominent heterochromatic knobs, which greatly aid in the recognition of specific chromosomes. Maize stocks differ in the number and location as well as in the size of the knobs. Twenty-two possible knob positions have been described among the 10 chromosomes; these are shown in the composite diagram (Figure 66.1). A particular stock possesses a specific constellation of knobs, which are constant features

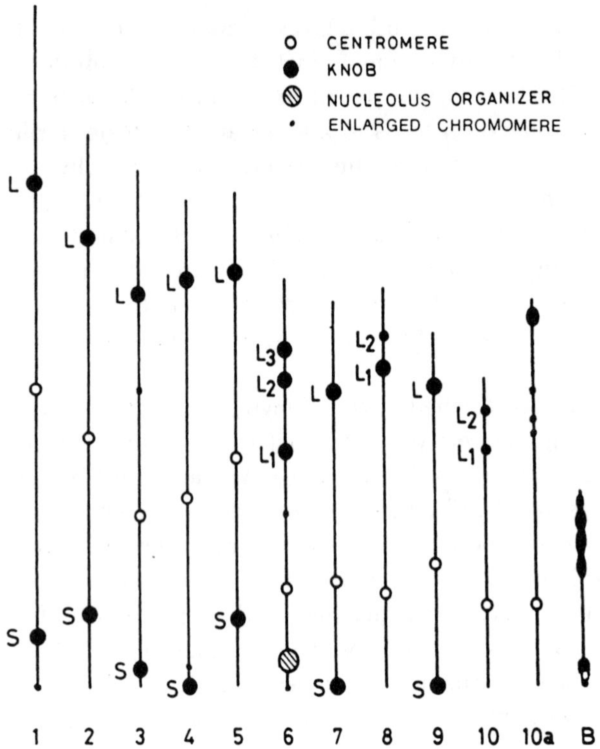

FIGURE 66.1 Idiogram of the 10 pachytene chromosomes of maize showing a compilation of the knob positions observed in different races. L refers to long arm and S to short arm. In addition to the normal A chromosomes, the abnormal chromosome (l0a) and the B type chromosome (B) are represented. Enlarged chromomeres on chromosomes 1, 3, 4, 6, and abnormal chromosome 10 are indicated. (Modified from Figure 1 of McClintock et al. 1981.)

of its genome. Chromosome length is increased by the presence of a knob. Each knob may be homozygous or heterozygous, depending on the degree of inbreeding within the strain. Large knobs may vary in shape, from a thin, attenuated condition to a shorter, rounded state. When a large internal knob is opposed by a knobless homolog, there is often asynapsis in the knob region, and recombination between marker genes spanning the knob site is reduced. Knobs do not affect the plant phenotype but do play a role in chromosome mechanics within the cell. [See studies on preferential segregation (Rhoades 1952) and high-loss (Rhoades and Dempsey 1971).] The knob on abnormal chromosome 10, known as K10-I, has properties unlike the other knobs, inducing neocentromeres, causing preferential segregation, and increasing crossing over in many parts of the genome (Rhoades and Dempsey 1966). A second abnormal chromosome 10 knob (K10-II), found by Kato in teosinte, differs somewhat in structure from K10-I but has the same unusual properties. K10-II has been introduced into maize (Rhoades and Dempsey 1985).

One of the first studies involving maize knobs was the demonstration by Creighton and McClintock (1931) that crossing over in maize involves a physical exchange of chromatin. Crossovers between two genes in the short arm of chromosome 9 were found to be associated with exchange between two physical landmarks, a terminal knob and a translocated segment, each represented in only one of the homologs. Knobs have been useful in characterization of maize races (Ghatnakar 1965; Longley and Kato 1965; Suto

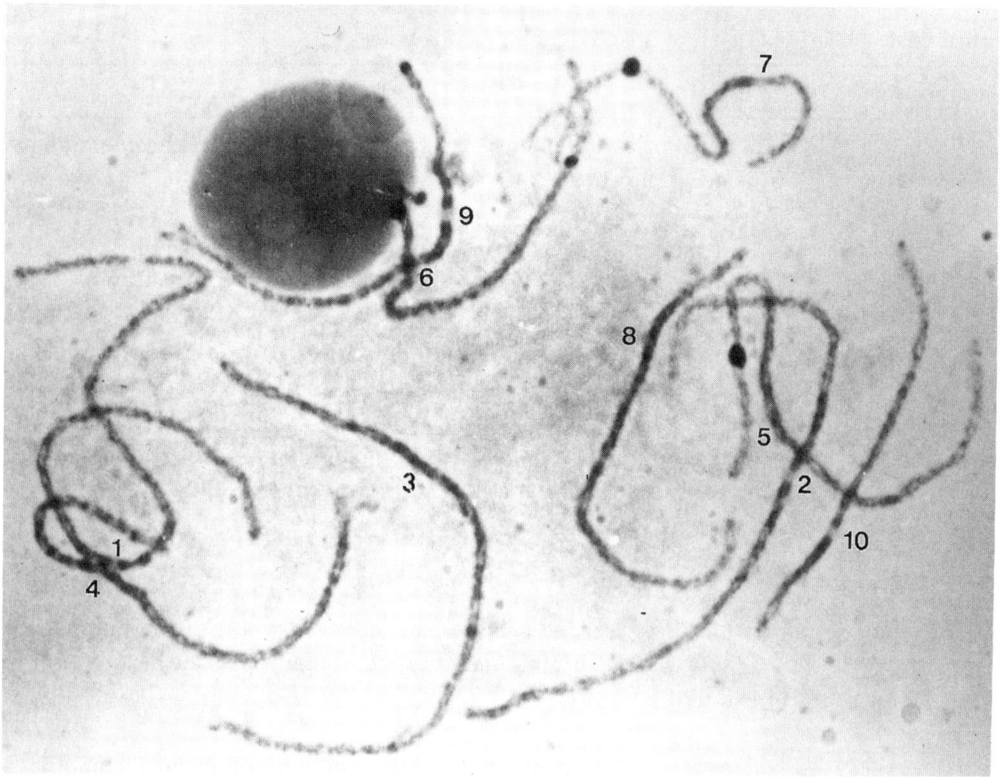

FIGURE 66.2 Acetocarmine squash of KYS pollen mother cell in pachytene stage of meiosis. Slide prepared by Janice Lovett. Chromosomes identified by numbers placed near centromeres. Chromosomes 5 and 7 have large knobs, while chromosomes 6 and 9 have small knobs. Centromeres of chromosomes 1 and 4 are fused in this cell.

et al. 1979; McClintock 1978); from cytological studies such as these, information has been derived about interrelationships and possible sites of origin of native races. In 1981, a 185-bp segment of DNA, identified in highly heterochromatic maize stocks, was shown to hybridize in situ preferentially to the heterochromatin of knobs. This sequence was estimated to vary from 10^4 to 10^6 copies in knobs of different sizes (Peacock et al. 1981).

Both centromeres and knobs show nonspecific fusions that may persist into early diplonema. Studies by S.R. Peterson (U. of Illinois masters thesis, 1955) and by J. Gurgel (1958) indicate that in the case of centromeres, the frequency of nonhomologous fusion is correlated with chromosome length, except for chromosome 5, which participates more frequently than expected. In the case of knobs, nonspecific fusions occur most frequently between large knobs. Fusions between centromeres and knobs have been reported but occur only rarely.

The excellence of the pachytene stage in maize and the availability of several trans-

TABLE 66.1. Features of maize pachytene chromosomes

Chromosome	Description
1	Longest of complement; arm ratio 1.3:1; enlarged chromomere at end of short arm
2	Arm ratio 1.25:1; prominent centromere with adjacent regions very pyknotic
3	Arm ratio 2:1; enlarged chromomere at 0.4 from the centromere in the long arm
4	Arm ratio 1.6:1; enlarged chromomere near end of short arm; dark-staining chromatin near end of long arm (0.7–0.9).
5	Arm ratio 1.1:1
6	Arm ratio 3.1:1; attached to nucleolus at end of heterochromatic NOR region in 6S; enlarged chromomeres at 0.25 in long arm and at end of 6S
7	Arm ratio 2.8:1; prominent centromere; region of long arm adjacent to centromere very pyknotic
8	Arm ratio 3.2:1; regions on both sides of centromere equally pyknotic
9	Arm ratio 1.8:1; prominent centromere; pyknotic region occupying proximal third of short arm
10	Shortest of complement; arm ratio 2.8:1; prominent centromere; centric heterochromatin deep staining in short arm and lighter in long arm.

locations with one breakpoint within the nucleolus organizer region made possible a detailed study of the function of that region (Phillips 1978). The NOR region of different maize stocks shows considerable variation in size and in content of ribosomal RNA genes, as well as in the site of attachment to the nucleolus. The heterochromatic portion of the NOR containing 90% of the redundant ribosomal DNA is proximal to the secondary constriction containing about 10% of the ribosomal DNA. Studies of nucleolar size at pachynema in plants with homozygous translocations where the two NOR segments often form separate nucleoli indicated that the secondary constriction is the important functional region in nucleolus formation, but two secondary sites in the heterochromatic portion are also functional, especially in the absence of the primary region.

Pachynema is the meiotic stage when pairing of homologous chromosomes has been completed, and early pachynema has been reported to be the time when crossing over takes place (Rhoades 1968). The two homologs are closely appressed along their length although the ends occasionally diverge. Since replication has already occurred, each homolog consists of two chromatids. These cannot usually be distinguished in squash preparations but occasionally the chromatid split can be seen in an unpaired segment. As early as 1933, it was recognized that two-by-two associations, normally found between homologs, may sometimes involve nonhomologous segments when the normal sequence is altered in one member of the pair (inversions, translocations) or when plants are monosomic or trisomic for a particular chromosome (McClintock 1933). Foldbacks within a single homolog are common in asynapsed regions. There has been a suggestion that some of the pairing observed in haploids may represent synapsis of remnant homologies in an allopolyploid where the basic number is five. Rare crossovers in such regions have led to formation of translocations (Weber and Alexander 1972).

Initiation of homologous pairing is most likely to occur in a region near the ends of the chromosome arms (0.8–0.9 from the centromere) with pairing at secondary sites also favoring more distal regions (Burnham et al. 1972). This conclusion was reached in an exhaustive study of the pachytene stage in plants combining two different translocations involving chromosomes 1 and 5. In addition to quadrivalents, two types of bivalents were theoretically possible at pachynema (those with homologous ends and nonhomologous centromere regions and those with homologous centromere regions and nonhomologous ends). Only the former were observed, even in translocations with distal breakpoints where the terminal segments are short and the internal segments long. Initiation of pairing in distal regions may account, at least partially, for the higher recombination frequencies observed in terminal segments.

Analysis of pachytene chromosomes has made possible the physical location of a number of mutant genes. Genes have been placed proximal or distal to various cytological markers, such as knobs or breakpoints of translocations and inversions. Translocations between members of the regular complement (A chromosomes) and supernumerary B chromosomes have been particularly useful because sperm cells lacking the segment distal to the breakpoint (hypoploid gametes) are transmissible. In certain other translocations, unbalanced gametes arising by adjacent segregation are functional. In both cases, crosses are made between the translocation stock and the recessive mutant tester, and the appearance of the mutant phenotype in the hemizygous offspring indicates the gene in question is contained in the missing segment. Precise location of three genes in the short arm of chromosome 9 (*Pyd*, *Wd*, and *Yg2*) was accomplished by careful study of a series of terminal deficiencies (McClintock 1944). *Pyd* is located in the stalk connecting the terminal knob and first chromomere, while *Wd* and *Yg2* are in the distal half of the first chromomere. In another study (McClintock 1941) several mutant genes were localized in the four chromomeres adjacent to the centromere in the short arm of chromosome 5.

The summary given above describes only a few of the findings based on analysis of pachytene chromosomes and is intended to display the range of problems and the types of solutions provided by such studies. Complete reviews of cytogenetic research on maize are found in Rhoades (1955), Burnham (1962), Carlson (1988), and Sharma and Sarma (1988). A description and discussion of the B chromosome, omitted in this narrative, are given in another chapter (see Carlson) of this volume.

REFERENCES

Belling J (1926) The iron-acetocarmine method of fixing and staining chromosomes. Biol Bull 50: 160–162

Buck JB (1935) Permanent acetocarmine preparations. Science 81: 75

Burnham CR (1962) Discussions in Cytogenetics. Burgess, Minneapolis, MN, 375 pp

Burnham CR (1982) Details of the smear technique for studying chromosomes in maize. In Sheridan WF (ed) Maize for Biological Research, Plant Molec Biol Assoc, Charlottesville, VA, pp 107–118

Burnham CR, Stout JT, Weinheimer WH, Kowles RV, Phillips RL (1972) Chromosome pairing in maize. Genetics 71: 111–126

Carlson WR (1988) The cytogenetics of corn. In Sprague GF, Dudley JW (eds) Corn and Corn Improvement, Agronomy Monograph No. 18, Am Soc Agronomy, Madison, WI, pp 259–343

Creighton HB, McClintock B (1931) A correlation of cytological and genetical crossing-over in *Zea mays.* Proc Natl Acad Sci USA 17: 492–497

Dempsey E (1957) "KYS" male sterility. Maize Genetics Cooperation News Letter 31: 81–83

Ghatnakar MV (1965) Heterochromatic knobs in Italian maize population and evolution of maize in Italy. Cytologia 30: 402–425

Gurgel JTA (1958) The non-homologous associations of centromeres and knobs of maize chromosomes at meiosis. Proc X Int Congr Genetics Montreal: 107

Lima-de-Faria A (1958) Compound structure of the kinetochore in maize. J Hered 49: 299–302

Longley AE, Kato TA (1965) Chromosome morphology of certain races of maize in Latin America. CIMMYT, Mexico Res Bulletin 1: 1–112

McClintock B (1929a) A method for making acetocarmin smears permanent. Stain Technol 4: 53–56

McClintock B (1929b) Chromosome morphology in *Zea mays.* Science 69: 629

McClintock B (1933) The association of non-homologous parts of chromosomes in the midprophase of meiosis in *Zea mays.* Z Zellforsch Mikrosk Anat 19: 191–237

McClintock B (1941) The association of mutants with homozygous deficiencies in *Zea mays.* Genetics 26: 542–571

McClintock B (1944) The relation of homozygous deficiencies to mutations and allelic series in maize. Genetics 29: 478–502

McClintock B (1978) Significance of chromosome constitutions in tracing the origin and migration of races of maize in the Americas. In Walden DB (ed) Maize Breeding and Genetics, John Wiley and Sons, New York, pp 159–184

McClintock B, Kato TA, Blumenschein A (1981) Chromosome constitution of races of maize. Colegio de Postgraduados, Chapingo, Mexico, 517 pp

Peacock WJ, Dennis ES, Rhoades MM, Pryor AJ (1981) Highly repeated DNA sequence limited to knob heterochromatin in maize. Proc Natl Acad Sci USA 78: 4490–4494

Phillips KL (1978) Molecular cytogenetics of the

nucleolus organizer region. In Walden DB (ed) Maize Breeding and Genetics, John Wiley and Sons, New York, pp 711–741

Rhoades MM (1952) Preferential segregation in maize. In Gowen JW (ed) Heterosis, Iowa State College Press, Ames, IA, pp 66–80

Rhoades MM (1955) The cytogenetics of maize. In Sprague GF (ed) Corn and Corn Improvement, Academic Press, New York, pp 123–219

Rhoades MM (1968) Studies on the cytological basis of crossing over. In Peacock WJ and Brock RD (eds) Replication and Recombination of Genetic Material, Australian Acad Sci, Canberra, pp 229–241

Rhoades MM, Dempsey E (1966) The effect of abnormal chromosome 10 on preferential segregation and crossing over in maize. Genetics 53: 989–1020

Rhoades MM, Dempsey E (1971) On the mechanism of chromatin loss induced by the B chromosome of maize. Genetics 71: 73–96

Rhoades MM, Dempsey E (1985) Structural heterogeneity of chromosome 10 in races of maize and teosinte. In Freeling M (ed) Plant Genetics, Vol 35 UCLA Symp Molec Cell Biol, Alan R Liss, New York, pp 1–18

Sharma AK, Sarma JSP (1988) Chromosome structure, rearrangements and genome relationship in Maydeae. Feddes Repertorium 99: 291–337

Schwartz D (1951) The interaction of nuclear and cytoplasmic factors in the inheritance of male sterility in maize. Genetics 36: 676–696

Suto T, Murakami K, Shiga T, Yoshida Y, Sugiyama S, Toyama M, Mochizuki M, Yamada M, Takayanagi K (1979) Collection and characteristics of races of maize in Japan. Misc Publ Natl Inst Agric Sci (Tokyo) Series D, No. 3, pp 1–210

Weber DF, Alexander DE (1972) Redundant segments in *Zea mays* detected by translocations of monoploid origin. Chromosoma 39: 27–42

67

Techniques for Preparing Whole-Mount Spreads of Maize Pachytene Chromosome Complements for Electron-Microscopic Visualization of Synaptonemal Complex Structures

MARJORIE P. MAGUIRE

The following describes a successful spreading procedure for maize which combines parts of techniques set forth separately by Stack, Holm, Gillies, Jones, and collaborators (mostly for other organisms) and contains a few additional elements. It can be used with silver, phosphotungstic acid (PTA) or uranyl acetate–lead citrate (UP) staining. Electron-microscopic (EM) viewing of silver-stained spread preparations allows very clear visualization of synaptonemal complex (SC) lateral elements, either as cores in advance of synapsis or in fully synapsed configurations (Figure 67.1). However, only the lateral elements of the SC stain appropriately with silver, and central elements, recombination nodules (RNs), and centromeres remain invisible. The other stains, PTA or UP, must be used to stain the additional features (Figure 67.2). PTA and UP also stain chromatin and other nuclear components, and this obscures viewing of the SC structures and RNs. Procedures must be adapted for making spread preparations that effectively remove the undesired components while keeping the SC structures and RNs. Unfortunately each organism may require a unique adaptation. The procedure described here allows visualization of maize SC components and RNs and also displays centromeres as rather large spheres when used with PTA or UP staining. It is equally suitable for the much less demanding silver staining.[1]

New microscope slides are coated in advance with plastic by dipping them in a

[1]A new technique for silver staining has now been devised which allows visualization of central elements, RNs, and centromeres: (Sherman, J.D., Herickhoff, L., and Stack, S.M. 1992. Genome 35: 907–915).

The Maize Handbook—M. Freeling, V. Walbot, eds.

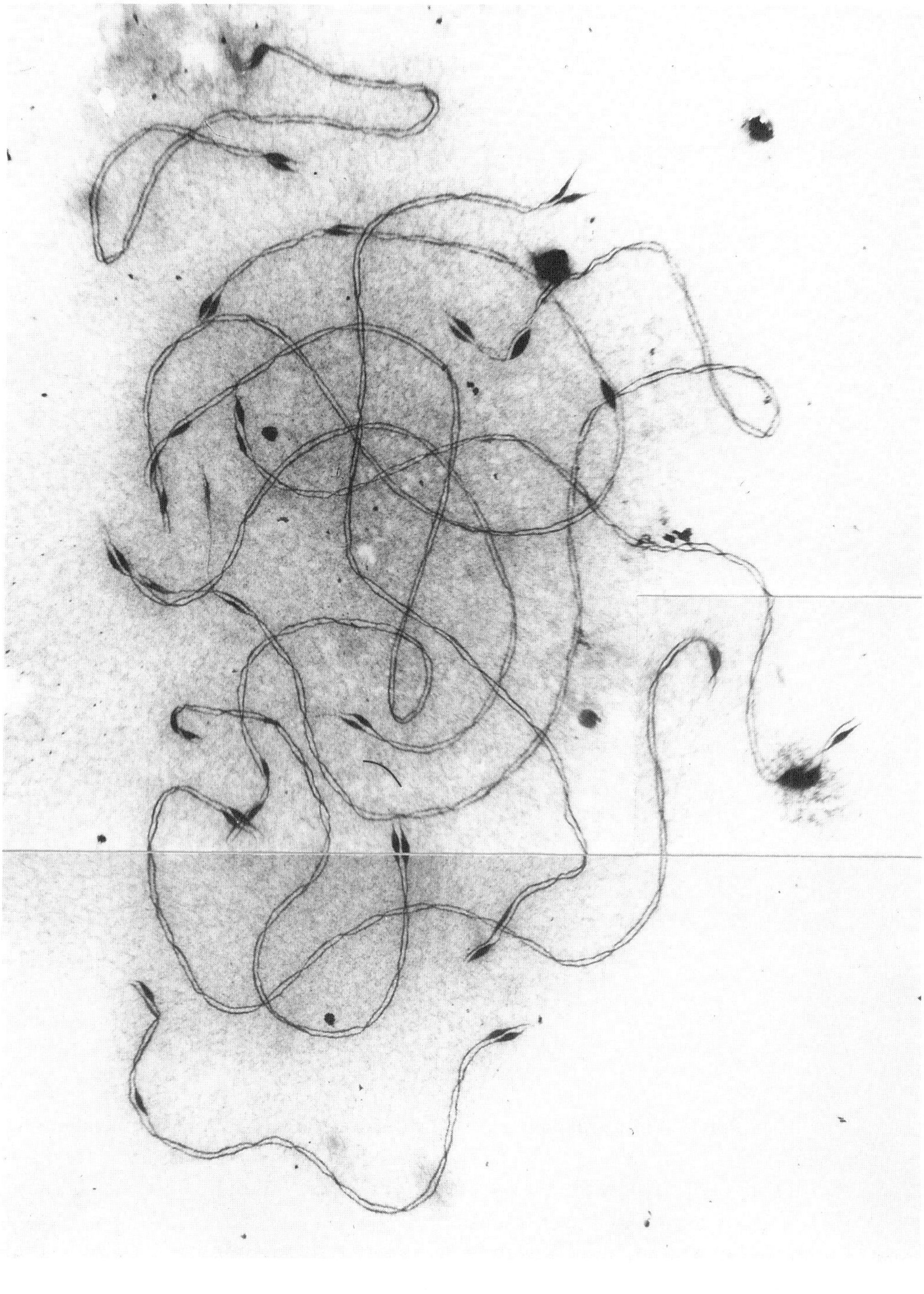

FIGURE 67.1 Silver-stained pachytene SC complement from a normal maize stock (KYS). The dark thickenings seen here appear only sporadically in such preparations. (From Maguire et al. 1991.)

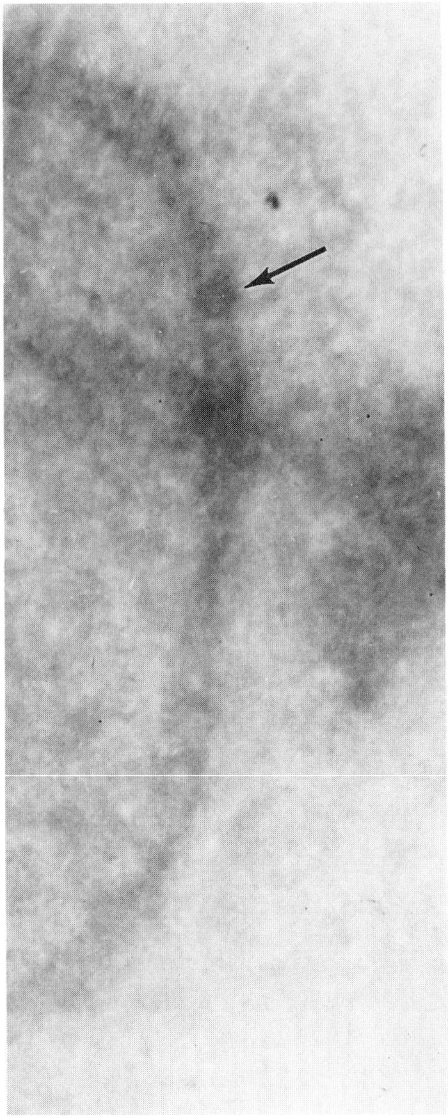

FIGURE 67.2 A portion of a pachytene SC complement from a trisome 9 stock (otherwise genetically normal), stained with 4% PTA. An RN is indicated by the arrow.

chloroform plastic solution (4-g broken Falcon petri dishes: 400 ml chloroform) and standing them on end in a test tube rack to dry. For this procedure, it is not necessary to treat the plastic coated slides with a glow discharge unit (to render the surface hydrophilic), because a relatively large amount of detergent is used.

At the outset, measured anthers from tassel branches thought to be at meiotic stage are briefly checked for stage in acetocarmine smears to determine the length of pachytene stage anthers. This length differs with the stock but is usually slightly in excess of 2 mm. See figures in the chapter by Chang and Neuffer (this volume) for guidance.

Fresh anthers at pachytene stage are macerated in a deep depression slide (kept over ice) in 5 μl of a freshly prepared ice cold medium: 1.5% sucrose, 1% polyvinylpyrrolidone, and 2.5 mM EDTA, adjusted to pH 4.7–4.8 with KOH. The maceration consists of watching under a dissecting scope while anthers (usually three at a time) are cut in half and their contents are pressed out. The suspension of microsporocytes is then transferred by pipette to the surface of a 5-μl drop of 0.1% Nonidet P40 in another deep depression slide (over ice) where it is left for 5 minutes. Then 60 μl of an ice cold fix-detergent mixture is added. The fix-detergent mixture consists of 6% paraformaldehyde, 1.5% sucrose solution adjusted to pH 8.6 to which SDS has been added to a final concentration of 0.03%. The depression slide is then covered and placed over an ice bath.

After 30 minutes to 1 hour, the depression slide is removed from over the ice bath and quickly warmed to room temperature on a laboratory bench. The contents are then micropipetted to plastic coated slides, sucking out the liquid from around the anther remnants. The plastic-coated slides are then vibrated for 10 seconds by touching an electric vibrating engraver to the surface of the ground glass end, and these slides are then allowed to air dry: they are left overnight at room temperature. Then the dried preparations on the slides are rimmed with nail polish (to prevent loss of plastic during fixation and staining), and the drying process is completed by placing the slides for 3–4 hours on a slide warmer at 37°C. These slides can be stored for several weeks before fixation and staining.

Immediately before staining, slides are fixed by treating with an ice cold, freshly prepared solution (4% paraformaldehyde, 1.5% sucrose, adjusted to pH 8.6) for 10 minutes (changed for fresh solution after the first 2 minutes). Then slides are briefly washed in 0.4% Photoflo and air dried.

Slides can be effectively stained with silver by the procedure of Stack and Anderson (1987), with 4% PTA aqueous solution by the general procedure of Albini and Jones (1987), or with UP by the procedure of Stack and Anderson (1986).

Slides are scanned with phase contrast microscopy (with at least a 20× to 25× objective) to locate complements to be studied further with EM; the positions of complements on the slides should be recorded. To facilitate transfer of preparations from slides to grids, using a plastic floating method, store the slides overnight in a humid chamber (such as atop a piece of bent glass tubing in a closed petri dish to which water has been added to a depth which does not wet the slide directly). Copper grids (Pelco IGC 50) are made slightly sticky by dipping them in dichloroethane (in which a short piece of Scotch tape has been briefly swished) and drying them on parafilm. Such grids are carefully positioned on good synaptonemal complex configurations. (It is a good idea to tilt the slides slightly at this stage to determine whether the grids are securely positioned so that they will not slip during additional manipulations.) The plastic film is then scored in a circle around and about 2 mm from the grids. Next the grids on their plastic rafts (one at a time) are floated on a drop or two of distilled water. (The water is placed at the edge of the scoring and with luck will creep under the plastic.) Then the slide is carefully immersed at a slant under the surface of distilled water in a bowl in such a way that the rafts carrying their grids are floated on the surface of the water, and the slide is then withdrawn. (The slide can then be dried and additional grids can be placed and floated in sequence. With acquisition of skill several grids per slide can be floated at once, or several close together grids can be included on a single raft.) Plastic rafts with grids are then picked up on lens paper. (Push down on the rafts from above, pushing them momentarily below the surface, and deftly invert and withdraw the lens paper so that the plastic is on top of the grids on top of the lens paper.) The lens paper is laid on a paper towel for drying. Later, after careful removal of the dried grids from the lens paper, it helps to examine them on a microscope slide with phase microscopy and map the positions of the configurations to be observed. Store the grids in grid boxes for future EM viewing.

Some cautionary notes: Coating slides with plastic film works best in relatively low humidity (below 60%). In higher humidity a cloudy plastic film is produced.

Floating plastic rafts (with grids) off of slides works best at relatively high humidity (above 60%). In low humidity, apparently, static electricity frequently inspires grids just floated to make incredible flip dives below the water surface back to the slide from which they have just been removed, before the slide can be withdrawn. With practice, rafts can be manipulated away from slides with a gentle jet of distilled water from a dropper so that none of this happens.

Additional directions: Plastic petri dish pieces are quickly dissolved in chloroform under sonication to generate the slide coating solution; the solution can be stored indefinitely.

Maceration medium pH adjustment is performed with 0.1 N and 0.01 N KOH and 0.1 N HCl.

A 25 mM acid EDTA stock solution is prepared in advance (and stored indefinitely). Dissolution of the EDTA is accomplished by adding 4 or 5 pellets of KOH; pH is then adjusted to 4.1 with HCl and KOH.

Paraformaldehyde is dissolved by addition of 1 N NaOH (2 ml to every 100 ml of solution) and heating in a hood while stirring to no hotter than 50°C. The solution is then cooled to about 4°C in an ice bath, and the pH is adjusted to 8.6 with 90% formic acid, 1 N NaOH, and 0.05 M sodium borate.

REFERENCES

Albini SM, Jones GH (1987) Synaptonemal complex spreading in Allium cepa and A. fistulosum I. The initiation and sequence of pairing. Chromosoma 95: 324–328

Maguire MP, Paredes AM, Riess RW (1991) The desynaptic mutant of maize as a combined defect of synaptonemal complex and chiasma maintenance. Genome 34: 879–887

Stack SM, Anderson LK (1986) Two-dimensional spreads of synaptonemal complexes from solanaceous plants II. Synapsis in Lycopersicum esculentum (tomato). Am J Bot 73: 264–281

Stack SM, Anderson LK (1987) Hypotonic bursting method for spreading synaptonemal complexes of Zea mays. J Hered 78: 178–182

A Smear Technique for the Study of Meiosis in Pollen Mother Cells of Maize

INNA GOLUBOVSKAYA

COLLECTION OF SPOROCYTE MATERIAL OF MAIZE

The microsporocyte material in the young developing tassel is located inside the rolled leaf sheaths. The readiness of material for study of meiosis can be checked by squeezing the upper part of plant stem between the fingers. When the region feels soft, the material is at or near meiosis. Also see Figure 72.37 in the chapter by Chang and Neuffer (this volume) for an orientation to tassel development.

FIXING OR KILLING MAIZE SPOROCYTES

The entire tassel, or only a portion of tassel, is placed in a vial of freshly mixed Farmer's killing solution consisting of 3 parts of 95% ethanol and 1 part glacial acetic acid for 24 hours at room temperature. For good fixation killer solution should be used in large excess (approximate proportion of tassel samples to fixative is 1:5).

RINSING FIXED MATERIAL

Rinse fixed material three times in 70% ethanol; store in 70% ethanol in the refrigerator until cytological study.

STAINING MICROSPOROCYTES FOR A STUDY OF MEIOSIS

It is necessary to stain microsporocytes in 2% acetocarmine or 2% propionic carmine made up in 100 ml of a 45% solution of acetic acid or 45% propionic acid. The stain

The Maize Handbook—M. Freeling, V. Walbot, eds.

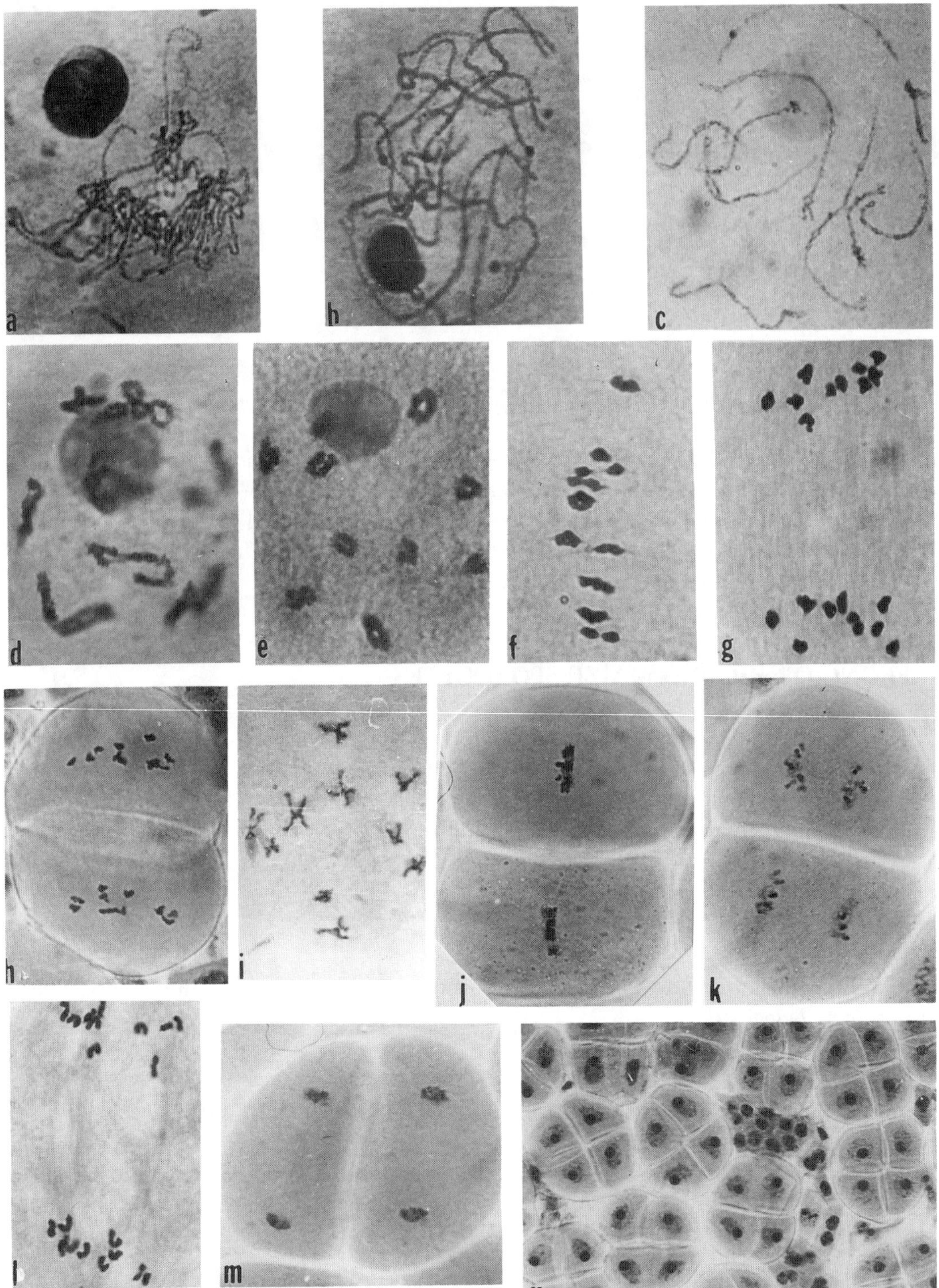
a
b
c
d
e
f
g
h
i
j
k
l
m
n

FIGURE 68.1 Normal meiosis in maize. a–g. The first meiotic division. a. Late zygotene; both paired and unpaired homologs are seen. b, c. Late pachytene; all 10 chromosome pairs are paired and 10 bivalents are evident. d, e. Late diplotene and diakinesis with 10 bivalents seen. f. Metaphase I; 10 bivalents form the metaphase plate. g. Anaphase I; 10 chromosomes pass to each of the opposite poles. h–n. The second meiotic division. h, i. Prometaphase II; 10 chromosomes (haploid number) are seen. Note each chromosome consists of two sister chromatids joined to each other in the centromere region only. j. Metaphase II; metaphase plates are seen in each daughter cell resulting from the first meiotic division. k, l. Anaphase II; normal separation; sister chromatids are seen. m. Telophase II. n. Tetrads of haploid microspores as a result of normal meiosis. Magnification: a–m 90× objective; 7× ocular; n 20× objective; 7× ocular.

should be mixed in a flask and simmered for 5–6 hours with a reflux condenser. Cool, filter, and keep in the dark. For pachytene preparations, 4.5–5% carmine is best.

PREPARING SLIDES FOR ANALYSIS OF MEIOSIS

An anther is placed in a drop of aceto-carmine (or propio-carmine). Cut off the ends of the anther and, using dissection needles, gently squeeze microsporocytes out into the stain. Remove anther pieces. Examine the slide in a low power light microscope. If the pollen mother cells (PMCs) are at the desired stage of meiosis (Figure 68.1) and are stained intensely, the coverslip is added and heated over an alcohol flame. Try to prevent boiling the stain, and examine the intensity of staining under the microscope. When PMCs are stained intensely enough, apply a temporary seal to prevent drying out at the perimeter of the coverslip (Burnham 1982). If all stages of meiosis are necessary to study, cells may be stained overnight. Toward this end, anthers of some tassel branches are removed from florets and placed in a small vial of 2% aceto-carmine, heated (but not boiled) three to five times over an alcohol flame, and left overnight in stain that has been carefully capped to prevent evaporating. For preparing the smear, a drop of 45% acetic acid is prepared. The resulting slides are very clear and a lot of meiocytes at the same stage are observed and scored. (Golubovskaya, 1989)

REFERENCES

Burnham C (1982) Details of smear technique for studying chromosomes in maize. In Sheridan WF (ed) Maize for Biological Research, Plant Molecular Biology Association, Charlottesville VA 22905, pp 107–118

Golubovskaya I (1989) Meiosis in maize: mei-genes and conception of genetic control of meiosis. Adv Genet 26: 149–192

Protocol for Preparing Maize Macrospore Mother Cells for the Study of Female Meiosis and Embryo-sac Development

INNA GOLUBOVSKAYA AND N. A. AVALKINA

Dissect young ears (2–4 cm long) from husk leaves and fix for 24 hours in Chemberlen's fixative: 50% ethanol:40% formalin:glacial acetic acid in proportion 90:5:5, and rinse several times in 95% ethanol. Store in 70% ethanol in the refrigerator.

Before staining, the ears are rinsed in distilled water, and the cell walls are digested by adding weak aqueous pectinase solution (2%) for 12–24 hours (Jongedijk 1987). Then the ear is treated with cold hydrolysis (room temperature) in hydrochloric acid (HCl) as follows: 1.5 hours in 1 N solution HCl, 2 hours in 50% HCl, followed by 1 hour in 1 N HCl. The material is then rinsed in distilled water three times for 3 minutes and stained with Feulgen's reagent for 3 hours. The material can be kept in Feulgen's fluid overnight (at 4°C) (Orlova and Avalkina 1985).

An ear floret is placed on a slide in a drop of glycerol and 45% acetic acid is mixed in a 1:1 proportion. The floret tissue is carefully dissected under a dissecting microscope (magnification 14 × 2) with sharp needles, and a single ovary is isolated. The ovary is positioned beneath the silk; attachment is close to the nucellus epidermis. Place the isolated ovary on a slide in a drop of the glycerol and 45% acetic acid mixture. Very carefully, radially dissect the ovary from surrounding tissue and isolate the central group of cells, including the megaspore mother cell (MMC). It is possible to distinguish the megaspore mother cell under high-power magnification. It is large with an elongate shape and has a thick callose envelope at prophase I.

The somatic cells surrounding the MMC are removed with a dissecting needle and filter paper. The dividing MMC is isolated in a free area of the slide, a drop of glycerol

The Maize Handbook—M. Freeling, V. Walbot, eds.

and acetic acid is added, and the preparation is then covered by a cover glass. Sometimes we add 2% aceto-carmine for more intense staining of MMC cytoplasm.

This technique allows us to obtain well-stained megaspore mother cells and makes it possible to analyze all stages of female meiosis and the developing embryo sac in both normal (Figures 69.1, 69.2) and meiotic mutants (Avalkina et al., 1992.)

CHEMBERLEN'S FIXING SOLUTION

50% ethanol	90 ml
40% formalin	5 ml
glacial acetic acid	5 ml

HYDROLYZING SOLUTION

Hydrochloric acid (HCl)	1 N and 50%

Staining Fluids

Feulgen's Reagent

Preparation: 1 g basic fucsin is dissolved in 200 ml boiled distilled water and simmered for 5 minutes over a flame and then filtered. Cool at 50°C and add 20 ml 1 N hydrochloric acid. Cool at 25°C and add 1 g dry $NaHSO_3$ or $Na_2S_2O_5$. Then, pour into a brown flask with a ground glass stopper and place in the dark for 24 hours. The stain must be light yellow in color.

2% Acetocarmine

See the preceding chapter in this volume.

Digesting Reagent

2% pectinase solution.

Mixture for Making Slides

Glycerol and 45% acetic acid mixed in 1:1 proportion.

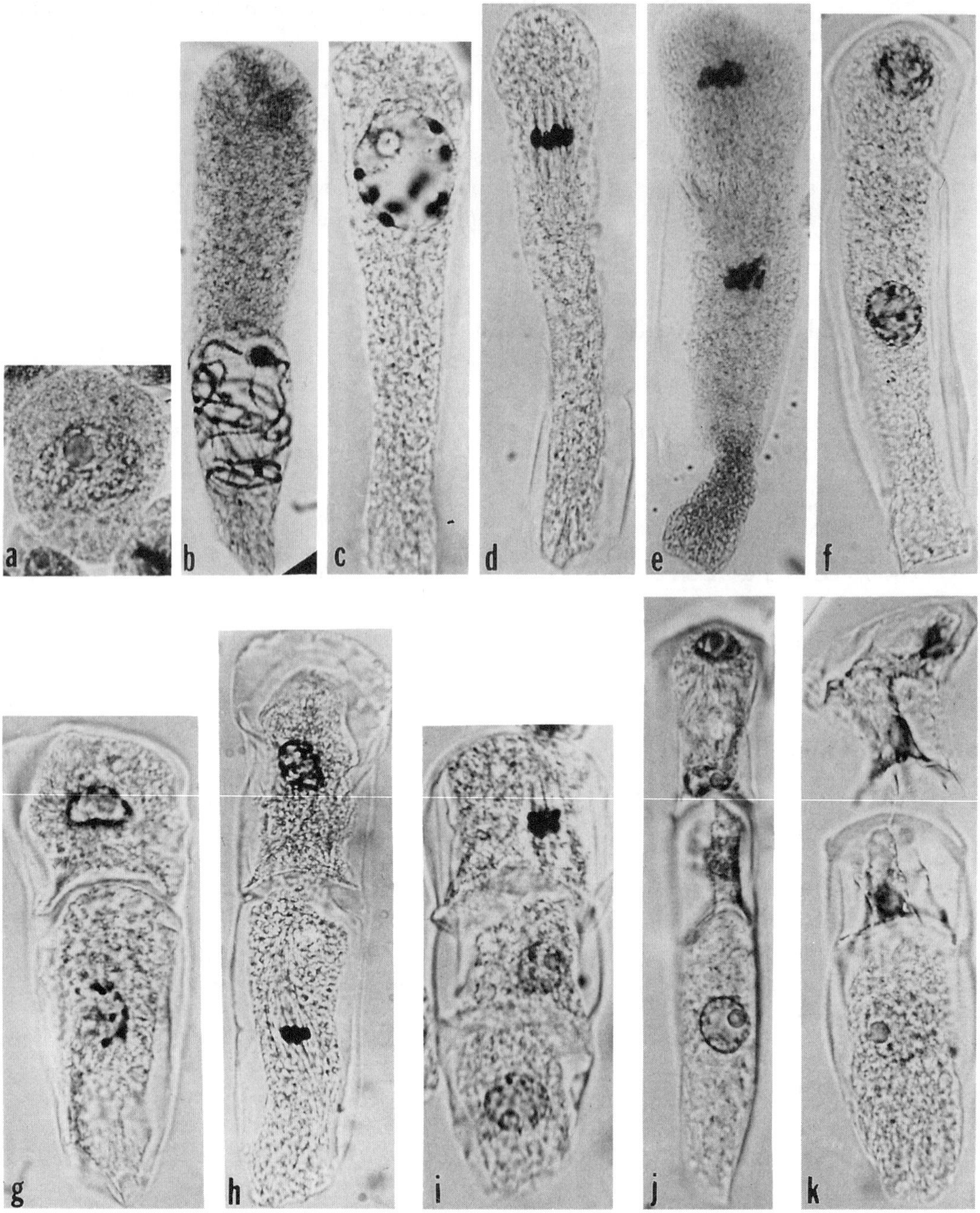

Figure 69.1 Female Meiosis in a Normal Maize Plant. a. Archeosporic cell meiotically dividing or macrospore mother cell. b. Pachytene, nucleus located at bottom of cell. c. Diakinesis, 10 bivalents present. d. Metaphase I. e. Late anaphase I. f. Dyad cells. g, h, i. Second meiotic division: prophase II and metaphase II at the lower cells (g, h). Complete second division at lower cell but metaphase II at upper cell (i). j. Tetrad stage, all products of normal female meiosis are seen. k. Tetrad, three upper products of meiotic division are degradated; embryo sac will develop from chalazal haploid macrospore (lower cell). Magnification: 60× objective; 7× ocular.

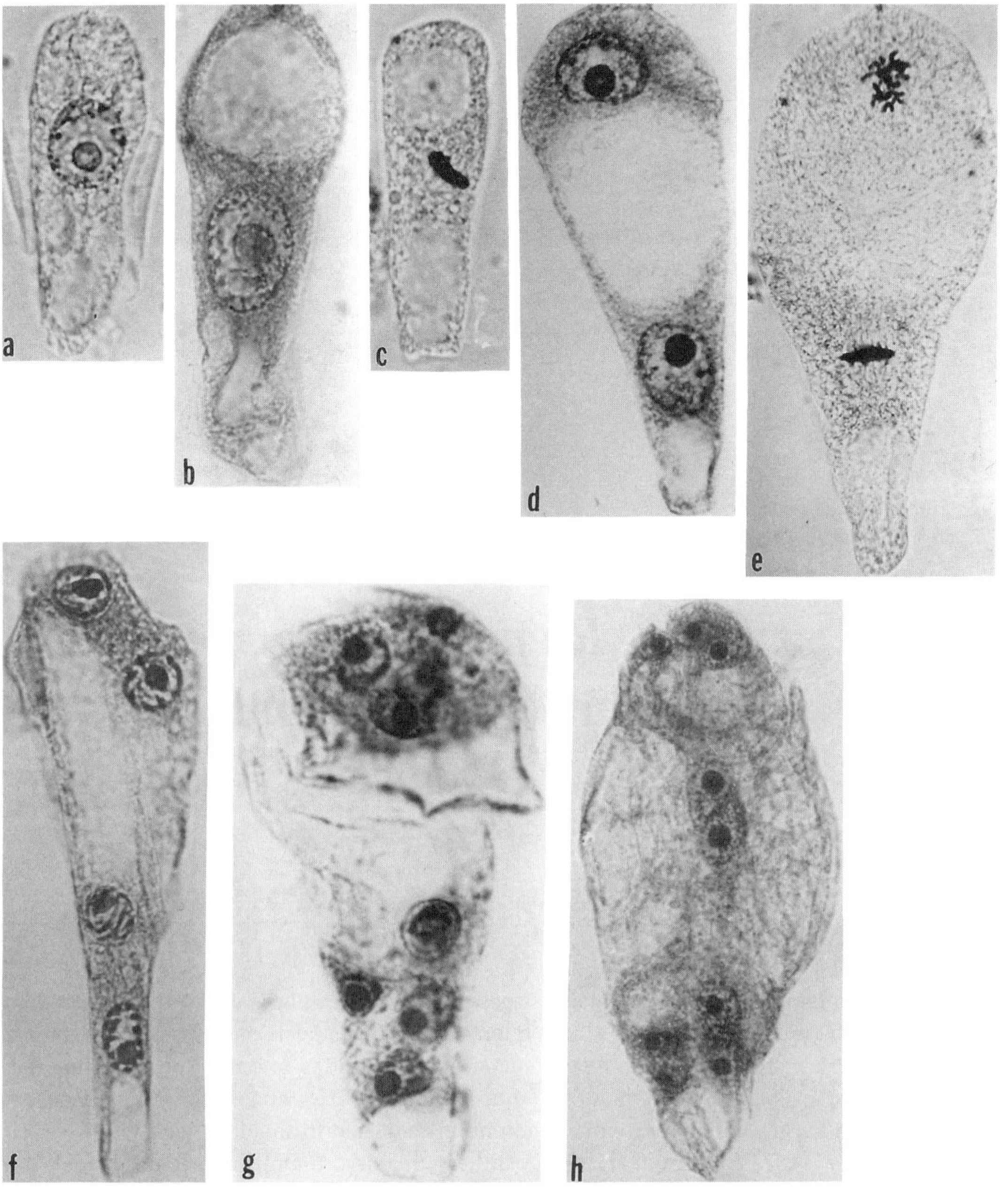

Figure 69.2 Embryo-sac Development in Normal Maize Plant. a. Haploid macrospore as a living product of complete normal female meiosis. b. One-nucleus embryo sac; note one nucleus in center cell and two visible vacuoles. c. First round mitotic division of nucleus (metaphase) in one-nucleus embryo sac. d. Two-nuclei embryo sac as a product of complete first round of mitosis. e. Second round of mitosis in two-nuclei cell—synchronous metaphase stages in both nuclei of embryo sac. f. Four-nuclei embryo sac as a product of complete second mitotic round. g. Eight-nuclei embryo sac as a product of complete third (last) mitotic division. h. Embryo sac is almost ready for fertilization: two polar nuclei in center of embryo sac, ovule and two synergids at upper space, and also antipodal cells at bottom are seen.
Magnification: 60× objective; 7× ocular.

REFERENCES

Avalkina NA, Golubovskaya I, Peremislova E (1992) Mei-genes and female meiosis in maize. Genetics (Russian) 28 (8): 130–141

Jongedijk E (1987) A quick enzyme squash technique for detailed studies of female meiosis in Solanum. Stain Technol 62: 135–142

Orlova IN, Avalkina NA (1985) An isolation of embryo-sacs of cereals without digestion of ovary tissue. Methodical Handbook, L. 12 pp (Russian)

70

Preparing a Suspension of Microsporocytes for Spreading and Electron Microscopy

INNA GOLUBOVSKAYA AND Z. K. GZEBENNIKOVA

1. Ten anthers at the same prophase stages of meiosis (prophase stage anthers are identified by examination under a light microscope using smears of fresh material stained with 2% aceto-carmine) are placed on siliconized slides in a drop of spreading medium consisting of 0.1% bovine serum albunin (BSA) and 2 mM EDTA (disodium salt) in Eagle's minimal essential medium (pH adjusted to about 7.7 with 0.5 N NaOH) (Gillies 1981; Fedotova et al. 1989). Anthers are cut at the top, and their contents are squeezed using siliconized cover glass. The anther walls and cover glass are removed. The spreading medium containing the meiotic cells is taken up in a pipette and suspended by repeated pipetting of the cells.
2. Several drops of a 0.2 M solution of sucrose are placed on a waxy plate and one drop of suspension of meiocytes is touched to the surface of the drops of sucrose; the

The Maize Handbook—M. Freeling, V. Walbot, eds.

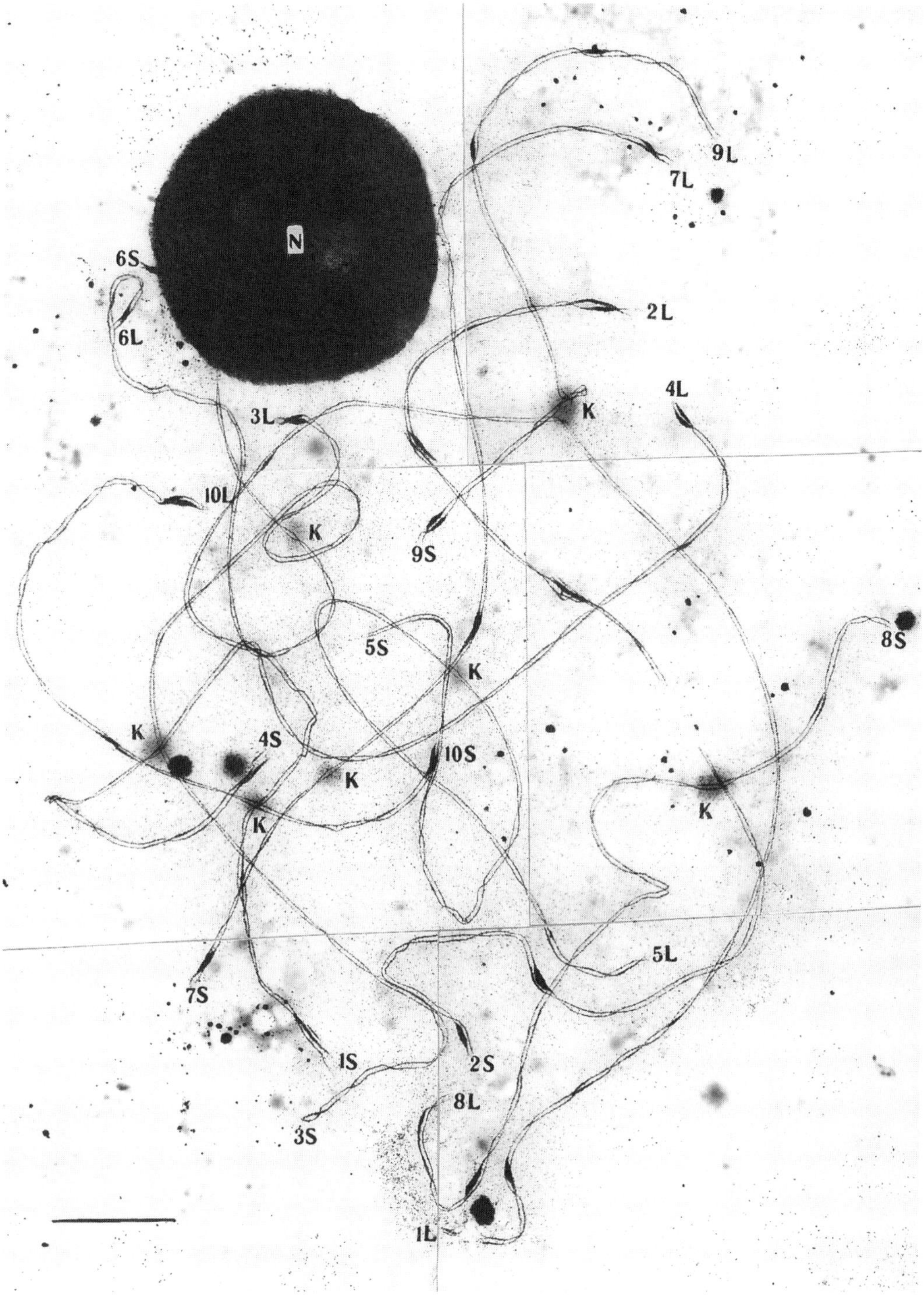

Figure 70.1 Electron Micrograph of Spread Nucleus Stained with Silver Nitrate. The ten synaptonemal complexes are numbered at their short arm ends (S), long arm ends (L), and kinetochore (K). Note the fusion of kinetochore regions of the 9th and 1st chromosomes, 2nd and 8th chromosomes, and 3rd and 4th chromosomes. The lateral element thickenings are also visible. Bar = 4 μm.

spread nuclei are then picked up by touching the surface with a plastic-coated slide. (For a method of plastic coating slides, see the chapter by Maguire.)

3. The slide with spread nuclei is fixed by exposure to the vapor of a 40% formalin solution in a covered petri dish from 4 hours to overnight at 4°C.
4. Fixed slides are dried for 3–4 days in the air, but covered to keep clean.
5. Stain in 70% silver nitrate solution in a petri dish under a cover glass in an incubator at a temperature of 55°C for 2–5 hours. Examine the slides under a light microscope to be sure that the spread materials are stained dark brown. Dry stained slides for 24 hours.
6. Score the plastic film around a marked nucleus and then float off the plastic raft on a clean drop of water that is added from a pipette; pick up the raft on a grid covered by stick mount and then air dry.

The results of this procedure are seen in the electron micrograph (Figure 70.1). The lateral element thickenings of the synaptonemal complexes are visible.

REFERENCES

Gillies CB (1981) Electron microscopy of spread maize pachytene synaptonemol complexes. Chromosoma 83: 575–591

Fedotova Yu S, Kolomiets OL and Bogdanov Yu F (1989) Synaptonemal complex transformation in rye microsporocytes at diplotene stage of meiosis. Genome 32: 816–823

71

Three-dimensional Fluorescence Microscopy of Maize Chromosomes

R. Kelly Dawe

A three-dimensional analysis of chromosome architecture requires the ability to analyze a series of in-focus sections. This can be done by physical sectioning of fixed material, or, more simply, by taking optical sections. Unfortunately, three-dimensional objects viewed under a light microscope contain out-of-focus information from other focal planes. One way to overcome this problem is by using the confocal microscope, which scans the image with a fine beam of light. (See Ruzin and Sylvester, "Light Microscopy II: Observation, Photomicrography, and Image Analysis"; section on confocal microscopy, this volume). A second approach is to use a standard light microscope to collect all the information in three dimensions and use computational methods to reconstitute the in-focus image (Agard et al. 1989). A major difference in these two treatments is that mathematical deconvolution *returns* the out-of-focus information to its source on the chromosomes. This approach is better for weakly stained objects, and when the data are collected with a charge-coupled device (CCD), the resolution is superior to the confocal microscope. Combining confocal microscopy with mathematical deconvolution will ultimately provide the highest resolution.

In order to preserve the spatial arrangement of chromosomes, it is best to minimize the handling of cells prior to fixation. Thus, entire anthers are fixed before the meiocytes are extruded from the anthers. Because fixation can be a slow process, a buffer that preserves chromatin structure is used while the anthers are in fixative; one such chromatin-stabilizing buffer is Buffer A (Belmont et al. 1987). Buffer A can also be used for visualizing microtubules (Staiger C.J., "Indirect immunofluorescence: Localization of the Cytoskeleton," this volume) and is compatible with fluorescent DNA dyes, such as DAPI, Hoechst 33258, and propidium iodide.

For three-dimensional analysis of microscopic images, an interactive computer program is used to automatically move the stage, collect, and store the information. The data are corrected for variations in source light intensity and for the fading of fluorescent dyes. After deconvolution, in-focus projections and stereo pairs can be used to view the data

The Maize Handbook—M. Freeling, V. Walbot, eds.

as rotating three-dimensional images. Chromosome paths can be modeled and the chromosomes can be computationally straightened to simplify analysis (Chen et al. 1989). After paraformaldehyde fixation and staining with DAPI, detailed cytological features were visible (Figure 71.1) in maize chromosomes (Dawe RK, Agard DA, Sedat JW, Cande WZ, *Maize Genetics Cooperation News Letter* 66: 1992).

MATERIALS

Buffer A

80 mM KCl, 20 mM NaCl, 0.5 mM EGTA, 2 mM EDTA, 15 mM PIPES buffer (pH 7.0), 15 mM beta-mercaptoethanol, 0.5 mM spermidine, 0.2 mM spermine.

Paraformaldehyde

Paraformaldehyde solution, 16% (Electron Microscopy Sciences, Fort Washington PA).

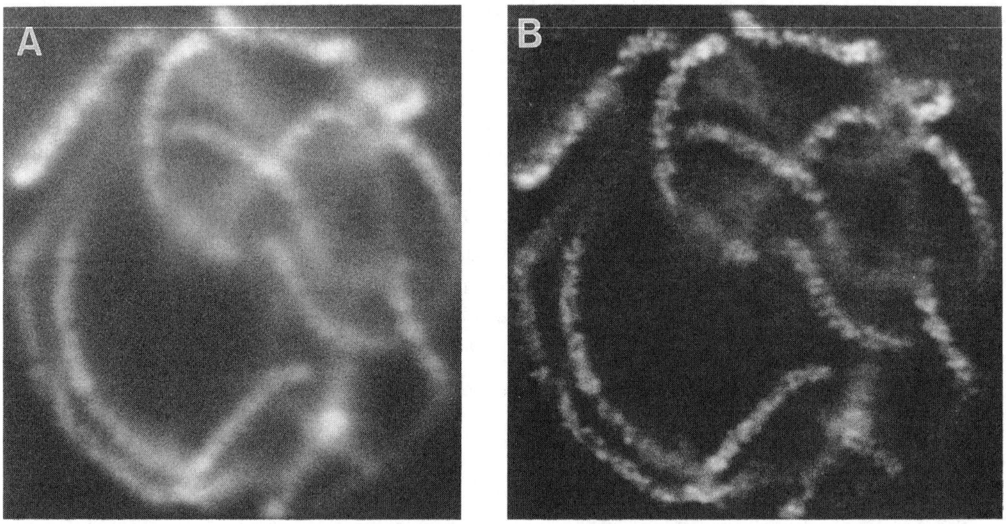

FIGURE 71.1 A. Fluorescent image of DAPI-stained chromosomes from the inbred KYS. This image was recorded using a 60× objective with an additional 1.5× magnification. B. Same image after three-dimensional deconvolution of a series of optical sections taken at 0.25-μm intervals. These images were produced in the laboratories of John W. Sedat and David A. Agard, Department of Biochemistry and Biophysics, U.C. San Francisco and Howard Hughes Medical Institute.

DAPI

(4,6-diamidine-2-phenylindole dihydrochloride; Sigma) 1 mg/ml stock in water.

Antifade Mounting Solution

100 mg/ml stock DABCO (1,4-diazobicyclo-(2,2,2)-octane; Ernest F. Fullam, Schenectady, NY) in 1 part PBS (137 mM NaCl, 2.7 mM KCl, 8 mM $Na_2HPO_4.7H_2O$, 1.47 mM KH_2PO_4, pH 7.0) and 9 parts glycerol. An alternative to DABCO is 2% *n*-propyl-gallate in the same medium.

METHODS

1. Dissect appropriately staged anthers (Burnham 1982; and see chapter by Chang and Neuffer, this volume) from the florets into Buffer A. Add an equal volume of 8% paraformaldehyde in Buffer A (final concentration 4% paraformaldehyde) and shake or rock the anthers for approximately 2 hours at room temperature.
2. Use a fine scalpel to cut off an end of each anther. Using a forceps or scalpel, gently press the meiocytes out the cut end.
3. A 200 μl micropipetter can be used to transfer the meiocytes directly to a microslide or to acid-washed coverslips treated with 1 mg/ml polylysine. Cells can be spun down (100× *g*) onto polylysine-coated coverslips to increase cell adhesion.
4. Stain the meiocytes with 0.05 μg/ml DAPI for 10 minutes. Mount the cells in antifade mounting medium and seal the edges of the coverslip with nail polish.
5. For three-dimensional analysis of nuclei, consult recent reviews (Agard et al. 1989; Chen et al. 1989)

REFERENCES

Agard DA, Hiraoka Y, Shaw P, Sedat JW (1989) Fluorescence microscopy in three dimensions. Meth Cell Biol 30: 353–377

Burnham CR (1982) Details of the smear technique for studying chromosomes in maize. In Sheridan W (ed) Maize for Biological Research, Plant Molecular Biology Association, Charlottesville, VA, pp 107–118

Belmont AS, Sedat JW, Agard DA (1987) A three dimensional approach to mitotic chromosome structure: evidence for a complex hierarchical organization. J Cell Biol 105: 77–92

Chen H, Sedat JW, Agard DA (1989). Manipulation, display, and analysis of three dimensional biological images. In Pawley J (ed) The Handbook of Biological Confocal Microscopy, IMP Press, Madison, WI, pp 127–135

Dawe, RK, Agard, DA, Sedat, JW, and Cande, WZ (1992) Pachytene DAPI map. Maize Genet Coop News Lett 66, 23.

72

Chromosomal Behavior During Microsporogenesis

MING T. CHANG AND M. GERALD NEUFFER

The developmental stages of maize microsporogenesis have been examined to determine the time interval of floret maturation and stage duration (Hsu and Peterson 1981; Hsu et al. 1988), to identify meiotic mutants and genetic male sterility (Albertsen and Phillips 1981; Curtis 1985; Golubovskay and Khristolyubova 1985; Staiger and Cande 1991), to study chromosomal behavior (McClintock 1938, 1984; Rhoades 1950; Rhoades and Dempsey 1966, 1972, 1973; Kindiger 1986; Rhoades et al. 1986; Zhao and Weber 1989; Kindiger et al. 1991), to assist anther and microspore culture (Hu 1985; Petolino and Jones 1986; Pechan and Keller 1988; Pescitelli and Petolino 1988; Gaillard et al. 1991), and to facilitate the biochemical analysis of gene products (Bedinger and Edgerton 1990; Mandaron et al.

→

FIGURE 72.1 Premeiotic interphase. The irregularly shaped pollen mother cell has dense protoplasm, no vacuoles, no clear cell wall structure, and an undifferentiated nucleus. **FIGURES 72.2–22** Meiosis. **FIGURE 72.2** Leptotene. Cell becomes round with dense protoplasm. The chromatin threads are greatly extended and coiled around the nucleolus. Synapsis is initiated. Single- and double-strand configuration is evident. The chromomeres (→) are visible. **FIGURE 72.3** Late zygotene–early pachytene. The pairing of the homologous chromosomes is complete. The condensed chromosomes show details of heterochromatin and knobs. The nucleolus and nucleolar-organizing region (→) of chromosome 6 are visible. **FIGURE 72.4** Pachytene. The paired chromosomes are further condensed to become a very thick thread. Individual chromosomes can be identified by their relative lengths, distinctive chromomere patterns, position of knobs, and other recognizable characteristics. The nucleolar-organizing region (→) of chromosome 6 is clearly attached to the nucleolus. **FIGURE 72.5** Diplotene. The chromosomes continue to condense into short, thick threads. The paired chromosomes appear to be repulsing one another, except in regions where an actual crossover took place. The chiasmata are frequently seen as X-shaped (→) and looped chromosome configurations. **FIGURE 72.6** Late diplotene. The chiasmata are terminalized and the very short condensed chromosome pairs are separated from each other. The X-shaped and looped chromosome configurations are still shown. The nucleolar-organizing (→) region of chromosome 6 is firmly attached to the nucleolus.

The Maize Handbook—M. Freeling, V. Walbot, eds.

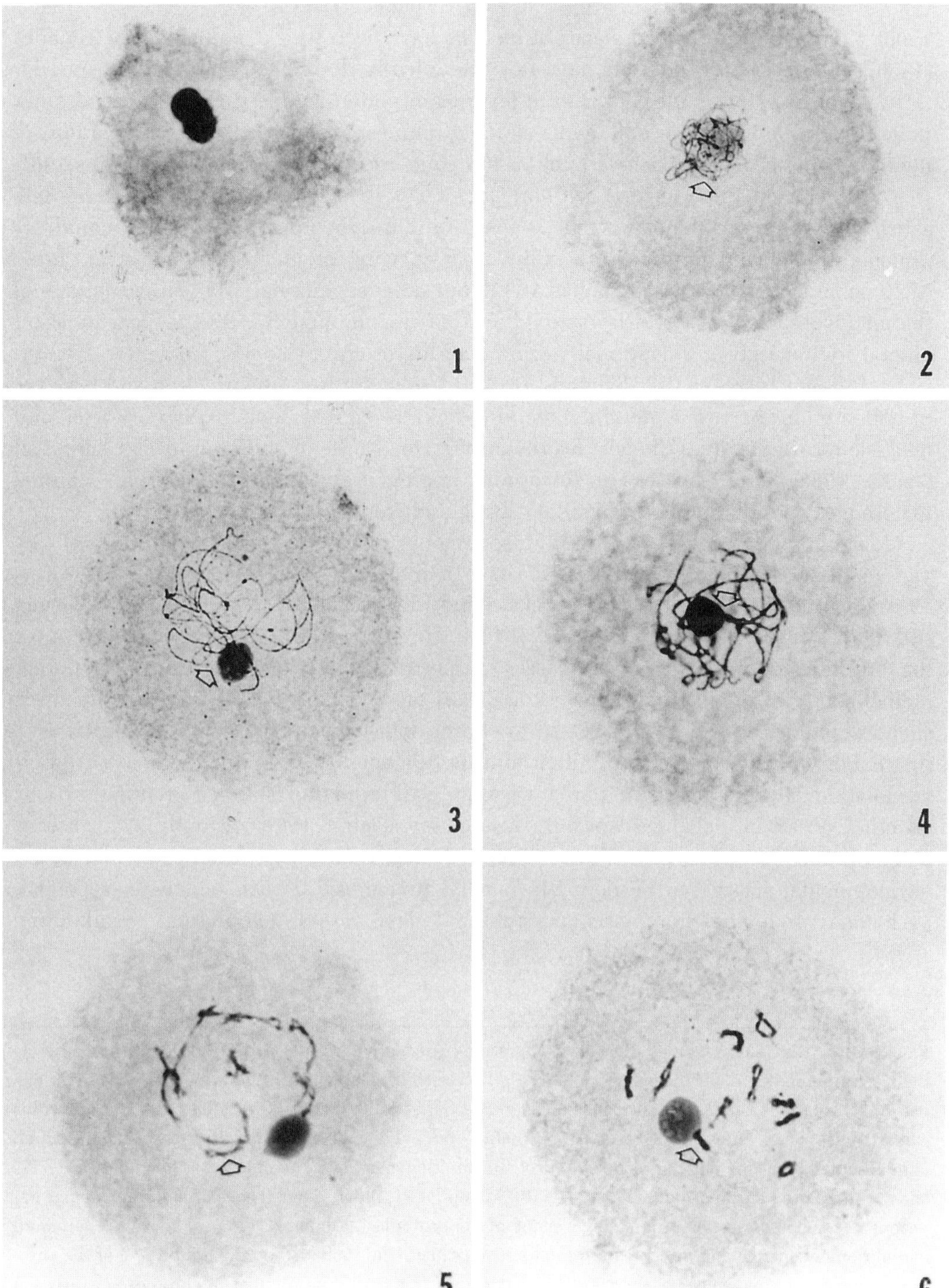

FIGURES 72.1–36 Maize microsporogenesis. Stages are transitory and represent a momentary expression, which may not be a good representation of the complete events. They are explained in sequence as follows, with an arrow (→) indicating the event.

1990). Complete features and stages of meiosis have been well documented by Rhoades (1950), but stages after quartets have not been clearly described. The main reasons for this lack of description are that certain features of pollen grains, such as the rigid outer pollen wall, exine, and abundant starch granules, prevent direct observation of developmental changes. Aceto-carmine or propiono-carmine (Albertsen and Phillips 1981; Hsu and Peterson 1981; Burnham, 1982) can effectively stain meiotic stages, but are less satisfactory for stages thereafter. The DNA-specific fluorochrome DAPI (4', 6-diamidino-2-phenylindole) and mithramycin allow clear viewing of nuclei (Vergne et al. 1987; McConchie et al. 1987; Mascarenhas 1989), but detailed internal and wall structures of the microspores are obscure. Hematoxylin, used in conjunction with the clearing agent chloral hydrate, gives exceptionally good staining of chromosomes, nuclei, and sperm cells of the microspores (Kindiger and Beckett 1985), but iron precipitation and bursting of pollen grains occurs. A modified staining mixture of 18% hematoxylin, 2% iron, and 80% aceto-carmine gives the best results for viewing stages of pollen mitosis (Chang and Neuffer 1989a), and a sucrose aceto-carmine treatment gives the best result for staining mature and germinating pollen grains (Chang and Neuffer 1989b).

Developmental stages during microsporogenesis have been defined differently by various researchers (Laser and Lersten 1972; Bennett et al. 1973; Albertsen and Phillips 1981; Hsu and Peterson 1981; Golubovskaya and Khristolyubova 1985; Curtis 1985; Chang and Neuffer 1989a; Bedinger and Edgerton 1990; Mandaron et al. 1990). In general, however the term microsporogenesis encompasses the entire developmental process from pollen mother cell (PMC) to mature pollen grain. Stages can be characterized by distinct features such as cytological events with regard to chromosome structure in meiosis, cytokinesis, first pollen mitosis, and second pollen mitosis (Rhoades 1950; Bennett et al. 1973). Cell vacuolation, enlargement, germ pore formation, wall formation, starch accumulation, and all other observable changes are quite useful for staging. Here we outline the use of chromosomal events to define developmental stages. By these criteria, a total of 30 developmental stages can be defined: namely, premeiosis (PMC), leptotene, zygotene, pachytene, diplotene, diakinesis, metaphase I, anaphase I, telophase I, cytokinesis,

FIGURE 72.7 Diakinesis. The condensed chromosome pairs are separated from each other and become thick staining bodies. The chiasmata (→) and the X-shaped and looped configurations are still seen. **FIGURE 72.8** Late diakinesis. The chromosome pairs are dark round staining (→) bodies and the nucleolus starts to disappear. **FIGURE 72.9** Metaphase I (side view). The nucleolus has disappeared. The paired chromosomes lie at the equatorial plate (→) of the spindle structure. The chiasmata have moved to the ends of the paired chromosomes. **FIGURE 72.10** Metaphase I (polar view). The paired chromosomes appear as dense bodies scattered on a single plane of the protoplast. **FIGURE 72.11** Anaphase I. The paired chromosomes separate and move toward the opposite poles. The V-shaped configuration (→) of the chromosome is due to movement of the centromere ahead of the arms. The number of chromosomes at each pole now is reduced to half. **FIGURE 72.12** Telophase I. The chromosomes at each pole are now extended. The nucleolus reappears and the cytoplasm divides (→) (cytokinesis) to form two half-mooned cells.

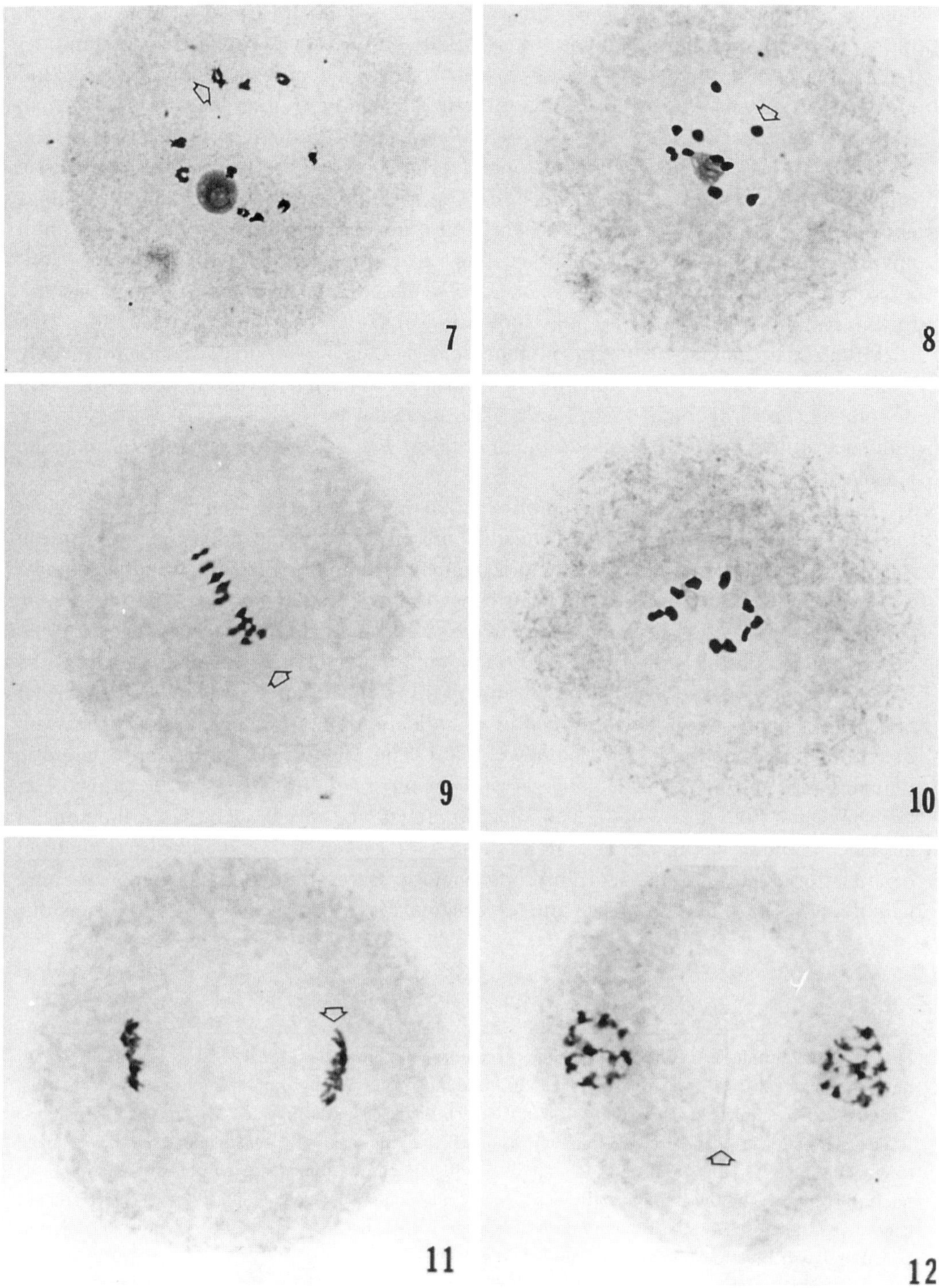
7
8
9
10
11
12

interphase II (dyad), prophase II, metaphase II, anaphase II, telophase II, cytokinesis, tetrad, uninucleate, late uninucleate, interphase, 1st pollen mitosis (1st prophase, 1st metaphase, 1st anaphase, and 1st telophase), interphase, 2nd pollen mitosis (2nd prophase, 2nd metaphase, 2nd anaphase, and 2nd telophase), and mature pollen (Figures 72.1–72.36) (Rhoades 1950; Carlson 1988; Pescitelli and Petolino 1988; Chang and Neuffer 1989a).

Figure 72.37 illustrates the macroscopic development of anthers, florets, and the tassel during microsporogenesis. The average duration of complete maize microsporogenesis is about 23 days, but is quite variable and is influenced by genotype, tassel size, and environmental conditions. The meiotic stages—from PMC to quartet—account for about 28% of the time for microsporogenesis; microspore maturation and mitotic divisions account for the remaining 72% of the time as shown in Table 72.1 (Hsu and Peterson 1981; Hsu et al. 1988; Mascarenhas 1989). The onset of meiosis of the upper floret starts approximately 4 days earlier than the lower floret of the same spikelet (Hsu et al. 1988), but the position of spikelet, tassel age, and many other environmental factors could have a significant influence in determining the developmental difference between florets within the same spikelet.

Stages can be further classified. For example, maize meiotic mutants can be assigned to one of seven key developmental events that are presumably under specific gene control: namely, the initiation of meiosis, homologous pairing, meiotic recombination, chiasmata formation, microtubule formation, and segregation of homologues, cytokinesis, and the initiation of the second division (Golubovskaya and Khristolyubova 1985; Curtis 1985; Staiger and Cande 1991). The classification of mutant phenotypes was mainly determined by cytological observations (Albertsen and Phillips 1981; Curtis 1985). The genes that affect the phenotypes were determined by F_2 data and their chromosomal locations were determined by genetic means (Coe et al. 1988). The phenotypic expression of meiotic mutant genes and male sterile genes is well documented (Albertsen and Phillips 1981; Carlson 1988; Coe et al. 1988). The specific effects of individual mutants include (1) absence of meiosis, (2) absence or disruption of synapsis, (3) changes in structural organization of chromosomes, (4) improper meiotic segregation and/or defective meiotic spindle, (5) failure of cytokinesis and/or irregularities of cell shape, (6) extra divisions

FIGURE 72.13 Prophase II. The chromosomes condense into short thick threads surrounding the small nucleolus. **FIGURE 72.14** Metaphase II. The chromosomes (each chromosome has two sister chromatids) lie at the equatorial plate of the spindle structure. Nucleoli have again disappeared. **FIGURE 72.15** Anaphase II. The two sister chromatids seen collectively as a dark staining mass are now separated and have moved toward the opposite poles. **FIGURE 72.16** Telophase II. The chromosomes at each pole are extended, the nucleoli reappear, and the cytoplasm divides to form four cone-shaped cells. **FIGURE 72.17** Quartets. Four cone-shaped microspores are formed and are enclosed inside the maternal wall (→), which is being digested and will thus release the four microspores. **FIGURE 72.18** Free cell from quartet. The newly released free microspores are undifferentiated and cone shaped and appear to have no distinct cell wall (Chang and Neuffer 1989a).

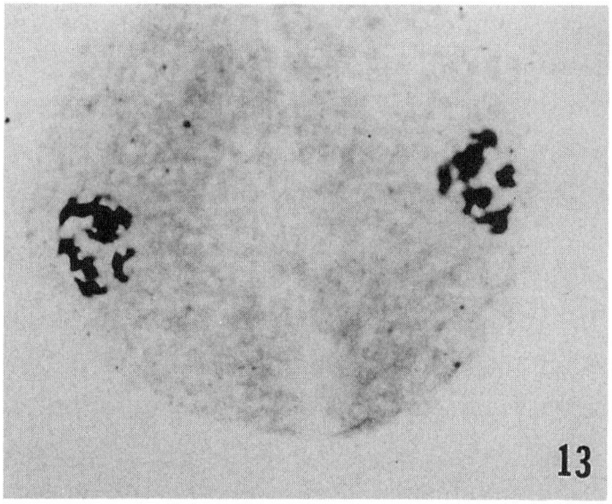

13

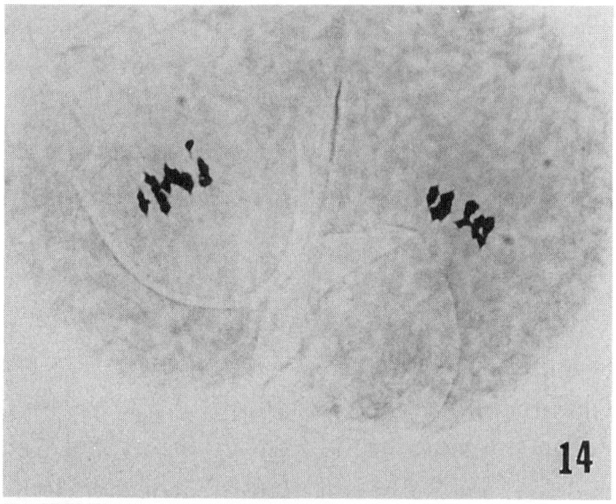

14

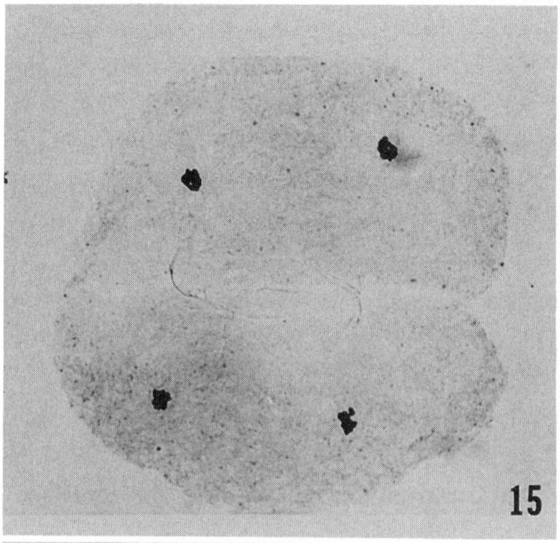

15

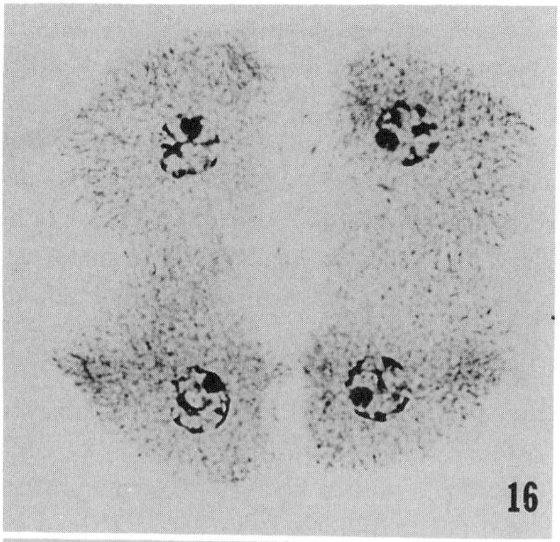

16

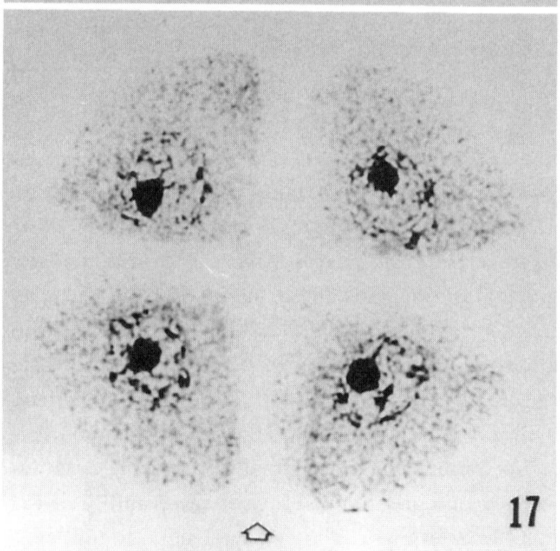

17

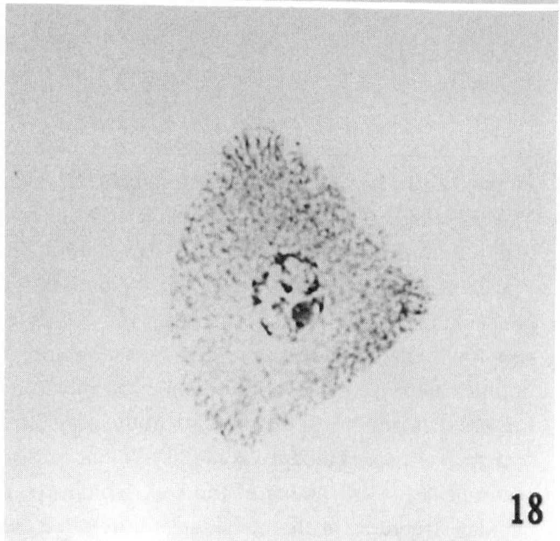

18

following meiosis, (7) a combination of effects on meiosis, (8) failure of microspore differentiation or vacuolation or wall formation, (9) microspore degeneration, and (10) absence of pollen mitosis. A summary of these mutant genes and of their effect in blocking the developmental stages is shown in Table 72.1 (Carlson 1988; Coe et al. 1988). The male sterile mutations listed are all sporophytic and thus represent genes expressed in diploid cells.

The effects of chromosomal deficiencies on the development of the maize pollen grain were studied by B-A translocations (TB-A) (Kindiger et al. 1991). Generally, loss of part of a chromosome arm caused abnormal microspore development, a slowing of the normal mitotic or developmental processes in the male gametophyte, or a termination of development (Table 72.1).

The effect of cytoplasmic male sterility (*cms*) on microspore degeneration is influenced by genotype and environment. For use in hybrid production, the *cms* microspore should degenerate as early as meiosis, but abortion has been reported to occur during almost every stage of pollen development (Laser and Lersten 1972; Kaul 1988). The cell degeneration effects of *cms-T* and *cms-C* are more stable in a fluctuating environment and complete than *cms-S* (Warmke and Lee 1977; Coe et al. 1988; Walbot and Messing 1988; Fauron et al. 1990; Levings and Williams 1990). Developmental effects of *cms*-T and *cms-C* have been shown to occur during cell vacuolation of mid- to late-uninucleate stage (Laser and Lersten 1972; Warmke and Lee 1977). For comparison purposes, the impact of *cms* is approximately determined (Table 72.1) according to the author's best estimate.

The aceto-carmine procedure to stain meiotic stages will not be described here. See other chapters in this volume for protocols. Procedures that Chang and Neuffer (1989a,b) developed to stain microspore mitosis, mature and germinating pollen grains are described in detail as follows.

FIGURE 72.19 Early uninucleate stage cell. The shape of the microspores is round with dense cytoplasm. The nucleus is located near the center and the cells are undifferentiated with no vacuoles and no clear wall structure. **FIGURE 72.20** Later early uninucleate stage cell. The microspores start to differentiate. The exine and intine structures are being formed. The cytoplasm remains dense, but many small vacuoles are being formed. The nucleus is still near the center of the protoplast. **FIGURE 72.21** Middle uninucleate stage cell. A large vacuole is forming in the protoplast, pushing the nucleus to one side. **FIGURE 72.22** Late uninucleate stage cell. The differentiation of exine and intine, germ pore, and annulus (→) is complete. Creases seen are due to the pressure of the cover slip on rigid spherical pollen wall. Cell volume increases four to six times. **FIGURES 72.23–36** Pollen mitosis. Illustrations show a gradually increasing accumulation of the starch grains in the cell, which progressively obscure the visibility or resolvability of the cellular structures of the male gametophyte. **FIGURE 72.23** Interphase. The cell nucleus is round, condensed, and nondifferentiated (→). **FIGURE 72.24** First prophase. The chromosomes condense into short thick threads (→) surrounding the nucleolus.

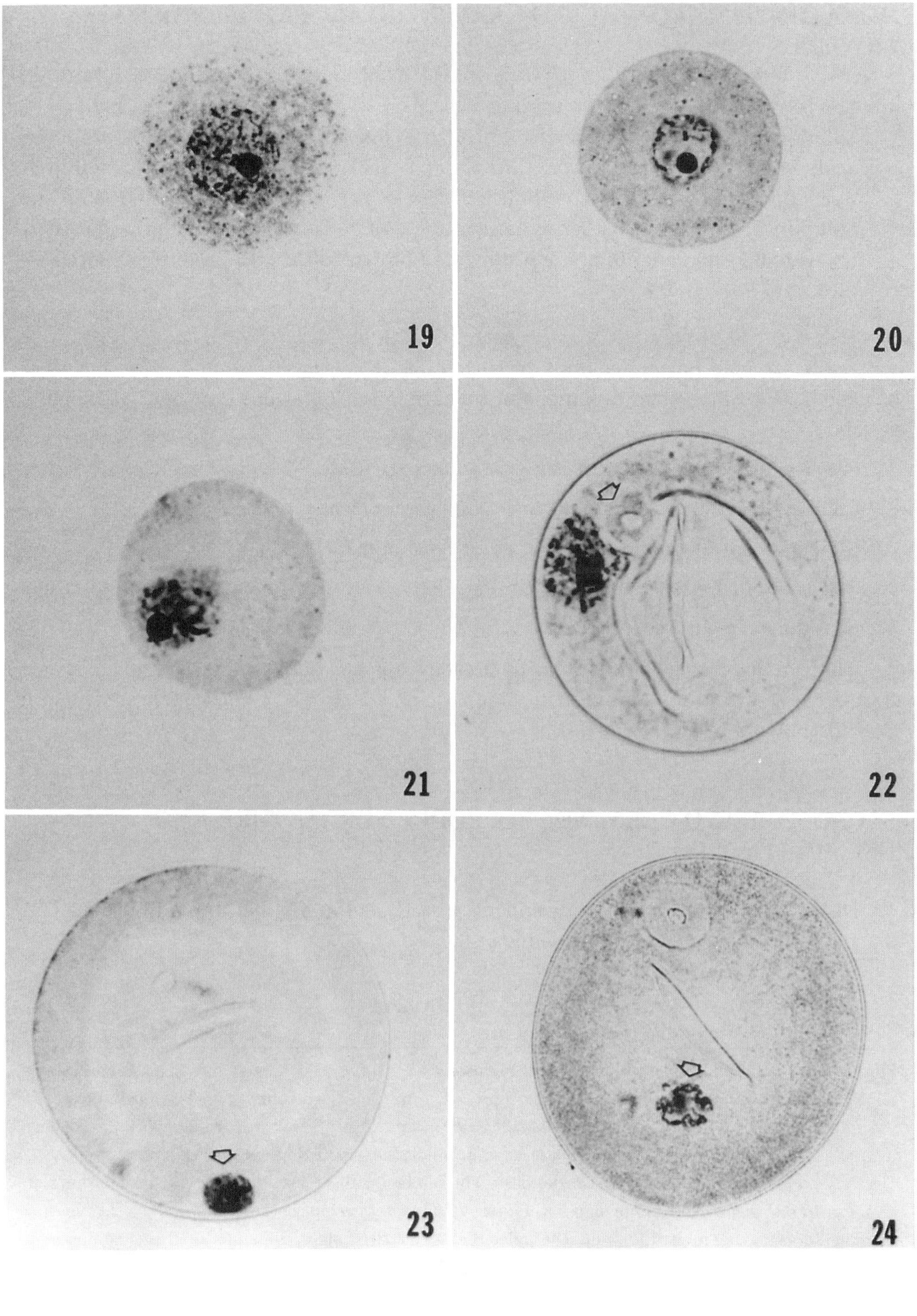
19
20
21
22
23
24

HEMATOXYLIN–IRON–ACETO-CARMINE STAIN FOR MICROSPORE MITOSIS

1. The hematoxylin stain is prepared as described by Henderson and Lu (1968): Solution A: Dissolve 2 g hematoxylin (Fisher, lot 701073) in 100 ml 50% propionic acid. Allow the solution to stand for about 1 week. The solution keeps indefinitely in a stoppered brown bottle without refrigeration (Kindiger and Beckett 1985). Solution B: Dissolve 0.5 g ferric ammonium sulfate in 100 ml 50% propionic acid. Solution B keeps indefinitely in a stoppered brown bottle without refrigeration (Kindiger and Beckett 1985).
2. Mix 18% solution A, 2% solution B, and 80% aceto-carmine.
3. Fresh anthers from upper floret of each spikelet are used. cut anthers in half, then add a drop of the stain mixture, and squeeze microspores out from anthers. Remove debris.
4. Cover slide with a coverslip and gently heat the slide.
5. Cells are greatly expanded and the cytoplasm is clear.
6. Slides are left overnight for better color development.
7. Stages are identifed under a light microscope.
8. Seal the coverslip with Permount, and slides can be kept for a short time.

The main advantage of this staining procedure is that slides are clean and color contrast between cytoplasm and chromosome/nuclei is sharp; as a consequence, staging is easy (Chang and Neuffer 1989a).

SUCROSE–ACETO-CARMINE PROCEDURE FOR MATURE AND GERMINATING POLLEN

1. Dissolve 3.6 g calcium chloride and 1.2 g boric acid in distilled water to make a 100-ml stock solution (Cook and Walden 1965).

FIGURE 72.25 Middle first prophase. The chromosomes continue to condense into short thick threads, allowing identification of individual chromosomes. The nucleolus and the nucleolar-organizing region (→) of chromosome 6 are visible. FIGURE 72.26 Late first prophase. The chromosomes further condense to become short thick rods. The nucleolus and the nucleolar-organizing regions of chromosome 6 are clearly seen (→). FIGURE 72.27 First metaphase. The nucleolus disappears and the 10 chromosomes are arranged in one plane close to one another. FIGURE 72.28 First anaphase. The sister chromatids are now separated and moving toward the opposite poles. FIGURE 72.29 Late first anaphase. The separated chromosomes have reached the opposite poles and formed two chromosome clusters. FIGURE 72.30 First telophase. The chromosomes at each pole are now extended and surround the nucleolus.

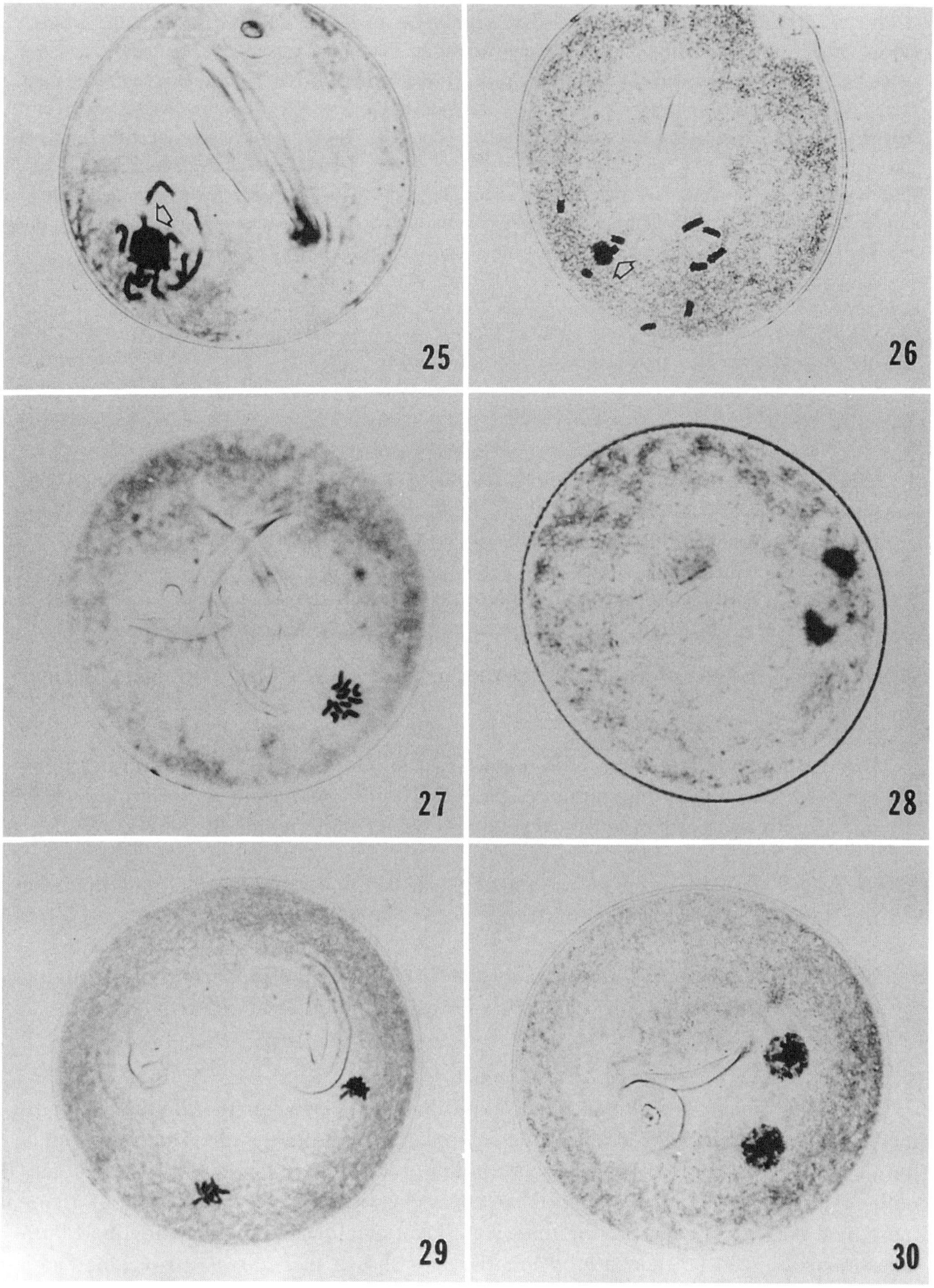
25
26
27
28
29
30

FIGURE 72.31 Binucleate stage. Two nuclei are formed at the opposite poles. The generative nucleus (→)(bottom) is usually located near the germ pore. It will further divide to form two sperms. **FIGURE 72.32** Second prophase. The generative nucleus underneath the surface of the intine wall becomes cup shaped (→) and will proceed to the second nuclear division. The vegetative nucleus is not at resting stage and appears to continue its metabolic activities. **FIGURE 72.33** Second metaphase. The nucleolus disappears and the 10 chromosomes are arranged in one single plane (→). The vegetative nucleus is stained dark. **FIGURE 72.34** Second anaphase. The sister chromatids are now separated and moving towards the opposite poles (→). The vegetative nucleus remains large and clear. **FIGURE 72.35** Second telophase. The chromosomes at each pole are now extended and surround the nucleolus. The vegetative nucleus remains large and clear (→). **FIGURE 72.36** Mature pollen. The mature pollen grain now has three nuclei. The top two condensed, crescent-shaped nuclei, surrounding the germ pore, are the sperms. The large one at the bottom (→) is the vegetative nucleus (Chang and Neuffer 1989).

2. Dissolve 15 g sucrose in 80 ml distilled water; add 1 ml of the above solution; and add distilled water to make a total of 100 ml solution. This solution should contain 15% sucrose, 360 ppm calcium chloride, and 120 ppm boric acid.

3. Three to five milligrams of fresh pollen are added to the solution and the beaker is gently shaken until the pollen grains form a uniform suspension.

4. Incubate at 25°C for at least 5 hours. After incubation the pollen suspension becomes a light syrup.

5. The mature and germinated pollen grains settle out, so most of the solution can be removed. The remaining volume is about 5–10 ml. One part of the solution is then mixed with two parts of aceto-carmine and left to stain overnight.

6. Slides are made using the conventional smear technique and then left overnight for development of better color contrast.

7. The intensity of staining can be controlled by varying the proportion of syrup and stain. Such adjustment may be needed for maize pollen from different sources.

The main advantages of this method are that the sucrose in the mixture helps to increase color contrast between the pollen cytoplasm (light pink) and the nuclei (reddish purple), decreases the frequency of burst pollen, increases pollen expansion, stabilizes pollen figures, and automatically seals the cover glass (Chang and Neuffer 1989b). Using these new staining procedures, the most significant cytological events during maize microsporogenesis can be sequentially identified as shown in Figures 72.1–72.36. These events can be used as a means for separating maize microsporogenesis into different developmental stages and for analysis of mutants of this process.

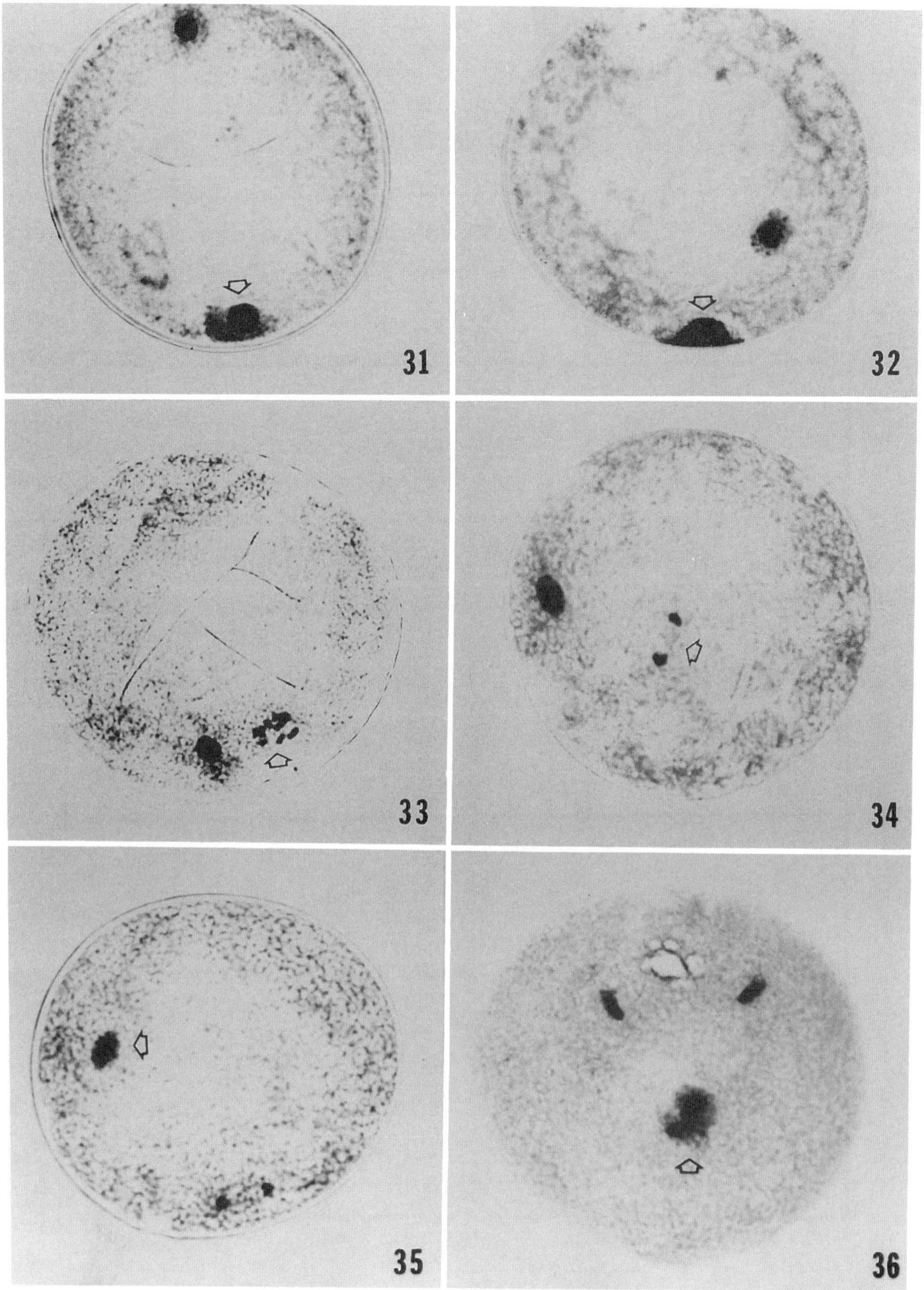
31
32
33
34
35
36

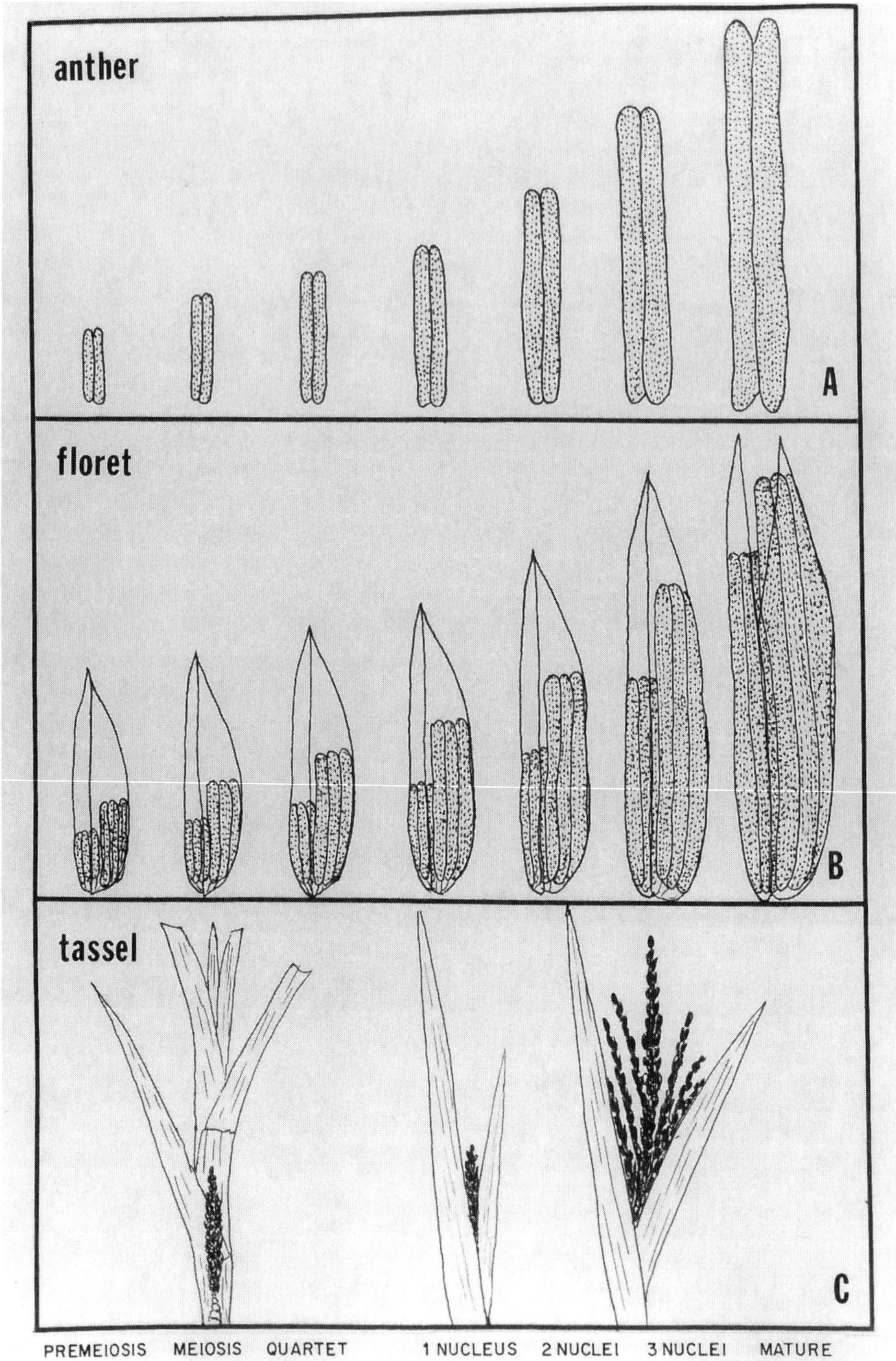

FIGURE 72.37 Developmental stages in anther maturation (A), floret maturation (B), and tassel maturation (C).

TABLE 72.1. Mutant genes, chromosomal deficiencies (TB) and cytoplasmic male sterility (cms) and their blocking effects. (Albertsen et al. 1981; Coe et al. 1988; Carlson, 1988; Hsu et al. 1988; Kindiger et al. 1991).

	Duration		
Developmental Stage	Days	Percent	Genes/Effects Identified
PMC			
Meiosis	5	21	*afd, as, dv, dy, el, dsy1, dsy2, pam1, pam2, st*
Meiosis I		20	*ms3, ms8, ms9, ms22, ms23, ms28, ms43, mei025, am1, am2*
Meiosis II		2	
Cytokinésis			*va1, va2*
————			*po=ms4, ms6, ms17*
Quartet	1.5	7	TB-1La
————			
Uninucleate	9	38	TB-1Sb, TB-3La, TB-3Lc, TB-3Ld
Vacuolation		25	*ms1, ms2, ms7, ms10, ms12, ms13*, cms-C, cms-T
Differentiation		13	
Microspore			cms-s, TB-3Sb, TB-5Sc,TB-7Lb
————			
First Pollen Mitosis	3.5	15	*ms5, ms11, ms14 Ms41*, TB-9Sd
Binucleate			TB-4Sa, TB-4Lc, TB-4Lf, TB-5La, 6Lb, 6Lc, TB-8Lc, TB-9Sb, TB-9La, TB-9Lc, TB-10Sc, TB-10L9, TB-10L19
————			
Second Pollen Mitosis	4	19	*ms24*, cms-S, TB-8La
Trinucleate			TB-6Sa, TB-7Sc, TB-10L32
Engorged Mature Pollen			

REFERENCES

Albertsen MC, Phillips RL (1981) Developmental cytology of 13 genetic male sterile loci in maize. Can J Genet Cytol 23: 195–208

Bedinger PA, Edgerton MD (1990) Developmental staging of maize microspores reveals a transition in developing microspore proteins. Plant Physiol 92: 474–479

Bennett MD, Rao MK, Smith JB, Bayliss MW (1973) Cell development in the anther, the ovule, and the young seed of Triticum aestivum L. var. Chinese Spring. Phil Trans R Soc Lond B 226 (875): 39–81

Burnham CR (1982) Details of the smear technique for studying chromosomes in maize. In Sheridan WF (ed) Maize for Biological Research, Plant Molecular Biology Association, Charlottesville, VA, pp 107–118

Carlson WR (1988) The cytogenetics of corn. In Sprague GF, Dudley JW (eds) Corn and Corn Improvement: Agronomy 18, American Society of Agronomy, Inc, Madison, WI, pp 259–344

Chang MT, Neuffer MG (1989a) Maize microsporogenesis. Genome 32: 232–244

Chang MT, Neuffer MG (1989b) A simple method for staining nuclei of mature and germinated maize pollen. Stain Technol 64: 181–184

Coe EH Jr, Neuffer MG, Hoisington DA (1988) The genetics of corn. In Sprague GF, Dudley JW (eds) Corn and Corn Improvement: Agronomy 18, American Society of Agronomy, Inc, Madison, WI, pp 81–258

Cook FS, Walden DB (1965) The male gametophyte of Zea mays L. II. In vitro germination. Can J Bot 45: 605–613

Curtis CA (1985) Meiotic mutants of maize. PhD dissertation, Univ Missouri, Columbia (Diss. Abstr. 47: 914-B)

Fauron CM-R, Havlik M, Hafezi S, Brettell RIS, Albertsen M (1990) Study of two different recombination events in maize *cmsT*-regenerated plants during reversion to fertility. Theor Appl Genet 79: 593–599

Gaillard A, Vergne P, Beckert M (1991) Optimization of maize microspore isolation and culture conditions for reliable plant regeneration. Plant Cell Reports 10: 55–58

Golubovskaya IN, Khristolyubova NB (1985) The cytogenetic evidence of the gene control of meiosis. In Freeling M (ed) Plant Genetics, UCLA Symposia on Molecular and Cellular Biology, New Series, Vol 35, Alan R Liss, New York, pp 723–738

Henderson SA, Lu BC (1968) The use of hematoxylin for squash preparations of chromosomes. Stain Technol 43: 233–236

Hsu S-Y, Peterson PA (1981) Relative stage duration of microsporogenesis in maize. Iowa State J Res 55(4): 351–373

Hsu S-Y, Huang Y-C, Peterson PA (1988) Development pattern of microspores in Zea mays L., the maturation of upper and lower florets of spikelets among an assortment of genotypes. Maydica XXXIII: 77–98

Hu H (1985) Use of haploids for crop improvement in China. In Swaminanthan MS, Hu H, Shao QQ (eds) Genetic Manipulation in Crops Newsletter 1: Chinese Agricultural Science and Technology Press, Beijing, China, pp 11–23

Kaul MLH (1988) Male sterility in higher plants. Monographs on Theoretical and Applied Genetics, Vol 10, Springer-Verlag, Berlin, 1005 pages

Kindiger B, Beckett JB (1985) A hematoxylin staining procedure for maize pollen grain chromosomes. Stain Technol 60: 265–270

Kindiger B (1986) Developmental abnormalities in hypoploid pollen grains of *Zea mays* L. PhD dissertation, University of Missouri, Columbia, MO

Kindiger B, Beckett JB, Coe EH Jr (1991) Differential effects of specific chromosomal deficiencies on the development of the maize pollen grain. Genome 34: 579–594

Laser KD, Lersten NR (1972) Anatomy and cytology of microsporogenesis in cytoplasmic male sterile angiosperms. Bot Rev 38: 425–454

Levings CS III, Williams ME (1990) Developments in cytoplasmic male sterility in corn. Proc Annu Corn Sorghum Res Conf 44: 76–86

Mandaron P, Niogret MF, Mache R, Moneger F (1990) In vitro protein synthesis in isolated microspores of *Zea mays* at several stages of development. Theor Appl Genet 80: 134–138

Mascarenhas JP (1989) The male gametophyte of flowering plants. The Plant Cell 1: 657–664

McClintock B (1938) The fusion of broken ends of sister half-chromatids following chromatid breakage at meiotic anaphase. Research Bull 290, University of Missouri, College of Agriculture, Columbia, MO, 47 pp

McClintock B (1984) The significance of responses of the genome to challenge. Science 226: 792–801

McConchie CA, Hough T, Knox RB (1987) Ultrastructural analysis of the sperm cells of mature pollen of maize, Zea mays. Protoplasma 139: 9–19

Pechan PM and Keller WA (1988) Identification of potentially embryogenic microspores in Brassica napus. Physiol Plant 74: 377–384

Pescitelli SM, Petolino JF (1988) Microspore development in cultured maize anthers. Plant Cell Reports 7: 441–444

Petolino JF, Jones AM (1986) Anther culture of elite genotypes of maize. Crop Sci 26: 1072–1074

Rhoades CA, Phillips RL, Green CE (1986) Cytogenetic stability of aneuploid maize tissue cultures. Can J Genet Cytol 28: 374–384

Rhoades MM (1950) Meiosis in maize. J Hered 41: 58–67

Rhoades MM, Dempsey E (1966) Induction of chromosome doubling at meiosis by the elongate gene in maize. Genetics 54: 505–522

Rhoades MM, Dempsey E (1972) On the mechanism of chromatin loss induced by the B chromosome of maize. Genetics 71: 73–96

Rhoades MM, Dempsey E (1973) Chromatin elimination induced by the B chromosome of maize. I. Mechanism of loss and the pattern of endosperm varigation. J Hered 64: 12–18

Staiger CJ, Cande WZ (1991) Microfilament distribution in maize meiotic mutants correlates with microtubule organization. Plant Cell 3: 637–644

Vergne P, Delvallee I, Dumas C (1987) Rapid assessment of microspore and pollen development stage in wheat and maize using DAPI and membrane permeabilization. Stain Technol 62: 299–304

Walbot V, Messing J (1988) Molecular genetics of corn. In Sprague GF, Dudley JW (eds) Corn and Corn Improvement: Agronomy 18, American Society of Agronomy, Madison, WI, pp 389–430

Warmke HE, Lee S-LJ (1977) Mitochondrial degeneration in Texas cytoplasmic male-sterile corn anthers. J Hered 68: 213–222

Zhao Z-Y, Weber DF (1989) Male gametophyte development in monosomics of maize. Genome 32: 155–164

A Staining Procedure for Pollen Grain Chromosomes of Maize

BRYAN KINDIGER

Iron acetocarmine has long been used to visualize the chromosomes in maize pollen (Belling 1921, 1928; Burnham 1982) (Figure 73.1). The staining procedure is cumbersome and often provides unsatisfactory results. Maize pollen compounds the difficulty by presenting special problems for cytological studies. The thick exine and abundant starch grains prevent clear observation of the chromosomes and nuclei. The —addition of chloral hydrate or ferric hydrate to the iron acetocarmine stain can clear some of the starch (Satina and Blakeslee 1935; Belling 1921; Maheshwari 1937); however, satisfactory results can remain elusive.

Hematoxylin stains have been successfully used for many years to stain thin sections (Darlington and LaCour 1947) and squashes of plant, animal, and fungal material (Henderson and Lu 1968; Wittman 1965). Recent studies utilizing a hematoxylin stain have provided desirable results in the cytological studies of the maize pollen grain (Kindiger and Beckett 1985; Chang and Neuffer 1989; Zhao and Weber 1989; Kindiger et al. 1991).

Morrison (1953) demonstrated that by using a saturated monobromo-napthalene solution as a prefixative, inhibition of the mitotic spindle can contract the pollen grain chromosomes of several grass species. The use of dimethyl sulfoxide (DMSO) in squash preparations has been shown by Sallee (1982) to facilitate the movement of monobromo-napthalene into root tip cells of wheat and maize.

The successful utilization of monobromo-napthalene + DMSO in the prefixative step, in conjunction with the use of a hematoxylin stain, provides excellent staining of the maize microspore and pollen grain chromosomes. Addition of chloral hydrate enhances the procedure by clearing the starch and providing for excellent resolution of the stained chromosomes.

The Maize Handbook—M. Freeling, V. Walbot, eds.

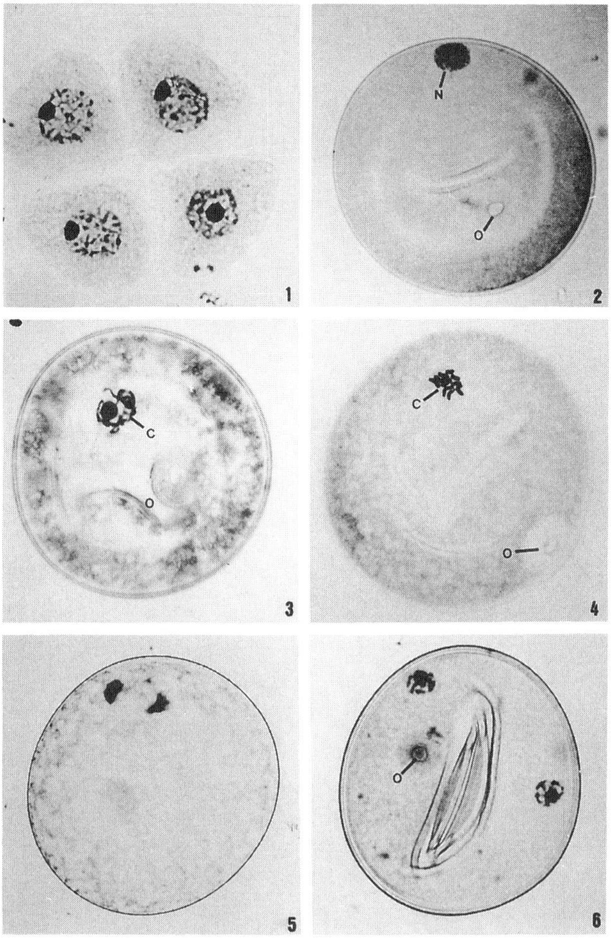

FIGURE 73.1 Maize pollen grain divisions: (1) Quartet stage, (2) single nucleus (n=nucleus; o=pollen pore), (3) prophase I (c=chromosomes), (4) metaphase I, (5) anaphase I, (6) telophase I (Continued)

HEMATOXYLIN STAIN (AS DESCRIBED BY HENDERSON AND LU 1968)

Solution A

Dissolve 2 g hematoxylin in 100 ml 50% propionic acid. Allow the solution to incubate for about 1 week. The solution keeps indefinitely in a stoppered brown bottle without refrigeration.

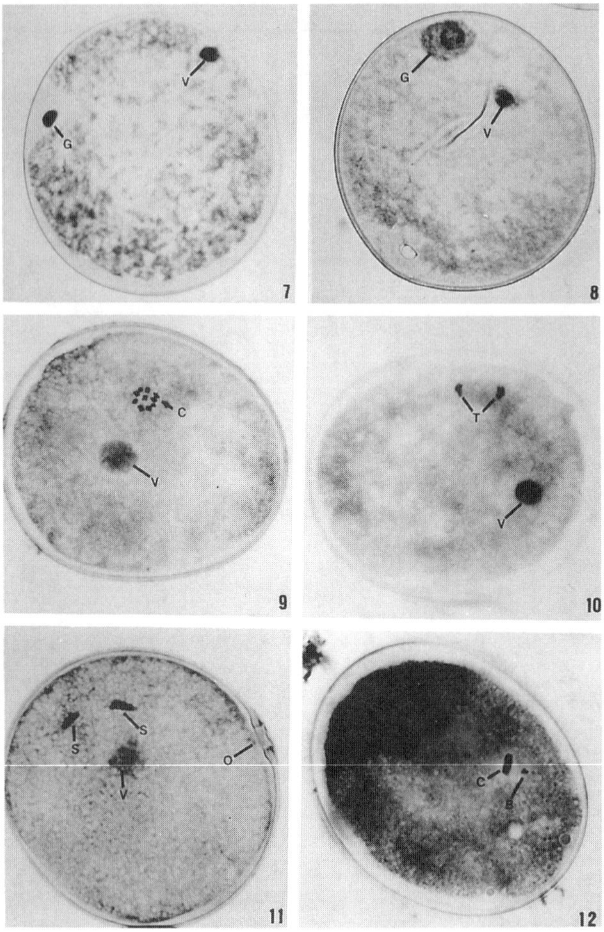

FIGURE 73.1 (Continued) (7) two nuclei (g=germinal; v=vegetative), (8) two nuclei (advanced) prophase II, (9) metaphase II, (10) anaphase II (t=dividing generative nucleus), (11) mature pollen (s=sperm; v=vegetative nucleus; o=pollen pore), (12) two-nuclei stage showing B chromosome (b=B chromosome).

Solution B

Dissolve 0.5 g ferric ammonium sulfate ($FeNH_4(SO_4)_2$) in 100 ml 50% propionic acid. Solution B keeps indefinitely in an unrefrigerated, stoppered brown bottle.

Mix equal volumes of solutions A and B. The mixture, which turns dark brown, is ready for use immediately. The stain remains good for about 2 weeks.

COLLECTION AND PREPARATION OF POLLEN GRAINS

Collect maize tassels when 3 cm of the tassel becomes visible above the leaf whorl. Fresh material can be collected, placed in a waxed shoot bag, and stored in the refrigerator for about 7 days. Material that is to be retained for longer periods should be placed in a 14:1 mixture of 70% ethyl alcohol:formaldehyde and refrigerated.

PREFIXATION

Prefixation is not necessary, but can be used prior to staining to provide shortened, well-defined mitotic chromosomes. Dissect out anthers from fresh material and prefix for 30–45 minutes in a solution consisting of 100 ml tap water, four drops monobromo-napthalene, and one drop DMSO. After prefixation, remove anthers and fix in glacial acetic acid for 30 minutes.

PROCEDURE FOR STAINING

Dissect out anthers and place in a drop of 45% acetic acid on a slide. Add a crystal of chloral hydrate. Macerate the anther with a dissecting needle to liberate the pollen grains and dissolve the chloral hydrate. Use of an iron dissecting needle is not necessary; a stainless steel dissecting needle is adequate. Remove the debris. Add a drop of stain and mix on the slide. For fixed material, mix the stain with the pollen grains on a slide and wait approximately 1 minute. Apply a glass coverslip. Once the coverslip is applied, its weight is often sufficient to gently squash the pollen grains. After applying the coverslip, blot off excess stain by placing the slide between sheets of bibulous paper and applying gentle pressure. Gently heat the slide until the stain just begins to boil. Heating will alter the color of the stain to a light golden brown. Heat must be applied if the nuclei and chromosomes are to take up the stain. Upon cooling, chromosomes and nuclei will darken and the cytoplasm will clear. Successive trials of alternate heating and cooling will increase the contrast between the stained material and the cytoplasm. Continue heating and cooling until the desired contrast is obtained. Usually only one or two cycles are required to successfully stain fresh material—three or four cycles for fixed material.

To stain fresh or mature pollen preserved in 70% ethyl alcohol:formaldehyde, allow the pollen grains to remain in the stain for 1 or 2 minutes before applying the coverslip. Heat slides of mature pollen grains gently. Excessive heat will cause the mature pollen grains to break open and extrude their contents. Successive trials of gentle heating and cooling will eventually increase the contrast between nuclei and cytoplasm.

ACK--Mention of a trademark of proprietary product does not constitute a guarantee or warranty of the product by the USDA and does not imply its approval to the exclusion of other products that may also be suitable.

REFERENCES

Belling J (1921) On counting chromosomes in pollen mother cells. Am Nat 55: 573–574

Belling J (1928) A method for the study of chromosomes in pollen mother cells. Univ Calif Publ Bot 14: 293–299

Burnham CR (1982) Details of the smear technique for studying chromosomes in maize. In Sheridan W (ed) Maize for Biological Research, Plant Molecular Biology Assn, Charlottesville, VA pp 107–118

Chang MT, Neuffer MG (1989) Maize microsporogenesis. Genome 32: 232–244

Darlington CD, LaCour LF (1947) The Handling of Chromosomes, Second Edition. Udwin Brothers, Woking, England

Henderson SA, Lu BC (1968) The use of haematoxylin for squash preparations of chromosomes. Stain Technol 43: 233–236

Kindiger B, Beckett JB (1985) A hematoxylin staining procedure for maize pollen grain chromosomes. Stain Technol 60: 265–269

Kindiger B, Beckett JB, Coe EH Jr (1991) Differential effects of specific chromosomal deficiencies on the development of the maize pollen grain. Genome 34: 579–594

Maheshwari P (1937) Recent advances in microtechnic. I. Methods of studying the development of the male gametophyte in angiosperms. Stain Technol 12: 61–70

McClintock B (1929) Chromosome morphology in Zea mays. Science 69: 629

Morrison JW (1953) A new technique for pollen grain study in the Gramineae. Can J Agric Sci 33: 399–401

Sallee PJ (1982) Prefixation and staining of the somatic chromosomes of corn. In Sheridan W (ed) Maize for Biological Research, Plant Molecular Biology Assn, Charlottesville, VA pp 19–20

Satina S, Blakeslee AF (1935) Fertilisation in the incompatible cross Datura Stramonium x D. Metel Bull Torrey Bot Club 62: 301–310

Wittmann W (1965) Aceto-iron-haematoxylin-chloral hydrate for chromosome staining. Stain Technol 40: 161–164

Zhao Z-Y, Weber DF (1989) Male gametophyte development in monosomics of maize. Genome 32: 155–164

74

A Technique for the Preparation of Somatic Chromosomes of Maize

BRYAN KINDIGER

The staining and counting of the somatic chromosomes of *Zea mays* constitute a valuable technique for studies in maize genetics. The Feulgen staining technique has been successful in cytological studies of a wide range of somatic plant chromosomes. The Feulgen technique is a nucleal-type reaction and was originally developed by Feulgen and Rossenbeck as a microchemical test (1924). The technique was later refined by DeTomasi (1936), Heitz (1936), Margolena (1932) and Whitaker (1939). Numerous early botanists/cytogeneticists (a few of whom are cited below), in a refinement of the technique, demonstrated that the mitotic inhibitor (colchicine) could be applied as a pretreatment to shorten plant chromosomes and thus allow superior visualization of the somatic chromosomes prior to staining (Burrell 1939; Nichols 1941; O'Mara 1939). Minor modifications in the pretreatment step have allowed further improvements in the resolution of the somatic chromosomes (Sallee 1981, 1982; Kindiger 1983).

Success of the technique depends upon the use of a reduced or colorless form of basic fuchsin which, upon contact with an aldehyde, develops a specific purple-red color on the chromosomes. Feulgen stain is also known as Schiff's reagent. Prepared Schiff's reagent can be readily obtained from several chemical supply companies. The strength of over-the-counter Schiff's reagent tends to be weak, but its staining capacity is adequate. If a stronger solution of Schiff's reagent is desired, the protocol for preparation of classical Schiff's reagent is presented below:

Schiff's reagent
800 ml distilled H_2O
4 g basic fuchsin
12 g potassium bisulfite
120 ml 1N HCL
Activated charcoal (carbon)

Boil 800 ml H_2O and then add 4 g basic fuchsin. Mix until all basic fuchsin is dis-

The Maize Handbook—M. Freeling, V. Walbot, eds.

solved. Allow to cool, then filter. To the solution, add 12 g potassium bisulfite and 120 ml 1 N HCl. Mix and allow to sit overnight in the dark. The following day, add activated carbon and allow the solution to clear. The clearing usually takes 1–4 hours. Filter the cleared solution. The resulting solution should be as clear as water. Store the reagent in a sealed brown bottle and keep refrigerated.

ROOT TIP PREPARATION

Germination

In a 30° C incubator, germinate maize seeds in a drainable container of moist, course sand until roots are 3–4 cm long (usually 36–48 hours).

PRETREATMENT

Pretreatment Solution

Thoroughly dissolve approximately 70 mg cycloheximide in 100 ml warm tap water. (A precise quantity of cycloheximide is not critical; experience has shown that a "pinch" is all that is required.) Four drops of monobromo-napthalene and 1–2 drops of dimethyl sulfoxide (DMSO) are added to the solution. Thoroughly mix the solution by vigorously squirting back and forth with an eyedropper. The purpose for each chemical is as follows: (1) cycloheximide will allow condensation of prophase chromosomes into a type of pseudo-metaphase conformation; (2) monobromo-napthalene, a spindle inhibitor, allows visualization of chromosomes at metaphase; and (3) dimethyl sulfoxide (DMSO), a carrier molecule, allows enhanced penetration of the above chemicals into plant tissue.

1. Collect root tips (approximately 1 cm long) and pretreat in glass vials for 4.5 hours at room temperature.
2. Pour off prefixative and replace with glacial acetic acid. Roots should remain in the glacial acetic acid from 1 to 24 hours.
3. Pour off all the glacial acetic acid and replace with hot (60° C) 1 N HCl. Place in a 60° C incubator and hydrolyze for 8–12 minutes. The relative size of the root tips should be taken into account when determining the hydrolysis time. Thin and small root tips will take less hydrolysis than thick, large tips.
4. Pour off all the 1 N HCl and rinse in tap water.

STAINING AND SLIDE PREPARATION

1. Pour off the tap water and stain in the Schiff's reagent for 10–15 minutes or until the tips become bright purple. If the hydrolysis time is correct, the margin between

the purple-stained tips and the nonstained white root tissue should be sharp and clear. If the hydrolysis time is too long, the whole excised root will stain purple. If hydrolysis time is too short, the root tip will stain a faint purple color.

2. To prepare the squash, remove about 1 mm of the stained tip with a blade and place on a glass slide. Add to the tip a drop of propionic orcein. Place a plastic cover slip adjacent to the pool of stain. Then, cover the root tip with a second plastic cover slip so that one edge rests on the edge of the first slip. Tap the second slip many times with a pencil eraser or dull needle to break up the tip and disperse the cells. If the hydrolysis time is correct, the root tip should be firm but friable, and the cells should easily disperse by gentle tapping of the coverslip. After the tip is broken up and the cells dispersed, slide the first coverslip out from under the second coverslip. Blot excess stain with a sheet of bibulous paper. Invert the slide on a sheet of bibulous paper with coverslip facing down and apply gentle pressure with your thumb. Be sure to prevent movement of the coverslip because such motion will roll the cells into useless half moon configurations.

If it is necessary to keep the root tips for more than 2 days, pour off the Schiff's reagent and replace with distilled water. Cap and refrigerate the vial. Eventually, both the distilled water and the entire root will turn purple. No deterioration in the quality of the chromosome preparations was observed after 3 weeks of storage.

ACK--Mention of a trademark of proprietary product does not constitute a guarantee of warranty of the product by the USDA and does not imply its approval to the exclusion of other products that may also be suitable.

REFERENCES

Burrell PC (1939) Root tip smear method for difficult material. Stain Technol 14: 147–149

DeTomasi JT (1936) Improving the technic of the Feulgen stain. Stain Technol 11: 137–144

Feulgen R and Rossenbeck H (1924) Mikroscopisch-chemischer Nachweis einer Nucleinsaure vom Typus der Thymonucleinsaure. Hoppe-Seyl Z. 135: 203–248

Heitz E (1936) Die Nucleal Quetschmethode. Ber Deut Bot Gesellsch 53: 870–888.

Kindiger B, Beckett JB (1983) A modified root tip squash technique. Maize Genetics Cooperation News Letter pp 32–33

Margolena LA (1932) Feulgen's reaction and some of its applications for botanical material. Stain Technol 7: 9–16

Nichols C (1941) Spontaneous chromosome aberrations in *Allium*. Genetics 26: 89–100

O'Mara JG (1939) Observations on the immediate effects of colchicine. J Hered 30: 35–37

Sallee PJ, Kimber G (1981) The use of DMSO in the prefixation of somatic chromosomes. Cereal Res Commun 9: 199–203

Sallee PJ (1982) Prefixation and staining of the somatic chromosomes of corn. In Sheridan W (ed) Maize for Biological Research, Plant Mol Biology Assoc, Charlottesville, VA, pp 119

Whitaker TW (1939) The use of the feulgen technic with certain plant materials. Stain Technol 14: 13–16

A Technique for Somatic Chromosome Preparation and C-banding of Maize

DAVID C. JEWELL AND NURUL ISLAM-FARIDI

Cytogenetic studies of both plants and animals have been considerably enhanced by differential staining techniques developed since the pioneering studies of Casperson et al. with fluorescent stains (1968, 1969a,b). The first non-fluorescent banding technique (employing the Giemsa stain) was developed by Pardue and Gall (1970). The Giemsa stain has an obvious advantage over fluorescent stains in that the researcher can utilize ordinary light microscopy and permanent preparations. Many researchers have applied chromosome banding techniques in research on maize and its wild relatives (Horn and Walden 1971; Vosa and Marchi 1972; Hadlaczky and Kalman 1975; Sachan and Tanaka 1976; Ward 1980; Mastenbroek and de Wet 1983; Aguiar-Perecin and Vosa 1985; Bernard and Jewell 1985; Gu et al. 1985; Laurie and Bennett 1985; Rayburn et al. 1985; Kakeda et al. 1990; Porter and Rayburn 1990). In brief, these studies have shown that heterochromatin patterns observed in pachytene chromosomes of maize can be detected in mitotic chromosomes by C-banding. In other preparations, the knobs on the maize chromosomes are not visible at mitotic metaphase where the chromosomes have been estimated to be approximately 13× shorter than at meiotic midpachynema (Filion and Walden 1973). The C-banding technique of mitotic chromosomes apparently detects the heterochromatin present in the knobs visible at pachynema in meiosis; the knobs contain a highly repeated 185-bp DNA sequence (Peacock et al. 1981). The C-banding technique has proved useful for chromosome characterization and identification because the C-banding patterns are heritable and remain discernible throughout the mitotic and meiotic cell cycles.

The use of banding techniques in studies of plant chromosomes has been hampered by problems such as the difficulties in preparing whole-cell metaphase plates without the normal aid of hydrolysis, the need to remove a coverslip (required for squashing the preparation), the existence of natural variation in the banding patterns of different cultivars, and a lack of understanding of the mechanism(s) that underlies chromosome banding. Two techniques can partially ameliorate the problems involved in obtaining good preparations of plant chromosomes for banding. The first technique uses the application

The Maize Handbook—M. Freeling, V. Walbot, eds.

of a quick-freeze method to remove the coverslip (Cogner and Fairchild 1953). The second uses enzymes to soften the cell wall prior to squashing the preparations (Gill and Hornby 1973; Cataldo et al. 1974).

The technique presented here involves complete digestion of the cell wall with enzymes and spreading the cells without squashing. This technique results in preparations that are relatively free of cell debris; most of the cells in the meristematic tissue are separated from each other, and there are no overlapping chromosomes in approximately 80% of the metaphase cells. Preparations that are free of cell wall material are required for in situ hybridization, as the cell wall can interfere with the hybridization procedure (Ambros et al. 1986; Mouras et al. 1987; Simpson et al. 1988).

What follows is a detailed description of a technique for the C-banding of somatic maize chromosomes. Figure 75.1 illustrates the pachytene chromosomes of the maize line W23. Somatic preparations of the same genotype (seed kindly supplied by Dr. R.L. Phillips) were prepared with this protocol. A somatic cell photographed using phase contrast microscopy is presented in Figure 75.2 and, following C-banding, in Figure 75.3. The contraction of chromosomes at mitotic metaphase leads to the coalescence of bands that

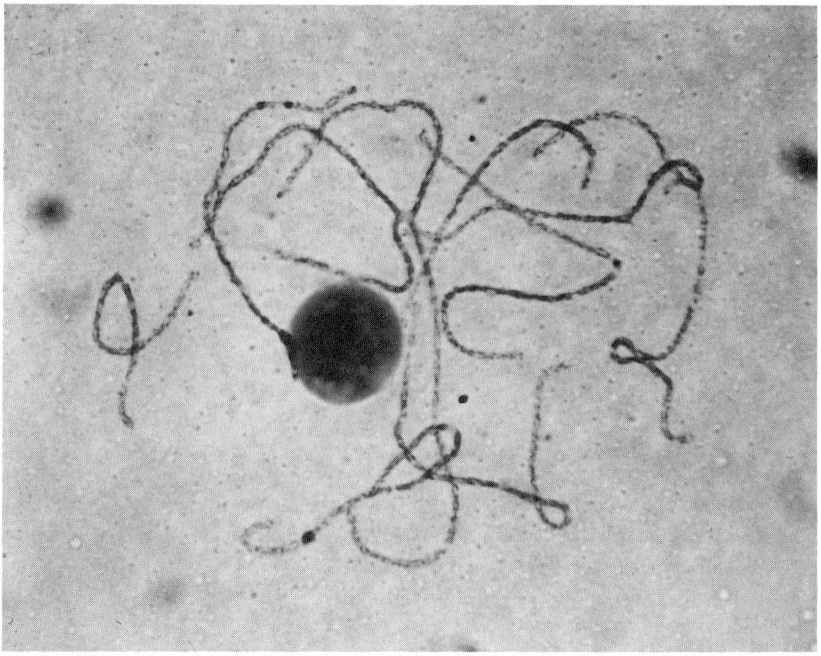

FIGURE 75.1 Pachytene chromosomes of the maize line W23. Chromosome 6 has a heterochromatic block on the short arm (located on the proximal side of the nucleolus) and has two knobs on the distal part of the long arm. Chromosome 7 has a single knob (that is larger than the knobs present on chromosome 6) on the long arm. The pachytene spread was prepared and photographed by John T. Stout and the photograph kindly provided by Dr. Ronald L. Phillips.

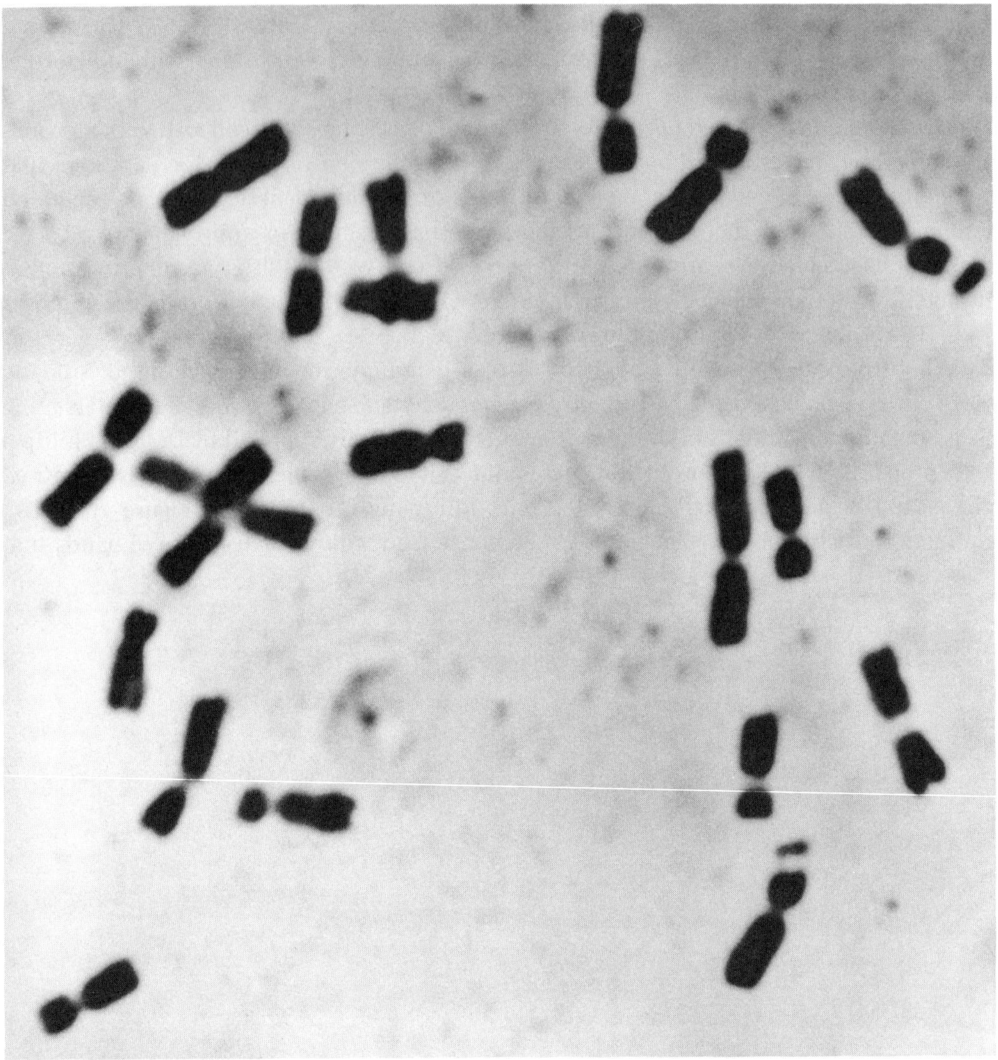

FIGURE 75.2 Mitotic metaphase chromosomes of W23. This photograph of an unstained preparation was taken using phase contrast microscopy. Centromere positions (and the secondary constriction on the short arm of chromosome 6) are clearly defined.

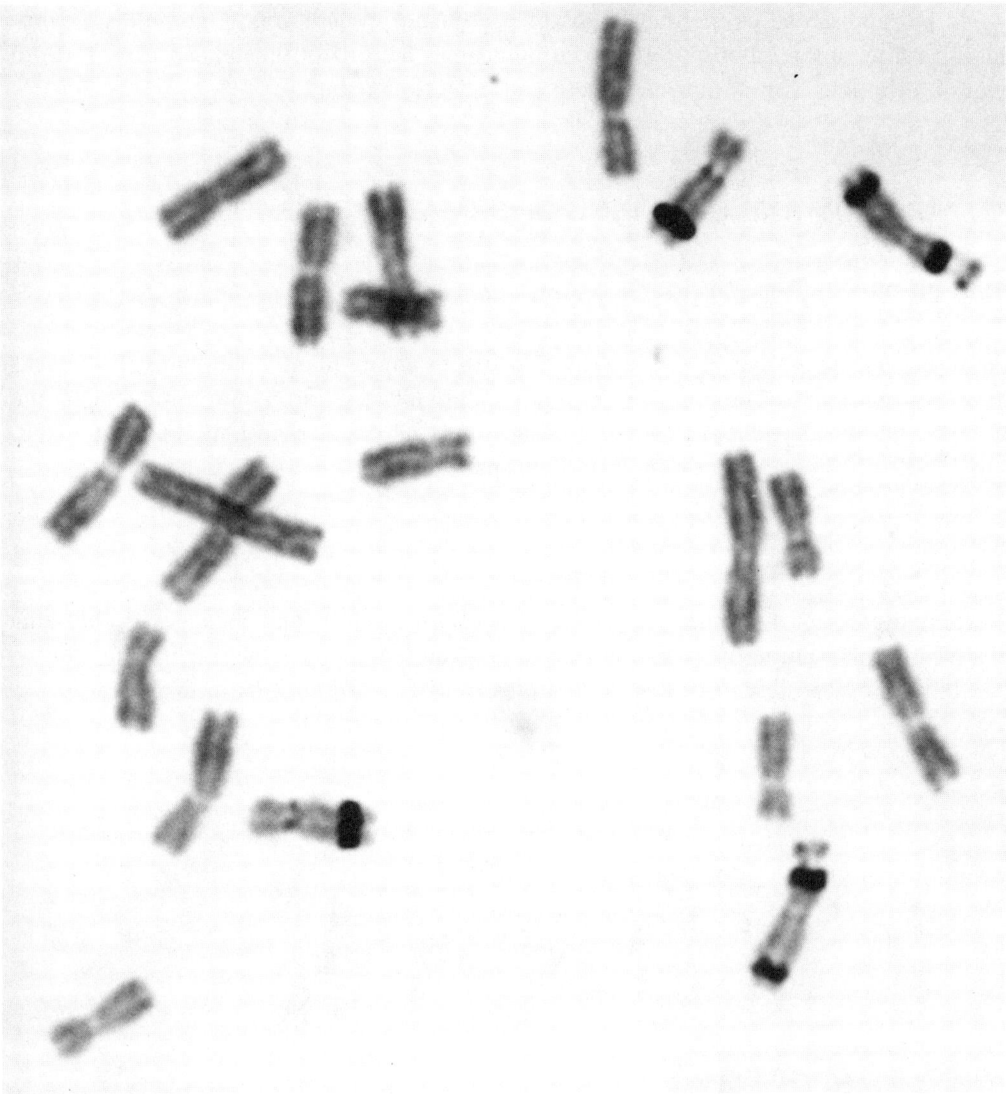

Figure 75.3 A C-banded preparation of W23 (the same cell presented in Figure 75.2). Chromosome 6 has two major C-bands (one on the short arm on the proximal side of the nucleolus and one on the distal part of the long arm). Chromosome 7 has a single prominent band on the long arm.

would be represented as distinct knobs in pachytene preparations. Figure 75.4 depicts the C-banding of the mitotic prophase chromosomes 6 and 7 of the genotype W23. (Both knobs on the long arm of chromosome 6 can be distinguished.)

PREPARATION AND C-BANDING OF SOMATIC CHROMOSOMES

Pretreatment

Excise the terminal 1 cm of the growing roots from young plants, preferably 3–4 weeks after planting in pots. The root tips are pretreated in 0.04% hydroxyquinoline (colchicine or other pretreatment chemicals can be used) for 3 hours at room temperature to accumulate cell divisions and inhibit spindle formation.

Fixation

Fix the samples in 3:1 (ethanol:glacial acetic acid) followed by one change with the same fixative after 5–10 minutes (optional); leave at room temperature overnight. The fixative should be prepared just prior to use. Samples can be stored at 4° C up to several weeks

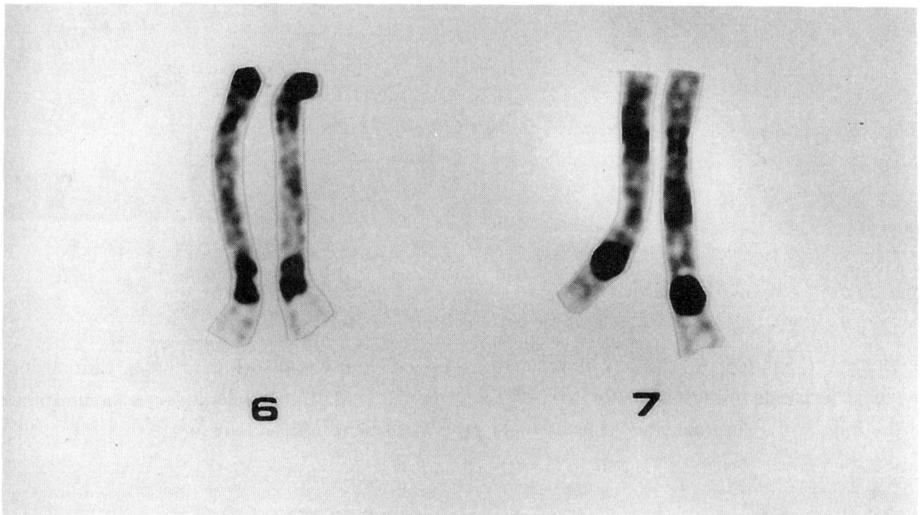

FIGURE 75.4 C-banding of the mitotic prophase chromosomes 6 and 7 of W23. Two distinct bands are evident on the long arm of chromosome 6 and represent the two knobs seen at pachynema (Figure 75.1). Only a single prominent C-band is present on the long arm of chromosome 7. The relative size of the C-bands corresponds well with the size of the knobs seen at pachynema (Figure 75.1).

before chromosome preparations are made. The more recent the sample, the better the preparation for chromosome banding.

Enzyme Treatment

Wash root samples with distilled water 2–3 times (5–10 minutes each). Wash with cold (4° C) 0.01 M citrate buffer for 10–20 minutes (preferably one change after 5 minutes). Transfer roots to a glass slide and cut off the meristematic portion (solid white tip) and discard the rest of the root tissue. Remove the excess buffer near the meristematic portions and transfer them into a small eppendorf tube (0.5 ml) containing about 100 μl of enzyme solution. (The volume depends on the number of root tips.) Incubate the root tips at 37° C for 60–120 minutes depending on the healthiness/thickness of roots and the fixation period. (If the root tips have been in fixative longer than a week, the time required for digestion is increased). Agitate the root tips very gently with a Pasteur pipet once or twice during incubation. To check digestion, transfer a single tip from the enzyme solution into a small watch glass containing distilled water for 2–3 minutes and prepare a slide as described below.

Slide Preparation

After digestion replace the enzyme solution with 0.01 M citrate buffer and change the buffer once (optional) using a Pasteur pipet. Transfer the tips to a small watch glass containing distilled water and leave for 10–20 minutes before slide preparation. Transfer a single tip onto an ethanol-washed glass slide using a Pasteur pipet. Remove excess water using a narrow-tip Pasteur pipet (homemade by drawing out the tip in a flame). Place a drop (or two) of 3:1 (ethanol:acetic acid) onto the tip and macerate the tissue. Immediately spread the cell suspension over the slide using very fine-pointed forceps and remove debris. If needed, one or two more drops of 3:1 can be added. Allow the slide to air dry, or dry it by gently warming the slide over an alcohol flame. (It should not boil.)

Checking the Preparation

Examine the slide under phase contrast microscopy and record the position of the cells for further study (for example: C-banding or in situ hybridization). If the slide is just for screening, a drop or two of 0.5% or 1% aceto-carmine or propionic carmine can be added and the preparation can be covered with a coverslip prior to examination.

A phase-contrast photograph of the cell(s) can be taken for comparison with photographs of the same cell following chromosome banding or in situ hybridization. To prepare the sample place one or two drops of distilled water onto the glass slide and cover with the correct-size coverslip. Use filter paper to soak up excess water. Following photography, simply remove the cover glass using a razor blade. There is no loss or damage of cells during this procedure.

Regardless of whether the cells are photographed, the preparations are then immersed in absolute ethanol and stored at 4° C overnight.

C-Banding

The following morning prepare 0.2 N HCl, 2×SSC, and 5% barium hydroxide. The barium hydroxide solution should be prepared fresh.

Remove the slides from the ethanol and air dry for about 30 minutes. Immerse the slides in 0.2 N HCl at 60° C for 90–100 seconds. Wash the slides in distilled water twice and transfer them to 5% barium hydroxide solution and treat for 8–10 minutes at room temperature (21° C). Wash the slides three times in distilled water and immerse them in 2×SSC at 60° C for 60 minutes. Remove the slides from 2×SSC and immerse in 5% Giemsa solution and stain for 40–60 minutes (staining period may vary from 20 to 120 minutes). One should check the slide under the microscope to optimize the staining time.

Wash the slides in distilled water and air dry overnight. (The preparations can be used after about 1 hour.) To make the slides permanent, the authors use Cytoseal 60, low viscosity (Stephens Scientific, Thomas Scientific Catalog 6705-A05).

Stock Solutions

Citrate Buffer

Prepare 500 ml of 0.01 M citrate buffer (pH 4.5). Add 1.47 g of trisodium citrate–dihydrate ($Na_3C_6H_50_7.2H_22O$) and 1.05 g of citric acid–monohydrate ($C_6H_8O_77.H_2O$) to distilled water and completely dissolve and adjust volume to 500 ml. This solution is approximately pH 4.5 and need not be adjusted.

Phosphate Buffer

Prepare 2,000 ml of buffer I and II. For buffer I add 18.92 g of Na_2HPO_4 to distilled water and completely dissolve and adjust volume to 2,000 ml. For buffer II add 18.14 g of KH_2PO_4 to distilled water (completely dissolve) and adjust the volume to 2,000 ml. Store these buffers at 4° C. A more concentrated stock solution can be made.

2×SSC (saline/sodium citrate buffer)

Prepare a stock solution of 20×SSC (17.43 g of NaCl and 8.82 g of $Na_3C_6H_5O_7.2H_2O$ in 100 ml of distilled water and adjust the pH to 7.0) that can be stored indefinitely at -15° C. The stock solution is diluted to 2×SSC for use.

Enzyme Solution

Prepare 10 ml or more enzyme solution. Add 0.5 g (5%) of cellulase Onozuka R-10 (Yakult Honsha Co. Ltd., Japan) and 0.1 g (1%) of pectolyase Y-23 (Seishin Pharmaceutical Ltd., Japan) to 10 ml of 0.01 M citrate buffer. (The concentration of enzyme solution is not critical.) Store the enzyme solution in 1-ml eppendorf tubes at -20° C or -70° C.

Barium Hydroxide

Prepare 100 ml of 5% solution. Add 5 g of $Ba(OH)_2.8H_2O$ (J.T. Baker) to 100 ml of distilled water and stir for 30–40 minutes. Filter this solution twice using Whatman No. 4 filter paper. This solution should be prepared immediately before use.

Giemsa Solution

A stock solution of Giemsa stain is prepared using 50 ml of glycerol ($C_3H_8O_3$) to which 1 g of Giemsa powder [Sigma Chemical Co. (G4507)] is added and stirred at 40–45° C for 20–30 minutes. Then add 50 ml of methanol and stir for another 20–30 minutes. Filter the solution using Whatman No. 4 filter paper and store it in a sealed brown bottle. (Keep it refrigerated.)

To prepare the staining solution for chromosomes, mix 5 ml of Giemsa stock solution with 95 ml phosphate solution (59 ml buffer I, 1/15 M Na_2HPO_4 and 36 ml buffer II, 1/15 M KH_2P_4).

General Notes: Root tips can be collected from plants at any stage prior to flowering. The number of dividing cells depends on soil moisture; soil in the pots should not be too wet. (Water the plant 1–3 days before collecting root tips.)

Although the actual concentration of the enzyme solution is not critical, it is important that the concentration of cellulase be higher than that of pectolyase. (A 6:1, 4:1, or 3:2 mix of the enzymes is adequate.)

Several authors have noted different staining reactions for Romanovsky stains used on different species and have observed changes in staining reaction following the application of different C-banding protocols (Seal and Bennett 1982). The type of stain used (and the brand) appears to be important in obtaining good C-banding. Giemsa stain made from powder (Sigma—see above) and Improved R66 Giemsa (Gurrs) have been used successfully in our laboratory.

The relative position of the C-bands (knobs) on the mitotic metaphase chromosome arms is not entirely in agreement with the positions determined by pachytene analysis. Chromosome contraction differences between meiosis and mitosis have been assumed to account for these differences (Ward 1980). In general the bands on mitotic chromosomes are in more distal positions than the knob positions determined by pachytene analysis.

It should be noted that the deeply stained C-bands have the appearance of being swollen (Mastenbroek and de Wet 1983) and can sometimes make the determination of chromosome length difficult, especially when the C-band is terminal or nearly so. Also, following C-banding it is often difficult to determine the exact position of the centromere. In our laboratory, karyotyping is done by utilizing the arm ratios determined from phase contrast photographs of unstained cells and combining this information with that of C-banding of the same cell.

REFERENCES

Aguiar-Perecin MLR de, Vosa CG (1985) C-banding in maize II. Identification of somatic chromosomes. Heredity 54: 37–42

Ambros PF, Matzke MA, Matzke AJA (1986) Detection of a 17kb unique sequence (T-DNA) in plant chromosomes by *in situ* hybridization. Chromosoma 94: 11–18

Bernard S, Jewell DC (1985) Crossing maize with sorghum, *Tripsacum* and millet: the products and their level of development following pollination. Theor Appl Genet 70: 474–483

Casperson T, Farber S, Foley GE, Kudynowski J, Modest EJ, Simonsson E, Wagh U, Zech L (1968) Chemical differentiation along metaphase chromosomes. Exp Cell Res 49: 219–222

Casperson T, Zech L, Modest EJ, Foley GE, Wagh U, Simonsson E (1969a) Chemical differentiation with fluorescent alkylating agents in Vicia faba metaphase chromosomes. Exp Cell Res 58: 128–140

Casperson T, Zech L, Modest EJ, Foley GE, Wagh U, Simonsson E (1969b) DNA-binding fluorochromes for the study of the organization of the metaphase nucleus. Exp Cell Res 58: 141–152

Cataldo FL, Miller MW, Brown JA (1974) Cellulase facilitated squash preparations of Vicia faba chromosomes for Quinacrine fluorescent staining. Stain Techn 49: 47–48

Cogner AD, Fairchild LM (1953) A quick-freeze method for making smear slides permanent. Stain Technol 28: 281–283

Gill JJB, Hornby C (1973) The use of snail gut cytase in fluorescent banding studies on plant chromosomes. Stain Technol 48: 251–253

Gu MG, Zhang XA, Huang DN, Ting YC (1985) Giemsa banding of mitotic chromosomes in diploid perennial teosinte, maize and their hybrids. Maydica XXX: 97–106

Hadlaczky GY, Kalman L (1975) Discrimination of homologous chromosomes of maize with Giemsa staining. Heredity 35: (3)37–374

Horn JD, Walden DB (1971) Fluorescent staining of euchromatin and heterochromatin in maize (*Zea mays*). Can J Genet Cytol 13: 811–815

Kakeda K, Yamagata H, Fukui K, Ohno M, Fukui K, Wei ZZ, Zhu FS (1990) High resolution bands in maize chromosomes by G-banding methods. Theor Appl Genet 80: 265–272

Laurie DA, Bennett MD (1985) Nuclear DNA content in the genera *Zea* and *Sorghum*. Intergeneric, interspecific and intraspecific variation. Heredity 55: 307–313

Mastenbroek I, de Wet JMJ (1983) Chromosome banding of maize and its closest relatives. Can J Genet Cytol 25: 203–209

Mouras A, Saul MW, Essad S, Potrykus I (1987) Localization by *in situ* hybridization of a low copy chimaeric resistance gene introduced into plants by direct gene transfer. Mol Gen Genet 207: 204–209

Pardue ML, Gall JG (1970) Chromosomal localization of mouse satellite DNA. Science 168: 1356–1358

Peacock WJ, Dennis ES, Rhoades MM, Pryor AJ (1981) Highly repeated DNA sequence limited to knob heterochromatin in maize. Proc Natl Acad Sci USA 78: 4490–4494

Porter HL, Rayburn AL (1990) B-chromosome and C-band heterochromatin variation in Arizona maize populations adapted to different altitudes. Genome 33: 659–662

Rayburn AL, Price HJ, Smith JD, Gold JR (1985) C-band heterochromatin and DNA content in maize mitotic chromosomes. Can J Genet Cytol 22: 61–67

Sachan JKS, Tanaka R (1976) A banding method for *Zea* chromosomes. Jpn J Genet 51: (2)139–141

Seal AG, Bennett MD (1982) Preferential C-banding of wheat and rye chromosomes. Theor Appl Genet 63: 227–233

Simpson PR, Newman MA, Davies DR (1988) Detection of legumin gene sequences in pea by *in situ* hybridization. Chromosoma 96: 454–458

Vosa CG, Marchi P (1972) Quinacrine fluorescence and Giemsa staining in plants. Nature 237: 191–192

Ward EJ (1980) Banding patterns of maize mitotic chromosomes. Can J Genet Cytol 22: 61–67

76

Duplications

JAMES A. BIRCHLER

Tandem duplications served as the basis for some of the central concepts in genetics that emerged from scutiny of maize. The work of L.J. Stadler on the *R* locus contributed to current thinking on spontaneous mutation, recombination, and divergence of tissue-specific expression (e.g., Stadler 1946; Stadler and Emmerling 1956). Work on the *A* locus by J. Laughnan gave rise to the concept of intrachromosomal recombination involving tandem duplications (Laughnan 1949).

Tandem duplications appear to be quite common, having been documented as a polymorphism at several loci. These include *R*, *A*, and *Alcohol dehydrogenase* (Schwartz and Endo 1966; Birchler and Schwartz 1979). In all cases, the duplication and simplex both exist in maize populations, and the members of the duplication are diverged. A popular idea to explain the high degree of tandem duplications is that there is unequal exchange between transposable elements of like type. Molecular analysis could reveal whether the boundaries involve a transposon or whether the duplication originated in another way. Given the number of duplications and their importance in providing raw material for evolution, the mechanism of duplication formation is an important problem.

In addition to tandem duplications, there are many duplicate loci in maize. This fact has been used to support the contention of an allotetraploid past (Rhoades 1951). Recent analysis using RFLP loci indicates that blocks of certain chromosomes do carry homeologous sequences (Helentjaris et al. 1988), although no sets of homeologous chromosomes are obvious from their data.

The Maize Handbook—M. Freeling, V. Walbot, eds.

REFERENCES

Birchler JA, Schwartz D (1979) Mutational study of the alcohol dehydrogenase-1 F Cm duplication in maize. Biochem Genet 17: 1173–1180

Helentjaris T, Weber DF, Wright S (1988) Duplicate sequences in maize and identification of their genomic locations through restriction fragment length polymorphisms. Genetics 118: 353–363

Laughnan JR(1949) The action of allelic forms of the gene A in maize. II. The relation of crossing over to mutation of A-b. Proc Natl Acad Sci USA 35: 167–178

Rhoades MM (1951) Duplicate genes in maize. Am Nat 85: 105–110

Schwartz D, Endo T (1966) Alcohol dehydrogenase polymorphism in maize-simple and compound loci. Genetics 53: 709–715

Stadler LJ (1946) Spontaneous mutation at the R locus in maize. I. The aleurone-color and plant-color effects. Genetics 31: 377–394

Stadler LJ, Emmerling MH (1956) Relation of unequal crossing over to the interdependence of R-r elements (P) and (S). Genetics 41: 124–137

77

Deficiency Analysis

JAMES A. BIRCHLER

Simple deficiencies that can be transmitted from one generation to the next are rare. The haploid gametophyte generation will eliminate any deficiency that removes a vital gene required for cell viability or gametophyte function. Fortunately in maize there are two techniques for the generation of deficient gametes at stages of the gametophytes that will permit recovery of the deficient chromosome in the next generation. First, the B-A translocations, described elsewhere in this volume, routinely nondisjoin at the second microspore division, producing genetically distinct sperm. Second, the *r-x1* deficiency that conditions nondisjunction late in the divisions of the megagametophyte gives rise to deficient eggs. These two tools, not available in other species, provide a powerful advantage to maize genetics.

The Maize Handbook—M. Freeling, V. Walbot, eds.

Of the simple deficiencies that can be passed from one generation to the next, it is often the case that the recovery through the female side is much greater or the only mode of transmission. (See references.) A chromosome that will pass through the female side but that has reduced or no recovery from the male fulfills the genetic criterion for carrying a deficiency. The *r-x1* deficiency, for example, does not transmit through the pollen.

The reason for this disparity involving deficiencies has been the subject of speculation over the years. One possibility might be that the megagametophyte remains enclosed in the sporophyte and potentially could receive metabolites from the surrounding tissue while the pollen must be free of the sporophyte for at least some time. Also, pollen tubes that are placed at a competitive disadvantage will not achieve fertilization in competition with normal ones; hence, pollen with a deficiency may be eliminated by competition. On the female side, there is no competition and debilitated but viable gametophytes might still give rise to progeny. An alternative view is that there are considerably more genes expressed in the microgametophyte than in the megagametophyte; thus there would be a higher likelihood that a given deficiency will eliminate a gene vital to microgametophyte functions.

REFERENCES

Baker RL, Morgan DT Jr (1969) Control of pairing and meiotic interchromosomal effects of deficiencies in chromosome 1. Genetics 61: 91–106

McClintock B (1941) The association of mutants with homozygous deficiencies in *Zea mays*. Genetics 26: 542–571

McClintock B (1944) The relation of homozygous deficiencies to mutations and allelic series in maize. Genetics 29: 478–502

Mottinger JP (1970) The effects of X-rays on the bronze and shrunken loci in maize. Genetics 64: 259–271

Neuffer MG (1957) Additional evidence on the effect of X-ray and ultraviolet radiation on mutation in maize. Genetics 42: 273–282

Phillips RL, Burnahm CR, Patterson EB (1971) Advantages of chromosomal interchanges that generate haplo-viable deficiency-duplications. Crop Sci 11: 525–528

Rhoades MM, Dempsey E (1973) Cytogenetic studies on a transmissible deficiency in chromosome 3 of maize. J Hered 64: 125–128

Rhoades MM, Dempsey E (1953) Cytogenetic studies of deficient-duplicate chromosomes derived from inversion heterozygotes in maize. Am J Botany 40: 405–424

Stadler LJ, Roman HL (1948) The effect of X-rays upon mutation of the gene A in maize. Genetics 33: 273–303

The Gametophyte Factors of Maize

Oliver E. Nelson

While there are a number of loci in maize that are specifically expressed in the gametophytic generation (e.g., *lethal ovule1* and *2*, *small pollen*, etc.) that are gametophyte factors in the broad sense of the term, maize geneticists use the term to refer specifically to the numerous loci designated as *gametophyte factor* (*ga*). At all but one of these loci, pollen grains carrying a *Ga* allele have a pronounced competitive advantage in effecting fertilization over pollen carrying the *ga* allele on the silks of plants that are *Ga/Ga* or *Ga/ga* but not on the silks of *ga/ga* plants. In some instances, the competitive advantage may be nearly complete. The exception to this generalization is *ga7* on chromosome 3, where the competitive advantage of *Ga* gametes over *ga* is only slightly influenced by the genotype of the silks in which they are growing (Rhoades 1948). The *ga7* locus thus differs from the other *ga* factors to be discussed, and the *ga7* allele may be a hypomorphic mutation at a locus whose product is required for pollen tube growth.

In the absence of a mutation conditioning a visible phenotype linked to one of the alleles in a *ga/Ga* heterozyote, there is obviously no evidence that one class of pollen grains is predominant in fertilization. When such a mutation is present, the distortion of the expected segregation ratio reflects the heterozygosity at the *ga* locus. Such a departure from an expected ratio was reported as early as 1902 by Correns in the F_2 progeny of a cross between a Rice Popcorn and a sweet corn (*sugary1*), where 16% *su1/su1* kernels (rather than 25%) were produced. The tentative explanation was that the differential fertilization was caused by the effect of the *su1* mutation on the male gametophyte. Emerson (1925) suggested, on the basis of his observation that in advanced progenies of a cross between a Rice Popcorn and a *su1/su1* strain there were some plants producing high ratios (ca. 35%) of *su1/su1* kernels, that selective fertilization related to the presence of a linked gametophyte factor was involved. Mangelsdorf and Jones (1926) reported a more complete investigation of a Rice Popcorn × *su1/su1* cross in which not only the F_2 progeny was generated but F_1 plants were backcrossed both as male and female parents to the Rice Pop and the *su1/su1* stocks. The resultant progenies from the backcrosses to the Rice Pop (*Su1/Su1*) were selfed to ascertain what proportion of fertilization events involved *su1* gametes. The percentages of *su1/su1* kernels (or *su1* gametes functioning in the

The Maize Handbook—M. Freeling, V. Walbot, eds.

backcrosses to the *Su1/Su1* parent) observed are summarized in Figure 78.1, which is taken from their paper. On the basis of these data showing that there was competition between *Su1* and *su1* gametes only when *Su1/su1* heterozygotes were selfed or backcrossed as the male parent to the Rice Pop (*Su1/Su1*) stock, they concluded that *Su1* was linked to a dominant allele (*Ga1*) at a gametophyte factor locus and that this conferred a competitive advantage on the *Su1* gametes but only on the silks of plants that were *ga1/Ga1* or *Ga1/Ga1*. The existence of this fourth chromosome locus, *gametophyte factor1* (*ga1*), has been confirmed in numerous subsequent investigations.

As noted, there are numerous *gametophyte factor* loci in maize. Such factors have been reported on chromosomes 1 (*ga4* and *ga6*), 2, 3 (*ga7*), 4 (*ga1*), 5 (*ga2* and *ga10*), 6, 7 (*ga3*), and 9 (*ga8*). In addition, there are data suggesting a second gametophyte factor on chromosome 9. Most data on gametophyte factors appear in short notes in the *Maize Genetics Cooperation News Letter (MNL),* and they can be located by reference to the index to the *MNL* compiled by E.H. Coe as the appendix to Vol. 36 (1962), the symbol index for Vols. 36–53 that appeared in Vol. 53 (1979), and the symbol indices that have appeared in each volume from Vols. 54 to 65 (1991).

The first gametophyte locus identified, *ga1*, has been the subject of numerous publications. The *ga1* locus was not only the first gametophyte factor locus reported, but it is unique in that the *Ga1-s* allele (Schwartz 1950) conditions nonreciprocal cross-sterility. A *Ga1-s/Ga1-s* plant is not receptive to *ga* pollen, but *Ga1-s* pollen induces a full seed set on *ga/ga* plants. In addition, many popcorns and some Central American races of dent and flint corn have been identified as being *Ga/Ga* or *Ga1-s/Ga1-s* while all North American flint and dent corns are *ga/ga* (Nelson 1952, 1954, 1960). Since it is not possible to derive a *ga1/ga1* race from one that is *Ga1/Ga1* or *Ga1-s/Ga1-s* owing to the great competitive advantage of *Ga1* or *Ga1-s* pollen over *ga1* pollen, allelic constitution at this locus offers a useful marker of evolutionary descent.

There are interesting aspects to the determination of recombination percentage between

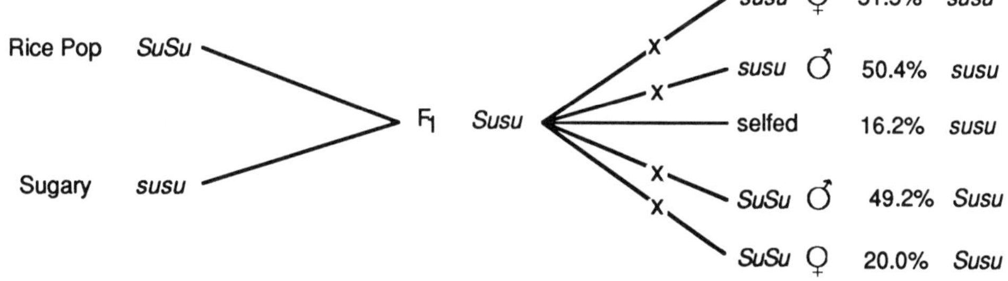

FIGURE 78.1 The results of self-pollinating the F_1 hybrid between Rice Popcorn and a *su1/su1* variety and of backcrossing the F_1 as a male and as a female to both parents. For the backcrosses to the Rice Popcorn (*Su1/Su1*) parent, it was necessary to go to the next generation in order to determine what proportion of the plants were *Su1/su1*. The figure is taken from Mangelsdorf and Jones (1926).

a mutant conditioning a visible phenotype and a linked *ga* locus. If *Ga* pollen were to effect fertilization to the complete exclusion of *ga* pollen, it would be simple to calculate the recombination from the excess or deficiency of the visible marker in an F_2 progeny. Such complete exclusion of *ga* gametes seems rarely to happen in self-pollinations of *ga*/*Ga* heterozygotes. In theory, an observed percentage of individuals with the mutant phenotype in an F_2 progeny could result from a wide range of combinations of *ga* gametes functioning and recombination percentages between the *ga* locus and the marker locus. For example, 15% sugary (*su1/su1*) kernels in the F_2 progeny of a cross between a *ga1 su1/ga1 su1* and a *Ga1 Su1/Ga1 Su1* stock could result from zero recombination between *ga1* and *su1* with 30% *ga1* gametes functioning or 30% recombination and no *ga1* gametes functioning or the whole range of combinations between these limits. In such an instance, the most expeditious way to estimate the recombination percentage is to backcross the F_1 as a female parent by the *ga su/ga su* parent so that no gametic competition is involved. Self-pollination of the plants grown from nonsugary seeds to ascertain the proportion of plants (*ga Su/ga su*) that give ratios grouped around 25% *sugary* kernels indicates the extent of recombination between *ga* and *su*.

Mangelsdorf and Jones (1926), in considering this problem, showed that if there were two marker loci (*a* and *b*) linked to the *ga* locus and if one knew the recombination percentage between these loci, it would be possible to obtain approximate solutions graphically to the percentage of *ga* gametes effecting fertilization as well as the recombination between *a* and *ga* and between *b* and *ga*. They illustrated this with two F_2 progenies. One was an F_2 giving an excess (7.8%) of a chromosome 4 *defective seed1* (*de1*) trait while the other was an F_2 giving a deficiency (8.8%) of *sugary1* kernels from the expected 25% mutant kernels. It was known that the recombination percentage between *de1* and *su1* was 39, so it is obvious that the *ga1* locus must be located between the two marker loci and that *de1* is linked to a *Ga1* allele while *su1* is linked to a *ga1* allele. On a graph on which the Y axis was the percentage of crossing-over between *ga1* and the marker locus and the X axis was the ratio of *Ga1* to *ga1* gametes effecting fertilization, one curve was drawn through the points marking the various combinations of crossing-over and relative effectiveness of *Ga1* vs. *ga1* gametes that would result in the observed deficiency of *su1/su1* kernels. Similarly, a curve was drawn through the set of points that would result in the observed excess of *de1/de1* kernels. Along these curves, an ordinate can be erected that crosses the two curves at points whose recombination percentages converted into map distances and added equal the map distance between *su1* and *de1*. Applying this method allowed the tentative conclusion that *ga1* gametes effected fertilization 20% of the times and that the crossing-over percentage between *ga1* and *su1* was 21.2 and between *ga1* and *de1* 24.3. It should be clear that one is on firmer ground if the two marker loci are segregating in the same F_2 progeny and the double-recessive seeds can be distinguished. It should also be pointed out that the method is not restricted to instances in which the *ga* locus is located between the marker loci as was the case investigated by Mangelsdorf and Jones, and as illustrated by them, the *Ga* allele may be in coupling or repulsion with the mutant allele at the marker locus.

Emerson (1934) calculated recombination percentages between the fourth chromosome mutants *su1* and *Tu* and *ga1* by a three-point method and by the graphical method of Mangelsdorf and Jones and showed that both methods gave essentially the same results. Brieger (1937) also considered this problem. With data from the F_2 progenies of crosses of *Bt1 Ga2 Pr* × *bt1 ga2 pr* on chromosome 5, he derived equations by which one could calculate the frequency of *ga2* gametes functioning and the recombination between *ga2* and *bt1* or *pr* for both the case in which one could distinguish the double-recessive phenotype and the case in which the double recessive was not distinguishable.

The allele (*Ga1-s*) conditioning nonreceptivity to *ga1* pollen has already been noted. Demerec (1929) reported the first case of the inability of one variety of corn (a popcorn) to set any seed when pollinated by another variety. Brunson (1937) made the same observation with another variety of popcorn, and Schwartz (1950), investigating a third case of cross-sterility, showed that the genetic basis of this nonreceptivity to foreign pollen was homozygosity for an allele, *Ga1-s* (s for strong), of the *ga1* locus, and that while *ga1* pollen could not fertilize *Ga1-s/Ga1-s* plants, *Ga1* pollen was effective. He demonstrated also that *ga1* pollen could germinate on the silks of *Ga1-s/Ga1-s* plants and grow into the stylar tissue. This observation still left open one of the possibilities that Demerec had advanced as the mechanism of cross-sterility, which is that the recessive pollen tubes could not reach the ovules. To this conjecture, Schwartz added the possibility of an abnormal reaction within the ovule. House and Nelson (1958) labeled pollen (both *ga1* and *Ga1-s*) with ^{32}P by injecting $^{32}PO_4$ into the terminal internode before pollen shed. At various times after the pollination of *Ga1-s/Ga1-s* plants with the labeled pollen, the silks were removed and dried on glass plates, and autoradiographs were made. The results showed that while the *ga1* pollen germinated and the pollen tubes grew into the stylar canals, the growth was progressively slower than that of *Ga1-s* pollen tubes until growth finally ceased altogether well short of the ovules. Thus there is clearly an interaction between style and pollen that results in a cessation of growth.

The inability of *Ga1-s/Ga1-s* plants to set seed with *ga* pollen can be exploited to protect maize being grown for special uses (popcorn, waxy, high amylose, sweet corn) from contamination by dent (*ga*) pollen both during the production of hybrid seed and the production of the crop itself (Nelson 1953). This system is being used to isolate seed production fields of popcorn hybrids that, owing to their original constitution at the *ga1* locus, would be susceptible to outcrossing with dent corn.

Nelson (1952) investigated the incidence of cross-sterility among the popcorns by a series of reciprocal crosses in all combinations between 10 popcorn inbreds from a number of varieties and the dent inbred, Hy, that was known not to be able to effect fertilization on several popcorn inbreds. Schwartz's stocks, *Ga1/Ga1* and *Ga1-s/Ga1-s*, were added in a second year. Six of these inbreds from different varieties reacted in the same manner as the *Ga1-s/Ga1-s* stock. Pollination with Hy pollen (*ga1*) produced no seed set or a very scanty seed set, but pollen from these inbreds induced full seed set in every other inbred in the test. One inbred from Rice Pop, 4519–4, behaved like the *Ga1/Ga1* stock in that it set seed with every source of pollen and was capable of inducing seed set in every other inbred. Three inbreds (the two from the Supergold variety of popcorn and the one from the Baby Golden variety) behaved like Hy (*ga1/ga1*). They were not capable of inducing seed set in the *Ga1-s/Ga1-s* stock nor most popcorns but could accept pollen from any other inbred or variety with which they were tested. These data are compatible with Schwartz' conclusion that there are three alleles, *ga*, *Ga*, and *Ga1-s*, at the *ga1* locus that condition the types of responses described above. There is, however, a reasonable doubt as to whether there is a *Ga1* allele (Nelson 1954). The observation, which raised this doubt initially, was that when the presumed *Ga1* allele was placed in the dent inbred, Hy, by a series of backcrosses, the plants homozygous for the transferred allele behaved as though they were *Ga1-s/Ga1-s* and could not set seed with *ga1* pollen. Kermicle (personal communication) has also found that when Maize Genetics Co-op's stock of *Ga1/ga1* is selfed, many of the progeny are cross-sterile, and this should not be a consequence of homozygosity for *Ga1*.

Two other observations bolster the doubts as to whether there is a *Ga1* allele. Jimenez and Nelson (1965) showed that the White Rice inbred, 4519–4, which reacted as though it were *Ga1/Ga1*, contained a *ga1* factor that had male-only action. Plants that are homozygous for this allele can fertilize plants homozygous for the *Ga1-s* allele, but do not select for *Ga1-s* pollen over *ga1* pollen nor for *Ga1-m* pollen over *ga1* pollen. In the tests described above (the ability to fertilize a *Ga1-s/Ga1-s* tester but be fertilized by *ga1* pollen) the effects of such an allele cannot be distinguished from those of a presumed *Ga1* allele. However, one finds 25% *su1/su1* kernels in the F_2 progeny of a cross between *Ga1-m Su1* and *ga1 su1* stocks. Jimenez and Nelson erroneously assigned this allele to *ga9*, a new fourth chromosome locus, which was the same distance from *su1* as *ga1* but on the other side of *su1*. Ashman (1981) showed that this was another allele of *ga1*, and my own unpublished data agree with this correction.

A third reason for suspecting that *Ga1* may not be a valid allele is that it has been shown that the *Ga1-s* alleles in certain genetic backgrounds condition the type of reactions described as being diagnostic of *Ga1* alleles. All Supergold popcorn inbreds are *ga1/ga1*. Attempts have been made to convert these inbreds to cross-sterility to eliminate contamination by dent (*ga1*) pollen when used as female parents in hybrid seed production fields. They have been crossed to a *Ga1-s/Ga1-s* inbred and backcrossed for a number of generations to the recurrent parent following the retention of the *Ga1-s* allele by testcrossing plants to a *Ga1-s/Ga1-s* tester. These procedures have resulted in converted inbreds that behave as though they were *Ga1/Ga1* (Ashman 1975). The Supergold inbreds clearly have a complex of modifying factors that are capable of attentuating the action of *Ga1-s* alleles. Ashman also demonstrated that no cytoplasmic entity was involved in the Supergold inbreds' modification of *Ga1-s* action.

The cited instances—showing that the *Ga1* alleles tested behave as *Ga1-s* alleles in the dent inbred backgrounds in which they have been placed or have a male-only action and that some popcorns possess factors that modify *Ga1-s* alleles so that they behave as the presumed *Ga1* alleles—do not demonstrate conclusively that there is not a *Ga1* allele. That demonstration can be made only by transferring the allele present in all the popcorn varieties that behave as *Ga1/Ga1* into the background of Hy, which has revealed that the *Ga1* alleles tested to date are *Ga1-s* in fact. There is reason to doubt the existence of the *Ga1* allele, but since the designation of an inbred or a variety as *Ga1/Ga1* indicates a unique set of reactions in the tests employed, I have used the designation in a descriptive sense throughout this paper.

There is a nonrandom distribution of the different alleles at the *ga1* locus. The *Ga1-s* alleles abound in the popcorns, which are considered to be primitive races of corn. North American dents and flints that have been tested have all been *ga1* (Nelson 1952), and the only North American corn tested that has been found to be other than *ga1/ga1* was Papago Indian corn from our Southwest. An extensive test of races of maize from Central America showed that many races were either *Ga1-s/Ga1-s* or *Ga1/Ga1* (Nelson 1960), but a few primitive races such as Palmero Toluqueno and Harinoso de Ocho were *ga1/ga1*. Some races had *ga1/ga1* plants as well as plants that were either *Ga1/*− or *Ga1-s/*−. These represent instances in which a race that was initially *ga1/ga1* is undergoing introgression of the *Ga1-s* allele. The observation that all North American corns tested were *ga1/ga1* requires that there be races of primitive maize of the same genotype for the reason noted earlier.

The observations that most popcorns and many Central American races of dent and flint maize are *Ga1* or *Ga1-s* has prompted the question of whether the teosintes share these alleles. There are several reports in the *MNL* of teosintes having *Ga1* or *Ga1-s* alleles. However, the most thorough investigation was that of Kermicle and Allen (1990). They found that five of six accessions of *Zea mays* ssp. *mexicana* and the one accession of ssp. *huehuetenangensis* tested were unable to set seed when pollinated with dent (*ga1*) pollen. Most plants in six accessions of *Z.m.* ssp. *parviglumis* had good seed sets with *ga1* pollen. The results with the Guatemalan teosinte, *Z. luxurians*, were mixed. Plants in three accessions set seed poorly with *ga* pollen while in three other accessions some plants set seed well. All the ssp. in which *ga* pollen did not effect fertilization or did so poorly are teosintes that grow in or on the edges of maize fields. The fact that one accession of ssp. *mexicana* set seed with *ga* pollen as did some plants of *Z. luxurians* suggests that their genotype is *ga1/ga1*. This suggests further that the original genotype of the species was *ga/ga* and that an allele similar to *Ga1-s* is introgressing into the population from the sympatric maize. In the case of the *mexicana* accessions, the introgression is complete while the process is ongoing in the *Z. luxurians* populations.

The genetic basis of the inability to set seed with *ga1* pollen was investigated for two of the ssp. *mexicana* accessions following incorporation of the factors conditioning cross-sterility into the dent inbred, W22, by successive backcrosses. The cross-sterility of the Chalco accession was shown to be conditioned by a *Ga1-s* allele. The basis of the cross-sterility of the Central Plateau 48703 accession is more complicated. It involves a complex of chromosome 4 factors designated as TIC-CP (teosinte incompatibility complex–Central Plateau) that render maize plants heterozygous for the complex incapable of setting seed either with *ga1* or *Ga1-s* pollen. One factor is an allele of *ga1*, *Ga1-m:CPT*, that has male-only action as described by Jimenez and Nelson (1965) for the *Ga1-m* allele found in the White Rice popcorn inbred, 4519–4. This component of the complex permits plants that are heterozygous or homozygous for TIC-CP to fertilize *Ga1-s/Ga1-s* plants. A second factor, CP2, is located 4 map units distal to the *su1* locus (the *ga1* locus is 36 map units distal to *su1*), and its presence permits pollen to function on silks carrying TIC-CP. A third linked factor is apparently required (either alone or in combination with CP2) to constitute the stylar barrier in TIC-CP. These incompatibility factors and displacement of the flowering periods apparently account for the low frequency of maize-teosinte hybrids in fields where the two ssps. are growing together. To be certain of this conjecture one would need to know also the incompatibility factors present in the sympatric maize.

Rashid and Peterson (1992) have identified another case of nonreciprocal cross-sterility that is also under multifactorial control but is a sporophytic incompatibility system. Plants that are *cif/cif* (*cross-incompatible female*) have very poor seed set (<25 kernels per ear) with pollen from plants that are *cim1/cim1; cim2/cim2* (*cross-incompatible male*).

The *gametophyte factors* of maize (especially *ga1*) have raised interesting questions that have not been answered. The basis(es) of the pollen tube/stylar interactions that are crucial in cross-incompatibility reactions or selective fertilization should be accessible for the various *ga*'s by molecular techniques. Whether the *Ga1-s* allele and other systems providing unilateral isolating mechanisms have played a role in the evolution of the *Zea* ssps. or whether they and the *gametophyte factors* conditioning selective fertilization are simply prime examples of selfish DNA may be more difficult to discern.

REFERENCES

Ashman RB (1975) Modification of cross-sterility in maize. J Hered 66: 5–9

Ashman RB (1981) Failure to verify the *Ga9* locus on chromosome 4. Maize Genetics Cooperation News Letter 55: 50–51

Brieger FG (1937) Genetic control of gametophyte development in maize II. A gametophyte character in chromosome five. J Genet 54: 57–80

Brunson AM (1937) Popcorn breeding. USDA Yearbook of Agriculture: 395–404

Correns C (1902) Scheinbare Ausnahmen von der Mendel'schen Spaltungsregel fuer Bastarde Ber. Dtsch Bot Ges 20: 157–159

Demerec M (1929) Cross-sterility in maize. Z Indukt Abstamm Verebungsl 50: 281–291

Emerson RA (1925) A possible case of selective fertilization in maize hybrids. An Rec 29: 136 (Abstract)

Emerson RA (1934) Relation of the differential fertilization genes, *Ga ga*, to certain other genes of the *Su-Tu* linkage group of maize. Genetics 19: 137–156

House LR, Nelson OE (1958) Tracer study of pollen-tube growth in cross-sterile maize. J Hered 49: 18–21

Jimenez JR, Nelson OE (1965) A new fourth chromosome gametophyte locus in maize. J Hered 56: 256–263

Kermicle JL, Allen JO (1990) Cross-incompatibility between maize and teosinte. Maydica 35: 399–408

Mangelsdorf PC, Jones DF (1926) The expression of Mendelian factors in the gametophyte of maize. Genetics 11: 423–455

Nelson OE (1952) Non-reciprocal cross-sterility in maize. Genetics 37: 101–124

Nelson OE (1953) A genic substitute for isolation in hybrid corn seed production. Econ Bot 7: 382–384

Nelson OE (1954) Gametophyte factors in the "standard exotics." Maize Genetics Cooperation News Letter 28: 36–37

Nelson OE (1960) The fourth chromosome factor in some Central and South American races. Maize Genetics Cooperation News Letter 34: 114–116

Rashid A, Peterson, PA (1992) The RSS system of

unidirectional cross-incompatibility in maize: I. Genetics. J. Hered. 83: 130–134.
Rhoades MM (1948) Gametophyte factor in chromosome 3. Maize Genetics Cooperation News Letter 22: 9–10
Schwartz D (1950) The analysis of a case of cross-sterility in maize. Proc Natl Acad Sci USA 36: 719–724

79

Ring Chromosomes

JAMES A. BIRCHLER

Ring chromosomes were first documented in maize by McClintock (1932). Because they undergo sister chromatid exchange and exist as a circular molecule, they often form dicentric chromosomes that are fractured by the cell wall formation between daughter cells. Consequently, they are unstable in size and chromatin content. It is this property that makes them so useful in genetic mosaic studies. Whenever a mutant is heterozygous with a ring chromosome, the instability makes sectors of tissue in which the ring or at least the normal allele is lost. Observations of such plants can reveal the autonomy of the mutant during development as well as the pattern of cell divisions.

The instability requires that the full size ring be selected repeatedly by the investigator. This apparently is difficult when particularly large rings are the starting material (G. Neuffer, personal communication). Also, for reasons that are not clear, there are apparently some small rings that become stable and therefore will no longer be suitable for mosaic analysis.

Ring chromosomes have typically been induced by ionizing irradiation (Schwartz 1953). There are also two reports noting that ring chromosomes form spontaneously from B-A translocations (Ghidoni 1973; Carlson 1973). The situation under which they were observed was when a tertiary trisomic (A A B-A) was used as a male parent onto a tester stock. Kernels exhibiting a mosaic endosperm phenotype were subsequently shown to possess a ring chromosome for the respective chromosome arm. The basis for this behavior is unknown as is whether it also occurs with other B-A translocation genotypes. The pollen tubes of A B-A genotype typically grow too slowly to achieve fertilization in

The Maize Handbook—M. Freeling, V. Walbot, eds.

competition with the normal tubes that will also be present. The ring chromosomes formed may delete some genetic material, leading to a more rapid rate of pollen tube growth and fertilization success. Rings might therefore be formed at a very low frequency, but would be selected for, when they do occur. Such an explanation might account for the recovery of ring chromosomes in the progeny of tertiary trisomics but not in other genotypes.

REFERENCES

Carlson WR (1973) Instability of the maize B chromosome. Theor Appl Genet 43: 147–150

Ghidoni A (1973) Changes in the structure of B-A chromosomes in maize. Theor Appl Genet 43: 151–161

McClintock B (1932) A correlation of ring-shaped chromosomes with variegation in *Zea mays*. Proc Natl Acad Sci USA 18: 677–681

Schwartz D (1953) The behavior of an X-ray induced ring chromosome in maize. Am Nat 87: 19–28

80

In situ Hybridization of DNA and RNA Probes to Maize Chromosomes

S.M. LIVINGSTON AND R.L PHILLIPS

The cytogenetic mapping method of in situ hybridization involves fixation of tissue samples, DNA denaturation in situ, incubation of the tissue with radioisotope-labeled DNA or RNA probes, removal of nonhybridized radioactive material in order to limit background signal, and finally, detection of the radioactive molecular hybrids using autoradiography.

The *Zea mays* L. meiotic cell stages of diakinesis and pachynema are particularly useful. In diakinesis, the chromosomes are so condensed that it is fairly simple to determine the general site of hybridization and the approximate size of the involved chromosome. In pachynema, maize chromosomes can be identified by morphological features such as

The Maize Handbook—M. Freeling, V. Walbot, eds.

total length, arm ratio, centromere placement, heterochromatic knobs, nucleolus organizer, and satellite regions. Using these "landmarks," a probe can be localized to a specific area of a known chromosome. Translocations can be used effectively to confirm the chromosome location of a specific sequence (Mascia et al. 1981).

PROCEDURES

1. Fix freshly collected microsporocyte tissue in 95% ethanol:glacial acetic acid (3:1). Maintain tissue at room temperature for 2 days; then store at −20°C.
   Some researchers change the fixative to 70% ethanol after a few days to weeks, but we have noticed increased tissue brittleness and degradation of cell morphology as well as decreased specific hybridization using this technique.
   It is important to use the tissue as soon as possible after fixation. Long exposure of tissue to fixative has been shown to have adverse effects on hybridization (Brigati et al. 1983).
2. Use one of the primary anthers in a floret to identify the appropriate meiotic stage. Maintain the other anthers in fixative until used.
3. Squash (3–5) anthers in 15 μl 45% acetic acid on each acid-cleaned slide. Remove as much excess anther tissue as possible with tweezers. Apply an acid-cleaned coverslip. Place slides in petri dishes on acetic-acid-dampened filter paper. Store overnight at 4°C.
   Acetic acid in the fixative and squash solution helps remove basic proteins that can interfere with hybridization (Dick and Johns, 1967; Gillespie and Spiegelman, 1965).
4. Apply light pressure to coverslips. Place slides, coverslip-side up, on dry ice for 10–15 minutes.
5. Pop off coverslips with a razor blade. Store slides in 2 × SSC buffer (0.3 M NaCl, 0.03 M sodium citrate, pH 7.4).
6. Incubate slides in 0.2 mg/ml RNAse A in 2 × SSC buffer for 1 hour at 37°C.
   The RNAse treatment helps rid cell preparations of endogenous RNA and reduces nonspecific binding of the input probe.
   First, a concentrated RNAse stock (20 mg/ml) made with double distilled water should be heated to 80°C for 10 minutes. This removes any contaminating DNAse. The stock is frozen in aliquots and diluted with sterile 2 × SSC just prior to use.
7. Wash slides in 2 × SSC for 5 minutes (three times).
8. Denature DNA in situ in either 0.2 N HCl for 20 minutes or in 0.07 N NaOH for 2 minutes.
   For accurate localization of a probe sequence, chromosome morphology must be preserved; consequently, the denaturation step is a critical one. Denaturation can be accomplished in a variety of ways: treatment with NaOH (Gall and Pardue 1969),

HCl (Macgregor and Kezer 1971), heat (John et al. 1969), high concentrations of formamide, or combinations of both heat and formamide (Jones and Robertson 1970; Wimber and Steffenson 1970).

In our experience, HCl denaturation maintains chromosomal distinctiveness better than high heat and formamide, and acid is a less-damaging procedure than using NaOH. Some researchers have concluded that using HCl as a denaturant limits the efficiency of hybridization (Jones 1973), but we have not found this to be the case using maize microsporocyte tissue.

9. Serially wash slides in 30%, 60%, 70%, 95%, and absolute ethanol (3 minutes per wash at 30% and 60%; 5 minutes per wash at 70%, 95%, and 100%). Air dry.

10. Apply 25 μl per slide of prehybridization mixture: 50% formamide, 50 μg denatured salmon sperm (50 μg yeast tRNA if using an RNA probe) and buffer to obtain a final concentration of 2 × SSC. Apply coverslip and incubate in a humid chamber at hybridization temperature for 2–3 hours. Flooding squashes with large amounts of nonhomologous DNA or RNA reduces nonspecific binding of the radioactive probe and thereby lessens background.

    We maintain the same incubation temperature during prehybridization and hybridization. Generally, this is about 42°C for DNA probes and about 60°C for RNA probes. Optimum hybridization temperatures vary with different probes and target tissues. Usually, DNA/DNA reassociation (Tr) occurs about 25°C below the melting temperature (Tm) of the duplex (Marmur and Doty 1962; Wetmur and Davidson 1968) while the Tr and Tm of an RNA/DNA hybrid are similar (Nygaard and Hall 1964).

11. Serially ethanol wash slides. Air dry.

12. Apply 15 μl hybridization mixture per slide: 10 ng radiolabeled RNA probe, or 10–15 ng heat-denatured, radiolabeled DNA probe, 50% formamide, 40 μg yeast tRNA for an RNA probe, or 40 μg heat-denatured salmon sperm DNA for a DNA probe, and buffer to a final concentration of 2 × SSC. Dextran sulfate (10%) can also be used to increase probe network formation; this allows visualization of hybridization sites in a shorter time (Harper and Saunders 1981; Gerhard et al. 1981). On the other hand, dextran sulfate can increase background. Apply coverslips and incubate slides in a humid chamber overnight at the hybridization temperature.

13. Allow slides to reach room temperature and wash them in 2 × SSC for 5 minutes (two times). Discard these washes in a radioactive waste container.

14. If an RNA probe is used, incubate slides in 0.2 mg/ml RNAse A in 2 × SSC for 1 hour at 37°C. This treatment will eliminate any nonhybridized radioactive probe.

15. Wash slides in 50% formamide, 50% 4 × SSC at 37° C for 15–30 minutes.

    The stringency of this wash will vary depending upon probe homology and copy number of target sequence. Higher percent formamide, lower salt concentration, and higher temperature all contribute to increased stringency. Tm is reduced 0.72°C per 1% formamide in solution (McConaughy et al. 1969) and may vary by 30°C over a salt concentration range of 0.0 to 1.0 M according to Marmur and Doty (1962).

16. Wash slides in 2 × SSC for 5 minutes (three times) at room temperature.

17. Volume wash slides in 2 × SSC at room temperature for 1–2 hours. This helps reduce background.

18. Heat Kodak NTB2 Nuclear Track Emulsion (previously diluted 2 parts sterile distilled water: 1 part emulsion) at 45°C for 1 hour. The emulsion must be thoroughly melted so a thin coat can be applied to the slides. Note: The emulsion should never be shaken or overheated. This can cause formation of background silver grains.

19. Dip slides into emulsion in a darkroom equipped with a Kodak No. 2 Wratten filter safelight with a 15-W bulb. Let slides dry upright for 20 minutes before storing them in a light-tight slidebox with desiccant. Put black electrical tape on the seams of the slidebox before wrapping it in aluminum foil. Store at 4°C until time for development.
    Exposure time can vary from a few hours to months depending upon the size and copy number of the target sequence as well as the specific activity and homology of the probe.

20. Develop slides in Kodak D19 developer for about 45 seconds. The development time varies according to specific activity of the probe and the time of exposure. In our experience, development time can vary from 5 seconds up to about 4 minutes.

21. Dip slides in stop solution (1.5% glacial acetic acid) for 30 seconds.

22. Soak slides in Kodak fixer for 2 minutes.

23. Wash slides in running tap water for 5 minutes.

24. Stain slides with freshly made 4% Giemsa in 0.01 M sodium phosphate buffer, pH 6.8, for 8–10 minutes. Destain slides with the same phosphate buffer. Air dry. Add a drop of Permount or Euparol mounting solution and cover with an acid-cleaned coverslip.

From the time of Gall and Pardue's descriptive report in 1969, the techniques of in situ hybridization have been increasingly refined to allow greater sensitivity in hybrid detection.

In 1981, Mascia and co-workers localized the high copy number 5S ribosomal RNA genes to the long arm of chromosome 2. The same year, Peacock et al. isolated a highly repeated 185 bp sequence homologous to knob heterochromatin in different maize stocks. In our laboratory, the methods used by Phillips and Wang (1982) enabled the mapping of the repetitive 17S-26S ribosomal RNA genes to the nucleolar organizing region of chromosome 6S. Hybridization efficiency has reached the point where it is now possible to visualize a single-copy gene on a maize pachytene chromosome (Shen et al. 1987) using ^{3}H as the radionuclide.

Several researchers have developed nonradioactive procedures as well to identify single copy gene sequences (Leary et al. 1983; Landegent et al. 1985).

REFERENCES

Brigati DJ, Myerson D, Leary JJ, Spalholz B, Travis SZ, Fong CKY, Hsiung GD, Ward DC (1983) Detection of viral genomes in cultured cells and paraffin-embedded tissue sections using biotin labeled hybridization probes. Virology 126 (1): 32–50

Dick C, Johns EW (1967) The removal of histones from calf thymus deoxyribonucleoprotein and calf thymus tissue with acetic acid-containing fixatives. Biochem J 105: 46P

Gall JG, Pardue ML (1969) Formation and detection of RNA-DNA hybrid molecules in cytological preparations. Proc Natl Acad Sci USA 63: 378–383

Gerhard DS, Kawasaki ES, Bancroft FC, Szabo P (1981) Localization of a unique gene by direct hybridization *in situ*. Proc Natl Acad Sci USA 78: 3755–3759

Gillespie D, Spiegelman S (1965) A quantitative assay for DNA-RNA hybrids with DNA immobilized in a membrane. J Mol Biol 12: 829–842

Harper ME, Saunders GF (1981) Localization of single copy DNA sequences on G-banded human chromosomes by in situ hybridization. Chromosoma 83(3): 431–439

John HA, Birnstiel ML, Jones KW (1969) RNA-DNA hybrids at the cytological level. Nature 223: 582–587

Jones KW (1973) The method of *in situ* hybridization. In Pain RH, Smith BJ (eds) New Techniques in Biophysics and Cell Biology, Vol 1, John Wiley and Sons, London, pp 29–66

Jones KW, Robertson FW (1970) Localisation of reiterated nucleotide sequences in Drosophila and mouse by *in situ* hybridisation by complementary RNA. Chromosoma 31(3): 331

Landegent JE, Jansen in de Wal N, van Ommen G-JB, Baas F, de Vijlder JJM, van Duijn P, van der Ploeg M (1985) Chromosomal localization of a unique gene by non-autoradiographic *in situ* hybridization. Nature 317: 175–177

Leary JJ, Brigati DJ, Ward DC (1983) Rapid and sensitive colorimetric method for visualizing biotin-labeled DNA probes hybridized to DNA or RNA immobilized on nitrocellulose: Bioblots. Proc Natl Acad Sci USA 80: 4045–4049

Macgregor HC, Kezer J (1971) Chromosomal localization of a heavy satellite DNA in the testis of Plethodon c. cinereus. Chromosoma 33(2): 167

Marmur J, Doty P (1962) Determination of the base composition of DNA from its thermal denaturation temperature. J Mol Biol 5: 109–118

Mascia PN, Rubenstein I, Phillips RL, Wang AS, Xiang LZ (1981) Localization of the 5S ribosomal RNA genes and evidence for diversity in the 5S RDNA region of maize. Gene 15 (1): 7–20

McConaughy BL, Laird CD, McCarthy BJ (1969) Nucleic acid reassociation in formamide. Biochemistry 8 (2): 3289–3295

Nygaard AP, Hall BD (1964) Formation and properties of RNA-DNA complexes. J Mol Biol 9: 125–142

Peacock WJ, Dennis ES, Rhoades MM, Pryor AJ (1981) Highly repeated DNA sequence limited to knob heterochromatin in maize. Proc Natl Acad Sci USA 78 (7): 4490–4494

Phillips RL, Wang AS (1982) In situ hybridization with meiotic cells. In Sheridan WF (ed) Maize for Biological Research, Plant Mol Biol Assoc, Charlottesville, VA, pp 121–122

Shen DL, Wang ZF, Wu M (1987) Gene mapping on maize pachytene chromosomes by *in situ* hybridization. Chromosoma 95 (5): 311–314

Wetmur JG, Davidson N (1968) Kinetics of renaturation of DNA. J Mol Biol 31: 349–370

Wimber DE, Steffenson DM (1970) Localization of 5S RNA genes on Drosophila chromosomes by RNA-DNA hybridization. Science 170: 639

Analysis of Traits With Complex Inheritance in Maize Using Molecular Markers

Tim Helentjaris and Manfred Heun

A recent outgrowth of the efforts to develop RFLPs and other molecular marker systems has been their use to analyze traits with inheritance patterns indicative of a multigenic nature. "QTL (quantitative trait locus) mapping," as it is often referred to, has provided new insight into a branch of genetics that has otherwise missed out on the advances of the molecular genetic revolution. This omission has been unfortunate as many of our most important agronomic or scientifically interesting traits, such as plant height, seed set, stress tolerance, maturity, disease resistance, and yield, require the products of a large number of genes to create the eventual phenotype. Phenotypic variation for each of these traits is probably controlled by a much smaller subset of these genes and molecular marker analysis can be used to determine both the genomic locations of the major factors and their effective impact upon this variation.

While the statistical methods usually used to analyze these types of traits may seem somewhat foreign to both "Mendelian" and molecular geneticists, in fact, the general approach to mapping QTLs is not significantly different from mapping phenotypic variants that result from the action of single genes. A population is constructed that is segregating for the trait of interest, progeny are assayed for both phenotypic values and for molecular marker constitution at different chromosomal sites, and finally the two are compared for correlation. The primary difference is that with single gene traits most of the variation can be attributed to a single marker locus (the remaining value may be attributed to recombination between the gene and the marker locus); in contrast, with quantitative traits, the genetic variance must be partitioned among several marker loci. To achieve this requires that one must use more segregating progenies and collect replicated phenotypic data. Hence the experiments are larger and the statistical analyses more complicated than with single gene traits.

Nevertheless, a standard approach can be defined. First, a population that is segregating

The Maize Handbook—M. Freeling, V. Walbot, eds.

for the trait of interest must be selected. Several population types are appropriate: F_2 plants, bulked F_3 families, backcross plants, recombinant inbred lines (RIs), etc. All of these types of segregating populations contain approximately the same amount of information and can be used; however, some experimental designs (F_3s, RIs) are preferable to others (F_2s, BCs) as they are more amenable to replication of the phenotypic analysis. The desire for replication can also be satisfied with the latter group by testcrossing the individual progeny to an unrelated tester line, which can then provide important data on the combining ability of the genotypes being examined. A much more important consideration is the choice of the parents used to create the population. With single gene traits it is not usually appreciated that one is essentially mapping the variation between the two parental phenotypes, as one parent is "mutant" and the other is any "wild-type" line. With quantitative trait analysis, the choice of parents is much more important, as they are usually both considered "wild type," differing metrically for the trait of interest. Hence, it should be remembered that one is not mapping all of the genes that are important for the expression of a phenotype but simply a much smaller class of loci whose alleles differ between the parents and where that difference has a significant impact on phenotypes of the progeny. Two parents that differ as much as possible for the trait of interest allow determination of the locations of major genes in that particular cross, but these loci may or may not be of significance in other crosses. Because of the ease of maize RFLP analysis, it has proven practical in most cases to actually analyze those crosses that are of most economic or scientific interest and to not resort to very extreme materials.

The number of progeny needed for analysis can be determined by how complex the trait is (i.e., how many loci will individually make a measurable contribution to phenotypic variation) and how accurately the phenotype of individuals can be determined. There are published protocols for estimating the number of individuals needed to detect loci accounting for a given percentage of total variance (Lander and Botstein 1989). These methods are moderately useful as guidelines, but it is probably not necessary to design an experimental plan based upon these types of estimates. For most analyses where reliable phenotypic values can be measured, 100 individual genotypes would be considered an acceptable minimum population, with more progeny providing a more reliable estimate of the locations and phenotypic values of important QTLs. Population sizes above a few hundred progeny probably do not provide much additional insight for the effort required for their analysis. If the phenotypic analysis is not too difficult in relation to the genotypic analysis, it has been pointed out that most of the information within a population is contained in those members exhibiting the most extreme phenotypes (Lander and Botstein 1989). Consequently one can first "phenotype" a population and then select only the "tails" for subsequent genotypic analysis. This can significantly reduce the amount of genotypic analysis required to locate accurately the most important QTLs. The penalty is that a less-reliable estimate of the actual value of any one region is obtained, but if necessary this can be remedied by subsequently including more individuals from the remaining segments of the populations.

As a practical guide we recommend the following: (1) construct a population of 100 to a few hundred members, (2) phenotype them, (3) select the extreme members (bottom and top 10%) for genotypic analysis, (4) determine if this particular population will provide reasonable information as to the locations and numbers of QTLs, and then (5) genotype a subset of the remaining members to gain a more accurate estimate of the phenotypic value of individual QTLs. This strategy has the added advantage of allowing the investigator to use only those markers in the subsequent analysis that were determined to be important in the first.

Once the population has been constructed and phenotypically analyzed, the next step is to analyze the members genotypically with molecular markers. Because of the intrinsically high frequency of polymorphism within maize, identifying informative markers is trivial unless the two parents selected are highly related by pedigree. Hybridization of probes against DNAs prepared from both parents and digested with three restriction enzymes is usually sufficient to define a set of informative markers. The number of markers to be used can be determined mathematically, but again, a practical guide will often suffice. It is not usually appreciated that higher numbers of markers do not return much for the effort expended for their analysis. In most populations, use of only a few markers per chromosome arm in an initial analysis will reveal whether important factors exist on that arm. Subsequent use of markers no closer than 20 cM in these regions will probably then be sufficient to locate and to assess the value of these loci. Increased numbers of markers will not provide any more resolution as to the location of QTLs, as this is essentially limited by the amount of recombination within the population (i.e., number of individuals tested). If higher resolution of the positions of QTLs is required, more individuals can be added, but the effort required increases dramatically. Perhaps more practical in some instances is to use individuals where recombination has occurred for several generations. For example, Paterson et al. (1990) have demonstrated that "progeny of progeny" can be used in this strategy. Many more markers targeted at the region of interest are required in this instance, but a much better assignment of a QTL position can be obtained.

Once both the phenotypic and genotypic data are obtained, the next step is to compare the two for correlations. There has been quite a bit of discussion regarding the best method for analyzing the data, but for the most part the distinctions between the methods are irrelevant. As these experiments are so structured, in that one usually utilizes homozygous parents and replicated progeny, all of the methods of analyses appear to yield essentially the same result, when they have been compared. The most straightforward approach is to simply run an analysis of variance (ANOVA), using statistical packages such as SAS or SYSTAT, to determine if the phenotypic means of the classes defined by each marker genotype are significantly different. If they are, above some specified level of significance, then those particular markers can be presumed to be located near a gene contributing to the phenotypic variance. Accepted minimum levels of significance should probably be set at least at the 0.01 level or even better at the 0.001 level. A recently developed program using multipoint analysis and maximum likelihood, Mapmaker-QTL (Lincoln and Lander 1989), represents an improved method for analyzing this type of data, but it does suffer in that it cannot handle certain population types and also does not allow for replication.

The culmination of the analysis is a table or plot displaying the effects of allelic substitution at areas of the genome on the phenotypes of progeny created from two specific parents. A few cautions about analyzing these data are probably in order. First, the quality of the final result is no better than the phenotypic data; consequently, this aspect of the analysis cannot be overstressed. Investigators often overlook this consideration in their concern over the lab aspects of RFLP analysis, but the problems of phenotypic analysis do not become apparent until the end of the analysis, if at all. Second, the model derived from one experiment should not be considered absolute, as a model is fitted to the data obtained. Consequently these models often seem to fit the data set "too well." A second data set should be used to verify that the model was not specific for the first set alone but can be generalized across replications. Past experience has shown that for the most part, the major loci defined in one replication will emerge as significant in replications, but their relative values may change somewhat. Third, it should be remembered that one has not defined all of the genes that are important for a particular trait but simply those regions where the two parents differ and that also possess the potential for creating significant impacts upon a phenotype. Other crosses should be tested to identify additional genomic regions that exert meaningful effects. Fourth, the use of the term "gene" when defining a QTL by this method is inappropriate, as one is more accurately defining a genomic region, and a broad one at that, that exerts an impact upon the phenotypic variation. Confirmation of the involvement of a single gene within that region will require significantly more analyses; it may be appropriate to consider a single gene as a starting hypothesis during further examination. Finally, the assignment of precise locations of QTLs should be considered as very approximate given the usual experimental design. With the random segregation of the rest of the genome and the effect of multiple loci on the final phenotype, the location of any one QTL in a single experiment should be viewed as having a variance of perhaps as much as ±20 cM. More precise localization or even definition to a single locus requires that a large part of the remaining genome be fixed such that the exact effect of a single region can be assayed more precisely. The use of near-isogenic lines (NILs) offers the best prospect in this regard.

Some additional considerations are worth mentioning. Although this discussion has centered on RFLPs, there is no significant difference between these markers and use of other molecular markers such as isozymes (Stuber et al. 1982) or PCR with random primers (Williams et al. 1991). Data generated from any of these markers can be easily integrated together; the investigator should be aware of and utilize the most appropriate methods in their own application. Additionally, the experimental designs described here are just a guide; other types of populations may be more appropriate for particular traits and offer significant advantages. Analysis of groups of inbreds only partially related by pedigree or of diallels may provide much better insight into traits such as agronomic yield or combining ability than the use of segregating progeny.

One of the exciting prospects of this type of analysis in maize is the possibility of linking up major QTLs with known genes. As maize is blessed with an excellent set of mapped mutant loci, which is likely to grow with current research efforts, it is possible as pointed out by Robertson (1985), that we may be able to correlate quantitative and qualitative variation at single loci. As it seems extremely difficult to envision cloning a gene contributing only 10% of the variance to a phenotype (although it might be quite important, scientifically or economically), relating important quantitative alleles with those of more extreme effects would seem to address this dilemma. Consequently, it may not be so far-fetched to think about cloning QTLs or even using those clones in genetic engineering strategies for crop improvement. The possibility of linking up the results of previous quantitative studies with molecular, Mendelian, and cytogenetic efforts holds great potential not only for the quantitative geneticist but also for the rest of the maize genetics community.

REFERENCES

Lander ES, Botstein D (1989) Mapping mendelian factors underlying quantitative traits using RFLP linkage maps. Genetics 121: 185–199

Lincoln SE, Lander E (1989) Mapping genes controlling quantitative traits with mapmaker/QTL. Whitehead Institute for Biomedical Research Technical Report, Cambridge, MA

Paterson AH, DeVerna JW, Lanini B, Tanksley SD (1990) Fine mapping of quantitative trait loci using selected overlapping recombinant chromosomes in an interspecies cross of tomato. Genetics 124: 735–742

Robertson DS (1985) A possible technique for isolating genic DNA for quantitative traits in plants. J Theor Biol 117: 1–10

Stuber CW, Goodman MM, Moll RH (1982) Improvement of yield and ear number resulting from selection at allozyme loci in a maize population. Crop Science 22: 737–740

Williams JGK, Kubelik AR, Livak KJ, Rafalski JA, Tingey SV (1991) DNA polymorphisms amplified by arbitrary primers are useful as genetic markers. Nucl Acids Res 18: 6531–6535

82

Heterofertilization

JAMES A. BIRCHLER

Ordinarily, in the process of double fertilization, the pollen tube enters the synergid and releases two sperm. One of these will join with the egg cell to form the zygote, which develops into the embryo and associated scutellum. The other sperm joins with the polar nuclei to form the triploid primary endosperm nucleus, which develops into the endosperm. This results in a concordance of genotype between the embryo and endosperm; this allows many characters expressed in the endosperm to be scored without growth of the F_1. For this reason, the persistent endosperm provides a major advantage to maize as a genetic organism—it eliminates a generation time of growth in many analyses and allows manipulation and foresight before planting.

It should be noted, however, that heterofertilization occurs in approximately 1–2% of all fertilizations (Sarkar and Coe 1971). Heterofertilization occurs when two pollen grains contribute to the double fertilization; the egg and polar nuclei are fertilized by sperm from different pollen grains (Sprague 1932). If the male parent is heterozygous, this could present a problem if one is mapping extremely closely linked endosperm traits relative to seedling or plant characteristics. Perhaps the most crucial case involves testing recombination between alleles. Heterofertilization would mimic a low level of recombination, whereas in fact there may be much lower levels or none at all if one allele is an intragenic deletion. Generally it will not matter for more loosely linked factors, where a 1% misclassification will not affect the results. Of course, using the heterozygote as a female will eliminate the problem altogether.

Another instance in which heterofertilization should be kept in mind involves classification of dosage in crosses with B-A translocations. Certain expected classes such as hyperploid heterozygotes have nonconcordant phenotypes. If, however, the male parent that contributes the B-A translocation is itself a hyperploid that carries two different markers, nonconcordant phenotypes might occasionally result from heterofertilization. Because the translocation must be used as a male, some very low frequency of kernels might mimic the hyperploid heterozygote phenotype but in fact be some other genotype. The frequency of this problem is so low as to be trivial for most concerns, but if the absolute accuracy of classification is required, other characteristics besides the phenotype should be used.

The Maize Handbook—M. Freeling, V. Walbot, eds.

[Editors' Note: A third instance of nonconcordance is not from heterofertilization. If transposable element insertion or excision events (in a reporter allele or gain or loss of the regulatory element) occur after the mitotic division of the generative cell, the two sperm will be of different genotypes. As with B-A translocations, heterofertilization could be confused with events in individual sperm of a single pollen grain. Using the transposable element stock as female will eliminate heterofertilization and allow detection of events after separation of the egg and polar nuclei lineages. The frequency of transposon events in the mega- and microgametophyte may not be equivalent.]

REFERENCES

Sarkar KR, Coe EH Jr (1971) Analysis of events leading to heterofertilization in maize. J Hered 62: 118–120

Sprague GF (1932) The nature and extent of hetero-fertilization in maize. Genetics 17: 358–368

IV
Molecular Biology Protocols

Isolation of Small Nuclear RNAs

TAMÁS KISS AND WITOLD FILIPOWICZ

For isolation of small nuclear (sn)RNAs from maize leaf tissue, we use a slightly modified procedure described originally by Kiss et al. (1985) for dicot plants. This procedure consists of two basic steps: (1) purification of maize nuclei and (2) extraction of snRNAs.

PURIFICATION OF NUCLEI FROM MAIZE LEAVES

Endonucleolytic degradation of the abundant high-molecular-weight RNAs (mainly ribosomal RNAs) during the nuclei isolation procedure can result in formation of small RNA fragments that can interfere with the detection of authentic snRNAs. The method given below describes how to obtain maize nuclear preparations with minimal RNA degradation.

1. Soak maize seeds in deionized water for 5–6 hours; sow and grow them under ordinary greenhouse conditions, until the leaves reach the size of about 15 cm.
2. Harvest the leaves (100–300 g) and rinse with cold distilled water. Do all the following steps at 4°C.
3. Homogenize the tissues in 10 volumes of ice-cold homogenization buffer containing 5% (w/v) citric acid, 250 mM sucrose, and 0.25% (v/v) Triton X-100 using a Waring blender at medium speed with 5 pulses of 20 seconds each. The homogenizer should be fitted with newly sharpened blades. For smaller-scale preparations, the Ultra-Turrax and Polytron homogenizers work fine as well.
4. Filter the homogenate through 4–5 layers of cheesecloth and two nylon filters (Züricher Beuteltuchfabric AG, Rüschlikon, Switzerland) of decreasing pore size (67 μm and 20 μm).
5. Centrifuge the filtrate in a Sorvall GSA-3 rotor at 5,000 rpm (4,000*g*) for 10 minutes.
6. Gently resuspend the pellet with a soft paintbrush in 100 ml of homogenization buffer.
7. Sediment the nuclei at 4,500 rpm (3,500*g*) in a Sorvall HS-4 rotor for 10 minutes.

The Maize Handbook—M. Freeling, V. Walbot, eds.

8. Gently resuspend the nuclei with a paintbrush in 40 ml of homogenization buffer and centrifuge as in step 7. At this stage the nuclear pellet should be completely free of contaminating chloroplasts (green color). If not, repeat steps 6 and 7.

Nuclei obtained by the above procedure are suitable for direct extraction of nuclear RNAs. Because of the extremely high amount of contaminating starch grains found in crude nuclear preparations from some dicot plants, it is necessary to further purify the nuclei by an additional Percoll density gradient centrifugation (Kiss et al. 1985). In the case of maize leaves we find this step to be unnecessary. Nuclei obtained by the procedure described in this chapter, although containing slightly more starch grains, do not appear to be more contaminated with cell debris or other organelles, as indicated by iron-acetocarmine staining and light microscopy.

EXTRACTION OF RNA FROM PURIFIED MAIZE NUCLEI

This procedure is a modification of the method described by Steele et al. (1965) for the extraction of RNA from isolated nuclei of rat liver.

1. Add to the nuclear pellet 10 volumes of RNA extraction buffer containing 50 mM sodium acetate, 140 mM sodium chloride, 0.3% sodium dodecyl sulfate, pH 5.1, at room temperature (RT). Lyse nuclei by homogenization with a loosely fitting Teflon pestle.
2. Add 1 volume of water-saturated phenol (not pH-adjusted) to the highly viscous nuclear homogenate. Vortex for 1 minute at RT and transfer to a 65°C water bath and shake vigorously for an additional 10 minutes.
3. Chill in an ice/water bath for 2–3 minutes and centrifuge at 8,000*g* for 10 minutes at 4°C.
4. Transfer the water phase to a tube containing a half volume of water-saturated phenol (not pH-adjusted). Vortex for 5 minutes at room temperature. Centrifuge at 8,000*g* for 10 minutes at 4°C.
5. Transfer water phase to a fresh tube, add 2.5 volumes of ice-cold ethanol, and mix thoroughly. Leave at −20°C for at least 1 hour.
6. Recover nucleic acids by centrifugation at 10,000*g* for 10 minutes at 4°C. Discard the supernatant and let the pellet of nucleic acids dry at room temperature for 5–10 minutes.
7. Dissolve the pellet in 200–400 μl of 0.3 M sodium acetate, pH 5.1, and transfer to a 1.5-ml eppendorf reaction tube. Centrifuge at 12,000*g* for 15 minutes.
8. Transfer the supernatant to a fresh tube and add 2.5 volumes of cold ethanol. Mix and chill at −20°C for 10 minutes. Recover RNA by centrifugation at 12,000*g* for 5 minutes in a microfuge.

9. Wash the pellet with 70% ethanol, recentrifuge for 5 minutes, and allow RNA pellet to dry at room temperature.
10. Redissolve the RNA in a small volume of water. Store at −20°C. The quality of RNA preparation can be assayed by electrophoresis on a denaturing 10% polyacrylamide–8 M urea gel, followed by ethidium bromide staining (Kiss et al. 1985). RNA preparations obtained from maize leaves by the above procedure contain all the major snRNA species that have been characterized in mammalian and plant cells. (See for reviews: Reddy and Busch 1988; Goodall et al. 1991.) RNAs are intact and suitable for further molecular characterization by northern blot analysis, cDNA synthesis, and direct RNA sequencing by either chemical or enzymatic methods.

REFERENCES

Goodall G, Kiss T, Filipowicz W (1991) Nuclear RNA splicing and small nuclear RNAs and their genes in higher plants. Oxford Surveys of Plant Molecular and Cell Biology 7: 255–296

Kiss T, Tóth M, Solymosy F (1985) Plant small nuclear RNAs. Nucleolar U3 snRNA is present in plants: partial characterization. Eur J Biochem 152: 259–266

Reddy R, Busch H (1988) Small nuclear RNAs: RNA sequences, structure, and modifications. In Birnstiel ML (ed) Structure and Function of Major and Minor Small Nuclear Ribonucleoprotein Particles, Springer-Verlag, Berlin, pp 1–37

Steele WJ, Okamura N, Busch H (1965) Effects of thioacetamide on the composition and biosynthesis of nucleolar and nuclear ribonucleic acid in rat liver. J Biol Chem 240: 1742–1749

Plant DNA Miniprep and Microprep: Versions 2.1–2.3*

Stephen Dellaporta

The original maize DNA miniprep protocol (version 1) and its successor, version 2, are used extensively for many plant species and different tissues. We (S.L. Dellaporta, J. Wood, J.B. Hicks) prefer this slightly modified version (2.1) for most large (ca. 50–100 μg) DNA extractions. This procedure has the advantage of speed, use of inexpensive reagents, and environmental soundness because neither phenol nor chloroform waste is generated.

MATERIALS

Extraction buffer (EB)
- 50 mM Tris, pH 8
- 10 mM EDTA, pH 8
- 100 mM sodium chloride
- 1% SDS
- 10 mM beta-mercaptoethanol

Liquid nitrogen
3M potassium acetate
Isopropanol
80% ethanol
3 M sodium acetate
Miracloth (Calbiochem 475855) filters cut about 4 cm^2
Falcon 2059 tubes or equivalent
Corning 8441 centrifuge adapters cut three-quarter length
1.5-ml microfuge tubes
Mortar and pestles

*Slightly modified by S. Dellaporta from the published procedure of Dellaporta SL, Wood J, Hicks, JB (1983) "A Plant DNA Minipreparation: Version 2." *Plant Molecular Biology Reporter* 1: 19–22).

The Maize Handbook—M. Freeling, V. Walbot, eds.

Powder funnel (15-mm stem)
65°C temperature block or water bath

METHOD I (MINIPREP)

1. Dispense 7 ml EB in Falcon 2059 tubes for the number of minipreps required.
2. Place 0.5–2 g (1 g is ideal) of plant tissue in mortar and add an excess of liquid nitrogen. When the nitrogen evaporates, grind the tissue thoroughly into a fine powder.
3. Transfer the powdered tissue from the mortar into the Falcon tube containing EB solution. (This can easily be done using a metal spatula and powder funnel whose stem fits snugly into the Falcon tube.) Quickly mix the powdered tissue and EB with the spatula.
4. Cap the tube, shake vigorously, and incubate at 65°C while the next tissue sample is ground and extracted.
5. After the tubes have incubated at 65°C for at least 5 minutes, add 2.5 ml of 5 M potassium acetate. Cap, shake vigorously, and incubate tubes on ice for at least 15 minutes.
6. Spin the tubes at 8,000*g* at 4°C for 10 minutes. [The Falcon tubes can be centrifuged in a fixed angle rotor (e.g., JA-17 or equivalent) using rubber adapters (e.g., Corning 8441) that are cut three-fourths their length to accommodate the Falcon tube and cap. Alternatively, the tubes can be spun in a swinging bucket rotor.]
7. With a 5-ml pipette, carefully remove the supernatant and filter through Miracloth into a Falcon tube containing 5 ml isopropanol. (This is done by forming a Miracloth funnel by pressing the center of the square into the Falcon tube with the pipette tip. Slowly filter the supernatant through the Miracloth funnel. This filtering step removes any suspended plant debris that may not have pelleted in the previous step.)
8. Cap the tubes and invert several times to precipitate the DNA. Spin the tubes at 8,000*g* to pellet DNA. (See note 1.)
9. Rinse the pellet with 80% ethanol and if necessary respin to pellet the DNA. Air dry the pellet by inverting the tubes on paper towels for several minutes; then redissolve DNA in 500 μl TE (10 mM Tris, pH 8; 1 mM EDTA). Transfer the DNA solution to a microfuge tube.
10. Reprecipitate the DNA with 50 μl 3 M sodium acetate and 500 μl isopropanol. Pellet the DNA by brief (30 seconds) centrifugation in the microfuge, wash the pellet with 80% ethanol, dry the pellet, and redissolve in 200–300 μl TE.
11. Before using the DNA for restriction enzyme digestion or PCR analysis, spin the DNA solution for 5 minutes in the microfuge to pellet insoluble materials.

Notes

1. An even quicker version (2.2) of the above procedure can be done when the DNA forms a single clump (which it usually does) after the first precipitation. Remove the DNA from the Falcon tubes with a glass hook (made by flaming the end of a Pasteur pipette) before the pelleting step. While the DNA is still on the hook, carefully rinsed with 80% ethanol from a squirt bottle and transfer directly into the microfuge tube. (The glass hooks can be reused by soaking in bleach, cleaning and sterilizing.). The DNA is air dried and redissolved in 200–300 μl TE and used directly for restriction or PCR analysis.
2. For restriction analysis, 7 μl of miniprep DNA (approximately 3 μg) is adequate for single-copy Southern analysis of maize (2 pg/haploid genome). Include DNase-free RNase in the restriction reaction to remove residual RNA. I also include 4 mM spermidine as this facilitates cleavage of DNA.
3. For PCR analysis, 1 μl of miniprep DNA is sufficient to amplify single-copy genomic sequences. Occasionally, inhibitors of the PCR reactions are present in some DNA minipreps from some species. These inhibitors can be removed by including PVP in the extraction buffer (Jychian Chen, personal communication) or by passing the final miniprep DNA sample through a spin dialysis column (see enclosed protocol) before PCR analysis.
4. For genomic cloning, further purification of miniprep DNA is sometimes necessary for maximum ligation and packaging efficiencies. Size fractionation by glycerol gradient centrifugation or by gel electrophoresis gives DNA that ligates and packages extremely well. Recombinant lambda libraries ($>10^6$ recombinants per microgram DNA) are routinely obtained using size-fractionated miniprep DNA.
5. For critical work such as PCR and cloning, mortars and pestles are best used a single time and then soaked in a dilute bleach solution immediately after use until cleaning. After cleaning and thorough rinsing, cover the mortar and pestle with aluminum foil and bake to sterilize. For standard Southern analysis, the same mortar and pestle can be reused several times to grind multiple tissue samples by carefully wiping the surfaces clean between samples. Falcon tubes can also be reused by cleaning and sterilization.

METHOD II (MICROPREP VERSION 2.3)

A scaled down version of the miniprep (microprep) can be performed very efficiently in microfuge tubes.

1. This procedure requires a hand drill; better yet, a table top drill press with variable speed control works best for maximum efficiency. Connect a microfuge pestle (Kontes Sci., Vineland, NJ; item number 749520) to the drill. Place a small amount of leaf

tissue (1–2 leaves will do) directly in a microfuge tube. Add liquid N_2 to freeze tissue and when the N_2 evaporates, quickly grind to pulverize the tissue to a fine powder. (You may break tissue up with the pestle while the N_2 is evaporating—this precools the pestle.)

2. Add EB to the tissue to bring the volume to about 750 μl. Vortex to suspend tissue in buffer. Incubate at 65°C for 10 minutes.
3. Add 150 μl 5 M potassium acetate, vortex briefly, and incubate for 20 minutes on ice. Spin in the microfuge at 4°C for 10 minutes; then transfer supernatant to a new microfuge tube.
4. Precipitate DNA with an equal volume of isopropanol. (Just squirt it into the tube.) Immediately spin out genomic DNA for 1 minute in microfuge, wash with 80% ethanol, and respin if necessary. Dry pellets and redissolve in 25–50 μl TE. (The volume the DNA is dissolved in depends on the size of the pellet.)

Notes

1. I recommend passing the DNA through CL-6B spin columns before critical PCR work but this is usually unnecessary. Restriction digest and Southerns can be done immediately at noted above. Use one-tenth volume of 40 mM spermidine in the restriction digestion reaction if partial cleavage is a problem.
2. Put micropipette tips over the ends of the squirt bottles and change them whenever they get contaminated. This precaution will save time in the long run, especially with PCR work. We always use a separate set of pipetman for genomic DNA work and microfuge tips with filters when preparing oligos and solutions for PCR. Believe me, it's necessary!

Urea-based Plant DNA Miniprep

JYCHIAN CHEN AND STEPHEN DELLAPORTA

This procedure is sometimes used with recalcitrant species or older plant tissues. Because the procedure employs a phenol:chloroform extraction it is slightly more time-consuming than the method described in the previous chapter and generates organic wastes. However, the recovered DNA usually digests well with restriction enzymes and stores for long periods without signs of degradation.

MATERIALS

- Urea extraction buffer (UEB)
  - Urea
  - 300 mM NaCl
  - 50 mM Tris, pH 8 at 25°C
  - 20 mM EDTA
  - 1% sarkosine
  - (Do not autoclave urea—make solution from autoclaved stocks of 1 M Tris, 0.5 M EDTA, 20% sarkosine, and sterile water to which urea crystals are added.)
- Liquid nitrogen
- 5 M ammonium acetate
- Isopropanol
- TE (10 mM Tris, pH 8 at 25°C; 1 mM EDTA)
- 80% ethanol
- Mortar and pestle
- Miracloth (Calbiochem 475855) filters cut about 4 cm^2
- Falcon 2059 tubes or equivalent
- Corning 8441 centrifuge adaptors cut three-quarter length
- 1.5 ml microfuge tubes

Jychian Chen's modifications of the published procedure of Shure M, Wessler S, Fedoroff N (1983) "Molecular Identification and Isolation of the *Waxy* Locus in Maize." *Cell* 35: 235–242

The Maize Handbook—M. Freeling, V. Walbot, eds.

Powder funnel (15-mm stem)
65°C temperature block or water bath

1. Dispense 6 ml UEB in Falcon 2059 tubes for the number of minipreps required.
2. Place 0.5–2 g (1 g is ideal) of plant tissue in mortar and add an excess of liquid nitrogen. When the nitrogen evaporates, grind the tissue thoroughly into a fine powder.
3. Transfer the powdered tissue from the mortar into the Falcon tube containing UEB solution. (This can easily be done using a metal spatula and powder funnel whose stem fits snugly into the Falcon tube.) Quickly mix the powdered tissue and UEB with the spatula.
4. Add 5 ml phenol:chloroform and firmly cap the tube; shake *vigorously.* Leave the tube at room temperature while going on to grind the next tissue sample. (*CAUTION*: Be sure the tubes are tightly capped before shaking. Steps 4 and 5 should be done under the fumehood wearing safety glasses and a lab coat.)
5. Continue until all the tissue is ground and extracted. After the phenol:chloroform and aqueous phases partially separate, reshake the tubes vigorously.
6. Spin the tubes at 8,000*g* at 4°C for 10 minutes. [The Falcon tubes can be centrifuged in a fixed-angle rotor (e.g., JA-17 or equivalent) using rubber adapters (e.g., Corning 8441) that are cut three-quarter their length to accommodate the Falcon tube and cap. Alternatively, the tubes can be spun in a swinging bucket rotor.]
7. With a 5-ml pipet, carefully remove the supernatant and filter through Miracloth into a Falcon tube containing 5 ml isopropanol. (This is done by forming a Miracloth funnel by pressing the center of the square into the Falcon tube with the tip of the pipet. Slowly filter the supernatant through the Miracloth funnel.)
8. Cap the tubes and invert several times to precipitate the DNA. Spin the tubes at 8,000*g* for 3 minutes to pellet DNA then carefully pour off the supernatant.
9. Rinse the pellet with 80% ethanol and if necessary respin to pellet the DNA. Air dry the pellet by inverting the tubes on paper towels for several minutes; then redissolve DNA in 500 μl TE (10 mM Tris, pH 8; 1 mM EDTA). Transfer the DNA solution to a microfuge tube.
10. Precipitate the DNA with 50 μl 3 M sodium acetate and 500 μl isopropanol. Pellet the DNA by brief (30 seconds) centrifugation in the microfuge, carefully pour off the supernatant, wash the pellet with 80% ethanol, dry the pellet, and redissolve in 200–300 μl TE.
11. Before using the DNA for restriction or PCR analysis, spin the DNA solution for 5 minutes in the microfuge to pellet insoluble materials.

Notes See notes 1–5 in the Preceding Chapter.

High-Molecular-Weight Plant DNA Preparation for CHEF Gel Analysis

ELSBETH WALKER

This method routinely gives good yields of very-high-molecular-weight (MW) DNA (> 2 Mb) from young plant tissue. Unlike other methods, this DNA is suitable for digestion with restriction enzymes. Other procedures employing protoplasts or less-rigorous washing procedures usually give a bimodal distribution of DNA sizes (high-MW and lower-MW peaks) or DNA that is difficult to cut with restriction enzymes. Additional information is available in the next chapter.

MATERIALS

- Liquid nitrogen
- Mortar and pestle
- PK solution (100 ml)
  - 10 mM Tris, pH 8 at 25°C
  - 1% *N*-lauryl sarcosine
  - 500 mM EDTA, pH 8
  - 1 mg/ml proteinase K
- 50 mM EDTA
- TE (10 mM Tris, pH 8; 1 mM EDTA)
- Agarose solution
  - 0.7% Imbed agarose (New England Biolabs)
  - 10 mM Tris, pH 8 at 25°C
  - 15 mM sodium chloride
  - 200 mM EDTA

The Maize Handbook—M. Freeling, V. Walbot, eds.

METHOD

1. In a mortar, freeze 0.2 g young leaf tissue (see note 1) with an excess of liquid nitrogen. When the nitrogen evaporates, grind the tissue to a fine powder.
2. Transfer the frozen tissue powder to tube containing 1 ml molten agarose solution at 50°C.
3. Quickly mix with a spatula and pour into a mold (see note 2); let agarose set for 15 minutes at room temperature.
4. Transfer solid agarose plugs to 5 ml PK solution; incubate overnight at 50°C.
5. Replace PK solution with fresh PK and incubate for 18–20 hours.
6. In 15-ml tubes (e.g., Falcon 2051), wash plugs ten times with a large excess (>10 ml) of 50 mM EDTA (10 minutes per wash, invert tubes periodically). The plugs can be stored for months at 4°C in the EDTA solution.
7. Cut plugs into well-size pieces of about 100 μl each.
8. In a microfuge tube, wash plugs ten times in 1 ml of TE (10 minutes per wash).
9. Wash plugs in 1 ml of appropriate restriction enzyme buffer.
10. Digest overnight in 200–500 μl of RE buffer plus three-fold excess of restriction enzyme.

Notes

1. It is imperative to use only fresh, young tissue for extraction. Older tissue will not give high MW DNA or will not digest well with restriction enzymes. Do not use >0.2 g of tissue for each extraction. Larger quantities of tissue cannot be resuspended well in the molten agarose.
2. Special "gel syringes" are commerically available (New England Biolabs) that work exceptionally well as gel molds.
3. All this washing is time consuming and boring. However, if you skimp, the DNA will not digest well with restriction enzymes.
4. This procedure routinely yields about 3–5 μg per 100 μl gel of high-molecular-weight DNA that is larger than chromosome 1 of yeast (> 2 Mb). This DNA is suitable for restriction enzyme digestion and CHEF analysis using enzymes certified for "in gel" DNA digestion.

87

Preparation of High-Molecular-Weight Maize DNA and Analysis by Pulsed-Field Gel Electrophoresis

AVRAHAM A. LEVY

Maize has a large haploid genome of about 3×10^6 kb organized on 10 chromosomes and resolved into 2,000 cM. The average physical distance to genetic distance is approximately 1,500 kb/cM. Large DNA molecules (up to 10 megabases—Mb) can be resolved with pulsed field gel electrophoresis (PFGE). This technique, together with yeast artificial chromosome vectors (YACs) and RFLP mapping, opens the prospect of isolating genes on the basis of their known genetic location. Protocols for high-molecular-weight (HMW) DNA isolation and resolution have been published for several plant species: Guzman and Ecker (1988) for *Arabidopsis*; Ganal and Tanksley (1989) for tomato; Cheung and Gale (1990) for wheat, barley, and rye; Sobral et al. (1990) for rice. A protocol for maize HMW DNA is described here and in the previous chapter.

CHOICE OF PLANT MATERIAL

Double-stranded maize DNA prepared by conventional techniques involving tissue grinding is ~100 kb on average and is not amenable to analysis by PFGE. Big DNA is typically prepared by embedding and lysing protoplasts in a low-melting-temperature (LMT) agarose plug.

Protoplasts are prepared by incubating the tissue with cell wall-digesting enzymes in the appropriate protoplast isolation buffer (PIB). See references in Table 87.1 for PIB composition and protoplast preparation with various types of maize tissues.

DNA gets quickly degraded in lysed protoplasts, therefore, it is important to have high quality protoplasts and a minimal digestion time (usually between 2 and 5 hours). Conditions need to be optimized for each type of material and each batch of cell-wall-digesting enzymes. Average DNA size in Table 87.1 was determined for each type of

The Maize Handbook—M. Freeling, V. Walbot, eds.

TABLE 87.1. Material for big DNA preparation

Protoplast source	Reference	DNA average size
Cell suspension	Fromm et al. 1987	Larger than 12 Mb
Immature embryo calli	Planckaert and Walbot 1989	10 Mb
Young ears (at silk emmergence)	Same conditions as in Planckaert and Walbot 1989	10 Mb
Etiolated seedlings	Suprasanna et al. 1986	10 Mb

protoplast by comparison to molecular weight markers (A.A. Levy unpublished). The biggest DNA we obtained was with the Black Mexican Sweet (BMS) cell suspension, the material that also yielded the best looking protoplasts.

Nuclei from etiolated seedlings or from young ears can also be embedded in agarose and used for HMW DNA extraction. DNA obtained from nuclei is of about 1 Mb average size; it is not big enough for most PFGE applications but can be used to analyze fragments up to 100 kb in Southern blots, or for cloning in YACs. Similarly, HMW DNA can be isolated from mitochondria (Levy et al. 1991).

PREPARATION OF HMW DNA

1. Check that cells have yielded protoplasts by microscopic observation.
2. Filter protoplasts through a 60 μM mesh; collect by centrifugation at 250*g* in a clinical centrifuge.
3. Wash twice in PIB without enzymes and without BSA if it is included in PIB composition.
4. Resuspend in PIB at room temperature to a final concentration of 2×10^7 cells per ml. This concentration is required to obtain 2–3 μg DNA per lane (50-μl plug) based on a genome size of 5.7 pg per diploid nucleus (Galbraith et al. 1983). It was found that band resolution decreased when larger amounts of DNA were loaded on a lane. Smaller amounts make it difficult to detect a single copy band by Southern blot analysis.
5. Mix protoplasts with an equal volume of 1.2% LMT agarose dissolved in PIB and kept at 42°C. At this point work quickly to keep the agarose from solidifying.
6. Pour the mixture into a prechilled (on ice) mold and let solidify for 5 minutes. The mold should be of the same width as the well of the gel; it can be purchased or be homemade with Teflon or Plexiglas.

7. Transfer the agarose plugs into the lysis solution (LS= 1% SDS, 0.5 M EDTA pH 8.0, 0.5 mg/ml proteinase K). Use 5 ml LS per ml of plug. Incubate overnight at 50°C with gentle shaking. Discard the LS and replace it with fresh LS for another 16–24 hours at 50°C.
8. Wash the plugs in 50 ml TE pH 7.6 at 50°C for 1 hour. Wash twice (1 hour each time) at 50°C in 50 ml TE with 40 μg/ml PMSF to inhibit the proteinase K in LS. PMSF has a short half life in aqueous solutions; it should be made fresh or stored as a stock solution of 4 mg/ml in 100% isopropanol at −20°C. Wash a last time in TE without PMSF.
9. Store plugs at 4°C in 0.5 M EDTA pH 8.0 for up to several months.

DIGESTION OF DNA IN AGAROSE PLUGS

Restriction enzymes with 8- or 7-bp recognition sites and enzymes that are sensitive to cytosine methylation will produce large DNA fragments [Table 87.2]. Information on the region studied such as G+C % or C-methylation should therefore be taken into consideration before choosing an enzyme for PFGE analysis. The G+C % in maize was estimated to be 50% but is variable for different regions, e.g., the rDNA cluster is G+C rich. An updated analysis on known maize sequences should be done to determine the frequency of all possible di- or trinucleotides and thus indicate whether some restriction sites are expected to be found more frequently than others.

1. Cut a piece from the plug that corresponds to about 50 μl (2–3 μg DNA).
2. Wash the samples to be digested in 50 volumes of TE, twice at room temperature, and once in 10 volumes of 1× restriction enzyme buffer.

TABLE 87.2. Rare cutter enzymes—1991

Enzyme	Recognition sequence	Manufacturer
*Asc*I	GGCGCGCC	New England Biolab
*Pac*I	TTAATTAA	New England Biolab
Pme	GTTTAAAC	New England Biolab
*Rsr*II	CGG(A/T)CCG	New England Biolab
*Not*I	GCGGCCGC	(and others)
*Sfi*I	GGCCNNNNNGGCC	(and others)
*Sgr*AI	C(A/G)CCGG(T/C)G	Boehringer Mannheim
*Swa*I	ATTTAAAT	Boehringer Mannheim
*Sse*8387I	CCTGCAGG	Takara Biochemical

3. Remove the buffer and replace it with 1× buffer, just enough to cover the sample (about 100 μl).
4. Add 30–50 units of the enzyme and incubate for 5 hours at the optimal temperature recommended by the supplier.

RESOLUTION AND ANALYSIS

The digested plug can be stored in a 0.5 M EDTA solution or loaded on a pulsed-field gel. Prior to loading, the restriction buffer is removed, and the plug is washed for at least 15 minutes with 10 volumes of the electrophoresis running buffer (usually 0.5× TBE). It is convenient to add loading dye to the TBE to stain the transparent plug.

Resolution conditions are as recommended by the PFGE apparatus manufacturer. Southern blot analysis of pulsed field gels is similar to that of standard agarose gels. UV nicking, or HCl depurination, however, should be included because of the large size of the fragments to be transferred.

REFERENCES

Cheung WY, Gale MD (1990) The isolation of high molecular weight DNA from wheat, barley and rye for analysis by pulse-field gel electrophoresis. Plant Mol Biol 14: 881–888

Fromm M, Callis J, Taylor LP, Walbot V (1987) Electroporation of DNA and RNA into plant protoplasts. Meth Enzymol 153: 351–366

Galbraith DW, Harkins KR, Maddox JM, Ayres NM, Sharma DP, Firoozabady E (1983) Rapid flow cytometric analysis of the cell cycle in intact plant tissues. Science 220: 1049–1051

Ganal MW, Tanksley SD (1989) Analysis of tomato DNA by pulsed field gel electrophoresis. Plant Mol Bio Rep 7: 17–27

Guzman P, Ecker JR (1988) Development of large DNA methods for plants: molecular cloning of large segments of Arabidopsis and carrot DNA into yeast. Nucleic Acids Res 16: 11091–11105

Levy AA, André CP, Walbot V (1991) Analysis of a 120 kilobase mitochondrial chromosome in maize. Genetics 128: 417–424

Planckaert F, Walbot V (1989) Transient gene expression after electroporation of protoplasts derived from embryogenic maize callus. Plant Cell Rep 8: 144–147

Sobral BWS, J. HR, Atherly AG, McClelland M (1990) Analysis of rice (Oryza sativa L.) genome using pulsed-field gel electrophoresis and rare-cutting restriction endonucleases. Plant Mol Biol Rep 8: 253–275

Suprasanna P, Rao KV, Reddy GM (1986) Isolation of protoplasts from seedlings. Maize Genetics Cooperation News Letter 60: 66

Isolation of Genomic DNA from Calli*

CHRISTINE A. WARREN

MATERIALS

Liquid nitrogen
Polytron
Homogenization buffer

25 mM MOPS (free acid)	Add MOPS to distilled H_2O (two-thirds of final volume); adjust to pH 7.8 with NaOH; add next three reagents, mix, and readjust to pH 7.8; bring volume up with H_2O, filter, autoclave, and store at 4°C; add beta-mercaptoethanol and BSA before use.
0.4 M mannitol	
1 mM EGTA	
0.1% cysteine	
0.1% BSA	
0.3% beta-mercaptoethanol	

20:10 TE
- 20 mM Tris pH 7.5
- 10 mM EDTA

10% SDS
7.5 M ammonium acetate
Isopropanol
75% ethanol
10:1 TE
- 10 mM Tris pH 8.0
- 1 mM EDTA

METHODS

1. Freeze calli in liquid nitrogen and store at -80°C in a sealed plastic bag. Pulverize in the bag with a rubber mallet, transfer to a cold mortar, and grind in homogenization

*Technique devoloped by Frederique Planckaert and Christine Warren

The Maize Handbook—M. Freeling, V. Walbot, eds.

buffer, keeping the volume small (5 ml/g tissue). To obtain a uniform suspension, grind with a polytron at low to medium speed for 10–20 seconds.

2. Centrifuge sample at 3,000*g* for 5 minutes. Remove supernatant with pipette and discard. Resuspend the pellet in 400 *ml* of 20:10 TE; then add 60 *ml* of 10% SDS. Add 300 *ml* of 7.5 M ammonium acetate and mix vigorously. Leave on ice for 30 minutes.
3. Sediment the SDS precipitate at 12,000*g* for 20 minutes. Remove supernatant to a microfuge tube and add 750 *ml* isopropanol. Let sit for 15 minutes at room temperature. You should see the DNA precipitate.
4. Pellet the DNA in a microfuge at room temperature for 5 minutes. Wash pellet with 75% EtOH and dry. Add 100 *ml* of TE (10 mM Tris, 1 mM EDTA) to the tube and allow the DNA to rehydrate overnight. Check DNA concentration on a gel.

Note

1. For large preparations of callus DNA, use the protocol for cob DNA isolation in the next chapter.

Isolation of DNA from Immature Cobs*

CHRISTINE A. WARREN

MATERIALS

Liquid nitrogen
Rubber mallet
Polytron
Miracloth (Calbiochem)
50-ml Oakridge tubes
16 gauge needles
3 ml syringes
Homogenization buffer

25 mM MOPS (free acid) 0.4 M mannitol 1 mM EGTA 0.1% cysteine 0.1% BSA 0.3% beta-mercaptoethanol	Add MOPS to distilled H_2O (two-thirds final volume); adjust to pH 7.8 with NaOH; add next three reagents, mix, and readjust to pH 7.8; bring volume up with H_2O, filter, autoclave, and store at 4°C. Add beta-mercaptoethanol and BSA before use

Lysis buffer
  50 mM Tris pH 8.0
  20 mM EDTA
  1% NaSarkosyl
Solid CsCl
10:1 TE
  10 mM Tris pH 8.0
  1 mM EDTA

*This is a modified version of the method of Rivin et al. (1982) Sheridan WF (ed) In *Maize for Biological Research*, Plant Molecular Biology Association, pp 161–164

The Maize Handbook—M. Freeling, V. Walbot, eds.

CsCl, r=1.6
  0.816 g CsCl/ml in TE
10 mg/ml ethidium bromide (EtBr)
Isopropanol equilibrated with CsCl-saturated TE
Isopropanol
70% ethanol

METHODS

1. Harvest immature ears before silks emerge and freeze in liquid nitrogen. Store in seal-a-meal bags at -80°C.
2. Pulverize the cobs in the bag using a rubber mallet and transfer to a mortar. Grind in 20 ml cold homogenization buffer. Transfer to a 50 ml disposable plastic tube and pulse 1–3 times for 30 seconds in a polytron at low to medium speed. Care must be taken to keep the suspension cold. Filter suspension through Miracloth into 50-ml Oakridge tubes.
3. Centrifuge at 2,000*g* for 15 minutes. Discard supernatant and quickly resuspend pellet in 8 ml lysis buffer. Add 8.2 g solid CsCl, and when dissolved, add 300 μg of 10 mg/ml EtBr. Measure the CsCl density on a refractometer and adjust if necessary to $\rho = 1.6$. Transfer to an ultracentrifuge tube; top off with CsCl in TE, $\rho = 1.6$. Centrifuge 40 hours in Ti75 rotor at 40,000 rpm.
4. Collect the DNA band with a 16 gauge needle attached to a 3-ml syringe. Extract the EtBr by repeatedly mixing with an equal volume of isopropanol equilibrated with CsCl-saturated TE until the DNA solution is no longer pink.
5. Measure the volume; add 2 volumes of water and 3 volumes of isopropanol. Spool out the high-molecular-weight DNA if visible; otherwise spin at 8,000*g* for 20 minutes at 10°C. Decant the isopropanol, wash the pellet in 70% EtOH, and dry briefly. Resuspend the DNA in TE, measure the OD_{260}, and check the quality of the DNA on a minigel.

Note

1. Grinding with a mortar and pestle can be omitted; however, the polytron can shear high molecular weight DNA and its use should be kept to a minimum. Use it at 25–50% speed and grind only until a uniform suspension is obtained.
2. The yield is routinely 100–300 mg DNA/g tissue. The younger cobs (2–5 cm) give the best yield/g tissue and are much easier to work with.

Preparation of Nucleic Acids from Maize Microspores and Pollen

ANNE H. BROADWATER AND PATRICIA BEDINGER

This protocol allows the separation of microspores and young pollen at several developmental stages with as much as 98% purity utilizing continuous sucrose gradients. Sucrose fractions are collected, assessed for stage and purity, and processed to isolate RNA or DNA.

MATERIALS

One large beaker of sterile, distilled, deionized water at 4°C
One large funnel with four layers of cheesecloth on top of one layer of fine mesh nylon cloth, autoclaved and chilled
Vanadyl Ribonucleoside Complex (VRC) 200 mM, purchased from BRL
20% and 60% sucrose solutions (DEPC-treated)
French press
Homogenization buffer (DEPC-treated)
  0.5 M mannitol
  1 mM EGTA
  20 mM HEPES pH 7
  0.1 mM DTT (added just prior to use)
Lysis buffer (DEPC-treated)
  50 mM EGTA
  100 mM NaCl
  1% SDS
  100 mM Tris HCl pH 7.6 (to be added after DEPC treatment)
  50 mM 2-mercaptoethanol (added just prior to use)
1:1 phenol:chloroform (use 24:1 chloroform:isoamyl alcohol)
Absolute ethanol

The Maize Handbook—M. Freeling, V. Walbot, eds.

3 M sodium acetate (DEPC-treated)
5 M LiCl (DEPC-treated)
Distilled, deionized water (DEPC-treated)

METHODS

Keep sample at 4°C for entire procedure. Wear gloves for entire procedure to prevent RNase contamination. To collect nucleic acids from mature pollen, start at step 14.

1. Sample anthers and examine microscopically to select plants of similar developmental stage. Stage plants at oldest part of main spike; examine both older and younger flowers for initial sample selection.
2. Pour 20–60% weight/volume continuous sucrose gradients; store at 4°C until needed. We use an SW41 rotor and pour 9.5-ml gradients (1.5 ml/minute) with a 0.5-ml pad of 60% sucrose at the bottom of Beckman Ultra-clear tubes.
3. Harvest upper portion of plants, remove leaves, and submerge tassels in the beaker of sterile, chilled water.
4. Strip spikelets off individual spikes into chilled Waring blender containing just enough homogenization buffer (HB) to submerge the spikelets.
5. Blend at high speed in 3-second pulses to make a uniform homogenate (typically three pulses).
6. Pour the homogenate into the prepared funnel and filter through cheesecloth to remove large debris, squeezing through all the HB.
7. Gather edges of fine mesh cloth and squeeze through all but 2–10 ml HB. The microspores and young pollen are retained on the cloth surface.
8. Replace the cloth in the funnel and, using a micropipeter and the retained HB, wash the sample down the sides of the cloth.
9. Transfer sample into a calibrated RNase-free tube and add 1/20th volume VRC.
10. Load 2–3 ml onto each sucrose gradient and balance tubes.
11. Centrifuge gradients at 4°C for 30 minutes at 30,000 rpm. Microspores with a single large vacuole do not enter the sucrose; microspores with multiple small vacuoles band at 35–40% sucrose; young prevaculated microspores band at 50% sucrose.
12. Collect bands separately into calibrated RNase-free tubes.
13. Remove a 15-μl sample from each fraction and assess the developmental stage and purity microscopically.
14. Add 3 volumes lysis buffer plus 1/20th volume VRC and French press ≥10,000 psi. Assess breakage microscopically; repeat French pressing if incomplete.
15. Extract the pressed sample with an equal volume of phenol:chloroform, centrifuge

10 minutes at 10,000 rpm, and collect the aqueous phase. Repeat extraction until there is no detectable interphase.

16. Precipitate final aqueous phase with 1/10th volume of DEPC-treated 3 M sodium acetate and 2.5 volumes absolute ethanol for several hours to overnight at −20°C.
17. Pellet nucleic acids by centrifuging for 30 minutes at 10,000 rpm. Remove supernatant carefully as pellet may be loose.
18. Resuspend pellet in 50–100 μl DEPC-treated water and transfer to an RNase-free microfuge tube.
19. Add an equal volume DEPC-treated 5 M LiCl and place on ice for 1 hour.
20. Centrifuge in 4°C microfuge for 20 minutes. RNA will pellet and DNA will remain in solution.

DNA Isolation

a. Transfer supernatant to a new tube and add 1/10th volume 3 M sodium acetate and 2.5 volumes 95% ethanol.
b. Precipitate for several hours to overnight at −20°C.
c. Centrifuge in microfuge for 20 minutes.
d. Resuspend DNA pellet in water or TE buffer.

RNA Isolation

a. Resuspend LiCl pellet in ≤50 μl DEPC-treated water.
b. Repeat LiCl precipitation.
c. Resuspend RNA pellet in ≤50 μl DEPC-treated water.

21. Examine by agarose gel electrophoresis to assess purity, quality, and quantity.

Note: The most problematic developmental stage to collect is that of prevacuolate young microspores because yields tend to be low and inviable older cells often contaminate the sample. Therefore, to isolate young microspores, we only use plants with prevaculate microspores or microspores with multiple small vacuoles in the oldest flowers. In addition, we keep the sample volume as small as possible, loading material from as many as six plants on a single sucrose gradient. Separate collection of upper and lower portions of the visible band at 50% sucrose may improve the purity of young prevaculate microspore samples.

Preparation of DNA and RNA from Leaves: Expanded Blades and Separated Bundle Sheath and Mesophyll Cells

TIMOTHY NELSON

Isolation of Mesophyll Protoplasts From Leaves

This preparation results in homogeneous intact mesophyll protoplasts suitable for nucleic acid isolation, assay of activities, or gene expression studies. The reader may wish to compare similar protocols used by Sheen (1990) for expression studies and by Kanai and Edwards (1973) for photosynthetic studies. The reader should be forewarned that the digestive enzymes used are expensive reagents.

Materials

Clean razor blades
Sidearm vacuum flask
Large petri dish
60-μm nylon net
135-μm filter mesh
Cellulase buffer (CB)

20 mM MES pH 5.5	2 ml 1 M
1 mM $MgCl_2$	100 μl 1 M
0.6 M sorbitol	30 ml 2 M
2% cellulase	2 g (Onozuka cellulase R10, Gallard Schlesinger)
0.1% pectinase	0.1 g (Macerase pectinase, Calbiochem)
Water	to 100 ml

The Maize Handbook—M. Freeling, V. Walbot, eds.

Wash buffer (WB)	
50 mM Tris-HCl, pH 8.0	2.5 ml 2 M
0.6 M sorbitol	30 ml 2 M
1 mM $MgCl_2$	100 μl 1 M
100 mM beta-mercaptoethanol	700 μl
Water	to 100 ml

Method

1. Crosscut 5 g of leaves (perpendicular to midrib) into 0.5 to 1-mm strips.
2. Add to 80 ml CB in sidearm flask. Apply vacuum until all sections are infiltrated.
3. Transfer to large petri dish and continue digestion at (RT) room-temperature, 3–5 hours.
4. Discard broken cells by filtration through 135-μm mesh. Resuspend residual, partially digested segments in 50 ml WB in the same petri dish.
5. Press the leaf strips gently with a spatula to release protoplasts.
6. Filter through 60-μm nylon net. Pellet cells by spinning 300*g* for 5 minutes. Gently resuspend in 25 ml WB.
7. Repeat spin and resuspend steps. Freeze in liquid nitrogen and store at −70°C.

ISOLATION OF BUNDLE SHEATH STRANDS FROM LEAVES

It is relatively simple to isolate homogeneous strands consisting of intact bundle sheath cells surrounding vascular strands. In this method, mesophyll cells are quantitatively eliminated by mechanical disruption and washes, while the walls of bundle sheath cells resist homogenizing forces. The isolation of intact bundle sheath cells or protoplasts is much more difficult than the isolation of strands and is usually unnecessary.

Materials

60-μm nylon net

Disruption buffer (DB):	
50 mM Tris-HCl pH 8.0	2.5 ml 2 M
0.6 M sorbitol	30 ml 2 M
1 mM $MgCl_2$	100 μl 1 M
100 mM beta-mercaptoethanol	700 μl
Water	to 100 ml

Method

1. Cut 3 g leaves into 2 × 2-mm squares; transfer to small beaker with 25 ml DB.
2. Treat with Polytron homogenizer, 40 seconds at speed 6; remove foam with a Kimwipe.
3. Filter through 60-μm nylon net; resuspend residue in fresh 25-ml DB.
4. Repeat steps 2 and 3 twice.
5. Wash bundle sheath strands retained on third net with 25 ml DB.
6. Proceed with RNA prep or freeze cells (and net) directly in liquid nitrogen. Store cells at −70°C.

PREPARATION OF RNA AND DNA FROM LEAF BLADES OR ISOLATED LEAF CELLS

This preparation yields high-quality RNA and/or DNA from a single sample of leaf tissue or from preparations of separated leaf cells. This serves well as a general RNA prep, and is based on the guanidinium method devised for RNase-rich animal tissue (Chirgwin et al. 1979). Diethyldithiocarbamic acid (DECA) is added to inhibit polyphenol oxidase—the accumulating brown polyphenolic products can greatly reduce RNA yields from leaf cells. The combined RNA/DNA version is of particular value for analysis of separated cells, for which materials are often limiting and for experiments in which it is critical to monitor the state of RNA and DNA from the same sample (e.g., Langdale et al. 1991). For a simpler method of DNA preparation from leaf tissue, see the chapter on DNA minipreps.

Materials

Lysis buffer (LB)
- 100 mM Tris-HCl, pH 8.6
- 1% Sarkosyl
- 4 M guanidinium thiocyanate
- 25 mM EDTA pH 8.0
- 25 mM EGTA pH 8.0
- 100 mM beta-mercaptoethanol
- add 20 mM diethyldithiocarbamic acid fresh (35 mg/10 ml)

Phenol
Chloroform:isoamyl alcohol (24:1)
Phenol:chloroform:isoamyl alcohol 25:24:1
Concentrated acetic acid
Isopropanol

DEPC-treated water
8 M LiCl
3 M NaAc pH 4.5
Ethanol

Method

1. *For leaves:* grind up to 10 g of tissue in liquid nitrogen with precooled mortar and pestle. *For separated cells:* grind 5 g frozen mesophyll cells or 3 g frozen bundle sheath cells in liquid nitrogen.
2. Transfer to 10 ml of LB (15 ml for cell preps) in a 30 ml tube.
3. Add 0.5 volume phenol and shake vigorously.
4. Add 0.5 volume chloroform:isoamyl alcohol (24:1). Mix well. Spin at 10,000 rpm for 10 minutes at 4°C.
5. Remove supernatant. Extract once with phenol:chloroform:isoamyl alcohol 25:24:1. Spin at 10,000 rpm for 5 minutes.
6. Decant supernatant. Add 50 μl (one drop) concentrated acetic acid (solution should turn colorless).
7. Add 0.6 volumes isopropanol. Freeze at −20°C (about 2 hours).
8. Thaw and spin at 10,000 rpm for 10 minutes. Discard supernatant.
9. Transfer pellet in 600 μl DEPC-treated water + 0.1% SDS to an eppendorf tube. Pipette up and down to resuspend. Incubate at 37°C for 10 minutes. Pipette once more and put on ice for 15 minutes.
10. Clear remaining insoluble material (mostly not nucleic acid) by a brief spin in the microfuge. Transfer supernatant to a fresh microtube.
11. Add 0.25 volumes of 8 M LiCl. Incubate overnight at 4°C.
12. Spin 10 minutes in microfuge. DNA remains in supernatant; RNA pellets.
13. *For DNA* (LiCl supernatant): Remove supernatant to fresh tube. Add 1/10th volume 3 M NaAc pH 4.5 and 2 volumes ethanol. Mix well. Spin at 10,000 rpm for 10 minutes. Wash pellet with 70% ethanol. (You can transfer pellet to microtube.) Dry down and resuspend in TE overnight at 4°C. DNA may contain some RNA, but you can RNase-treat during restriction digests.
14. *For RNA* (LiCl pellet): Redissolve pellet from LiCl precipitation in 400 μl DEPC-treated water + 0.1% SDS. Precipitate RNA with 1/10th volume 3 M NaCl and 2 volumes ethanol at −70°C. Wash pellets with 70% ethanol, dry, and resuspend in DEPC-treated water. (You can also add 0.1% SDS.) Store at −70°C.

REFERENCES

Chirgwin JM, Przybyla AE , MacDonald RJ, Rutter WJ (1979) Isolation of biologically active RNA from sources enriched in RNase. Biochemistry 18: 5294–5299

Kanai R, Edwards GE (1973) Separation of mesophyll protoplasts and bundle sheath cells from maize leaves for photosynthetic studies. Plant Physiol 51: 1133–1137

Langdale JA, Taylor WC, Nelson T (1991) Cell-specific accumulation of maize phosphoenolpyruvate carboxylase is correlated with demethylation at a specific site >3 kb upstream of the gene. Mol Gen Genet 225: 49–55

Sheen J-Y (1990) Metabolic repression of transcription in higher plants. Plant Cell 2: 1027–1038

92

Isolation of RNA from Wx and wx Endosperms

SUSAN R. WESSLER

ISOLATION FROM *Wx* ENDOSPERMS

Materials

Lysis buffer

42 g/100 ml (w/v) urea (ultrapure grade)
0.35 M NaCl
20 mM Tris-Cl pH 7.5
1 mM EDTA pH 8

The Maize Handbook—M. Freeling, V. Walbot, eds.

2% Sarkosyl

Keep this buffer at 4°C.

Phenol:CHCl$_3$:Isoamyl Alcohol (25:24:1)Saturated With 10 mM Tris-Cl pH 7.5, 1 mM EDTA

Keep this solution at room temperature.

Methods

1. Add 8 g of frozen powdered tissue to 50 ml cold lysis buffer and 50 ml phenol solution in a 250-ml beaker. Stir until uniformly mixed.
2. Polytron in a fumehood at setting 8 for 20 seconds; repeat three times.
3. Pour into 50-ml disposable, capped, conical bottom centrifuge tubes. Centrifuge 15 minutes in Sorvall RT 6000 refrigerated table top machine at 4°C and 3,500 rpm.
4. Reextract aqueous layer twice with equal volume of phenol solution.
5. Add 1/10th volume 3 M NaOAc and 2.5 volumes ethanol; precipitate overnight.

ISOLATION FROM *WAXY* ENDOSPERMS

The hydrophilic starch from *wx* mutants dissolves in the aqueous phase and results in a gel from which RNA can not be extracted. The gelation of *wx* starch can be prevented by substituting 0.35 M Na_2S0_4 for 0.35 M NaCl in the lysis buffer and by equilibrating the phenol solution with 0.35 M Na_2S0_4 (in addition to Tris-EDTA). Because Na_2S0_4 has limited solubility in ethanol, the RNA must be dialyzed prior to ethanol precipitation.

REFERENCES

Fedoroff NV, Mauvais J, Chaleff D (1983) Molecular studies on mutations at the *Shrunken* locus in maize caused by the controlling element *Ds*. J Mol Appl Genet 2: 11–29

Shure M, Wessler S, Fedoroff N (1983) Molecular identification and isolation of the *waxy* locus in maize. Cell 35: 225–233

RNA Isolation From Electroporated Protoplasts

KENNETH R. LUEHRSEN AND JANE HERSHBERGER

MATERIALS

Guanidinium thiocyanate solution
- 4 M guanidinium thiocyanate
- 0.5% N-lauroylsarcosine
- 0.05 M sodium citrate pH 7
- 0.3% beta-mercaptoethanol (add just before use)

CsCl cushion
- 5.7 M CsCl, 0.1 M EDTA pH 7, treated with DEPC

1 × SET
- 10 mM Tris-HCl pH 7.5
- 5 mM EDTA
- 1% SDS

Phenol/chloroform
- 1:1 mixture of TE-equilibrated phenol and chloroform

TE
- 10 mM Tris-HCl pH 8
- 0.1 mM EDTA

5 M NaCl (DEPC-treated)
Ethanol (100%)
Water (DEPC-treated)

METHOD

1. Electroporate two batches of 5×10^6 protoplasts with 50 μg expression plasmid and 100 μg salmon sperm DNA. Incubate at 26°C for 6–24 hours.

The Maize Handbook—M. Freeling, V. Walbot, eds.

2. Pool the protoplasts and pellet at 100*g*. There is no need to wash them prior to lysis.
3. Add 3 ml of the guanidinium thiocyanate solution; the protoplasts should lyse immediately. Vortex at high speed for 15 seconds to shear the DNA. Spin at 2,000*g* for 5 minutes to pellet the debris.
4. Layer the cleared solution over a 1.1-ml CsCl cushion in a Beckman SW60 tube. Spin at 41,000 rpm overnight (>14 hours) at 20°C.
5. The RNA will be a clear pellet at the bottom of the tube. Aspirate off all of the supernatant; the interface at the top of the cesium cushion will be cloudy with DNA and other debris. Cut off the top of the tube with a razor blade, leaving 1–2 cm at the bottom.
6. Resuspend the pellet in 200 μl 1 × SET. Extract once with an equal volume of phenol/chloroform. Back extract the organic phase with an additional 75 μl 1 × SET.
7. To the extracted aqueous phase, add 15 μl DEPC-treated 5 M NaCl and 900 ml 100% EtOH; place on ice for at least 30 minutes. Spin for at least 20 minutes in a microfuge. Resuspend the RNA in 50 μl H_2O and read the OD_{260} on 2 μl.

Notes

1. The yield for 10^7 BMS protoplasts should be 75–100 μg of total RNA. Maize RNA purified by this procedure is relatively free of contaminating polysaccharides and DNA.
2. See Sambrook et al. (1989) for instructions on how to prepare phenol/chloroform and the DEPC-treated solutions.
3. See Berger and Kimmel (1987) "Guide to Molecular Cloning Techniques" p 225 for using other ultracentrifuge rotors to pellet the RNA.

REFERENCES

Berger SL, Kimmel AR (eds) (1987) Guide to Molecular Cloning Techniques. Academic Press, Orlando, 812 pp

Sambrook J, Fritsch EF, Maniatis T (1989) Molecular Cloning: A Laboratory Manual. Second Edition, 3 vol. Cold Spring Harbor Laboratory Press, New York

Procedures for Isolating Mitochondria and Mitochondrial DNA and RNA

KATHLEEN J. NEWTON

STARTING MATERIALS AND GENERAL CONSIDERATIONS

Mitochondrial yield is greatly influenced by the source of the plant parts used for the preparation. Cobs, young unpollinated ear shoots (within 2 days after the silks emerge from husk leaves), are excellent source materials for both mtDNA and RNA, as well as for mitochondria for *in organello* protein synthesis experiments. Good mtDNA/RNA yields can also be obtained from tassels removed from the stalks at approximately the time that meiosis is taking place in the anthers. Mesocotyl and epicotyl regions of 3–5-day-old etiolated seedlings are reasonable sources, although the relative yields tend to be lower and extreme care must be taken to avoid bacterial and fungal contamination if the seedling mitochondria are to be used for *in organello* protein synthesis studies. The procedures can also be easily adapted for tissue cultures, but the reader is referred to the excellent compilation of Hanson et al. (1985) for detailed protocols for cultured cells.

Yields are greatest when care is taken to keep mitochondria as intact as possible, when preparations are kept cold, and when the procedures are completed rapidly. Some researchers prefer to work in the cold room for the initial steps, but good results are obtained by keeping everything on ice. For RNA preparations, breakdown by nucleases can be largely avoided if sterile glassware and DEPC-treated autoclaved solutions and, where indicated, nuclease inhibitors are used.

If enough starting material is available (e.g., two or more ear shoots), the preparation may be divided after the differential centrifugation steps to obtain both DNA and RNA or DNA and mitochondria for protein synthesis studies. In such cases, the *MOPS* or mtRNA homogenization buffer can be used for grinding.

Solutions

Note: Add BSA (Sigma # A4503) 0.1–0.2% to homogenization buffers (H.B) just before use.

The Maize Handbook—M. Freeling, V. Walbot, eds.

mtDNA H.B.
0.4 M mannitol
10 mM TES pH 7.5
5 mM EGTA
0.05% cysteine

mtRNA H.B.
0.35 M sorbitol
50 mM Tris-HCl pH 8
5 mM EDTA
20 mM 2-mercaptoethanol
(Added just before use)

MOPS (protein) H.B.
0.4 M mannitol
25 mM MOPS pH 7.8
1 mM EGTA
4 mM cysteine

Sucrose Pad
20% sucrose
10 mM TES pH 7.3
20 mM EDTA

DNA Wash Buffer
10% sucrose
20 mM Tris-HCl pH 8.0
25 mM EDTA

mtRNA Wash
0.35 M sorbitol
50 mM Tris-HCl pH 8.0
20 mM EDTA

RNA lysis buffer
10% "sarkosyl"
(N-lauroyl sarcosine)
20 mM Na_2EDTA
25 mM Tris-HCl pH 8.0

Mitochondrial dilution
0.2 M mannitol
10 mM tricine
1 mM EGTA
pH to 7.3 with KOH

2X label buffer
0.5 M mannitol
20 mM tricine
20 mM Na succinate
0.18 M KCl
20 mM $MgCl_2$
2 mM EGTA; pH to 7.2

KPO Solution pH 7.2
Acidic KH_2PO_4 1 M stock
Basic KH_2PO_4 1 M stock
For pH 7.2, mix about 95 ml of the basic solution
with about 50 ml of the acidic solution. Filter sterilize.

100 mM ATA stock solution
0.95 g aurintricarboxylic acid
1 ml 1 M Tris-HCI pH 8.0
19 ml sterile water. Filter sterilize the solution
(Keep covered at 4°C.
Should make new monthly.)

Amino Acid Mix (Minus Methionine)

Make up 19 amino acids from the Sigma L-amino acid set (kit LAA-21) at 2 mM each. Dissolve them in 10 ml of sterile water, add 50 μl 0.1M NaOH, sterilize the mix with a syringe filter, and aliquot it into microfuge tubes. Quick-freeze on dry ice and store at −80°C. The amino acid mix can be stored for years under these conditions. Each tube can be thawed and refrozen at least three times.

0.1 M GTP

To a 25-mg bottle (Sigma G8877), add 478 μl sterile water. Aliquot, quick-freeze, and store at −80°C.

0.26 M ADP

To a 100-mg bottle (Sigma A6521), add 390 μl sterile water. To pH to 6.8, add approximately 280 μl 1 M NaOH and then 130 μl 0.1 M NaOH. Aliquot, quick-freeze, and store at −80°C.

MITOCHONDRIAL DNA

The following protocol is based on the procedure of Kemble et al. (1980) and is for 2–3 ear shoots (usually 20–50 g) of starting material.

1. Add BSA to homogenization buffer to 0.1–0.2% just before use. Note: keep everything cold on ice and work quickly.
2. Harvest tissue, weigh, and briefly surface sterilize in 5% bleach; then rinse several times in sterile cold water.
3. Homogenization: Cut up ear shoots (or seedling shoots) into Waring blender jar. Add a minimum of 3 (ears and immature tassels) or 5 (seedlings) volumes of DNA homogenization buffer. Homogenize at low speed for 3 seconds and at high speed for 3 seconds. Repeat. Do not "overgrind" or yield will be low. Note: Small amounts of tissue (1–3 g) can be ground in a mortar and pestle. Pour through a funnel lined with four layers cheesecloth and one layer Miracloth into a centrifuge bottle. Squeeze hard to obtain all liquid possible. (Change gloves between samples and clean blender jar with water, 10% bleach, and ice-cold sterile water.)
4. Differential centrifugation: Centrifuge at 1,000*g* (2,500 rpm in GSA rotor) for 5 minutes to remove nuclei and cell debris. Transfer supernatant to new bottle and centrifuge at 2,000*g* (3,500 rpm in GSA rotor) for 10 minutes. Transfer supernatant again and centrifuge at 10,000*g* (8,000 rpm in GSA rotor) for 15 minutes to pellet mitochondria. Pour off supernatant and place bottle with pellet on ice. (Prepare the DNase solution, 10 mg/ml in 10 mM Tris pH 7.5, and mix gently by inversion during the 10,000*g* centrifugation.)
5. DNase treatment: Resuspend pellet in a final volume of 5–10 ml homogenization buffer using a sterile paintbrush or by pipetting with a cotton-plugged sterile plastic pipette. Transfer to 30 ml Corex tube. Add 100 μl 1.5 M $MgCl_2$ and 200 μl of 10 mg/ml DNase to each sample. Mix gently (by swirling/inverting tubes). Place on ice for 30 minutes, mixing gently every 5 minutes or so. Add a second aliquot of $MgCl_2$ and DNase and repeat procedure for another 30 minutes.
6. Inactivate DNase: At the end of the incubation, add 50 μl of 0.5 M EDTA to stop the DNase reaction. Carefully underlay each with 10 ml of the sucrose pad solution (20% sucrose) and centrifuge at 10,000*g* for 15 minutes (9,200 rpm in SS34 rotor). Resuspend the pellet in 5 ml of sucrose pad solution and then add 5 ml DNA wash buffer. Centrifuge at 10,000*g* for 10 minutes. Resuspend in a total of 10 ml of DNA

wash buffer, transferring to 15 ml Corex tube. Centrifuge at 10,000*g* for 10 minutes. (Weigh out proteinase K: 10 mg/ml dissolved in sterile water and set at 37°C during centrifugation.)

7. Lyse mitochondria: Resuspend washed mitochondrial pellet in 2 ml of TE buffer. Add 0.1 ml proteinase K solution, swirl, and let sit at 37°C for 5–10 minutes. Add 0.2 ml of 10% N-lauroyl sarcosine (sarkosyl) solution, swirl and let sit at 37°C for up to 30 minutes. Note: suspension should clear as mitochondria lyse.
8. Extractions to remove proteins: Add 2 ml of phenol equilibrated in 10× TE; vortex and centrifuge at 7,000 rpm (SS34 rotor) for 10 minutes. Transfer upper aqueous layer to a new Corex tube. Add 2 ml of phenol:chloroform (1:1), vortex and centrifuge at 5000 rpm 5 minutes. Transfer upper aqueous layer to a new corex tube. Add 2 ml chloroform, vortex, and centrifuge at 5,000 rpm 5 minutes. Transfer upper aqueous layer to a new Corex tube.
9. Ethanol precipitate: Add 1/10th volume 3 M NaOAc (pH 5.6) and 2–3 volumes 95% ethanol (ice-cold). Precipitate at −20°C overnight.
10. On following day, pellet mtDNA by centrifuging at 10,000*g* for 30 minutes. Rinse pellet with 70% ethanol; centrifuge 10 minutes. Invert tube and allow ethanol to evaporate (approximately 30 minutes). Resuspend near-dry DNA in 1× TE, using approximately 5 μl per gram of starting weight.

The DNA prepared in this way still contains RNA. The RNA is usually degraded during restriction digestion of the DNA. To ensure this, a small amount of RNase (e.g., 1 μl of 10 mg/ml RNase A) may be added at the end of the digests, prior to loading the DNA on the gel. The mtDNA from ear shoots is more contaminated with chloroplast DNA than is mtDNA from etiolated seedlings. However, the chloroplast DNA bands can be distinguished in many digests if lanes containing plastid DNA digests are run next to lanes with the mtDNA on the gels. (See Newton and Coe 1986.) The most serious problem encountered in these preparations is the partial degradation of nuclear DNA. This is usually caused by clumps present in the mitochondrial resuspensions and/or inadequate DNase treatment.

MITOCHONDRIAL RNA

The RNA isolation protocol below is slightly modified from the procedure published by Stern and Newton (1985). The RNA prepared from mitochondria lysed in the presence of aurintricarboxylic acid (ATA) shows very little degradation and is suitable for northern blot analysis. The inhibitor, which is difficult to remove from the RNA, inhibits many enzymes and is not recommended for RNA that will be used for cDNA cloning or primer extension. However, most mtRNA preparations from maize ear shoots are quite good, even when no ATA is added, if isolations are done as quickly as possible and care is taken to avoid nuclease contamination.

For the same amount of ear shoots, the grinding and differential centrifugation steps (1–4) are the same as for DNA preps, although a different homogenization buffer can be used (RNA homogenization buffer). When enough material is available, mitochondria are purified on gradients, prior to extracting the RNA.

Preparing the Gradients

Sucrose step gradients are made with sterile 60%, 47%, 35%, and 20% sucrose solutions (in 10 mM tricine, 1 mM EGTA, pH 7.5) in ultraclear tubes prior to starting the prep. The solutions are layered smoothly but rapidly enough to cause slight distortions at the interfaces. The tubes are covered with Parafilm and placed at 4°C prior to use. If the SW41 rotor is used, 1.0 ml 60% sucrose, 2.5 ml 47% sucrose, 3.0 ml 35% sucrose, and 1.5 ml 20% sucrose is added to each tube. If larger or smaller gradients are to be run (e.g., SW27 or SW60), the amounts of the sucrose solutions are adjusted proportionately.

Steps 1–4. See DNA Procedure Above.

5. Gradient purification of mitochondria: Resuspend the crude mitochondrial pellet from the 10,000*g* centrifugation in a total of 2–3 ml cold RNA homogenization buffer (or RNA wash buffer). Layer the mitochondria carefully onto the sucrose gradients and centrifuge at 30,000 rpm in a SW41 rotor for 60 minutes at 4°C (with brake off for the last 800 rpm of deceleration). If larger amounts of starting material are used, mitochondria can be banded using the SW28 rotor (25,000 rpm for 60–90 minutes). Mitochondria band at the 35%/47% interface.
6. Diluting mitochondria: Remove the 20% and part of the 35% sucrose layer. Collect the mitochondria from the 35–47% interface with a sterile plastic dispo-pipette and transfer them to a 30-ml Corex tube on ice. Over 15–20 minutes, dilute the mitochondria with a total of 3 volumes of RNA wash buffer. Start by adding single drops and gradually increase the number of drops. Note: Rapid dilution will lyse the mitochondria. Pellet mitochondria by centrifuging at 10,000*g* (9,200 rpm in SS 34 rotor) for 15 minutes at 4°C.
7. Lyse mitochondria: Add 2–5 ml of RNA wash buffer to the drained pellets and resuspend. If ATA is to be used, add $1/100^{th}$ volume of 100 mM ATA stock. Lyse pellets by adding one-fourth volume RNA lysis buffer. Keep on ice 5 minutes, swirling until turbidity of solution clears.
8. Extractions: Extract at least twice with an equal volume of 1:1 (v/v) mix of chloroform and redistilled phenol (equilibrated with 10× TE pH 8.0 and containing 0.1% 8-hydroxyquinoline). Extract until there is a neglible interface. Then extract once with chloroform. If ATA is used, some loss to the organic phase will occur and more ATA should be added ($1/200^{th}$ volume) after every two extractions.
9. Precipitate RNA with lithium chloride: Estimate volume of lysate and add one-fifth volume of ice-cold 12 M LiCl (final concentration 2 M LiCl). Place at 4°C overnight

(minimum 6 hours). Pellet RNA by centrifugation at 10,000g (9,200 rpm in SS34 rotor) for 40 minutes at 4°C. DNA and tRNAs remain in the supernatant. Resuspend pellets in 1× TE, pH 8.0 (containing 50 μM ATA if ATA+ prep), approximately 5 μl/g starting material.

10. Second lithium chloride precipitation: Transfer RNA to microfuge tubes. Add cold 12 M LiCl to a final concentration of 2 M to reprecipitate RNA (4°C for 6 hours). Pellet RNA by full-speed centrifugation in a microcentrifuge (eppendorf) at 4°C, 20 minutes. Carefully remove supernatant from the microfuge tube. Keep the RNA pellets on ice.
11. Rinse RNA pellets with 300 μl 100% ethanol and centrifuge 10 minutes at 4°C, full speed in the microfuge. Carefully remove supernatant. Dry pellet briefly with tubes laid on sides (not inverted) as pellets may slide.
12. Resuspend RNA pellets in 1× TE, pH 8.0 (with 50 μM ATA, freshly diluted from the stock solution, if used). Use 2 μl/g starting material. Aliquot and quick-freeze aliquots on dry ice. Store at −80°C.

PROTEIN SYNTHESIS IN ISOLATED MITOCHONDRIA

In organello protein synthesis experiments work only when materials are freshly harvested. The procedures described follow those of Leaver et al. (1983) with only minor modifications. The young ears are excellent source materials. With seedling shoots, contamination by bacteria is common. It is imperative that a "minus energy" control be included in all *in organello* studies. (See Newton and Walbot 1985.)

For the same amount of ear shoots, the grinding and differential centrifugation steps (1–4) are the same as for DNA preps; a different homogenization buffer can be used. (MOPS homogenization buffer.) For protein synthesis studies the mitochondria are purified on sucrose gradients; the gradients are prepared as described above for RNA.

Steps 1–4. See DNA Procedure Above.

5. Gradient purification of mitochondria: Resuspend the crude mitochondrial pellet from the 10,000g centrifugation in a total of 1.5–3 ml cold MOPS homogenization solution (+0.2% BSA). Layer the mitochondria carefully onto the sucrose gradients and centrifuge at 30,000 rpm in a SW41 rotor for 60 minutes at 4°C (with brake off for the last 800 rpm of deceleration). If smaller amounts of starting material (e.g., single ear shoots) are used, mitochondria can be banded using the SW60 rotor (35,000 rpm for 50–60 minutes). Mitochondria band at the 35%/47% interface.
6. Diluting mitochondria: Remove the 20% and part of the 35% sucrose layer. Collect the mitochondria from the 35–47% interface with a sterile plastic dispo-pipette and transfer them to a 30 ml Corex tube on ice. Over 15–30 minutes, dilute the mito-

chondria with a total of 3 volumes of mitochondrial dilution solution. Start by adding single drops and gradually increase the number of drops. Note: Rapid dilution will lyse the mitochondria. Pellet mitochondria by centrifugation 10,000*g* (9,200 rpm in SS 34 rotor) for 15 minutes at 4°C.

Inhibitors can be prepared during this centrifugation. For cycloheximide and streptomycin, 5 mg of each is dissolved in 500 μl cold sterile water. Five milligrams of erythromycin is dissolved in 250 μl of ethanol; then 250 μl of cold sterile water is added. From the −80°C freezer thaw [^{35}S]-methionine, the amino acid (minus methionine) mix and, on ice, the 1 M DTT, 0.26 M ADP, and the 0.1 M GTP stock solutions.

7. Resuspend the mitochondrial pellet in 5 ml of 0.4 M mannitol, 10 mM tricine pH 7.2, and 1 mM EGTA buffer, and transfer to a 15-ml Corex tube. Pellet the mitochondria again at 10,000*g* for 15 minutes. Make up 1× labeling buffer: 5 ml cold 2× label buffer, 4.5 ml cold, sterile water, 50 μl KPO_4, and 20 μl 1 M DTT.

8. Drain the final mitochondrial pellets. Resuspend in approximately 1 ml of 1× labeling buffer for each 20–40 g starting material (if ear shoots), using pipetter with wider-bore (cut off) 1 ml tips.

9. Labeling mitochondrial proteins: Add 230 μl of resuspended mitochondria to each of four microfuge tubes. Add 3 μl of amino acid mix (minus methionine) to all four tubes. The first tube will be a minus energy control and will only receive label in addition to the amino acid mix. To each of the other 3 tubes, add 3 μl 0.1 M GTP and 7 μl 0.26 M ADP (pH 6.8–7). To each of the +energy tubes, add 1 μl of a different inhibitor solution: cyclohexamide (inhibits cytosolic protein synthesis), erythromycin (inhibits bacterial and chloroplast protein synthesis), chloramphenicol (inhibits organellar and bacterial protein synthesis). Mix and add 5–7 μl (50–70 μCi) of [^{35}S]-methionine (>1,000 Ci/mmol). Mix, cover with foil, and shake (200 rpm) at room temperature for 60–90 minutes.

10. Estimating radioactive incorporation: Pipet 5 μl of each reaction onto a small piece of Whatman 3MM paper (numbered with a lead pencil) and let air dry for 1–2 minutes. Place the filters in a beaker with approximately 30 ml of 10% ice-cold TCA for 10 minutes. To remove the unincorporated radioactivity, wash filters with hot (approximately 90°C) 5% TCA (10 minutes) and then with 5% TCA at room temperature three times (5 minutes each). Rinse the filters at room temperature with 95% ethanol (approximately 20 ml), 5 minutes, ethanol:ether for 5 minutes, and ether for 5 minutes. Air dry the filters and estimate radioactivity in a scintillation counter.

11. Aliquot mitochondria: After the reactions are completed, aliquot each reaction into microfuge tubes (usually 50-μl aliquots); centrifuge for 4 minutes at 4°C at full speed in a microfuge. Remove the supernatant, quick freeze the pellets on dry ice, and store at −80°C.

REFERENCES

Hanson MR, Boeshore ML, McClean PE, O'Connell MA, Nivison HT (1985) The isolation of mitochondria and mitochondrial DNA. Methods Enzymol 118: 437–453

Kemble RJ, Gunn RE, Flavell RB (1980) Classification of normal and male-sterile cytoplasms in maize II. Electrophoretic analysis of DNA species in mitochondria. Genetics 95: 451–458

Newton KJ, Coe EH (1986) Mitochondrial DNA changes in abnormal growth (nonchromosomal stripe) mutants of maize. Proc Natl Acad Sci USA 83: 7363–7366

Stern DB, Newton KJ (1985) Isolation of plant mitochondrial RNA. Methods Enzymol 118: 488–496

Leaver CJ, Hack E, Forde B (1983) Protein synthesis by isolated plant mitochondria. Methods Enzymol 97: 476–484

Newton KJ, Walbot V (1985) Maize mitochondria synthesize organ-specific polypeptides. Proc Natl Acad Sci USA 82: 6879–6883

95

Isolation of Maize Chloroplasts and Chloroplast DNA

STEVEN RODERMEL

This procedure is for the isolation of chloroplasts and highly purified chloroplast DNA from 1 kg of maize seedling leaves. Because starch reduces the yield of intact chloroplasts, light-grown seedlings are placed into darkness for 1–2 days prior to DNA isolation. All manipulations should be performed at 4°C, unless otherwise noted.

MATERIALS

Buffer A: 0.3 M mannitol, 50 mM Tris-HCl, 3 mM disodium EDTA, 1 mM 2-mercaptoethanol, 0.1% bovine serum albumin, pH 8.0

The Maize Handbook—M. Freeling, V. Walbot, eds.

Buffer B: 0.3 M sucrose, 50 mM Tris-HCl, 20 mM disodium EDTA, pH 8.0
Sucrose pad solution: 0.5 M sucrose, 50 mM Tris-HCl, 20 mM disodium EDTA, pH 8.0
DNase solution: 5 mg/ml DNAse (Sigma DN-100), 1× SSC (0.15 M NaCl, 0.015 M sodium citrate), 5 mM $MgCl_2$
Sucrose gradient solutions: 5% (or 30%) sucrose, 1 M NaCl, 10 mM disodium EDTA, pH 8.0
Resuspension buffer: 50 mM Tris-HCl, 20 mM disodium EDTA, pH 8.0

Sodium Sarkosyl-NL 97 (ICN Pharmaceuticals)
Pronase (Calbiochem grade B), 1.5 mg in 0.3 ml distilled water, self-digested for 2 hours at 37°C
Miracloth (Calbiochem)
SS-34, GSA, and GS-3 centrifuge tubes and rotors (Sorvall)
SW-40 centrifuge tubes and rotor (Beckman)
Waring blender (or food processor)
Tissue homogenizer (Wheaton, 60 ml capacity, type "A")

METHOD

1. Homogenize leaves in a Waring blender (or food processor) in a total of 2.5 liters of ice-cold buffer A. Pour the homogenates through four layers of Miracloth, transfer the filtrates to 500 ml GS-3 bottles, and centrifuge at 3,000 rpm for 20 minutes at 4°C.

2. Gently resuspend the pellets in a small volume of buffer A (~15 ml), pool the samples into one GSA bottle (~200 ml total), and centrifuge as above. (Gentle resuspensions are conveniently performed using a camel hair brush.)

3. Aspirate off the supernatant and gently resuspend the pellet in 60 ml of buffer A. Disperse the plastids by three gentle passes through a chilled tissue homogenizer; then transfer the suspension to a flask on ice. Rinse the GSA bottle with 40 ml of buffer A, pass the rinse through the homogenizer three times, and then add it to the flask (100 ml total plastid suspension).

4. To remove contaminating nuclear and mitochondrial DNAs, add 2 ml of the DNase solution and 1 ml of 1 M $MgCl_2$ to the 100 ml plastid suspension. Incubate 1 hour on ice; then add 2 ml of 1 M disodium EDTA (pH 8.0).

5. Layer the plastid suspension onto a 100 ml sucrose pad solution in a GSA tube and centrifuge at 6,000 rpm for 20 minutes at 4°C. Aspirate off the supernatant; then wash the plastids by gently resuspending the pellet in 200 ml of buffer B. Centrifuge at 6,000 rpm for 20 minutes at 4°C. Repeat the washing step. Aspirate off the final supernatant and gently resuspend the pellet in 20 ml of resuspension buffer. Transfer the suspension to a flask.

6. Add 5 ml of 10% Sarkosyl (2% final w/v) and 0.3 ml of self-digested Pronase (final concentration of ~60 μg/ml) to the flask and incubate for 60 minutes at 37°C.
7. Transfer the lysed chloroplast suspension to two SS-34 tubes and extract three times with equal volumes of 0.1 M Tris-HCl (pH 8.0)-saturated phenol. Centrifuge at 7,000 rpm for 10 minutes (SS 34 rotor) to separate the organic and aqueous phases. Precipitate the nucleic acids from the final aqueous phase by adding 2.5 volumes of 100% ethanol (at −80°C for 30 minutes or at −20°C overnight); then collect by centrifugation at 10,000 rpm for 30 minutes at 4°C.
8. Wash the pellets with 70% ethanol; invert the tubes to drain excess ethanol; then resuspend the pellets in a total of 2 ml of resuspension buffer.
9. To obtain highly purified chloroplast DNA (free from contaminating RNA), load the suspension onto four 12.8-ml 5–30% sucrose gradients and centrifuge at 21,000 rpm in a Beckman SW40 rotor for 15 hours at 4°C.
10. Remove all but the bottom 4 cm of each gradient, fill the tubes with resuspension buffer, and centrifuge at 35,000 rpm for 16 hours at 4°C.
11. Rapidly pour off the supernatants, invert the tubes to drain excess sucrose, and then resuspend the pellets in a total of 0.5 ml of resuspension buffer. Determine the absorbance at 260 nm. (A_{260} of a 1 mg/ml solution is 20.) This procedure typically yields 100–200 μg of chloroplast DNA.

FURTHER READING

Bogorad L, Gubbins EJ, Krebbers E, Larrinua IM, Mulligan BJ, Muskavitch KMT, Orr EA, Rodermel SR, Schantz R, Steinmetz AA, DeVos G, Ye YK (1983) Cloning and physical mapping of maize plastid genes. Methods Enzymol 97: 524–554

Kolodner R, Tewari KK (1975) The molecular size and conformation of the chloroplast DNA from higher plants. Biochim Biophys Acta 402: 372–390

Kolodner R, Tewari KK, Warner RC (1976) Physical studies on the size and structure of the covalently closed circular chloroplast DNA from higher plants. Biochim Biophys Acta 477: 144–155

96

In vitro Capping of Maize Mitochondrial RNA and Transcription Initiation Site Characterization by RNase Protection

R. MICHAEL MULLIGAN

Unlike nuclear mRNAs, transcripts from prokaryotes and organelles are not capped at the 5′ terminus with a guanine residue. They can be capped in vitro with guanylyltransferase. Guanylyltransferase (also known as GTP transferase) from vaccinia virus has been extensively studied by Moss and colleagues (Moss 1981) and utilizes GTP to incorporate GMP into a 5′ to 5′ linkage (Fig. 96.1A). Guanylyltransferase is specific for di- or triphosphorylated transcripts (Martin and Moss 1975); thus, reaction with α-^{32}P-GTP may be used to specifically radiolabel primary, unprocessed transcripts. Transcription initiation sites for maize mitochondrial genes may be identified with an RNase protection assay by digestion of capped RNA/single-stranded DNA heteroduplexes with single-strand–specific ribonucleases (Mulligan et al. 1988a,b, 1991).

MATERIALS

in vitro Capping Reaction

- 10× capping buffer (0.5 M Tris-HCl, pH 7.9; 12.5 mM $MgCl_2$)
- 50 mM DTT
- α- ^{32}P-rGTP (3,000 Ci/mmol) in neutral buffer or ethanol
- 5 mM *S*-adenosylmethionine—prepared in 5 mM sodium acetate buffer (pH 5.0)
- Human placental RNase inhibitor (RNasin, Promega)
- DEPC-treated water
- 3.0 M sodium acetate (pH 5.2)
- 7.5 M ammonium acetate
- 100% ethanol (stored at −20°C)

The Maize Handbook—M. Freeling, V. Walbot, eds.

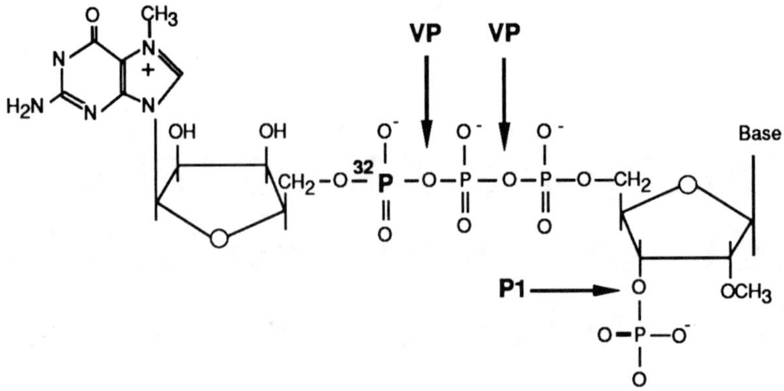

FIGURE 96.1A Enzymatic digestion of in vitro capped RNA demonstrates that guanylytransferase incorporates radiolabel from α-^{32}P-GTP into a cap structure. A. Structure of the cap showing cleavage sites by nuclease P1 (P1), and snake venom phosphodiesterase (VP).

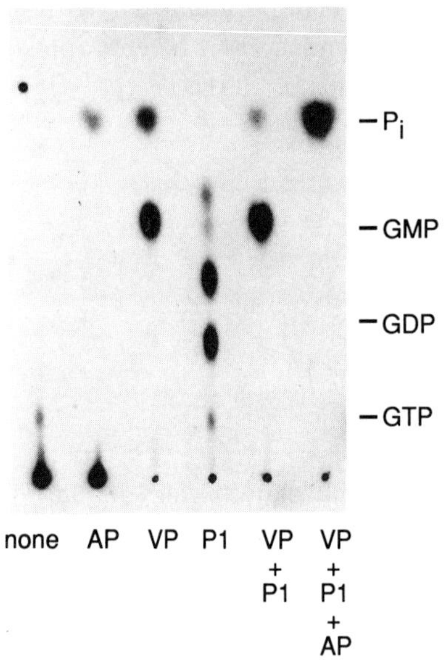

FIGURE 96.1B Enzymatic digestion of in vitro capped RNA demonstrates that guanylytransferase incorporates radiolabel from α-^{32}P-GTP into a cap structure. B. Maize mtRNA was capped as described, but *S*-adenosylmethionine was omitted from the reaction so that methylation of the cap would not affect the mobility on TLC; 10 μg of capped mtRNA was passed twice over a Biogel P-10 column to remove unincorporated α-^{32}P-GTP. Capped RNA (40 ng, 7,000 cpm) was digested with various enzymes. Digestions were buffered at pH 5.5 with 50 mM sodium acetate for nuclease P1 treatment, or at pH 9.0 with 50 mM Tris-HCl, 10 mM $MgCl_2$, and 10 mM spermidine for treatment with calf intestinal alkaline phosphatase or snake venom phosphodiesterase; 10 ng of the digested capped RNA was applied to a PEI-cellulose TLC plate and chromatographed with 0.6 M ammonium sulfate. Capped RNA samples were treated with the following enzymes and loaded in the corresponding lanes: none, no enzyme; AP, 50 units calf intestinal alkaline phosphatase; VP, 3 μg of snake venom phosphodiesterase; P1, 20 units nuclease P1; VP+P1, nuclease P1 digestion followed by snake venom phosphodiesterase treatment; AP+VP+AP, nuclease P1 digestion followed by snake venom phosphodiesterase and alkaline phosphatase treatment.

70% ethanol (stored at −20°C)
Guanylyltransferase (BRL # 8024SA)
0.5 M EDTA
10 mg/ml proteinase K
1% SDS
Phenol:chloroform (1:1)

RNase Protection Assay

Ribonuclease A (DNase-free, RNA sequencing grade)
Ribonuclease T1 (DNase-free, RNA sequencing grade)
Yeast tRNA
5× hybridization buffer (200 mM PIPES-NaOH, pH 6.4; 2 M NaCl; 5 mM EDTA)
Recrystalized formamide
RNase digestion buffer (10 mM Tris-HCl, pH 7.7; 300 mM NaCl; 5 mM EDTA)
Glycogen (optional)
Isopropanol
Electrophoresis sample buffer (90% formamide, 1× TBE, 0.1% bromphenol blue, 0.1% xylene cyanole)

Labeling Reaction

1. Transfer 250 μCi of α-^{32}P-rGTP to a microfuge tube and dry in a Speed-Vac.
2. Precipitate 100 μg of total mitochondrial RNA with sodium acetate and ethanol. Collect the RNA by centrifugation for 10 minutes, wash the pellet with 70% ethanol, and dry in a Speed-Vac.
3. Add the following reagents to the dry RNA:

Stock reagent	*Vol (μl)*
10× capping buffer	2
50 mM DTT	1
RNasin (80 units)	2
5 mM *S*-adenosylmethionine	2
DEPC-treated water	11

4. Dissolve the RNA with the reaction mixture and transfer the entire aliquot into the tube containing the dry α-^{32}P-rGTP. Some of the 11 μl of water may be reserved and used to rinse the tube to obtain a quantitative transfer. Dissolve the α-^{32}P-rGTP with the reaction mixture by repeated pipeting with a Pipetman.

5. Add 2 μl (14 units) of guanylyltransferase and incubate for 30 minutes at 37°C.
6. Terminate the reaction with 1 μl of 0.5 M EDTA. Add 2 μl of 10 μg/μl proteinase K and 2.3 μl of 1% SDS and incubate for 15 minutes at 37°C.
7. Extract with 25 μl phenol:chloroform (1:1). Precipitate the capped RNA with 15 μl of 7.5 M ammonium acetate and 120 μl of 100% ethanol at −80°C for 1 hour. Two precipitations with ammonium acetate will remove most of the unincorporated nucleotide. Store the capped RNA as an ethanol precipitate at −80°C.

The products of the capping reaction may be tested by probing the sensitivity of the cap structure to digestion with several enzymes (Figure 96.1B). The radioactive phosphate residue (Figure 96.1A) is in a 5′ to 5′ linkage and is inaccessible to alkaline phosphatase, a phosphomonoesterase. Digestion of capped RNA with alkaline phosphatase did not release radiolabel (Figure 96.1B, lane AP; the small amount of inorganic phosphate is from residual α-^{32}P-GTP in the preparation; note the GTP in the undigested lane ("none"). Snake-venom phosphodiesterase should digest and release the radiolabel as GMP (Figure 96.1A), and TLC of this sample confirmed this prediction (lane VP). Nuclease P1 does not digest the 5′ to 5′ linkage of the cap structure (Banerjee 1980), and TLC of capped RNA digested with nuclease P1 indicated that two major products are generated with chromatographic properties consistent with the presence of the dinucleotide product (lane P1). This is further corroborated by the ability of phosphodiesterase to convert these compounds to GMP (lane VP+P1). Finally, digestion with all three enzymes releases the radiolabel as inorganic phosphate (lane VP+P1+AP).

Notes

1. The reaction should be done in a minimal volume. The K_m for GTP is about 20 μM (Moss 1981), but the labeling reaction contains only ~4 μM GTP to avoid handling excessive amounts of radioisotope. *S*-adenosylmethionine is included to help drive the reaction toward completion. The amount of RNA may be reduced to as little as 10 μg; however, the volume should be kept to an absolute minimum.
2. Polypropylene is brittle at low temperatures and must be allowed to warm before centrifugation.
3. Inhibitory activity of RNasin dependends on DTT.
4. Dissolution of the radiolabel can be checked by removal of the reaction mixture (for labeling reaction) or coprecipitated nucleic acids (for RNase protection) into a pipet tip and passed in front of a Geiger counter.
5. Microfuge tubes may not seal well during phenol extractions. It is advisable to use microfuge tubes that seal very tightly or lock when extracting radioactive samples.

6. The radiospecific activity of the RNA can be checked by passing the capped RNA through a spun column packed with Biogel P10 resin (BioRad) and counting an aliquot with a scintillation counter. Typical incorporation of radiolabel is about 100,000 cpm/μg of RNA.
7. A simple and reliable test for effective capping is to probe a DNA gel blot of mitochondrial genes with the capped RNA. (See Mulligan et al. 1988b).

RNase PROTECTION ASSAY

1. Add to a microfuge tube 200 ng of radiolabeled capped mtRNA, 200 ng of antisense single-stranded DNA, and 5 μg of yeast tRNA. Precipitate with 0.1 volume of 3.0 M sodium acetate and 3 volumes of 100% ethanol at −20°C overnight or at −80°C for 1 hour. Control reactions should include substitution of 200 ng of the sense DNA, omission of DNA, and an extra reaction that will have RNase omitted.
2. Collect the nucleic acids by centrifugation for 10 minutes in a microfuge and wash the pellet with ice-cold 70% ethanol. Dry the pellet briefly with a Speed-Vac.
3. Prepare 1× hybridization solution (10 μl/digestion) by mixing 1 volume of 5× hybridization buffer with 4 volumes of recrystalized formamide (Favaloro et al. 1980). Dissolve the precipitated nucleic acids in 10 μl of 1× hybridization solution by repeated pipeting in and out of a disposable pipet tip. Check for dissolution of the radioactivity.
4. Denature the nucleic acids at 75°C for 5 minutes and transfer directly to a 42°C water bath. The optimal temperature will vary depending on nucleic acid sequence, length of heteroduplex, etc. Allow the nucleic acids to hybridize for at least 4 hours to overnight.
5. Prepare the RNase digestion solution. Typical conditions for RNase digestion are 250 ng of RNase A plus 200 units of RNase T1 per 300 μl digestion. The appropriate volume of RNase digestion buffer (number of reactions × 0.33 ml) should be equilibrated to 0°C in an ice-water bath. Immediately prior to use, RNase A is added to a concentration of 833 ng/ml and RNase T1 is added to a concentration of 666 units/ml. Add 0.3 ml of RNase digestion solution directly to the renatured nucleic acids while still in the water bath at the hybridization temperature, and immediately remove the microfuge tube from the water bath. Rapidly mix the contents with a flick of your finger and collect the solution at the bottom of the tube with a snap of your wrist. Place the microfuge tube in an ice water bath and move on to the next sample. The important principle at this step is that the renaturation occurs at the specified annealing temperature and at high concentration of nucleic acids, and hybridization is terminated by 30-fold dilution of the nucleic acids and rapid decrease of the temperature to 0°C. Thus, it is very important to rapidly dilute and cool the sample.
6. After the RNase digestion solution has been added to all the samples, move them to a 30°C water bath and incubate 1 hour.

7. Terminate the reaction with the addition of 30 μl of 1% SDS and 5 μl of 10 μg/μl proteinase K. Incubate at 37°C for 15 minutes.
8. Extract the nucleic acids with 300 μl phenol:chloroform (1:1). Add 1 μg of carrier tRNA or 1 μg of glycogen. Add 400 μl of isopropanol to precipitate the nucleic acids. (Do not add salt—there is already 0.27 M NaCl present from the RNase digestion buffer.) Incubate at −20°C overnight or at −80°C for 1 hour. Collect the nucleic acids by centrifugation in a microfuge for 10 minutes. Wash the pellet with 70% ethanol; this is a critical wash to remove the isopropanol that will not dry in a Speed-Vac. Dry the pellets in a Speed-Vac and dissolve the sample in 10 μl of electrophoresis sample buffer. Denature sample at 75°C for 5 minutes and electrophorese 3–5 μl of sample on a denaturing polyacrylamide gel and analyze by autoradiography.

Notes

1. The protocol for nuclease S1 protection assays by Favalaro et al. (1980) provides excellent background information on transcript mapping. An excellent discussion of temperature-related artifacts in nuclease S1 protection analyses may be found in Sazer and Schimke (1986).
2. Examples of RNase protection analysis may be found in Figure 2 and Figure 6 of Mulligan et al. (1988a) and Figure 4 of Mulligan et al. (1988b).
3. Sequencing gels or 15 × 25-cm gels with 0.5-mm spacers may be used to obtain good resolution for transcript mapping. Size standards can be prepared by phosphorylation of DNA size standards with γ-^{32}P-ATP and T_4 polynucleotide kinase; however, DNA fragments migrate about 10% further than similar-length RNAs. Thus, sizing of the protected RNA fragments is difficult. Precise identification of the transcription initiation site should be confirmed by primer extension analysis or direct sequence analysis of the capped RNA (Covello and Gray 1991).

REFERENCES

Banerjee AK (1980) 5′-terminal cap structure in eucaryotic messenger ribonucleic acids. Microbiol Rev 44: 175–205

Covello PS, Gray MW (1991) Sequence analysis of wheat mitochondrial transcripts capped in vitro: definitive identification of transcription initiation sites. Curr Genet 20: 245–251

Favaloro J, Treisman R, Kamen R (1980) Transcription maps of polyoma virus-specific RNA: analysis by two-dimensional nuclease S1 gel mapping. Methods Enzymol 65: 718–749

Moss B (1981) 5′ end labeling of RNA with capping and methylating enzymes. In Chirikjian JG, Papas TS (eds), Gene Amplification and Analysis, Elsevier, New York, pp 253–266

Martin SA, Moss B (1975) Modification of RNA by mRNA guanylyltransferase and mRNA (guanine-7) methyltransferase from vaccinia virions. J Biol Chem 250: 9330–9335

Mulligan RM, Lau GT, Walbot V (1988a) Numerous transcription initiation sites exist for the maize mitochondrial genes for subunit 9 of the ATP synthase and subunit 3 of cytochrome oxidase. Proc Natl Acad Sci USA 85: 7998–8002

Mulligan RM, Maloney AP, Walbot V (1988b) RNA processing and multiple transciption initiation sites result in transcript size heterogeneity in maize mitochrondria. Mol Gen Genet 211: 373–380

Mulligan RM, Leon P, Walbot V (1991) Transcriptional and posttranscriptional regulation of maize mitochrondrial gene expression. Mol Cell Biol 11: 533–543

Sazer S, Schimke RT (1986) A re-examination of the 5′ termini of mouse dihydrofolate reductase RNA. J Biol Chem 261: 4685–4690

97

Editors' Note

Many successful variations on the Southern blotting procedure are described in existing molecular biology manuals. Two groups working with maize genomic DNA have contributed their lab protocols to illustrate some of the "tricks of the trade" that can be useful. For many applications, quantitative information is required from a Southern blot, i.e., distinguishing diploid from haploid copy number or quantifying the copy number of a transposable element. To provide an internal control, reconstruction lanes, which contain a mixture of known amounts of plasmid DNA (i.e., 1 ng, 3 ng, 10 ng) and known amounts of genomic DNA (1 μg, 3 μg, 10 μg), are usually included as separate lanes on the gel, and the intensity of hybridization in sample lanes (which may vary in DNA concentration) is compared to the standards. For laser densitometry of blots, a light exposure is required to ensure that the film was exposed within its linear range of sensitivity (where silver grain density is proportional to radioactive decay from the blot). Somewhat less than one order of magnitude of intensity differences can be reliably measured from a single film; two film exposures of the same blot can be used to evaluate reconstruction lanes over 1–1.5 orders of magnitude of intensity. Qualitative comparisons (i.e., RFLP) can be made without reconstruction lanes and when sample lanes contain different amounts of DNA; however, it is easier to prepare useful exposures when the amount of DNA loaded is similar per lane.

The Maize Handbook—M. Freeling, V. Walbot, eds.

Southern Blots of Maize Genomic DNA

CHRISTINE A. WARREN AND JANE HERSHBERGER

MATERIALS

DNA size markers
10× gel loading buffer
  0.25% bromophenol blue
  0.25% xylene cyanol
  50% glycerol

10 mg/ml ethidium bromide
1× TBE
  89 mM Tris-OH
  89 mM boric acid
  2.5 mM EDTA

Make as a 20× stock; adjust pH to 8.3 using boric acid; filter through a 0.45 μM nitro-cellulose membrane to prevent precipitation

Denaturing solution
  1.5 M NaCl
  0.5 M NaOH
Neutralizing solution
  3 M NaCl
  0.5 M Tris-HCl pH 7.5
20× SSPE
  20 mM Na_4EDTA
  0.2 M $NaH_2PO_4 \cdot H_2O$
  0.16 M NaOH
  3.6 M NaCl
  pH 7.5
20 X SSC
  3 M NaCl
  0.3 M Na citrate ($Na_3C_6H_5O_7 \cdot 2H_2O$)

The Maize Handbook—M. Freeling, V. Walbot, eds.

Nylon membrane filter (see note 2)
Blotting paper (Schleicher & Schuell 6B002)
Pre-Prehybridizing solution
  0.1 × SSPE
  0.5% SDS
Prehybridizing/hybridizing solution
  50% deionized formamide
  3× SSPE
  5× Denhardt's
  100 μg/ml denatured salmon sperm DNA
  1% SDS
Blot washing solution
  0.1× SSPE
  0.1% SDS
0.4 M NaOH
Stripping buffer
  0.2 M Tris pH 7.5
  0.1 × SSPE
  0.5% SDS
Equipment for restriction digests, agarose gel electrophoresis, and autoradiography

METHODS

1. Digest 2.5–5 μg of DNA in the greatest volume possible for the well size. Add ~4 units enzyme/μg DNA, digest for 2 hours, add another aliquot of enzyme, and digest for 2–4 more hours. When digestion is complete, add 10% by volume of 10× gel loading buffer.
2. In a 13.5 × 20-cm gel box, pour a 300-ml, 0.8% agarose gel in 1× TBE with 10 μg/ml ethidium bromide, using a 12- to 18-tooth comb. Load samples (up to 70 μl) and a lane of size markers, and run at 30–40 V overnight.
3. Photograph the gel with a transparent ruler laid alongside the size markers and transfer the gel to a shallow dish. (Optional: To enhance the transfer efficiency for fragments greater than 5 kb, depurinate the DNA by covering the gel with 0.2 N HCl and rocking gently for 10 minutes. Rinse the gel briefly in H_2O twice before proceeding.) Add denaturing solution to cover the gel. Denature with gentle agitation for 45 minutes, replacing the denaturing solution every 15 minutes; then neutralize for 45 minutes, replacing the neutralizing solution every 15 minutes.
4. Put several hundred milliliters of 10× SSC into a shallow baking dish and set a plastic bridge across the sides of the dish. Lay a 25 × 40-cm wick of blotting paper across the bridge with both ends in the 10× SSC reservoir. Cut the nylon membrane to the size of the gel, wet it in H_2O for 2 minutes, and then transfer it to 2× SSC for 2

minutes. In addition, cut two pieces of blotting paper to the size of the gel. To assemble the transfer setup, invert the gel onto the wick and remove any air bubbles by smoothing with a 10-ml pipette. Place the nylon membrane on top of the gel, remove the air bubbles, and place two pieces of blotting paper on top of the membrane. Surround the gel with strips of old X-ray film to prevent the transfer buffer from wicking around the gel. Finally, lay 2 inches of single-fold paper towels onto the blotting paper and place a book on top of the stack to weight it evenly. Allow the transfer to proceed overnight.

5. To disassemble, remove the weight, paper towels, and blotting paper. Invert the filter with the gel still attached onto Saran wrap. Mark through the wells onto the nylon membrane with a pencil, remove the gel, and trace over the pencil marks with an indelible VWR pen. Float the filter on 2× SSPE for 5 minutes, dry, and bake for 2 hours at 80°C between two pieces of blotting paper.
6. For Genetran, "pre-prehyb" the filter in 0.1× SSPE, 0.5% SDS at 65°C for 2 hours. For Hybond-N, wet the filter in 2× SSPE. Transfer the filter to a hybridization tube or seal-a-meal bag and prehybridize for 2–24 hours at 42°C in 8–10 ml of prehybridization solution. Add denatured probe ($\sim 10^7$ cpm) to the tube/bag and hybridize overnight at 42°C.
7. Rinse the filter twice for 15 minutes with shaking in blot wash; then wash for 2 hours at the desired temperature (up to 70°C) with two changes of buffer. Blot the filter dry, wrap in Saran wrap, and expose at −80°C with screens. For best results, "double film" the filter by sandwiching it between two pieces of X-ray film, securing it to the lower one, and placing this between two intensifying screens. Develop the upper film after 1–3 days, and depending on the intensity of the first exposure, develop the lower film after 3–10 days.

Notes

1. The filter can be stripped by washing first in 0.4 M NaOH at 30°C for 30 minutes and then in stripping buffer at 45°C for 30 minutes. To store, blot the stripped filter dry, wrap it in Saran wrap, and place in a sealed plastic container at −20°C. To reprobe, wet it in 2× SSPE, prehybridize for 2 hours, and probe as usual.
2. There are many nylon membranes on the market. We have found that Genetran (Plasco) and Hybond-N (Amersham) give the greatest sensitivity.

99

Southern Blot Hybridization*

STEPHEN L. DELLAPORTA AND MARIA A. MORENO

MATERIALS

20× electrophoresis buffer
  0.8 M Sigma 7–9 (cheap Tris base)
  20 mM disodium EDTA
  0.1 N NaOH
  80 ml per liter of glacial acetic acid
  (pH should be about 7.8)
  Autoclave.

0.4 N NaOH

20× SCP
  2 M NaCl
  0.6 M disodium phosphate
  20 mM disodium EDTA
  12 ml per liter HCl
  (pH should be around 6.8)
  Autoclave.
2× "Gold Juice" hybridization buffer
To a 1 liter beaker, add in the following order:
  100 ml dH_2O
  300 ml 20× SCP
  100 g dextran sulfate (MW >500,000, Phamacia 17–0340–02)

*This is a modified version of procedure published in Chomet PS, Wessler S, SL Dellaporta (1987: *EMBO J* 6: 295–302). The changes we have incorporated are to directly transfer the DNA from agarose gels to Zetaprobe membranes (BioRad) in 0.4 N NaOH. These membranes can be reused many times with a good signal-to-noise ratio.

The Maize Handbook—M. Freeling, V. Walbot, eds.

Warm the solution to 37°C and stir this solution with a spatula until the dextran is completely dissolved. Add 40 ml 20% *N*-lauryl-sarcosine and bring volume to approximately 540 ml. (Premark a 540 ml line on the beaker and bring the final volume to this mark.) Filter sterilize the solution through 0.45 μM nitrocellulose filter unit (Nalgene). Add 540 mg heparin (Sigma H7005) and store at room temperature.

Nonstringent wash solution

2× SCP

1% SDS

1. Gel electrophoresis: 7 μl (several micrograms) of miniprep DNA is digested with 10 units of restriction enzymes in a 20-μl reaction volume according to manufacturer's conditions. Two microliters 10× stop buffer is added to the reaction and then loaded directly into the gel well (7 mm). For standard conditions, a 0.8% agarose gel is used. Dilute concentration of ethidium bromide can be added directly to the agarose solution and running buffer (1× E buffer) to avoid staining the gel after electrophoresis. Electrophoresis is carried out for 8–15 hours (usually overnight).
2. Photograph the gel under UV light. We do not acid depurinate the DNA after electrophoresis. If high-molecular-weight DNA transfer is a problem, then leave the gel on the transilluminator for an additional 1 minute to nick the DNA.
3. Pretreat the gel for 15 minutes in 250 ml 0.4 N NaOH in a small Pyrex dish with slow shaking so that the solution passes over the top surface of the gel.
4. In a large Pyrex dish, add approximately 2 liters of 0.4 N NaOH solution. Saturate a large sponge with the solution. Bring the 0.4 N NaOH solution level to just below the top of the sponge. To the top of the sponge, carefully place two sheets of blotting paper (saturated with 0.4 N NaOH), avoiding trapping any air bubbles between the two sheets of paper.
5. Place the gel on top of the sponge/blotting paper, carefully avoiding trapping air bubbles between the gel and paper.
6. Prewet a sheet of Zetaprobe membrane with 0.4 N NaOH. Be sure to mark a corner of the membrane with a pencil so that you can orient the autoradiogram later on. Place the membrane on top of the gel, carefully avoiding air bubbles. Roll a pipet over the membrane to remove trapped air. Place two sheets of saturated of blotting paper over the membrane, carefully avoiding bubbles.
7. To avoid short-circuiting the transfer of DNA from the gel to the membrane, cover the remaining surface of the sponge with strips of plastic. (These can be made from old X-ray film by first removing the emulsion. Wash and reuse them). Place the plastic strips right up to the edge of the gel.
8. Place a 4-cm stack of brown paper towels (single-fold 9.5 × 10.6 sheets; Fort Howard Corp., Green Bay, WI, product #235-04) over the gel. The stack of towels fits over

two of our standard 10 cm × 14 cm pieces of gel. Put a glass plate on top of the towels and a 1-kg weight on top of the plate.

9. Blot the gel 8–15 hours (usually overnight). Carefully remove the membrane from the gel and rinse it for 15 seconds in 0.1 M Tris, pH 7. Place the membrane (DNA side up) on a dry blotting paper. While it is still damp, UV crosslink the DNA to the membrane.

10. Place the membrane inside a hybridization vessel (e.g., large roller glass tubes) with the DNA side facing in. Several filters can be hybridized in a single vessel. Add 15 ml 1× Gold Juice to the vessel. Prehybridize the filter for 15–30 minutes at 65°C. Without removing the prehybridization solution, directly add the probe to the prehybridization solution. Hybridize at 65°C for 8–15 hours (usually overnight).

11. Pour off the hybridization solution. (Save this solution; it can be reused several times). Add 25-ml wash solution and place the tube back on the roller at 65°C for 15 minutes. Rewash the filter 1–2× with wash solution. The final wash should be done under stringent conditions (dilute wash solution 1:10 with dH_2O = 0.2× SCP, 0.1% SDS) for 15 minutes at 65°C.

12. Blot excess liquid from membrane with paper towels and place the damp membrane between two sheets of Saran wrap and expose to film. (Drying out the membrane will make it impossible to effectively strip the probe from the filter for rehybridization).

Notes

1. After blotting, cover the sponge unit with Saran wrap until the next use. It is not necessary to wash out the unit before each experiment. We "reuse" the sponge many times, periodically rinsing out the NaOH solution, which increases in concentration from evaporation.

2. Large sponges can often be found in an auto supply store or can be made from sheets of foam rubber purchased from an upholstery shop.

3. Probe solutions can be reused 2–3 times for up to 2 weeks. Store the solution in disposable polypropylene tubes at room temperature in a shielded box. To reuse the solution, place the loosely capped tube in a boiling water bath for 5 minutes. Cool on ice and add to prehybridized membrane.

4. Strip the probe from membranes by shaking the filters in 0.4 N NaOH for 15 minutes at room temperature. Neutralize the filters in 0.1 M Tris, pH 7.

5. For plaque hybridization filters (nitrocellulose), follow a standard protocol for lifts and fixing DNA to filters. In a Tupperware box (or similar sealable plastic container), add 50 ml of Gold Juice WITHOUT the dextran sulfate. Carefully layer each filter into the solution, covering each one with hybridization mix. Prehy-

bridize at 65°C for 15 minutes. In another box, add 50 ml Gold Juice WITHOUT dextran sulfate containing the radiolabeled probe. Carefully layer the prehybridized filters into the hybridization solution, covering each filter with the solution. Add some of the prehybridization solution to the hybridization box if additional solution is needed. Hybridize and wash as above.

100

Northern Blotting

KENNETH R. LUEHRSEN

MATERIALS

Formamide

Deionize by adding 5 g of a mixed bed resin (i.e., Amberlite MB-1; Sigma) to 100 ml formamide and stir for 30 minutes. Remove the resin by filtration through Whatmans #1 filter paper. Store at −20°C.

37% formaldehyde

10× MOPS buffer

- 0.2 M MOPS (4-morpholinepropanesulfonic acid)
- 10 mM EDTA
- 10 mM sodium acetate pH 7

400 μg/ml ethidium bromide (EtBr)

10× RNA gel loading buffer

- 0.2% xylene cyanol
- 0.2% bromphenol blue
- 10 mM EDTA pH 7
- 50% glycerol

20× SSC

The Maize Handbook—M. Freeling, V. Walbot, eds.

1× SSC is 0.15 M NaCl, 15 mM sodium citrate pH 7

20× SSPE

1× SSPE is 0.18 M NaCl, 10 mM sodium phosphate pH 7.7, 1 mM EDTA

Hybridization solution

50% formamide

5× Denhardt's reagent (1× is 0.02% Ficoll, 0.02% polyvinylpyrrolidone, 0.02% BSA)

5× SSPE

0.1% SDS

100 μg/ml denatured salmon sperm DNA

50 μg/ml yeast tRNA

METHOD

The mRNA is first separated through a denaturing agarose gel before transfer to the blotting membrane. The amount of RNA loaded will depend on the transcript abundance for the desired gene. Generally, 15 μg of total RNA or 2 μg polyA$^+$-selected mRNA is an appropriate amount. The percent agarose gel will depend on the size(s) of the transcripts to be resolved; a 1% agarose gel works well for most purposes. See Kroczek and Siebert (1990) for additional information about this procedure.

1. Dry ~15 μg total cellular RNA in a vacuum dessicator.
2. Prepare a 1% agarose solution in 1× MOPS buffer and boil to dissolve. Cool the solution to about 50°C and, working in a fumehood, add 1/20th volume of 37% formaldehyde. Pour into a gel mold and let solidify. Immerse in running buffer (1× MOPS without formaldehyde) just before use.
3. Resupend the RNA in 2 μl DEPC-treated water. Add 5 μl formamide, 2 μl formaldehyde, 1 μl 10× MOPS buffer, 1 μl 400 μg/ml EtBr. Heat at 65°C for 10 minutes. Add 1 μl 10× loading buffer.
4. Load the samples and run the gel at ≤5 V/cm until the bromphenol blue marker dye reaches the end of the gel. Recirculate the buffer during the run.
5. Remove the gel from the electrophoresis chamber and soak for 20 minutes in 10× SSC. Place on a UV light (302 nM) box and photograph the gel. If total RNA was run, two major bands at ~3.4 kb and ~1.6 kb will be evident (corresponding to the large and small rRNAs), with a light smear of mRNA in the background.
6. Transfer the RNA to a nylon membrane by capillary action using 20× SSC.
7. Place the nylon membrane on a UV light box; all of the RNA should be transferred and be visible on the membrane. Expose the membrane with a UV transilluminator for 5 minutes; this helps covalently crosslink the RNA to the membrane. Bake at 80°C until dry.
8. Prehybridize the blot in hybridization solution without probe for ~2 hours. Add

10^6 to 10^7 cpm/ml of ^{32}P-labeled probe and hybridize overnight at 43°C for DNA probes and at 55°C for RNA probes.

9. Wash the membrane twice for 5 minutes with 2× SSC/0.1% SDS at room temperature.
10. Do a stringent washing twice for >30 minutes in 0.1× SSC/0.1% SDS. For DNA probes wash at ~65°C and for RNA probes at ~75°C. The optimal wash temperature will differ according to the GC content of the probe.
11. Autoradiograph the membrane, using intensifying screens if necessary. If the blot is to be reprobed, do not let the membrane dry completely; simply blot off the excess wash buffer and sheath it in saran wrap.

Notes

1. I use Seakem agarose from FMC BioProducts, although other brands may also work well.
2. HybondN gives a stronger hybridization signal than Nytran in my hands.
3. I generally use ^{32}P-labelled single-stranded RNA probes; try to keep the probe size <1000 nt. Larger probes will give higher backgrounds even with prehybridization.
4. Most probes can be stripped from the membrane by washing three times for 10 minutes with 0.1XSSPE, 0.1% SDS at 100°C. RNA probes may require longer washing times; those that are especially GC-rich (greater than 70%) cannot be removed.
5. RNA size standards should be run alongside with the cellular RNAs. BRL sells RNA standards covering different size ranges.

REFERENCE

Kroczek RA, Siebert E (1990) Optimization of northern analysis by vacuum-blotting, RNA-transfer visualization, and ultraviolet fixation. Anal Bioch 184: 90–95

101

RNase Protection Assay

KENNETH R. LUEHRSEN

MATERIALS

5× SP6 RNA polymerase buffer
- 200 mM Tris-HCl pH 7.9
- 30 mM $MgCl_2$
- 10 mM spermidine

5× T3/T7 RNA polymerase buffer
- 200 mM Tris-HCl pH 8
- 40 mM $MgCl_2$
- 10 mM spermidine
- 125 mM NaCl

Formamide/hybridization buffer
- 80% deionized formamide
- 20% 5× hybridization buffer

5× hybridization buffer
- 200 mM PIPES pH 6.7
- 2 M NaCl
- 5 mM EDTA

RNase H buffer
- 10 mM Tris-HCl pH 7.5
- 0.3 M NaCl
- 10 mM $MgCl_2$
- Add RNase H to 10 units/ml before use

RNase mix solution
- 10 mM Tris pH 7.5
- 0.3 M NaCl
- 5 mM EDTA
- 2× RNase from a 50× stock (add fresh)

50× RNase mix

The Maize Handbook—M. Freeling, V. Walbot, eds.

2 mg/ml RNase A
0.1 mg/ml RNase T1
Gel loading buffer
95% formamide
20 mM EDTA
0.05% bromphenol blue
0.05% xylene cyanol FF

METHOD

The probe is prepared by first cloning a DNA fragment corresponding to the probe sequence in a vector, such as pBluescript (Stratagene), capable of generating single-stranded RNAs; these vectors have T3, T7, or SP6 phage RNA polymerase promoters abutting the polylinker region. The optimal insert/probe size is between 200 and 1,000 bp; because the plasmid must be linearized before probe synthesis (see below) the insert can be as long as you wish, as long as the final probe size is approximately what is suggested above. The plasmid is linearized either downstream of the insert or within the insert, if all of the insert is not required as the probe. For linearization, use a restriction enzyme leaving either a 5′ overhang or a blunt end; 3′ overhangs have been shown to act as initiation points for the phage polymerases. Also, you can linearize with an enzyme that cleaves the plasmid multiple times as long as the phage promoter and insert remain together. Remember to choose the appropriate polymerase/RE digest to generate the *antisense* RNA. After linearization, the plasmid is purified by phenol/chloroform extraction and ethanol precipitation. The cut plasmid is resuspended in DEPC-treated water at a concentration of 1 mg/ml. See Krieg and Melton (1987) for additional information about *in vitro* transcription of RNA.

Probe Preparation

1. In a 0.5 ml microfuge tube

5× phage polymerase buffer	2 μl
ATP, UTP, GTP each	1 μl
10 mM and pH 7.5	
5 mM DTT	0.5 μl
DEPC-treated water	2 μl
^{32}P-CTP	3 μl
20 mCi/ml 800 Ci/mM	
Mix	
Linearized plasmid	0.5 μl
1 mg/ml	
Mix	

RNasin 20–40 units/μl	0.5 μl
Phage polymerase 10–50 u/μl	0.5 μl

Incubate at 40°C for 30 minutes.

2. Add 0.5 μl RQ1 DNase (Promega Corp) @ 1 units/μl and incubate an additional 15 minutes at 37°C. The DNase treatment removes the template DNA.
3. Remove 10 μl and transfer to a new tube containing 90 μl of DEPC-treated water and extract once with an equal volume of phenol/chloroform. The unincorporated nucleotides must be removed, and this can be done by either multiple (≥3) ethanol precipitations or by gel filtration over P10 (BioRad).

Notes

1. UTP and GTP can also be used as the labeled nucleotide; ATP is not recommended.
2. Generally expect about 100 ng of labeled RNA to be synthesized. To determine the approximate specific activity, count 2 μl in a scintillation counter and divide by 100 ng.

RNase Protection

1. The following protocol is slightly modified from that described by Goodall et al. (1990). To set up the hybridization use a 0.5 ml eppendorf tube and add ~10^5 cpm (Cerenkov) of probe RNA to 10–20 μg purified RNA (i.e., after centrifugation through CsCl). Bring the volume up to 50 μl with DEPC-water. Add 50 μl 2.5 M ammonium acetate and 300 μl of 100% ethanol. Place on ice for at least 30 minutes and microfuge >15 minutes to collect the precipitate. Wash the pellet with 100% EtOH and dry.
2. Resuspend the pellet in 10 μl formamide/hybridization buffer and heat at 80°C for 2 minutes. Place the tubes at 45°C overnight.
3. Add 50 μl RNase H solution and incubate at 30°C for 30 minutes. RNase H will degrade RNA from an RNA/DNA duplex and will thus remove any probe protected by DNA contaminating the RNA preparation.
4. Add 50 μl RNase mix solution and incubate at 22–30°C for 30 minutes. The mixture contains both RNase A (pyrimidine-specific) and RNase T1 (G-specific). This will degrade any probe that remains single-stranded; duplex RNA will remain undegraded.
5. Remove 110 μl of the reaction and place in a fresh tube containing 2 μl of 10% SDS. Add 2 μl proteinase K (20 mg/ml in water) and incubate at 37°C for 15 minutes.
6. Phenol/chloroform extract, transfer the upper phase to a fresh tube containing 5 μg

tRNA and add 300 μl 100% ethanol. Place on ice for at least 30 minutes and microfuge >15 minutes to collect the precipitate. Wash (do not mix) the pellet with 0.5 ml 70% and then 0.5 ml 100% EtOH and dry.

7. Resuspend the entire pellet in 3 μl gel loading buffer, heat at 80°C for 2 minutes, and separate on a 6% sequencing gel. Fix the gel with 10% methanol/10% acetic acid. Dry the gel and autoradiograph with two screens for the appropriate length of time.

Notes

1. After the RNase digestions it is important to transfer the reaction to a new tube: in this way any undegraded probe adhering to the sides of the eppendorf tube will not be carried through the procedure.
2. The temperature of the RNase mix reaction can be altered to accommodate the Tm of the duplex RNA. Generally 22–30°C works well, but keep in mind that a run of seven contiguous AT base pairs can be cleaved at these temperatures. Temperatures as low as 15°C allow for sufficient RNase activity; >30°C is not recommended.
3. More than one probe can be added to each protection assay as long as the protected fragments will be resolved on the gel.
4. Always run marker RNAs (^{32}P-labeled) and a sample of the probe. (1 μl of a 1:200 dilution works well.) You must be sure that the synthesized probe is full length.
5. If DNA contaminates the cellular RNA, it can be treated with RQ1 DNase prior to the mapping procedure to remove the DNA.
6. The size of the protected fragments should be between 75 and 400 bp. Fragments smaller than 75 bp are hard to see and to distinguish from spurious digestion products; fragments greater than 400 bp become difficult to size accurately on a sequencing gel.
7. When interpreting the results, bear in mind that a visualized fragment means that some portion of the probe is protected but you cannot determine *a priori* what part of the probe that is. Unlike S1 mapping in which the probe is end-labeled, the RNA probe is continuously labeled and thus any part of it can be represented in the protected fragment.

REFERENCES

Goodall GJ, Wiebauer K, Filopowicz W (1990) Analysis of pre-mRNA processing in transfected plant protoplasts. Methods Enzymol 181: 148–161

Krieg PA, Melton DA (1987) *In vitro* RNA synthesis with SP6 RNA polymerase. Methods Enzymol 155: 397–415

102

Genomic Sequencing in Maize

ANNA-LISA PAUL AND ROBERT J. FERL

Genomic sequencing is a method for generating a genomic DNA blot from a sequencing gel. The technique was first described by Church and Gilbert (1984) as a means of visualizing the state of individual nucleotides in mouse chromosomes. The power of genomic sequencing is twofold. First, it enables a researcher to characterize genomic DNA directly; thus changes in chromatin structure, methylation and protein interactions can be followed in their native state (without cloning or PCR steps). Second, the position of each reactive nucleotide can be identified precisely through the use of indirect end-labeling and strand-specific probes.

The first application of genomic sequencing to plants was an investigation of cytosine methylation in maize (Nick et al. 1986). The primary use of genomic sequencing has been for footprinting the in vivo interactions of DNA binding factors in maize (Ferl and Nick 1987; Paul and Ferl 1991), *Arabidopsis* (Ferl and Laughner 1989), and parsley (Schulze-Lefert et al. 1989; Lois et al. 1989). The procedures outlined in the following sections deal with in vivo footprinting, but the basic techniques of sequencing-gel electrophoresis and electroblotting of genomic DNA can be utilized for a wide variety of applications.

IN VIVO FOOTPRINTING

The primary reagent used to modify DNA for footprinting in vivo is dimethyl sulfate (DMS). DMS readily passes through cell walls and membranes; it methylates guanine and (to a lesser extent) adenine. The access of DMS to a particular nucleotide can be altered by proteins bound to the DNA. The subsequent cleavage of the DNA molecule will either be enhanced (resulting in a darker band than the control) or repressed (a lighter band) if a protein is in close association with the DNA near that position.

METHOD

1. Prepare at least two cell suspensions (one to be treated with DMS and the other for the naked DNA control) that contain 3–5 g of cells in a volume of 50 ml.

The Maize Handbook—M. Freeling, V. Walbot, eds.

2. Add 100 μl DMS (99+% from Aldrich) directly to the flask containing the 50 ml of cell suspension and gently swirl the cells for 2 minutes in a fumehood. Volumes can be adjusted as long as the concentration of DMS is kept to 0.2%. Do not treat the control flask with DMS.
3. Filter the cells promptly through a Buchner funnel lined with Miracloth (Calbiochem), and wash the cells with 1 liter of water. Place 20 ml of 10 N NaOH in the filter flask to inactivate the highly toxic DMS as it is flushed from the cells. The cells for the naked DNA control should also be collected by filtering, but only one wash with about 200 ml of water (to remove excess culture medium) is necessary.
4. Freeze both sets of recovered cells with liquid nitrogen and store at −80°C until the DNA is to be extracted. For best results, extract the DNA by direct lysis of frozen, powdered tissue and then purify on CsCl gradients (Paul and Ferl 1991). (See other chapters in this section for DNA preparation methods.) DNA isolated by methods that circumvent CsCl gradients typically shows an increased level of in-lane background over that DNA purified with CsCl.

Note Whole plant organs can be treated with DMS as well (McKendree et al. 1990); organs with a simple tissue composition and gene expression profile are better suited to this type of analysis than are more complex organs.

PREPARING THE DNA FOR SEQUENCING

Before genomic DNA can be resolved on a sequencing gel, the molecules must be cut to generate a common end to facilitate indirect end-labeling (Wu 1980). The restriction site that defines the "end" should be about 300 bases away from the center of the region of interest (Figure 102.1).

1. Set up microfuge tubes containing 20 μg of DNA in 200 μl of 1× TE for each lane. There should be at least two tubes, one containing the naked (unmodified) genomic DNA from the control flask of cells, and the other from the in vivo DMS treatments. Digest each sample to completion with the restriction enzyme that defines the common end.
2. Add 0.5 μl of DMS to the naked genomic control DNA (in 200 μl 1× TE) and vortex. Let the reaction proceed for 2 minutes, then add 200 μl of phenol:chloroform:isoamyl (25:24:1). Take care not to accidentally "re-DMS" the DNAs from the in vivo treatments; this mistake would completely destroy any chromatin-influenced variations. (See note below.)
3. Phenol-extract the DNA from the in vivo treatments as well, and then extract both the naked genomic controls and the in vivo treatments with 200 μl of chloroform:isoamyl alcohol (24:1).

4. Precipitate the aqueous phases with 80 μl of 7.5 M ammonium acetate and 600 μl of 95% ethanol on ice for 15 minutes.
5. Centrifuge at top microfuge speed for 10 minutes, discard supernatant, wash pellets with 70% ethanol and then with 95% ethanol, and air dry.
6. Resuspend the pellets in 50 μl of 10% piperidine and heat at 90°C, 20 minutes.
7. Add 250 μl of distilled water to each sample, freeze in liquid nitrogen, and lyophilize until dry. Resuspend pellets in another 50 μl of water and lyophilize again.
8. Resuspend the pellets in 5 μl of standard sequencing dye. The samples can be electrophoresed on a standard sequencing-gel apparatus, but several parameters should be taken into account to achieve optimum resolution. The gel should be 0.75 mm thick to facilitate loading 20 μg of DNA and to give the gel strength to hold up to the rigors of electroblotting. The percentage of acrylamide and the length of the run depend upon the size of the targeted sequences. For example, a set of bands found between 300 and 400 bp away from the restricted end are well resolved on a 6% acrylamide gel where the xylene cyanol dye has traversed 60 cm. It is not necessary to run these gels hot; in fact they should be cool to the touch before the gel is prepared for transfer.

Note: The treatment of naked genomic DNA with DMS produces a guanine ladder that has not been influenced by any aspects of in vivo DNA/protein interaction. A guanine ladder from a plasmid containing the region of interest is not a suitable control. Variations in guanine sensitivity to DMS and endogenous nicks in genomic DNA are not necessarily reflected in a plasmid sequence.

TRANSFER OF THE DNA TO A NYLON MEMBRANE

Transferring the electrophoresed DNA fragments to a nylon membrane allows the DNA to be probed for indirect end labeling in a strand-specific manner. After hybridizing with one strand, the blot can be stripped and reprobed with the other strand.

1. Carefully pry open sequencing plates to leave the gel entirely on the bottom plate and overlay with plastic wrap.
2. Cut a piece of GeneScreen (DuPont) slightly larger than the set of lanes to be transferred and a piece of Whatman 3-mm paper slightly larger than that.
3. Mark the strip of GeneScreen to designate the DNA side, then place in a tray of 1×TBE (89 mM Tris/89 mM boric acid/2.6 mM EDTA) for 10 minutes.
4. Use one of the strips of protective sheeting from the GeneScreen as a template to trace a rectangle on the plastic wrap over the section of the gel to be transferred. Follow the traced lines with a sharp scalpel, cutting through the plastic wrap and gel.

5. Carefully pull back the plastic wrap that covers the cut gel section and overlay the gel with the dry piece of 3-mm paper to pick the cut section of gel up off the glass.
6. Lay the gel section/3-mm paper (gel side up) on to a Scotch-Brite support pad. Wet down the gel with 5–10 ml of 1× TBE (having the buffer in a squeeze bottle is helpful) and overlay the pre-wet GeneScreen, marked side down, with a rolling motion. Do not introduce any bubbles between the membrane and the gel.
7. Place a second Scotch-Brite pad and a support grid on top, flip the sandwich over, remove the pad, and ease off the filter paper by wetting with more TBE. Reapply the paper with a rolling motion to exclude all bubbles and assemble the sandwich again.
8. Electroblot in 1× TBE at 1.6–1.8 A for 2 hours.
9. Peel the GeneScreen away from the gel, cover with plastic wrap, and UV crosslink (marked side up). The UV crosslinking times and distances should be determined empirically for each UV source. A bank of four germicidal UV bulbs (GE-G158T) placed at a distance of 30 cm for 6–10 minutes works well.

PRODUCTION OF THE PROBE AND HYBRIDIZATION

It is critical that the probe be of a very high specific activity (10^9–10^{10} cpm/μg) because the individual bands to be visualized will be present in minute amounts (ca. 5×10^{-15}g per 20 μg total DNA) in a typical reaction from a single-copy maize gene.

Generation of a Single-Stranded Probe in M13:

1. Choose a restriction fragment of ca. 200 bp that has one end in common with the restriction site chosen for the genomic digests, and clone this fragment into both M13mp18 and M13mp19 to produce the orientations that will be specific for either the top or bottom strand (Sambrook et al. 1989) (see Figure 102.1).
2. Anneal the standard 17mer primer to the M13 phage DNA.
3. Synthesize a radiolabeled single strand by incubating:
   4.5 μg annealed phage/primer (in 10 μl)
   1 μl each dGAT, dATP, and dTTP (2 mM)
   4 μl 10× Taq polymerase buffer
   1 μl Taq polymerase
   25 μl (250 μ Ci) ^{32}P-CTP
4. Incubate at 65°C for 30 minutes. Add 80 μl of standard sequencing dye to the reaction and boil.
5. Separate the single-stranded labeled DNA from the vector on a short sequencing gel. Cast the gel with a well large enough to accommodate 120 μl. Pre-run the gel until it is hot to the touch; then load the labeling reaction mix (see Figure 102.2).

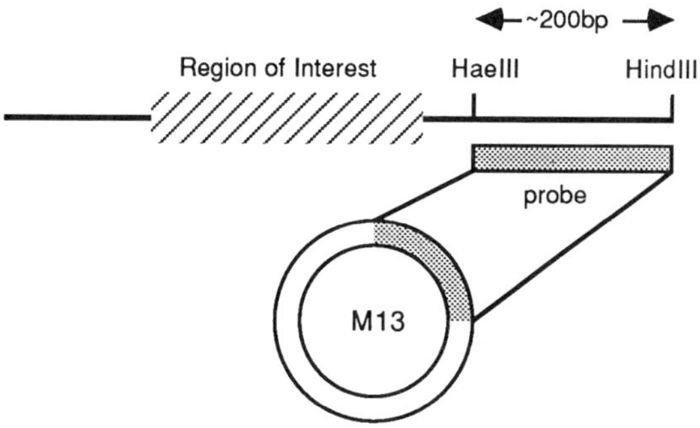

FIGURE 102.1 Choosing a probe for indirect end labeling. The cartoon illustrates the "end" restriction site to facilitate indirect end labeling (*Hind* III) and the second restriction site to generate a suitable size DNA fragment for the probe (*Hae* III), with respect to the region of the promoter to be footprinted (cross-hatched area). The M13 vector containing the cloned fragment is shown below. The single stranded probe synthesized from the M13 clone (shaded bar) will hybridize to genomic DNA in the position indicated.

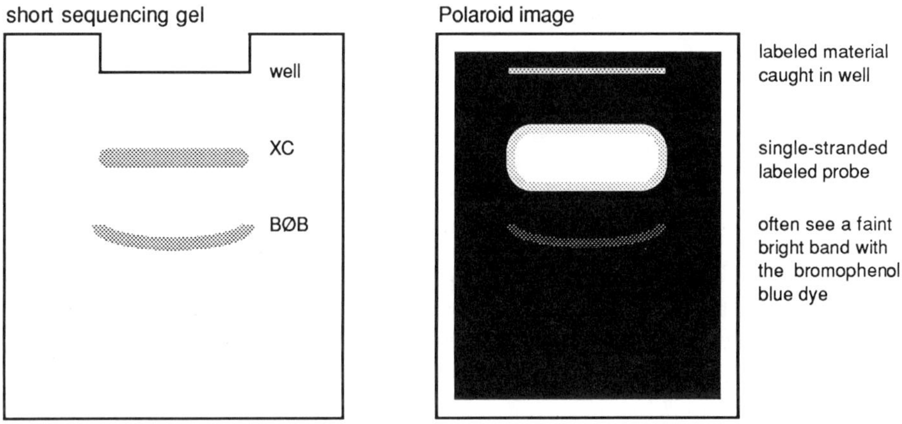

FIGURE 102.2 Preparing single-stranded labeled probe. The left panel shows the migration of the xylene cyanol (XC) and bromophenol blue (B Ø B) dye fronts on a preparative sequencing gel. The right panel represents the Polaroid image generated by placing Polaroid type 57 film over the preparative gel for several minutes. The position of the radioactively labeled fragment corresponds to the thick bright band on the film.

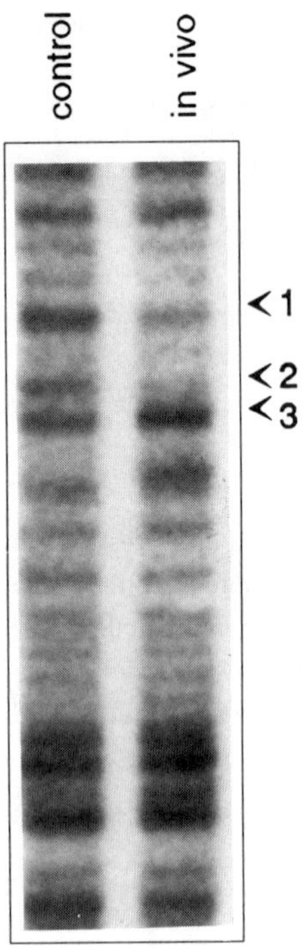

FIGURE 102.3 An example of in vivo footprinting within the maize *Adh2* promoter. The control lane shows the banding pattern generated by treating protein-free ("naked") genomic DNA with dimethyl sulfate (DMS) as described in the text. Protein interactions with the same region of the promoter in vivo influence the ability of DMS to modify guanines, which is reflected as a change in the intensity of the corresponding band. Proteins associated with the promoter DNA can either enhance (band 3) or inhibit (bands 1 and 2) the ability of DMS to modify the guanine in that position.

6. Allow the xylene cyanol to migrate about 3 cm from the well. Separate the plates to leave the gel on the bottom, then cover with plastic wrap.
7. Place a piece of Polaroid type 57 instant film over the gel, pierce in two places for later alignment, and apply a light pressure for 3–5 minutes. Develop the film and place it under the glass (such that the pierced spots match) and cut out the bright band of labeled DNA. The probe DNA should be between 150 and 200 bp in length. To increase the *length* of the labeled DNA, increase the ratio of ^{32}P-CTP to phage DNA.
8. Cut the acrylamide slice up into small (2–3 mm) pieces and elute in 5–10 ml of hybridization buffer at 65°C while the genomic blot is prehybridizing. Adding a small scrap of GeneScreen to the elution reaction will reduce nonspecific spots.

HYBRIDIZATION CONDITIONS

1. Prehybridize the blot for at least 1 hour at 65°C. Rolling the blots around Plexiglas rods and hybridizing inside rotating Plexiglas tubes is a convenient method for handling these long blots in a minimum volume of hybridization solution.

2. Discard the pre hybe solution and add the contents (buffer plus the gel pieces) of the tube of eluting probe mixture and incubate 8–16 hours at 65°C.
3. Wash the blot with at least three changes, 20 minutes each, of wash buffer at 65°C.
4. Pat the blot dry, wrap in plastic, and autoradiograph. The necessary exposure time will vary, but 2–3 days on Kodak XAR with two intensifying screens is typical. An example of an in vivo footprint is shown in Figure 102.3. Note: The phosphate stock solutions must be pH adjusted with H_3PO_4.

Hybridization solution	Wash solution
0.5 M Na_2HPO_4 (pH 7.2)	10 mM Na_2PO_4 (pH 7.2)
7% SDS	1 mM EDTA
1% BSA (Sigma A-4378)	0.1% SDS
	40 mM NaCl

REFERENCES

Church GM, Gilbert W (1984) Genomic sequencing. Proc Natl Acad Sci 81: 1991–1995

Ferl RJ, Laughner BH (1989) *In vivo* detection of regulatory factor binding sites of *Arabidopsis thaliana Adh*. Plant Mol Biol 12: 357–366

Ferl RJ, Nick HS (1987) *In vivo* detection of regulatory factor binding sites in the 5' flanking region of maize *Adh1*. J Biol Chem 262: 7947–7950

Lois R, Dietrich A, Hahlbrock K, Schulz W (1989) A phenylalanine ammonia-lyase gene from parsley: structure, regulation and identification of elicitor and light responsive cis-elements. EMBO J 8: 1641–1648

McKendree WL, Paul A-L, DeLisle AJ, Ferl RJ (1990) *In vivo* and *in vitro* characterization of protein interactions with the G-box of the *Arabidopsis Adh* gene. Plant Cell 2: 207–214

Nick HS, Bowen B, Ferl RJ, Gilbert W (1986) Detection of cytosine methylation in the maize *Adh1* gene by genomic sequencing. Nature 319: 243–246

Paul A-L, Ferl RJ (1991) *In vivo* footprinting reveals unique cis-elements and different modes of hypoxic induction in maize *Adh1* and *Adh2*. Plant Cell 3: 159–168

Sambrook J, Fritsch EF, Maniatis T (1989) Molecular Cloning, a Laboratory Manual, Second Edition. Cold Spring Harbor Laboratory Press, New York

Schulze-Lefert P, Dangl JL, Becker-Andre M, Hahlbrock K, Schulz W (1989) Inducible *in vivo* DNA footprints define sequences necessary for UV light activation of the parsley chalcone synthase gene. EMBO J 8: 651–656

Wu C (1980) The 5' ends of *Drosophila* heat shock genes in chromatin are hypersensitive to DNase I. Nature 286: 854–860

103

The Polymerase Chain Reaction: Applications to Maize Transposable Elements

ANNE B. BRITT AND DAVID J. EARP

The polymerase chain reaction serves to amplify specifically the DNA sequence located between two primers selected by the investigator. This enables the investigator to, among other things, directly clone new alleles of previously cloned genes without constructing and screening a library. For an overview of the theory and application of PCR techniques we direct the reader to *PCR Protocols* (Innes et al. 1990) and a recent review (Bej et al. 1991). This chapter is devoted to techniques designed to overcome the difficulties that are commonly encountered in the application of PCR to maize DNA.

MAIZE-RELATED DIFFICULTIES

Three technical difficulties frequently arise when attempting to amplify maize sequences. The first of these is template switching during amplification. No PCR reaction is 100% efficient; all reactions generate some incomplete products. These incomplete products can, in subsequent annealing steps, hybridize to related but nonidentical template molecules, thereby generating artifactual recombinant sequences. When amplifying unique regions of DNA (such as mutant alleles of *Bronze-1*) this problem can be overcome simply by using a homozygous (or mutant/deletion) line for the preparation of template DNA. Greater difficulties will arise if the investigator attempts to clone repetitive sequences such as transposable elements. While the products of such a reaction will include many sequences that are actually present in the template genome, some scrambling of related sequences will inevitably occur. For this reason the products of such an amplification cannot be regarded as a faithful representation of the actual genomic sequences from which they are derived.

A second technical difficulty can arise when attempting to amplify sequences (such

The Maize Handbook—M. Freeling, V. Walbot, eds.

as *Mu* elements) that include large inverted repeats. These sequences will form hairpin structures that may inhibit the progress of the polymerase. The solution in this case is to change the amplification strategy and design the primers so that only one of the inverted repeats is amplified per reaction tube. In this way the PCR-generated "short product" template carries only one repeat, cannot form a hairpin, and is efficiently amplified. Alternatively, the inclusion of deaza dGTP (discussed below) in the nucleotide mix may destabilize hairpin structures in the short product.

The most commonly encountered difficulty in the amplification of maize sequences is the poor efficiency of amplification of GC-rich sequences; such sequences frequently occur within the coding regions of maize genes. Two approaches have been employed to overcome this problem. One laboratory (B. Burr, personal communication) has achieved the successful amplification of the entire GC-rich *Bronze-1* coding region by the addition of the GC base-pair destabilizing nucleotide 7-deaza-2'-deoxyguanosine 5'-triphosphate, in a 3:1 ratio of deaza dGTP to dGTP, to an otherwise-typical PCR reaction mix (McConlogue et al. 1988). Another method for increasing the specificity and the final concentration of product is to undergo a second round of amplification with one or two nested primers. In addition, if the product of either the primary or, in the case of a nested reaction, the secondary round of amplification is nonspecific the annealing temperature should be raised and/or the annealing time should be reduced. The complete elimination of the annealing step, while reducing the overall efficiency of some reactions, often produces a highly specific amplification product.

The specificity and yield of a reaction can also be improved by gel-purifying the product of a primary reaction and adding a few microliters of the purified product (in molten agarose) to a secondary amplification reaction. Before cutting the primary product out of the gel, the investigator should destain the gel extensively against water to remove all of the gel running buffer and as much of the ethidium bromide as possible. The gel slice should then be melted and mixed thoroughly. Gel purification should be undertaken only as a last resort as it can easily lead to contamination of the secondary reaction with undesired template DNAs.

THE CLONING OF TRANSPOSON-TAGGED LOCI BY THE INVERSE POLYMERASE CHAIN REACTION (IPCR)

IPCR utilizes routine DNA manipulation methods to create a novel template from which the genomic sequences flanking a transposon are amplified by PCR. The strategy is outlined in Figure 103.1. In the illustrated scheme the amplified product contains sequences originally flanking both ends of the transposon. A variation on this approach involves the utilization of a restriction enzyme for the first digestion that cleaves within the element; in conjunction with appropriate primers, amplification of a template created in this way yields sequences flanking only one end of the transposon. The availability of suitable restriction sites will determine the appropriate approach. Both methods have

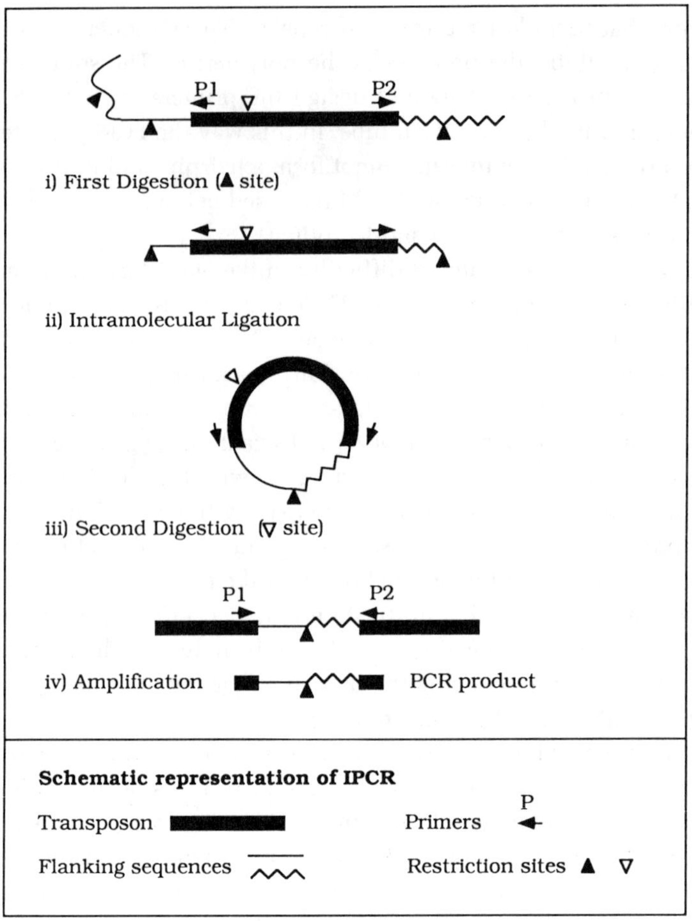

FIGURE 103.1 Amplification of target sit DNA by inverse PCR (IPCR). This figure illustrates the four steps required for PCR amplification of the transposable element's insertion site: primary digestion, circularization, secondary digestion, and amplification.

been used to isolate maize transposable elements from tobacco (Earp et al. 1990; Masson et al. 1991; Osborne et al. 1991).

MAIZE-SPECIFIC CONSIDERATIONS

While IPCR should be well suited to cloning transposons and their flanking sequences in maize, published examples of the successful application of the technique are limited to date. Some basic considerations specifically relevant to maize are given below. The appropriate use of the described controls at all stages of the process will indicate which steps need attention if problems arise.

Template Preparation

The complex families of transposon-related sequences in most maize lines complicate amplification of a specific element. Enrichment of the substrate DNA for the target element—for example, by restriction digestion and size fractionation on glycerol gradients (Earp et al. 1990)—may reduce the number of "background" products obtained following amplification. This restriction digestion may also serve as the first digestion for IPCR template generation. Southern analysis can be used to determine suitable restriction sites. When dealing with the *Ac* or *Spm* transposon families, methylation-senstive restriction enzymes may be useful to release an active transposon from genomic DNA and may facilitate the enrichment of a specific transposon in a size-selection protocol. The size of an IPCR product can also be determined by Southern analysis; as the efficiency of amplification decreases with increasing product size, it is advisable to choose a restriction enzyme that will generate a template of no more than 3–4 kb.

The efficiency of the template preparation can be assessed by removing samples after each step and analyzing them by Southern blots; using a transposon probe, it should be possible to see the mobility of the target fragment change following ligation and linearization. Alternatively, a small amount of circular pUC DNA may be added to the uncut maize DNA. Small aliquots of the preparation can then be removed after each stage and used to transform *E. coli*. The number of transformants should drop substantially (compared to a sample removed from the uncut starting material) following the first digestion and then recover following ligation. If the cells are plated onto XGAL/IPTG plates, the percentage of white colonies will give an indication of the frequency of (undesirable) intermolecular ligation; if this is a problem, it may be addressed by decreasing the concentration of DNA in the ligation mix. (See below.)

Primers

The enzymes chosen for restriction digestion will dictate the locations of primers for the amplification step. Some primers used for the amplification of *Ac* and *Spm* are listed below. The *Ac* primers have been used successfully for IPCR in tobacco and tomato; the *Spm* primers for IPCR in maize (Earp et al. 1990). Nested primer pairs (for a second round of amplification) are listed for *Ac*. The numbers following each primer give the location of the primer on the Genbank sequence accession noted for each transposon. The primers initiate replication away from the center of the transposons, through the nearest end into the flanking genomic sequence.

Ac *Genbank Accession Code: MZEAC1*

5' end	(1st round)	5'TCGGGTTCGAAATCGATCGGGAT3'	(bp 68–90)
	(2nd round)	5'ATACGATAACGGTCGGTACGGG3'	(bp 27–48)

3' end	(1st round)	5'TTATCCCGTTCGTTTTCGTTACCG3'	(bp 4447–4470)
	(2nd round)	5'CGTTTTCGTTTCCGTCCCGCAAG3'	(bp 4479–4501)

Spm *Genbank Accession Code: MZETNSPM*

5' end	(3' of SalI site)	5'ACGCCGCTGGCTAGACTGGAGAGA3'	(bp 280–303)
3' end		5'TCGGCTTATTTCAGTAAGAGTGTG3'	(bp 8225–8248)

More *Spm* primers are listed in Masson et al. (1991). A particular primer may be tested for function prior to IPCR using a purified clone of the transposon type under study—amplification being assessed using a primer located at one end of the element in conjunction with a primer located within the vector sequence, oriented toward the transposon primer.

Mu Elements

Because of the long inverted terminal repeats of the Mu element family, amplification of sequences containing *Mu* can be difficult. Successful IPCR amplification of *Mu* and flanking sequences has not yet been reported. The primer below functions at both ends of the *Mu* element. It has been used in maize to amplify, by direct PCR, one end of a *Mu1* element and the sequence flanking it, in conjunction with a primer located in the flanking genomic sequence. The efficiency of these PCR amplifications can be increased significantly by selecting a restriction enzyme that cleaves within the transposon and by replacing the 200 μM dGTP in the amplification mixture (see below) with 100 μM dGTP plus 100 μM deaza dGTP (S. Hake, personal communication). This primer also functions on *Mu8* (the primer contains a 1-base mismatch on *Mu8*) and may function with varying efficiency on other *Mu* elements (S. Hake, personal communication).

Mu1 *Genbank Accession Code: MZETNMU1*

Dual end primer 5'CCATAATGGCAATTATCTC3'

The primer is located at the very ends of the transposon; the Genbank entry (placing the primer at bp 110–128 and 1467–1485) contains additional flanking sequence.

PROTOCOL FOR IPCR

1. Preparation of genomic DNA. Any DNA extraction procedure that yields DNA of sufficient quality for restriction enzyme digestion and ligation will suffice. The DNA does not have to be purified on a CsCl gradient.
2. Template preparation. Following size fractionation to enrich for the specific transposon, 3–5 μg of DNA is required for the template generation. If the template preparation requires digestion with an enzyme other than that used in the fractionation step, this digestion is performed now. Note that any partial digestion products will

create alternative template molecules and should be avoided. Following digestion/ enrichment and standard extractions with phenol and phenol/chloroform and ethanol precipitation, the DNA is ligated at a DNA concentration of approximately 2 μg/ml with 1 unit/μl T4 DNA ligase. These conditions should result in intramolecular rather then intermolecular ligation. The latter can cause problems on amplification (Bej et al. 1991). The ligated mixture is extracted with phenol, phenol/chloroform, and precipitated with 0.1 volume of 3 M sodium acetate pH 5.6 and 3 volumes of ethanol and 2 μg/ml tRNA to enhance recovery. The material is then linearized by digestion with a suitable restriction enzyme, followed by a further round of extractions and ethanol precipitation. This final linearization step may not always be necessary but should be included to maximize efficiency of the PCR amplification.

3. Amplification conditions. The amplification mixture should contain ca. 0.2 μM of each primer and 200 μM of each dNTP, along with buffer constituents as described by the manufacturer for the particular Taq polymerase. Conditions for amplification will vary depending on the primer annealing temperatures and size of the expected product. A typical amplification protocol (for a product of up to 1 kb) might be: 1 minute denaturation at 95°C, 2 minutes annealing at 55°C, and 1 minute extension at 72°C, repeated for 35 cycles. If no product is detectable, 10% of the reaction mix from the first amplification should be used as a template for a second round of amplification. Nested primer pairs may be used for this second round of amplification to increase specificity.

REFERENCES

Bej A, Mahbubani MH, Atlas RM (1991) Amplification of nucleic acids by polymerase chain reaction (PCR) and other methods and their applications. Crit Rev Biochem Mol Bio 26(3/ 4): 301–334

Britt AB, Walbot V (1991) Germinal and somatic products of *Mu* excision from the *Bronze-1* gene of *Zea mays*. Mol Gen Genetics 227: 267–276

Clark JM (1988) Novel non-templated nucleotide addition reactions catalyzed by procaryotic and eucaryotic DNA polymerases. Nucl Acids Res 16(20): 9677–9687

Doseff A, Martienssen R, Sundaresan V (1991) Somatic excision of the *Mu1* transposable element of maize. Nucl Acids Res 19(3): 579–584

Earp DJ, Lowe B, Baker B (1990) Amplification of genomic sequences flanking transposable elements in host and heterologous plants: a tool for transposon tagging and genome characterization. Nucl Acids Res 18(11): 3271–3279

Innes MA, Gelfand DH, Sninsky JJ, White TJ (1990) PCR Protocols: A Guide to Methods and Applications. Academic Press, San Diego, CA, 482 pp

Loh EY, Elliott JF, Cwirla S, Lanier LL, Davis MM (1989) Polymerase chain reaction with single-sided specificity: analysis of T-cell receptor delta chain. Science 13: 217–220

Masson P, Strem M, Fedoroff N (1991) The *tnpA* and *tnpD* gene products of the *Spm* element are required for transposition in tobacco. Plant Cell 3: 73–85

McConlogue L, Brow MAD, Innis MA (1988) Structure-independent DNA amplification by PCR using 7-deaza-2'-deoxyguanosine. Nucl Acids Res 16(20): 9869

Nash J, Luehrsen KR, Walbot V (1990) *Bronze-2* gene of maize: reconstruction of a wild-type allele and analysis of transcription and splicing. Plant Cell 2: 1039–1049

Ochman H, Gerber AS, Hartl DL (1988) Genetic applications of an inverse PCR reaction. Genetics 120: 621–623

Osborne BI, Corr CA, Prince JP, Hehl R, Tanksley SD, McCormick S, Baker B (1991) *Ac* transposition from a T-DNA can generate linked and unlinked clusters of insertions in the tomato genome. Genetics 129: 833–844

Ralston E, English JJ, Dooner HK (1988) Sequence of three *bronze* alleles of maize and correlation with the genetic fine structure. Genetics 119: 185–197

Shyamala V, Ames GF (1989) Genome walking by single-specific-primer polymerase chain reaction: SSP-PCR. Gene 84: 1–8

Sullivan TD, Schiefelbein JW, Nelson OE (1989) Tissue-specific effects of maize *Bronze* gene promoter mutations induced by *Ds1* insertion and excision. Dev Gen 10: 412–424

104

In vitro Synthesis of Capped and Polyadenylated mRNA for Translational Studies *in vitro* and *in vivo*

DANIEL R. GALLIE

The use of vectors containing a bacteriophage promoter (e.g., T7, SP6, T3) allows in vitro synthesis of uniformly capped and polyadenylated mRNA free from contaminating RNA (Yisraeli and Melton 1989). Such specific mRNAs are useful for in vivo studies of RNA translation and stability after introduction of the mRNA into cells.

In vitro synthesized mRNAs have been introduced into maize cells by electroporation (Gallie et al. 1989). One such vector, pT7-A_{50} (2857 bp; Figure 104.1), a hybrid between pBluescript and pUC19, contains a T7 promoter immediately upstream of a polylinker region. Introduction of a gene at the *Stu I* site, which is situated at the transcriptional

The Maize Handbook—M. Freeling, V. Walbot, eds.

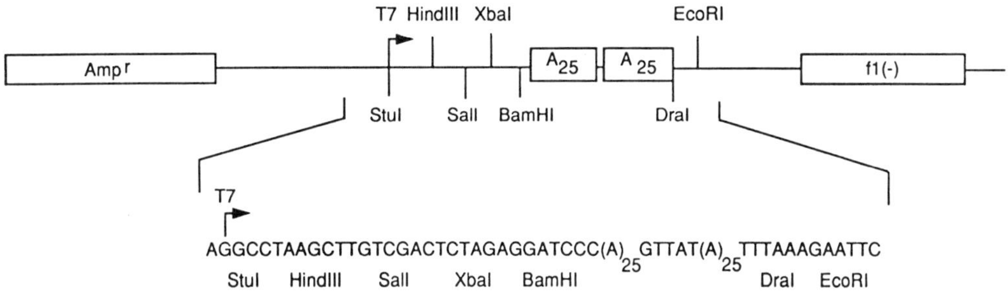

FIGURE 104.1. Schematic map of the in vitro transcription vector, pT7-A_{50} (2,857 bp). The plasmid map is shown at top (not drawn to scale). The sequence of the T7 transcription start site (indicated by the arrow), the multiple cloning site, and the poly(A) region is shown below.

start site, results in the addition of only two G residues to the 5'-terminus of the transcript. A poly$(A)_{50}$ tract downstream of the polylinker region allows the production of polyadenylated mRNA with a uniform tail length. The presence of the f1 phage origin allows the production of single-stranded DNA (the strand complementary to the "coding" strand) for use in site-directed mutagenesis or sequencing of the target gene.

Capping of the transcript is carried out concomitant with transcription using a cap structure such as m^7GpppG. As T7 RNA polymerase initiates transcription at a G, the penultimate base of the cap must be a G. mRNA capped with the monomethylated form of the cap is more translationally active than mRNA with a nonmethylated cap (Gallie et al. 1989).

SOURCES OF MATERIALS

Bovine serum albumin (RNase and DNase free), Boehringer Mannheim; m^7GpppG and T7 RNA polymerase, New England BioLabs; RNasin, Promega; all other materials, Sigma. 10× T7 buffer: 400 mM Tris-HCl, pH 7.5, 60 mM $MgCl_2$, 20 mM spermidine.

IN VITRO SYNTHESIS AND PURIFICATION OF mRNA

Assemble the following reagents and add in the order given.

Linear DNA (0.5 mg/ml)	10 μl
H_2O	14.6 μl
BSA	3.0 μl
5 mM ATP	7.5 μl
5 mM GTP	2.5 μl

5 mM CTP	7.5 μl
5 mM UTP	7.5 μl
10 mM m^7GpppG	6.3 μl
10x T7 buffer	7.5 μl
100 mM DTT	7.5 μl
RNasin (40 units/ml)	0.5 μl
T7 RNA polymerase (50 units/ml)	0.75 μl
Total	75.0 μl

1. Add in the order listed in a microfuge tube and incubate 2–3 hours at 37°C.
2. Stop the reaction by adding 1/10th volume 100 mM EDTA pH 8.0.
3. Extract once with 1 volume of phenol/chloroform (1:1) and centrifuge 10 minutes at 12,000*g*.
4. Extract the aqueous phase (top phase) once with 1 volume ether. Discard the ether (top) phase.
5. Add 1/10th volume 7.5 M ammonium acetate followed by 1 volume isopropanol. After mixing, incubate on ice 60 minutes and centrifuge 15 minutes at 12,000*g*.
6. Wash the pelleted RNA several times with 80% ethanol. The RNA can be stored under ethanol at −70°C indefinitely.
7. To resuspend the mRNA for use, remove all traces of ethanol (do not dry the pellet), and add sterile, RNase-free H_2O. Keep on ice for 30 minutes with intermittent vortexing.

NOTES

1. Linearize the vector before mRNA synthesis. The site of linearization will define the 3' terminus of the resulting mRNA. In order to generate polyadenylated mRNA from pT7-A_{50}, for example, *Dra* I is used. Restriction enzymes that leave a 3' overhang may promote strand-switching by the T7 RNA polymerase resulting in additional sequence at the 3' terminus. Those enzymes that produce a 5' overhang or a blunt end are more suitable.
2. The nucleotide stocks should be brought to pH 7.5 before use.
3. Following transcription, the DNA template can be removed with DNase (RNase-free) or the RNA purified on preparative agarose gels; however, the presence of the vector does not pose a problem for most applications as it is not transcribed either in vitro or in vivo in the absence of T7 RNA polymerase.
4. For efficient translation in vivo, the mRNA must contain both a cap and a polyA tail. Because of the synergism between the cap and the polyA tail, neither regulatory element will function efficiently in the absence of the other (Gallie 1991).

REFERENCES

Gallie DR, Lucas WJ, Walbot V (1989) Visualizing mRNA expression in plant protoplasts: factors influencing efficient mRNA uptake and translation. Plant Cell 1: 301–311

Gallie DR (1991) The cap and poly(A) tail function synergistically to regulate mRNA translational efficiency. Genes Devel 5: 2108–2116

Yisraeli JK, Melton DA (1989) Synthesis of long, capped transcripts *in vitro* by SP6 and T7 RNA polymerases. Methods Enzymol 180: 42–50

105

Construction of a Genomic Library in Lambda Phage

CLIFFORD F. WEIL AND THOMAS E. BUREAU

In most respects, preparation of maize genomic libraries in lambda phage is similar to preparing any other genomic library. We follow the procedures for cloning with lambda phage vectors detailed by Sambrook et al. (1989) with the variations described below. We focus here on those aspects that may require special attention when using maize DNA. Please note the "Other Considerations" before beginning to clone.

PPEPARATION OF MAIZE INSERT DNA

Following digestion with the appropriate restriction enzyme, the maize DNA is size-selected by electrophoresis through a 0.7% (w/v) low-melting-point agarose gel. A convenient protocol used in our lab is as follows:

Materials

50 μg of digested genomic DNA

The Maize Handbook—M. Freeling, V. Walbot, eds.

Low melting point agarose [SeaPlaque (FMC Corp)]
$1\times$ Tris-acetate-EDTA (TAE) electrophoresis buffer
10 mg/ml ethidium bromide (EtBr)
TE-saturated phenol (TE=10 mM Tris, pH 8, 1 mM EDTA)
3 M sodium acetate pH 5.2
95% ethanol
70% ethanol
TE buffer (pH 8.0)
SM buffer (for 1 liter)
- 5.8 g NaCl
- 2.0 g Mg_2SO_4
- 50 ml 1 M Tris-HCl pH 7.5
- 5 ml 2% gelatin

Tris-acetate-EDTA (TAE) electrophoresis buffer ($50\times$)
- 242 g Tris base
- 57.1 ml glacial acetic acid
- 100 ml 0.5 M EDTA (pH 8.0)
- Distilled water to 1 l

1. Following digestion, precipitate the DNA by adding 1/10th volume of 3 M sodium acetate and 2.5 volumes cold 95% ethanol. Mix well (do not vortex!) and spin 15 minutes in a microfuge. Wash the pellet with 70% ethanol, spin 5 minutes, and dry in a Speed-Vac. Resuspend the pellet in approximately 50 μl TE overnight to ensure complete hydration. (Note: Maize DNA is notoriously difficult to rehydrate at high concentrations. If the sample has not rehydrated completely, heating to 68°C for 10 minutes can help.)
2. Cast a 0.7% (w/v) low melting point agarose gel made with $1\times$ TAE buffer including ethidium bromide at a final concentration of 1 μg/ml. Allow to harden approximately 1 hour at 4°C to insure that the wells set completely. We have used gels of 10.5×14 cm and 20.5×24 cm with equally good results.
3. Add loading dye to sample and load sample onto gel. Be sure to load a size standard leaving a gap of one or two wells between the standard and the sample(s). Electrophoresis conditions should be at or below 50 V for best results.
4. When the gel has finished running, examine it under long-wave UV (366 nm) illumination ONLY. (Short-wave UV will damage the DNA you are trying to isolate.) Using a clean razor blade or scalpel, cut out the desired size range and remove to a clean glass plate. Mince gel slice and place in microfuge tube. Add 500 μl of TE-saturated phenol. Do not expose the gel slice to phenol/chloroform or chloroform.
5. Vortex briefly and freeze at −70°C for 10 minutes. Allow to thaw completely at room temperature (NOT at temperatures that may melt the low-melting-point agarose). Repeat the freeze/thaw cycle once more or until no particulate matter remains.
6. Spin 10 minutes in microfuge. Remove aqueous (top) layer staying well away from

interface and place in a fresh microfuge tube. Extract once more with 500 μl of TE-saturated phenol. Spin and remove aqueous layer to fresh tube.

7. Add 1/10th volume 3 M sodium acetate (pH 5.2) and 2.5 volumes cold 95% ethanol. Spin 15 minutes. Wash pellet with 70% ethanol. Spin 5 minutes. Dry pellet in Speed-Vac. Resuspend pellet in 10 μl of TE.

OTHER CONSIDERATIONS

1. Most of the vectors available are described in detail by Sambrook et al. (1989) and can be obtained commercially along with complete instructions.
2. It is essential that the lambda arms not be overdigested (e.g., no more than 3–4 hours using 1–3 units enzyme/μg DNA) because this reduces the efficiency of ligation. We have found this especially important for LambdaZAP phage when cutting with *Xho*I, where digestion should be ≤ 2 hours.
3. To reduce the possibility of rearrangements when cloning repeated sequences, use an *E. coli* host strain that is multiply recombination-deficient. Even though some lambda strains require a *recA*$^+$ host, *recBC* , *recJ*, *uvrC*, *umuC* , *sbcC* mutants are available [e.g., Epicurean Coli SURE (Stratagene)].
4. Maize DNA is highly methylated, and we find it is most easily cloned using host strains that are *mcrA*, Δ(*mcrBC-hsdRMS-mrr*)171 [e.g., Epicurean Coli SURE (Stratagene)].

Instructions on preparation of and ligation to phage arms, in vitro packaging, infection of *E. coli* cells, and plating are provided by most commercial sources. Plaque lifts are described in Sambrook et al. (1989). We find that 20,000–30,000 pfu per 150-mm petri plate rather than the 50,000 pfu per plate mentioned in Sambrook et al. is a more desirable plaque density. Duplicate lifts are made from each 150-mm plate.

HYBRIDIZATION OF PLAQUES TO NITROCELLULOSE FILTERS

The following hybridization buffers and conditions are used routinely with excellent results. One advantage to these buffers is that they do not utilize toxic compounds such as formamide.

MATERIALS

Prehybridization buffer	*Hybridization buffer*
6× SSC	6× SSC
1% SDS	1% SDS
250 μg/ml heparin	250 μg/ml heparin
	10 mM EDTA
	10% (w/v) dextran sulfate

METHODS

1. Heat the prehybridization buffer at 68°C for 10 minutes to dissolve SDS. Add to filters. Prehybridize filters at 68°C for at least 30 minutes. Use ~2 ml prehybridization buffer for each 150 mm filter. Up to 12 filters can be incubated in a single seal-a-meal bag or other sealed container.
2. Add the probe to the hybridization buffer. Use the same volume of hybridization buffer as prehybridization buffer. Denature the probe by boiling the probe/hybridization buffer mixture for 10 minutes. Drain the prehybridization buffer from the seal-a-meal bag, add the denatured probe and reseal. Incubate ≥ 16 hours at 68°C with mild shaking.
3. Remove the probe solution AND SAVE FOR SUBSEQUENT ROUNDS OF PLAQUE SCREENING. Wash the filters for 5 minutes in 0.1 × SSC, 0.5% SDS at room temperature to remove debris.
4. Wash the filters in 0.1 × SSC, 0.5% SDS at 68°C for 45 minutes. Repeat wash.
5. Air dry filters. Mark with ^{35}S-labeled ink for orientation and autoradiograph.
6. Positive plaques found on both duplicate lifts are then purified to homogeneity.

NOTES

1. In subsequent isolation of phage DNA, we recommend preparing phage particles by the plate lysate method; we obtain inconsistent results from growing lambda phage in liquid culture. Centrifugation of the plate lysate to remove cell debris followed by RNase/DNase treatment, PEG/NaCl precipitation, or simply pelleting the phage through SM buffer in an ultracentrifuge yields adequately clean preparations of phage particles.
2. The library of in vitro packaged phage can be stored in SM buffer containing 0.4% chloroform at 4°C. Longer-term storage is improved by plating out the library and storing the plate lysate that results; however, this will increase the representation of faster-growing phage in the library.

REFERENCE

Sambrook J, Fritsch EF, Maniatis T (1989) Molecular Cloning: A Laboratory Manual, Second Edition. Cold Spring Harbor Press, Cold Spring Harbor, NY

106

Agroinfection

JESUS ESCUDERO AND BARBARA HOHN

The plant–*Agrobacterium* interaction represents an unusual kind of sharing of genetic information: bacterial genes are transferred to and expressed in the plant (Winans 1992 and reviews cited therein). Although many dicotyledonous species and a few monocot species have been described as hosts for *Agrobacterium*, the scientifically and agronomically important grasses have not been demonstrated as hosts. T-DNA delivery from *Agrobacterium* to maize was detected, however, by using the technique called "agroinfection" (Grimsley et al. 1987). This consists of the inoculation of the plant with an agrobacterial strain carrying an artificial T-DNA with viral sequences that allow propagation of the introduced DNA. In this chapter, we describe a simple protocol for agroinfection in maize using Maize Streak Virus (MSV), a single stranded DNA monopartite geminivirus as part of the T-DNA construct. The agroinfection technique can be used to study different aspects of maize, bacterial, and viral biology.

DNA CONSTRUCTIONS

The original plasmid used in agroinfection contained an MSV dimer in the T-DNA of a binary vector, but some other DNA constructions carrying an MSV 1.6 mer or 1.4 mer have been shown to be as efficient as the duplicated MSV construction. On the other hand, there is no difference in efficiency between the two possible orientations of the MSV genome with respect to the T-DNA borders. Figure 106.1a simplifies the plasmid DNA construction: LB and RB correspond to the left and right borders, respectively, on the T-DNA of a binary vector; 1.X means the duplicated MSV genome, where "X" can vary from 0.4 mer to 1.0 mer.

BACTERIAL STRAINS

A. tumefaciens strains C58 or LBA4301 were grown in YEB medium at 28°C. Derivatives of these strains were made chromosomically resistant to the antibiotic rifampicin, and

The Maize Handbook—M. Freeling, V. Walbot, eds.

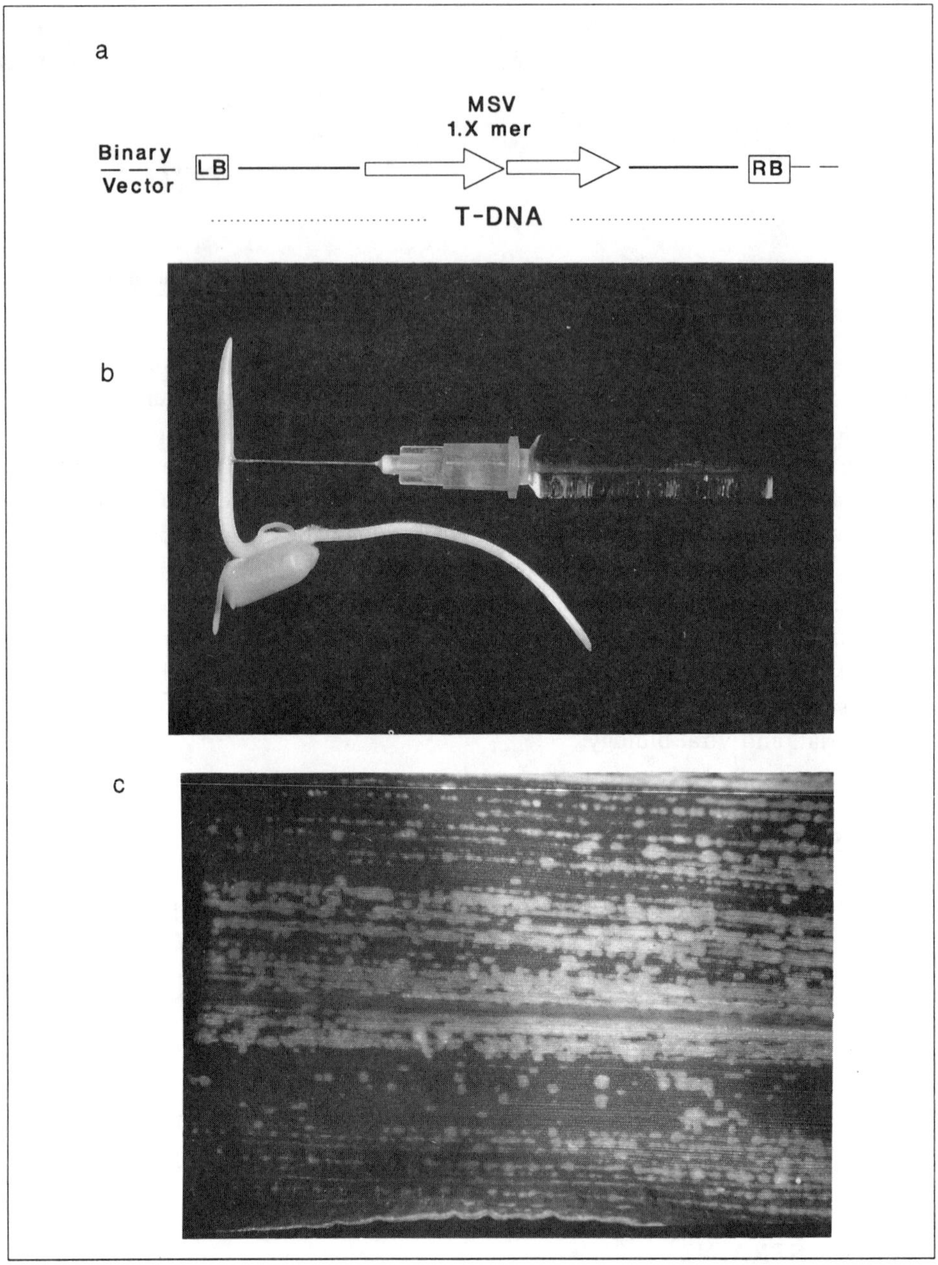

FIGURE 106.1. a. Schematic DNA construct. b. Seedling inoculation. c. Maize leaf showing MSV symptoms 10–14 days after inoculation.

plasmids carrying the MSV DNA constructions were mobilized into agrobacteria using the helper strain HB101 (pRK2013) (An et al. 1988). Routinely, bacteria were taken from a glycerol stock, grown overnight in liquid YEB medium with antibiotics, harvested, and resuspended in 10 mM $MgSO_4$ without antibiotics at OD600 = 1.0 prior to plant inoculation; this corresponds to about 10^5–10^6 bacteria per μl.

It is necessary to use nopaline strains, as octopine strains are inactive in agroinfection of maize. It is important to be aware of biosafety requirements when manipulating MSV.

MAIZE LINES

Three- or 4-day-old seedlings of the lines A188, Golden Bantam, B73, and W23 were grown and inoculated at the coleoptilar node (see Figure 106.1a), giving a survival rate after the injection with bacteria of usually >90%. Also, immature embryos ranging between 14 and 18 days after pollination have been shown to be good recipients for agroinfection, although the efficiency depends on the maize line used (Schläppi and Hohn 1992).

MATERIALS AND SOLUTIONS

1. Sterile equipment: glass bottles, magnetic stirrers, pipettes, filter papers, petri dishes, eppendorf tubes, Hamilton syringe, and needles.
2. Soil and pots or convenient medium containers.
3. 1% Na hypochlorite, 0.05% SDS aqueous solution.
4. Sterile liquids: water, 10 mM $MgSO_4$ aqueous solution, YEB medium.
5. Appropriate antibiotics, depending on the agrobacterial strain and binary vector. (*rifampicin*, kanamycin, carbenicillin, and chloramphenicol are the most frequently used.)

METHOD

1. Put kernels (10–100) in a sterile bottle with a magnetic stirrer and add 2 volumes of the 1% Na hypochlorite, 0.05% SDS solution. Disinfect by stirring for 10–15 minutes.
2. Using a vacuum device, remove the used bleach solution and add 2 volumes of sterile water. Again stir for 10–15 minutes; repeat this step.
3. After rinsing the kernels twice with water, place them on a filter paper inside a petri dish. (Sterility is recommended.) Wet the filter paper with 5 ml of sterile water and cover the dish. Normally, 12–15 kernels are enough for a standard plate (9-cm diameter).

4. Germinate the kernels inside the plates (in the dark or in the light) at 28°C for 3–4 days, making sure they remain wet.
5. Inoculate the agrobacterial strain harboring the construction with the viral DNA when shoots are approximately 2 cm in size. When the maize shoots are pushing the lid of the petri dish trying to open it (by the third or fourth day) they are ready to be injected with the agrobacteria.
6. Collect the bacteria from 4 ml of the overnight culture by spinning them down for 3 minutes, 5,000*g* in a microfuge. Resuspend the cell pellet in the same volume of a 10 mM $MgSO_4$ solution. The optical density at 600 nm should be around 1, corresponding to about 10^5–10^6 bacteria per μl for most of the *Agrobacterium* strains.
7. Load the bacterial suspension into a clean Hamilton syringe and inoculate the maize seedlings with 10–20 μl by using a sterile fine needle (0.45×10 mm or 0.40×20mm), avoiding air bubbles. The point of inoculation on the seedling is important (Grimsley et al. 1988). To get the best results, injection must be at the meristematic ring in the coleoptilar node. (See Figure 106.1b.)
8. After inoculation, place the seedlings on soil or the desired culture medium in a growth chamber and grow as usual, checking the viability of the plantlets periodically. (For most of the maize lines tested the viability is higher than 90%.)
9. After 6–7 days postinoculation, expect to see the first viral symptoms as chlorotic spots, normally on the second or third leaf. The spots will extend to streaks as the leaf grows. Most of the infected plants will develop symptoms within the first 2 weeks (see Figure 106.1c) and usually the final scoring can be done after 3–4 weeks.
10. DNA can be extracted from leaves showing symptoms, and the presence of viral genomes can be confirmed by Southern blot hybridization using an internal probe for the virus.

NOTES

1. Excessive disinfection may diminish the germination rate of the kernels. Do not overdo this first step.
2. It is practical to keep a −80°C frozen aliquot of the agrobacterial cells to check the DNA constructions if necessary.
3. We recommend growing a fresh overnight culture of the agrobacteria before the plant inoculation. In addition, the bacterial culture should be started from a glycerol stock made with cells harboring the right DNA construction. For most of the plasmids used as binary vectors the presence of antibiotics is always required to maintain selection pressure.
4. If the agrobacteria overgrow when using in vitro tissue culture medium, the antibiotic cefotaxim (at 500 μg/ml) can be used.
5. Biosafety: The described manipulations with MSV are currently done under a

BL3 containment system in Switzerland. Researchers should consult their respective guidelines prior to initiating work with MSV.

REFERENCES

An G, Ebert PR, Mitra A, Ha SB (1988) Binary vectors. In Gelvin SB, Schilperoort RA (eds) Plant Molecular Biology Manual, Dordrecht: Kluwer Academic Publishers, pp 1–19

Grimsley N, Hohn T, Davies JW, Hohn B (1987) *Agrobacterium*-mediated delivery of infectious maize streak virus into maize plants. Nature 325: 177–179

Grimsley N, Ramos C, Hein T, Hohn B (1988) Meristematic tissues of maize plants are most susceptible to Agroinfection with maize streak virus. Bio/Technology 6: 185–189

Schläppi M, Hohn B (1992) Competence of immature maize embryos for *Agrobacterium*-mediated gene transfer. Plant Cell 4: 7–16

Winans SC (1992) Two-way chemical signaling in *Agrobacterium*-plant interactions. Microbiol. Reviews 56: 12–31

107

Polyethylene Glycol-mediated DNA Uptake into Maize Protoplasts

Leszek A. Lyznik and Thomas K. Hodges

Polyethylene glycol (PEG)-mediated DNA uptake is an alternative to electroporation in the transformation of maize protoplasts. Both methods produce comparable results and have been successful for the generation of stably transformed maize callus tissue. About 10–30 transformed calli are obtained from 10^5 protoplasts after PEG treatment (Lyznik et al. 1989; Armstrong et al. 1990) compared with approximately 1–8 transformed calli per 10^5 electroporated protoplasts (Fromm et al. 1986; Huang and Dennis 1989). Polycation Polybrene and cationic lipid Lipofectin have also been successfully used for maize protoplast transformation (Antonelli and Stadler 1990).

Two different transformation procedures are described here, and the one chosen

The Maize Handbook—M. Freeling, V. Walbot, eds.

depends on the final goal—transient expression assays or stable transformation experiments (Lyznik et al. 1991; Lyznik et al. 1989). The initial steps of protoplast preparation and PEG incubation are the same. Stable transformation experiments require more complex conditions after PEG incubation to assure protoplast growth.

The protocols presented are for protoplast isolation, PEG-mediated transformation, and selection of protoplasts using A188×BMS maize suspension cultures. The information provided in the notes may be useful in adapting the methods to other genotypes.

REAGENTS AND MEDIA

Protoplast Incubation Medium—PIM

MS salts	4.3 g/liter	*Stock solutions*	
Thiamine stock	10 ml/liter	Thiamine · HCl stock	0.025 g/500 ml H_2O
2,4-D stock	20 ml/liter	2,4-D stock	1 mg/10 ml H_2O
Mannitol	36.4 g/liter	2 M $CaCl_2 \cdot 2H_2O$ stock	29.4 g/100 ml
$CaCl_2 \cdot 2H_2O$ stock	40 ml/liter		

Adjust pH to 5.9 with 0.1 N KOH.

Protoplast Culture Medium—PCM

MS salts	4.3 g/liter
2,4-D stock	20 ml/liter
Glucose	0.25 g/liter
Thiamine stock	10 ml/liter
Sucrose	20 g/liter
Coconut water	20 ml
Mannitol	36.4 g/liter

Adjust pH to 5.9 with 0.1 N KOH. Add 8 g/liter low-melting-point (LMP) agarose (Gibco BRL/Life Technologies, Gaithersburg, MD) to make PCM-LMP agarose.

Postfeeder Culture Medium—MS 2D

MS salts	4.3 g/liter	*Modified White Vitamins*	
Modified White Vitamins	10 ml/liter	Glycine	200 mg/liter
Myo-inositol	100 mg/liter	Nicotinic acid	50 mg/liter
Sucrose	20 g/liter	Pyridoxine HCl	50 mg/liter
2,4-D stock	20 ml/liter	Thiamine HCl	50 mg/liter

Adjust pH to 5.9 with 0.1 N KOH. Add 4 g/liter agarose (Gibco BRL/Life Technologies, Gaithersburg, MD), autoclave for 15 minutes, and dispense medium into tissue culture dishes. Add 2 ml/liter filter-sterilized kanamycin sulfate stock solution (50 mg/ml) to the medium used for selection of transgenic callus.

Transformation Medium—TM

MES 2(*N*-morpholino)ethanesulfonic acid	19.5 g/liter
Mannitol	36.4 g/liter
$CaCl_2 \cdot 2H_2O$ stock	40 ml/liter

Adjust pH to 5.5 with 0.1 N KOH.

Krens' F Medium—F-Medium

NaCl	8.2 g/liter
KCl	372 mg/liter
Na_2HPO_4	105 mg/liter
Glucose	900 mg/liter
$CaCl_2 \cdot 2H_2O$ stock	65 ml/liter

Adjust pH to 7.2.

50% (W/V) PEG Transformation Solution

Add 50 g PEG [polyethylene glycol, average molecular weight (MW): 8,000; Sigma Chemical Co., St. Louis) to 50 ml 2× F-medium. Heat solution to solubilize PEG and adjust volume to 100 ml with water. Autoclave for 15 minutes.

Fluoroscein Diacetate Stock Solution—50 mg/100 ml of Acetone

The modifications to the standard MS and N6 media are marked in the text. All culture media contained 2% sucrose.

PROTOCOLS

Protoplast Isolation

1. Cycle suspension cultures in N_6 medium (supplemented with 6 mM proline, 1 g/liter casein hydrolysate, and 3.5 mg/liter 2,4-D) by weekly subculture of 5 ml packed cell volume (pcv) into 35 ml of fresh N_6 medium in 250 ml Erlenmeyer flasks. Rotate the flasks containing suspension cells on a gyratory shaker at 120 rpm, 26°C, in the dark. The suspension culture medium is replaced with 35 ml of fresh medium at 3–4 days after each subculture.

2. One week prior to protoplast isolation, transfer 2 ml pcv into 35 ml of MS medium (supplemented with 3.5 mg/l 2,4-D).
   This subculture into MS medium improved the protoplast yield; however, if the yield is 1×10^6 protoplasts or more per 1 ml pcv, this treatment may be omitted.

3. Remove old medium from flasks and wash cells with PIM. Add 20 ml of the PIM containing 2% cellulase and 0.25% pectinase (both enzymes from Worthington Biochemical, Freehold, NJ) to 5 ml (pcv) of cells. Incubate for 3 hours at room temperature on a rotary shaker set at 40 rpm.
   Selection of a suitable suspension culture composed of small clusters is a prerequisite for a high protoplast yield. Cell wall digestion is never complete. We do not recommend extending the incubation time for yield improvement, because the viability of protoplasts may decrease. The yield can be improved by using other enzymes. Pectolyase Y-23 (Seishin Pharmaceutical, Tokyo, Japan), at a concentration of 0.1%, proved to be quite effective, permitting an even shorter incubation time of about 1 hour instead of the standard 3-hour incubation. It is always prudent, however, to verify the quality of protoplast preparations, because some hydrolytic enzymes cause extensive damage to protoplasts (Ye and Earle 1991). The same cell culture line can yield more or fewer protoplasts depending on the physiological status of the suspension. Cells at the late logarithmic phase of growth tend to produce higher protoplast yields and higher transient gene activity signals.

4. Filter the digestion mixture through a 50 μm nylon screen, wash cell clusters on the screen with PCM, and pellet protoplasts by centrifugation at 50*g* for 15 minutes.

5. Remove supernatant and suspend protoplast pellet in PCM containing 9% Ficoll (Ficoll 400, Sigma Chemical, St. Louis, MO).
   The 9% Ficoll solution may not be optimal for separation of protoplasts from all suspension cultures, and other concentrations of Ficoll solution should be tested.

6. Dispense the protoplast suspension into a new centrifuge tube and overlay with TM. Centrifuge at 75*g* for 10 minutes and collect the band of protoplasts formed at the interface between Ficoll and TM.
   The volume of Ficoll solution and density of the protoplast suspension should be adjusted experimentally based on the size of tubes used for centrifugation and the size of the protoplast band collected. The pellet and the band of protoplasts should be separated by a clear Ficoll solution. Protoplasts form a compact band that can be removed by applying a gentle suction to the pipette and pulling the entire protoplast band through the upper solution. Small amounts of protoplasts do not form a compact band making it difficult to remove without Ficoll contamination.

7. Dilute a sample of the protoplast suspension in TM containing 2.5 μg/ml fluoroscein diacetate (200× diluted FDA stock solution with PCM). Place the protoplasts onto a hemacytometer slide and count the number of viable protoplasts; adjust the final concentration of viable protoplasts to 1×10^7 protoplasts per ml with TM.
   A high-quality protoplast preparation is essential for efficient transformation. Un-

usually low yields of protoplasts, heavy contamination with undigested cells, many broken protoplasts, or poor viability should disqualify the preparation. In such cases, it is better to abort the experiment at this step than to invest the time and effort required for the transformation procedures.

TRANSIENT TRANSFORMATION

1. Fill the wells of a 12-well microculture plate (Corning, NY) with a 2 ml solution of PCM-LMP. Leave plates (with lid on) in the laminar flow transfer hood to let agarose solidify inside the wells.
   It is quite likely that LMP-agarose can be substituted by other gel-forming reagents. If stable transformation experiments are planned, the same LMP-agarose PCM solution has to be used for the preparation of feeder cell layers.
2. Transfer 0.1 ml of protoplast suspension into each tube (Falcon 205412×75-mm-style tubes, Becton Dickinson, NJ 07035). Add 5 μl of plasmid DNA (standard concentration; 1 mg/ml) and then 0.1 ml of 50% PEG in F-medium. Slowly mix solutions by rotating tubes. Incubate at room temperature for 20 minutes.
   Any small tube could be used for the protoplast incubation. It is difficult to mix the PEG solution with the protoplast suspension in a cone-shaped 1.5 ml Eppendorf tubes. Plasmid DNA preparations should be as pure as possible; cesium chloride gradient centrifugation of plasmid DNA is highly recommended.
3. Transfer all of the incubation mixture from each tube onto separate agarose blocks in the 12-well microculture plate. Rotate the plate to disperse protoplast suspensions. Wrap plates with parafilm and incubate overnight at 26°C without shaking.
   During the initial experiments, it is important to verify that at least 50% protoplasts survive the overnight incubation.
4. Harvest protoplasts by transfering all liquid into 1.5 ml eppendorf tubes (Fremont, CA). Wash agarose blocks with an additional 0.2 ml of PCM. Pellet protoplasts by centrifugation, remove supernatant, and suspend pellet in about 0.2–0.5 ml extraction buffer.
   Protoplasts can be destroyed by sonication or detergent treatment. For GUS protein extraction, the protoplast pellet can be suspended in buffer containing 0.1% Triton X-100, vortexed for about 5 seconds, and centrifuged for 5 minutes at 16,000 *g* (the maximum speed of a minicentrifuge e.g., Centrifuge 5414, Eppendorf). The collected supernatant can be used directly for GUS activity assays.

STABLE TRANSFORMATION

1. Use tissue culture dishes 60×15 mm to prepare feeder plates. Pipette 3 ml PCM-LMP into dishes and let it solidify. Add additional 3 ml of PCM-LMP containing 0.2 ml

packed cell volume of feeder cells (selected small clusters of the same suspension culture used for protoplast isolation). PCM-LMP should be cooled to 40–45°C before mixing with the feeder cells. Put sterile 0.8-μm Millipore (Bedford, MA) filters on the top of the solidified feeder cell layer. The use of "conditioned medium" for protoplast growth may be beneficial for some cell lines (Armstrong et al. 1990).
The feeder plates can be prepared during the protoplast digestion. They may be prepared the day before transformation, but then the plated protoplasts dry very fast. This rapid drying can have a detrimental effect on plating efficiency.

2. Transfer 1 ml of the protoplast suspension into a 50-ml tube (Corning polypropylene 25330, Corning, NY 14831) for each treatment. Add 50 μl of plasmid DNA (standard concentration; 1 mg/ml) and then 1 ml of 50% PEG in F-medium (the same as for transient transformation). Slowly mix solutions by rotating tubes. Incubate at room temperature for 30 minutes.
   Add plasmid DNA containing the neomycin phosophotransferase (*neo*) gene or hygromycin-resistance gene (*hpt*) for selection of transgenic calli. Another plasmid DNA containing a gene of interest (if not present on the same plasmid) may be added for co-transformation. The efficiency of co-transformation is generally in the range of 40–80%.

3. Slowly dilute the transformation mixture with 30 ml of F-medium over 30 minutes (e.g., 2×0.5 ml and 2×1 ml at 2-minute intervals and 2 ml every 2 minutes thereafter). Pellet protoplasts by centrifugation at 50*g* for 10 minutes.
   Protoplasts tend to aggregate during incubation with PEG. A fine suspension should form upon dilution with F-medium.

4. Suspend the protoplast pellet in 2 ml of PCM containing 9% Ficoll, transfer the suspension into sterile 15 ml Falcon tubes, overlay the protoplast suspension with 2 ml of PCM, and centrifuge tubes at 75*g* for 10 minutes.

5. Remove the protoplast band, count the number of viable protoplasts, and adjust the concentration to 1×10^6 protoplasts per 1 ml of PCM.
   About 50% of the protoplasts should survive the PEG treatment. Small clusters of 2–5 viable protoplasts are acceptable.

6. Plate 0.2 ml of the protoplast suspension onto the 0.8 μm Millipore filters that overlay the feeder cells. Leave the closed feeder plates in the hood to let the protoplast suspensions dry on filters. (Usually it takes overnight). Wrap plates with Parafilm and incubate at 25°C in the dark.

PROTOPLAST GROWTH AND SELECTION

1. One week after transformation and protoplast culture, prepare new feeder plates as before, but add 100 μg/ml kanamycin sulfate for selection of transgenic calli (if hy-

gromycin-resistance gene was used, the selection protocol of Huang and Dennis 1989 could be used.)

There are several other selectable markers available (e.g., methotrexate-, phosphinothricin-, glyphosate-, chlorosulfuron-resistance genes); however, applications for maize protoplast selection still need to be tested. The kanamycin-based selection protocols in association with the *neo* gene work quite well for maize protoplasts providing kanamycin is applied at the right time and concentration. Our selection protocol requires 100 μg/ml kanamycin sulfate in the culture medium applied 1 week after protoplast plating. The time of application of kanamycin selection may vary depending on the rate of protoplast growth. Application of kanamycin too soon after transformation will kill transformed cells, but application of kanamycin too late results in many "escapes."

2. Transfer Millipore filters with protoplasts and/or cells onto fresh feeder plates 1 week after transformation. Always include the protoplast growth control treatment (protoplasts and/or cells transferred onto fresh feeder plates without kanamycin sulfate in the medium) and the selection control treatment (mock-transformed protoplasts, i.e., protoplasts carried through the transformation protocol in the absence of foreign DNA and plated on feeder plates containing kanamycin sulfate). Incubate in the same conditions for 1 additional week.
3. After 2 weeks on feeder plates, transfer the callus onto MS 2D medium supplemented with 100 μg/ml kanamycin sulfate for further selection of the transgenic calli.
4. After an additional 2 weeks identify the fast growing and putatively transgenic calli. Continue to grow the transgenic calli on MS 2D medium containing the kanamycin sulfate with subcultures every 2 weeks until the callus is analyzed for the presence and expression of the foreign genes or for some other characteristics.

REFERENCES

Antonelli NM, Stadler J (1990) Genomic DNA can be used with cationic methods for highly efficient transformation of maize protoplasts. Theor Appl Genet 80: 395–401

Armstrong CL, Petersen WL, Buchholz WG, Bowen BA, Sulc SL (1990) Factors affecting PEG-mediated stable transformation of maize protoplasts. Plant Cell Rep 9: 335–339

Fromm ME, Taylor LP, Walbot V (1986) Stable transformation of maize after gene transfer by electroporation. Nature 319: 791–793

Huang YW, Dennis ES (1989) Factors influencing stable transformation of maize protoplasts by electroporation. Plant Cell Tissue Organ Culture 18: 281–296

Lyznik LA, Peng J, Hodges TK (1991) Simplified procedure for transient transformation of plant protoplasts using polyethylene glycol treatment. BioTechniques 10: 294–299

Lyznik LA, Ryan RD, Ritchie SW, Hodges TK (1989) Stable transformation of maize protoplasts with *gusA* and *neo* genes. Plant Mol Biol 13: 151–161

Ye GN, Earle ED (1991) Effect of cellulases on spontaneous fusion of maize protoplasts. Plant Cell Rep 10: 213–216

DNA Delivery into Maize Cell Cultures Using Silicon Carbide Fibers

H.F. KAEPPLER AND D.A. SOMERS

Silicon carbide fibers can be used to deliver foreign DNA into maize tissue culture cells for transient expression studies and genetic transformation (Kaeppler et al. 1990, 1992). The fibers are easily handled and are very inexpensive. At its current level of development, frequencies of DNA delivery and stable transformation using the silicon carbide fiber method are somewhat lower than can be achieved with microprojectile bombardment (Klein et al. 1989; Spencer et al. 1990; Fromm et al. 1990; Gordon-Kamm et al. 1990). Further improvement of the silicon carbide fiber procedure will likely lead to increased DNA delivery frequencies. Materials and methods are given below for delivery of plasmid DNA into nonregenerable and regenerable maize cell cultures using silicon carbide fibers.

MATERIALS

Liquid cell culture medium [MS2D (Green 1977) or N6 (Armstrong and Green 1985)]
Plasmid DNA (1 μg/μl) suspended in TE buffer (10 mM Tris, pH 8.0, 1 mM EDTA)
Silicon carbide fibers (Silar Whiskers SC-9, and Material Safety Data Sheet (MSDS), Advanced Composite Materials, 1525 S. Buncombe Rd., Greer, SC 29651–9208, 1–800–272–6638)
Disposable petri plates (20 mm × 60 mm)
1.5 ml eppendorf tubes
Wig-L-Bug machine and 1.5 ml reusable Wig-L-Bug containers (Crescent Dental Manufacturing Co., Chicago, IL)
Millipore 47-mm Sterifil aseptic filter system
47-mm Millipore MF support pads

The Maize Handbook—M. Freeling, V. Walbot, eds.

METHODS

Preparation of Sterile Suspensions of Silicon Carbide Fibers

Silicon carbide fibers must be handled with care to avoid inhalation and possible lung damage (consult MSDS). Dry fibers should be handled in a vented exhaust hood. It is best to work with the fibers in a liquid suspension.

1. In an exhaust hood, transfer approximately 50 mg dry silicon carbide fibers into pre-weighed eppendorf tubes.
2. Close tubes and reweigh to determine weight of silicon carbide fibers.
3. Puncture a hole in the top of the tube with a needle; then cover the top of the tube with two layers of aluminum foil.
4. Autoclave the tube containing silicon carbide fibers on the "fast exhaust and dry" cycle for 25 minutes at 121°C and 20 psi.
5. Add sterile, deionized H_2O to make up a 5% (w/v) suspension of fibers.

Notes

1. Fresh silicon carbide fiber suspensions have been observed to transfer DNA at much higher frequencies than older suspensions; therefore, it is recommended that new fiber suspensions be made for each experiment.
2. Tubes containing silicon carbide fiber suspensions must be vortexed immediately before use to resuspend the fibers.
3. Proper disposal of silicon-carbide-fiber-contaminated materials should be discussed with an institutional agency responsible for hazardous waste disposal.

DNA Delivery Into Nonregenerable Black Mexican Sweet (BMS) Suspension Culture Cells

1. All steps are conducted using aseptic conditions.
2. Black Mexican Sweet suspension culture cells (Green 1977) are used for DNA delivery 3–6 days after subculture. Cells are collected on filter support pads using the Millipore Sterifil vacuum filtration apparatus and rinsed with 35 ml MS2D medium.
3. Add 20 μl plasmid DNA solution and 40 μl of a sterile 5% suspension of silicon carbide fibers to a 1.5 ml eppendorf centrifuge tube. To pipet the silicon carbide fiber suspension, use a sterile 200 μl pipet tip with the end (1 cm) cut off.
4. Cap the tube and vortex (Vortex Genie 2, Scientific Industries, Inc.) on medium speed for 10 seconds to thoroughly mix the DNA and silicon carbide fiber suspensions.
5. Add approximately 300 μl packed volume of BMS cells with a spatula followed by 50 μl MS2D medium to tubes containing DNA/silicon carbide fiber suspensions.

6. Invert tube and vortex on high setting to mix cells with silicon carbide fibers (about 10 seconds).
7. Return the tube to the upright position and vortex for 1 minute on high setting.
8. Transfer contents of each tube into petri plates (20 × 60 mm) by adding 500 μl MS2D culture medium and pouring contents into a petri dish. Remove remaining cells from tube with two more 500 μl rinses of culture medium into the same petri dish.
9. Transient expression assays and selection of stable transformants are described in Kaeppler et al. (1990, 1992). Also see protocols in this volume.

DNA Delivery into Regenerable Embryogenic Suspension Culture Cells

Regenerable suspension culture cells apparently require more force for silicon-carbide-fiber-mediated DNA delivery. In this procedure a Wig-L-Bug macerator is used to provide harder and more frequent intercellular collisions.

1. Add 20 μl DNA suspension and 40 μl of a 5% (w/v) suspension of silicon carbide fibers to an autoclaved 1.5 ml Wig-L-Bug capsule.
2. Mix DNA and silicon carbide fibers with the end of the pipet tip used to transfer the silicon carbide fiber suspension.
3. Add 300 μl packed volume of regenerable suspension culture cells, which have been prepared as described above and rinsed with N6 medium, plus 0 to 50 μl N6 medium to a capsule containing mixed DNA and fibers. (Note: do not add the metal Wig-L-Bug macerating bar to the capsule.)
4. Close capsule and agitate with Wig-L-Bug machine for 30 seconds.
5. Use N6 liquid culture medium (1–2 ml) to rinse cells into 20 × 60 mm sterile, disposable petri plates.
6. Procedures for selection of stable transformants are described in the chapters on cell culture and transformation in this volume.

REFERENCES

Armstrong C, Green CE (1985) Establishment and maintenance of friable, embryogenic maize callus and the involvement of L-proline. Planta 164: 207–214

Fromm ME, Morrish F, Armstrong C, Williams R, Thomas J, Klein TM (1990) Inheritance and expression of chimeric genes in the progeny of transgenic maize plants. Bio/Technology 8: 833–839

Gordon-Kamm WJ, Spencer TM, Mangano ML, Adams TR, Daines RJ, Start WG, O'Brien JV, Chambers SA, Adams WR, Willetts NG, Rice TB, Mackey CJ, Kreuger RW, Kausch AP, Lemaux PG (1990) Transformation of maize cells and regeneration of fertile transgenic plants. Plant Cell 2: 603–618

Green CE (1977) Prospects for crop improvement in the field of cell culture. HortScience 12: 131–134

Kaeppler HF, Gu W, Somers DA, Rines HW, Cockburn AF (1990) Silicon carbide fiber-mediated DNA delivery into plant cells. Plant Cell Rep 9: 415–418

Kaeppler HF, Somers DA, Rines HW, Cockburn AF (1992) Silicon carbide fiber-mediated stable transformation of plant cells. Theor Applied Genet, 84: 560–566

Klein TM, Kornstein L, Sanford JC, Fromm ME (1989) Genetic transformation of maize cells by particle bombardment. Plant Physiol 91: 440–444

Spencer TM, Gordon-Kamm WJ, Daines RJ, Start WG, Lemaux PG (1990) Bialaphos selection of stable transformants from maize cell cultures. Theor Applied Genet 79: 625–631

109

Transient Gene Expression Assay by Electroporation of Maize Protoplasts

KENNETH R. LUEHRSEN AND JEFFREY R. DE WET

MATERIALS

BMS media for maize suspension
- 1 package Murashige and Skoog salts (Gibco BRL)
- 1 ml 1000× vitamins (1.3 mg/ml niacin, 0.25 mg/ml thiamine, 0.25 mg/ml pyroxidine, 0.25 mg/ml pantothenate)
- 0.13 g asparagine
- 2 ml 2,4 D (2,4-dichlorophenoxyacetic acid @ 1 mg/ml in 100% EtOH)
- 20 g sucrose
- 0.2 g inositol
- (per liter
- pH 5.8 [with 0.1 N NaOH])

Protoplast isolation medium (PIM)
- 0.25 M mannitol

The Maize Handbook—M. Freeling, V. Walbot, eds.

50 mM $CaCl_2$
10 mM sodium acetate pH 5.8
PIM+enzymes
0.3% Cellulase (CELF/Worthington)
1.0% Cytolyase (Genencor)
0.02% pectolyase Y23 (Seishin)
0.5% BSA
β-mercaptoethanol (add 25 μl per 50 ml of solution)
(Prepare using the PIM solution. Spin out the particulates at 10,000*g* for 10 minutes. Filter-sterilize through a 0.45 μM filter.)
Poration solution (PS)
200 mM mannitol
120 mM KCl
10 mM NaCl
4 mM $CaCl_2$
10 mM HEPES pH 7.2 (with 0.1 N NaOH)
Plating out medium for transients (POMT)
0.3 M mannitol in 800 ml BMS medium
add 200 ml conditioned medium (see step 1 under Method)

METHOD

This protocol was derived from that described in Fromm et al. (1987) and Luehrsen et al (1992).

1. Routinely culture BMS protoplasts in ~40 ml BMS media in 100-ml flasks; shake at 100 rpm at 26°C. Split the culture 1:1 into fresh medium every 3–to 4 days. Harvest when the protoplasts are in log-phase growth (1–2 days after splitting). The yield is approximately 10^7 protoplasts/40 ml. To harvest, spin for 2 minutes at ~500 rpm at room temperature. Save and filter sterilize the conditioned media for the plating out media (POMT).
2. Wash the cells with 50 ml of protoplast isolation medium (PIM).
3. Resuspend the cells in an equal volume of PIM+enzyme solution. Distribute 10 ml of the cell suspension into individual 100 × 15 mm petri plates. Incubate at 26°C on a bed shaker (50 rpm) for 2–5 hours.
4. *Gently* transfer the protoplasts to a 50-ml conical screw-cap tube. Spin to pellet the cells as before.
5. Wash the protoplasts twice with 50 ml PIM.
6. Wash the protoplasts once in 50 ml electroporation solution (PS). Spin to pellet the cells as before.

7. Resuspend in PS (about 3 ml for each 40 ml culture starting volume) for a final cell density of ~3×10^6 per ml.
8. *Optional.* Heat shock the protoplasts for 10 minutes at 45°C. Chill the protoplasts on ice for >30 minutes.
9. Use sterile plastic microcuvettes with a 0.4-cm opening. Add 0.5 ml cells and 0.5 ml of the expression plasmid DNA solution (@10–50 μg/ml in PS) to the cuvette; mix gently.
10. Set the capacitance to 1550 μF. For machines with permanent electrodes, sterilize the electrodes by dipping in water and then in ethanol; flame and cool in a sterile petri dish. Insert the electrodes into the protoplast/DNA suspension. Charge the capacitors to 175 V (450 V/cm final field strength) and discharge using a 12-msec time cutoff.
11. Transfer the electroporated protoplasts directly into 7.5 ml POMT in a 100 × 15-mm petri plate.
12. Incubate at 26°C in the dark for 24–48 hours. Harvest the protoplasts for either RNA recovery or preparation of a reporter gene enzyme extract.

Notes

1. Using this procedure we have assayed firefly luciferase, β-glucuronidase (GUS), and chloramphenicol acetyltransferase (CAT) activities.
2. At step 8, the heat shock is optional. Using the heat shock, we have observed an approximately two- to fivefold increase in both transcript abundance and reporter gene enzyme activity.
3. For electroporation, we use aluminum plate electrodes spaced 0.4 cm apart.

REFERENCES

Fromm ME, Callis J, Taylor LP, Walbot V (1987) Electroporation of DNA and RNA into plant protoplasts. Methods Enzymol 153: 351–366

Luehrsen KR, de Wet JR, Walbot V (1992) Transient expression analysis in plants using the firefly luciferase reporter gene. Methods Enzymol, 216: 397–414

110

Assay of Bacterial Chloramphenicol Acetyl Transferase in Transformed Maize Tissues*

STEVE GOFF

This protocol describes a method for detecting bacterial CAT enzyme activity in homogenates of maize cells that have been transiently transformed with plasmid constructs capable of expressing the bacterial CAT gene. A variety of intact plant tissues or cultured maize cell lines may be used. Several transformation techniques are described in other chapters. The method described in this protocol uses the conversion of ^{14}C-acetyl coenzyme A to ethyl acetate soluble radioactivity followed by radioactive determination in a liquid scintillation counter. The general approach is to heat-inactivate activities that may cause background, allow the bacterial enzyme to generate organic-soluble product for 1 hour, extract the product, back-extract the organic phase, and count the resulting product in a scintillation vial. The ^{14}C-acetyl coenzyme A is very expensive, and the CAT activity may be marginally detectable if the promoter driving CAT is weak or if the number of cells transformed is low. This assay requires the use of radioactivity in an volatile organic solvent, and several manipulations of the sample are necessary. It has been reported that the use of firefly luciferase as a reporter is approximately 1,000 times more sensitive than is bacterial CAT.

MATERIALS

Tissues expressing bacterial CAT
Control tissue that does not express CAT
1.5-ml microfuge tubes (three per sample)
Scintillation vials (one per sample)

*From Sleigh M.J. (1986) A nonchromatographic assay for expression of chloramphenicol acetyl-transferase-gene in eycariptic cells. Anal Biochem 156: 251–256.

The Maize Handbook—M. Freeling, V. Walbot, eds.

Organic scintillation fluor (4 ml per sample)
Ethyl acetate (about 400 μl per sample)
250 mM Tris-HCl (pH 7.8)
Homogenization buffer
100 mM potassium phosphate pH 7.8
1 mM dithiotheitol

Reaction solution (per sample, make up extra to account for pipeting losses)

Stock Solution	*Amount per sample*	*For 30 samples*
5 mM chloramphenicol	32 μl	960 μl
1 M Tris-HCl pH 7.8	15 μl	450 μl
10 mM acetyl coenzyme A	0.9 μl	27 μl
0.1 μCi ^{14}C-acetyl coenzyme A	1.0 μl	30 μl
Water	26.1 μl	783 μl

Equipment

0°C mortar and pestle
0°C water bath
65°C water bath
37°C water bath
Fumehood
Vortex
Pipetman (200 μl)
Scintillation counter
Microcentrifuge at 4°C

METHOD

1. Prepare three sets of appropriately labeled 1.5 ml microfuge tubes. The microfuge tubes used for steps involving radioactive ^{14}C-acetyl coenzyme A should be very tight sealing (not the flex-top variety) as the ethyl acetate will leak from poor-sealing tubes.
2. Homogenize the tissue on ice in a mortar and pestle using 0.5–2 ml of homogenization buffer per sample. (Use as little homogenization buffer as possible; 1 ml buffer works well for a 3-cm-diameter embryogenic callus culture, and use the same amount of homogenization buffer per sample in a given experiment or set of related experiments.) Homogenize the sample very well to release as much bacterial CAT activity as possible. Pipet the extract into a 1.5 ml microfuge tube, and store on ice until all extracts have been prepared.
3. Centrifuge the extracts at full speed in a microcentrifuge at 4°C for 5 minutes and return the sample to ice.

4. Aliquot 25 μl of the extract supernatant into a fresh 1.5-ml tight-sealing microfuge tube.
5. Incubate the 25-μl cell-extract aliquots at 65°C for 10 minutes. This step reduces the non-CAT conversion of the substrate to product, and should be done at exactly 65°C for 10 minutes to allow valid comparisons between repetitions of the experiment.
6. Remove the extract aliquots to room temperature. Add 75 μl of the reaction solution to each tube and incubate in a 37°C water bath for exactly 60 minutes.
7. While the reaction proceeds, add 100 μl of 250 mM Tris-HCl (pH 7.8) to the final set of microfuge tubes. This will allow the tubes to be closed immediately following addition of the ethyl acetate from the first extraction of the reaction mixture. (See step 11.)
8. Remove the reactions to room temperature in a fumehood. Add 200 μl ethyl acetate to each reaction mixture, capping each tube immediately after adding the ethyl acetate to prevent liquid loss.
9. Vortex for 1 minute at room temperature using the highest vortexing speed.
10. Centrifuge for 2 minutes in a room-temperature microfuge at highest speed setting.
11. In a fumehood, transfer 180 μl of the upper ethyl acetate layer to a fresh microfuge tube with the 100 μl 250 mM Tris-HCl (pH 7.8) as prepared in step 7. Cap the tube immediately after transferring the volatile ethyl acetate extract to prevent liquid loss.
12. Vortex for 1 minute at room temperature using the highest vortexing speed.
13. Centrifuge for 2 minutes in a room-temperature microfuge at the highest speed setting.
14. In a fumehood, transfer 150 μl of the upper ethyl acetate layer to a labeled scintillation vial. Add 4 ml of scintillation fluid able to mix freely with the organic ethyl acetate. Cap the vial and count the radioactivity for 1–5 minutes in a scintillation counter.

Notes: The volume of the homogenization buffer will depend on the amount of sample that is being used. Because only 25 μl of extract is required for the assay, the lower the volume of homogenization buffer used, the higher the activity of CAT will be. Soft tissues such as callus can be homogenized with a great deal less effort using a ground-glass tissue grinder (Potter-Elvehjem device) attached to a motorized grinder. A low-speed bench-top drill press is an inexpensive alternative to the motorized grinders available from scientific supply companies, although it may be a bit more dangerous to use. Care must be taken throughout the protocol to avoid loss of ethyl acetate from the samples. Loss of the volatile organic solvent will make it difficult to pipette the required amounts in steps 11 and 14. Rapid capping of the microfuge tubes limits the loss of ethyl acetate.

111

Assay of Firefly Luciferase in Transformed Maize Tissues*

STEVE GOFF

This protocol describes the assay of firefly luciferase enzyme activity in homogenates of maize cells transiently transformed with plasmid constructs able to express the firefly luciferase gene. A variety of intact plant tissues or cultured maize cell lines may be used for the detection of firefly luciferase activity following gene transfer. Transformation techniques are described in other chapters. The general approach is to homogenize the tissue, centrifuge down the debris, and determine the amount of luciferase by measuring the light output when the extract is mixed with the substrates luciferin and ATP in a luminometer. This assay is very sensitive and has been used to detect luciferase expression in a very low percentage of the total cell population.

MATERIALS

Tissues expressing firefly luciferase
Control tissue that does not express luciferase
1.5-ml microfuge tubes (one per sample)
Luminometer cuvettes (one per sample)
0.5 mM Luciferin (from a 5.0 mM stock solution prepared in water and stored at −20°C)
(It is useful to have some luciferase (Sigma) as a positive control.)
Homogenization buffer
100 mM potassium phosphate pH 7.8
1 mM dithiotheitol
Assay buffer
25 mM tricine
15 mM $MgCl_2$

*From Callis, J., M.E. Fromm and V. Walbot (1987), in *Genes & Development* 1: 1183–1200.

The Maize Handbook—M. Freeling, V. Walbot, eds.

5 mM ATP
500 μg/ml BSA

Equipment
0°C mortar and pestle
0°C water bath
Pipetman
Microcentrifuge at 4°C
Luminometer

METHOD

1. Warm up the luminometer, and thaw all reagents required. Place the luciferase assay buffer on ice.

2. Prepare a set of appropriately labeled 1.5-ml microfuge tubes.

3. Homogenize the tissue on ice in an ice-cold mortar and pestle using 0.5–2 ml of homogenization buffer per sample. Use as little homogenization buffer as possible; 1 ml buffer works well for 2 kernels or a 3-cm-diameter embryogenic callus culture; use the same amount of homogenization buffer per sample in a given experiment or set of related experiments. It is crucial to grind the sample very well in order to release as much luciferase as possible. Pipet the extract into a 1.5-ml microfuge tube, and store on ice until all extracts have been prepared.

4. Centrifuge the extracts at full speed in a microcentrifuge at 4°C for 5 minutes, and replace on ice.

5. Aliquot 200 μl of luciferase assay buffer into a set of luminometer cuvettes at room temperature. Prepare an extra six or more cuvettes to determine the machine background.

6. Determine the machine noise by activating the sensor for three rounds of measurements without any sample present. Place the 0.5 mM luciferin in the machine substrate reservoir for luminometers equipped with autoinjection pumps. Fill the lines by activating the injector pump several times, and check to be certain substrate solution is being delivered to an empty cuvette. Determination of the background (dark current) through the photomultiplier provides the operator with an indication of the stability of the machine and allows detection of any light leaks.

7. Allow the cell extract to equilibrate to room temperature. Aliquot 100 μl of cell extract supernatant into one of the luminometer cuvettes and place it in the machine. Activate the automatic pump/timer/photomultiplier and wait for the reading to stop (either 10 or 30 seconds for most luminometers). Record the result. Process the remaining samples in an identical manner.

Notes: The homogenization buffer can be prepared as a 10× concentrated stock and stored at −80°C. The luciferase assay buffer salts (without ATP) can also be prepared as a 10× concentrated stock and stored at −80°C. The volume of the homogenization buffer will depend on the amount of sample that is being used. Because only 100 μl of extract is required for the assay, the lower the volume of homogenization buffer used, the higher the activity will be. Soft tissues such as callus can be homogenized with a great deal less effort using a ground glass tissue grinder (Potter-Elvehjem device) attached to a motorized grinder. A low-speed drill press is an inexpensive alternative to the motorized grinders available from scientific supply companies.

Luminometers are very different in sensitivity and background current. They are also very different in price, all being fairly expensive. In an effort to reduce the cost of this assay, I have had a luminometer designed and built using a very sensitive photomultiplier. This machine is now commercially available, and I can provide further information upon request.

Several recent developments should improve the already high sensitivity of the luciferase assay. It is thought that luciferase is conformationally unstable, and that this leads to a rapid loss of activity within cells during the incubation period. (I generally allow 48 hours for expression of the enzyme prior to tissue homogenization.) It was found that substrate inhibitors would increase luciferase activity, presumably by increasing the stability of the enzyme. It was also found that coenzyme A increased the activity by promoting release of the product from the enzyme. Promega now sells a Luciferase assay reagent kit that incorporates Coenzyme A, and the use of these reagents can increase the detected luciferase activity substantially.

Storage of Frozen Maize Tissue

SUSAN R. WESSLER

Most protocols for the isolation of DNA and RNA require grinding frozen tissue to a fine powder prior to extraction. Several of the protocols in this book include a short description of how maize tissue should be ground for particular purposes. In this section an efficient system for the storage of frozen ground maize tissue is described.

CONSIDERATIONS IN THE STORAGE OF FROZEN TISSUES

Long-term storage of frozen maize tissue is of critical importance to labs involved in the routine isolation of nucleic acids. It is often impractical to grow fresh material every time it is needed. Although isolation of milligram quantities of DNA from a single maize plant is feasible, over a month of growth is necessary to have sufficient tissue. Similarly, mRNA is readily obtained from all tissues. If your gene is expressed in the flowers or the kernel, for example, you will have to wait months for these tissues to develop. Limited availability of greenhouse space or seasonal availability of field space could also be important considerations in determining when fresh tissue will be available.

These problems are easily solved by planning ahead and storing adequate supplies of frozen tissue in a −80°C freezer. In my lab we have frozen stocks from most of the 40 or so *waxy* mutants and from a similar number of *R* mutants. These stocks include young seedlings for DNA isolation and other tissues at various developmental stages for the isolation of mRNA. Below is a summary of the procedures we follow to ensure that frozen tissues are well organized, accessible, and occupy a minimum amount of valuable freezer space.

METHODS

1. Tissues are harvested and frozen in liquid N_2 as soon as possible.
2. The following steps are taken to reduce bulk prior to storage:

The Maize Handbook—M. Freeling, V. Walbot, eds.

a. Leaves, tassels, roots stems etc. are ground to a fine powder using a mortar and pestle that has been pre-chilled in liquid N_2.
b. Frozen ears are shelled and the kernels stored. In preparation for shelling, ears are immersed in liquid N_2 until they stop bubbling. These ultra-frozen ears can be shelled in much the same way as mature, dried ears. To reduce bulk even further, kernels can be ground or dissected prior to long-term storage.

Note 1: It is critical that ears be immersed in liquid N_2. Failure to do so results in damaged kernels upon shelling.
Note 2: Older kernels (over 30 DAP) are difficult to grind in a mortar and pestle. These can be crudely ground in a chilled coffee-bean mill prior to use of the mortar and pestle.

3. All tissues are stored in capped, 50-ml (114 × 29 mm), conical-bottom disposable test tubes. Information regarding genotype and age can be recorded on the cap with a permanent marker. This information is also recorded on a tag kept with the tissue inside the tube.
4. Tubes are stored in 20 × 17-cm Styrofoam racks (25 tubes/rack; many manufacturers supply these racks with the tubes). If you have a chest-model −80°C freezer, these racks can be stored in stainless steel towers similar to those used to store small cardboard freezer boxes. Our towers have four open shelves with a handle at the top. They can be custom made by any competent machine shop.
5. The position of frozen material can be recorded (e.g., tower 4, shelf 3, row 2) on a master list to facilitate rapid retrieval.

113

Maize Methods—Starch Biosynthetic Genes

L. CURTIS HANNAH, EDWIN R. DUKE, KAREN E. KOCH, AND B. GREGORY COBB

Starch biosynthetic genes are of interest to maize biologists for a number of reasons. These genes have been used to follow chromosomal behavior during endosperm development, as traps for transposable elements and other mutations, and as marker genes to follow tissue-specific patterns of expression. The economic importance of their substrates (sugars) and products (starches) has provided another reason for inquiry.

The cloning of the starch synthetic genes *waxy* (*wx*) *shrunken-1* (*sh1*), *shrunken-2* (*sh2*), and *brittle-2* (*bt2*) has been reported. Isolation of other starch synthetic genes is in progress. In many experiments, enzymic assay of the gene products is advantageous or required. A description of these assays is given below. Furthermore, we outline an in vitro kernel development scheme that can be used to alter the external stimuli that affect gene expression and to study sugar metabolism.

ADP-GLUCOSE PYROPHOSPHORYLASE REACTION: GLUCOSE-1-PO_4 + ATP = ADP-GLUCOSE + PP

Assay employs a technique developed by Jack Preiss (see Dickinson and Preiss 1969 for example) and subsequently modified. A 100-μl reaction mixture contains 8 μmol HEPES buffer, pH 8.0, 2 μmol $MgCl_2$, 0.2 μmol glucose-1-PO_4, 100,000 cpm C-14 glucose-1-PO_4, 0.2 μmol ATP, 50 μg bovine serum albumin, 2 μmol 3-phosphoglycerate and enzyme. Stock solutions (10×) of the reactants can be frozen for several months. We routinely make up a solution containing buffer, glucose-1-PO_4, BSA, $MgCl_2$, and sometimes 3 PGA. With these components mixed together as a stock solution, a typical reaction involves the simple addition of the enzyme and ATP to an aliquot of the mixture above. Water is substituted for ATP in typical controls. After reaction times of 10–60 minutes, tubes are boiled and cooled, and 0.03 mg alkaline phosphatase (Worthington—BAPSF) is added

The Maize Handbook—M. Freeling, V. Walbot, eds.

for an additional 1–2 hours. We have used various brands and grades of alkaline phosphatase but none leads to as complete removal of glucose-1-PO_4 as observed with the Worthington enzyme. This factor is critical for the detection of very low levels of enzymic activity. Note that the definition of a unit of phosphatase differs among companies.

A 20-μl aliquot is then placed on a DEAE-cellulose paper disk (DE-81) that fits in the bottom of a scintillation vial (usually 16 mm). A straight pin inserted into the numbered disk allows for easy handling of the paper. Marking of the disk is best done with a pencil because some other labeling materials (ink, for example) can leach from the paper and subsequently cause quenching of radioactivity. The papers are then washed in distilled, deionized water. The product of the reaction, ADP-glucose, is charged; glucose arising by the action of phosphatase on the remaining glucose-1-PO_4 does not contain a charge and thus is washed from the DE paper.

Various washing chambers can be used. An inexpensive chamber consists of a typical laboratory plastic squeeze bottle. The top is removed with a knife and 20 or so holes are made with a hot punch. A glass rod inserted through holes at the top of the bottle allows the chamber to be placed in a beaker of water without the bottom of the chamber resting on the bottom of the beaker. Water is mixed with a stirring rod. We routinely use three 15-minute washes. After washing, the samples are dried with a heat lamp, placed in vials, and quantified with a scintillation counter. Because the radiolabel binds to the paper, the vials and scintillation solution can be reused. We routinely monitor the vials after disk removal with the scintillation counter.

Various enzyme extraction buffers can be used. Buffers containing phosphate have the advantage that phosphate, an inhibitor of the enzyme, does seem to stabilize activity. Because the reaction mixture usually contains high levels of the activator 3-PGA, which overcomes this inhibition, phosphate inhibition is effectively eliminated.

It is imperative that enzyme activity be demonstrated to be proportional to time and to dilution. Enzyme extracted from endosperms 22 days post-pollination at 1 g/ml buffer must be diluted 10–100-fold in order to remove the effects of an inhibitor(s) and thus have linearity with protein content. This is also true for activity extracted from *sh2* and *bt2* mutants that contain only a small percentage of wild type enzyme. Furthermore, crude preparations lose activity rapidly, and the enzyme sticks to glass.

Several controls are required when setting up the assay. The washing procedure can be checked by counting reactions not treated with phosphatase or enzyme preparation before and after washing. This is critical because some laboratory water contains concentrations of salts high enough to remove phosphorylated sugars from the DEAE-cellulose paper. By the same rationale, significant phosphatase activity in the enzyme preparation can be detected by monitoring preparations before and after washing in which the Worthington phosphatase has not been added. It is also important that the radioactivity detected on the paper be proportional to the amount of reaction mixture added to the paper. The addition of charged molecules such as 3-PGA to the reaction mixture will compete with ADP-glucose for binding to the paper.

Once the assay procedure is established, controls of zero time of incubation or ones lacking ATP can be run. We prefer the latter because they can, in theory, detect the presence of enzymes that convert glucose-1-PO_4 to a phosphatase-resistant charged product other than ADP-glucose.

SUCROSE SYNTHASE REACTION: SUCROSE + UDP = UDP-GLUCOSE + FRUCTOSE

Several assays are used for this enzyme; the most common one being a nonradioactive assay that involves the measurement of reducing sugar arising from the reaction. We have found that a radioactive assay is superior, because it has greater sensitivity and, since reducing sugars in the enzyme preparation need not be removed before assay, dialysis preceding assay can be avoided. This is important because significant enzyme activity is lost in this step when the enzyme is extracted from certain maize tissues. The assay was developed independently here and elsewhere (Delmer 1972; Su and Preiss 1978). The assay measures the production of ^{14}C-labeled UDP-glucose from ^{14}C-labeled sucrose. Because the product is charged and the labeled substrate is not, DE-52 paper can be used, as in the ADP-glucose pyrophosphorylase assay above, to separate the two reactants.

Reaction mixtures include 80 mM MES, pH 5.5, 5 mM NaF, 100 mM sucrose, 100,000 cpm ^{14}C-sucrose, and 5 mM UDP plus enzyme in a volume of 50 μl. Reactions are terminated by boiling. Controls lack UDP. Aliquots are spotted on DE-52 paper; the samples are washed, dried, and counted as in the pyrophosphorylase assay. Again, it is imperative that the amount of product produced be linear with time and with protein content, and that good enzymological practices be followed. The specific activity of the sucrose can be adjusted to the sensitivity needed in the experiment.

We have found that commercially available ^{14}C sucrose contains traces of a radioactive contaminants that bind to DE paper. This was initially detected as high levels of apparent incorporation in reactions lacking UDP. Because the K_m for sucrose is much higher than that for UDP, high levels of sucrose (and the contaminant) are needed relative to the level of UDP. Because the reaction should never be allowed to approach equilibrium in experiments designed to measure enzymic activity, the amount of UDP-glucose produced must be neglible compared to the remaining amount of sucrose. The amount of UDP-glucose produced is usually comparable to the amount of the radioactive contaminant. Because of this, commercially available ^{14}C sucrose must be purified before use in the assay. An easy procedure consists of the following: Commercial ^{14}C sucrose is streaked onto a strip (7 cm × 20 cm) of DE-52 paper in a line parallel to the short side of the paper some 5–6 cm from one edge. The paper is then placed in a chromatography jar and eluted in a descending direction with deionized, distilled water. The bottom of the paper is cut into a V-shape and a tube is placed at the bottom of the V. Almost all ^{14}C sucrose is collected in the first drop of water while the contaminant is held on the paper. This easy procedure can be completed in just a few hours and enough sucrose for hundreds of reactions can be prepared at one time.

STARCH-BOUND ADP-GLUCOSE-STARCH GLUCOSLYTRANSFERASE REACTION: (AMYLOSE)n + ADP-GLUCOSE = (AMYLOSE)n+1 + ADP

Because the product of the reaction is insoluble while the substrate ADP-glucose is soluble, it is quite simple to separate the reactants following incubation. The procedure given below was taken from Tsai et al. (1970) and modified slightly to incorporate the use of a scintillation counter.

A 50-μl reaction mixture contains 6 μmol HEPES, pH 8.0, 0.25 μmol labeled ADP-glucose (5,000–100,000 cpm), and variable (usually 5 mg) amounts of starch granules. Following incubation at 37°C, usually for 30 minutes, 0.5 ml of 0.1 N NaOH is added. Starch is then precipitated with methanol to a final concentration of 75%. The suspension can then be passed through a conventional filter apparatus (for example Millipore); the resulting pellet can be dried on the filter disk and then counted in a scintillation counter.

The use of NaOH and methanol is not essential; the important points are to stop the reaction at the specified time (this can be done simply by diluting the reaction mixture with buffer) and quantitatively recovering the starch granules on the filter. The substitution of commercially available amylopectin and a conventional enzyme preparation for the starch granules can be used to assay the soluble ADP-glucose-starch glucosyltransferases.

Sound enzymological practices concerning linearity with time, with amount of starch granules, etc., should always be followed.

Starch granules containing the Wx enzyme can be prepared easily and the enzyme is quite stable. Developing kernels are ground at 1 g/ml of buffer and centrifuged at 31,000*g* for 20 minutes. (Lower speeds and lesser times can clearly be used.) A typical buffer for this is 10 mM Tris-maleate, pH 7.0. (Probably any reasonable buffer will work.) The supernatant is discarded, and the yellow layer on top of the white starch pellet is removed with a spatula. Additional buffer is added, the pellet is mixed into the buffer, and the tube is respun. Removal of the carotenoid layer, remixing, and recentrifugation are repeated although they are probably not essential. Ice-cold acetone is then mixed with the starch pellet, and the tube is recentrifuged. Following decanting, the starch is removed with a spatula, placed in a petri dish, and allowed to dry in a refrigerator. Enzymic activity is quite stable to refrigerator temperatures. Furthermore, starch granules from *sh2* mutants are approximately twice as active as wild type, most probably because of the reduction in the amount of starch synthesized in this mutant.

IN VITRO KERNEL DEVELOPMENT

To study the regulation of the starch synthetic genes and to determine the types and amounts of sugars used in starch biosynthesis, we have used and modified (for example, Cobb and Hannah 1983) an in vitro kernel development scheme originally described by Gengenbach (1977). A step-by-step protocol is given below. (Also consult other chapters of this volume.)

1. Harvest ears 5–7 days postpollination. Enough time should elapse between pollination and harvest to allow visual determination of ovule enlargement (discussed below). Ears are transported to the lab and held at 4°C until placed in culture. We have held ears at this temperature for 5 days before culture.
2. Cut the tips and the base from the ears and remove the outer husk. (From this point on all procedures are carried out using sterile technique).
3. Spray the husk lightly with 95% ethanol and flame.
4. Cut along the long axis of the ear from the tip to approximately 1 cm from the base. Make a second cut opposite (180°) from the first cut in a similar manner. Connect the two cuts with a third cut at the base and gently remove the flap of husks.
5. Determine the degree of fertilization. The fertilized ovules should be larger than unfertilized ovules by 5 dpp. Additionally, the silks will no longer be connected to fertilized ovules. If a majority of ovules are not fertilized, discard the ear.
6. Make transverse cuts through the ear at 10–12-ovule increments to produce cylinders of ear tissue. It may be necessary to use a knife with a serrated blade if the cob tissue is fibrous.
7. Insert the tip of a probe into the pith tissue of the ear cylinder to ease manipulation and rotate the cylinder on end.
8. With the cylinder of tissue on end carefully remove kernels with a scalpel leaving three pairs of kernel rows separated approximately 120° from each other. (At this point the cylinder will roughly look like a radiation sign when viewed from the end).
9. Cut into the pith tissue along each side of the kernel rows to separate the three explants. As much as possible make the cuts into the pith at an angle to maximize the amount of cob tissue attached to each block of kernels.
10. Place the kernel blocks into the medium described below. Culture the blocks at 29°C.

MURASHIGE-SKOOG STOCK SOLUTIONS

Liter	*Stock*		*Amt. to add to liter*
M.S. nitrates	NH_4NO_3	82.5 g/liter	20 ml
	KNO_3	95.0 g/liter	
M.S. sulfates	$MgSO_47 \cdot H_2O$	37.0 g/liter	10 ml
	$MnSO_41 \cdot H_2O$	1.69 g/liter	
	$ZnSO_47 \cdot H_2O$	1.06 g/liter	
	$CuSO_45 \cdot H_2O$	2.5 mg/liter	
M.S. halides	$CaCl_2 \cdot 2H_2O$	4.4 g/liter	10 ml

	KI	8.3 mg/liter	
	$CaCl_2 \cdot 6H_2O$	2.5 mg/liter	
M.S. P-B-Mo	KH_2PO_4	17.0 g/liter	10 ml
	H_3BO_3	620 mg/liter	
	$Na_2MoO_4 \cdot 2H_2O$	25.0 mg/liter	
FeEDTA	Na_2-EDTA	1.863 g/500 ml	10 ml
	Boil vigorously for 2 minutes, then add		
	$FeSO_4 \cdot 7H_2O$	1.393 g/500 mliter	
Thiamine-HCl	50 mg/50 ml		100 μl
Myo-inositol			100 mg
Nicotinic acid	50 mg/50 ml		500 μl
Pyridoxin HCl	25 mg/50 ml		1 ml
2,4-D	50 mg/50 ml		1 ml
Casein hydrolysate			1 g
Sucrose			150 g
pH	Adjust to 5.8 w/1 N HCl or KOH		
Agar			4 g

After mixing, the medium is sterilized by autoclaving for 15 minutes at 15 psi. A total of 50 ml is added to each container (Phytatray, Sigma).

REFERENCES

Cobb BG, Hannah LC (1983) Development of wild type, *shrunken-1* and *shrunken-2* maize seeds grown *in vitro*. Theor Appl Genet 65: 47–51

Delmer DP (1972) The regulatory properties of purified *Phaseolus aureus* sucrose synthetase. Plant Physiol 50: 389–393

Dickinson DB, Preiss J (1969) ADP-glucose pyrophosphorylase from maize endosperm. Arch Biochem Biophys 130: 119–128

Gengenbach BG (1977) Development of maize caryopses resulting from in vitro pollination. Planta 134: 91–93.

Su J-C, Preiss J (1978) Purification and properties of sucrose synthase from maize kernels. Plant Physiol 61: 389–393

Tsai CY, Salamini F, Nelson OE (1970) Enzymes of carbohydrate metabolism in the developing endosperm of maize. Plant Physiol 46: 299–306

114

Identifying and Characterizing the TATA Box Promoter Sequence Element in a Maize Nuclear Gene

JULIE M. VOGEL

Nuclear genes in plants that are transcribed by RNA polymerase II are likely to contain an AT-rich sequence, the TATA box, in the proximal region of the promoter. Based on a survey of plant promoter sequences, most plant genes, including those of maize, have a fairly recognizable TATA element (the consensus sequence is TATA(T/A)AT) and this element is located within a stringent distance of 32 ± 7 nucleotides upstream of the mapped site for transcript initiation (Joshi 1987). Like its demonstrated function in other eukaryotes, this TATA box sequence element in plants appears to be required to direct transcript initiation from a single, specific site downstream. The TATA element serves as the binding site for the evolutionarily conserved transcription initiation factor, TFIID, and this protein-DNA binding event appears to be the first step in the formation of an active transcription intiation complex (reviewed by Greenblatt 1991).

STRATEGIES

As part of the process of understanding how a newly cloned maize gene may be regulated at the transcriptional level, it is necessary to identify the TATA sequence element. The easiest way to predict the location of this element is first to identify accurately the transcript inititation site using standard RNA-mapping techniques (primer extension, S1 analysis, and/or RNase protection). Then, the probable TATA element should be identifiable as an AT-rich stretch of 5–10 bp located ~25–40 nucleotides upstream of this start site. If such an AT-rich sequence is difficult to verify, or if it is desirable to know more about the sequence requirements for TATA function in a given gene, then a more rigorous structure-function analysis of this region of the promoter can be performed. Several strategies are available, depending on information already at hand and on the level

The Maize Handbook—M. Freeling, V. Walbot, eds.

of specific information desired. All involve mutagenesis, followed by quantification of the mutation's effect on expression of a linked downstream reporter gene, typically by transient assay in electroporated maize protoplasts from an appropriate cell type. However, particle bombardment methods of gene transfer also are likely to yield suitable results. If the TATA element cannot be identified by eye, then a series of small, interstitial deletions, or linker-scan mutations, can be made in the general region predicted for the TATA box. The mutation that abolishes or severely reduces constitutive expression is likely to be that overlapping the TATA-functioning element. Alternatively, if a presumptive TATA element can be predicted by eye, then its actual function can be verified by specific mutagenesis directed at this sequence. To accomplish this, either site-directed mutagenesis to introduce systematic, defined changes or undirected, random mutations can be employed. Regardless of the method used, each mutation must be tested for its effect on reporter gene expression in a transient expression assay.

Below, I outline a general method for introducing a specific, desired mutation into a known TATA element; this protocol uses the oligonucleotide-directed site-specific mutagenesis method of Kunkel (1987). Protocols for mutagenesis can also be found in Sambrook et al. (1989) or in *Current Protocols in Molecular Biology* (Ausubel et al. 1989). Transient expression assays in maize cells and tissues are described elsewhere in this book.

MATERIALS

Single-stranded bacteriophage (or phagemid that can be induced for single-stranded growth) containing the TATA-sequence on the insert.

E. coli host CJ236 [*dut*1 *ung*1 *thi*1 *rel*A1/pCJ105 (cam^r F′)].

Mutagenic oligonucleotide (single-stranded, 17–40 nucleotides, with sequence complementary to the single-strand DNA to be mutagenized); this should be phosphorylated (kinased) at its 3′ end.

Sequenase v2.0 (modified T7 DNA polymerase) (U.S. Biochemical).

T4 DNA ligase (any supplier).

10× annealing buffer (1×=20 mM Tris-HCl pH 7.4, 10 mM $MgCl_2$, 50 mM NaCl, 10 mM DTT).

10× synthesis buffer (1×=0.5 mM each dATP, dCTP, dGTP, TTP, 1 mM ATP, 10 mM Tris-HCl pH 7.4, 5 mM $MgCl_2$, 5 mM DTT).

METHOD

Briefly, the single-stranded (+) template containing the wild type TATA box is grown in the *E. coli* host CJ236; this strain misincorporates UTP in place of some of the TTP residues in the DNA and is also incapable of correcting these misincorporations. After annealing the phosphorylated (−) strand mutagenic oligonucleotide to this U-containing single-stranded DNA, a primer extension reaction (carried out by the Sequenase enzyme) and a ligation reaction are carried out simultaneously to produce the mutant complementary strand. This U-containing wild type strand :: T-containing mutant strand heteroduplex is transformed into a normal *E. coli* strain (such as HB101) capable of recognizing and eliminating U-containing DNA. Thus, only the mutant T-bearing strands are allowed to replicate and produce transformants. These transformants are picked, and the mutant TATA sequence is verified by DNA sequence analysis. Often, this mutagenized sequence cassette must then be subcloned into an appropriate expression vector (i.e., upstream of GUS or luciferase reporter genes).

1. Combine in a microfuge tube, in a total volume of 10 μl:
   0.3–0.5 pmol ss U-containing template DNA (grown in CJ236 host)
   10 pmol 3′-phosphorylated oligonucleotide
   1 μl 10× annealing buffer
   H_2O to 10 μl
   Heat to 90–95°C, 5 minutes; then slowly cool to room temperature (1–2 hours).

2. With tubes on ice, add in this order:
   1 μl 10× synthesis buffer
   5 Weiss units T4 DNA ligase
   2.5 units Sequenase v2.0
   Incubate on ice 5 minutes, then at room temp 5 minutes, and finally at 37°C 1.5–2 hours.

3. Add 90 μl stop buffer (10 mM Tris-HCl pH8, 1 mM EDTA). Store reactions at −20°C.

4. Transform 10 μl (at a time) into competent *E. coli* HB101, DH5, or equivalent.

5. Pick colonies from the transformation plates, inoculate into 2–3 ml medium, and isolate plasmid DNA.

6. Check the plasmid by DNA sequencing to verify the mutant sequence.

7. Subclone an appropriate portion of the mutagenized cassette into an expression vector. (I usually reverify the sequence of the mutagenized region after this subcloning—just to be sure that the actual recombinant molecule that is to be tested for expression contains mutant sequence.)

REFERENCES

Ausubel F et al (eds) (1989) Current Protocols in Molecular Biology. John Wiley & Sons, New York

Greenblatt J (1991) Roles of TFIID in transcriptional initiation by RNA polymerase II. Cell 66: 1067–1070

Joshi CP (1987) An inspection of the domain between the putative TATA box and translation start site in 79 plant genes. Nucl Acids Res 15 (16): 6643–6653

Kunkel TA, Roberts JD, Zakour RA (1987) Rapid and efficient mutagenesis without phenotypic selection. Meth Enzymol 154: 367–382

Sambrook J, Fritsch EF, Maniatis T (1989) Molecular Cloning: A Laboratory Manual, Second Edition. Cold Spring Harbor Laboratory Press, New York

115

Promoters

Kenneth R. Luehrsen

In recent years, the expression of chimeric genes in maize has become routine through the use of gene transfer techniques such as electroporation and particle bombardment. It is now easy to study the regulatory sequences directing transcription by joining them to reporter genes such as chloramphenicol acetyltransferase (CAT), firefly luciferase, or β-glucuronidase (GUS). Alternatively, the RNA and protein products of a gene can by studied by linking the coding region to a high activity promoter.

Maize promoters transcribed by RNA polymerase II are, in general, about 500 bp in length and contain several conserved motifs involved in transcription regulation. The most conserved of these is the TATA box, an 8-bp, AT-rich conserved motif that lies about 25–40 bp upstream of the transcription start site; this is the binding site for the transcription factor TFIID. (See preceding chapter.) There is no well-conserved CAAT box as is found in animal promoters; however, several short, conserved protein-binding motifs are found upstream of the TATA box. These include motifs for the *trans*-acting transcription factors involved in light regulation, anaerobic induction, hormonal regulation, or anthocyanin biosynthesis, as appropriate for each gene.

When constructing chimeric expression cassettes, choosing which promoter to use depends on the circumstances in which expression is to be studied. An initial consideration is whether or not the construct will need to be expressed in a plant besides maize. One promoter, cauliflower mosaic virus (CaMV) 35S, is expressed in most cell types and is strongly active in plant tissues and in cultured cells. Even though the normal host of the

The Maize Handbook—M. Freeling, V. Walbot, eds.

CaMV virus is a dicot, genes driven by the 35S promoter are expressed well in most monocots, with wheat being an exception (Last et al. 1991). The maize *Alcohol dehydrogenase-1* (*Adh1*) promoter has been extensively used in maize gene expression studies; it is also active in *Panicum maximum* and *Triticum monococcum* but has low activity in rice (Kyozuka et al. 1990; McElroy et al. 1990), carrot (Hauptmann et al. 1988), and tobacco (Ellis et al. 1987). The maize *Shrunken-1* promoter is active in *Panicum maximum* and *Pennisetum purpureum* as well as in corn (Maas et al. 1991; Vasil et al. 1989), but not in carrot (Hauptmann et al. 1988). It is, however, difficult to quantify promoter strength from the existing literature, because expression constructs often differ in several features, i.e., inclusion of an intron, intron position; composition of the 5′ and 3′ untranslated regions (UTRs); and ATG context. Also, while there must be similarities in the binding sites for *trans*-acting transcription factors between monocots and dicots, subtle differences may compromise promoter strength in heterologous species.

A second consideration is whether to use a constitutive or an inducible promoter; Table 115.1 summarizes some of the available promoters. In most studies in which a

Table 115.1 Promoters active in maize

		Notes[a]	Reference
Constitutive			
CaMV 35S		1	Callis et al. (1987), Last et al. (1991)
Nos		1	Callis et al. (1987)
Sh1		2	Hauptmann et al. (1988), Vasil et al. (1989)
Adh1		2	Callis et al. (1987)
pEmu		2	Last et al. (1991)
Inducible	*Inducible by:*		
Adh1	hypoxia	3	Olive et al. (1990)
	cold stress		Christie et al. (1991)
Hsp70	heat stress		Callis et al. (1988)
PPDK	light		Sheen (1991)
Anthocyanin:			
A1 and *Bz1*	coexpressed *R* (or *B*) and *C1* genes		Klein et al. (1989), Goff et al. (1990)
Bz2	coexpressed *R* and *C1* genes		J. Bodeau, personal communication

[a]Notes: (1) also active in other monocots and dicots; (2) also active in other monocots; (3) also inducible in other monocots.

strong constitutive promoter is required, the CaMV 35S promoter is used. The nopaline synthase (*Nos*) promoter from *Agrobacterium tumefaciens* has also been successfully used in maize protoplasts (Callis et al. 1987). Alternatively, the maize *Adh1* promoter is active in tissue culture materials and has the added property of being further inducible under hypoxic (Walker et al. 1987) and cold stress conditions (Christie et al. 1991). Recently, a promoter chimera (pEMu) containing a truncated maize *Adh1* promoter, six copies of the 40 bp ARE (anaerobic responsive element), and four copies of the octopine synthase (OCS) enhancer element was shown to express at ~20-fold greater levels than than the CaMV 35S promoter in maize protoplasts; the promoter did not need to be induced by hypoxia for full activity (Last et al. 1991). The pEMu promoter is also active in several other monocots. It is thus possible to add binding sites of *trans*-acting transcription factors to boost expression from a weak promoter.

There are a number of promoters inducible by either environmental stimuli or by coexpressed transcription factors. A CAT gene under the control of the maize *hsp70* promoter was induced 200-fold at 42°C in electroporated maize protoplasts, indicating that heat stress can be used to modulate expression of a reporter gene (Callis et al. 1988). It is possible to induce expression of the maize *Adh1* promoter about 2.5-fold with hypoxic stress (Oliver et al. 1990; Walker et al. 1987). The pyruvate, orthophosphate dikinase (PPDK) promoter can be induced by light in maize leaf protoplasts (Sheen 1991). The *R* (and *B*) and *C1* genes encode transcription factors and regulate expression of the anthocyanin biosynthetic genes *A1*, *Bz1*, and *Bz2*. With the cloning of *R* (and *B*) and *C1*, chimeric gene constructs in which these regulators are constitutively expressed have been made. By using a reporter gene driven by an anthocyanin structural gene promoter, it is possible to study promoter induction by coexpressing *R* (or *B*) and *C1* in maize cells (Goff et al. 1990; Klein et al. 1989; Roth et al. 1991).

REFERENCES

Callis J, Fromm M, Walbot V (1987) Introns increase gene expression in cultured maize cells. Genes Dev 1: 1183–1200

Callis J, Fromm M, Walbot V (1988) Heat inducible expression of a chimeric maize hsp70CAT gene in maize protoplasts. Plant Physiol 88: 965–968

Christie PJ, Hahn M, Walbot V (1991) Low-temperature accumulation of *Alcohol dehydrogenase-1* mRNA and protein activity in maize and rice seedlings. Plant Physiol 95: 699–706

Ellis JG, Llewellyn DJ, Dennis ES, Peacock WJ (1987) Maize Adh-1 promoter sequences control anaerobic regulation: Addition of upstream promoter elements from constitutive genes is necessary for expression in tobacco. EMBO J 6: 11–16

Goff SA, Klein TM, Roth BA, Fromm ME, Cone KC, Radicella JP, Chandler VL (1990) Transactivation of anthocyanin biosynthetic genes following transfer of *B* regulatory genes into maize tissues. EMBO J 9: 2517–2522

Hauptmann RM, Ashraf M, Vasil V, Hannah LC, Vasil IK, Ferl R (1988) Promoter strength comparisons of maize *Shrunken 1* and *Alcohol dehydrogenase 1* and *2* promoters in mono- and dicotyledonous species. Plant Physiol 88: 1063–1066

Klein TM, Roth BA, Fromm ME (1989) Regulation of anthocyanin biosynthetic genes introduced into intact maize tissues by microprojectiles. Proc Natl Acad Sci USA 86: 6681–6685

Kyozuka J, Izawa T, Nakajima M, Shimamoto K (1990) Effect of the promoter and first intron of maize *Adh1* on foreign gene expression in rice. Maydica 35: 353–357

Last DI, Brettell RIS, Chamberlain DA, Chaudhury AM, Larkin PJ, Marsh EL, Peacock WJ, Dennis ES (1991) pEmu: an improved promoter for gene expression in cereal cells. Theor Appl Genet 81: 581–588

Maas C, Laufs J, Grant S, Korfhage C, Werr W (1991) The combination of a novel stimulatory element in the first exon of the maize *Shrunken-1* gene with the following intron 1 enhances reporter gene expression up to 1000-fold. Plant Mol Biol 16: 199–207

McElroy D, Zhang W, Cao J, Wu R (1990) Isolation of an efficient actin promoter for use in rice transformation. Plant Cell 2: 163–171

Olive MR, Walker JC, Singh K, Dennis ES, Peacock WJ (1990) Functional properties of the anaerobic response element of the maize Adh1 gene. Plant Mol Biol 15: 593–604

Roth BA, Goff SA, Klein TM, Fromm ME (1991) *C1*- and *R*-dependent expression of the maize *Bz1* gene requires sequences with homology to mammalian *myb* and *myc* binding sites. Plant Cell 3: 317–325

Sheen J (1991) Molecular mechanisms underlying the differential expression of the maize pyruvate, orthophosphate dikinase genes. Plant Cell 3: 225–245

Vasil V, Clancy M, Ferl RJ, Vasil IK, Hannah LC (1989) Increased gene expression by the first intron of maize *shrunken-1* locus in grass species. Plant Physiol 91: 1575–1579

Walker JC, Howard EA, Dennis ES, Peacock WJ (1987) DNA sequences required for anaerobic expression of the maize alcohol dehydrogenase 1 gene. Proc Natl Acad Sci USA 84: 6624–6628

116

Introns

KENNETH R. LUEHRSEN

In the design of expression constructs, it is generally desirable to engineer the highest possible level of expression. Including a spliceable intron in the transcription unit of both plant and animal chimeric expression constructs (or cDNAs driven by their native promotor) has been shown to increase gene expression at both the mRNA and protein levels from two- to 1,000-fold (Buchman and Berg 1988; Callis et al. 1987). This phenomenon has been termed *intron enhancement* of gene expression. The molecular mechanisms responsible for the enhancement are not well understood. Although some

The Maize Handbook—M. Freeling, V. Walbot, eds.

animal introns contain transcriptional enhancers and thus increase initiation at the proximal promoter, other introns do not stimulate transcription but appear to mediate enhancement posttranscriptionally. Studies in mammalian cells suggest that entry of the pre-mRNA into the splicing pathway improves the nuclear stability of the transcript, possibly by more efficient capping (Inoue et al. 1989) or polyadenylation (Huang and Gorman 1990) or by protection of the transcript from nuclease attack while in the spliceosome particle.

Several studies have shown that introns can enhance gene expression in both monocots (Callis et al. 1987; Luehrsen and Walbot 1991; McElroy et al. 1990; Oard et al. 1989) and dicots (Dean et al. 1989; Leon et al. 1991; McCullough et al. 1991). In the first reported study, Callis et al. (1987) showed that a high level of expression of the maize *Alcohol dehydrogenase-1* gene in BMS was dependent on one or more introns being included in the transcription unit. The level of enhancement was up to 50-fold depending on the intron(s) used and its position in the transcription unit. (See below.) Enhancement was seen for both transient assays and stable transformants expressing the chimeric genes. In this and subsequent studies, it was also shown that *Adh1-S* intron 1 could enhance the expression of chimeric gene constructs driven by the CaMV 35S and *Nos* promoters having the CAT, luciferase, neo, or GUS reporter genes (Callis et al. 1987; Luehrsen and Walbot 1991). In addition to *Adh1-S* intron 1, several other maize introns have been reported to enhance expression. *Adh1* introns 2 and 6 (Mascarenhas et al. 1990), *Shrunken-1* intron 1 (Maas et al. 1991; Vasil et al. 1989), the single *Bronze-1* intron (Callis et al. 1987), and actin intron 3 (Luehrsen and Walbot 1991) enhanced expression from two- to 1,000-fold when placed upstream of a reporter gene. Introns can also function across species barriers; *Adh1-S* intron 1 has been shown to enhance expression in other monocots such as rice and wheat (Last et al. 1991) and the *Sh1* intron 1 enhances expression in guinea grass and napier grass (Vasil et al. 1989).

Intron enhancement results in both an increase in the steady-state level of mRNA and in the amount of translated reporter enzyme (Luehrsen and Walbot 1991). The placement of *Adh1-S* intron 1 and *Sh1* intron 1 up- or downstream of the promoter had little effect on expression (Callis et al. 1987; Maas et al. 1991), suggesting that these introns do not function as transcriptional enhancers. Using *Adh1-S* intron 1, the levels of mRNA can be enhanced as much as 30-fold over an intronless control (Luehrsen and Walbot 1991). Surprisingly, in transient assays using constructs containing *Adh1-S* intron 1 or actin intron 3, not all of the pre-mRNA was spliced and significant amounts of unspliced RNA accumulated (Luehrsen and Walbot 1991).

When incorporating an intron into expression constructs, several considerations must be addressed. The placement in the transcription unit is critical: enhancement is greatest when the intron is placed near the 5′ end of the transcription unit. Callis et al. (1987) found that *Adh1-S* intron 1 was able to stimulate CAT expression 110-fold when placed in the 5′ untranslated region (UTR) but only fivefold when placed in the 3′ UTR. Similarly, Mascarenhas et al. (1990) found that *Adh1* introns 2 and 6 placed in the 5′ UTR enhanced expression of a CAT-containing chimeric gene from 12- to 20-fold, but had no effect when placed in the 3′ UTR. If the intron is placed in the 5′ UTR, there should be no ATG sequences in the exon sequences flanking the intron, as these would likely be efficient translation start sequences and result in nonfunctional fusion proteins or a truncated protein product. Also, the introduced intron will be surrounded by different exon sequences for each chimeric construct. As exon sequences flanking the intron do affect splicing efficiency (Luehrsen and Walbot 1991), the processing of the intron should be assessed by northern blotting or RNase mapping.

It is recommended that maize introns be used in the expression constructs. While some dicot introns are spliced in monocots (Peterhans et al. 1990; Tanaka et al. 1991), their processing in maize needs to be assessed on an individual basis. The use of mammalian introns is not recommended as the different splicing requirements for plants and animals will likely preclude efficient splicing of mammalian introns in maize. Also, the use of heterologous introns will increase the chance of fortuitous alternative splicing or polyadenylation events.

REFERENCES

Buchman AR, Berg P (1988) Comparison of intron-dependent and intron-independent gene expression. Mol Cell Biol 8: 4395–4405

Callis J, Fromm M, Walbot V (1987) Introns increase gene expression in cultured maize cells. Genes Dev 1: 1183–1200

Dean C, Favreau M, Bond-Nutter D, Bedbrook J, Dunsmuir P (1989) Sequences downstream of translation start regulate quantitative expression of Petunia *rbcS* genes. Plant Cell 1: 201–208

Huang MTF, Gorman CM (1990) Intervening sequences increase efficiency of RNA 3′ processing and accumulation of cytoplasmic RNA. Nucl Acids Res 18: 937–947

Inoue K, Ohno M, Sakamoto H, Shimura Y (1989) Effect of the cap structure on pre-mRNA splicing in *Xenopus* oocyte nuclei. Genes Dev 3: 1472–1479

Last DI, Brettell RIS, Chamberlain DA, Chaudhury AM, Larkin PJ, Marsh EL, Peacock WJ, Dennis ES (1991) pEmu: an improved promoter for gene expression in cereal cells. Theor Appl Genet 81: 581–588

Leon P, Planckaert F, Walbot V (1991) Transient gene expression in protoplasts of *Phaseolus vulgaris* isolated from a cell suspension culture. Plant Physiol 95: 968–972

Luehrsen KR, Walbot V (1991) Intron enhancement of gene expression and the splicing efficiency of introns in maize cells. Mol Gen Genet 225: 81–93

Maas C, Laufs J, Grant S, Korfhage C, Werr W (1991) The combination of a novel stimulatory element in the first exon of the maize *Shrunken-1* gene with the following intron 1 enhances reporter gene expression up to 1000-fold. Plant Mol Biol 16: 199–207

Mascarenhas D, Mettler IJ, Pierce DA, Lowe HW (1990) Intron-mediated enhancement of heterologous gene expression in maize. Plant Mol Biol 15: 913–920

McCullough AJ, Lou H, Schuler MA (1991) *In vivo* analysis of plant pre-mRNA splicing using an autonomously replicating vector. Nucl Acids Res 19: 3001–3009

McElroy D, Zhang W, Cao J, Wu R (1990) Isolation of an efficient actin promoter for use in rice transformation. Plant Cell 2: 163–171

Oard JH, Paige D, Dvorak J (1989) Chimeric gene expression using maize intron in cultured cells of breadwheat. Plant Cell Rep 8: 156–160

Peterhans A, Datta SK, Datta K, Goodall GJ, Potrykus I, Paszkowski J (1990) Recognition efficiency of *Dicotyledoneae*-specific promoter and RNA processing signals in rice. Mol Gen Genet 222: 361–368

Tanaka A, Mita S, Ohta S, Kyozuka J, Shimamoto K, Nakamura K (1991) Enhancement of foreign gene expression by a dicot intron in rice but not in tobacco is correlated with an increased level of mRNA and an efficient splicing of the intron. Nucl. Acids Res 18: 6767–6770

Vasil V, Clancy M, Ferl RJ, Vasil IK, Hannah LC (1989) Increased gene expression by the first intron of maize *shrunken-1* locus in grass species. Plant Physiol 91: 1575–1579

117

Characterization of Zein Genes and Their Regulation in Maize Endosperm

GARY A. THOMPSON AND BRIAN A. LARKINS

The prolamine proteins of maize seed are traditionally called zeins (Osborne 1987). Zein synthesis initiates in the endosperm following the period of free nuclear division and continues until desiccation (Murphy and Dalby 1971). Zeins are made on rough endoplasmic reticulum (RER) membranes (Larkins and Dalby 1975) and form accretions (protein bodies) within the lumen of the RER (Burr and Burr 1976; Larkins and Hurkman 1978). Analysis of these proteins by SDS-PAGE typically reveals a mixture of polypeptides ranging in size from M_r 27—kD to M_r 10—kD. Although the separation pattern of zeins is reproducible among various laboratories, assignment of molecular weights is not. This is largely a consequence of the unusual structures of these proteins, which cause them to migrate faster or slower than their molecular weights predict. In addition, the differences in solubility of zeins in alcoholic solutions causes them to partition into several fractions, depending on the nature of the solvent. Both of these factors have made zein nomenclature complex (Esen 1987; Rubenstein and Geraghty 1989). Recently, we suggested a classification system based on protein structure (Larkins et al. 1989), rather than on differences in mobility on SDS-PAGE (Wilson 1991) or solubility (Esen 1986). With our classification system the M_r 22-kD and 19-kD proteins, which are typically the most abundant, are

The Maize Handbook—M. Freeling, V. Walbot, eds.

called alpha-zeins, the M_r 14-kD protein is called the beta-zein, the M_r 27-kD and 16-kD proteins are called gamma-zeins, and the M_r 10-kD protein is called the delta-zein (Table 117.1).

The alpha-zein proteins range in size from 210 to 245 amino acids. They all have high contents of glutamine (25%), leucine (20%), alanine (15%), and proline (11%); none have been found that contain lysine. A distinguishing feature of alpha-zeins is the presence of tandemly repeated peptides of approximately 20 amino acids in the central region of the protein (Geraghty et al. 1981; Pedersen et al. 1982; Spena et al. 1983). Based on circular dichroism analysis and computer-generated models (Argos et al. 1982), it was proposed that the repeated peptides are alpha-helices that interact through hydrogen bonding to fold the proteins into rod-shaped molecules.

The beta-zein protein, which is 160 amino acids long, contains less glutamine (16%), leucine (10%), and proline (9%) than the alpha-zeins, but has significantly more of the sulfur amino acids, methionine (7%) and cysteine (4%). Unlike the alpha-zeins, the beta-zein contains no alpha-helical structure and is composed mostly of beta-strand and turn structure (Pedersen et al. 1986).

The M_r 27-kD gamma-zein is a cysteine-rich (7%) protein of 180 amino acids with an exceptionally high content of proline (25%) (Prat et al. 1985; Wang and Esen 1986). This high proline content is a consequence of a series of hexapeptide repeats (PPPVHL) at the NH_2 terminus. There are eight tandem copies of this repeat in the M_r 27-kD gamma-zein, but only three copies of a homologous repeat in the M_r 16-kD gamma-zein; this accounts, in part, for the difference in their molecular weight.

The delta-zein is a small protein of 130 amino acids (Kirihara et al. 1988a). It is exceptionally rich in the sulfur amino acids, methionine (23%) and cysteine (4%), and undoubtedly constitutes an important source of sulfur storage for the seed.

Immunocytochemical studies revealed a temporal pattern of zein association into protein bodies (Lending et al. 1988). The least developmentally mature cells of the endosperm, which are just beneath the aleurone layer, contain small protein bodies that consist predominantly of beta- and gamma-zeins. In older, more centrally located endosperm cells, protein bodies are much larger and contain primarily alpha-zeins. In these cells, the alpha-zeins fill the center of the protein body, while the beta- and gamma-zeins are peripheral. Recently, we have found that the delta-zein is also in the central part of the protein body (Lending and Larkins, unpublished). These differences in the temporal appearance and spatial distribution of zeins in protein bodies suggest that the cysteine-rich beta- and gamma-zeins play a role in organizing zeins into protein bodies (Lending and Larkins 1989). There is as yet no experimental evidence to support this hypothesis.

Zein synthesis is affected by a number of genes, some of which are defined by mutant alleles (Table 117.2). The development of *high-lysine* corn was predicated on the use of these mutations, because they inhibit zein synthesis, which consequently elevates the percent of lysine in the grain. Among the first mutations characterized were *opaque-2* (Mertz et al. 1964) and *floury-2* (Nelson 1981). The floury phenotype of these mutants served as the basis for the identification of other mutations that affect zein accumulation. Many of these mutant genes have been mapped and have had their effect on zein synthesis

TABLE 117.1 Summary of zein characteristics

Class	M_r	% Total zein	Mature protein (amino acids)	Gene No.	Genomic clones	Variety	Comments	Reference
Alpha	22-kD	20	242–245	25	gZ22.8	W64A	L	Thompson et al. (1992)
					ψgZ22.8	W64A	ψ[a] L	Thompson et al. (1992)
					Z7	W22	ψ	Kridl et al. (1984)
					Za1	W64A	ψ	Spena et al. (1982)
					pML1	A619	ψ	Langridge & Feix (1983), Wandelt & Feix (1989)
Alpha	19- kD	40	210–245	50	gZ19ab1	BMS*		Pedersen et al. (1982)
					gZ19ab11	W64A	L	Kriz et al. (1987)
					pMS1	A619		Brown et al. (1986)
					pMS2	A619		Langridge et al. (1985)
					zE19	W64A	ψ L	Spena et al. (1983)
					zE25	W64A	ψ L P	Spena et al. (1983)
					Z4	W22		Hu et al. (1982)
Beta	14-kD	10	160	1 or 2	gZ15A	W64A		Pedersen et al. (1986)

(Continued)

TABLE 117.1 (Continued)

Class	M_r	% Total zein	Mature protein (amino acids)	Gene No.	Genomic clones	Variety	Comments	Reference
Gamma	27-kD	20	180	1 or 2	p268c	W64A		Boronat et al. (1986), Reina et al. (1990)
					W22-A	W22	L	Das and Messing (1987), Das et al. (1991)
					W22-B	W22	L	Das and Messing (1987), Das et al. (1991)
					A188-A	A188	L	Das and Messing (1987), Das et al. (1991)
					A188-B	A188	L	Das and Messing (1987), Das et al. (1991)
Gamma	16-kD	<5	164	1 or 2	pZ3	W64A		Reina et al. (1990)
Delta	10-kD	<5	129	1 or 2	pZ10B	BSSS53		Kirihara et al. (1988a)

[a]ψ, pseudogene or contains premature translation termination codons;
L, linked to another zein gene;
P, partial or incomplete coding region;
*, Black Mexican Sweet cell culture.

Table 117.2 Genes affecting zein accumulation[a]

Locus	Inheritance	Location	Zein accumulation	Genetic interaction
Opaque-1 (o1)	recessive	chromosome 4 long arm		
Opaque-2 (o2)	recessive	chromosome 7 short arm	22-kD elimination 19-kD reduction	*additive with o7,o6,De-B30,Mc* *epistatic over fl2*
Opaque-5 (o5)	recessive	chromosome 7 long arm		
Opaque-6 (o6)	recessive		general reduction	*additive with o2,o7*
Opaque-7 (o7)	recessive	chromosome 10 long arm	19-kD reduction	*additive with o2* *epistatic over fl2*
Opaque-9 (o9)	recessive			
Opaque-10 (o10)	recessive			
Opaque-11 (o11)	recessive			
Opaque-12 (o12)	recessive			
Opaque-13 (o13)	recessive			
Opaque-2 modifiers	semidominant		27-kD overproduction	*independent of o2,fl2,fl2/o2*
Floury-1 (fl1)	semidominant	chromosome 2 short arm		
Floury-2 (fl2)	semidominant	chromosome 4 short arm	general reduction	*hypostatic to o2,o7*
Floury-3 (fl3)	semidominant	chromosome 8 long arm		
Defective Endosperm-B30 (De-B30)	semidominant	chromosome 7 short arm	22-kD reduction	*additive with o2*
Mucronate (Mc1)	semidominant		general reduction	*additive with o2*
Zpr10/(22)		chromosome 10 long arm	10-kD overproduction	

[a]Adapted from Motto et al. 1989. Also see Benner et al. (1989), Coe et al. (1988), Nelson (1981), and Soave and Salamini (1984).

described (Motto et al. 1989; Shotwell and Larkins 1989), but their mechanisms of action are unknown. Only in the case of *opaque-2*, where the gene was first tagged with a transposable element and then cloned (Motto et al. 1988; Schmidt et al. 1987), has an explicit function been assigned to a gene. *Opaque-2* encodes a leucine zipper-type transcriptional regulatory protein that binds to the promoters of the 22-kD alpha-zein genes and is required for their transcription (Lohmer et al. 1991; Schmidt et al. 1990).

The impact of these mutations on zein synthesis reduces the size and sometimes alters the shape of protein bodies (Geetha et al. 1991; Wolf et al. 1969). It has been postulated that these changes lead to the formation of air spaces around the starch grains in desiccated endosperm, which are thought to be responsible for the soft, floury texture of the mature seed (Robutti et al. 1974). Genes called *opaque-2 modifiers* alter the phenotype of floury mutants, giving rise to a hard, vitreous endosperm. The *opaque-2* modifiers have been used to develop a high-lysine corn (Quality Protein Maize) with enhanced seed quality (Vasal et al. 1980). These genes have not been mapped, but they appear to function by increasing the synthesis of the 27-kD gamma-zein protein (Geetha et al. 1991; Lopes and Larkins 1991). A different gene that causes high levels of expression of the delta-zein gene has also been partially characterized (Benner et al. 1989); however, as with the *opaque-2 modifiers*, its mechanism of action is unknown.

Interest in understanding the molecular mechanisms that regulate zein gene expression led to the isolation and characterization of genes encoding the various types of zein proteins. A number of genes encoding the alpha-zeins have been cloned and sequenced (Table 117.1). Because of the number and complexity of these genes, multiple systems of nomenclature have been developed to classify them (Heidecker and Messing 1986; Marks and Larkins 1982; Rubenstein and Geraghty 1989; Wilson 1991). A uniform system for gene designation has not been adopted. Gene classification is less complex for the beta-, gamma-, and delta-zeins, because they are encoded by genes present in only one or two copies per genome (Boronat et al. 1986; Das and Messing 1987; Kirihara et al. 1988b; Prat et al. 1987; Wilson and Larkins 1984).

Attempts have been made to identify transcriptional regulatory sequences that are common among the various types of zein genes by comparing their nucleotide sequence. Consensus sequences in the 5′ and 3′ noncoding regions of these genes have been identified (Heidecker et al. 1991; Thompson and Larkins 1989), but with the exception of the *Opaque-2* binding region, the functional role of these sequences is unknown. Recent advances in maize transformation and regeneration will undoubtedly lead to a more complete characterization of DNA regulatory elements.

REFERENCES

Argos P, Pedersen K, Marks MD, Larkins BA (1982) A structural model for maize zein proteins. J Biol Chem 17: 9984–9990

Benner MS, Phillips RL, Kirihara JA, Messing JW (1989) Genetic analysis of methionine-rich storage protein accumulation in maize. Theor Appl Genet 78: 761–767

Boronat A, Martinez MC, Reina M, Puigdomenech

P, Palau J (1986) Isolation and sequencing of a 28 kD glutelin-2 gene from maize. Common elements in the 5′ flanking regions among zein and glutelin genes. Plant Sci 47: 95–102

Brown JWS, Wandelt C, Feix G, Neuhaus G, Schweiger HG (1986) The upstream regions of zein genes. Sequence analysis and expression in the unicellular alga *Acetabularia*. Eur J Cell Biol 42: 161–170

Burr B, Burr FA (1976) Zein synthesis in maize endosperm by polyribosomes attached to protein bodies. Proc Natl Acad Sci USA 73: 515–519

Coe EH, Neuffer MG, Hoisington DA (1988) The genetics of corn. In Sprague GF, Dudley JW (eds) Corn and Corn Improvement, Third Edition, American Society of Agronomy, Crop Science Society of America, Soil Science Society of America, Madison, pp 81–236

Das OP, Messing JW (1987) Allelic variation and differential expression at the 27-kilodalton zein locus in maize. Mol Cell Biol 7: 4490–4497

Das OP, Poliak E, Ward K, Messing J (1991) A new allele of the duplicated 27 kD zein locus of maize generated by homologous recombination. Nucl Acids Res 19: 3325–3330

Esen A (1986) Separation of alcohol-soluble proteins (zeins) from maize into three fractions by differential solubility. Plant Physiol 80: 623–627

Esen A (1987) A proposed nomenclature for the alcohol-soluble proteins (zeins) of maize (*Zea mays* L.). J Cereal Sci 5: 117–128

Geetha KB, Lending CR, Lopes MA, Wallace JC, Larkins BA (1991) *Opaque-2* modifiers increase gamma-zein synthesis and alter its spatial distribution in maize endosperm. Plant Cell 3: 1207–1219

Geraghty D, Peifer MA, Rubenstein I, Messing J (1981) The primary structure of a plant storage protein: zein. Nucleic Acids Res 9: 5163–5174

Heidecker G, Messing J (1986) Structural analysis of plant genes. Annu Rev Plant Physiol 37: 439–466

Heidecker G, Chaudhuri S, Messing J (1991) Highly clustered zein gene sequences reveal evolutionary history of the multigene family. Genomics 10: 719–732

Hu N-T, Peifer MA, Heidecker G, Messing J, Rubenstein I (1982) The primary structure of a genomic zein sequence of maize. EMBO J 1: 1337–1342

Kirihara JA, Petri JB, Messing J (1988a) Isolation and sequence of a gene encoding a methionine-rich 10 kD zein protein from maize. Gene 71: 359–370

Kirihara JA, Hunsperger JP, Mahoney WC, Messing JW (1988b) Differential expression of a gene for a methionine-rich storage protein in maize. Mol Gen Genet 211: 477–484

Kridl JC, Vieira J, Rubenstein I, Messing J (1984) Nucleotide sequence analysis of a zein genomic clone with a short open reading frame. Gene 28: 113–118

Kriz AL, Boston RS, Larkins BA (1987) Structural and transcriptional analysis of DNA sequences flanking genes that encode 19 kilodalton zeins. Mol Gen Genet 207: 90–98

Langridge P, Feix G (1983) A zein gene of maize is transcribed from two widely separated promoter regions. Cell 34: 1015–1022

Langridge P, Brown JWS, Pintor-Toro JA, Feix G (1985) Expression of zein genes in *Acetabularia* mediterranea. Eur J Cell Biol 39: 257–264

Larkins BA, Dalby A (1975) *In vitro* synthesis of a zein-like protein by maize polyribosomes. Biochem Biophys Res Commun 66: 1048–1054

Larkins BA, Hurkman WJ (1978) Synthesis and deposition of zein in protein bodies of maize endosperm. Plant Physiol 62: 256–263

Larkins BA, Wallace JC, Galili G, Lending CR, Kawata EE (1989) Structural analysis and modification of maize storage proteins. Dev Indus Micro 30: 203–209

Lending CR, Larkins BA (1989) Changes in the zein composition of protein bodies during maize endosperm development. Plant Cell 1: 1011–1023

Lending CR, Kriz AK, Larkins BA, Bracker CE (1988) Structure of maize protein bodies and immunocytochemical localization of zeins. Protoplasma 143: 51–62

Lohmer S, Maddaloni M, Motto M, Di Fonzo N, Hartings H, Salamini F, Thompson RD (1991) The maize regulatory locus *opaque-2* encodes a DNA-binding protein which activates transcription of the b-32 gene. EMBO J 10: 617–624

Lopes MA, Larkins BA (1991) Gamma-zein content is related to endosperm modification in Quality Protein Maize. Crop Sci 31: 1655–1662

Marks MD, Larkins BA (1982) Analysis of sequence microheterogeneity among zein messenger RNAs. J Biol Chem 257: 9976–9983

Mertz ET, Bates LS, Nelson OE (1964) Mutant gene that changes protein composition and increases lysine content of maize endosperm. Science 145: 279–280

Motto M, Maddaloni M, Ponziani G, Brembilla M, Marotta R, Di Fonzo N, Soave C, Thompson R, Salamini F (1988) Molecular cloning of the o2-*m5* allele of *Zea* mays using transposon marking. Mol Gen Genet 212: 488–494

Motto M, Di Fonzo N, Hartings H, Maddaloni M, Salamini F, Soave C, Thompson R (1989) Regulatory genes affecting maize storage protein synthesis. In Oxford Surveys of Plant Molecular and Cell Biology, Vol 6, Clarendon Press, Oxford, pp 87–114

Murphy JJ, Dalby A (1971) Changes in the protein fractions of developing normal and *opaque-2* maize endosperm. Cereal Chem 48: 336–349

Nelson O (1981) The mutants *opaque-9* through *opaque-13*. Maize Genetics Cooperation News Letter 55: 68

Osborne TB (1987) The amount and properties of the proteins of the maize kernel. J Am Chem Soc 19: 525–528

Pedersen K, Devereaux J, Wilson DR, Sheldon E, Larkins BA (1982) Cloning and sequence analysis reveal structural variation among related zein genes in maize. Cell 29: 1015–1026

Pedersen K, Argos P, Naravana SVL, Larkins BA (1986) Sequence analysis and characterization of a maize gene encoding a high-sulfur zein protein of Mr 15,000. J Biol Chem 261: 6279–6284

Prat S, Cortadas J, Puigdomenech P, Palau J (1985) Nucleic acid (cDNA) and amino acid sequences of the maize endosperm protein glutelin-2. Nucl Acids Res 13: 1493–1504

Prat S, Perez-Grau L, Puigdomenech P (1987) Multiple variability in the sequence of a family of maize endosperm proteins. Gene 52: 41–49

Reina M, Guillen P, Ponte I, Boronat A, Palau J (1990) DNA sequence of the gene encoding the Zc1 protein from *Zea* mays W64A. Nucl Acids Res 18: 6425

Reina M, Ponte I, Guillen P, Boronat A, Palau J (1990) Sequence analysis of a genomic clone encoding a Zc2 protein from *Zea* mays W64A. Nucl Acids Res 18: 6426

Robutti JL, Hosenei RC, Wasson CW (1974) Modified *opaque-2* corn endosperms. II. Structure viewed with scanning electron microscope. Cereal Chem 51: 173–180

Rubenstein I, Geraghty DE (1989) The genetic organization of zein. In Pomeranz Y (ed) Advances in Cereal Science and Technology, Vol VIII, American Association of Cereal Chemists, Inc, St Paul, MN, pp 297–315

Schmidt RJ, Burr FA, Aukerman MJ, Burr B (1990) Maize regulatory gene *opaque-2* encodes a protein with a "leucine zipper" motif that binds to zein DNA. Proc Natl Acad Sci USA 87: 46–50

Schmidt RJ, Burr FA, Burr B (1987) Transposon tagging and molecular analysis of the maize regulatory locus *opaque-2*. Science 238: 960–963

Shotwell MA, Larkins BA (1989) The biochemistry and molecular biology of seed storage proteins. In Marcus A (ed) The Biochemistry of Plants: A Comprehensive Treatise, Vol 15, Academic Press, New York, pp 296–345

Soave C, Salamini F (1984) Organization and regulation of zein genes in maize endosperm. Phil Trans R Soc Lond B304: 341–347

Spena A, Viotti A, Pirrotta V (1982) A homologous repetitive block structure underlies the heterogeneity of heavy and light chain zein genes. EMBO J 1: 1589–1594

Spena A, Viotti A, Pirrotta V (1983) Two adjacent

genomic zein sequences: Structure, organization and tissue-specific restriction pattern. J Mol Biol 169: 779–811

Thompson GA, Larkins BA (1989) Structural elements regulating zein gene expression. Bioessays 10: 108–113

Thompson GA, Siemieniak DR, Sieu LC, Slightom JL, Larkins BA (1992) Sequence analysis of linked maize 22-kD α-zein genes. Plant Mol Biol 18: 827–833

Vasal SK, Villegas E, Bjarnason M, Gelan B, Goertz P (1980) Genetic modifiers and breeding strategies in developing hard endosperm *opaque-2* materials. In Pollmer WG, Phillips RH (eds) Improvement of Quality Traits of Maize for Grain and Silage Use, Martinus Nijhoff, London, pp 37–73

Wandelt C, Feix G (1989) Sequence of a 21 kD zein gene from maize containing an in-frame stop codon. Nucl Acids Res 17: 2354

Wang SZ, Esen A (1986) Primary structure of a proline-rich zein and its cDNA. Plant Physiol 81: 70–74

Wilson CM (1991) Multiple zeins from maize endosperms characterized by reversed-phase high performance liquid chromatography. Plant Physiol 95: 777–786

Wilson DR, Larkins BA (1984) Zein gene organization in maize and related grasses. J Mol Evol 29: 330–340

Wolf MJ, Khoo U, Seckinger HL (1969) Distribution and subcellular structure of endosperm protein in varieties of ordinary and high-lysine maize. Cereal Chem 46: 253–263

118

Overview of Cloning Genes Using Transposon Tagging

VICKI CHANDLER

Genes in which a transposable element has inserted can be cloned using hybridization probes specific for the adjacent transposable element. Three families of transposable elements, *Ac/Ds*, *Spm*(*En*), and *Mu*, have been used successfully to clone maize genes. Below I briefly summarize the strategies that have been utilized once an insertion mutation has been isolated and genetically characterized. See the appropriate section in Section III, Genetics Protocols, for a brief review of these transposable elements, to learn how to

The Maize Handbook—M. Freeling, V. Walbot, eds.

isolate mutations using the different element systems, and to confirm that the mutation is in fact the result of the desired element.

The most general method for identifying the element responsible for the mutation is to look for an element-homologous restriction fragment that cosegregates with the mutant phenotype. This is done by examining the DNA from mutant progeny arising from outcrosses or self-pollinations, using DNA blot analyses. I will describe this method first, using *Mu* elements as an example, and then follow with special considerations for *Ac/Ds* and *Spm*(*En*) elements. I will then describe two other approaches that have been successful, and end with a discussion of the criteria used to establish that the cloned DNA sequences represent the gene of interest.

Mu

Mu elements are the most heterogeneous family of elements described in maize. All *Mu* elements share very similar ~220-bp terminal inverted repeats, but different classes of *Mu* elements have completely unrelated internal sequences. A total of six classes of *Mu* elements have been described to date, with each class defined by homologies within the internal sequences. (See Chandler and Hardeman 1992 for a review of *Mu* elements.) *Mu1* is frequently the most abundant element found in Mutator stocks, and to date is responsible for generating the largest number of mutations. Thus, when one isolates a mutation from Mutator stocks, a reasonable first step is to probe a blot containing DNA from mutants with the *Mu1* internal probe. Frequently one will observe 20–40 copies of *Mu1* elements in the stocks. The use of different restriction enzymes can often allow adequate resolution of the fragments. However, subsequent outcrosses may be necessary to reduce the number of *Mu1* elements to a tractable number. Outcrossing does not reduce the number of elements significantly if the *Mu* system remains active, and thus, several generations may be required. (See below.) Therefore, it is recommended that the blots be hybridized with internal probes for the other classes of elements (*Mu3*, *Mu4*, *Mu7*, *Mu8*, and *MuR*, which can be obtained by writing to the Chandler laboratory). Most of the other *Mu* elements are not found in high copy number in Mutator stocks. Thus, frequently with just one or two DNA blots, it is possible to either eliminate a particular class of *Mu* element as being responsible for the mutation of interest or to identify a candidate restriction fragment cosegregating with the mutation. If the latter case is observed, additional individuals can then be analyzed to verify cosegregation. Multiple alleles of *Mu*-induced mutations in particular genes have been isolated and examined for the nature of the insertion within them. Some genes, such as *bz1*, have contained a high frequency of *Mu1*-related elements (Taylor et al. 1986; Brown et al. 1989; Hardeman and Chandler 1989), while other genes such as *Kn* and *Sh1* also have contained numerous insertions of other classes of *Mu* elements, such as *Mu7* and *Mu8* (Viet et al. 1990) or *MuR* (Chomet et al. 1991; Hardeman, unpublished data), respectively. Thus, when working with a new gene, it is not safe to assume that *Mu1* will have generated the mutations being characterized.

To generate the Mutator stocks to use for cosegregation analyses, outcrossing the mutant with different non-Mutator backgrounds and then self-pollinating (if working with recessive alleles) to generate several families, each of which can then be scored for cosegregating fragments, is recommended. (See next chapter for RFLP analyses.) One must keep in mind that sequential selfing of Mutator plants frequently leads to a shutdown of the system. This can be useful or deleterious depending on the nature of the mutation being characterized. If the mutant phenotype is easy to follow (i.e., an easily scored phenotype, tightly linked to a kernel marker), shutting down the *Mu* system enables quicker dilution of the *Mu* element copy number upon subsequent outcrosses. The discovery of suppressible alleles that produce the wild-type phenotype when the *Mu* system becomes inactive (Martienssen et al. 1989, 1990) could complicate cosegregation analyses, however. For this reason, using DNA from mutant individuals to screen for the presence of a cosegregating fragment is recommended. Looking for the absence of a particular fragment in individuals with the wild-type phenotype could result in erroneously concluding that a particular fragment is not cosegregating with the mutation because it is also in normal sibs. Another potential problem is that an endogenous *Mu* element may be tightly linked to the gene of interest and cosegregate with the mutant phenotype. This can be readily eliminated as a concern by examining the parental lines used to generate the original mutation (assuming they still exist).

Multiple genes have been cloned using the *Mu* system, including *a1* (O'Reilly et al. 1985), *bz2* (McLaughlin and Walbot 1987), *vp1* (McCarty et al. 1989), *hcf106* (Martienssen et al. 1989), *y1* (Buckner, Kelson and Robertson 1990), *a2* (Menssen et al. 1990), *iojap* (Han et al. 1992), and *hm1* (S. Briggs, personal communication).

Ac/*Ds* AND *Spm*(*En*)

If the element responsible for the insertion is *Ac*, then a similar strategy to that described above for *Mu* elements is usually adequate for cloning the gene of interest. If one uses internal *Ac* probes, there are usually ~10 cross-hybridizing restriction fragments in most maize plants (Fedoroff et al. 1983), making it relatively straightforward to identify the *Ac*-homologous restriction fragment that cosegregates with the mutant phenotype. The *bz1* (Fedoroff et al. 1984) and *r* (Dellaporta et al. 1988) genes were cloned using this strategy. By employing several different *Ds* probes, the *bz2* (Theres et al. 1987) and *Kn* (Hake et al. 1989) genes were cloned using *Ds*. Using a similar strategy, *c1* (Paz-Ares et al. 1986) and *c2* (Wienand et al. 1986) genes were cloned using *Spm*(*En*). However, sequences that cross-hybridize with *Ds* elements and *Spm*(*En*) elements exist in very high copy numbers in essentially all maize lines, making this a laborious approach. A strategy that simplifies the search for a restriction fragment cosegregating with a *Spm*(*En*)-induced mutation is to use methylation sensitive enzymes on DNA prepared from the mutants (Schmidt et al. 1987; Cone et al. 1988). The observation is that the active elements, which represent a small number of the total cross-hybridizing sequences in the genome, are undermethylated relative to most of the sequences in the genome. Thus, most of the sequences remain in large DNA fragments at the top of the gel, and only the DNA within or surrounding the smaller number of active elements are cut, producing restriction fragments distinguishable by DNA blot analyses. *c1* (Cone et al. 1986) and *opaque2* (Schmidt et al. 1987) were cloned using this strategy. A similar phenomenon is also observed with *Ac* elements (Schwartz and Dennis 1986; Chomet et al. 1987), which enabled the efficient cloning of the *P* locus containing an *Ac* element using methylation sensitive enzymes (Chen et al. 1987). Given that inactive *Mu1* elements are also hypermethylated (Chandler and Walbot 1986; Bennetzen 1987), a similar strategy might be successful with this system as well.

Two other strategies have been successfully used to clone maize genes. The *a1* gene was cloned by preparing libraries from maize plants with *a1* mutations caused by *En* or *Mu*. The libraries were then screened with *En* or *Mu* probes, and the resulting phages that contained *En* and *Mu* sequences were then cross-screened to identify the phage in both libraries carrying *a1* sequences (O'Reilly et al. 1985). The *bz2* gene was cloned by combining transposon tagging and differential hybridization (McLaughlin and Walbot 1987). A library was prepared from a plant with a *Mu*-induced mutation in *bz2*, and several *Mu1*-homologous phage were isolated. To identify the *Mu1*-homologous phage that also contain *bz2* sequences, the phage were screened for differential hybridization with RNA isolated from wild-type and *bz2* deletion strains. Both of these strategies assumed that the *Mu*-induced mutations were caused by the *Mu1* element, and fortunately that was the case. A more general approach would be to screen with the *Mu* termini.

Once a candidate clone has been isolated, one must then prove that the adjacent sequences are in fact part of the gene of interest. The fact that these sequences cosegregate with the gene of interest is a necessary prerequisite, but by itself does not constitute proof. If something about the protein encoded by the gene of interest is known, then using the cloned sequences for hybrid selection, in vitro translation, and immune precipitation is one approach. However, for most genes isolated by transposon tagging, very little if anything is known about the gene product. Complementation of the mutant phenotype by transformation is commonly used with other organisms. With the advent of maize transformation, this may become routine in the future, but at present it is not a method readily available to most maize researchers, although somatic complementation has been used with the easy-to-score anthocyanin pathway (Nash et al. 1990). What has been commonly done is to use DNA blot analyses to compare restriction patterns of independent insertion mutations to the progenitor wild-type allele and revertants (if available). Independent insertion mutations should contain restriction-fragment differences relative to the wild-type allele due to the insertion of the element within the gene. The revertant may or may not be distinguishable from the wild-type allele, depending on whether the element excised precisely. It is also possible that the element did not excise from the gene, but underwent a genetic change that restored the gene's function (suppression phenomenon described by Martienssen et al. 1989, 1990 is one example). With *Ac* and *Spm*, revertants are usually readily isolated. This is not the case with many *Mu*-induced mutations. However, most *Mu*-induced mutations do undergo somatic reversion at fairly high frequencies. If revertants or multiple, independent insertions are not available for the gene of interest, DNA from somatic sectors can be examined and compared to mutant tissue. Martienssen et al. (1989) prepared DNA from adjacent revertant and mutant sectors, demonstrating a concomitant restriction-fragment-length polymorphism, and establishing that their cloned sequences did in fact represent the *hcf106* gene.

REFERENCES

Bennetzen JL (1987) Covalent DNA modification and the regulation of *Mutator* element transposition in maize. Mol Gen Genet 208: 45–51

Brown WE, Robertson DS, Bennetzen JL (1989) Molecular analysis of multiple *Mutator*-derived alleles of the *Bronze* locus of maize. Genetics 122: 439–445

Buckner B, Kelson TL, Robertson DS (1990) Cloning of the *y1* locus of maize, a gene involved in the biosynthesis of carotenoids. Plant Cell 2: 867–876

Chandler VL, Walbot V (1986) DNA modification of a maize transposable element correlates with loss of activity. Proc Natl Acad Sci USA 83: 1767–1771

Chandler VL, Hardeman KJ (1992) The *Mu* elements of *Zea* mays. In Scandalios J (ed) Advances in Genetics 30: 77–122, Academic Press, New York

Chomet PS, Wessler S, Dellaporta S (1987) Inactivation of the maize transposable element *Activator* (*Ac*) is associated with its DNA modification. EMBO J 6: 295–302

Chomet P, Lisch D, Hardeman KJ, Chandler VL, Freeling M (1991) Identification of a regulatory transposon that controls the Mutator transposable element system in maize. Genetics 129: 261–270

Cone KC, Burr FA, Burr B (1986) Molecular analysis of the maize anthocyanin regulatory locus *C1*. Proc Natl Acad Sci USA 83: 9631–9635

Cone KC, Schmidt RJ, Burr B, Burr FA (1988) Advantages and limitations of using *Spm* as a transposon tag. In Nelson OE (ed) Plant Transposable Elements, Plenum Press, New York, pp 149–160

Dellaporta SL, Greenblatt I, Kermicle J, Hicks JB, Wessler SR (1988) Molecular cloning of the maize *R-nj* allele by transposon-tagging with *Ac*. In Gustafson JP, Appels R (eds) Chromosome Structure and Function: Impact of New Concepts, 18th Stadler Genetics Symposium, Plenum Press, New York, pp 263–282

Fedoroff N, Wessler S, Shure M (1983) Isolation of the transposable maize controlling elements *Ac* and *Ds*. Cell 35: 235–242

Fedoroff NV, Furtek DB, Nelson OE (1984) Cloning of the *bronze* locus in maize by a simple and generalizable procedure using the transposable element *Activator* (*Ac*). Proc Natl Acad Sci USA 81: 3825–3829

Hake S, Vollbrecht E, Freeling M (1989) Cloning *Knotted*, the dominant morphological mutant in maize using *Ds2* as a transposon tag. EMBO J 8: 15–22

Han, C.-D., Coe, EH Jr, and Martienssen RA (1992) Molecular cloning and characterization of iojap (ij), a pattern striping gene of maize. EMBO J. 11: 4037–4046

Hardeman KJ, Chandler VL (1989) Characterization of *bz1* mutants isolated from Mutator stocks with high and low numbers of *Mu1* elements. Dev Genetics 10: 460–472

Martienssen RA, Barkan A, Freeling M, Taylor WC (1989) Molecular cloning of a maize gene involved in photosynthetic membrane organization that is regulated by Robertson's Mutator. EMBO J 8: 1633–1639

Martienssen RA, Barkan A, Taylor WC, Freeling M (1990) Somatically heritable switches in the DNA modification of *Mu* transposable elements monitored with a suppressible mutant in maize. Genes Dev 4: 331–343

McCarty DR, Carson CB, Stinard PS, Robertson DS (1989) Molecular analysis of *viviparous1*: an abscisic acid-insensitive mutant of maize. Plant Cell 1: 523–532

McLaughlin M, Walbot V (1987) Cloning of a mutable *bz2* allele of maize by transposon tagging and differential hybridization. Genetics 117: 771–776

Menssen A, Hohmann S, Martin W, Schnable PS, Peterson PA, Saedler H, Gierl A (1990) The *En/Spm* transposable element of *Zea* mays contains splice sites at the termini generating a novel intron from a *dSpm* element in the *A2* gene. EMBO J 9: 3051–3057

Nash J, Luehrsen KR, Walbot V (1990) Bronze-2 gene of maize: reconstruction of a wild-type allele and analysis of transcription and splicing. Plant Cell 2: 1039–1049

O'Reilly C, Shepherd NS, Pereira A, Schwarz-Sommer Z, Bertran I, Robertson DS, Peterson PA, Saedler H (1985) Molecular cloning of the *a1* locus of *Zea* mays using the transposable elements *En* and *Mu1*. EMBO J 4: 877–882

Paz-Ares J, Ghosal D, Wienand U, Peterson PA, Saedler H (1986) Molecular cloning of the *c* locus of *Zea* mays: a locus regulating the anthocyanin pathway. EMBO J 5: 829–833

Schmidt RJ, Burr FA, Burr B (1987) Transposon tagging and molecular analysis of the maize regulatory locus *opaque2*. Science 238: 960–963

Schwartz D, Dennis E (1986) Transposase activity of the *Ac* controlling element in maize is regulated by its degree of methylation. Mol Gen Genet 205: 476–482

Taylor LP, Chandler VL, Walbot V (1986) Insertion of 1.4 kb and 1.7 kb *Mu* elements into the *bronze1* gene of *Zea mays* L. Maydica XXXI: 31–45

Theres N, Scheele T, Starlinger P (1987) Cloning of the *Bz2* locus of *Zea* mays using the transposable element *Ds* as a gene tag. Mol Gen Genet 209: 193–197

Wienand U, Weydemann U, Niesbach-Klosgen U, Peterson PA, Saedler H (1986) Molecular cloning of the *c2* locus of *Zea* mays, the gene coding for chalcone synthase. Mol Gen Genet 203: 202–207

Viet B, Vollbrecht E, Mathern J, Hake S (1990) A tandem duplication causes the *Kn1–0* allele of *Knotted*, a dominant morphological mutant of maize. Genetics 125: 623–631

How RFLP Loci Can Be Used to Assist Transposon-Tagging Efforts

STEVEN P. BRIGGS AND WILLIAM D. BEAVIS

The purpose of this chapter is to enable the investigator to obtain and utilize restriction fragment-length polymorphism (RFLP) probes for the ultimate goal of cloning genes by transposon tagging. Probes for transposable elements often yield a complex RFLP pattern in maize DNA, and this chapter presents methods for interpreting such patterns. While purely genetic methods can be used to establish an association between an autonomous transposable element and a mutant locus (McClintock 1965), molecular methods such as those presented here are often required to perform a similar analysis with nonautonomous elements. The latter elements cause far more mutations in maize than do their autonomous counterparts.

METHODS

Determining Linkage Between Mutant Loci and Transposable Elements

Restricted Mutagenesis

Detecting mutants in the F_1, by crossing wild-type, transposable element-containing plants with plants that carry a standard recessive allele of the target gene is termed *restricted mutagenesis* because only mutations in the target locus will be observed (with the exception of new dominant mutations). If the mutagenized allele and the standard recessive allele have the same phenotype, flanking RFLP markers may provide the only means to distinguish them. Some mutant phenotypes are suppressible by transposable elements, uncoupling the cosegregation of the mutant phenotype from the locus (McClintock 1965; Martienssen et al. 1990). Furthermore, certain mutations or traits can be poorly penetrant, thus confounding the classification of progeny. The use of flanking RFLP loci rather than trait scoring to classify the progeny alleviates this difficulty. (Of course, one must rely

The Maize Handbook—M. Freeling, V. Walbot, eds.

upon trait scoring to initially isolate the mutant alleles.) Therefore, the first step is to place the target locus on an RFLP map. Because the map distances between RFLP loci are known, it is generally sufficient to place the target locus between RFLP loci (i.e., to determine the linear order) rather than to determine the map distances between the target locus and the flanking RFLPs (which would require a larger population). To bracket a target locus, it must fall within an interval bordered by RFLP probes. This interval must be no greater than about 20 cM so that the number of double crossovers between the flanking RFLP probes is negligible within a mapping population of less than 100 progeny. Genetic linkage of about 1 cM to a phenotypic marker would be equally valid but would place constraints on the opportunities to make mapping crosses.

The RFLP probes are used to classify the progeny of any population in which the mutagenized allele is segregating; it is not necessary to observe a mutant phenotype. Progeny in which the RFLP probes have recombined with each other are discarded. Restriction enzymes are used that do not digest the element suspected to have caused the mutation. Each of these elements will thus generate a different sized fragment. To begin the analysis, DNAs from progeny are run on a gel by class, as illustrated in Figure 119.1. To establish linkage between the mutagenized allele and a particular transposable element, a Southern blot of the gel is hybridized with a probe for the element. The observation of a restriction fragment that is always present in progeny that inherit the mutagenized allele tentatively identifies the element as a cause of the mutation. The observation of recombination between the element and the region that is bounded by the RFLP loci eliminates the element as a candidate for having caused the mutation.

Consider that most, if not all, of the transposable elements under observation will

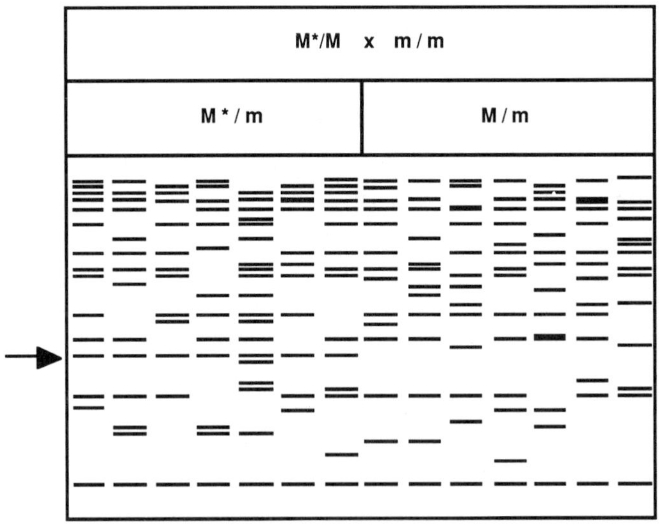

FIGURE 119.1 Illustrative Southern blot of transposable elements that are segregating in a backcross.

be unlinked to any given mutant locus. Even when there are as many as 36 elements, the probability of any one of them being within 25 cM of the mutant locus is only 50% (Table 119.2).[1] Therefore, the first and simplest test is to determine whether or not all of the elements are unlinked to the mutation. In a backcross population, four progeny are sufficient to detect recombination between unlinked loci with 95% confidence; seven progeny will provide 99% confidence. Examining only progeny that inherit the mutant allele permits candidate elements to be more easily identified as ubiquitous restriction fragments on a Southern blot, but does not affect the confidence level. Inclusion of a few progeny that inherit the alternative allele can be used to confirm that the ubiquitous fragment is cosegregating with the mutant locus. If the ubiquitous element is linked to but not at the locus, then the number of progeny required to observe at least one recombinant is given in Table 119.1.[2] For example, a total of 98 backcross progeny would reveal a recombinant if the transposable element were at least 3 cM from either edge of the

Table 119.1 Number of progeny required to be 95% certain of observing recombination between the mutant locus and a transposable element restriction fragment

Recombination frequency (cM)	Backcross (F_1)	Self-pollinated (F_2)	
		Homozygous mutant	Homozygous nonmutant
1	298	29,956	149
2	148	7,488	74
3	98	3,327	49
4	73	1,871	37
5	58	1,197	29
6	48	831	24
7	41	610	21
8	36	467	18
9	32	368	16
10	28	298	14

[1]The probability of detecting linkage between a transposable element and a mutant locus is calculated by determining the proportion of the maize genome that is linked to the loci where the elements reside and using this information in an approximation derived by Elston (1975).

[2]The number of progeny needed to detect a recombinant in the restricted and unrestricted mutagenesis experiments is calculated by determining the expected frequency of recombinants in the mutant phenotypic class, r, setting the probability of detecting no recombinants in n progeny to 0.05, and solving for n from the binomial distribution, i.e., $n = \ln(0.05)/\ln(1 - r)$.

bounded region. Success at this stage (i.e., failure to detect a recombinant) should be confirmed by testing for cosegregation between the element and the target locus directly, rather than between the element and the bounded region. This can be done either (1) by cloning the element and utilizing the DNA that flanks the cloned element as a probe on Southern blots to examine DNA from phenotypically classified progeny in which the gene is segregating or (2) by observing the pattern of segregation of the element in progeny that can be scored phenotypically for inheritance of the mutant allele. The number of progeny required for each type of test is determined as for the bounded case above.

Similar tests can be done with F_2 progeny. In this case, it is best to examine only the two homozygous classes of progeny. A sample of 16 homozygous mutant progeny will provide 95% confidence that a recombinant will be observed that lacks the unlinked element. If a ubiquitous restriction fragment is observed, then the most efficient way to test further the linkage between the fragment and the mutant gene is to examine the homozygous, nonmutant class (i.e., the class that did not inherit the mutant allele); only recombinants will carry the fragment. Table 119.1, column 3, shows that only 49 of these progeny would be required to observe a recombinant if the gene (or either edge of the bounded region) and the fragment are 3 cM or more apart. If only the homozygous, mutant class can be unambiguously identified (e.g., if flanking RFLP loci are unavailable) then 3,327 progeny would be required to have the same level of confidence of detecting a recombinant (i.e., homozygous mutant progeny that lack the fragment; Table 119.1, column 2). Alternatively, the fragment can be cloned so that the DNA that flanks the element is used as a probe of the homozygous mutant class; if the cloned locus is homozygous in 49 of these progeny then, again, the fragment and the gene map within 3 cM of each other. If there are 20 transposable elements present, then there is less than a 5% chance that one of them will be within 3 cM of a given gene, if they are randomly distributed in the genome (Table 119.2).

If both cloning and RFLP mapping are unattractive alternatives, then there is another

TABLE 119.2 Probability of detecting linkage between transposable elements randomly placed in the genome and a mutant locus

No. of transposable elements	Genetic distance (cM) between a given locus and the nearest randomly distributed transposable element					
	3	5	10	15	20	25
5	0.00	0.02	0.04	0.06	0.08	0.09
10	0.01	0.04	0.08	0.12	0.15	0.17
15	0.01	0.06	0.12	0.17	0.21	0.25
20	0.02	0.08	0.15	0.22	0.27	0.32
25	0.02	0.10	0.19	0.26	0.33	0.38
50	0.04	0.19	0.34	0.46	0.54	0.61

test that is less efficient but requires only Southern blots and the transposable element probe. First, a ubiquitous restriction fragment is identified among 16 F_2 homozygous mutant progeny. Second, a total of 49 such progeny are grown and crossed to an inbred that lacks a similar-sized fragment. Finally, four progeny from each cross are examined to determine whether or not the ubiquitous fragment was homozygous in each of the 49 parent plants; if it was, then all progeny from every cross will contain the fragment. (There is a 95% probability that the fragment will be missing in one of the four progeny if the parent was heterozygous.) These results from 212 progeny provide the same information (i.e., that the ubiquitous fragment is less than 3 cM from the mutant locus) and level of confidence (95%) as the preceding methods.

Unrestricted Mutagenesis

Crossing a mutagenized, wild-type, plant with another wild-type individual and then self-pollinating the progeny produces an F_2 in which new mutations will be segregating. All mutations that produce a scorable phenotype can be observed in this F_2, thus, the method is termed unrestricted mutagenesis. The advantage of unrestricted mutagenesis is that all of the genes that affect the trait of interest are potential targets. The disadvantage of unrestricted mutagenesis is that each mutation is detected as a segregating family rather than as an individual (in contrast to restricted mutagenesis). Approximately 20 times more progeny are required to observe the same number of mutations in a given gene. The choice of whether to use restricted or unrestricted mutagenesis should depend upon whether the target is a particular locus or a trait, respectively.

Because there is no standard recessive allele in unrestricted mutagenesis, there is no need for flanking RFLP markers to identify the mutant allele (assuming that there are no problems with expression of the gene). Such markers still have great value, though, for making tentative assignments of allelism and for demonstrating cosegregation of the mutant allele with a transposable element restriction fragment. Normally, however, identification of a restriction fragment that cosegregates with the mutant allele is based upon scores of phenotype rather than use of RFLPs. RFLP loci are used to identify the reciprocal homozygous class among the wild-type siblings. Evaluation of linkage is performed as described above.

Confirmation

The preceding experiments may identify elements that are linked to the mutant locus—but these tests do not prove that the transposon is inserted into the gene. The challenge at this stage is to detect and discard false positives. Recombination can be used to rule out those elements that are located at least a few centimorgans away from the mutant locus; the methods to do this are described above. It is more fruitful to perform other types of experiments to verify the identity of a "tagged" locus rather than to conduct extremely high resolution linkage tests. After all, if the element caused the mutation, then it is part of the locus, by definition, and cannot be removed by recombination. Where an investigator draws the line between doing further linkage tests vs. other types of tests is a matter of judgment and depends upon the particulars of each case.

Multiple forward mutations are useful for demonstrating that other genetic events at the locus are associated with DNA rearrangements in the vicinity of the linked element.

Revertants are often used to verify a causal relationship between an insertion element and a mutation. Either germinal or somatic reversions are sufficient. "Reversion" to normal function may be caused by cycling of an element system rather than by excision but this is equally useful as long as an association between a change in the DNA (methylation or length) and a change in the phenotype can be established. Reversion is the only conclusive test that can be performed without first cloning the suspect fragment. Typically, however, the fragment is cloned and the flanking DNA is used as a probe to confirm that excision of the element has occurred.

Intragenic recombination between different mutant alleles to restore wild-type function, if accompanied by loss of both insertion elements, establishes that at least part of the gene lies between the two insertions. If cloning of the mutant alleles indicates that they are inserted into a gene that spans their sites of insertion, then the gene is almost certainly the target locus.

Transformation of mutant plants (or antisense expression in wild-type plants) with the cloned, wild-type allele should restore or suppress the wild-type phenotype.

The pattern of expression of the cloned gene may confirm the expected phenotype, e.g., a floral mutation may be associated with a gene that is only expresed in flowers.

Cosegregation of the cloned gene with a mutant allele that conditions the same phenotype in another organism would show that the clone was for the gene of interest, provided that the other organism lacked general colinearity with the maize genome (e.g., *Arabidopsis* rather than sorghum).

SUMMARY

1. Progeny from a self-pollinated heterozygote

$$M^*/M \rightarrow M^*/M^* \; M^*/M \; M/M$$

   a. Use the transposable element to probe a Southern blot containing DNA from 16 M*/M* progeny.
   b. If a common restriction fragment is observed, then repeat the experiment using 49 M/M progeny.
   c. If the common fragment observed in a is not seen in b, then the transposable element is within 3 cM of M* and confirming experiments should be conducted.

2. Progeny from a testcross of a heterozygote

$$M^*/M \times m/m \rightarrow M^*/m \; M/m$$

   a. Use the transposable element to probe a Southern blot containing DNA from 7 M*/m progeny.

b. If a common restriction fragment is observed, then repeat the experiment using 91 more progeny (can be selected at random but must be genotyped as either M*/m or M/m).
c. If the common fragment observed in a always cosegregates with M*/m, then the transposable element is within 3 cM of M* and confirming experiments should be conducted.

Any means can be used to classify the progeny. If only the M*/M* class can be unambiguously identified in the first case, then more than 3,000 M*/M* progeny would need to be examined for absence of the common restriction fragment (compared to examining 49 of the M/M class for presence of the fragment). Clearly, having codominant markers such as RFLP loci can save work. If RFLP loci are unavailable, then the same thing can be accomplished by cloning the restriction fragment and using DNA that flanks the transposable element as a codominant probe of the M*/M* class. If only the M*-linked allele of the probe is present in 49 M*/M* progeny (controls must be done to insure that the M-linked allele of the probe can also be detected), then the common restriction fragment maps within 3 cM of M*.

REFERENCES

Burr B personal communication: Matz EC, Burr FA, Burr B (1991) Mapping new mutations using RFLPs. Maize Genetics Cooperation News Letter 65: 104–105

Elston RC (1975) An approximation for the prior probability of autosomal linkage. Cytogen Cell Genet 14: 290–292

McClintock B (1965) The control of gene action in maize, In Genetic Control of Differentiation, Brookhaven Symposia in Biology 18: 162–184

Martienssen R, Barkan A, Freeling M, Taylor W (1990) Somatically heritable switches in the DNA modification of Mu transposable elements monitored with a suppressible mutant in maize. Genes Dev 4: 331–343

Ott J (1985) Analysis of Human Genetic Linkage. John Hopkins, Baltimore, MD, 302 pp

V
Cell Culture Protocols

120

Regeneration of Plants from Somatic Cell Cultures: Applications for in vitro Genetic Manipulation

CHARLES L. ARMSTRONG

Plant regeneration from in vitro cultures of maize was first reported in 1975 by Green and Phillips. Steady improvements have been made since then in the culture systems and in applying them in genetic studies. Protoplasts have been regenerated into fertile plants by several groups (Prioli and Sondahl 1989; Shillito et al. 1989; Morocz et al. 1990). Selection experiments have yielded disease- and herbicide-resistant plants (Gengenbach et al. 1977; Shaner and Anderson 1985). Fertile transgenic plants are now reproducibly obtained in some laboratories through either direct DNA delivery into protoplasts or particle bombardment of intact, cultured cells. Successful applications of corn tissue culture technology have been largely limited to large research groups. The goal of this chapter is to present a simple *how-to* guide for successful establishment of maize somatic cell tissue cultures suitable for in vitro genetic manipulation. Attention is focused on *model genotypes* and also on *minor* technical details that are important, yet difficult to find elsewhere.

COMMENT ON TERMINOLOGY

Regenerable maize cultures are generally classified as *type I* or *type II*. The culture types are distinguished by (1) friability (degree of physical association between cells), (2) mode of regeneration (embryogenesis or organogenesis), and (3) degree of differentiation. Type I cultures are nonfriable (they must be subcultured with a scalpel and forceps), regenerate through both somatic embryogenesis and organogenesis, and are highly differentiated (often contain leaflike structures and advanced-stage somatic embryos on maintenance me-

The Maize Handbook—M. Freeling, V. Walbot, eds.

dium). In contrast, type II cultures are very friable (easily subcultured with a spatula), regenerate almost exclusively via somatic embryogenesis, and are relatively undifferentiated. (Numerous globular-stage somatic embryos are present that do not develop further on maintenance medium.) The classifications are not absolute, and many cultures fall between the two extremes. Type II cultures are preferred for many applications, primarily because of reduced differentiation and enhanced friability. These characteristics make in vitro selection and establishment of liquid suspension cultures much easier.

ESTABLISHMENT OF TYPE II CULTURES

Choice of Genotype

Maize type II culture establishment is very genotype-dependent. The most commonly used, publicly available genotypes are the inbred lines A188 and B73 and the F_1 hybrid of these two lines. A188 was one of the first genotypes successfully regenerated, and it has been used extensively. B73 cultures are significantly more difficult to establish and maintain. The hybrid is generally preferred over either inbred because of the enhanced vigor of both the cultures and the regenerated plants. A188 matures significantly earlier than B73, so several delayed plantings of A188 should be made to ensure ease of crossing. Crosses can be made in either direction with satisfactory results. A188 × B84 F_1 cultures have also been used quite successfully (Gordon-Kamm et al. 1990). Genetic stocks with a very high frequency and vigor of type II culture initiation have been selected out of an A188 × B73 cross, and this *Hi-II* germplasm is also publicly available (*Maize Genetics Cooperation News Letter* 65: 92–93).

Explant Tissue and Growth of Donor Plants

While regenerable cultures have been initiated from many meristematic tissues of the maize plant, the preferred explant is the immature embryo. The developmental stage of the embryo is extremely important and should be determined by embryo length (base to tip of scutellum) and not by days postpollination. Optimum embryo length is 1.0–2.0-mm. The growing conditions of the donor plants are also important—poor plants will result in poor culture response. One good measure of donor plant health is the number of days required to form a 1.0–2.0-mm embryo. Normally, this should occur within 8–12 days of pollination. If longer is required, growing conditions should be improved. [Editors' Note: in northern California, and elsewhere with cooler weather than the Midwest, the 1.0–2.0-mm stage is reached after 12–15 days without impairing culture establishment.]

Embryo size is determined by peeling back the husks of an ear still attached to the plant, slicing off the tip of several kernels in the center portion of the ear using a scalpel or pocketknife, and then scooping out the developing endosperm and embryo with a spatula. The embryo, which is located on the side of the kernel facing the tip of the ear, is removed and placed onto a ruler with millimeter grid markings to measure. (More accurate measurements can be obtained using a dissecting microscope with appropriate eyepiece reticle, if desired.) With a little practice, the proper size range can be easily determined visually without actual measurement. A few points to note are: (1) Embryo size is not uniform for the entire ear; embryos from kernels at the very base or tip of the ear will be significantly smaller than those from the center; (2) the immature embryo is growing extremely rapidly, and the size should be checked at least daily once the proper developmental stage is near (start checking at 8-days postpollination until you know how long it will take with your genotype under your specific growing conditions); (3) under field conditions, insects and fungi can rapidly invade ears that have been sampled to check embryo size. Avoid checking much earlier than necessary. Overplanting to allow some plants to be sacrificed strictly for rough developmental staging can be useful. (4) F_1 embryos developing on an inbred plant generally grow significantly faster than corresponding embryos of the inbred parents.

When the embryos are at the proper developmental stage, the ear is removed from the plant. If seed is desired from the same plant, the upper part of the ear can be cut off with pruning shears, and the cut surface of the remaining ear can be treated with fungicide (e.g., Captan) and allowed to mature. The harvested ear can be used either immediately for embryo dissection and culture initiation or else can be stored at 4°C for later use. For storage, the ears are dehusked and placed in Ziploc storage bags. Overnight storage at 4°C has no known detrimental effect on culture response and can be very useful when ears are harvested from a remote nursery. Reasonably good culture response has been obtained even after 1–2 weeks of refrigeration, but storage for longer than several days is not recommended as a general practice.

Sterilization

Ears must be surface-sterilized prior to embryo isolation. This is accomplished using half-strength liquid bleach (1:1 dilution of household bleach with water) plus a surfactant (e.g., a few drops of Tween 20). The tip of the ear is removed; then the ear is cut in half (using pruning shears), and both halves are placed into an autoclaved 100 × 80-mm Pyrex storage dish containing half-strength bleach plus Tween 20. (For large ears, several storage dishes or a large beaker can be used.) All work should be performed in a laminar flow hood or other aseptic work area. After 10 minutes, the floating ears are rotated using a sterile instrument to ensure adequate exposure of all surface tissues to the bleach. After an additional 10 minutes, the ears are transferred to another autoclaved storage dish containing sterile water. At least one and preferably two additional transfers to fresh sterile water are made, with at least 10 minutes for each rinse. Ears can be left floating in the final rinse for at least several hours. Typically, a number of ears are sterilized in the morning, and embryos are isolated from these for the remainder of the day.

As an additional precaution against contamination, brief vacuum infiltration during the bleach step will help to remove trapped air from between kernels and permit more complete sterilization. This can be extremely helpful for field-grown material but is generally unnecessary for greenhouse-grown plants. Care must be taken not to infiltrate for too long, as the embryo can be killed if bleach enters the kernel.

Embryo Isolation

An ear section is removed from the last rinse with a large (10 inch) sterile forceps, and placed onto a sterile surface (e.g., 100 × 15-mm disposable petri dish). A scalpel handle or straight forceps is placed into the pith of the cob at the upper cut surface to provide a convenient handle for manipulating the ear. The tips of several rows of kernels are cut off with a scalpel, cutting close to but not touching the embryo. The endosperm is then scooped out of the kernel using a spatula. The embryo will normally remain associated with the endosperm but sometimes will remain inside the kernel. Using the spatula, the embryo is carefully removed from the detached endosperm or kernel and placed scutellum-side up onto the surface of the culture medium. (The scutellum is the rounded side of the embryo that was embedded in the endosperm.) It is essential that the embryo is not cut or otherwise damaged during this process.

For the beginner, immature embryo isolation can be a painstaking task. With a little practice, however, one can easily isolate several hundred embryos in a few hours. A few hints: (1) A dissecting microscope or illuminated magnifier may help when first learning, but will just slow you down after the first ear or two. It is faster to isolate embryos without magnification and then check the plates afterward under a dissecting scope for injured or upside-down embryos. (2) A properly maintained spatula is essential—find several that work well for you (generally narrow, somewhat sharpened end, free of nicks) and clean off accumulated dried starch frequently. (3) The flat (embryonic axis) side of the embryo adheres tightly to both the spatula and the surface of the medium. Remove the embryo such that the scutellum rests on the spatula and then lightly touch the embryonic axis to the surface of the culture medium.

Culture Medium

A type II culture initiation medium used successfully in several laboratories is given below. It is a modification of the N6 medium developed for rice anther culture by Chu et al. (1975), and is described by Armstrong and Green (1985) with further modifications by Songstad et al. (1991).

N6 1-100-25-Ag Medium

Component	Concentration	Source
N6 salts (Sigma)	4.0 g/liter	All media components
L-proline	2.88 g/liter	except casamino acids:
Glycine*	2.0 mg/liter	Sigma Chemical Co.
Thiamine HCl*	1.0 mg/liter	P.O. Box 14508

Pyridoxine HCl*	0.5 mg/liter	St. Louis, MO 63178
Nicotinic acid*	0.5 mg/liter	1-800-325-3010
Casamino acids*	100 mg/liter	
Sucrose	20.0 g/liter	Difco Bacto Vitamin
2,4 -D	1.0 mg/liter	Assay casamino acids:
Silver nitrate	10 μM	Fisher Scientific
Phytagel	2 g /liter	catalog # DF0288-02-1
		711 Forbes Avenue
		Pittsburgh, PA 15219
		1-412-562-8300

*Eriksson's Powder Vitamin Mixture, available through Sigma, can be used to provide these components, but an additional 0.5 mg/liter thiamine HCl should still be added.

The pH is adjusted to 5.8 with 1 N potassium hydroxide prior to autoclaving. A 1 mg/ml stock solution of 2,4-D is made by dissolving in a small amount of 1 N potassium hydroxide and bringing to the appropriate volume with double-deionized water. All components are added prior to autoclaving except for silver nitrate. Silver nitrate is made as a 10 mM aqueous stock solution and is stored in a light-proof container. It is filter-sterilized and added after autoclaving. After autoclaving and cooling to 50°C, the medium is poured into disposable petri dishes (100 × 25 mm work well). After the media has solidified, the plates are stored in black plastic bags to prevent photodegradation.

Culture Incubation Conditions

Petri dishes can be conveniently stored in plastic sweater boxes. When plates are stored inside clean boxes, it is not necessary to seal the plates; if desired, either Parafilm or masking tape (3M Scotch brand #202) can be used to seal the plates and protect against microbial contamination. Cultures are incubated in the dark at 28°C for callus initiation and maintenance. Green light filters are useful for *working lights* in a culture room, providing enough light for the researcher without adversely affecting culture response.

PLANT REGENERATION FROM TYPE II CULTURES

Regeneration from type II cultures can be accomplished in a three-step process. (See media recipes below.) Tissue is incubated on each medium for about 2 weeks. *Regen 1* promotes somatic embryo differentiation (progression to approximately *stage 1*; see Abbe and Stein 1954). *Regen 2* promotes embryo enlargement and maturation, and *Regen 3* promotes *germination.* The first two steps are carried out in the dark at 28°C, and the final step under a 16:8-hour photoperiod, about 70 $\mu E\ m^{-2}\ sec^{-1}$ provided by cool-white fluorescent bulbs, at about 25°C. For the final step, plates are placed in a single layer on shelves in a lighted incubator and are sealed with 3M Scotch brand #394 venting tape.

(This tape "breathes" better than Parafilm or masking tape and helps reduce condensation problems.) Small green shoots formed on Regen 3 in the 100 × 25-mm petri plates are transferred to Regen 3 medium in 200 × 25-mm Pyrex tubes or Phytatrays to permit further plantlet development and root formation. Once a good root system has developed, the plants are carefully removed from the medium, the root system is washed thoroughly under running water, and the plants are then placed into 2.5-inch square pots containing Metromix 350 growing medium. Watering with the systemic fungicide benomyl (100 mg/l Benlate) will reduce the risk of fungal contamination. It is critical to maintain the freshly transplanted plants in a high-humidity environment for several days. Placing flats full of well-watered plants into large, sealed, transparent plastic bags is a simple, economical way to accomplish this. Condensation should be visible on the inside of the plastic bags. After several days, the humidity should be gradually reduced to harden off the plants. (Cut a few holes in the bag; also, water the plants again at this point if needed.) Several days later (about 1 week since taking to soil), the plants should be hardened off well enough to survive typical growth chamber or greenhouse conditions for corn. Transplanting from 2.5-inch square to 6-inch diameter round and finally to 10–12-inch diameter round pots works well. Metromix 350 or a similar growing medium will work well for all stages—but do not forget to fertilize! Note: During the hardening-off stage only, temperature and light intensity should both be lower than for normal corn greenhouse conditions. (23°C and 225 μE m^{-2} sec^{-1} works well for hardening off.)

Regen 1	Regen 2	Regen 3
MS salts (Sigma; 4.4 g/liter)	N6 salts (Sigma; 4.0 g/liter)	MS salts (Sigma; 4.4 g/liter)
1.30 mg/liter nicotinic acid	0.5 mg/liter nicotinic acid	1.30 mg/liter nicotinic acid
0.25 mg/liter pyridoxine HCl	0.5 mg/liter pyridoxine HCl	0.25 mg/liter pyridoxine HCl
0.25 mg/liter thiamine HCl	1.0 mg/liter thiamine HCl	0.25 mg/liter thiamine HCl
0.25 mg/liter Ca-pantothenate	2.0 mg/liter glycine	0.25 mg/liter Ca-pantothenate
100 mg/liter myo-inositol	60 g/liter sucrose	100 mg/liter myo-inositol
1mM asparagine	2.0 g/liter Phytagel	1mM asparagine
0.1 mg/liter 2,4-D	pH 5.8	20 g/liter sucrose
0.1 μM ABA		2.0 g/liter Phytagel™
20 g/liter sucrose		pH 5.8
2.0 g/liter Phytagel™		
pH 5.8		

All ingredients can be obtained from Sigma Chemical Co., P.O. Box 14508, St. Louis, MO 63178.

CULTURE SYSTEMS FOR OTHER GENOTYPES

Often it is necessary or desirable to work with a specific genotype, which may or may not culture well with the conditions described above. Several strategies may be useful: (1) Alternative media. Several media formulations have been developed that support regenerable culture formation from a wide range of corn genotypes (e.g., Duncan et al. 1985; Close and Ludeman 1987). "Type I" cultures are generally produced on these media, however, and they may be difficult to use for some applications. (2) "Brute force." Some genotypes will form type II cultures on N6 1-100-25-Ag or similar medium, but at a very low frequency. By isolating hundreds (if not thousands) of embryos, you might get lucky. Success with this approach is most likely with Iowa Stiff Stalk Synthetic–related lines. (3) Crossing to highly regenerable lines. Culture response is highly heritable. F_1 and/or F_2 embryos from crosses of many genotypes with highly culturable lines (e.g., Hi-II or A188; Hi-II generally much better) will generally form quite usable type II cultures.

SOURCES OF MATERIALS

Supplies	Supplier address and phone
Pyrex storage dishes (catalog # 08-782) Spatulas (catalog # 21-401-10 for embryo isolation) Plastic vacuum desiccator (catalog # 08-642-5) 200 × 25-mm tubes	Fisher Scientific 711 Forbes Avenue Pittsburgh, PA 15219 412-562-8300
Kaputs (caps) for 200 × 25-mm tubes	Bellco Glass, Inc. P.O. Box B, Edrudo Road Vineland, NJ 08360 1-800-257-7043
Plastic bags for media storage (7 × 18 inches; 0.002 poly bags, black)	Crown Packaging 8514 Eager Road St. Louis, MO 63144 314-968-9400
Pruning shears, Captan (Orthocide), pots and flats for regenerated plants	A.H. Hummert Co. 2746 Chouteau St. Louis, MO 63103 314-771-0646 (or local hardware store/garden center)
Petri dishes (100 × 25 and 100 × 15 mm)	Nunc, Inc. 2000 Aurora Road Naperville, IL 60563-1796 1-800-288-6862

Scotch brand 3M #202 masking tape, 10-inch forceps	Baxter Healthcare Corp. 1430 Waukegan Road McGaw Park, IL 60085-6787 312-689-8410
Scotch brand 3M #394 venting tape	Ray Engel Packaging Supply 2326 Centerline Industrial Drive St. Louis, MO 63146 314-567-3388
Media ingredients, Phytatrays	Sigma Chemical Co. P.O. Box 14508 St. Louis, MO 63178 1-800-325-3010
Metromix 350	Grace Horticultural Products W.R. Grace and Co. Cambridge, MA 02140

REFERENCES

Abbe EC, Stein OL (1954) The origin of the shoot apex in maize: embryogeny. Am J Bot 41: 285–293

Armstrong CL, Green CE (1985) Establishment and maintenance of friable embryogenic maize callus and the involvement of L-proline. Planta 164: 207–214

Chu CC, Wang CC, Sun CS, Hsu C, Yin KC, Chu CY, Bi FY (1975) Establishment of an efficient medium for rice anther culture through comparative experiments on the nitrogen sources. Sci Sin (Peking) 18: 659–668

Close KR, Ludeman LA (1987) The effect of auxin-like plant growth regulators and osmotic regulation on induction of somatic embryogenesis from elite maize inbreds. Plant Science 52: 81–89

Duncan DR, Williams ME, Zehr BE, Widholm JM (1985) The production of callus capable of plant regeneration from immature embryos of numerous *Zea mays* L. genotypes. Planta 165: 322–332

Gengenbach BG, Green CE, Donovan CM (1977) Inheritance of selected pathotoxin resistance in maize plants regenerated from cell cultures. Proc Natl Acad Sci USA 74: 5113–5117

Gordon-Kamm WJ, Spencer TM, Mangano ML, Adams TR, Daines RJ, Start WG, O'Brien JV, Chambers SA, Adams WR, Jr, Willets NG, Rice TB, Mackey CJ, Krueger RW, Kausch AP, Lemaux PG (1990) Transformation of maize cells and regeneration of fertile transgenic plants. Plant Cell 2: 603–618

Green CE, Phillips RL (1975) Plant regeneration from tissue cultures of maize. Crop Sci 15: 417–420

Morocz S, Donn G, Nemeth J, Dudits D (1990) An improved system to obtain fertile regenerants via maize protoplasts isolated from a highly embryogenic suspension culture. Theor Appl Genet 80: 721–726

Prioli LM, Sondahl MR (1989) Plant regeneration and recovery of fertile plants from protoplasts of maize (*Zea mays* L.). Bio/Technology 7: 589–594

Shaner DL, Anderson PC (1985) Mechanism of action of the imidazolinones and cell culture selection of tolerant maize. In Zaitlin M, Day

P, Hollaender A, Wilson CM (eds) Biotechnology in Plant Science: Relevance to Agriculture in the Eighties, Academic Press, New York, pp 287–299

Shillito RD, Carswell GK, Johnson CM, DiMaio JJ, Harms CT (1989) Regeneration of fertile plants from protoplasts of elite inbred maize. Bio/Technology 7: 581–587

Songstad DD, Armstrong CL, Petersen WL (1991) $AgNO_3$ increases type-II callus production from immature embryos of maize inbred B73 and its derivatives. Plant Cell Rep 9: 699–702

121

Initiation, Maintenance, and Plant Regeneration of Type II Callus and Suspension Cells

J.C. SELLMER, S.W. RITCHIE, I.S. KIM, AND T.K. HODGES

The following steps for intiating, selecting, maintaining, and regenerating plants from type II callus and type II callus-derived suspension cell cultures represent a general format that has proven successful for inbred A188 and A188 derivatives (Armstrong and Green 1985; Kamo and Hodges 1986). The following protocols and the timing of specific steps should be used primarily as a guideline; they will require fine-tuning for different maize lines. In addition, we stress that visual selection and testing for plant regeneration from the type II callus and suspension cells must be done continuously.

TYPE II CALLUS

Immature Embryo Isolation

1. Prepare petri plates containing 25 ml of type II callus initiation medium ($3SN62D+AgNO_3$). This medium contains N6 salts and vitamins (Chu et al. 1975),

The Maize Handbook—M. Freeling, V. Walbot, eds.

3% sucrose, 2.3 g/liter proline (Armstrong and Green 1985), 0.1g/liter enzymatic casein hydrolysate, 2 mg/liter 2,4-dichlorophenoxy-acetic acid (2,4-D), 15.3 mg/liter $AgNO_3$ (Songstad et al. 1991), and 0.8% Bacto-agar, and was adjusted to pH 6.0 before autoclaving.

2. At 9–11 days after pollination, select an ear with immature embryos measuring 1.25–1.75 mm in length. Because embryo size is critical, randomly check kernels for the proper embryo length prior to harvesting the entire ear. This is accomplished by first cutting away the top of the kernel with a scalpel. Second, insert a thin spatula down along the side opposite of the silk scar to the base of the kernel and lift the endosperm and embryo out of the kernel. If embryos are of the proper size, remove the ear from the plant. Conversely, if embryos are too small, replace the husks around the ear and return the ear bag to hold the husks in place. Recheck these ears in 1–2 days.
3. To sterilize the freshly harvested ears, place a large magnetic stir bar into a 1-liter beaker, cover with aluminum foil, and sterilize in an autoclave. Within a laminar flow hood or other sterile environment, pour 500–700 ml of a solution containing 10% Clorox (2.5% sodium hypochlorite) and 20 drops/liter Tween 20 surfactant into the beaker.
4. Remove the husks and silks from the harvested ear, break the ear into halves (if longer than 10 cm), and place into the Clorox/Tween 20 sterilizing solution. After 30 minutes of vigorous stirring, carefully decant the Clorox/Tween 20 solution and rinse the ear three separate times with 500–700 ml of sterile deionized water (5 minutes of vigorous stirring with each rinse).
5. Remove the ear from the beaker by spearing up through the pith at either end with a sterile dissecting needle. Place the ear onto a sterile glass petri plate, and place the petri plate with ear and dissecting needle onto a dissecting microscope that is in a laminar flow hood.
6. Cut away the top of the kernels and extract each embryo using the technique described in step 2 for determining embryo size. Separate the embryo from the endosperm and place the intact embryos onto the type II callus initiation medium. The embryo axis side should be in contact with the medium (scutellar side up). Plate 10 embryos onto each plate, wrap the plates with Parafilm, and incubate in the dark at 25°C.

Type II Callus Selection and Maintenance

Type II callus is distinguishable as seen with the dissecting microscope as soft and friable with numerous individual embryoids protruding from the callus surface. Type II callus may show heterogeneity in embryoid structure, stage of embryoid development, or in color from translucent to light yellow. Type II callus must be continually selected at each subculturing to maintain the highly embryogenic, soft, and friable callus. Hard callus tissue not showing embryoid structures or tissue showing highly advanced embryoid development should be discarded.

1. The time required for callus growth and embryoid formation is genotype-dependent. The A188-derived cultures form transferrable type II callus in approximately 1.5–3 weeks.
2. Examine the embryos for the presence of type II callus using a dissecting microscope. Once type II callus is formed, select and transfer that callus to type II callus maintenance medium (3SN62D without $AgNO_3$). Although $AgNO_3$ is beneficial for type II callus induction, we have found it detrimental for type II callus maintenance. Give each selected callus clump a number to distinguish its point of origin (the genotype, embryo, and date of isolation).
3. Following the initial selection for type II callus, continue to select and transfer the friable, soft, embryogenic callus every 7–14 days. During the first 4–6 weeks of this process, the amount of type II callus from each embryo is increased. However, callus that is slow growing or that produces relatively few embryoids can be discarded. Depending on the genotypes used for developing type II callus and suspension cells, a higher concentration of 2,4-D or a shorter subculture cycle may be required to prevent embryoid maturation and to maintain a friable, soft type II callus.
4. Once a type II callus line has been established (approximately 4–6 weeks after embryo isolation), initiation of suspension cells and testing for plant regeneration is undertaken. Because of the time and expense required for type II callus maintenance, it is usually desirable to maintain only the most embryogenic lines, i.e., those that regenerate the most plants. Thus, to insure the usefulness of each type II callus line, regularly test for the level of plant regeneration. In our laboratory, suspension cultures are most readily developed using type II callus during the first 6 months after embryo isolation.

Plant Regeneration From Type II Callus

Regeneration of plants from type II callus is based upon allowing the embryoids on the surface of the type II callus to mature and germinate. The protocol used in our laboratory is as follows:

1. Collect and weigh the callus in preweighed petri plates.
2. Evenly distribute 1–2 g fresh weight of the soft, friable type II callus containing numerous embryoids over the surface of a 100 × 15-mm petri plate containing 25 ml of regeneration medium. Regeneration medium consists of Murashige and Skoog (MS) basal salts (Murashige and Skoog 1962), modified White's vitamins (0.2 g/liter glycine, and 0.5 g/liter of each of thiamine-HCl, pyridoxine-HCl, and nicotinic acid), supplemented with 6% sucrose, 0.1 g/liter myo-inositol, and 0.8% Bacto-agar (6SMS0D). Wrap the plates with Parafilm and place in the dark. After 1 week, move the plates to a lighted growth chamber with a 16-hour light (75 $\mu E\ m^{-2}\ sec^{-1}$) and an 8-hour dark photoperiod (Kamo and Hodges 1986).
3. Three weeks after plating the type II callus to 6SMS0D examine the calli for shoot

formation. Transfer the calli and shoots to fresh 6SMS0D plates for another 2 weeks.

4. Select and transfer calli with shoots to petri plates with reduced sucrose (3SMS0D). The callus without shoots can be left on the 6SMS0D for a longer period if the callus is slow in embryo development.
5. Upon distinct formation of a shoot and root (some may be ready for transfer after the first 3 weeks on regeneration medium), transfer the newly developed green plantlets to Magenta GA-7 (Magenta Corp, Chicago, IL) containers containing 60 ml of 3SMS0D medium solidified with 0.6% Bacto-agar. This will allow good root development and plant elongation. In addition, the lower agar concentration reduces the chance of root damage when transplanting the plantlets to soil.
6. When the plant has developed a strong root system (10–15 days after transfer into the Magenta boxes), gently remove the plant from the agar, wash the remaining agar from the roots and shoot, and carefully transplant into a 4-inch pot containing a moist soil. Place the pots in a high humidity chamber and over a period of 10 days slowly reduce the humidity to approximately that of the greenhouse. Once the plants are adapted to a lower humidity, move them to the greenhouse and treat them like seedlings.

SUSPENSION CELL CULTURES

Suspension Cell Culture Initiation

Suspension cultures are initiated from the friable type II callus cultures. Attempts to establish suspension cell cultures from hard embryogenic callus have been unsuccessful.

1. For the 2–3 weeks before initiating a suspension culture, the type II callus should be subcultured on no more than a 7-day cycle. This short subculture cycle reduces embryoid development and aids in the dispersion of the type II callus in liquid medium.
2. Prepare 2SN61.5D (or 3.5D) liquid suspension medium containing N6 salts and vitamins (Armstrong and Green 1985), 2% sucrose, 0.7 g/liter proline, 0.1 g/liter enzymatic casein hydrolysate, and 1.5 or 3.5 mg/liter 2,4-D. Before autoclaving, adjust the medium to pH 6.0 and dispense 20 or 40 ml of medium into 125 or 250 ml erlenmeyer flasks, respectively. In addition, 100–125 ml of medium can be dispensed into 250-ml flasks for washing purposes. (See step 5.) Media flasks should be covered with two layers of aluminum foil or other appropriate material to ensure sterility and yet allow air exchange.
3. Select and dispense 1–2 g fresh weight of type II callus (weigh in a sterile preweighed petri plate) into the 125- or 250-ml flasks of prepared medium.
4. Place the flask in the dark at 25°C on a gyratory shaker at 120 rpm.

5. After 1 week, begin a 7-day wash cycle with a complete washing of the cells every third and seventh day. Washing consists of removing all of the old medium and cellular debris produced by the cells with a sterile 10- or 25-ml-wide-bore pipet and replacing with fresh medium. Care is taken to minimize removal of cells. This wash procedure is repeated two more times. Washing continues for 2–3 weeks.
6. After 2–3 weeks of washing, the cultures will have grown to such an extent that subculturing is required. At this time, examine the cells under a microscope to determine the types of cells that make up the suspension culture. The goal in the development of a suspension culture is to produce small aggregates of highly cytoplasmic and actively dividing cells. Large, highly vacuolated, and irregularly shaped cells are undesirable, although these types of cells will be prevalent in young cultures. It is also common at this time for some cultures to have several to many very large cell clusters or aggregates. The purpose of subculturing is to amplify the number of small active cell aggregates and reduce the presence of undesirable cells and large cell clusters.
7. Once subculturing is initiated, a standard 7-day culture/selection cycle is followed. The subculture/selection process begins after washing the cells once with fresh suspension cell medium (as described in step 5). Cells are then subcultured by *selectively* pipetting a 4–5-ml packed cell volume (PCV) of small cell aggregates with a 10—ml cutoff pipet. This is accomplished by placing the mouth of the cutoff pipet at the bottom of the flask and drawing cells up while, at the same time, moving the pipet mouth along the bottom of the flask in a back-and-forth motion. This is continued until a 4–5-ml PCV is attained. The PCV is measured by placing the mouth of the cutoff pipet *squarely* on the bottom of the flask while expelling the liquid. Dispense the selected cells into a new 250-ml flask containing 40 ml of medium. The PCV used for subculturing depends on the speed at which the cells grow and should be adjusted to fit the response of each callus line placed into culture.
8. On the third or fourth day of the 7-day culture/selection cycle, refresh the cells by removing the old medium and replacing with fresh medium one time. (This is equivalent to a single washing.)

Suspension Cell Culture Establishment and Maintenance

1. During the next several months, continue the 7-day culture/selection cycle in conjunction with periodic examinations of the culture cells under a microscope. With continued culture/selection, the number of large, highly vacuolated cells should diminish while small clusters of highly cytoplasmic cells should become more prevalent. Once established, suspension cultures are maintained using the same 7-day culture/selection cycle. Continuation of the single washing step during the cell culture cycle will depend on the growth rate and amount of debris secreted from the suspension cells. If the cultures remain actively dividing and produce a small amount of debris, then the washing step may be omitted from the cell culture cycle.

2. Approximately 3 months after suspension culture initiation, the culture should be relatively stabilized with respect to growth rate. Tests for plant regeneration should be done every 3–6 months. Testing for plant regeneration is important as suspension cultures of this type often lose the ability to regenerate large numbers of plants after 1 year. Because of the loss of plant regeneration capacity, some researchers have utilized cryopreservation of established suspension lines in an attempt to reduce the need and expense of continually developing new lines (Gordon-Kamm et al. 1990; Shillito et al. 1989).

Plant Regeneration From Suspension Cell Cultures

The initiation of suspension cultures from type II callus requires the promotion and selection of actively dividing and highly cytoplasmic cells within clusters averaging about 0.7 mm in diameter (Kamo and Hodges 1986). Once a suspension cell line is established (approximately 2 months) it should be tested for regeneration. Plant regeneration from the actively dividing suspension cells relies on the reinitiation of callus.

1. To begin the regeneration process, use a 5-ml cutoff pipet to transfer and evenly distribute a 0.5-ml PCV of a 3–6-day-old suspension cell culture across the surface of a 100 × 15-mm petri plate containing 3SN61D solidified with 0.8% Bacto-agar. The reduced concentration of 2,4-D is used to promote cell growth without inhibiting embryo development.
2. Remove the excess liquid from around the calli by blotting with sterile filter paper or use a sterile syringe and needle. Seal the plates with Parafilm and place in the dark at 25°C.
3. After 3 weeks, examine the plates for the presence of embryogenic callus. Transfer the callus to fresh plates containing 3SN61D.
4. Reexamine the plates, 3 weeks later, for the presence of embryoids. Select and transfer the embryogenic callus to regeneration medium 6SMS0D (described above).
5. To complete the regeneration process for embryogenic callus derived from suspension cells, follow the procedures outlined in the regeneration of type II callus described above.

REFERENCES

Armstrong CL, Green CE (1985) Establishment of friable, embryogenic maize callus and the involvement of L-proline. Planta 164: 207–214

Chu CC, Wang CC, Sun CS, Hus C, Yin KC, Chu CY (1975) Establishment of an efficient medium for anther culture of rice through comparative experiments on the nitrogen sources. Scientia Sinica 18: 659–668

Gordon-Kamm WJ, Spencer TM, Mangano ML, Adams TR, Daines RJ, Start WG, O'Brien JV, Chambers SA, Adams WR, Jr, Willetts NG, Rice TB, Mackey CJ, Krueger RW, Kausch AP, Lemaux PG (1990) Transformation of maize cells and regeneration of fertile transgenic plants. Plant Cell 2: 603–618

Kamo KK, Hodges TK (1986) Establishment and characterization of embryogenic maize callus and cell suspension cultures. Plant Sci 45: 111–117

Murashige T, Skoog F (1962) A revised medium for rapid growth and bioassays with tobacco tissue cultures. Physiol Plant 15: 473–497

Shillito RD, Carswell GK, Johnson CM, DiMaio JJ, Harms CT (1989) Regeneration of fertile plants from protoplasts of elite inbred maize. Bio/Technology 7: 581–587

Songstad DD, Armstrong CL, Petersen WL (1991) $AgNO_3$ increases type II callus production from immature embryos of maize inbred B73 and its derivatives. Plant Cell Rep 9: 699–702

122

Production of Transgenic Maize Plants via Microprojectile-Mediated Gene Transfer

MICHAEL FROMM

This chapter describes the rationale and techniques for producing transgenic maize plants. The emphasis will be on a basic understanding of how to develop a system for producing transgenic plants rather than a specific protocol. (Table 122.1 lists references for various protocols.) This is because the specific details are changing rapidly and depend on the availability to the investigator of the specific reagents. However, the basic principles remain the same and a good understanding of these will allow the investigator to effectively apply the latest technical improvements.

The technique used to transfer DNA into plant cells is the high velocity microprojectiles method. The DNA-coated, 1-μm tungsten or gold particles are accelerated by a microprojectile gun, of which there are several designs available. The critical aspect is that the particle gun transfers particles and DNA efficiently into the target cells, as measured by the transient and stable transformation assays described below. Any particle acceleration

The Maize Handbook—M. Freeling, V. Walbot, eds.

TABLE 122.1. References for maize transformation protocols

Topic	References, sources, components[a]
Particle gun designs	
Air rifle	13
Electric discharge	2, 11
Gunpowder	6–10
Helium	15
Microprojectile/DNA preparations	2, 6–11, 13–15, 17
Plasmids	
C1 and B anthocyanin genes	5
GUS genes for corn	7, 12
Bar gene selectable marker	3, 4, 6, 14, 16
EG5 BMS cell line	1
BMS transformation	10, 14
Transgenic plant production	4, 6
Herbicide sources	
BASTA herbicide formulation	18
Pure PPT	19
Herbiace herbicide formulation	20
Pure Bialaphos	20
PAT enzyme and leaf assay	3, 6, 14, 16
Media	
BMS media	21
N6 media	22

[a]References, sources, components:

1. Chourey PS, Zurawski DB (1981) Callus formation from protoplasts of a maize culture. Theor Appl Genet 59: 341–344.

2. Christou P, Swain WF, Yang, N-S, McCabe DE (1989) Inheritance and expression of foreign genes in transgenic soybean plants. Proc Natl Acad Sci USA 86: 7500–7504.

3. De Block M, Botterman J, Vandewiele M, Dockx J, Thoen C, Gossele V, Movva NR, Thompson C, van Montagu M, Leemans J (1987) Engineering herbicide resistance in plants by expression of a detoxifying enzyme. EMBO J 6: 2513–2518.

4. Fromm ME, Morrish F, Armstrong C, Williams R, Thomas J, Klein TM (1990) Inheritance and expression of chimeric genes in the progeny of transgenic maize plants. Bio/Technology 8: 833–839.

5. Goff SA, Klein TA, Roth BA, Fromm ME, Cone KC, Radicella JP, Chandler VL (1990) Transactivation of anthocyanin biosynthetic genes following transfer of the *B* regulatory genes into maize tissues. EMBO J 9: 2517–2522.

6. Gordon-Kamm WJ et al (1990) Transformation of maize cells and regeneration of fertile transgenic plants. Plant Cell 2: 603–618.

Table 122.1. (Continued)

7. Klein TM, Gradziel T, Fromm ME, Sanford JC (1988) Factors influencing gene delivery into Zea mays cells by high-velocity microprojectiles. Bio/Technology 6: 559–563.
8. Klein TM, Fromm ME, Weissinger A, Tomes D, Schaaf S, Sletten M, Sanford JC (1988) Transfer of foreign genes into intact maize cells with high-velocity microprojectiles. Proc Natl Acad Sci USA 85: 4305–4309.
9. Klein TM, Harper EC, Svab Z, Sanford JC, Fromm ME, Maliga P (1988) Stable genetic transformation of intact Nicotiana cells by the particle bombardment process. Proc Natl Acad Sci USA 85: 8502–8505.
10. Klein TM, Kornstein L, Sanford JC, Fromm ME (1989) Genetic transformation of maize cells by particle bombardment. Plant Physiol 91: 440–444.
11. McCabe DE, Swain WF, Marinell BJ, Christou P (1988) Stable transformation of soybean Glycine max by particle acceleration. Bio/Technology 6: 923–926.
12. McElroy D, Zhang W, Cao J, Wu R (1990) Isolation of an efficient actin promoter for use in rice transformation. Plant Cell 2: 163–172.
13. Oard JH, Paige DF, Simmonds JA, Gradziel TM (1990) Transient gene expression in maize, rice, and wheat cells using an airgun apparatus. Plant Physiol 92: 334–339.
14. Spencer TM, Gordon-Kamm WJ, Daines RJ, Start WG, Lemaux PG (1990) Bialaphos selection of stable transformants from maize cell culture. Theor Appl Genet 79: 625–631.
15. Sanford JC et al (1991) An improved helium-driven biolistic device. Technique J Meth Cell Molec Biol 3: 3–16.
16. Thompson CJ, Movva NR, Tizard R, Crameri R, Davies JE, Lauwereys M, Botterman J (1987) Characterization of the herbicide-resistance gene bar from *Streptomyces hygroscopicus.* EMBO J 6: 2519–2523.
17. Tomes D, Weissinger AK, Ross M, Higgins R, Drimmond BJ, Schaaf S, Malone-Schoneberg J, Staebell M, Flynn P, Anderson J, Howard J (1990) Transgenic tobacco plants and their progeny derived by microprojectile bombardment of tobacco leaves. Plant Molec Biol 14: 261–268.
18. Hoechst Celanese Corp, Somerville NJ.
19. Cresent Chemical Co. Inc, 1324 Motor Parkway, Hauppauge, New York 11788, phone (516) 348-0333; catalog #35605 glufosinate ammonium (CAS #7718-82-2, 1 g $89.00).
20. Meija Seika Kaisha, Ltd. Pharmaceutical Research Center, Morooka-CHO, Kohoku-ku, Yokohama 222, Japan.
21. BMS media: 4.4 g MS salts (Gibco); 0.25 mg/liter each of thiamine HCl, pyridoxine, and calcium pantothenate and 1.3 mg/liter niacin; 2 mg/liter 2, 4-D; 0.13 g/liter L-asparagine; 0.1 g/liter inositol; 20 g/liter sucrose; 0.2% GELRITE or 0.6% agarose; adjust pH to 5.8 before autoclaving.
22. N6 medium: 4.0 g/liter Chu (N_6) basal salts (Sigma C-1416), 1.0 ml/liter Eriksson's Vitamin Mix (1,000× stock made from Sigma E-1511 Powder), 0.5 mg/liter thiamine HCl, 20 g/liter sucrose, 1 mg/liter 2, 4-D, 2.88 g/liter L-proline (only with Bialaphos selection, not with PPT selection), 0.1 g/liter vitamin-free casamino acids, 0.2% GELRITE or 0.6 % agarose, adjust pH to 5.8 before autoclaving. (Add sterile herbicides after autoclaving and cooling to 55°C.)

device that meets these efficiency criteria is suitable. (The Dupont PDS 1000 and PDS 2000 work well in the author's experience.) The particle/DNA mixture is also an important parameter and several of the published protocols (see Table 122.1) should be evaluated by the rapid transient assays described below to determine which works best for you.

Basically, the protocol for the production of transgenic corn plants uses microprojectiles to transfer genes into embryogenic maize cells, a selectable marker gene system to recover these transgenic cells, and a regeneration system to convert the transgenic calli into fertile plants. This system is reproducible and is in use in a number of laboratories. However, it should be understood that this is still a difficult procedure that involves considerable tissue culture effort and experience with the microprojectile gun. Initially, one should expect to spend at least 40 hours of labor for each line of fertile transgenic plants produced.

There are two types of gene transfer assays: transient and stable transformation assays. After gene transfer, the majority of the introduced plasmid DNA persists as free plasmid DNA in the nucleus for a period of 7 days or so. During its transient existence in the cell, the introduced DNA is transcribed into mRNA and translated into protein given the appropriate regulatory signals. Transient gene expression is very useful to measure gene transfer efficiency because results can be obtained in 2 days and the efficiency can be quantified. Furthermore, the actual cells expressing the introduced DNA can be visualized with the 𝔅-glucuronidase (GUS) or anthocyanin genes described below.

In a small percentage of the cells, typically 1–5% of those transiently expressing the new DNA, the plasmid DNA will integrate into chromosomes. If these cells continue to divide, they grow into stably transformed calli containing the plasmid DNA. The bombarded cells of some cultures divide better than others, so the actual ratio of transiently expressing cells to stably transformed cells varies from 5% down to very small values, approaching zero in some cases. Thus, while transiently expressing cells are a good measure of the efficiency of gene transfer, they are not entirely predictive of stable transformation frequencies between different cultures. However, higher transient frequencies in a particular culture seem to correlate with better stable transformation frequencies in that culture system.

Two particularly useful reporter genes for transient assays are the *E. coli vidA* (GUS) gene and the anthocyanin regulatory genes *C1* and *B* (*Lc* or *R*), expressed from the CaMV 35S promoter. In the case of GUS, the cells can be stained with the substrate 5-bromo-4-chloro-3-indolyl B-D-glucuronic acid (XGLUC), which turns the GUS-expressing cells blue. The *B* and *C1* anthocyanin regulatory genes induce the endogenous maize anthocyanin structural genes, if present, and this results in purple-anthocyanin-pigmented cells. (This anthocyanin response is genotype-dependent, in that all the anthocyanin structural genes must be present. Most nonpigmented maize lines will respond, particularly the most common A188 or A188 × B73 cultures). [Editors' Note: Most modern inbred lines of maize are *c1* and *r-r* but contain all of the required enzyme-encoding genes. See Coe et al. (1988) for further details, in *Corn and Corn Improvement*.] Note that while the GUS reaction product diffuses out and colors multiple cells, the purple anthocyanin pigment is vacuolar and should be restricted to a single cell. In either case, these genes allow the efficiency of gene transfer to be measured as the number of pigmented cells 2 days after bombardment. This allows someone to optimize their skills with the particle gun quickly. Bombardment of embryogenic callus cultures should produce several hundred to several thousand spots. (Note: you need a dissecting microscope to see the anthocyanin-pigmented cells.) The number of spots reflects both the gene transfer efficiency and the expression level of the plasmid transferred—a poorly expressed plasmid will not produce very many colored cells.

Once you can use the particle gun efficiently as measured by transient assays, the next step is to use a nonregenerable Black Mexican Sweet (BMS) suspension culture as the target for stable transformation studies. This will require the use of a selectable marker plasmid (discussed below) and selection for a period of 4–8 weeks. The reason for using BMS cells for training is that BMS cells are at least 100-fold easier to transform relative to regenerable embryogenic cells. This allows even suboptimal procedures to produce positive transformation results while at the same time providing a useful training exercise in selection conditions that will be applicable to the regenerable cells. Additionally, learning to routinely transform BMS cells is a valuable skill as many gene expression questions can be answered in this system quickly and efficiently. Note that anthocyanin induction by expression of the *C1* and *B* cDNAs is very dependent on the BMS culture source, with the EG5 BMS culture working the best (Table 122.1).

There are a number of selectable markers available, preferably expressed from the vectors that give the highest expression. The ideal selectable marker allows the transgenic corn cells to grow rapidly in the presence of a metabolic inhibitor that prevents the growth of all the nontransgenic neighboring cells. The more efficient this process the quicker the growth of the transgenic calli can be observed and the less labor required to reach this stage. The essence of the selection dilemma is that initially the transgenic cell is surrounded by nontransgenic cells. Stringent selection pressure will cause these neighboring cells to die too quickly, and this will kill the rare transgenic cells, too. Therefore, the selection pressure and selective agent need to be mild enough, at least initially, so that the unhealthy nontransgenic cells do not inhibit the division of the transgenic cells. As the number of transgenic cells increases, this gets to be less of a problem. Because the stable transformation efficiency is low, quite a large number of cells are bombarded. These cells can rapidly grow into more plates of tissue than can be easily handled, so a selective agent that effectively inhibits growth quickly helps reduce the amount of tissue culture labor required to recover the transgenic calli.

The various selectable markers used successfully in monocots are genes encoding the neomycin phosphotransferase II, hygromycin phosphotransferase, acetolactate synthase (ALS), and phosphinothricin acetyltransferase (PAT) activities. Of these, I would recommend using the bar gene encoding the PAT enzyme. The ALS gene is a very efficient marker in tissue culture but appears to have a detrimental effect in some of the transgenic plants. The bar gene-encoded PAT enzyme detoxifies, via acetylation, the herbicide phosphinothricin (PPT), the active ingredient in the herbicide BASTA. An alternative active form of this herbicide is bialaphos: PPT with two alanyl groups attached. Bialaphos is activated to PPT in the plant cell and is then acetylated by the PAT activity. There are relatively convenient enzyme assays for PAT activity, and the plants can be scored phenotypically by painting a leaf with the herbicide.

In addition to the experience with transient and stable transformation systems above, one needs to learn how to establish embryogenic maize callus and suspension cultures. This is fully discussed in other chapters in this volume. The use of A188 and A188 hybrids, such as A188 × B73, greatly facilitates the embryogenic response, and success at establishing embryogenic callus cultures is likely on the first attempt. Establishing embryogenic suspension cultures is more difficult and takes several months.

The following protocol for producing transgenic corn plants assumes that embryogenic callus or a suspension culture has been established recently, within 1–6 months for callus cultures and 3–9 months for suspension cultures. (As the cultures get older, plant regeneration and fertility decrease). The protocol also assumes that transient assays have been used to obtain efficient gene transfer conditions and that BMS stable transformations have been carried out to demonstrate that the vector and selection system are functioning properly. This protocol assumes that the bar gene is being used as the selectable marker. The use of a selectable marker plasmid with a second gene that expresses GUS facilitates identifying the transgenic calli. Alternatively, co-transformation with a second plasmid encoding GUS can be used.

PROTOCOL FOR PRODUCING TRANSGENIC MAIZE PLANTS

Day 1. Bombardment of the Target Cells

1. Place embryogenic callus on N6 agarose medium or suspension cultures on paper filters on N6 agarose medium. Try to make a thin lawn of tissue in a circle of about 5 cm diameter. (Because gene transfer is on the surface of the tissue, the extra thickness just gives extra cells to select against.)
2. Bombard the tissues with the DNA-microprojectile preparation and particle gun parameters that gave the best transient assay results. Up to three bombardments of the same sample can often increase the transformation efficiency.

Days 1–14. Start Selection on Media Containing Bialaphos

1. Allow the bombarded cells to grow without selection pressure for 1–14 days, typically 2–7 days. This step is probably an important variable and several different times might be chosen to find out what works best for a given culture. Suspension culture cells can be placed back in suspension during this time or grown on solid media.
2. Transfer the cells to medium containing Bialaphos. (Note that PPT is not effective when the medium contains amino acids, while Bialaphos is effective even with amino acids present.) The recommended starting point is 1 mg/liter Bialaphos (1 mg/liter Bialaphos active ingredient if using a herbicide formulation). Usually the cells will

continue to grow the first time they are placed on selective media as they have a metabolic reserve pool and the Bialaphos takes some time to become effective. This growth is cell density-dependent, with higher densities giving more growth.

Week 3–4. Transfer Cells to Fresh Selective Media

When the cell density on the first selection plates becomes too much for further efficient selection (more than 25% of the tissue size of the maximum number of cells that can grow on the same media without Bialaphos), replate the cells by gently flattening and spreading the tissues on fresh selective plates (one original plate will end up on several new plates to reduce the tissue density) containing 1–3 mg/liter Bialaphos. This second plating will usually inhibit cell growth enough to observe growth of the transgenic calli above the background of "flat" nongrowing cell clusters.

Weeks 6–10. Isolate Individual Transgenic Calli

Transgenic calli become obvious by their faster growth and size at this stage. Pick obviously growing calli onto fresh selective plates. Observe the remaining calli on the selective plate to see if slower-growing transgenic calli appear later. Sometimes, if too much growth has occurred, all the cells must be replated to fresh selective plates to allow 12 weeks for transgenic calli to appear.

Weeks 8–12. Increase Amount of Individual Transgenic Calli

Continue to grow individual callus lines on selective media to increase the amount of material to initiate large-scale regeneration efforts. Once resistant calli have been obtained, it is important to verify that they are transgenic, at least when establishing the system, as it takes some time to regenerate the plants and analyze them. These tests should simply confirm that the calli are transgenic, with the more extensive analysis saved for the plants. The easiest first test is to measure enzyme activities, especially if the selectable marker plasmid also carried a second scorable marker like the GUS gene in pBARGUS, or was co-transformed with the GUS gene. However, the GUS gene might not always be present or expressed, so PAT assays of the bar gene selectable marker are also possible. Alternatively, the DNA can be analyzed by PCR or Southern Blot. (Note that the anthocyanin markers are still unproven as visual markers when expressed from promoters other than their own in stably transformed embryogenic cells—these should be used cautiously, if at all, until proven to be nondetrimental to the cells and plants.)

Weeks 10–14. Start the Regeneration Process

The regeneration process is identical to published procedures for nontransgenic calli except that selective pressure is usually maintained. The key concern is that some of the transgenic calli consist of a mixture of transgenic and nontransgenic cells. This can lead to a predominance of nontransgenic plants during regeneration. Fortunately, selection pressure during regeneration inhibits the nontransgenic shoots and facilitates the recovery of the transgenic plants. Some of the calli appear to consist entirely of transgenic cells and give rise to transgenic plants without selection pressure. Probably the best strategy is to regenerate both with and without selection, using PPT or Bialaphos levels similar to those used to originally recover the calli (1–3 mg/liter). If you obtain plants on selection, they most likely are transgenic. If you only obtain plants without selection, you might need to screen the plants. Fortunately, the BASTA leaf painting assay is easy and gives results in 3 days on which plants are transgenic. Alternatively, enzymatic or DNA analysis can be performed.

Weeks 14–18. Regeneration Produces Plantlets

The most difficult step is getting the plantlets into soil and into the greenhouse. The cuticle of the plants is very thin after the high-humidity conditions in tissue culture. Therefore, high-humidity conditions need to be used when first transplanted to soil. This is followed by a slow lowering of humidity before transfer to the greenhouse. Fortunately, humidity control can be achieved with simple closed plastic containers, with gradual opening of the system to the air by progressively propping open the lid over several days.

Months 6–7. Plants Reach Fertility

Quite frequently transgenic plants are male and female fertile. However, tissue-culture-derived plants are often male sterile. Even in the cases where the transgenic plants are male fertile, it is a good idea to cross the transgenic ear with pollen from a normal non-tissue-culture-derived plant to increase the vigor of the progeny. Self-pollinations can be done in the next generation.

A less-frequent problem encountered in plants regenerated from older tissue culture lines is that the endosperm does not develop fully, causing the kernels to abort 2–3 weeks after fertilization. This problem can be recognized by noticing that the seeds stop enlarging (it does not damage the ear to peel back the husk to look at the kernels, and then just cover the ear again) and that they seem hollow when pressed. Dissection will show that the endosperm has stopped growing or even shriveled. Fortunately, the embryo is normal and can be rescued by dissection and germination in the medium used for the regeneration process.

Months 7–9. Transgenic R1 Seeds Obtained

They can be stored and germinated as normal maize seeds.

ACKNOWLEDGMENTS

Many of the ideas and protocols developed for maize transformation were in collaboration with Chuck Armstrong, Arlene Barnason, Briana Dennehey, Beate Hairston, Ted Klein, Fionnuala Morrish, and Bill Petersen.

In vitro Selection

D.A. Somers and K.A. Hibberd

Direct selection of mutant (Gengenbach et al. 1977; Hibberd 1984; Miao et al. 1988; Anderson and Georgeson 1989; Parker et al. 1990; Diedrick et al. 1990) and transgenic (Gordon-Kamm et al. 1990; Fromm et al. 1990) maize tissue cultures is usually conducted using friable, embryogenic callus (Armstrong and Green 1985) or regenerable suspension cultures (Green et al. 1983). Regenerable haploid callus cultures (Coumans et al. 1989; Pescitelli et al. 1989) may allow direct selection of certain recessive mutations as well as negative selection protocols for selection and identification of auxotrophic mutants. The procedures described in this chapter outline direct selection protocols that have been used successfully to select diploid tissue cultures from which mutant and/or transgenic plants have been regenerated.

PILOT STUDIES

Successful tissue culture selection depends on evaluating and understanding the action of the selective agent on maize cell cultures. A selective agent is often chosen for investigation based on observations of specific effects on other organisms. Biochemical tests such as in vitro assays of enzyme inhibition and measurements of metabolite levels in response to the selective agent can help to determine if the selective agent is acting as predicted in maize cultures. In some cases, a metabolic block imposed by the selective agent(s) can be reversed by addition of a downstream metabolic product. Optimizing reversal of inhibition by the selective agent can aid in defining more efficient selection conditions.

Properties of the selective agent should be well defined before starting selection experiments. The stability of the selective agent is extremely important in defining the selection strategy because tissue culture selection often requires 3 or more months to complete. Culture conditions such as light or low pH will degrade some selective agents and can require changing growth conditions or length of culturing cycle. Variation in batches of the selective agents, often noted with antibiotics, should be explored before setting

The Maize Handbook—M. Freeling, V. Walbot, eds.

specific selection levels. Also solvents or carriers associated with the selective agents can greatly affect tissue culture growth and quality, requiring changes in selective agent source or presentation in the medium. In addition, media components can interfere with the action of some selective agents. For example, amino acids can reverse the inhibitory effects of selective agents that target amino acid biosynthesis.

Preliminary experiments will generally be needed to define a selection strategy. Specifically, these studies should examine the interactions between the selective agent and maize tissue cultures, the potency of the selective agent on maize cultures, and the response of specific cell culture lines to the selective agent. Agents that kill rapidly are generally more challenging to work with in maize cultures probably because rapid death of tissues can also kill rare resistant cells. Fine-tuning selection conditions is extremely important with this type of agent. With slower acting selective agents, there will be greater latitude in defining conditions for recovery of resistant lines. In some cases, slow-acting agents will require several cycles of selection to see the full effects on tissue growth and quality.

Maize tissue culture lines that are rapidly growing, highly regenerable, and stable under stress are preferred starting materials for selections. Determining interactions of specific maize culture lines with the selective agent can take the form of a series of growth inhibition tests. These involve placing tissue from several tissue culture lines on media containing four or more levels of the selective agent and observing tissue culture growth and quality for one or more tissue culture cycles on the selective agent. Problems with specific lines can be identified before selections are initiated; in addition, selective agent levels for initiation of selection can be identified.

Selection strategies can include on–off treatments and single-level and step-up enrichment selection procedures (Hibberd 1984). On–off treatments should be most useful for selective agents that rapidly kill maize cells. This should allow for some survival of neighboring tissues in the early period of the selection process. Single-level or lethal-dose selections are generally the simplest and least time consuming. However, conditions for the single-level selections are likely to be very narrow. The more general approach of step-up enrichment selection, which is applicable to many types of selective agents and under a wider range of conditions, will be described. Methods are also described for determining tissue culture tolerance to a selective agent; this provides a means of deducing selective agent concentrations to use to initiate selection and to characterize tolerance levels of selected tissue cultures.

MATERIALS

Friable, embryogenic callus or regenerable, suspension cultures

Solid N6 medium (Armstrong and Green 1985) poured in 100 × 25-mm sterile disposable petri plates

7-cm-diameter Whatman #1 filter papers

METHODS

Determination of Selective Agent Growth Inhibition Curve

1. Prepare callus maintenance medium containing a range of concentrations of the selective agent. Be sure to include a control containing the reagent used to dissolve or dilute the selective agent.
2. Pour 50 ml medium into 100 × 25-mm sterile disposable petri plates.
3. Determine dry weight (wt) of 7-cm Whatman #1 filter papers and label dry wt on each filter with #2 pencil before autoclaving.
4. Overlay solidified medium with autoclaved filter paper.
5. Evenly spread approximately 0.5 g fresh callus subsamples taken from a uniform callus or suspension culture onto the filter paper for each treatment. Determine dry wt of three replicate samples of 0.5 g fresh callus by drying at 60°C for 48 hours. Include 3–4 replications per selective agent concentration.
6. Depending on the rate of growth inhibition by the selective agent, after the appropriate period (usually 2 weeks), determine dry wt of callus and filter papers by removing from medium and drying on lid of petri plate for 48 hours at 60°C.
7. Determine dry wt increase of callus from each concentration of selective agent as follows:

   $$\text{Callus plus filter wt} - \text{filter wt} - \text{initial callus wt} = \text{Dry wt increase}$$

8. Tolerance level is often expressed as the concentration of selective agent that reduces growth by 50% as determined from the kill curve.

Notes

1. Some tissue culture lines react negatively to culturing on filter paper by growing slower and losing plant regeneration capacity. Screening of available tissue cultures should be conducted to identify lines that perform well on filter papers. Alternatively, kill curves and selection can be conducted directly on the culture medium by transferring callus tissue onto culture medium in small clumps (50 mg). Subcultures are conducted at 3-week intervals by moving each callus clump to fresh medium. It is recommended that this procedure be conducted in parallel with the filter paper method for determinations of kill curves to identify which culture procedure suits specific cell lines.
2. Cells can be removed from filter paper with a flat spatula for dry wt determinations; this eliminates weighing the filter paper.
3. Some investigators prefer to use fresh wt increase rather than dry wt determinations.

Enrichment Selection

In enrichment selection procedures, increasing selection pressure is gradually applied to cells by increasing the concentration of selective agent at each subculture. This procedure likely enhances recovery of mutants because the gradual increase in selection pressure may allow tolerant cells to proliferate without adverse effects of low cell population density imposed by the selective agent. Enrichment selection also allows visual selection for highly regenerable callus, ensuring plant regeneration.

1. Transfer samples of approximately 0.5 g cells onto 7-cm filter papers overlaying medium containing from 20% to 50% of the previously determined lethal dose of the selective agent. If filter papers are not used, plate several 50-mg clumps of tissue culture per petri plate directly onto selection medium.
2. After 2 or 3 weeks, transfer cells to an increased level of selective agent (usually a doubling series is used) by transferring the filter paper and cells to the fresh medium and so on until tissue cultures are isolated that grow on medium containing lethal concentrations of selective agent. During the selection process, callus sectors exhibiting rapid growth compared to the rest of the cell population are plated directly onto the next highest selective agent concentration.
3. Callus growing on lethal concentrations of selective agent is increased and subdivided onto callus maintenance medium with, and without, the selective agent and plant regeneration medium containing the selective agent.

Notes

1. To ensure selection of a mutant tissue culture, no less than 20–100 g callus or suspension cells should be placed on selection medium.
2. Use highly regenerable tissue cultures to ensure good plant regeneration following selection and bulk-up. Selection, callus maintenance, and plant regeneration may take as long as 4–6 months and could possibly reduce plant regeneration capacity.
3. Some investigators have utilized X-rays, sodium azide, or other mutagens to increase mutation frequency in tissue culture. Tissue culture per se results in high frequencies of mutations in plants regenerated from friable, embryogenic callus (Armstrong and Phillips 1988). The experience of the authors indicates that the inherent frequency of tissue-culture-induced genetic variation is sufficient for selection of dominant mutations at specific loci.

REFERENCES

Anderson PC, Georgeson M (1989) Herbicide-tolerant mutants of corn. Genome 31: 994–999

Armstrong CL, Green CE (1985) Establishment and maintenance of friable, embryogenic maize callus and the involvement of L-proline. Planta 164: 207–214

Armstrong CL, Phillips RL (1988) Genetic and cytogenetic variation in plants regenerated from organogenic and friable, embryogenic tissue cultures of maize. Crop Sci 28: 363–369

Coumans MP, Sohota S, Swanson EG (1989) Plant development from isolated microspores of *Zea mays* L. Plant Cell Rep 7: 618–621

Diedrick TJ, Frisch DA, Gengenbach BG (1990) Tissue culture isolation of a second mutant locus for increased threonine accumulation in maize. Theor Appl Genet 79: 209–215

Fromm ME, Morrish F, Armstrong C, Williams R, Thomas J, Klein RM (1990) Inheritance and expression of chimeric genes in the progeny of transgenic maize plants. Bio/Technology 8: 833–839

Gengenbach BG, Green CE, Donovan CM (1977) Inheritance of selected pathotoxin resistance in maize plants regenerated from cell cultures. Proc Natl Acad Sci USA 74: 5113–5117

Gordon-Kamm WJ, Spencer TM, Mangano ML, Adams TR, Daines RJ, Start WG, O'Brien JV, Chambers SA, Adams WR Jr, Willetts NG, Rice TB, Mackey CJ, Krueger RW, Kausch AP, Lemaux PG (1990) Transformation of maize cells and regeneration of fertile transgenic plants. Plant Cell 2: 603–618

Green CE, Armstrong CL, Anderson PC (1983) Somatic cell genetics systems in corn. In Downey K, Voellmy RW, Ahmad F, Schultz J (eds) Advances in Gene Technology: Molecular Genetics of Plants and Animals, Academic Press, New York, London, pp 147–157

Hibberd KA (1984) Induction, selection and characterization of mutants in maize cell cultures. In Vasil IK (ed) Cell Cultures and Somatic Cell Genetics of Plants, Vol I, Laboratory Procedures and Their Applications, Academic Press, New York, pp 571–576

Miao S, Duncan DR, Widholm JM (1988) Selection of regenerable maize callus cultures resistant to 5-methyl-DL typtophan, S-2-amino-ethylcysteine and high levels of L-lysine plus L-threonine. Plant Cell Tiss Org Cult 14: 3–14

Parker WB, Marshall LC, Burton JD, Somers DA, Wyse WL, Gronwald JW, Gengenbach BG (1990) Dominant mutations causing alterations in acetyl-coenzyme A carboxylase confer tolerance to cyclohexanedione and aryloxyphenoxypropionate herbicides in maize. Proc Natl Acad Sci USA 87: 7175–7179

Pescitelli SM, Mitchell JC, Jones AM, Pareddy DR, Petolino JF (1989) High frequency androgenesis from isolated microspores of maize. Plant Cell Rep 7: 673–676

124

Selection of Stable Transformants From Black Mexican Sweet Maize Suspension Cultures

JULIE ANDERSON KIRIHARA

Nonregenerable maize Black Mexican sweet (BMS) suspension cultures (Sheridan 1975, 1982) are finely divided and homogeneous in nature. Suspension cultures of BMS grow rapidly and are easily maintained. The characteristics of these suspension cultures permit their use in transformation and selection experiments employing protocols similar to those for microbial cells. The facility with which they can be genetically transformed by electroporation (Fromm et al. 1986) or microprojectiles (Klein et al. 1989) make BMS cultures useful as model systems for transformation and gene expression studies. This chapter outlines a plating and selection protocol that has reproducibly yielded large numbers of transformed cell lines by microprojectile bombardment and selection of plated BMS suspension cultures.

MATERIALS

Liquid MS2D medium (MS medium (Murashige and Skoog 1962) containing 2 mg/liter 2, 4-D)
Solid MS2D medium (MS2D medium solidified with 0.35% Gelrite in 100 × 25-cm plates)
Solid MS2D medium containing the selection agent, e.g., 30 nM chlorsulfuron (Chem Service Inc., West Chester, PA)
Sterile 7-cm Whatman #1 filter disks
Sterile Buchner funnel for 7-cm filters
Plasmid DNAs (1 mg/ml in 10 mM Tris-HCI, pH 8, 1 mM EDTA)
Microprojectile bombardment system

The Maize Handbook—M. Freeling, V. Walbot, eds.

METHODS

Culture Maintenance

Maize BMS suspension cultures are available from a number of laboratories, including the American Type Culture Collection (ATCC 54022); the author has not tested the ATCC culture. Maize BMS suspension cultures are maintained by weekly transfer of 12 ml of suspended cells to 125 ml of liquid MS2D medium in a 500 ml flask. The cultures are incubated at 26°C on a rotary shaker (150 rpm) in the dark. Cells are harvested for bombardment at 2–4 days postsubculture. For the selection method described in this chapter, it is important to use and maintain a very finely divided suspension culture. Selection of this type of tissue can be carried out by repeated selection of rapidly growing fine suspensions with the aid of a narrow-bore pipet. Once obtained, the use of a narrow-bore pipet during subculturing helps to maintain the finely divided tissue morphology.

Selectable Markers and Selection Agents

The use of different selectable marker genes in maize transformation is discussed elsewhere in this volume. In our laboratory we have had success using several selectable markers including genes encoding acetolactate synthase (ALS), hygromycin phosphotransferase II (HPT II), phosphinothricin acetyltransferase (PAT), and dihydrofolate reductase (DHFR), conferring resistance to chlorsulfuron, hygromycin, phosphinothricin, and methotrexate, respectively. All constructs have been driven by the CaMV 35S promoter. The use of maize introns has been beneficial in improving transformation frequency in several constructs. The construct (p35ALS) used for chlorsulfuron selection described in this chapter consists of a CaMV 35S promoter fused to the coding sequence and 3' noncoding sequence from a mutant *Arabidopsis thaliana* ALS gene (Olszewski et al. 1988).

Growth inhibition studies should be carried out for each BMS suspension culture line with the selection agent of choice. (See Sellmer et al. elsewhere in this volume.) A level of selective agent that results in approximately 90% inhibition of growth should be used for selection of transformants. In initial selection experiments, a range of concentrations should be tested to determine the optimal level of selection agent that allows maximum recovery of transformants with minimum "escapes" from control tissue. For example, 30 nM chlorsulfuron was found to be an effective level for selection of transformants using the p35ALS construct.

Transformation and Selection Protocol

1. Transfer approximately 1 ml settled-cell volume of cells (6–10 ml of suspended cells) to a 7-cm Whatman #1 filter disk by pipetting a suspension of cells onto the filter disk in a Buchner funnel under a very slight vacuum. Try to cover the filter with as even a layer of cells as possible. To aid in even dispersion of cells on the

filter, begin pipetting the cell suspension into the Buchner funnel before applying the vacuum.

2. Transfer the filter disk containing the cells to a plate of MS2D medium. The cells may be prepared a day in advance of bombardment and stored at 26°C until ready for use.
3. Repeat steps 1 and 2 to generate enough filters of cells for the entire experiment. Generally, prepare 2–3 filters of cells per DNA mixture to be bombarded; also include at least 2 filters of cells for nonbombarded or (−)DNA bombarded control treatments.
4. Precipitation of DNA onto microprojectiles may be carried out by one of the methods described by the manufacturer of the microprojectile bombardment system or by one of the protocols outlined in this volume. The method of precipitation depends on the type of apparatus used for the bombardment (e.g., gunpowder or helium). The DNA used for precipitation onto the microprojectiles generally consists

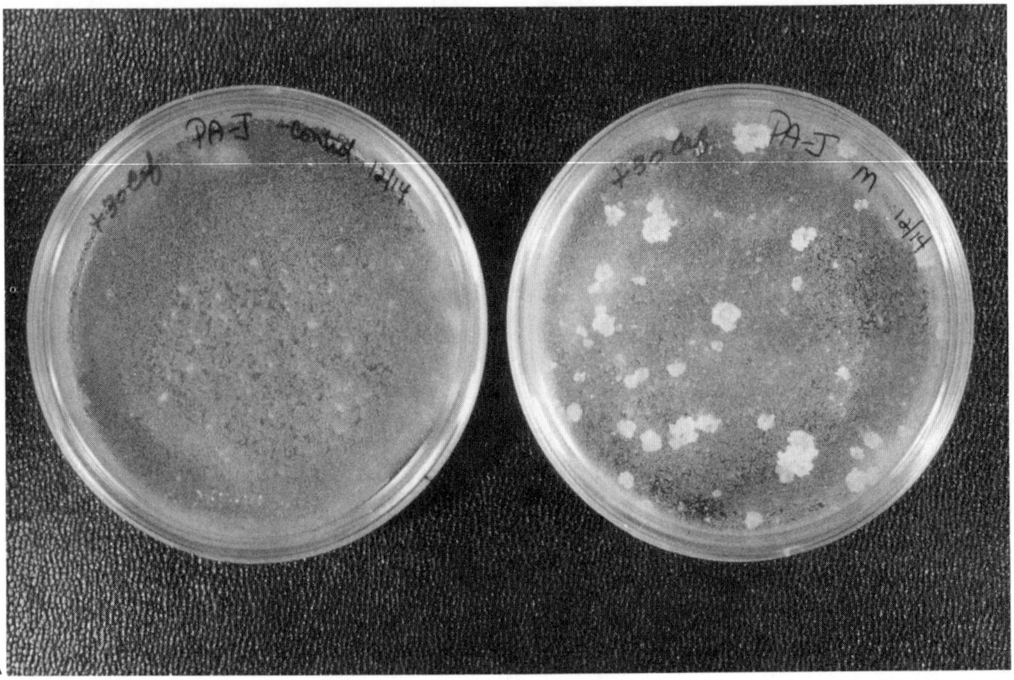

A B

FIGURE 124.1 Appearance of chlorsulfuron-resistant callus sectors at 5 weeks postbombardment. Black Mexican sweet suspension culture cells were plated onto 7-cm filter disks, bombarded with DNA-coated tungsten particles, and allowed to grow for 7 days on nonselective medium. The cells were then removed from the filter disks and plated as a thin lawn onto medium containing 30 nM chlorsulfuron. A. Nonbombarded control cells. B. Cells treated with p35ALS DNA.

TABLE 124.1. Effect of selectable marker DNA concentration on transformation frequency[a]

% p35ALS DNA	No. CSF-resistant sectors/bombardment
50	154 ± 17
25	146 ± 39
10	98 ± 30
5	31 ± 5
NBB	0

[a]Total DNA concentration maintained at 1 mg/ml in all treatments by addition of carrier DNA. Data are mean ±SD, based on six determinations per treatment. NBB: nonbombarded control treatment; CSF: chlorsulfuron.

of an equimolar mixture of separate plasmids. Transfer a filter disk containing the plant tissue to an empty 10-cm plastic petri dish and place the petri dish into the chamber of the microprojectile bombardment apparatus. Carry out the bombardment of the tissue according to the instructions for the particular instrument used.

6. Following bombardment, return the filter disk containing the plant tissue to solid MS2D medium and incubate at 26°C. Transient expression (e.g., transient GUS activity) can be assayed after 48 hours using tissue bombarded with the appropriate screenable marker constructs. For recovery of stable transformants, allow the tissue to grow for 7 days on nonselective media.
7. After 7 days on nonselective media, remove the tissue from the filter disk with a spatula and transfer the tissue to an empty petri dish.
8. Add 4–5 ml of liquid MS2D medium to the petri dish and mix the cells and medium together to generate a slurry of approximately 6 ml in volume.
9. Pipet approximately 2 ml of the cell mixture onto a plate of selective medium. Gently and slowly tip the plate and rotate it so that the cells form an even layer across the surface of the entire plate. Repeat the process with two more selective plates, for a total of three plates for each filter of tissue. Even plating of the cells is important for optimal results to be obtained in the selection. It may be helpful to practice with some extra plates of nonbombarded cells before plating critical experimental material.
10. The majority of the liquid used for plating the tissue will be absorbed by the media. It may be helpful to leave the plates in a laminar-flow hood for 15–30 minutes with the covers off to speed the drying process. The tissue should not move when the plate is tipped if properly dried.
11. Seal the edges of the plates with Parafilm and incubate the plates at 26°C for 4–5

weeks, at which point resistant sectors should be visible on plates of bombarded tissue (Figure 124.1).

12. Transfer the individual resistant cell clumps (approximately 0.5 cm in diameter) to fresh plates of selective medium at a density of 7–10 callus clumps per plate. After 2–3 weeks, resistant cell lines can be transferred to fresh selective medium for tissue increase and subsequent biochemical and molecular analyses.
13. Maintain the cells on selective medium (or on nonselective medium with occasional passes on selective medium) with a 2–3-week transfer period.

Notes

1. The use of cells at 2–4 days postsubculture is important. We have observed decreased frequency of transformation using older cultures.
2. An example of the transformation frequencies obtained with this system is shown in Table 124.1. In this experiment, a 1:1 mixture of p35ALS and a 35S-GUS construct was diluted to several final concentrations with carrier DNA (plasmid or salmon sperm DNA) before precipitation onto the microprojectiles. Total DNA concentrations were maintained at 1 mg/ml for all treatments. Following bombardment and selection, chlorsulfuron-resistant sectors were counted. The results indicated decreased transforrnation frequency with decreased p35ALS DNA used in the bombardment.
3. We have observed a 95% frequency of co-transformation and coexpression of two genes carried on separate plasmids (171/179 GUS+/chlorsulfuron-resistant lines).

REFERENCES

Fromm ME, Taylor LP, Walbot V (1986) Stable transformation of maize after gene transfer by electroporation. Nature 319: 791–793

Klein TM, Kornstein L, Sanford JC, Fromm ME (1989) Genetic transformation of maize cells by particle bombardment. Plant Physiol 91: 440–444

Murashige T, Skoog F (1962) A revised medium for rapid growth and bio assays with tobacco tissue cultures. Physiol Pl 15: 473–497

Olszewski NE, Martin FB, Ausubel FM (1988) Specialized binary vector for plant transformation: expression of the *Arabidopsis thaliana* AHAS gene in *Nicotiana tabacum*. Nucl Acids Res 16: 10765–10782

Sheridan WF (1975) Growth of corn cells in culture. J Cell Biol 67: 396a

Sheridan WF (1982) Black Mexican sweet corn: its uses for tissue cultures. In Sheridan WF (ed) Maize for Biological Research, Plant Mol Bio Assoc, Charlottesville, VA, pp 385–388

Maize Protoplast Culture

R.D. Shillito, G.K. Carswell, and C. Kramer

Protoplasts are the starting point for many genetic manipulations. These include genetic modification by incorporation of foreign DNA (direct gene transfer by PEG, electroporation, or other method), fusion with maize or other species, and generation of soma clonal variants. There have been few reports of the regeneration of fertile plants from maize protoplasts (Shillito et al. 1989; Prioli and Sondahl 1989; Morocz et al. 1990) although others have reported nonfertile plants (Cai et al. 1988; Rhodes et al. 1988a). While the methods that produce fertile plants from protoplasts differ in details of media composition and genotype, the basic procedure is based on obtaining a suitable embryogenic suspension culture as the source of the protoplasts. Although the preparation of protoplasts from callus cultures has been described in the literature (Kamo et al. 1987; Vasil and Vasil 1987), there are no reports of the successful culture of callus protoplasts to whole plants.

We describe the method we have used to initiate and select embryogenic suspension cultures for the isolation and culture of regenerable maize protoplasts (Shillito et al. 1989). In contrast to the methods described by Sellmer and co-workers elsewhere in this volume, the suspension culture method was developed with the express goal of obtaining protoplasts capable of forming fertile plants.

INITIATION OF CALLUS CULTURES

Embryogenic callus and suspension cultures of maize are maintained at 26°C in the dark.

Although a number of media have been described for initiating cultures, we typically use two formulations: N6 medium consisting of N6 salts and vitamins (Chu et al. 1975), 3% w/v sucrose, and 2 mg/l 2,4-D solidified with 0.24% w/v Gelrite {N6 maintenace medium} and a new medium, designated JMS, developed by Suttie et al. for initiation of embryogenic cultures from immature maize embryos. It consists of 1.69 g/liter NH_4NO_3; 1.82 g/liter KNO_3; 0.35 g/liter KH_2PO_4; 0.21 g/liter $CaCl_2 \cdot 2H_2O$; 0.4 g/liter $MgSO_4 \cdot 7H_2O$, minor elements and vitamins of the medium of Schenk and Hildebrant (1972), iron stock of Murashige and Skoog (1962), 2% w/v sucrose, with 10 g/liter myo-inositol,

The Maize Handbook—M. Freeling, V. Walbot, eds.

10 g/liter casein hydrolysate, 288 g/liter proline, and 1 mg/liter Dicamba (pH 5.8). The medium is solidified with 0.8% w/v Phytagar; 10 mg/liter fresh filter-sterilized $AgNO_3$ is added after autoclaving and cooling the medium. Use of JMS resulted, in many cases, in a higher percentage of embryos forming serially propagatable embryogenic callus. The resulting callus should be transferred to the N6 maintenance medium.

Explant immature embryos onto the initiation medium as described by Selmer and co-workers elsewhere in this volume.

SELECTION OF FRIABLE EMBRYOGENIC CALLUS SUITABLE FOR INITIATING SUSPENSION CULTURES

It is critical that the tissue used to initiate suspension cultures be of the correct morphology. Type II or friable callus arises predominantly on the distal end (toward the root pole of the embryo) of the scutellum 2 weeks after explanting the embryo. Remove the callus from the embryo, and subculture the tissue every 7–10 days on N6 maintenance medium, all the time selecting for yellow compact callus that is not watery. The callus to aim for becomes relatively non-mucilagenous, and it is granular and friable such that it separates into individual cell aggregates on placing into liquid medium. The time required to obtain such a callus can be from a few months to a year of subculture and is typically 3–6 months. When the callus reaches the desired morphology, it should be tested for its ability to regenerate fertile plants (see below), and one should attempt to initiate suspensions at monthly intervals.

PREPARATION OF SUSPENSION CULTURES

Place approximately 500-mg aliquots of the friable embryogenic callus into 25 ml of liquid N6 medium containing 2 mg/liter 2,4-D in 125 ml Delong flasks and culture them on a gyratory shaker (130 rpm, 2.5-cm throw). Initiate 10–12 cultures at any one time so as to be able to select the best cultures for further work. Subculture the suspension cultures weekly by transferring 0.5 ml packed cell volume (PCV: settled cell volume in a pipitte—see Sellmer et al. this volume) into 25 ml of fresh medium. Inspect the cultures after 2 weeks and retain those in which most of the cells consist of densely cytoplasmic dividing cell aggregates. Discard suspension cultures containing predominantly aggregates with large, expanded cells or many single elongated cells. Maintain the cultures by always choosing for subculture the flasks whose contents exhibit the best morphology. After 4–6 weeks of subculture in this fashion, the cultures increase two- to threefold per weekly subculture. After 3–4 weeks, or once a suspension culture is established, it can be transferred to 250-ml Delong flasks containing 50 ml N6 medium.

Periodic filtration through 630-μm-pore-size stainless-steel sieves (every 2 weeks) can be used in some cases to increase the dispersion of the cultures. This is not necessary if careful attention is given to developing the callus cultures and selecting strongly for the best suspension culture morphology. Regenerate a sample of the established suspension to check that it has retained its ability to regenerate fertile plants. The number of plants regenerated and the fertility of the plants are likely to be lower than those seen for the callus from which the suspension was derived.

Monitor the suspension every 2–3 weeks for release of protoplasts. When a yield of 5–10 million per g fresh weight is obtained, begin plating the protoplasts to monitor division and colony formation.

Cryopreservation can be used to maintain a supply of regenerable embryogenic suspension cultures with known behavior and to eliminate the need to continually initiate fresh culutres. Suspension cultures can be cryopreserved easily and will form new suspensions suitable as a source of protoplasts within 1–3 months of recovery from the freezer. Take samples of the established suspension culture every 4–6 weeks for cryopreservation (DiMaio and Shillito 1989). In this way, you will have a sample of the culture at the optimum stage in its development when it is subsequently needed to initiate new cultures. Callus can also be cryopreserved, with varying success, depending on its morphology. Good friable cultures cryopreserve well, whereas type II cultures are difficult to cryopreserve.

ISOLATION OF PROTOPLASTS

Incubate 1–1.5 ml PCV of the suspension culture cells in 10–15 ml of a mixture consisting of 4% w/v RS cellulase with 1% w/v Rhozyme in KMC (8.65 g/liter KCl, 16.47 g/liter $MgCl_2 \cdot 6H_2O$ and 12.5 g/liter $CaCl_2 \cdot 2H_2O$, 5 g/liter MES pH 5.6) salt solution. Two percent w/v macerase can also be added if desired and may increase the number of protoplasts released. Carry out digestion on a rocking table for a period of 3–4 hours. Monitor the release process under an inverted microscope. Young cultures will generally take longer to release protoplasts.

Collect the released protoplasts as follows: Filter the preparation through a 100-μm mesh sieve followed by a 50-μm mesh sieve. Wash the protoplasts through the sieves with a volume of KMC salt solution equal to the original volume of enzyme solution. Place 10 ml of the protoplast preparation in each of several plastic disposable centrifuge tubes and layer 1.5–2 ml of 0.6 M sucrose solution [buffered to pH 5.6 with 0.1% w/v morpholino-ethane sulfonic acid (MES) and KOH] underneath. Centrifuge at 60–100*g* for 10 minutes; collect the protoplasts banding at the interface and place them in a fresh tube.

Resuspend the protoplast preparation in 10 ml of fresh 13/14-strength KMC salt solution and centrifuge for 5 minutes at 60–100*g*. Remove the supernatant and resuspend the protoplasts gently in the drop of medium remaining, and then gradually add 10 ml of a 6/7-strength KMC solution. Remove an aliquot for counting and sediment the remaining protoplasts again by centrifugation. Resuspend the protoplasts at 2×10^7 per ml in KM culture medium (modified Kao and Michayluk 6p medium, containing double the concentration of organic acids, and with vitamins A and D, and ascorbic acid omitted) or, for transformation via electroporation or PEG treatment (Rhodes et al. 1988b; Fromm, this volume; Shillito et al. 1985; Kramer et al. 1992), in mannitol solution having the same osmolality as the KM medium and containing 6 mM $MgCl_2$, or any other suitable medium.

CULTURE OF MAIZE PROTOPLASTS

The protoplasts are cultured on filters placed on nurse cultures.

Number Durapore filters (Millipore catalog number GVWP04700, 0.22-μm pore size, 47-mm diameter) on the margin with a lead pencil on the upper surface as they are removed from the container. This is to insure that the hydrophilic side of the filter is facing up on the nurse cultures and that only one filter is placed on each. Autoclave the filters for 20 minutes. Soak the filters in liquid KM medium to thoroughly wet them.

Prepare nurse cultures from an embryogenic suspension culture. Increase the osmolality of N6 medium to ~530–540 mOs/kgH_2O by adding 6 % w/v glucose. Re-adjust the pH to 6.0 and filter-sterilize the medium through a 0.2-μm filter. Add 0.5 mg/liter 2,4-D and 100 mg/liter *O*-acetyl-salicylic acid (10 mg/ml in 10% v/v DMSO). Add the medium to sterile SeaPlaque agarose give a final agarose concentration of 1.2% w/v. Heat the medium to 60°C to melt the agarose, cool the medium to 44°C in a water bath, and add 1 ml PCV of the embryogenic suspension culture per 10 ml of medium. Distribute 5-ml aliquots into 60-mm-diameter petri dishes. When the medium has solidified, place the moistened sterile Durapore filters on the surface, numbered side up.

Dilute the protoplast preparation to 0.5–2 x 10^6 protoplasts/ml with cooled (44°C) KM medium containing 1.2% w/v SeaPlaque agarose, 0.5 mg/liter 2,4-D, and 100 mg/liter *O*-acetyl-salicylic acid. Pipette 0.5-ml aliquots of this preparation slowly onto the surface of each of the filters.

Place the cultures in the dark for 1–2 weeks (younger cultures may benefit from the longer time). Then transfer the filters to fresh N6 medium with 3 % w/v sucrose and 2 mg/l 2,4-D, solidified with Gelrite. After 1 more week, remove the agarose film containing the developing colonies from the filters and place them on fresh N6 maintenance medium.

REGENERATION OF PLANTS

As suspension or callus cultures age, they will become more difficult to regenerate. It is, therefore, important that all stages from initiation of the callus through to regeneration of plants be carried out with as little delay as possible. Extra stress due to delayed subculture or environmental factors may also reduce the capability of cells to subsequently regenerate and produce fertile plants.

Protoplast-derived callus may need to be subcultured and selected on N6 maintenance medium for 4–6 weeks to obtain a friable, non-mucilaginous morphology suitable for initiating regeneration.

For regeneration, incubate cultures in the light (10–50 μE m^{-2} sec^{-1}), and the plants that arise in 100–200 μE m^{-2} sec^{-1} light (12–16 hour days from an equal number of Gro-lux and cool white fluorescent lamps).

FROM CALLUS

Place callus (0.25–0.5 g/90-cm-diameter petri dish) on N6 medium without hormones, or MS medium containing 0.25 mg/liter 2,4-D and 1, 5, or 10/mg/liter kinetin, or 1, 5, or 10 mg/liter 6-benzylaminopurine to initiate regeneration.

Place the cultures in the light. Subculture every 10–14 days to fresh medium. Transfer the callus to fresh MS medium without hormones after 2 or 4 weeks. Plantlets take 4–8 weeks to appear. Callus derived from protoplasts or from older suspension cultures generally requires higher cytokinin levels to regenerate than freshly initiated callus.

FROM SUSPENSION

For suspension cultures, first acclimatize the cells to growing on solid medium. Plate the suspension culture cells on N6 maintenance medium and subculture every 10–14 days, selecting for the friable embryogenic callus morphology. After 4–8 weeks the callus will be ready to initiate regeneration.

PLANTLETS

Once the plantlets are at least 2 cm tall, transfer them to half-strength Murashige and Skoog (1962) (MS) salts, full-strength MS Vitamins, and 3% sucrose in Magenta GA-7 (Magenta Corp., Chicago IL) containers or other suitable culture vessels. Roots form in 2–4 weeks. Once the roots look to be formed well enough to support growth, transfer the plantlets to soil in peat pots, under a light shading (for example, two layers of cheesecloth) for the first 4–7 days. It is often helpful to maintain high humidity around the plants for the first week, either on a mist bed or by use of a humidifier.

REFERENCES

Cai Q, Kuo C, Qian Y, Jiong R, Zhou Y (1988) Somatic embryogenesis and plant regeneration from protoplast of maize (*Zea mays* L.). In Puite KJ et al (eds) Progress in Plant Protoplast Research, Proc 7th Int Protoplast Symp, Wageningen, The Netherlands, 1987, Kluwer, Dordrecht, p 120

Chu CC, Wang CC, Sun CS, Hus C, Yin KC, Chu CY (1975) Establishment of an efficient medium for anther culture of rice through comparative experiments on the nitrogen sources. Scientia Sinica 18: 659–668

DiMaio JJ, Shillito RD (1989) Cryopreservation technology for plant cell cultures. 1989. J Plant Tiss Cult Meth 12: 163–169

Kamo KK, Chang KL, Lymm ME, Hodges TK (1987) Embryogenic callus formation from maize protoplasts. Planta 172: 245–251

Kramer C, Shillito RD, DiMaio JJ, Carswell GK (1992) Selection of transformed maize on phosphinothricin and the pH indicator chlorophenol red. Planta 1993 in press

Morocz S, Donn G, Nemeth J, Dudits D (1990) An improved system to obtain fertile regenerants via maize protoplasts isolated from a highly embryogenic suspension culture. Theor Appl Genet 80: 721–726

Murashige T, Skoog F (1962) A revised medium for rapid growth and bioassays with tobacco tissue cultures. Physiol Plant 15: 473–497

Prioli LM, Sondahl MR (1989) Plant regeneration and recovery of fertile plants from protoplasts of maize (*Zea mavs* L.). Bio/Technology 7: 589–595

Rhodes CA, Lowe KS, Ruby KL (1988a) Plant regeneration from protoplasts isolated from embryogenic maize cell cultures. Bio/Technology 6: 56–60

Rhodes CA, Pierce DA, Mettler IJ, Mascarenhas D, Detmer JJ (1988b) Genetically transformed maize plants from protoplasts. Science 240: 204–207

Schenk RU, Hildebrandt AC (1972) Medium and techniques for induction and growth of monocotyledonous and dicotyledonous plant cell cultures. Can J Bot 50: 199–204

Shillito RD, Carswell GK, Johnson CM, DiMaio JJ, Harms CT (1989) Regeneration of fertile plants from protoplasts of elite inbred maize. Bio/Technology 7: 581–587

Shillito RD, Saul MW, Paszkowski J, Müller M, Potrykus I (1985) High efficiency direct gene transfer to plants. Bio/Technology 3: 1099–1103

Suttie J, Dunder E, Hill M, Pace G (1991) Improved media for establishing friable embryogenic cultures of inbred maize genotypes. Poster 905, 3rd International Congress of Plant Molecular Biology, Tucson, AZ, October 1991

Vasil V, Vasil IK (1987) Plant regeneration from friable embryogenic callus and cell suspension cultures of *Zea mays*. J Plant Physiol 124: 399–408

Anther and Microspore Culture

J.F. PETOLINO AND A. D. GENOVESI

The maize pollen grain represents the beginning of a short-lived gametophytic phase during which the two sperm are delivered to the embryo sac prior to fertilization. Although this stage of the life cycle normally consists of only a few cell divisions, under certain experimental conditions, gametophytes can be induced to undergo an altered development, leading to the production of haploid embryolike structures and/or callus, without an intervening fertilization. This remarkable process, referred to as androgenesis, is the biological basis for the in vitro techniques known as anther and microspore culture. This chapter describes some protocols associated with producing doubled haploid maize from cultured anthers and microspores.

GENOTYPE

In most in vitro techniques, donor plant genotype is a significant factor. Nowhere is this more evident than in maize androgenesis. Unlike most other systems (such as plant regeneration from immature embryo-derived callus) where some genotypes are more or less responsive than others, most genotypes of maize have been found to be unresponsive in anther and microspore culture. Indeed, after considerable screening, only a few genotypes have been found to undergo androgenesis to any measurable degree. The Chinese have identified a few genotypes, including Ching huang 13, Dan san 91, and Lai pin pai. U.S. and European scientists have also reported responsive genotypes such as Seneca 60, F492 × Rc525, (H99×FR16) × Pa91, and A665 × A634. More recently, special genotypes have been bred to be highly responsive to anther and microspore culture (i.e., 139/39, DH5×DH7). Attempting anther or microspore culture without a responsive genotype can be a frustrating and fruitless endeavor.

DONOR PLANT PHYSIOLOGY

Differences in response between field- and greenhouse-grown plants and between plants sown at various times during the year indicate that donor plant physiology is a critical

The Maize Handbook—M. Freeling, V. Walbot, eds.

factor in anther and microspore culture. Experience has taught that reproducible growth conditions are a prerequisite for obtaining consistent results. Maize can be grown in the greenhouse in 5-gallon pots in standard potting soil with appropriate fertilization (i.e., 14–14–14 Osmocote). Supplemental lighting in the form of metal halide and/or high-pressure sodium lamps is recommended, especially during the winter months. Healthy maize can be produced using a 13–16-hour photoperiod with day/night temperatures of about 30°/20°C. Although few systematic studies have been performed to address this issue, it appears that donor plant growth conditions should be near optimal to maximize anther and microspore culture response.

STAGE OF DEVELOPMENT

Tassels with anthers containing mid- to late-uninucleate microspores are most receptive to androgenic induction. Development continues during low-temperature pretreatment with the progression from mid/late-uninucleate to late-uninucleate/early binucleate. Thus, at the time of culture initiation, most microspores are somewhere between the late-uninucleate to early binucleate stage of development. Precise staging is accomplished using mithramycin staining followed by fluorescence DL microscopy. If fluorescence optics are not available, trypan blue can be used to determine developmental stage.

PRETREATMENT

Excised tassels containing uninucleate microspores are wrapped in paper towels or placed in plastic bags to prevent desiccation and incubated for 14–17 days at 8–10°C. It is believed that this low temperature pretreatment disrupts normal pollen development, stimulating androgenesis. Following pretreatment, tassels are surface sterilized by soaking for several minutes in a sodium hypochorite solution. An alternative pretreatment involves excising anthers from pretreated tassels after 1–4 days and floating them on a 0.3 M mannitol solution with the addition of 50 mg/liter ascorbic acid for 2–3 weeks at 10°C. The addition of 0.025% colchicine to the mannitol solution at this stage is an effective means of inducing chromosome doubling.

MICROSPORE ISOLATION

A variety of procedures have been used to isolate microspores from anthers or tassel segments. Excised anthers can be pressed against a 113-μm stainless-steel sieve using a glass rod so that the microspores pass through while the majority of the anther wall debris is trapped. Similarly, anthers can be chopped with a razor blade prior to being filtered through 130-μm mesh. Also, a specialized attachment on a commercial blender has been used effectively on anthers as well as tassel segments to liberate microspores from surrounding somatic tissue. The most important consideration for microspore isolation is to minimize stress. Refrigerating the apparatus and/or medium is advantageous presumably as a result of diminished chemical activities during this period.

Following maceration and separation from anther wall and/or tassel debris, microspores can be further purified by passing through a smaller sieve (i.e., 45–50 μm) or by centrifugation. The isolated microspores can then be cultured directly or be subjected to further enrichment (to remove nonviable microspores) via discontinuous density gradient centrifugation. For example, culture medium layered over 20% Percoll has been used to isolate viable microspores at the medium–Percoll interface. Similarly, microspores at different developmental stages have been separated using 20%, 30%, 40%, and 50% Percoll. Late-uninucleate to early binucleate microspores are captured at the 20–30% interface after centrifugation at 225*g* for 3 minutes.

CULTURE MEDIUM

The various media used for maize anther and microspore culture are similar to those used for other cereal tissue culture procedures. N6 and YP formulations, and various modifications thereof, have been most commonly used as basal medium for maize androgenesis (Pescitelli et al. 1989; Genovesi 1990). Some common features of anther and microspore culture media, however, can be identified. Androgenesis appears to require a rather high concentration of sucrose (6–12%). In addition, activated charcoal (0.5%) has been used as a component of most anther and microspore culture media. Other commonly found constituents include triiodobenzoic acid and proline.

In addition to culture medium, specific media have been developed for use during the isolation of microspores from anthers. For example, an isolation medium consisting of 6% sucrose, 50 mg/liter ascorbic acid, 400 mg/liter proline, 0.05 mg/liter biotin, and 10 mg/liter nicotinic acid has been found to maintain microspore viability during maceration.

CULTURE CONDITIONS

Anthers are typically cultured in 10–20 ml of medium in either 60×20 mm or 100×15 mm petri dishes (20–60 anthers per dish). The anthers are either floated on the surface of liquid medium or set on the surface of agar- or agarose-solidified medium. Postplating temperature treatments are occasionally used (i.e., 2 weeks at 30°C) before incubation at 25–28°C. Cultures are most typically grown in the dark, although a 16:8-hour photoperiod has also been used.

Isolated microspores are typically cultured in liquid medium at densities ranging from 7,500 to 250,000 per ml. In these stationary cultures, developing embyrolike structures and/or callus are submerged in the medium. An alternative approach involves culturing isolated microspores on the surface of a 10-μm nylon "raft." In this way, the medium is allowed to pass through the nylon while the developing embryolike structures and/or callus remain afloat. Additionally, this technique allows the tissue to be transferred to other media easily.

PLANT REGENERATION

After approximately 3–5 weeks, embryolike structures (0.5–2.0 mm) can be transferred directly to germination medium. This can be achieved using hormone-free compositions of standard media (i.e., N6, YP, MS). Alternatively, embryolike structures and callus can be transferred to a callus-induction medium for clonal propagation. This is particularly desirable if chromosome doubling was not performed earlier, because callus can be treated with colchicine (0.01–1.0%) prior to regeneration of plants. Media for the induction of embryogenic maize callus and for the regeneration of plants are well established (Phillips et al. 1988).

POLLINATION

Anther- and microspore-derived plants often exhibit developmental abnormalities that make successful self-pollination difficult. Lack of pollen production, terminal ear formation, asynchronous pollen shed/silk emergence, and premature senescence are commonly observed. If clonal propagation of callus is performed, it is desirable to regenerate several plants from each culture, thus allowing sib-pollinations to be made. Seed yield on primary regenerants is generally low and of questionable quality. Therefore, seed from regenerants oftentimes needs to be germinated under controlled conditions for successful establishment of doubled haploid lines.

REFERENCES

Genovesi AD (1990) Maize (*Zea mays* L.): *in vitro* production of haploids. In Bajaj YPS (ed) Biotechnology in Agriculture and Forestry, Vol 12, Springer-Verlag, Berlin, pp 176–203

Pescitelli SM, Mitchell JC, Jones AM, Pareddy DR, Petolino JF (1989) High frequency androgenesis from isolated microspores of maize. Plant Cell Rep 7: 441–444

Phillips RL, Sommers DA, Hibberd KA (1988) Cell/tissue culture and *in vitro* manipulation. In Sprague GF, Dudley JW (eds) Corn and Corn Improvement, American Society of Agronomy, Madison, WI pp 345–387

In vitro Culture of Maize Kernels

BURLE G. GENGENBACH AND ROBERT J. JONES

This method provides a means with which to study the development of intact maize kernels under conditions in which nutritional, hormonal, environmental and other factors can be modified and controlled. Kernels may be cultured from 2 days after pollination up to physiological maturity. Final kernel size may vary, but average kernel dry weight generally ranges from 70% to 100% of normal.

MATERIALS

Maize ears 2–4 days after pollination, intact with husks and shank attached
Sterile petri dishes (25 mm × 100 mm)
Aluminum foil (about 30 cm × 30 cm) folded in half two times to give a 15-cm square (Wrap squares in a large piece of foil and sterilize in 100°C oven for >1 hour or autoclave for 35 minutes)
Several types and sizes of forceps
Scalpels (#10- and 20-size blades)
Laminar-flow hood
Sterile modified culture medium

Glutamine	15.0 mM*	Glutamate	1797 mg/liter
KH_2PO_4	3.0 mM	Leucine	1030 mg/liter
$MgCI_2$	1.5 mM	Proline	634 mg/liter
Aspartate	510 mg/liter	$CaCI_2$	1.2 mM
H_3BO_3	0.1 mM	Alanine	486 mg/liter
$MnSO_4$	0.1 mM	Phenylalanine	395 mg/liter
Fe-EDTA	50.0 μM	Serine	354 mg/liter
$ZnS0_4$	10.0 μM	Valine	337 mg/liter
Kl	5.0 μM	Asparagine	306 mg/liter
Niacin	1.2 mg/liter	Glutamine	298 mg/liter
Thiamine-HCl	0.4 mg/liter	Tyrosine	267 mg/liter

The Maize Handbook—M. Freeling, V. Walbot, eds.

Folic acid	44.0 μg/liter	Isoleucine	242 mg/liter
Gentamycin sulfate	20 mg/liter	Tryptophan	208 mg/liter
Sucrose	150 g/liter	Threonine	202 mg/liter
Agar	5.5 g/liter	Glycine	171 mg/liter
$NaMoO_4$	1.0 μM	Histidine	167 mg/liter
$CuS0_4$	0.1 μM	Cysteine	163 mg/liter
$CoCI_2$	0.1 μM	Arginine	153 mg/liter
		Methionine	119 mg/liter
		Lysine	90 mg/liter

*The glutamine nitrogen source may be replaced by an amino acid mixture.
†Dissolve Na-EDTA in hot water; add FeSO4.$7H_2O$, stir, and cool.

METHOD

1. Grow plants under normal field or greenhouse conditions and make controlled pollinations.
2. Two to 4 days after pollination, remove the intact ear from the plant, taking care not to break off the ear shank. The ear shank serves as a convenient "handle" for holding the ear in subsequent steps in the lab.
3. Clean the laminar-flow hood work area with 95% ethanol. Place a sterile aluminum foil square in the hood and unfold one section. This provides a disposable sterile working area in the hood.
4. Before putting ear in hood, cut off the tip of the husks to remove protruding silks and remove the outer one or two layers of husks by hand.
5. Hold ear by the shank, and with the tip of ear down, use a large sterile forceps to remove the husks layer by layer. Flame sterilize the forceps after each husk is removed.
6. When all husks are removed, hold the ear horizontally and remove the silks by grasping them with the forceps.
7. Position tip of ear on unused area of foil and cut off the small part of the tip. The amount to be removed will vary depending on the age of the ear and the extent of the pollination.
8. Discard all husks and silks.
9. Using sterile forceps, completely unfold the aluminum foil and place ear only, not the shank, on a sterile section of foil.
10. Hold shank with one hand to steady the ear. Use a scalpel with #20 blade to cut about halfway into cob between two rows of kernels. Start about 1 or 2 cm from base (large end) of cob and cut toward the tip.
11. Rotate ear and make similar cuts between the rest of the paired rows of kernels.

The uncut base of the cob will help hold the cut strips of cob together until finished.

12. Cut off the base of the cob and discard it and the shank. Depending on the size of the ear, each ear should give five to seven wedge-shaped strips of double rows of kernels.
13. Lay each strip on an unused section of the sterile foil and cut the strips into blocks of 6 kernels (2 wide by 3 long).
14. Hold each block of 6 kernels with forceps and use a scalpel with #10 blade to remove 5 of the 6 kernels from the top of the cob. Remove the two from each end and then one of the middle pair. This block of cob tissue, which originally contained 6 kernels, and the single remaining kernel become the experimental unit.
15. Place the block into sterile agar-solidified medium (50 ml per 25 × 100 mm dish) such that the cob region is below and the kernels are above the medium surface. Five or 6 blocks/dish are convenient but a dish can accommodate more if needed.
16. Place top on dish, seal with masking tape or Parafilm.
17. Standard growth conditions for the kernels usually are 25° to 28°C in the dark. Kernels cultured under these conditions usually attain maximum weight in about 40 days.

Notes

1. Method works for most genotypes—some better than others. W64A and A619 inbreds and their hybrid are usually very good (Gengenbach 1977; 1984).
2. When plants are grown in cool or dry conditions, there usually is little microbial contamination of the immature ear. The ear and inner husks usually are sterile. Ear contamination (usually fungal) may be encountered especially in southern humid growing areas. In this case, surface sterilization of the inner husks with 95% ethanol may help. The best solution is to avoid contamination of the plant material; it may be necessary to grow plants in a greenhouse or growth chamber.
3. A reduced nitrogen source is required for kernel development and growth. Glutamine is a good source of reduced nitrogen for most experiments; however, a complete mixture of 20 amino acids usually results in slightly better kernel growth.
4. Although not tested by the authors, a synthetic gelling agent such as GELRITE might be advantageous in this culture system. Other researchers (Singletary and Below 1989) have used a liquid medium in which kernels are supported by a coil of aluminum screen overlaid with a 7-cm filter paper disk and inserted into 250-ml flasks.
5. Optimal kernel growth was found when kernels were removed from the cob segment in a ratio of five removed for one remaining. Many other configurations may be tried.
6. This method is amenable to many modifications (Culley et al. 1984) depending on the nature of the experiment and the ingenuity of the investigator.

7. For some purposes, pollination may be done in vitro as well (Gengenbach 1977). In this case the silks are left on the cob segments and none of the ovaries are removed. After placing the cob segments into the medium, the silks can be arranged over the edge of the petri dish so that they extend out under the lid. Pollen can be applied to the exposed silks and after 24–48 hours the silks can be removed and the dish can be sealed for kernel growth.

REFERENCES

Culley DE, Gengenbach BG, Smith JA, Rubenstein I, Connelly JA, Park WD (1984) Endosperm protein synthesis and L-[^{35}-S]methionine incorporation in maize kernels cultured *In vitro*. Plant Physiol 74: 389–394

Gengenbach BG (1977) Development of maize caryopses resulting from in-vitro pollination. Planta 134: 91–93

Gengenbach BG (1984) In vitro pollination, fertilization and development of maize kernels. In Vasil IK (ed) Cell Culture and Somatic Cell Genetics of Plants, Vol. 1, Academic Press, New York, pp 276–282

Singletary GW, Below FE (1989) Growth and composition of maize kernels cultured in vitro with varying supplies of carbon and nitrogen. Plant Physiol 89: 341–346

In vitro Ear Culture System

RICHARD V. KOWLES

An in vitro ear culture system is described in which the entire ear can be grown in a synthetic liquid medium under carefully controlled conditions. When sterile conditions can be maintained, the kernels will develop to maturity. The system allows for kernel uptake trials, pulse/chase experiments, and other chemical applications that might not be possible in field-grown plants.

MATERIALS

The medium, modified from Culley et al (1984), contains the following components:

Micro Stock Solution

	mg/liter
H_3BO_3	620
Kl	83
$Na_2MoO_4 \cdot 2H_2O$	25
$CoC1_2 \cdot 6H_2O$	2.5
$MnSO_4 \cdot H_2O$	1,690
$ZnSO_4 \cdot 7H_2O$	287
$CuSO_4 \cdot 5H_2O$	2.5

Iron Stock Solution

	mg/liter
Na_2EDTA	7,448
$FeSO_4 \cdot 7H_2O$	5,568

Dissolve the Na_2EDTA in 200 ml boiling distilled water. Also dissolve the $FeSO_4$ in

The Maize Handbook—M. Freeling, V. Walbot, eds.

200 ml boiling distilled water. While both are hot, slowly add the EDTA solution to the $FeSO_4$ solution. Mix the resultant solution on a magnetic stirrer until it comes to room temperature. Then bring the final volume up to 1 liter.

Other Stock Solutions

	mg/100 ml
Thiamine-HCl stock	8
Niacin stock	120
Folic acid stock	88

The final medium is made as follows:

0.51 g	KH_2PO_4
0.31 g	$MgC1_2 \cdot 6H_2O$
0.18 g	$CaCI_2 \cdot 2H_2O$

Add these ingredients to about 750 ml distilled water one at a time in this order; that is, thoroughly dissolve each before adding the next. An optional step includes the addition of 30 μl PBTC (phosphono-butane-1,2,4-tricarboxylic acid) to aid in the dissolving process. Then add the other stock solutions as follows:

10 ml	Micro stock solution
5 ml	Fe EDTA solution
5 ml	Thiamine-HCl
1 ml	Niacin
5 ml	Folic acid
2.2 g	Glutamine
15 g	Sucrose

Bring to 1 liter with distilled water and pH to 5.8. Autoclave at least 30 minutes.

METHOD

The ear is taken at about 8–10 days after pollination (dap) by severing the plant well above and below the ear. Quickly immerse the lower end into a large container of water. In a transfer hood, cut the stalk several inches below the ear with a sterile scalpel or knife. Submerge the lower third of the ear in 2.5% sodium hypochlorite (Chlorox) for 10 minutes. The severed upper end of the stalk is sealed with sterile lanolin. After surface sterilization, place the lower third of the ear into sterile distilled water to rinse for 10 minutes. Place the ear tightly into a sterile 250-ml side-arm Erlenmeyer flask containing the sterile medium. The juncture between the ear and the flask is sealed with sterile

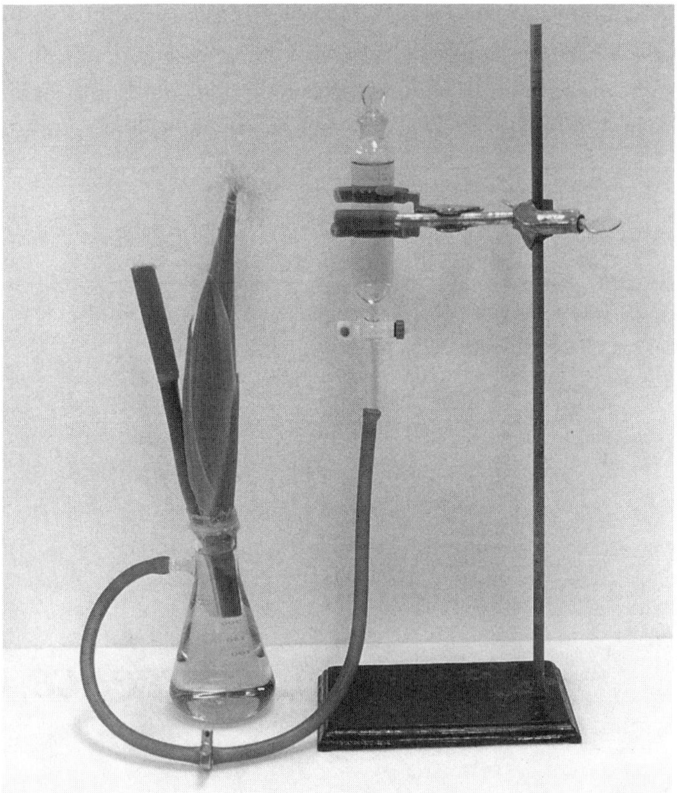

FIGURE 128.1 The in vitro ear culture system.

lanolin. The side-arm is connected by rubber tubing to a sterile buret or separatory funnel. This unit serves as a reservoir that holds additional medium and can be controlled with its stopcock and/or a clamp on the rubber tubing. The ear will initially take up large amounts of medium presumably as a result of transpiration activity. One can add more medium daily. The entire system is supported with an L-stand and clamps (Figure 128.1).

Notes

1. This system has been used successfully with the inbreds A188, W64a, and A619.
2. DNA content per nucleus and mitotic activity in the endosperm tissue are comparable to that of field- or greenhouse-grown materials.
3. Contamination over short periods does not seem to be a problem, but it can be difficult to eliminate over long periods.

4. When contamination can be minimized, A188 kernels grown in this way have had a 65% germination rate.
5. The system is sufficiently compact to enable placement into small growth chambers in which temperature and lighting can be completely controlled.

REFERENCES

Culley DE, Gegenbach BG, Smith JA, Rubenstein I, Connelly JA, Park WD. (1984) Endosperm protein synthesis and L-[^{35}Sl methionine incorporation in maize kernels cultured *in vitro*. Plant Physiol 74: 389–394

129

Maize Inflorescence Culture

R. I. Greyson

The objective of this protocol is to culture immature tassels or ears to observe and manipulate normal development. This is in contrast to other maize or grass infloresence culture protocols designed for somatic embryogenesis (Sharma et al. 1984; Vasil 1985; Rhodes et al. 1986; Pareddy and Petolino 1990) or to study the factors responsible for postfertilization grain filling (Donovan and Lee 1977, 1978; Singh and Jenner 1983; Nicholls 1986; Armstrong et al. 1987; Trione and Stockwell 1989).

The following protocol is applicable to the culture of tassels and ears of the sweet corn var. Seneca 60. It is a summary taken from Polowick and Greyson (1982), Pareddy and Greyson (1987, 1985), and Bommineni and Greyson (1987, 1990). The culture of inflorescences from other cultivars may require refinements of the medium.

GROWTH CONDITIONS FOR THE DONOR PLANTS

1. Plant one kernel per 20-cm plastic pot containing a standard greenhouse mixture of loam/sand/peat (8/5/4).

The Maize Handbook—M. Freeling, V. Walbot, eds.

2. Germinate and grow seedlings in greenhouse conditions with supplemental lighting: 18 hours/6 hours light/dark regime: temperature regime 24°C/20°C ± 2°C (day/night).
3. With variation, at 25–30 days after germination (DAG), Se60 seedlings yield tassels (0.5–1.0 cm). Ear primordia (0.5–1.0 cm) are available 45–50 DAG.

SURFACE STERILIZATION

4. For tassels, 5–8 cm stem sections containing tassel primordia surrounded by 2–4 leaf sheaths are surface sterilized in 20% Javex bleach for 15 minutes and rinsed 3–5 times in sterile H_2O. For ears, the uppermost axillary bud is sterilized in the same fashion.
5. Dissections are performed under sterile conditions in a laminar flow hood with a dissecting microscope, tweezers, and scalpels.

BASAL MEDIUM

6. Major and minor inorganic salts can be prepared from stock solutions (Murashige and Skoog 1962). We assume that commercially available freeze-dried formulations of MS salts would serve equally well.
7. Vitamins and glycine (White 1943).
8. I-inositol (100 mg/liter).
9. $FeSO_4 \cdot 7H_2O$ (27.8 mg/liter) and $Na_2 \cdot$EDTA (37.3 mg/liter).
10. Sucrose (1–6%).
11. Various plant growth substance additions, depending on the organ cultured and the objectives of the experiment. See 16 and 17 below.
12. Initial pH 5.8.
13. Media are dispensed 40 ml per 125 ml Erlenmeyer flask. Flask openings are covered with two layers of aluminum foil. Flasks are autoclaved.

CULTURE CONDITIONS

14. Tassels and ears are cultured individually.
15. Cultures are maintained for approximately 20 days in fluorescent-lighted incubators (18 hours light/6 hours dark; 28°C ± 2°C).

SPECIAL CONDITIONS FOR TASSEL CULTURE

16. Sucrose is adjusted to 0.3 M and kinetin is added at 10^{-7} M.

SPECIAL CONDITIONS FOR EAR CULTURE

17. Sucrose is adjusted to 0.175 M or 6% and kinetin is added at 10^{-7} M.
18. Ear primordia, depending on the initial size, respond to additions of cytokinin (kinetin) and gibberellin (gibberellic acid). (Refer to Bommineni and Greyson 1987, 1990.)

REFERENCES

Armstrong TA, Soong T-S, Hinchee MAW (1987) Culture of detached spikes and the early development of the fourth floret caryopsis in wheat. J Plant Physiol 131: 305–314

Bommineni VR, Greyson RI (1987) In vitro culture of ear shoots of *Zea mays* and the effect of kinetin on sex expression. Am J Bot 74: 883–890

Bommineni VR, Greyson RI (1990) Regulation of flower development in cultured ears of maize (*Zea mays* L.). Sex Plant Reprod 3: 109–115

Donovan GR, Lee JW (1977) The growth of detached wheat heads in liquid culture. Plant Sci Lett 9: 107–113

Donovan GR, Lee JW (1978) Effect of the nitrogen source on grain development in detached wheat heads in liquid culture. Aust J Plant Physiol 5: 81–87

Murashige T, Skoog F (1962) A revised medium for rapid growth and bioassays with tobacco tissue. Physiologia Plant 15: 473–497

Nicholls PB (1986) Induction of sensitivity to gibberellic acid in developing wheat caryopses: Effect of sugars in the culture medium. Aust J Plant Physiol 13: 795–801

Pareddy DR, Greyson RI (1985) In vitro culture of immature tassels of an inbred field variety of *Zea mays*, cv. Oh43. Plant Cell, Tissue Organ Culture 5: 119–128

Pareddy DR, Greyson RI, Walden DB (1987) Fertilization and seed production with pollen from in vitro cultured maize tassels. Planta 170: 140–143

Pareddy DR, Petolino JF (1990) Somatic embryogenesis and plant regeneration from immature inflorescences of several elite inbreds of maize. Plant Science 67: 211–219

Polowick PL, Greyson RI (1982) Anther development, meiosis and pollen formation in *Zea* tassels cultured in defined liquid medium. Plant Sci Lett 26: 139–145

Rhodes CA, Green CE, Phillips RL (1986) Factors affecting tissue culture initiation from maize tassels. Plant Sci 46: 225–232

Sharma HC, Gill BS, Sears RG (1984) Inflorescence culture of wheat-*Agropyron* hybrids: Callus induction, plant regeneration, and potential in overcoming sterility barriers. Plant Cell Tissue Organ Cult 3: 247–255

Singh BK, Jenner CF (1983) Culture of detached ears of wheat in liquid culture: modification and extension of the method. Aust J Plant Physiol 10: 227–236

Trione EJ, Stockwell VO (1989) Development of detached wheat spikelets in culture. Plant Cell Tissue Organ Cult 17: 161–170

Vasil IK (1985) Somatic embryogenesis and its consequences in the Gramineae. In Henke RR, Hughes KW, Constantin MJ, Hollaender A (eds) Tissue Culture in Forestry and Agriculture, Plenum Press, New York, pp 31–47

White PR (1943) A Handbook of Plant Tissue Culture. Jacques Cattell Press, Lancaster, PA, 277 pp

Shoot Meristem Culture

ERIN E. IRISH

This protocol provides techniques for the isolation, staging, and culture of the shoot apical meristem of maize (adapted from Irish and Nelson 1988, 1991). One of the interesting characteristics of the maize plant is the tight regulation of the number of nodes produced by the shoot apical meristem before it converts to a terminal reproductive structure, the tassel. Shoot apex (shoot apical meristem plus one or two of the youngest leaf primordia) culture can be used in developmental analyses to delay the conversion to a tassel by increasing the number of nodes the meristem produces. Isolation and culture of the meristem also allow surgical experiments to be performed on the exposed meristem.

MATERIALS

- Single-edged razor blades
- 70% ethanol
- Culture medium:
  - Murashige and Skoog salts
  - 1.0 ml/liter 1000× vitamins
    - *1,000× vitamin recipe*
    - 0.5 g/liter thiamine HCl
    - 0.1 g/liter pyridoxine HCl
    - 0.1 g/liter nicotinic acid
    - 3.0 g/liter glycine
  - 100 mg/liter myo-inositol
  - 30 g/liter sucrose, pH 5.8 w/1 N NaOH
  - 3.0 g/liter agar
- Fine-tipped forceps

The Maize Handbook—M. Freeling, V. Walbot, eds.

Dissecting needle* plus needle holder and/or micro dissecting knife (available from Roboz Surgical Instrument Co. Inc., 1000 Connecticut Avenue, N.W., Washington, D. C. 20036, 202-393-1234)
Alcohol lamp or bunsen burner
Counter if staging the meristem
Cell culture plates (24 wells, fill each well with approximately 2 ml medium)
Parafilm cut into strips for wrapping plates
Long forceps for transferring plantlets from plates to tubes
2.5 cm × 10-cm culture tubes + kaputs (caps from Bellco Glass)

Equipment

Stereomicroscope and light source
Laminar-flow hood
Plant growth chamber

METHOD

1. Remove plant from soil. Rinse off the shoot and roots in running water. Trim off most of the roots using a razor blade.
2. Remove the outermost, expanded leaf by slipping a blunt dissecting tool under the margin of the leaf or by catching the margin with the tip of your index finger near the level of the node of attachment (at the base of the shoot, Figure 130.1a and b, "B") and peeling away the leaf by rotating the shoot. Remove subsequent expanded leaves in a similar manner (Figure 130.1b, "C"). A counter may be used to keep track of the number of leaves removed in determining the stage of the meristem (total number of leaves initiated).
3. Trim the shoot to a length of 2–3 cm to facilitate removal of the thin, unexpanded leaves. Dip the shoot in 70% ethanol to surface sterilize.
4. Viewing through a stereomicroscope, remove leaf primordia with a sterile dissecting knife or needle. Sterilize the dissecting tool before removing each successive leaf by dipping it in 70% ethanol and flaming to evaporate the alcohol. The same technique of unrolling one leaf at a time that was used with the expanded leaves is sufficient to remove the primordia.

*Dissecting needles sharp enough for this procedure can be made by electrically etching tungsten wire. Cut approximately 3-cm lengths of 0.02- or 0.04-inch tungsten wire. Insert a length into a needle holder that has been insulated with a sleeve of Tygon tubing and wired to a rheostat. Extend an insulated wire from the other pole of the rheostat to a loop of exposed copper wire that is submerged into a beaker containing 10% NaOH. Wear safety glasses. Completing the circuit by dipping the needle into the solution will result in etching the submerged portions of the tungsten wire. Low amounts of current will slowly etch the needle; higher current will speed up the process but usually blows the fuse of the rheostat. The sharpened needle can be used in a needle holder.

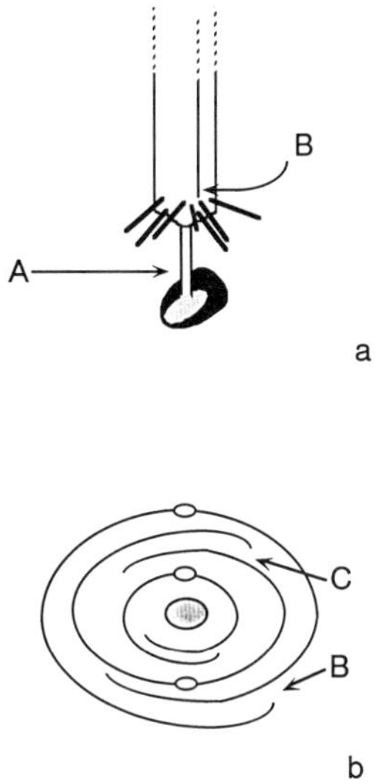

Figure 130.1 a. Side view of the base of a maize shoot with trimmed adventitious roots and seminal root removed. Cut at "A" to provide additional tissue for handling shoot during dissection. Insert blunt dissecting tool at "B" to remove leaf. b. Diagram of transverse section indicating relative orientation of outer leaf margins on successive leaves. After outermost leaf is removed (by insertion at "B") remove next leaf by inserting at "C," and so on.

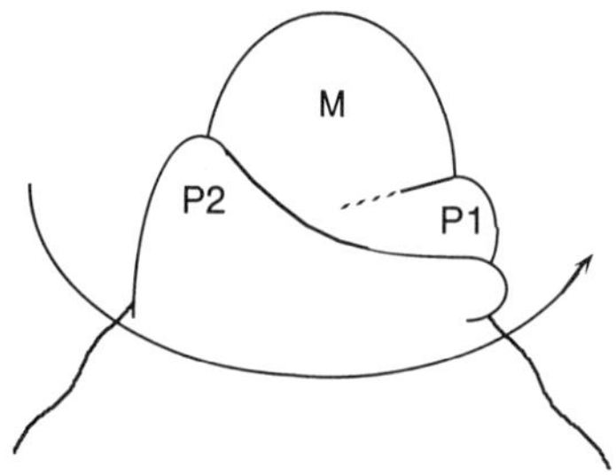

Figure 130.2 Diagram of side view of shoot apex. M = meristem, P1 = newest leaf primordium, P2 = next older primordium. Arrow indicates cut to remove shoot apex from the remaining plant.

5. Once the meristem is exposed, slice through the shoot apex with a dissecting needle, severing the meristem plus 1–2 leaf primordia from the rest of the plant (Figure 130.2).

6. Place the isolated shoot apex in approximately 2 ml of medium in a well of a 24-well culture plate. Wrap the plates with a strip of Parafilm to maintain humidity and reduce chances of contamination. Place in a plant growth chamber with at least 3,000 lux illumination.

7. When the apex has grown to a 1–2-cm-high plantlet, using long thin forceps transfer the plantlet to a culture tube containing 10 ml culture medium. Allow the plant to continue to grow in the growth chamber until adventitious roots are at least 2 cm. By this point the leaves are well developed enough photosynthetically to support the continued growth of the plant.

8. Transfer the plantlet to a small pot containing potting soil and place in a mist cham-

ber. Gradually transfer it to less humid conditions. Once sufficient cuticle is covering the shoot, allow the plant to grow in the greenhouse.

Notes:

Hormones are unnecessary for continued growth and function of cultured meristems of W23 and B73. Other genetic backgrounds may require the addition of hormones. As a starting point, you might try 0.04 mg/liter 2,4-D as an auxin or 0.10 μM kinetin for a cytokinin. These two hormones offer the advantage that they are not significantly altered by autoclaving.

1.* Multiple shoots can be processed at once if the the rinsed shoots are kept in a closed, humid container. The shoots can be stored this way for at least 2–3 hours at room temperature without affecting viability. For convenience, the shoots may be trimmed: leave enough of the shoot so that at least 2 cm of leaf tissue extends above the meristem.

2a. In order to manipulate the shoots easily during the dissection, leave attached enough tissue to be able to grip the shoot easily between thumb and forefinger. This can be just the base of the shoot for older plants. For seedlings, the stem is too short to be gripped easily, so the hypocotyl should be left attached also (Figure 130.1a, "A"). In order to have a good hypocotyl "handle," the seeds should be planted at least 1.5 cm deep.

2b. The leaves should pull off cleanly: If they do not, use a dissecting knife to remove whatever is left of the leaf at the point of attachment. This is critical: leaving leaf bases attached to the stem will completely obscure the meristem by the time you get there.

3. This is not necessary if the apex sticks to the needle (step 5), but dipping just in case is a good idea.

5a. This action generally results in the shoot apex sticking to the needle, which can be used to place the apex in the culture medium. If it does not stick, you will chase the apex with the needle around the shoot tip until it gets caught on the needle (hence, step 3).

5b. Explants of the meristem without any primordia attached from W23 or B73 do not live in culture, probably because a large percentage of the meristem is damaged during the removal of the newest primordia. This is probably also true in other lines of maize, but has not been tested.

7. The time required for a cultured apex to form a 2-cm plantlet and the time that the plantlet takes to form adventitious roots are both variable, from 2–3 weeks to that many months. There seems to be no difference between the fast and slow plants once they are established. Plantlets that take longer than 3 months to get out of culture usually have short, very leathery leaves and rarely survive to flower.

---

*numbers correspond to steps in Method.

REFERENCES

Irish EE, Nelson TM (1988) Development of maize plants from cultured shoot apices. Planta 175: 9–12

Irish EE, Nelson T (1991) Vegetative to floral conversion occurs in multiple steps in maize tassel development. Development 112: 891–898

131

Establishment and Culture of Maize Endosperm

JACK C. SHANNON

Straus and LaRue (1954) were the first to describe establishing endosperm cultures from sweet corn. Attempts to culture endosperm from normal were unsuccessful until Tabata and Motoyoshi (1965) reported the importance of genotype in the successful culture of endosperm from starchy genotypes. We confirmed this (Shannon and Batey 1973) in that only two inbreds (A636 and R168) out of 18 were capable of continuous active growth on the defined medium described by Straus (1960). Later we (Shannon and Liu 1977) reported that a simplified liquid medium containing the inorganic salts of the Murashige and Skoog (1962) medium plus sucrose, thiamine, and asparagine supported rapid growth of maize endosperm suspension cultures (Table 131.1). This simplified medium containing a gelling agent is also recommended for the initiation and maintenance of callus cultures. Note that the maize endosperm medium is free of any added hormones or growth regulators. The protocol used in my laboratory follows.

MATERIALS

For Callus Establishment

- A transfer hood
- Ear of A636 inbred, 9–11 days postpollination

The Maize Handbook—M. Freeling, V. Walbot, eds.

TABLE 131.1 Stock solutions (100×) and final quantity of components in a simplified medium for growth of maize endosperm[a]

Components	In Stocks (g/liter)	In medium (mg/liter)	Components	In Stocks (g/liter)	In medium (mg/liter)
Stock 1			Stock 4		
NH_4NO_3	165.0	1650.0	$CaCl_2 \cdot 2H_2O$	44.0	440.0
KNO_3	190.0	1900.0	KI	0.083	0.83
			$CoCl_2 \cdot 6H_2O$	0.0025	0.025
Stock 2					
$MgSO_4 \cdot 7H_2O$	37.0	370.0	Stock 5		
$MnSO_4 \cdot H_2O$	1.69	16.9	$FeSO_4 \cdot 7H_2O$	2.78	27.8
$ZnSO_4 \cdot 7H_2O$	0.86	8.6	$Na_2EDTA \cdot 2H_2O$	3.73	37.3
$CuSO_4 \cdot 5H_2O$	0.0025	0.025			
			Stock 6		
Stock 3			Thiamine	0.04	0.4
KH_2PO_4	17.0	170.0	Other		
H_3BO_3	0.62	6.2	Sucrose	—	30,000
$Na_2MoO_4 \cdot 2H_2O$	0.025	0.25	Asparagine	—	2,000

[a]For the preparation of 1 liter of liquid medium: to 800 ml of distilled water add, in order, 30 g of sucrose; and 10 ml each of stocks 1–6. Dissolve 2 g of asparagine in 100 ml of hot (about 80°C) water and then add to the medium. Adjust the pH of the medium to 5.6, bring to 1 liter volume with distilled water, distribute to the culture flask, and autoclave. For solid medium, before bringing to volume, add either 7 g of Noble agar or 1.5 g of GELRITE, adjust to volume, heat to dissolve the gelling agent, distribute to the culture tubes, and autoclave.

- Spray bottle of 95% ethanol
- Cylinder or tall beaker of sufficient size to completely immerse the ear
- 10% aqueous solution of a commercial chlorine bleach
- Two large beakers of autoclaved distilled water
- About 10 paper towels wrapped in brown Kraft paper and autoclaved
- A scalpel
- A micro spoon-spatula
- A container of 95% ethanol and burner for flaming the scalpel and spatula
- 2.5-cm × 9.5-cm culture vials containing 5–10 ml of sterile simplified medium (Shan-

non 1982) solidified with 0.7% Noble Agar or 0.15% GELRITE, capped with plastic closures

For Suspension Culture

- 250-ml large-mouth Erlenmeyer flasks containing 80 ml of liquid medium, closed with foam plastic plugs and covered with aluminum foil
- A transfer spoon fashioned from a 1-teaspoon-size stainless-steel measuring spoon attached to a 20-cm-long stainless-steel rod
- A container of 95% ethanol and burner for flaming the transfer spoon
- An orbital shaker

METHOD

Culture Initiation

1. Spray the outside of the ear with ethanol and then remove the husk.
2. Immerse the ear in the beaker of 10% bleach for 10 minutes.
3. Rinse the ear by dipping it in the two beakers of sterile water.
4. Place the ear on the sterile paper towels.
5. Slice off the top of several kernels with the sterile scalpel.
6. Scoop out the small endosperms with the sterile spatula.
7. Place 4–8 endosperms on the surface of the medium in each vial.
8. Move the vials to an unlighted culture room maintained at 29°C.
9. Subculture the callus to fresh medium every 3–4 weeks as needed.

Notes

1. The age of the ear is critical. If it is too old, the endosperm explant fills with starch and fails to initiate callus.
2. At the correct age the endosperm is easily removed essentially intact. The endosperm is a sphere 1–2 mm in diameter. It is more dense than the surrounding nucellar tissue and the two tissues are easily distinguished.
3. At the correct age the embryo is very small and it may also be transferred to the culture vials. If this occurs it will usually germinate and the seedling can be discarded at the time of subculture.
4. We observe a greater success of callus initiation when 4–8 endosperms are added

to each vial. With fewer endosperm explants per vial callus initiation occurs much later or not at all.

5. The medium may be prepared from six 100-fold-concentrated stock solutions (Table 131.1) or from a commercial Murashige and Skoog basal salt mixture (Sigma Chemical Co., St. Louis, MO), plus sucrose, thiamine, and asparagine.

Suspension Culture

1. Transfer actively growing callus from several vials to the Erlenmeyer flasks containing liquid medium. The size of the flask and quantity of medium may be varied depending on the quantity of callus available for inoculating the flasks.
2. Place the flasks on an orbital shaker in the unlighted 29°C culture room. Operate the shaker at about 120 rpm.
3. After 3–4 weeks transfer a teaspoon full of callus to fresh medium.
4. Repeat the subculture every 1–3 weeks as needed.

Notes

1. With subsequent transfers the growth rate should increase, and the size of the cell clusters should decline. As the growth rate increases the interval between transfers should be reduced. For example, we transfer our finely divided cultures weekly.
2. A large-bore transfer pipet may be used to subculture finely divided suspension cultures. However, care must be taken to insure that sufficient inoculum is transferred. The smaller the inoculum size the longer the lag phase.

REFERENCES

Murashige T, Skoog F (1962) A revised medium for rapid growth and bioassays with tobacco tissue cultures. Physiol Plant 15: 473–497

Shannon JC (1982) Maize endosperm cultures. In Sheridan WF (ed) Maize for Biological Research, Plant Mol Biol Assoc, Charlottesville, VA, pp 397–400

Shannon JC, Batey JW (1973) Inbred and hybrid effects on establishment of in vitro cultures of *Zea mays* L. endosperm. Crop Sci 13: 491–493

Shannon JC, Liu JW (1977) A simplified medium for the growth of maize (*Zea mays*) endosperm tissue in suspension culture. Physiol Plant 40: 285–291

Straus J (1960) Maize endosperm tissue grown in vitro. III. Development of a synthetic medium. Am J Bot 47: 641–647

Straus J, LaRue CD (1954) Maize endosperm tissue grown in vitro I. Culture requirements. Am J Bot 41: 687–694

Tabata M, Motoyoshi F (1965) Hereditary control of callus formation in maize endosperm cultures in vitro. Japan J Genet 40: 343–355

In vitro Pollen Germination

D. B. WALDEN

The ability to germinate fresh pollen on a "defined" medium offers opportunities for a number of areas of research. Applications, such as agrichemical assessment, gametophyte selection, etc., have been reported by a number of investigators over the last 25 years or more. Principal components of the technical strategy include:

1. Pretreatment of plants. Highest and consistent germination will be obtained from nonstressed, turgid plants. Watering 1–2 days before pollen collection is recommended if drought conditions prevail.
2. Harvest and collection of pollen. Collect the tassel bags with fresh pollen, within 1–2 hours of anther dehiscence. Debris should be removed by screening or other techniques.
3. Pretreatment of pollen. Pollen will be 65% (or higher) water content when shed. With high moisture pollen (often with a tendency to clump), it is useful to spread it out for an hour on a sheet of paper (tassel bag) to allow for the reduction of water content to 40–50%. This step may be bypassed by delaying collection.
4. Choice of genotype. Genotypes perform differentially on a specific medium. The medium may need to be altered to optimize germination for a specific genotype. Optimal conditions can produce consistent germination, 90% or higher.
5. The medium. Maize pollen will germinate (percentage low) on droplets of water, but tube growth will be stunted. Most workers report greater success on a solid medium with the additives listed below. Pollen of Seneca 60, Oh43 and a number of other inbreds do well on:

Noble agar	0.7%
Ca source, e.g., $CaCl_2 \cdot 2H_2O$	300 mg/liter
Carbohydrate source, e.g. sucrose	0.35 M
Boron source, e.g., H_3BO_3	100 mg/liter

The most critical component appears to be the use of Noble agar (agar that has been

The Maize Handbook—M. Freeling, V. Walbot, eds.

incubated with strong acids to remove a variety of impurities). The use of "non-Noblized" agar guarantees variability in the results from sample to sample. Medium can be prepared a few days in advance. Plates/dishes/containers should be poured and cooled just prior to use (1 hour) so that there is no accumulation of moisture on the surface of the agar.

6. Application of pollen to agar surface. Deliver small amounts of pollen and spread evenly in a monolayer. A camel's hair brush works well. Do not touch the brush to the surface of the agar. If pollen has received a pretreatment in a liquid, remove as much of the suspension liquid as possible before pipetting onto the agar surface (or silks).
7. Application of treatments to germinating pollen. Following inoculation of the agar surface, aerosol or physical environment (e.g., specific temperature control, air quality, etc.) conditions can be applied. Agrichemicals, metabolic enhancers/inhibitors, etc., have been assessed by several workers.
8. Incubation. Normal room temperature is adequate. Diffuse light is acceptable. The addition of a moistened (but not dripping) filter paper to the underside of the cover of the dish will help maintain a high humidity. The filter paper can be moistened 15–30 minutes before inoculation if convenient. Most germination takes place within 30 minutes.
9. Data collection. Enumerative data on germination or other detectable phenotypes can be obtained on a large scale. Sampling by selection of fields (10×) for photographic or video recording provides the opportunity to collect the data by visual or automatic means at a later time. Tube growth measurements can also be made directly from the plate or at a later date. Tubes will continue to elongate for several hours. We have measured tube elongation (several thousand microns) over 24-hour periods. Excellent experimental design/parametric statistical analyses can be employed.

REFERENCES

Blackmore S, Knox RB (eds) (1990) Microspores. Academic Press, New York, 347 pp

Greyson RI, Pareddy DR, Bommenini VR,Walden DB (1986) The use of *in vitro* methods in the production of pollen. In Mulcahy DL, Bergamini Mulcahy G, Ottaviano E (eds) Biotechnology and Ecology of Pollen, Springer-Verlag, New York, pp 135–137

Pfahler, PL (1978) Biology of the maize male gametophyte. In Walden DB (ed) Maize Breeding and Genetics, Wiley Interscience, New York, pp 517–530

Vasil IR (1987) Physiology and culture of pollen. Int Rev Cytol 107: 127–174

Walden DB, Greyson RI (1986) Maize pollen research: Preliminary reports from two projects investigating gamete selection. In Mulcahy DL, Bergamini Mulcahy G, Ottaviano E (eds) Biotechnology and Ecology of Pollen, Springer-Verlag, New York, pp 139–145

Axillary Bud in vitro Culture: Asexual Propagation of Maize

R. I. GREYSON AND D. B. WALDEN

Excised axillary buds may be grown on a defined medium and induced to produce sexually mature plants (Raman et al. 1980). Recently, the opportunity to cycle the proliferation of axillary bud cultures has led to the recovery of thousands of plants from an original culture (Walden et al. 1989). Such plants do not demonstrate examples of somaclonal variation, as may occur with plants regenerated from callus.

Axillary bud proliferation technology provides a mechanism to produce (asexually) a population of isogenic plants (e.g., genic male steriles). Hundreds to a few thousand plants can be produced in a 12-month period. Because each plant is sexually competent, large quantities of seed from a single original caryopsis are possible. The technique is equally useful as a means to recover a sexually mature plant from a sectored source (as might occur in plants regenerated from transformed callus).

METHODS

This protocol is for the maintenance of axillary bud cultures of Seneca 60 (Se60). Modifications may be required for the successful culture of axillary buds from other cultivars. The procedure is taken from Raman et al. (1980).

1. Growth conditions for donor plants are as described for inflorescence culture. (See Greyson, this volume.)
2. Cut stem segments from seedlings (20 days after planting) at the scutellar node and near the top of the first leaf sheath. Wash the segment and surface sterilize it in 20% Javex bleach for 15 minutes and rinse it several times in sterile H_2O.
3. Using sterile conditions, remove leaf blades and encircling sheaths, leaving the last two or three leaf primordia untouched.

The Maize Handbook—M. Freeling, V. Walbot, eds.

4. Cut the tip portion of the stem (10–20 mm) containing five to six axillary bud primordia, undissected leaf primordia, and the shoot apical meristem and culture it on appropriate shoot inducing agar medium (SIM). See item 8.
5. After approximately 10 days, dissect developing axillary buds from the cultured shoot tip and sub-culture each individually on SIM. Repeat this procedure at 10–14 day intervals to increase and maintain the clone. Maintain records and identifying labels on lines from individual seedlings.
6. Alternatively, sub culture axillary buds to appropriate root inducing medium (RIM). See item 9.

MEDIA

7. Basal medium (BM): A medium containing the following items was developed for Seneca 60. (a) Major and minor inorganic salts prepared from stock solutions (Murashige and Skoog 1962). We assume that commercially available freeze-dried formulations of M&S salts would serve equally well. (b) Vitamins and glycine (White 1943). (c) I-inositol (100 mg/liter). (d) $FeSO_4 \cdot 7H_2O$ (27.8 mg/liter) and $Na_2 \cdot EDTA$ (37.3 mg/liter). (e) Sucrose (30 g/liter). (f) Adenine sulfate dihydrate (120 mg/liter). (g) Monobasic sodium phosphate (170 mg/liter). (h) Agar (8 g/liter). pH 5.8 prior to autoclaving.
8. Shoot-inducing medium (SIM): To induce axillary bud development on Seneca 60 explants, kinetin (3 mg/liter) and indole butyric acid (1 mg/liter) are added to BM.
9. Root-inducing medium (RIM): to stimulate root production of excised axillary buds, indole acetic acid (3 mg/l) and kinetin (1 mg/l) are added to BM.
10. Rooted buds can be transplanted to soil and grown in pots or subsequently transplanted into the field nursery.

There is some variation in the morphology of the regenerated plants with this procedure. This variation is assumed to relate to the different states of determination of the individual axillary buds.

REFERENCES

Murashige T, Skoog F (1962) A revised medium for rapid growth and bioassays with tobacco tissue. Physiologia Plant 15: 473–497

Raman K, Walden DB, Greyson RI (1980) Propagation of *Zea mays* L. by shoot tip culture: A feasibility study. Ann Bot 45: 183–189

Walden DB, Greyson RI, Bommineni VR, Pareddy DR, Sanchez J-P, Banasikowska E, Kudirka DT (1989) Maize meristem culture and recovery of mature plants. Maydica 34: 263–275

Index

A, marker system for monosomic selection, 360
a1 gene, 649
A1
  cloned anthocyanin genes and their gene product, 283
  promoter active in maize, 634, 635
a2 gene, 649
  marker system for monosomic selection, 360
A2
  cloned anthocyanin genes and their gene product, 283
  marker system for monosomic selection, 360
A158 line, 393
A188 line, 344, 664, 682
  agroinfection, 601
  class and other zein characteristics, 642
  germination rate and in vitro ear culture system, 712
  hybrid lines, 682
A188-A (genomic clone), class and other zein characteristics, 642
A188-B (genomic clone), class and other zein characteristics, 642
A619 line, 344
  class and other zein characteristics, 641
  restoration patterns of inbred line of maize, 420
A632 line, 344, 392
  frequencies of androgenesis in crosses of three inbred lines as male to W23 *ig ig* females, 391
  restoration patterns of inbred line of maize, 420
A632/W23 hybrid, 15
A634 line, restoration patterns of inbred line of maize, 420
A636 inbred line, 719
a1 sh2 (3L) gene, 259
a2 (5S) gene, 259
A-A reciprocal translocation breakpoints, 365
A-A translocations
  breakpoints and stocks, listing, 364–376
  stable terminal tetrasomics, 380
A^B, B-A translocated chromosome, 308
A-B-A compound chromosomes, 334–335
Abaxial epidermis, 18, 22
Ac2, marker system for monosomic selection, 360
Acetocarmine squash technique, 432–434, 437
Acetolactate synthase (ALS) gene, 681, 691
Achromat, 96
A chromosome, interstitial deficiencies, segmental aneuploid analysis, 382
Ac—r-x1/r-m3, marker system for monosomic selection, 360
Activator (Ac) element, 238
Activator (Ac) tagging methods, 219–233
Activator 2 (*Ac2*), 359–360
Active Mutator lines, 243
Adaxial epidermis, 18, 19, 20, 21, 22
Adh1, promoter active in maize, 634
Aeciospores, 289–290
afd (gene), 474
Agroinfection, 599–603
Air drying, 109
Air rifle particle gun design, references, sources and components for maize transformation protocols, 678
Albino gene, 265
Albino marker, expression in maize tissues, 268
Albino mutation, 263, 265, 268
Alcohol dehydrogenase (Adh), 335
  allozymes, 378, 379
  locus, 493
Alcohol dehydrogenase 1 (*adh1*) gene, 301, 634, 637
Aleurone cells, 78–80
Aleurone development, stages of, 78–80
  cellularization, 79–80
  fertilization, 78
  growth as an epidermis, 80
  syncitial development, 78–80
Aleurone gene tagging, 222–228, 230
Alleles, completely null, 211
Allelism, 191–192
Allelism testing of lethal mutation, 407–412
  allelism tests with stocks bearing one ear, 408
  developing colored and colorless mutant stocks, 408–410
  increased efficiency by use of balanced lethals, 412
  preparation of the parent plants for double pollination, 410–412
Allopolyploid, 439
Allozyme, 377

Allozyme markers, 328, 329, 378, 379, 382
al mutation, 13
Alpha (class), zein characteristics
summarized, 641
Alpha-zeins, 639–640, 641, 644
Alternative combination of normal
chromosomes or their standard structures
(+), 361
a-m1, marker system for monosomic
selection, 360
Ameiotic (*am*) mutation, 418
Ameiotic 1 (*am1*) mutation, 474
maize meiotic/gametophytic mutant
behavior, name and reference, 57
Ameiotic 2 (*am2*) mutation, 474
maize meiotic/gametophytic mutant
behavior, name and reference, 57
American Type Culture Collection (ATCC
54022), 691
Amino acid mix, 550, 555
Amino acids, in vitro selection, 686
Ammonium nitrate (fertilizer), 200
Amylose extender mutations, 301
Analysis of variance (ANOVA), 511
Anastomose, 25, 26
Androgenesis, 390–392, 701–703
Andromonoecious dwarf (*d1*, *d2*, *d3*, *d5*, and
D8) mutations, 45
Andromonoecious dwarfing, 258
Aneuploid analysis, segmental, 377–382
Aneuploidy, 319
Anisotropy, 99
Annual teosinte, 67, 68
Anther and microspore culture, 701–704
culture conditions, 703–704
culture medium, 703
donor plant physiology, 702
genotype, 701
microspore isolation, 702–703
plant regeneration, 704
pollination, 704
pretreatment, 702
stage of development, 702
Anther-ear-1 (*an1*) mutation, 45, 46
Anthers, developmental stages in maturation,
473
Anthocyanin, 26, 27, 258
aleurone development, 78
gene for aleurone color, 259
locating recessive genes to chromosome
arm with B-A translocations, 315, 317,
321–322
not in *dek1* gene, 259
pigmentation conditioned by alleles of *R*
locus, 359
produced in shoot, 13
promoter active in maize, 634
use of indeterminate gametophyte, 386
Anthocyanin color markers, 328–329
Anthocyanin genes (*A1*, *A2*, *C2*, *B*, *Pl*, *Bz2*,
and *Bz1*), 265–266, 268, 680
Anthocyanin genes (cloned), and their
regulation, 282–283
anthocyanin synthesis requirements, 282
gene product and references of specific
genes, 283
Anthocyanin genetics, 192, 279–281
aleurone tissue colors, 280–281
anther colors, 281
anthocyanin color affected by differences,
280
cob colors, 281
embryo plumule colors, 281
pericarp colors, 281
plant colors, 280
pollen colors, 281
scutellum colors, 281
seedling colors, 281
silk colors, 281
Anthocyaninless
frequencies and parents of monosomic
types produced by *R*/*r-X1* x
Mangelsdorf's tester cross, 353
genetic marker and monosomics, 353, 355
Anthocyanin marker, expression in maize
tissues, 268
Anthocyanin production, 222
r-X1 deficiency, 350, 351
Anthocyanin synthesis, developing colored
and colorless mutant stocks, 408–410
Anthracnose resistance, trait distinguishing
juvenile and adult regions of the
shoot, 12
Antibodies
against tubulin, 143–144, 148
detection of hybrids, in situ hybridization,
177
immunocytochemistry, 152, 156–157
immunolocalization of nuclear proteins,
158, 161–162, 164
Antimorphic mutations, 271
Antisense RNA, 576
Antitubulin, 144
AO Cycloptic Binocular, 299
Apex, 21
Apical meristem, 6–8, 11–12, 13–14, 34
founder cells within, 23
shoot meristem culture, 715–718
structures, 11–12
Apochromat, 96

Araldite, 113–114
Archesporial cell, 52
Arm ratio, 347
Asc I (rare cutter enzyme-1991), recognition sequence and manufacturer, 532
Asexual propagation, axillary bud in vitro culture, 725–726
Asparagine, stock solutions (100x) and final quantity of components in a simplified medium for growth of maize endosperm, 720
Asynaptic (*as*) mutation, 394, 418
Asynaptic 1 (*as1*) mutation, 474
  maize meiotic/gametophytic mutant behavior, name and reference, 57
Auricles, 18, 19, 20, 21, 25, 26, 27
Aurintricarboxylic acid (ATA), 552
Autodiploids, 59
Autofluorescence, 100
Avermectin, 199
Avid (insecticide), 199
Axillary bud in vitro culture: asexual propagation, 725–726

b, marker system for monosomic selection, 360
B, cloned anthocyanin genes and their gene product, 283
b (2S) gene, 259
B14, restoration patterns of inbred line of maize, 420
B14A, restoration patterns of inbred line of maize, 420
B37
  restoration patterns of inbred line of maize, 420
  Rf1-Rf2 constitution of non-T-restoring inbred line, 421
B73 line, 664, 682
  agroinfection, 601
  restoration patterns of inbred line of maize, 420
  Rf1-Rf2 constitution of non-T-restoring inbred line, 421
  shoot meristem culture, hormones unnecessary, 718
B77 line, restoration patterns of inbred line of maize, 420
B84 line, restoration patterns of inbred line of maize, 420
B84F1 line, 664
B^A, B-A translocated chromosome, 308
Baby Golden popcorn, 499–500
Bacillus thuringensis, 199
Bacterial chloramphenicol acetyl transferase assay in transformed tissues, 616–619
Bacteriophage promoter, 592
Bagging, ear shoot, 200–201
Bar gene selectable marker (plasmids), references, sources and components for maize transformation protocols, 678
Barley seedlings
  electron micrographs of root tissue, 154–155
  immunocytochemistry, 153–156
Basic fuschin, counterstaining of sections, 163
Basidiospores, 289–290
Basipetal veins, 25
BASTA herbicide, 681, 684
  formulation, references, sources and components for maize transformation protocols, 678
B-A translocation manipulation, 308–314
  calculating nondisjunction and preferential fertilization, 313–314
  chromosome combinations, 309
  identifying chromosomal types, 309–310
  maintenance of stocks, methods, 311–313
  properties distinguishing them from standard (A-A) translocations, 308
  viability of genetically unbalanced microspores, megaspores, and gametes, 310–311
B-A translocations, use in locating recessive genes to chromosome arm, 276, 315–327
  additional details, 326–327
  centromeric overlaps, 380, 381, 382
  compound B-A translocations, 325–326, 327
  comprehensive list, 336–340
  definition, 315
  formation of ring chromosomes, 503
  list of B-A translocations, 317–318, 337–342
  locating mutant genes more precisely, 324
  locating recessive genes to chromosome arm, 318–319
  marker systems for, 330–331
  nondisjunction and deficiency analysis, 494
  procedure if mutant phenotype fails to segregate in F_1, 324–325
  stable terminal tetrasomics, 380, 381
  types of translocation stocks supplied, 319–324
  used in dosage analysis, 328–329
B chromosome, 275–276, 278
  A-B-A compound chromosome, 334, 335
  centromeric overlaps, 381
  construction of B-A translocations, 332
  dispensable nature of, 308–309, 310
  locating recessive genes to chromosome

B chromosome (*cont.*)
  arm with B-A translocations, 315, 324, 326
Beadex mutation, 271
Beadle's hypothesis (maize a domesticated teosinte), 66, 73, 74, 75
Beta (class), zein characteristics summarized, 641
Beta-glucanase, 48–50
Beta-glucuronidase (GUS), 615, 633, 680
Beta-1,3-glucan (callose) wall, 48
Beta-zeins, 639–640, 641, 644
Bf1 phenotype, 326
bf2 phenotype, 326
Bialaphos, 682–683, 684
Biased transmission of genes and chromosomes, 274–278
  constructing examples of biased transmission, 277–278
  interpreting abnormal ratios that arise in crosses, 278
  phenomena causing exceptions, 274
  types, in maize, 274–277
Binucleate, mutant genes, chromosomal deficiencies (TB) and cytoplasmic male sterility (cms) and their blocking effects, 474
Bipolaris maydis race T, 392
Birthday paradox statistic, 291
Black Mexican Sweet corn
  class and other zein charactersitics, 641
  DNA delivery into nonregenerable suspension culture cells, 611–612
  media, references, sources and components for maize transformation protocols, 678
  stable transformant selection from suspension cultures, 690–694
  suspension cells, 140
  suspension culture, 681
  transformation, references, sources and components for maize transformation protocols, 678
Blade, 18–21, 23, 25–27
B-Peru, marker system for monosomic selection, 359–360
B-Peru allele, 330
  b-m gene, 259
B-Peru gene, 322
Branched silkless (*bd*) mutation, 46
Brightfield illumination, 101–102
Brightfield microscopy, to view ^{35}S-labeled hybrids, 179
Brittle-2 (*bt 2*) gene, 624–629
Bronze-1 intron, 637
Bronze-1 mutant alleles, 586–587
Brown midrib-2
  frequencies and parents of monosomic types produced by *R/r-X1* x Mangelsdorf's tester cross, 353
  genetic marker and monosomics, 353, 355
BSSS53, class and other zein characteristics, 642
bt1 (5L) gene, 259
bt2 (4S) gene, 259
Bud, 11, 12
  cell lineage of the shoot, 14
Bundle sheath strands, isolation from leaves, 542–543
Bz, marker system for monosomic selection, 360
Bz1 gene
  cloned anthocyanin genes and their gene product, 283
  cloning, 649
  marker system for monosomic selection, 360
promoter active in maize, 634, 635
Bz2 gene, 649
  cloned anthocyanin genes and their gene product, 283
  cloning, 650
  marker system for monosomic selection, 360
  promoter active in maize, 634, 635
Bz2 bz2-m gene, 259
bz2 (1L) gene, 259
bz2-m, marker system for monosomic selection, 360

C1 gene
cloned anthocyanin genes and their gene product, 283
  cloning, 649, 650
  kernel specificity, 282
  molecular regulation, 282
C1 and B anthocyanin genes (plasmids), references, sources and components for maize transformation protocols, 678
c2 gene
  cloning, 649
  marker system for monosomic selection, 360
C2 gene
  cloned anthocyanin genes and their gene product, 283
  marker system for monosomic selection, 360
c2 (4L) gene, 259
C103 line, restoration patterns of inbred line of maize, 420
C123 line, restoration patterns of inbred line of maize, 420

$CaCl_2 \cdot 2H_2O$, stock solutions (100×) and final quantity of components in a simplified medium for growth of maize endosperm, 720
Callus cells, type II, initiation, maintenance and plant regeneration, 671–674
Callus cultures, 681
  initiation of, 695–696
Calyptrogen, 34–35, 36
CaMV 35S promoter, 634, 637, 691
CAT. *See* Chloramphenicol acetyltransferase
Cauliflower mosaic virus (CaMV), 633–634, 635, 637, 691
CB59G line, restoration patterns of inbred line of maize, 420
C-banding technique, 484–491
CE1, restoration patterns of inbred line of maize, 420
Cell cultures
  DNA delivery, silicon carbide fibers used, 610–612
  type I, 663–664, 669
  type I, sources of materials, 669–670
  type II, 663–669, 696
  type II, callus and suspension cell initiation, maintenance, and plant regeneration, 671–676
  type II, cryopreservation, 697
  type II, culture incubation conditions, 667
  type II, culture medium, 666
  type II, embryo isolation, 666
  type II, explant tissue and growth of donor plants, 664–665
  type II, genotype, 664
  type II, plant regeneration from, 667–668
  type II, sources of materials, 669–670
  type II, sterilization, 665–666
Cell lineage markers, 13
Cell shape (cross section), epidermal cells, trait distinguishing juvenile and adult regions of the shoot, 12, 22
Central cell, 54
Central pith, 34
Central Plateau 48703 accession, 501
Centromere, 304, 308, 315, 318, 324–325, 378–379
  B chromosome, 335
  B chromosome, construction of B-A translocations, 332, 333
  centromeric overlaps, 380–381
  and inversions, 346
  pachytene stage in maize, 435, 437, 440
  somatic chromosomes, 486
Centromeric overlaps, 380–381, 382
Centromeric trisomics, 377
Chalco accession, 501
Charge-coupled devices (CCD), 457
CHEF gel analysis, high molecular weight plant DNA prep, 528–529
Chemical (conventional) fixation, 118–121, 122, 128, 131
Chiasmata, 461, 462, 463, 464
Chimeras for genetic analysis, 258–261, 347
  cell autonomy, 259–260
  definition, 258
Chloral hydrate, 476, 479
Chloramphenicol acetyltransferase (CAT), 615, 633
  bacterial CAT enzyme activity in transformed tissues, 616–619
  genes, 637
Chlorophyll, produced in shoot, 13
Chlorophyll fluorescence, 104
Chlorophyll genes, 258, 260, 261
Chlorophyll pigmentation, 266–267, 269
Chloroplast mutations, 413, 415, 416
  isolation of, and chloroplast DNA, 556–558
Chlorosulfuron-resistance genes, 609
Chromatin, 146
  physical exchange of, 436
Chromatophore, 399
Chromomeres, 347, 435–436, 439, 461, 462
Chromosomal deficiencies (TB), 474
Chromosomal translocations involving the nucleolus organizer region or satellite of chromosome 6, 342–344
Chromosome doubling, 391, 392
Chromosome 4 complex, 75
Chromosome 6, chromosomal translocations involving the nucleolus organizer region (NOR) or satellite, 342–344
Chromosome 10 (abnormal), 275, 277, 278
  genetic map of rust resistance gene complex, 287
r-X1 deficiency, 351
Chromosome arms, specific types, B-A translocation marker systems
  chromosome arm 1L, 331
  chromosome arm 1S, 331
  chromosome arm 2S, 331
  chromosome arm 3L, 331
  chromosome arm 4L, 331
  chromosome arm 4S, 331
  chromosome arm 5L, 331
  chromosome arm 5S, 331
  chromosome arm 6L, 331
  chromosome arm 7L, 331
  chromosome arm 7S, 331
  chromosome arm 8L, 331
  chromosome arm 9S, 331

Chromosome arms, (*cont.*)
  chromosome arm 10L, 331
Chromosome doubling, 702
Chromosomes
  behavior during microsporogenesis, 460–474
  mapping of mutants, 264–265
  three-dimensional fluorescence microscopy, 457–459
CI21E line, restoration patterns of inbred line of maize, 420
cif/cif (cross-incompatible female), 502
cim1/cim1 (cross-incompatible male), 502
C1 gene, 259
Cl1 gene, 259
cl (3S) gene, 259
Classification of pollen abortion in the field, 297
Clonal analysis, 262–269
Clonal sector use for lineage and mutant analysis, 262–269
  chromosomes, mapping of mutants, 264–265
  complications, 269
  consideration of marker, 265–267
  experimental approaches for mutant analysis, 263–269
  expression of marker genes in maize tissues, 268
  genetic mutation, 263
  irradiation conditions, 267
  planting and observations, 267–269
Cloned anthocyanin genes and their regulation, 282–283
  anthocyanin synthesis requirements, 282
  gene product and references of specific genes, 283
Cloning, 247, 262–269
  of genes, 211
Cloning genes using transposon tagging, 647–651
cms-C (gene), 474
  analysis, diagnostic line (male parent), 422
  effect on microspore degeneration, 466
cms-s gene, 474
cms-S gene, 474
  analysis, diagnostic line (male parent), 422
  effect on microspore degeneration, 466
cms-T gene, 474
  analysis, diagnostic line (male parent), 422
  effect on microspore degeneration, 466
Coating of specimens, 115
$CoCli_2 \cdot 6H_2O$, stock solutions (100×) and final quantity of components in a simplified medium for growth of maize endosperm, 720
Coefficient of determination, 427, 428
Colchicine, 488
  as mitotic inhibitor to shorten plant chromosomes, 481
Cold-metal block freezing (CMBF), 124–126
Coleoptilar ring, 6
Coleoptile
  cryofixed, 129
  root systems, 30
Coleorhiza, 7, 29–30
Color genes, 265, 268; *see also* Anthocyanin
  Compilation of North American Maize Breeding Germ Plasm, 424
Complex locus, 304–305
Compound B-A translocations, construction of, 332–333
Compound light microscope
  degree of optical correction, 96–97
  desirable features, 96–97
  magnification, 96
  Numerical Aperture (NA), 96
  resolving power and lenses, 96
Computer hardware, 105–107
Computer software, 104, 105–107, 196
  INBRED program, 252–254
  Mapmaker-QTL, 511
  SAS (statistical package), 511
  SYSTAT (statistical package), 511
Confocal laser scanning microscopy, 103–105
Confocal scanning microscopy, ovule clearing and, 138
Contamination, 708
Contrast, 102–103
Corngrass (*Cg1*) gene, 72, 73
Corngrass (*Cg*) mutations, 13, 45–46
Cortex, 34
Cosegregation, 658
cp-1381* (9L) gene, 259
Critical-point drying, 109, 110
Cross-sterility, 497, 499, 500, 501, 502
Cross veins, 26
Cryofixation/freeze-substitution (technique), 122–125, 128–133
Cryopreservation, 697
Cryoprotectants, 123–124, 126
Cryptic (truncated) gene, 304–305
$CuSO_4 \cdot 5H_2O$, stock solutions (100×) and final quantity of components in a simplified medium for growth of maize endosperm, 720
Cuticle thickness, epidermal cells, trait distinguishing juvenile and adult regions of the shoot, 12
Cutting back, 201–202
Cycloheximide, 482

Cytogenetics to enhance transposon tagging with Ac throughout the maize genome, 234–239
  Ac and the P-vv allele, 238–239
  maize translocation and inversion stocks, 234–235
  modification strategy, 235–237
  reciprocal translocations and inversions between 1S and the other arms, 236–237
  source of Ac, the P alleles, and tester stocks, 235
Cytogenetic Working Map, 235
Cytokinesis, mutant genes, chromosomal deficiencies (TB) and cytoplasmic male sterility (cms) and their blocking effects, 474
Cytokinin, inflorescence culture, 714
Cytological analysis, monosomics identification, 352
Cytology, to survey molecular polymorphisms in nuclear genome of maize inbreds, 425
Cytometric measurements of nuclear DNA, 396–400
  DNA measurements, 398–400
  Feulgen cytophotometry, 396–397
  hydrolysis tests, 398
  spectral curve, 398
Cytoplasmically inherited mutants, analysis of, 413–416
  determining inheritance patterns, 414–415
  molecular analyses, 415–416
  nuclear-cytoplasmic interactions, 416
  recognizing a cytoplasmic mutation, 414
Cytoplasmic male sterility (cms), 418–422
  effect on microspore degeneration, 466
Cytoplasmic male sterility-Charrua (*cms-C*), 418, 421–422, 466, 474
Cytoplasmic male sterility-semisterile (*cms-S*), genetic nature of restorer genes, 419, 421, 422
Cytoplasmic male sterility-Texas (*cms-T*), 418, 420, 422
  analysis of restorer gene constitution, 421
  restorer genes in inbred lines F, 420–421
Cytoplasmic male sterility-USDA (*cms-USDA*), 418

D8 (dwarf) mutation, 263, 271
DAPI (4′,6-diamidino-2-phenylindole), 459, 460
Darkfield illumination, 97
Darkfield microscopy, to view 35S-labeled hybrids, 179
Database, germplasm, 424
Databases, of inbred lines and molecular markers, 425–426
7-Deaza-2′-deoxyguanosine 5′-triphosphate (deaza dGTP), 587
Defective endosperm-B30 (*De-B30*) locus, genes affecting zein accumulation, 643
Defective kernel (*dek*) mutations, 7–9, 407409, 412
Deficiency analysis, 494–495
Deficient-duplicate (Df-Dp) chromosome production, 347
Defective seed 1 (*de1*) mutation, 498
Dehydration, 90
dek1 (colorless floury defective mutant) gene, 259–261
dek1 (1S) gene, 259–261
dek5 (3S) gene, 259
dek14 (10S) gene, 259
dek28 gene, 322
Delta (class), zein characteristics summarized, 641
Delta-zeins, 639–640, 641, 644
Dent (*ga*) corn, 12, 497–499, 500, 501
Dental impression kits, 110–114
Desynaptic1 (*dsy1*) mutation, 474
  maize meiotic/gametophytic mutant behavior, name and reference, 57
Desynaptic1 (*dy1*) mutation, maize meiotic/gametophytic mutant behavior, name and reference, 57
Desynaptic2 (*dys2*) mutation, 474
  maize meiotic/gametophytic mutant behavior, name and reference, 57
Diakinesis, 452, 462, 463, 464
  in situ hybridization of DNA and RNA probes to chromosomes, 504–505
Diethyldithiocarbamic acid (DECA), 543
Differential-interference contrast microscopy, ovule clearing and, 138
Differential Interference Contrast (DIC, Nomarsky optics), 98–99
Differential RNA hybridization, 247
Differentiation, mutant genes, chromosomal deficiencies (TB) and cytoplasmic male sterility (cms) and their blocking effects, 474
Digoxigenin (DIG) labeling, 173–174, 177, 178–179
Dihydrofolate reductase (DHFR), 691
Dimethyl sulfate (DMS), in vivo footprinting, 579–585
Dimethyl sulfoxide (DMSO), 476, 479, 482
Dipel (insecticide), 199
Diploids, 394
Diplotene, 448, 449, 461, 462
Disease lesion mutants, 291–296
  factors causing lesion initiation in "sensitive" stage, 294–296

Disease lesion mutants (*cont.*)
  variations in phenotypic expression of dominant and recessive mutants, 292–293
Dissection, destructive, 83–84
Dissection, nondestructive, 84–85
Dissociation (Ds) element, 238
Dissociation (Ds) tagging methods, 220–224
Dithane (insecticide), 200
Divergent spindle (*dv1*) mutation, 474
  maize meiotic/gametophytic mutant behavior, name and reference, 57
DNA
  agroinfection, and DNA constructions, 599–603
  analysis, 247–248, 251
  chloroplast, 556–558
  construction of a genomic library in lambda phage, 595–598
  delivery into cell cultures using silicon carbide fibers, 610–612
  genomic sequencing in maize, 579–585
  high-molecular-weight (HMW) DNA preparation and analysis by pulsed-field gel electrophoresis, 530–533
  high molecular weight plant prep for CHEF gel analysis, 528–529
  in situ hybridization of probes to chromosomes, 504–507
  isolation from microspores and pollen, 540
  isolation of DNA from immature cobs, 536–537
  isolation of genomic DNA from Calli, 534–535
  measurements, 398–400
  microprep version 2.3, 524–525
  miniprep, 523–524
  miniprep and microprep: version 2.1, 522–525
  mitochondrial, 549–555
  polyethylene glycol (PEG)-mediated uptake into protoplasts, 603–609
  preparation from leaf blades or isolated leaf cells, 543–544
  preparation from leaves: expanded blades and separated bundle sheath and mesophyll cells, 541–544
  protoplast culture, 695–699
  southern blots of genomic DNA, 566–568
  transgenic plant production via microprojectile-mediated gene transfer, 677–684
  urea-based plant DNA miniprep, 526–527
Domains, 18
Double-mutant phenotypes, 210
Double pollination, 410–412
Dosage analysis using B-A translocations, 328–329
Dotted 2 (*Dt2*), 359, 360
Dotted 3 (*Dt3*), 359, 360
Doubling, of chromosome, 391, 392
Doubly monosomic plants, 360
Drosophila, 270–271, 272
Drying of tissue, 109–110
Ds-induced chromosome breaking, 258–259
Duplications, 493–494

Ear, 11–12, 37–40, 43–46
  branch (shank), 38
  cells giving rise to, 262
  cob, 38, 69–70
  development, 37–38
  disarticulation, 70–71
  glumes, 43–45, 69, 71, 73, 76
  gynoecium, 43, 44, 45
  husks, 38, 43, 45
  inflorescence culture, 712–714
  internode, 69–71
  in vitro culture system, 709–712
  irradiation and clonal analysis, 267
  lemma, 38, 43, 44, 45
  morphology, 38–40
  mutants, 45–46
  palea, 38, 43, 44, 45
  pedicellate spikelet, 70
  phase change, 61
  phyllotaxy, 71, 76
  prophyll, 38, 43
  reproductive and vegetative structures, 13
  sessile spikelet, 70
  shoot growth, 14
  silks, 38, 40, 44, 45, 51
  stylar canal, 43, 44, 45
  style, 38, 43
  tillers, 68, 73
  trichomes, 38
Editors' note, 565
EG5 BMS cell line, references, sources and components for maize transformation protocols, 678
Egg apparatus, 54
Electric discharge particle gun design, references, sources and components for maize transformation protocols, 678
Electron microscopy, plastic histology and, 136
Electrophoretic separation, 328
Electroporated protoplasts, RNA isolation, 547–548
Electroporation, 547–548, 690, 698
Elongate (*el*) mutation, 394–395, 474

maize meiotic/gametophytic mutant behavior, name and reference, 57
Elongate1 (*el1*) mutation, maize meiotic/gametophytic mutant behavior, name and reference, 57
Embedding (technique), 91, 118–119, 121, 127–128, 131–132
immunolocalization of nuclear proteins, 159–161
Embryogeny, 3, 4–9
classification systems of embryo stages, 4–9
coleoptilar stage embryo, 6–7
early embryogenesis in maize, 6
genes, role of, 7–9
late transition stage, 5–7
midtransition stage embryo, 5
phases, 4–9
proembryo, 5, 6, 7
recommended bibliography, 3, 4
scutellum, 6, 7, 9
stage 1, 5–7
stages 2–6, 7
transition-stage, 5, 6, 13
Embryonic callus, 7
Embryo sac, 3, 54, 55
development, macrospore mother cell preparation protocol for study, 450–453
eight-nucleate, immature, 56
four-nucleate, 55
mature, 55–59
techniques for histology of, 135–139
top wall, 54
two-nucleate, 55
Embryo-specific (*emb*) mutations, 8–9, 407–409
Empty magnification, 96
Endodermis, 34
Endoplasmic reticulum (RER) membranes, 639
Endosperm, zein gene characterization and regulation, 639–644
Endosperm nuclear DNA, flow cytometry for, 400–406
fixative, 402
methodology, 401–402
nuclear preparation, 403
running samples, 406
settings for the flow cytometer, 405
solutions for nuclei preparations, 402
staining with Hoechst 33258, 404–405
use of an internal standard, 403–404
Engorged mature pollen, mutant genes, chromosomal deficiencies (TB) and cytoplasmic male sterility (cms) and their blocking effects, 474
Enrichment selection, 688
Enstar 5E (insecticide), 199–200
Enzymes
electrophoretic variants, 425
starch gel electrophoresis, 425
Epicuticular wax, trait distinguishing juvenile and adult regions of the shoot, 12, 21, 23, 26
Epidermal hairs, trait distinguishing juvenile and adult regions of the shoot, 12
Epidermis, 18–21, 34
root system, 34
Epiflorescence illumination
cytoskeleton localization, 146
steps to set up, 100–101
Epiflorescence microscopy, 99–101, 103
Ethanol, for dehydration, 90
Ethylmethane sulfonate (EMS), 291
paraffin oil pollen treatment, 213–214
Excision assay, 219
to recover transposed Ac or Ds elements, 220–224
Exine, 49, 50

Female inflorescence. *See* Ear
Fertilization, 3–4
double, 4
embryo sac formation, 3
heterofertilization, 4
imprinting, 4
preferential, rate of, 314
recommended bibliography, 3, 4
$FeSO_4 \cdot 7H_2O$, stock solutions (100×) and final quantity of components in a simplified medium for growth of maize endosperm, 720
Feulgen cytophotometry, to measure nuclear DNA, 396–397
Feulgen staining technique, somatic plant chromosomes, 481–483
Filiform apparatus, 55
Filters, epiflorescence microscopy, 100
Firefly luciferase, 615, 616, 633
assay in transformed maize tissues, 619–621
First pollen mitosis, mutant genes, chromosomal deficiencies (TB) and cytoplasmic male sterility (cms) and their blocking effects, 474
Flatfield, 96
Flax-rust interaction, 286
Flint corn, 497, 501
Floret, developmental stages in maturation, 473
Fluorescence microscopy
paraffin histology and, 136

Fluorescence microscopy (*cont.*)
plastic histology and, 138
three-dimensional, of maize chromosomes, 457–459
Fluorescence photomicrography, 102
Fluorescent phallotoxins, 140
Fluorochromes, 100, 101, 104
Floury-1 (*fl1*) locus, genes affecting zein accumulation, 643
Floury-2 (*fl2*) locus, genes affecting zein accumulation, 643
Floury-2 mutation, 640
Floury-3 (*fl3*) locus, genes affecting zein accumulation, 643
Flow cytometry for endosperm nuclear DNA, 400–406
Focusing, 102
Forward-angle light scatter signal (FALS), 405
Founder cells, 23
Freeze drying, 109
Freeze-fracture, 132–133
Freeze-substitution, 122–132
Freon-22, 127
Functional megaspore, 53, 54
Fungicides, 199

Gametogenesis, 48–59
megagametogenesis, 51–52, 54–55
megasporogenesis, 51–54
microsporogenesis, 48–51
Gametophyte, 276–277
determines whether the pollen is normal or aborted, 418–419
isolation of, 138–139
placement of genes expressed in the male, 256–257
Gametophyte factors (ga) of maize, 496–502
Gamma (class), zein characteristics summarized, 641
Gamma-zeins, 639–640, 641, 644
Gaspe Flint line, 14
Genbank sequence accession, 589–590
Gene compensation, 356
Gene linkage, 309
Gene placement using waxy-marked reciprocal translocations, 255–257
genes expressed in the male gametophyte, 256–257
genes expressed in the sporophyte, 255–256
Genes, role in embryogeny, 7–9
Gene tagging with Activator (Ac)/Dissociation (Ds) elements in maize, 219–233
changes in Ac dosage to recover Ac transpositions, 224–226
excision assay to recover transposed Ac or Ds elements, 220–224
genetic tests to identify Ac-induced mutations, 230–231
logistical considerations, 228–229
molecular analysis, 231–232
mutational screens, 229–230
precautions, 226–228
propagating Ac lines, 226–227
remove off-type somatic sectors, 227
standard precautions, 227–228
utility of Ac-induced mutations, 232–233
Genetic experiments and mapping, 189–196
characterization of expressions, 192–193
establishment of inheritance pattern, 189
example, 190–191
mapping techniques, 191, 193–196
planning of experiments, 189–190
segregation analysis, 189–190
testing for allelism with existing mutants, 191–192
Genetic fine structure as revealed in pollen assays, 298–301
intragenic recombination in maize, 298–299, 300
order of mutant sites, 300
overlapping deletion method, 300
technique developed for collecting and assaying pollen samples, 299
Genetic fine structure from testcross progeny analysis, 303–305
analysis of complex loci, 304–305
construction of a genetic fine-structure map, 305
orientation of the physical map of the gene relative to the centromere, 304
physical and genetic distance correlated, 303–304
Genetic Maps, 196
Genetic markers, translocations as, 361–363
Genetics, 66–76
Genetran, nylon membrane, 568
Gene transfer assays, 680
Genic, analysis, diagnostic line (male parent), 422
Genomic library in lambda phage, construction of, 595–598
Genomic sequencing in maize, 579–585
Genotype, 701, 719
Germless (*gm*) mutations, 8
Germplasm, 424
Germplasm Resources Information Network (GRIN), germplasm database, 424
Gibberellin, inflorescence culture, 714
Giemsa stain, 484, 490, 491

gl15 mutation, 13
gl15 phenotype (glossy), 326
Glossy seedling
  frequencies and parents of monosomic types produced by *R/r-X1* x Mangelsdorf's tester cross, 353
  genetic marker, and monosomics, 353, 355
Glucose-6-phosphate dehydrogenase, 271
Glume factor, 75
Glycocalyx, 50
Glyphosate-resistance genes, 609
GMendel (program), 196
Golden Bantam line, agroinfection, 601
Golden stalk
  frequencies and parents of monosomic types produced by *R/r-X1* x Mangelsdorf's tester cross, 353
  genetic marker, and monosomics, 353, 355
Grassy tillers, 12
Grounding of specimens, 115
Ground tissue, 18
Growing maize for genetic studies, 197–208
  collecting pollen, 202
  contamination, 203–204
  controlled pollinations, 200–203
  cutting back, 201–202
  ear shoot bagging, 200–201
  field cultivation, 198–200
  greenhouse cultivation, 207
  harvesting, 204–205
  maintaining pedigrees, 205–207
  pollen viability, 204
  pollination, 202–203
  precautions, 203–204
  suppliers, list of, 208
Growth inhibition tests, 686
Guanidinium method, 543
Guanylyltransferase (GTP transferase), 559–562
Gunpowder particle gun design, references, sources and components for maize transformation protocols, 678
GUS genes for corn, 682, 683
  references, sources and components for maize transformation protocols, 678
  reporter genes, 637
GX122 hybrid, 15
gZ15A (genomic clone), class and other zein characteristics, 642
gZ19ab1 (genomic clone), class and other zein characteristics, 641
gZ19ab11 (genomic clone), class and other zein characteristics, 641
gZ22.8 (genomic clone), class and other zein characteristics, 641
H3 genes, 356
H4 genes, 356
H95 line, restoration patterns of inbred lines of maize, 420
Hairiness genes, 26
Hairy-sheath-frayed mutation, 13
"Half-vein rule," 25–26
"Halo effect," 98
Hand sectioning (technique)
  dried material, 88
  fixed material, 88
  live tissue (viewed live or fixed after sectioning), 87–88
Haploid, 394–395, 448–449, 452–453
  pollination, anther and microspore culture, 704
Haploid production, practical aspects of, 386–387
  indeterminate gametophyte, 386–387
  stock 6, 387
Harinoso de Ocho (race) corn, 501
H_3BO_3, stock solutions (100x) and final quantity of components in a simplified medium for growth of maize endosperm, 720
hcf106 gene, 649, 651
HCl depurination, 533
Helium particle gun design, references, sources and components for maize transformation protocols, 678
Hematoxylin, 460
Hematoxylin stains, 476, 477–478
Heptaploids, 394–395
Herbiace herbicide, references, sources and components for maize transformation protocols, 678
Herbicides, 199
Heteroalleles, 305
Heterofertilization, 4, 222, 322, 390, 514–515
Heterosis, 427–428
Heterozygotes
  A-B-A compound chromosomes, 335
  B-A translocations, 309, 312
  B-A translocation stocks, 319, 323, 324
  biased transmission of genes and chromosomes, 276, 277
  construction of compound B-A translocations, 333
  deficiency-duplication, translocations involving NOR or satellite of chromosome 6, 342–344
  mapping genes with recombinant inbreds, 249–254
  self-pollinated, progeny from, 659
  testcross, progeny from, 659–660

Hexadecene, 124, 130
Hexaploids, 394–395
High chlorophyll fluorescence (*hcf*) mutation, 413
High molecular weight plant DNA prep for CHEF gel analysis, 528–529
High pressure freezing (HPF), 110, 126
High velocity microprojectiles method, 677
Histology of maize megaspores and embryo sacs, techniques for, 135–139
  developmental staging, 135
  isolation of megaspores, embryo sacs, and gametes, 135, 138–139
  ovule clearing, 135, 136, 138
  paraffin histology, 135, 136
  plastic histology, 135, 136–138
  preparing material, 135–136
hm1 gene, 649
Homozygotes, 333
  B-A translocations, 309, 311–312
  B-A translocation stocks, 319
Hopi Indian varieties, 288
Hsp70, promoter active in maize, 634
Hybond-N, nylon membrane, 568, 574
Hybridization, 175–177, 179, 577–578, 583–585
  plaques to nitrocellulose filters, 597–598
  Southern blot, 569–572
Hydrolysis tests, 398
Hygromycin phosphotransferase, 681
Hygromycin phosphotransferase II (HPT II), 691
Hygromycin-resistance gene (*hpt*), 608
Hypermorphic mutations, 270–271
Hyperploids, 312–313
Hy pollen (*ga1*), 499–500
Hypomorphic mutations, 270, 271

I153 line, restoration patterns of inbred line of maize, 420
IGSS. *See* Silver-enhanced immunogold labeling
IllA line, restoration patterns of inbred line of maize, 420
Ils1–501B, variations in phenotypic expression of this recessive disease lesion mutant, 293
Image manipulations, 107
Immunocytochemistry for light and electron microscopy, protocol, 149–157
  immunolabeling procedure for maize kernels at the light microscope level, 150–153
  immunolabeling procedure for root tissue at the electron microscope level, 153–156
Immunofluorescence microscopy, cytoskeleton localization, 140–148
Immunogold localization, 118, 128–132
Immunolabelling procedure, maize kernels at the light microscope level, 150–153
Immunolocalization of nuclear proteins, 158–164
  antibody labeling, 161–162
  counterstaining sections, 163
  evaluating the results, 163–164
  permanent mounting of sections, 163
  slide preparation, 158–159
  tissue fixation, 159
  tissue infiltration, embedding, and sectioning, 159–161
Impression kits, 110–114
Imprinting, 4, 272, 390
Inbred lines and their molecular markers, 228, 423–430
  alleles per locus, 425
  chromosome map positions, 425
  data on agronomic traits of U.S. public inbreds, 424
  describing maize inbreds with molecular markers, 425–428
  development and origin of maize inbred lines, 423–424
  methods used to survey, 425
  some notes of caution, 428–430
  sources of seeds, 424
INBRED program (Keith Thompson's), 252–254
Inbreeding, 311–312, 319
Indeterminate gametophyte (ig) gene: biology and use, 388–393
  androgenesis and cytoplasmic substitution, 390–392
  embryological observations, 389
  haploid production, 386–387
  kernel abnormalities, 389–390
  male sterile cytoplasms in lines homozygous for ig, 387
  male sterility, 392–393
  ploidy series production, 395
  to derive diploids from tetraploids, 387
Indeterminate gametophyte (*ig*) mutation, 386–387, 389
Indeterminate gametophyte 1 (*ig1*) mutation, maize meiotic/gametophytic mutant behavior, name and reference, 58, 59
Indeterminate (*id*) mutation, 14
Indirect end labeling, 580–581
Indirect immunofluorescence, 100
Indirect immunofluorescence: localization of the cytoskeleton technique, 140–148

microfilaments, 140, 145–148
microtubules, 140, 141–148
Inducible, promoter active in maize, 634
Infiltration, technique, 90–91
Inflorescence culture, 712–714
Inflorescences, 11, 37–46, 61–62
maize vs. teosinte, 67–69, 76
sex, and plant architecture, 71–72
Inheritance patterns, traits with complex inheritance analyzed using molecular markers, 509–513
Insecticides, 199–200
In situ hybridization (protocol), 165–179
detection of hybrids, 177–178
digoxigenin (DIG) labeling, 173–174, 177, 178–179
fixation of leaf sections, 166–167
hybridization, 175–177, 179
paraffin embedding tissue, 167–168
preparation of poly-D-lysine-coated slides, 165–166
preparation of tissue sections, 165–170
pretreatment of sections, 174–175
riboprobe preparation, 170–174, 179
sectioning paraffin-embedded tissue, 168–169
^{35}S labeling, 170–173, 177–178, 179
staining sectioned tissue, 169–170
In situ hybridization of DNA and RNA probes to chromosomes, 504–507
Interference contrast microscopy, plastic histology and, 138
Internet, 252
Internode, 11, 12
cell lineage of the shoot, 14
root system, 30, 31
Interval mapping, 195
Intine, 49, 51
Intragenic recombination, 658
Intron enhancement, 636
Introns, 636–638
Inverse polymerase chain reaction (IPCR), 587–591
protocol, 590–591
"Inverse effect," 271, 272
Inversions
genetic markers and, 346
list available and, 346–349
paracentric, 346, 347
pericentric, 346, 347
uses of, 347–349
Inversions, specific types, breakpoints and source
Inv 1a SL, 348
Inv 1c SL, 348
Inv 1d L, 348
Inv 1f SL, 348
Inv 1g SL, 348
Inv 1h L, 348
Inv 1j SL, 348
Inv 1k L, 348
Inv 1l SL, 348
Inv 1m SL, 348
Inv 2a SL, 348
Inv 2b SL, 348
Inv 2d SL, 348
Inv 2e S, 348
Inv 2f SL, 348
Inv 2g SL, 348
Inv 2h L, 348
Inv 2i SL, 348
Inv 2l SL, 348
Inv 2m SL, 348
Inv 2o SL, 348
Inv 2p SL, 348
Inv 3a L, 348
Inv 3b L, 348
Inv 3c L, 348
Inv 3d SL, 348
Inv 3e SL, 348
Inv 3h L, 348
Inv 4a L, 348
Inv 4b SL, 348
Inv 4c SL, 348
Inv 4d L, 348
Inv 4e L, 348
Inv 4f L, 348
Inv 4g SL, 348
Inv 4h L, 348
Inv 4i L, 348
Inv 4j L, 348
Inv 4k SL, 348
Inv 5a SL, 348
Inv 5b SL, 348
Inv 5c SL, 348
Inv 5d SL, 348
Inv 5e SL, 348
Inv 5f SL, 348
Inv 5g L, 348
Inv 6a SL, 348
Inv 6b SL, 348
Inv 6c SL, 348
Inv 6d SL, 348
Inv 6e SL, 348
Inv 7a L, 348
Inv 7b SL, 348
Inv 7c L, 348
Inv 7d SL, 348
Inv 7e SL, 348
Inv 7f L, 348

Inversions, specific types, breakpoints and source (*cont.*)
Inv 8a SL, 348
Inv 8b L, 348
Inv 8c S, 348
Inv 9a SL, 348
Inv 9b SL, 348
Inv 9c SL, 348
Inv 10a SL, 348
In vitro capping of mitochondrial RNA and transcription initiation site characterization by RNase protection, 559–564
In vitro culture of axillary bud, asexual propagation, 725–726
In vitro culture of kernels, 705–708
In vitro ear culture system, 709–712
In vitro pollen germination, 723–724
In vitro selection, 685–688
In vitro synthesis of capped and polyadenylated mRNA for translational studies in vitro and in vivo, 592–594
In vivo footprinting, 579–585
iojap gene, 413, 416, 649
Ionizing irradiation, to induce ring chromosomes, 503
Iowa Stiff Stalk Synthetic-related lines, culture systems, 669
Iron-acetocarmine staining, 476, 520
Isolation, minimum distance in commercial seed fields, 305
Isolation of DNA from immature cobs, 536–537
Isolation of genomic DNA from Calli, 534–535
Isolation of small nuclear RNAs (snRNAs), 519–521
extraction of RNA from purified maize nuclei, 520–521
purification of nuclei from maize leaves, 519–520
Isozymes, 512
description of maize inbreds with molecular markers, 425, 426

Japonica
frequencies and parents of monosomic types produced by *R/r-X1* x Mangelsdorf's tester cross, 353
genetic marker, and monosomics, 353, 355

K55 line
as male parent of a newly arisen male sterile trait, 422
restoration patterns of inbred line of maize, 420
Karyotyping, 491
Kelvedon 33 cultivar, 36
Kernels
abnormalities, *ig* gene, 389–390
in vitro culture of, 705–708
in vitro development scheme, 627–628
KH_2PO_4, stock solutions (100×) and final quantity of components in a simplified medium for growth of maize endosperm, 720
KI, stock solutions (100×) and final quantity of components in a simplified medium for growth of maize endosperm, 720
Kinetochore (K), 455
Kinoprene (insecticide), 199–200
KNO_3, stock solutions (100×) and final quantity of components in a simplified medium for growth of maize endosperm, 720
Kn gene, 648
cloning, 649
Knob, 274–275, 277, 347, 484
pachytene chromosomes, 435–437, 485, 488
Knobbed chromosomes, 275
Knobless chromosome, 275
Knotted-1 (*Kn1*) mutation, 271, 272
Knotted-1 protein, 158, 163
Köhler illumination, 97, 98
Ky21 line, restoration patterns of inbred line of maize, 420
KYS line, 434–435
restoration patterns of inbred line of maize, 420
silver-stained pachytene SC complement, 443

L289 inbred line, 434
L317 line, restoration patterns of inbred line of maize, 420
Lambda phage, genomic library construction, 595–598
Lanthanum hexaboride filaments, 114, 115
Lasso/Atrazine combination (herbicides), 199
Lateral buds, trait distinguishing juvenile and adult regions of the shoot, 12
Lateral veins, 25, 26
Lc, cloned anthocyanin genes and their gene product, 283
Leaf, 11, 12, 17–27
adult, 21–23
cell lineage of the shoot, 14
cells giving rise to, 262
development from meristem founder cells to the primordium, 23–25
homologs, 26
juvenile, 21–23

lateral dimension, 18
longitudinal dimension, 18
marking transformations of one part into another, 26
postprimordial development, 25
shoot growth, 15
tissues and cells, mature adult, 18–21
transverse dimension, 18
vegetative, 17–18
veins during development, 25–26
Leaf number (L), 17
Leafy (*Lfy*) mutation, 13, 14
Leptotene, 461, 462
LERf line, restoration patterns of inbred line of maize, 420
Les mutations, 291–296
Les mutations, specific types, variations in phenotypic expression of this dominant disease lesion mutant
les 1–843, 292
les 2–845, 292
les 3-Ullstrup, 292
les 4–1375, 292
les 5–1449, 292
les 6–1451, 292
les 7–1461, 292
les 8–2005, 292
les 9–2008, 292
les 10-Kermicle, 292
les 11–1438, 292
les 12–1453, 292
les 13–2003, 292
les 14–2004, 292
les 15–2007, 292
les 16–2016, 292
les 17–2345 (A762), 292
les – 721, 293
les – 1378, 293
les – 1395, 293
les – 1442, 293
les – 1521C, 293
les – 2006, 293
les – 2012, 293
les – 2013, 293
les – 7145 (Beckett), 293
les –A467 Blanco, 293
Lethal ovule 2 (*lo2*) mutation, maize meiotic/gametophytic mutant behavior, name and reference, 58, 59
Lethal pollen 1 (*lp1*) mutation, maize meiotic/gametophytic mutant behavior, name and reference, 58
Lethal seedling stocks, 412
Light microscopy
confocal laser scanning microscopy, 103–105
destructive dissection, 83–84
hand-sectioning, 87–88
methods of observation, 95–101
microtechnique, 89–94
nondestructive dissection, 84–85
paraffin histology and, 137
recording the image: photomicrography, 101–103
small nuclear RNA contamination not indicated, 520
tissue clearing, 85–87
video microscopy and computer-mediated image manipulation, 105–107
Ligular region, 18
Ligule, 18–21, 23–25
Liguleless
frequencies and parents of monosomic types produced by *R*/*r-X1* x Mangelsdorf's tester cross, 353
genetic marker, and monosomics, 353, 355
Liguleless1–0 mutant sector, 25
Liguleless-2 (*lg2*) gene, 277
Liguleless 4–0 (*Lg4–0*) mutant heterozygote, 26, 27
Line test, 114
Lipid quantity, 357
lls1-Troyer, variations in phenotypic expression of this recessive disease lesion mutant, 293
Localization work, use of trisomics, 307
Locating recessive genes to chromosome arm with B-A translocations, 315–327
additional details, 326–327
compound B-A translocations, 325–326, 327
list of B-A translocations, 317–318, 337–342
locating mutant genes more precisely, 324
locating recessive genes to chromosome arm, 318–319
procedure if mutant phenotype fails to segregate in F1, 324–325
types of translocation stocks supplied, 319–324
Longley's culture numbers, 349
Lorsban insecticide, 199
Low-melting-temperature (LMT) agarose plug, 530
Luciferase genes, 637
lw mutation, 13

M14 line, 393
restoration pattern of inbred line of maize, 420

Macrospore mother cells (MMCs), protocol for preparation for study of female meiosis and embryo-sac development, 450–453
Maize and Puccinia sorghi: host-pathogen interactions, 286–290
  genetics of resistance to *Puccinia sorghi*, 286–287
  working with *Puccinia sorghi*, 288–290
Maize Genetics Cooperation News Letter, 196, 234, 265, 336, 497
  chromosome maps giving relative positions of gene loci and translocation breakpoints, 326
Maize Genetics Cooperation Stock Center, 234, 327, 386, 387, 408, 412
Maize Streak Virus (MSV), agroinfection protocol, 599–603
Male germ unit, 50
Male inflorescence. See Tassel
Male sterility and restorer genes, 418–422
  analysis of a newly arisen male sterile trait, 422
  characteristics of restorer-of-fertility genes in maize, 419
  genetic analysis of newly arisen male sterile traits, 421
  genetic nature of restorer genes, 418–419
  molecular analysis of cytoplasmic male sterility, 421–422
  restoration patterns of 42 inbred lines of maize, 419, 420
  restorer genes in inbred lines F, 419–421
Mangelsdorf's tester, 352–353, 355–356, 359
Mapping genes with recombinant inbreds (RI), 249–254
MAPMAKER, 196
Mapmaker-QTL (computer software), 511
Margin domains, 18
Marker linkage, 309
Marker systems for B-A translocations, 330–331
Marker systems for *r-X1*, 359–360
Megacytes, 59
Megagametophyte, deficiency analysis, 495
Megagametogenesis, 51–52, 54–55, 135
  definition, 51
Megaspore mother cell (MMC), 52
Megaspores, 52–53, 55
  techniques for histology of, 135–139
Megasporocyte, 52
Megasporogenesis, 51–54, 135, 389
  definition, 51
mei025 (gene), 474
Meiosis, 135
  macrospore mother cell preparation protocol for study, 450–453
  mutant genes, chromosomal deficiencies (TB) and cytoplasmic male sterility (cms) and their blocking effects, 474
  smear technique for study in pollen mother cells, 447–449
Meiosis I, mutant genes, chromosomal deficiencies (TB) and cytoplasmic male sterility (cms) and their blocking effects, 474
Meiosis II, mutant genes, chromosomal deficiencies (TB) and cytoplasmic male sterility (cms) and their blocking effects, 474
Mendelian ratio, 70, 74
  deviation from, 277
Merit cultivar, 36
Metacentric compound chromosomes, 334
Metal film replication methods, in freeze-fracture, 132–133
Metaxylem, 34
Methotrexate-resistance genes, 609
Methyl Cellosolve (ethylene glycol monomethyl ether), for dehydration, 90
Mexican annual teosinte, 66, 67, 69, 74
  architecture, 69
  ssp. *mexicana* Race Chalco, 74
Microcomputer image analysis systems, 105–107
Microcytes, 59
Microfilaments, 140, 145–148
Microgametophyte, deficiency analysis, 495
Microprojectile/DNA preparations, references, sources, and components for maize transformation protocols, 678
Microprojectiles, 690, 692–694
Micropylar region, 389, 390
Microspore, 48–50
  mutant genes, chromosomal deficiencies (TB) and cytoplasmic male sterility (cms) and their blocking effects, 474
  nucleic acid preparation from microspores and pollen, 538–540
Microspore culture, 701–704
Microspore mitosis, 49, 50
Microsporocytes, 41–43, 48, 140–143, 145, 147
  in situ hybridization of DNA and RNA probes to chromosomes, 504–507
  preparing a suspension for spreading and electron microscopy, 454–456
  staining, for meiosis study by smear technique, 447–449
Microsporogenesis and chromosomal behavior during, 48–51, 460–474
  chromosomal deficiencies, translocations,

mutant genes, cytoplasmic male sterility and blocking effects, 474
developmental stages defined, 461, 462–472
hematoxylin-iron-aceto-carmine stain for microspore mitosis, 466–468
mutant effects, specific, 464, 474
pachytene stage, 434–440
sucrose-aceto-carmine procedure for mature and germinating pollen, 468–470
time different stages accounted for, 462–464
Microtechnique
dehydration, 90
embedding, 91
fixation, 89–90
infiltration, 90–91
mounting coverslip, 94
mounting the sections, 92
sectioning, 91–92
staining, 92–94
Microtubules, 140–148
Middle blade, 18
Midrib, 18, 20, 21, 23–25, 26
Midvein, 18, 20, 21, 23–25, 26
Minimum sample method, 216
"Minus energy" control, 554, 555
Mitochondria
procedures for isolating, and mitochondrial DNA and RNA, 549–555
RNA, in vitro capping and transcription initiation site characterization by RNase protection, 559–564
Mitochondrial DNA (mtDNA), 421–422
Mithramycin, 460
Mithramycin A, 400
Mitotic index, 23, 24
Mitotic metaphase, analysis of, 310
$MgSO_4 \cdot 7H_2O$, stock solutions (100×) and final quantity of components in a simplified medium for growth of maize endosperm, 720
$MnSO_4 \cdot H_2O$, stock solutions (100×) and final quantity of components in a simplified medium for growth of maize endosperm, 720
MO17 line
restoration patterns of inbred line of maize, 420
Rf1-Rf2 constitution of non-T-restoring inbred line, 421
Modeling clay, 110, 112
Modulator (of pericarp) (Mp), 238
Molecular markers, 191, 192, 195
to analyze traits with complex inheritance, 509–513
Molecular markers of inbred lines, 423–430
Monobromo-naphthalene, 476, 479, 482
Monosomics, 335
aborted pollen, 351
characteristics produced by the *R/r-X1* x Mangelsdorf's tester cross, 355
for analysis of univalent chromosome behavior, 354
for mapping unplaced genes to chromosomes, 354–356
identification of, 352–354
production using the *r-X1* genetic system, 350–354
to alter the number of copies of known genetic loci, 356–357
to explore the genome for gene dosage effects, 357
use for gene localization and dosage studies, 350–357
Monosomy, 377
Monosporic Polygonum pattern, 53
Morphological evolution, maize and teosinte, 66–76
block inheritance of the key genes, 74–75
ear disarticulation, 70–71
genetic control of the key morphological traits, 70–72
number of genes, 74
role of known maize genes, 72–73
Morphological mutant characterization by a nine-step method, 209–211
Mounting coverslip, technique, 94
Mounting the sections, technique, 92
Mpl mutation, 271
mRNA
introns, 637
in vitro synthesis of capped and polyadenylated mRNA for translational studies in vitro and in vivo, 592–594
storage of frozen maize tissue, 622–623
ms1 (gene), 474
ms2 (gene), 474
ms3 (gene), 474
ms5 (gene), 474
ms6 (gene), 474
ms7 (gene), 474
ms8 (gene), 474
ms9 (gene), 474
ms10 (gene), 474
ms11 (gene), 474
ms12 (gene), 474
ms13 (gene), 474
ms14 (gene), 474
ms17 (gene), 474
ms22 (gene), 474

ms23 (gene), 474
ms24 (gene), 474
ms28 (gene), 474
Ms41 (gene), 474
ms43 (gene), 474
Mucronate (*Mc1*) locus, genes affecting zein accumulation, 643
Multiple forward mutations, 658
Multiranking (*mr*) gene, 64–65
Murashige-Skoog stock solutions, 628–629
Mutagenesis, 211, 212–218, 220, 232
  expected frequencies, 216
  general handling, 213
  handling M1, 215
  handling M2, 215
  harvest, 215
  paraffin oil pollen treatment using ethyl methane sulfonate (EMS), 213–214
  precautions, 215
  rationale, 212, 216–218
  restricted, 653–656
  seed (dry) treatment using EMS, 217, 218
  stocks, 213
  TATA box promoter sequence element, identifying and characterizing, 631–632
  transposon, 238–239
  treatment, 213–214
  treatment solution, 213
  unrestricted, 656–657
Mutant allele dosage effects, 209
Mutations, 57, 192–193, 195–196, 238
  activator (Ac)-induced, 228–233
  al, 13
  albino, 263, 265, 268
  ameiotic (*am*), 57, 418
  ameiotic 1 (*am1*), 57
  ameiotic 2 (*am2*), 57
  amylose extender, 301
  andromonoecious dwarf (*d1, d2, d3, d5,* and *D8*), 45
  anther-ear-1 (*an1*), 45, 46
  antimorphic, 271
  as1, 57
  asynaptic, 394, 418
  Beadex (*Drosophila*), 271
  branched silkless (*bd*), 46
  characterization of morphological mutants by a nine-step method, 209–211
  chloroplast, 413, 415, 416
  chromosomes, and mapping of maize mutants, 264–265
  Corngrass (*Cg*), 13, 45–46
  D8, 263, 271
  defective kernel (*dek*), 7–9, 407–409, 412
  defective seed 1 (*de1*), 498
  dsy 1, 57
  dsy 2, 57
  dv 1, 57
  dy 1, 57
  elongate, 394–395
  el1, 57
  embryo-specific (*emb*), 8–9, 407–409
  ethyl methane sulfonate (EMS)-induced, 8
  floury-2, 640
  germless (*gm*), 8
  gl15, 13
  Hairy-sheath-frayed, 13
  high chlorophyll fluorescence (hcf), 413
  hypermorphic, 270–271
  hypomorphic, 270, 271
  ig1, 58, 59
  indeterminate (*id*), 14, 386–387, 389
  induced, 220, 228–229
  in vitro selection and, 688
  Knotted-1 (*Kn1*), 271, 272
  Leafy (*Lfy*), 13, 14
  les, 291–296
  lo2, 58, 59
  lp1, 58
  lw, 13
  Mpl, 271
  Mu-induced, 650
  multiple forward, 658
  mutagenesis, 212–218
  neomorphic, 271
  nonchromosomal stripe (NCS), 414, 415, 416
  Notch (*Drosophila*), 270
  opaque-2, 640
  Papyrescent (*Pn*), 46
  Po1, 58, 59
  polytypic (*Pt*), 46
  prescreening stock lines, 227–228
  P-VV, 221
  ramosa-1 (*ra1*), 46
  ramosa-2 (*ra2*), 46
  ramosa-3 (*ra3*), 46
  R-sc, 222
  screens, 229–230
  segmental aneuploids for mutant analysis, 270–273
  segregating kernel, 228, 231
  silkless (*sk*), 46
  silky (*si*), 46
  specific effects of individual mutants during microsporogenesis, 462
  st1, 58
  sugary-1, 323
  tassel seed, 45, 46
  Teopod (*Tp1, Tp2, Tp3*), 13, 45–46, 271, 272

teosinte branched (*tb*), 45
terminal ear (*te*), 45, 46
transposon tagging with Mutator, 243–248
Tunicate (*Tu*), 46, 499
Vestigial glumes (*Vg*), 46
wd, 13
Wx, 298–301
Mutator (*Mu*) transposable element family, 211, 239, 243–248, 590, 650
cloning genes using transposon tagging, 648–651
Mutator transposons, 211

N6 line
restoration patterns of inbred line of maize, 420
Rf1-Rf2 constitution of non-T-restoring inbred line, 421
N6 media, references, sources and components for maize transformation protocols, 678
N28 line, restoration patterns of inbred line of maize, 420
$Na_2E\text{-}DTA \cdot 2H_2O$, stock solutions (100×) and final quantity of components in a simplified medium for growth of maize endosperm, 720
$Na_2M\text{-}oO_4 \cdot 2H2O$, stock solutions (100×) and final quantity of components in a simplified medium for growth of maize endosperm, 720
Near-isogenic lines (NILs), 512
Neocentromeres, 436
Neo gene, 637
Neomorphic mutations, 271
Neomycin phosophotransferase (neo) gene, 608
Neomycin phosphotransferase II, 681
NH_4NO_3, stock solutions (100x) and final quantity of components in a simplified medium for growth of maize endosperm, 720
"Nicking," 305
Nonchromosomal stripe (NCS) mutation, 414, 415, 416
Nondisjunction
B-A chromosome, 315–316, 319, 322, 324, 327
B chromosome, 315
B chromosome centromere, 328
capacity for, 308–314
factor for, 315
rate of, 313–314, 319, 322, 324
r-X1 deficiency and, 351
Nonring chromosome, 347
Northern blotting, 572–574
Nos, promoter active in maize, 634–635, 637
Not I (rare cutter enzymes-1991), recognition sequence and manufacturer, 532
Notch mutation, 270
N-propanol, surface application to leaf, stimulating lesion formation, 295
Nuclear DNA, cytometric measurements, 396–400
DNA measurements, 398–400
Feulgen cytophotometry, 396–397
hydrolysis tests, 398
spectral curve, 398
Nuclear proteins, immunolocalization of, 158–164
Nucleic acid preparation from microspores and pollen, 538–540
Nucleolar organizing region, 356, 461, 462
Nucleolus organizer region (NOR)
chromosome 6, chromosomal translocations, 342–344
mapping of RNA genes, 507
pachytene stage in maize, 435, 437–438
Nucleotide sequencing, 303
"Null stress" point, 53
NY821 line, restoration patterns of inbred line of maize, 420

Off-type progeny, 224, 226, 227
Oh07 line, restoration patterns of inbred line of maize, 420
Oh43 line
in vitro pollen germination, 723
restoration patterns of inbred line of maize, 420
Oh45 line, restoration patterns of inbred line of maize, 420
Oh51A line, restoration patterns of inbred line of maize, 420
Oh545 line, restoration patterns of inbred line of maize, 420
Opaque-1 (*o1*) locus, genes affecting zein accumulation, 643
Opaque-2 gene, 644
cloning, 650
Opaque-2 (*o2*) locus, genes affecting zein accumulation, 643
Opaque-2 modifiers, 644
Opaque-2 modifiers locus, genes affecting zein accumulation, 643
Opaque-2 mutation, 640
Opaque-5 (*o5*) locus, genes affecting zein accumulation, 643
o5 (7L) gene, 259
Opaque-6 (*o6*) locus, genes affecting zein accumulation, 643

Opaque-7 (*o7*) locus, genes affecting zein accumulation, 643
Opaque-9 (*o9*) locus, genes affecting zein accumulation, 643
Opaque-10 (*o10*) locus, genes affecting zein accumulation, 643
Opaque-11 (*o11*) locus, genes affecting zein accumulation, 643
Opaque-12 (*o12*) locus, genes affecting zein accumulation, 643
Opaque-13 (*o13*) locus, genes affecting zein accumulation, 643
Orbicular wall, 50
Orbicules, 50
Organizers, 21
Osmocote (fertilizer), 207
Othricin-resistance genes, 609
Oxalis sp., 289
Oxalis corniculata, 286, 289
Oxalis europaea, 286, 289

p35ALS construct, 691
p268c (genomic clone), class and other zein characteristics, 642
Pac I (rare cutter enzyme-1991), recognition sequence and manufacturer, 532
Pachynema, 434–440, 484, 485, 488
  in situ hybridization of DNA and RNA probes to chromosomes, 504–505
Pachytene, 448, 449, 452, 461, 462
Pachytene chromosome complements for electron-microscopic visualization of synaptonemal complex structures, preparation techniques of whole-mount spreads, 442–446
Pachytene chromosomes, heterochromatin patterns, 484, 485
Pachytene chromosomes, traditional analysis of, 432–440
  acetocarmine squash technique, 432–434, 437
  features of maize pachytene chromosomes, 438
  structure of maize chromosomes at pachynema, 434–440
Paired spikelet genes, 72, 74
Palmero Toluqueno (race) corn, 501
pam1 (gene), 474
pam2 (gene), 474
Papago Indian corn, gametophyte factors, 501
Papyrescent (*Pn*) mutation, 46
Paracentric inversion, 346, 347
Paraffin embedding, 158, 159–160, 161
Paraffin histology, 158
Parenchymatous cells, 20, 21
  root system, 34
Parental imprinting, 272
Parthenogenesis, 391
PAT enzyme and leaf assay, references, sources and components for maize transformation protocols, 678
Patroclinous offspring, 390–391
Pattern of expression, 658
Patterns of plant structures, 61–65
pBluescript, 592
  (Stratagene), 576
PCR analysis. *See* Polymerase chain reaction, applications to transposable elements
PCR Protocols, 586
Pedigrees, maintenance of (record keeping), 205–207
PEG treatment, protoplast isolation, 698
pEmu, promoter active in maize, 634
Pentaploids, 394–395
Pericarp cells, 221, 222, 223, 227, 228
Pericentric inversion, 346, 347
Pericycle, 34
Periodic acid-Schiff's reaction, 137
Peters Soluble (fertilizer), 207
Phase change, 61–62
Phase contrast microscopy, 97–98
  ovule clearing and, 138
Phloem, 34
Phosphinothricin acetyltransferase (PAT) enzyme, 681–682, 683, 691
Phosphinothricin (PPT) herbicide, 681–682, 684
Phosphin-resistance genes, 609
Photographing very-high-contrast objects, 103
Photography (color) with brightfield illumination, 101–102
Photography of maize, 180–185
  camera, its parts and settings, 181–182
  dodging, 184
  film, choice of, 183–184
  internegatives, 184
  printing B/W negatives, 184–185
  subject, and lighting, 182–183
Photography Through the Microscope (9th ed.), 95
Photoperiods, 15
Phragmoplast, 140, 147, 148
Phyllotaxy, 62–65
  ear, 71, 76
Phytochrome, 31
Phytomer, 11–13, 62–65
  cell lineage of the shoot, 13, 14
  manifestation in different regions of the maize plant, 64
Pl gene, cloned anthocyanin genes and their gene product, 283

Plant structure patterns, 61–65
Plasticene, 110
Plastochron number (P), 17, 23
Ploidy hybridization barrier, 394
Ploidy series, production of, 394–395
Plunge freezing, 110
Pme (rare cutter enzyme-1991), recognition sequence and manufacturer, 532
pML1 (genomic clone), class and other zein characteristics, 641
pMS1 (genomic clone), class and other zein characteristics, 641
pMS2 (genomic clone), class and other zein characteristics, 641
po=ms4 (gene), 474
Pocket microscope
  to follow inversion by pollen abortion examinations, 346
  to identify plants heterozygous for a particular aberration, 297
Polarization microscopy, 99
Polar nuclei, 54
Pollen, 49–51
  in vitro germination, 723–724
  nucleic acid preparation from microspores and pollen, 538–540
  related in lineage to the inner anther wall, 262–263
Pollen abortion
  in the field, classification of, 297
  and inversions, 346–347
  rate classification, 310
Pollen contamination, factors reducing, 305
Pollen grain chromosomes, staining procedure, 476–480
  collection and preparation of pollen grains, 479
  hematoxylin stains, 476, 477–478
  prefixation, 479
  procedure for staining, 479–480
Pollenkitt, 51
Pollen mother cells (PMCs), 41–43, 48, 49, 437
  microtubule staining, 142
  smear technique for study of meiosis, 447–449
Polydextran, 124, 130
Polyembryony, 390
Polyethylene glycol (PEG)-mediated DNA uptake into protoplasts, 603–609
  protocols, 605–607
  protoplast growth and selection, 608–609
  reagents and media, 604–605
  stable transformation, 607–608
  transient transformation, 607
Polymerase chain reaction (PCR): applications to transposable elements, 247, 250, 512, 586–591
  cloning of transposon-tagged loci by the inverse polymerase chain reaction (IPCR), 587–588
  DNA miniprep, 524
  maize-related difficulties, 586–587
  maize-specific considerations, 588–590
  protocol for IPCR, 590–591
Polymitotic 1 (*po1*) mutation
  maize meiotic/gametophytic mutant behavior, name and reference, 58, 59
  plants, 344
Polystichy, 72
Polytypic (*Pt*) mutation, 46
PPDK, promoter active in maize, 634
Preferential fertilization, rate of, 314
Preferential pairing in meiosis, analysis using trisomics, 307
Preligular region, 23, 24
Premeiosis (PMC), 461, 462
Preservation of tissue, 109, 110
Primary intermediate veins, 25
Primexine, 50
Primordium, 23, 24, 25
pro 1 (8L) gene, 259
Proembryo, 5, 6, 7
Promoters, 633–635
Propane jet freezing (PJF), 110, 126
Propane plunge-freezing, 122–123
Prophyll, 11, 12
Propidium iodide [PI], 104
Protein synthesis in isolated mitochondria, 554–555
Protoplast culture, 695–699
  culture of, 698
  from callus, 699
  from suspension, 699
  initiation of callus cultures, 695–696
  isolation of protoplasts, 697–698
  plantlets, 699
  preparation of suspension cultures, 696–697
  regeneration of plants, 698–699
  selection of friable embryogenic callus suitable for initiating suspension cultures, 696
Protoplast isolation buffer (PIB), 530
Protoplasts
  electroporated, RNA isolation, 547–548
  isolation, 605–607, 697–698
  isolation of mesophyll protoplasts from leaves, 541–542
  polyethylene glycol (PEG)-mediated DNA uptake into protoplasts, 603–609

Protoxylem, 34
psi gZ22.8 (genomic clone), class and other zein characteristics, 641
pUC19, 592
Puccinia sorghi Schw., 286–290
  axenic culture not successful, 286
  genetics of resistance to, 286–287
  identification of rust races, 288
  method of inoculation, 288–289
  sexual cycle, 289–290
  urediospore storage, 288
  working with, 288–290
Pulsed-field gel electrophoresis (PFGE)
  high-molecular-weight DNA preparation and analysis, 530–533
  choice of plant material, 530–531
  digestion of DNA in agarose plugs, 532–533
  preparation of HMW DNA, 531–532
  resolution and analysis, 533
Pure Bialaphos (herbicide sources), references, sources and components for maize transformation protocols, 678
Pure PPT (herbicide sources), references, sources and components for maize transformation protocols, 678
P-variegated allele (P-vv), 238
P-vv mutation, 221, 227
Pycniospores, 289–290
Pyd gene, 439
pZ3 (genomic clone), class and other zein characteristics, 642
pZ10B (genomic clone), class and other zein characteristics, 642

QTL (quantitative trait locus) mapping, 196, 509–513
Quantitative trait loci (QTLs), 196, 509–513
Quantitative trait locus (QTL) mapping, 196, 509–513
Quartet, mutant genes, chromosomal deficiencies (TB) and cytoplasmic male sterility (cms) and their blocking effects, 474

R138 line, 719
  restoration patterns of inbred line of maize, 420
R177 line, restoration patterns of inbred line of maize, 420
R213 line, Rf1-Rf2 constitution of non-T-restoring inbred line, 421
R802A line, restoration patterns of inbred line of maize, 420
Race Chalco teosinte, 74
Race Chapalote maize, 74
Radiation-induced chromosome breakage, 258
Radicle, 30
Ramosa-1 (*ra1*) mutation, 46
Ramosa-2 (*ra2*) mutation, 46
Ramosa-3 (*ra3*) mutation, 46
Randomly Amplified Primed DNAs (RAPDs), 191, 192
Reciprocal interchange, 361, 362–363
Reciprocal translocations, 361, 362–363
  listing of the current collection of 1003 translocations, address for obtaining, 363
Reciprocal translocations and inversions between 1S and the other arms, 236–237
Recombinant inbreds (RI), mapping genes with, 249–254
  derivation of recombinant inbreds, 250
  how mapping is done, 250–252
  potential problems, 252–254
  what is mapped, 250
Recombination nodules (RNs), 442, 444
Reconstruction lanes, 565
Red aleurone
  frequencies and parents of monosomic types produced by *R*/*r-X1* x Mangelsdorf's tester cross, 353
  genetic marker, and monosomics, 353, 355
Regeneration of plants from somatic cell cultures: applications for in vitro genetic manipulation, 663–670
Regions, 18
Replica technique, 110–114, 115
Restorer genes, 418–422
Restorer-of-fertility (Rf) genes, 256–257, 418–422
  Rf1, Rf2, Rf3, Rf4, Rf5 (specific genes), characteristics of restorer-of-fertility genes in maize, 419
Restricted mutagenesis, 653–656
Restriction digests, analysis of, 303
Restriction enzyme digestion, DNA miniprep, 524
Restriction Fragment Length Polymorphisms (RFLPs), 191–192, 195, 246, 252–253, 328, 651
  agreement with estimates on genetic relationship based on pedigree, 427–428, 429
  homeologous sequences of chromosomes, 493–494
  locating recessive genes to chromosome arm, 315, 318
  loci used to assist transposon-tagging efforts, 653–660
  maps, 211
  marker, 287
  marker, segmental aneuploid analysis, 377, 379, 380

molecular analyses, cytoplasmically inherited mutants, 415
monosomics identification, 352–353
quantitative trait analysis, 510, 512
reciprocal translocations and, 363
surveys of maize inbreds and loci, 425, 426–427
techniques, 210
to determine monosomic chromosome, 354, 355–356, 359
Reversion, 658
Revertants, 658
Rf. *See* Restorer-of-fertility
RFLP. *See* Restriction Fragment Length Polymorphisms
r-g, marker system for monosomic selection, 360
Rhodamine fluorescence, 104
Rhodamine-phalloidin, microfilament staining, 145–146
RI. *See* Recombinant inbreds
Rice Popcorn, 496–497, 499
Ring chromosomes, 258, 260, 503–504
paracentric inversions and, 347
r-m3, marker system for monosomic selection, 360
RN. *See* Recombination nodules
RNA
antisense, 576
in situ hybridization of probes to chromosomes, 504–507
in vitro synthesis of capped and polyadenylated mRNA for translational studies in vitro and in vivo, 592–594
isolation from electroporated protoplasts, 547–548
isolation from microspores and pollen, 540
isolation from Wx and wx endosperms, 545–546
mitochondrial, 549–555
mitochondrial, in vitro capping, 559–564
northern blotting, 572–574
preparation from leaf blades or isolated leaf cells, 543–544
preparation from leaves: expanded blades and separated bundle sheath and mesophyll cells, 541–544
small nuclear, isolation of, 519–521
RNase protection assay, 559, 563–564, 575–578
R-Navajo (*R-nj*) embryo anthocyanin marker, 390, 391–392
Robertson's Mutator, 232
Robertson's Mutator stocks, 8
Roguing, 424
Romanovsky stains, 491
Root cap, 34, 35–36
Root lodging, 200
Root meristematic cells, 140
Roots, 29–37
adventitious, 29–32, 34
anatomy, 33–37
apex, 36
brace, 29, 31, 32
cap, 34–36
elongation zone, 36–37
habit, 29–33
hilling, 32
lateral, 29, 32
primary, 29–31, 33–34
prop, 29, 31–33
proximal meristem, 36
quiescent center, 36
seminal, 29–31, 34
shoot-borne, 29
soil water content, 32–33
sterile liquid culture, 33
system, 29–31
tap, 29, 31
trait distinguishing juvenile and adult regions of the shoot, 12
Roundup (herbicide), 199
R-P (gene), cloned anthocyanin genes and their gene product, 283
Rp genes (resistant against *Puccinia sorghi*), 287, 288
R-r, marker system for monosomic selection, 360
R-S, cloned anthocyanin genes and their gene product, 283
R-sc gene, 259
R-scm2, marker system for monosomic selection, 360
R-scm2 allele, 333
R-scm3 allele, 330
R-sc mutation, 222
RsrII (rare cutter enzyme-1991), recognition sequence and manufacturer, 532
Rust diseases, 286–290
r-X1 deficiency, nondisjunction and deficiency analysis, 494–495
r-X1 genetic system
characteristics, 351–352
deficiency (*r-X1*), 350–351
of monosomics, 350
propagating the *r-X1* deficiency, 351
r-X1 marker system, 359–360
r-x1/R-r, marker system for monosomic selection, 360
r-x1/r-g, marker system for monosomic selection, 360

Safranin and fast green, 136, 137
SAS (computer software statistical package), 511
Sass's hemalum, 136, 137
Satellite, chromosome 6, chromosomal translocations, 342–344
Scanning electron microscopy (SEM), 108–116
grounding the specimen, 115
observations, techniques recommended, 115–116
plastic histology and, 137
specimen preparation methods, 109–114
stabilizing the specimen, 115
Schiff's reagent. *See* Feulgen staining
Scintillation counter, 627
Sclerenchyma cells, 20, 21
Scutella, colored, 321–322, 327
Scutellum, 6, 7, 9, 409
pigmentation, 330
root system, 30
SD10 line, restoration patterns of inbred line of maize, 420
Second pollen mitosis, mutant genes, chromosomal deficiencies (TB) and cytoplasmic male sterility (cms) and their blocking effects, 474
Sectioning
immunolocalization of nuclear proteins, 159–161
technique, 91–92
Segmental aneuploid analysis, 377–382
centromeric overlaps, 380–381, 382
interstitial deficiencies, 381–382
segmental tetrasomy, 377, 378–379
segmental trisomy, 377–378, 379
stable terminal tetrasomics, 379–382
Segmental aneuploids, use for mutant analysis, 270–273
classification scheme of dominant mutations, 270–271
Segmental transpositions, directed synthesis of, protocol, 383–385
Segmental trisomy, 377–378
Seedlings, irradiation and clonal analysis, 267
Selection, in vitro, 685–688
Selective agent growth inhibition curve, 687
Selfing, 215, 216, 228, 245–246
Self-pollination, 411–412
gametophyte factors, 497, 498
Semisterile hypoploids, 324, 325, 326
Semisterility, 311–312, 361–364, 419
Seneca 60 (Se 60)
axillary bud cultures, 725–726
inflorescence culture, 712–714
in vitro pollen germination, 723
Sequenase enzyme, 632
Sfi I (rare cutter enzymes-1991), recognition sequence and manufacturer, 532
*Sgr*A1 (rare cutter enzyme-1991), recognition sequence and manufacturer, 532
Sh bz-E2 Wx, *Bz* selections from *bz* heteroallelic plants pollinated with *sh bz-R wx*, 304
sh bz-m1 wx, *Bz* selections from *bz* heteroallelic plants pollinated with *sh bz-R wx*, 304
sh bz-m2 (*DI*) *wx*, *Bz* selections from *bz* heteroallelic plants pollinated with *sh bz-R wx*, 304
Sheath, 18–21, 26, 27
Shock absorbers, 124–126
Shoot, 6–7, 8, 11–15
adult phase of, 12–13
apical meristem, 6–8, 11–14
cell lineage, 13–14
growth of, 14–15
juvenile part of, 12–13
mutations, 13, 14
photosensitivity (photoperiod), 14–15
structure, 11–13
temperature sensitivity, 15
Shoot-bagging, 200–201
Shoot meristem, clonal analysis, 263
Shoot meristem culture, 84, 715–718
Shrunken-1 (*sh1*) gene, 624–629, 648
promoter active in maize, 634
Shrunken-1 intron, 637
Shrunken-2 (*sh2*) gene, 624–629
Sib-pollinations, 704
Silicon carbide fibers, DNA delivery into cell cultures using silicon carbide fibers, 610–612
Silkless (*sk*) mutation, 46
Silks, cutting back effect, 201–202
Silky (*si*) mutation, 46
Silver-enhanced immunogold (IGSS) labeling, 158–159, 163
Single-lens reflex cameras (SLR), 101, 103
SK2 line, restoration patterns of inbred line of maize, 420
^{35}S labeling, 170–173, 177–178, 179
Slime, 35–36
Smear technique, meiosis in pollen mother cells studied, 447–449
Sn (gene), cloned anthocyanin genes and their gene product, 283
Somatic cell cultures, regeneration of plants, in vitro genetic manipulation applications, 663–670

Somatic chromosome preparation and C-banding, technique, 484–491
Somatic chromosomes, technique for the preparation of, 481–483
Southern blot hybridization, 569–572
Southern blots, genomic DNA, 566–568
Spectral curve, 398
Spindle, 140
Splam freezing, 110
Spore abortion, 361
Sporocytes, collection fixing, or killing, for smear technique, 447
Sporophyte
  determining whether normal pollen is produced, 418
  placement of genes expressed in, 255–256
Sporopollenin, 49, 50
Spot metering, 103
spt 1 phenotype, 326
spt 2 phenotype, 326
Spurr's resin, 136
  kit, 113–114
*Sse*83871 (rare cutter enzyme-1991), recognition sequence and manufacturer, 532
st (gene), 474
Stable terminal tetrasomics, 379–382
Stable transformant selection from Black Mexican Sweet (BMS) maize suspension cultures, 690–694
Staining (technique), 92–94
  safranin and fast green (Johansen 1940), 92–93
  safranin and orange gold (Sharman 1943), 93–94
  toluidine blue (TBO), 94
Staining procedure for pollen grain chromosomes, 476–480
  collection and preparation of pollen grains, 479
  hematoxylin stains, 476, 477–478
  prefixation, 479
  procedure for staining, 479–480
Staining Procedures, Fourth Edition, 163
Starch biosynthetic genes, 624–629
Starch-bound ADP-glucose-starch glucosyl-transferase reaction, 627
Starch gel electrophoresis
  of enzymes, 425
  of isozymes, 425
Steedman's wax (SW) embedding, 158, 160–161, 162
Stereopairs, 116
Sticky chromosome 1 (*st1*) mutation, maize meiotic/gametophytic mutant behavior, name and reference, 58
Stock 6, haploid production, 386, 387
Storage of frozen maize tissue, 622–623
Structural genes, and anthocyanin synthesis, 282
Structures, patterns of plants, 61–65
S-type male-sterile cytoplasm (*cms-S*), 256–257
Sucrose, 123–124, 126, 127, 130
  stock solution (100x) and final quantity of components in a simplified medium for growth of maize endosperm, 720
Sucrose synthase reaction, 626
Sugary-1 mutation, 323, 496–499
Sugary endosperm
  frequencies and parents of monosomic types produced by *R*/*r*-*X1* x Mangelsdorf's tester cross, 353
  genetic marker, and monosomics, 353, 355
Supergold popcorn, 499–500
Suspension cells, type II, initiation maintenance and plant regeneration, 671, 674–676
Suspension cultures
  Black Mexican Sweet (BMS) maize, stable transformant selection, 690–694
  endosperm, *Zea mays* L., 719–722
Suspensor, 6, 7
Sweet corn (*sugary 1*), 323, 496–499
Sweet corn var. Seneca 60, inflorescence culture, 712–714
Swa I (rare cutter enzyme-1991), recognition sequence and manufacturer, 532
Symbol identifier (ID) letter or number for the aberration, 364
Synaptonemal complexes, preparation of a suspension of microsporocytes for spreading and electron microscopy, 454–456
Synaptonemal complex (SC) lateral elements, preparation techniques of whole-mount spreads of pachytene chromosome complements for electron-microscopic visualization, 442–446
SYSTAT (computer software statistical package), 511

T, 361
Tandem duplications, 493–494
Tapetal cells, 48–50
Tassel, 11–12, 37–43, 45–46
  androgenesis, 702–703
  cell lineage of the shoot, 13, 14
  cells giving rise to, 262
  development, 37–38, 41–43
  developmental stages in maturation, 473

differentiation, 41–43
inflorescence culture, 712–714
lemma, 38, 39
macroscopic development during microsporogenesis, 462, 473
microsporogenesis and, 48
morphology, 38
mutants, 45–46
palea, 38, 39
pedicellate spikelet, 38, 39, 41
phase change, 61
pollination and, 202–203
reproductive and vegetative structures, 13
sessile spikelet, 38, 39, 41
shoot growth, 14, 15
smear technique for the study of meiosis in pollen mother cells, 447–449
teosinte branched effect, 73
Tassel seed (*ts*) mutations, 45, 46
TATA box promoter sequence element in a nuclear gene, identifying and characterizing, 630–632
Teliospores, 289–290
Telomere, 304, 435
Telophase, 54
Teopod (*Tp1*, *Tp2*) gene, 72, 73
Teopod (*Tp1*, *Tp2*, *Tp3*) mutations, 13, 45–46, 271–272
Teosinte, 286
domestication of, 64
gametophyte factors, 501
morphological evolution, 66–76
photosensitivity, 70
Zea mays ssp. mexicana, 66
Zea mays ssp. parviglumis, 66
Zea spp. vs. maize, 66–68, 69–70, 73, 74
Teosinte branched (*tb1*) gene, 12, 72, 73, 272
Teosinte branched (*tb*) mutants, 45
Terminal ear (*te*) mutations, 45, 46
Terminal tetrasomics, 377
Testcrosses in coupling (cis), 194–195
Testcrosses in repulsion (trans), 194–195
Testcross progeny analysis, for genetic fine structure, 303–305
Tetrad stage, 452
Tetraploids, 394
Tetrasomics, 335
stable terminal, 379–380, 381, 382
Tetrasomy, 377, 378
segmental, 378–379
Thiamine, stock solutions (100×) and final quantity of components in a simplified medium for growth of maize endosperm, 720
38–11 line, restoration patterns of inbred line of maize, 420
Rf1-Rf2 constitution of non-T-restoring inbred line, 421
TIC-CP (teosinte incompatibility complex-Central Plateau), 501–502
Tillers, 12, 68, 73
Tissue clearing
with NaOH and chloral hydrate, technique (Arnott 1959), 85–86
without removing cytoplasmic components (Herr 1972), 86–87
Tissue infiltration, immunolocalization of nuclear proteins, 159–161
Toluidine blue, epidermal cells, traits distinguishing juvenile and adult regions of the shoot, 12, 20, 21
Tr
as male parent of a newly arisen male sterile trait, 422
restoration patterns of inbred line of maize, 420
Traits with complex inheritance, analysis using molecular markers, 509–513
Transcription initiation factor TFIID, 630, 633
Transcription initiation site characterization by RNase protection, 559–564
Transformation, 658
Transgenic plant production, references, sources and components for maize transformation protocols, 678
Transgenic plant production via microprojectile-mediated gene transfer, 677–684
Transgenic plants, protocol for production, 682–684
Transient gene expression assay by electroporation of protoplasts, 613–615
Translocations, specific types
1–6b, satellite interchanges and breakpoints, 343
1–6Li, NOR-interchanges and breakpoints, 343
1–6(4986), NOR-interchanges and breakpoints, 343
1–6(6189), NOR-interchanges and breakpoints, 343
1–6(8415), NOR-interchanges and breakpoints, 343
2–6(001–15), satellite-interchanges and breakpoints, 343
2–6(027–4), NOR-interchanges and breakpoints, 343
2–6(5419), NOR-interchanges and breakpoints, 343
2–6(8441), NOR-interchanges and breakpoints, 343

2–6(8786), NOR-interchanges and breakpoints, 343
3–6b, satellite-interchanges and breakpoints, 343
3–6(030–8), NOR-interchanges and breakpoints, 343
3–6(032–3), NOR-interchanges and breakpoints, 343
4–6c, satellite-interchanges and breakpoints, 343
4–6(003–16), satellite-interchanges and breakpoints, 343
4–6(4341), NOR-interchanges and breakpoints, 343
4–6(5227), satellite-interchanges and breakpoints, 343
4–6(7037), NOR-interchanges and breakpoints, 343
4–6(7328), satellite-interchanges and breakpoints, 343
5–6b, satellite-interchanges and breakpoints, 343
5–6d, satellite-interchanges and breakpoints, 343
5–6f, NOR-interchanges and breakpoints, 343
5–6(8219), satellite-interchanges and breakpoints, 343
5–6(8696), NOR-interchanges and breakpoints, 343
6–7(035–3), NOR-interchanges and breakpoints, 343
6–7(4964), NOR-interchanges and breakpoints, 343
6–7(5181), NOR-interchanges and breakpoints, 343
6–7(7036), satellite-interchanges and breakpoints, 343
6–9a, NOR-interchanges and breakpoints, 343
6–9d, NOR-interchanges and breakpoints, 343
6–9(017–14), satellite interchanges and breakpoints, 343
6–9(4778), NOR-interchanges and breakpoints, 343
6–10f, satellite-interchanges and breakpoints, 343
6–10(5253), NOR-interhcanges and breakpoints, 343
6–10(5519), NOR-interchanges and breakpoints, 343
6Lc, 474
6Lb, 474
B-3Ld, 392
B-10la, 391
TB-1La, 334, 474
TB-1La, arms uncovered by translocation and closest known genes (proximally or distally), 337
TB-1La, kernel and seedling factors for confirming the presence of the current basic set of B-A translocations, 320
TB-1La-3L4759–3, arms uncovered by translocation and closest known genes (proximally or distally), 337
TB-1La-3L5242, arms uncovered by translocation and closest known genes (proximally or distally), 337
TB-1La-3L5267, arms uncovered by translocation and closest known genes (proximally or distally), 337
TB-1La-3Le, arms uncovered by translocation and closest known genes (proximally or distally), 337
TB-1La-4L4692, arms uncovered by translocation and closest known genes (proximally or distally), 338
TB-1La-5S8041, arms uncovered by translocation and closest known genes (proximally or distally), 338
TB-1La-5S8041, kernel and seedling factors for confirming the presence of the current basic set of B-A translocations, 320
TB-1Lc, arms uncovered by translocation and closest known genes (proximally or distally), 337
TB-1Lc, homozygous B-A translocation stocks and their inbred backgrounds, 321
TB-1Sb, 474
TB-1Sb, arms uncovered by translocation and closest known genes (proximally or distally), 337
TB-1Sb, homozygous B-A translocation stocks and their inbred backgrounds, 321
TB-1Sb, kernel and seedling factors for confirming the presence of the current basic set of B-A translocations, 320
TB-1Sb-2Lc, arms uncovered by translocation and closest known genes (proximally or distally), 337
TB-1Sb-2L4464, arms uncovered by translocation and closest known genes (Proximally or distally), 337
TB-1Sb-2L4464, homozygous B-A translocation stocks and their inbred backgrounds, 321
Translocations, specific types (*cont.*)
TB-1Sb-2L4464, kernel and seedling factors for confirming the presence of the current basic set of B-A translocations, 320

TB-1Sb-2L4464–4Lf, arms uncovered by translocation and closest known genes (proximally or distally), 338
TB-2Sa, arms uncovered by translocation and closest known genes (proximally or distally), 337
TB-2Sa, kernel and seedling factors for confirming the presence of the current basic set of B-A translocations, 320
TB-2Sb, arms uncovered by translocation and closest known genes (proximally or distally), 337
TB-3La, 474
TB-3La, arms uncovered by translocation and closest known genes (proximally or distally), 337
TB-3La, homozygous B-A translocation stocks and their inbred backgrounds, 321
TB-3La, kernel and seedling factors for confirming the presence of the current basic set of B-A translocations, 320
TB-3La-2L7285, arms uncovered by translocation and closest known genes (Proximally or distally), 337
TB-3La-2S6270, 326
TB-3La-2S6270, arms uncovered by translocation and closest known genes (proximally or distally), 337
TB-3La-2S6270, kernel and seedling factors for confirming the presence of the current basic set of B-A translocations, 320
TB-3Lc, 474
TB-3Lc, arms uncovered by translocation and closest known genes (proximally or distally), 337
TB-3Ld, 474
TB-3Ld, arms uncovered by translocation and closest known genes (proximally or distally), 337
TB-3Ld, homozygous B-A translocation stocks and their inbred backgrounds, 321
TB-3Ld, indeterminate gametophyte and haploid production, 387
TB-3Lg, arms uncovered by translocation and closest known genes (proximally or distally), 337
TB-3Lg, homozygous B-A translocation stocks and their inbred backgrounds, 321
TB-3Lf, arms uncovered by translocation and closest known genes (proximally or distally), 337
TB-3Lh, arms uncovered by translocation and closest known genes (proximally or distally), 337
TB-3Li, arms uncovered by translocation and closest known genes (proximally or distally), 337
TB-3Lj, arms uncovered by translocation and closest known genes (proximally or distally), 337
TB-3Lk, arms uncovered by translocation and closest known genes (proximally or distally), 337
TB-3Ll, arms uncovered by translocation and closest known genes (proximally or distally), 337
TB-3Ll, homozygous B-A translocation stocks and their inbred backgounds, 321
TB-3Lm, arms uncovered by translocation and closest known genes (proximally or distally), 337
TB-3Sb, 474
TB-3Sb, arms uncovered by translocation and closest known genes (proximally or distally), 337
TB-3Sb, homozygous B-A translocation stocks and their inbred backgrounds, 321
TB-3Sb, kernel and seedling factors for for confirming the presence of the current basic set of B-A translocations, 320
TB-4Lb, arms uncovered by translocation and closest known genes (proximally or distally), 338
TB-4Lc, 474
TB-4Lc, arms uncovered by translocation and closest known genes (proximally or distally), 338
TB-4Lc, homozygous B-A translocation stocks and their inbred backgrounds, 321
TB-4Lc, kernel and seedling factors for confirming the presence of the current basic set of B-A translocations, 320
TB-4Ld, arms uncovered by translocation and closest known genes (proximally or distally), 338
TB-4Le, arms uncovered by translocation and closest known genes (proximally or distally), 338
TB-4Lf, 474
TB-4Lf, arms uncovered by translocation and closest known genes (proximally or distally), 338
TB-4Lf, homozygous B-A translocation stocks and their inbred backgrounds, 321
TB-4Lh, arms uncovered by translocation and closest known genes (proximally or distally), 338
TB-4Li, arms uncovered by translocation and closest known genes (proximally or distally), 338

TB-4Sa, 474
TB-4Sa, arms uncovered by translocation and closest known genes (proximally or distally), 337
TB-4Sa, homozygous B-A translocation stocks and their inbred backgrounds, 321
TB-4Sa, kernel and seedling factors for confirming the presence of the current basic set of B-A translocations, 320
TB-4Sg, arms uncovered by translocation and closest known genes (proximally or distally), 338
TB-5La, 474
TB-5La, arms uncovered by translocation and closest known genes (proximally or distally), 338
TB-5La, kernel and seedling factors for confirming the presence of the current basic set of B-A translocations, 320
TB-5La-3L5521, arms uncovered by translocation and closest known genes (proximally or distally), 337
TB-5La-3L7043, arms uncovered by translocation and closest known genes (proximally or distally), 337
TB-5La-3Lb, arms uncovered by translocation and closest known genes (proximally or distally), 337
TB-5Lb, arms uncovered by translocation and closest known genes (proximally or distally), 338
TB-5Lb, homozygous B-A translocation stocks and their inbred backgrounds, 321
TB-5Ld, arms uncovered by translocation and closest known genes (proximally or distally), 338
TB-5Sc, 474
TB-5Sc, arms uncovered by translocation and closest known genes (proximally or distally), 338
TB-5Sc, kernel and seedling factors for confirming the presence of the current basic set of B-A translocations, 320
TB-6Lb, arms uncovered by translocation and closest known genes (proximally or distally), 338
TB-6Lb, homozygous B-A translocation stocks and their inbred backgrounds, 321
TB-6Lc, arms uncovered by translocation and closest known genes (proximally or distally), 338
TB-6Lc, homozygous B-A translocation stocks and their inbred backgrounds, 321
TB-6Lc, kernel and seedling factors for confirming the presence of the current basic set of B-A translocations, 320
TB-6Ld, arms uncovered by translocation and closest known genes (proximally or distally), 338
TB-6Sa, 474
TB-6Sa, arms uncovered by translocation and closest known genes (proximally or distally), 338
TB-6Sa, homozygous B-A translocation stocks and their inbred backgrounds, 321
TB-6Sa, kernel and seedling factors for confirming the presence of the current basic set of B-A translocations, 320
TB-7Lb, 474
TB-7Lb, arms uncovered by translocation and closest known genes (proximally or distally), 338
TB-7Lb, homozygous B-A translocation stocks and their inbred backgrounds, 321
TB-7Lb, kernel and seedling factors for confirming the presence of the current basic set of B-A translocations, 320 TB-7Lb-4L4698, arms uncovered by translocation and closest known genes (proximally or distally), 338
TB-7Sc, 474
TB-7Sc, arms uncovered by translocation and closest known genes (proximally or distally), 338
TB-7Sc, kernel and seedling factors for confirming the presence of the current basic set of B-A translocations, 320
TB-8La, 474
TB-8La, arms uncovered by translocation and closest known genes (proximally or distally), 338
TB-8La, homozygous B-A translocation stocks and their inbred backgrounds, 321
TB-8Lb, arms uncovered by translocation and closest known genes (proximally or distally), 338
TB-8Lc, 474
TB-8Lc, arms uncovered by translocation and closest known genes (proximally or distally), 338
TB-8Lc, homozygous B-A translocation stocks and their inbred backgrounds, 321
TB-8Lc, kernel and seedling factors for confirming the presence of the current basic set of B-A translocations, 320
TB-9La, 474
TB-9La, arms uncovered by translocation and closest known genes (proximally or distally), 339
Translocations, specific types (*cont.*)
TB-9Lc, 474

TB-9Lc, arms uncovered by translocation and closest known genes (proximally or distally), 339
TB-9Lc, kernel and seedling factors for confirming the presence of the current basic set of B-A translocations, 320
TB-9Lc (Wc), homozygous B-A translocation stocks and their inbred backgrounds, 321
TB-9S (Saraiva), arms uncovered by translocation and closest known genes (proximally or distally), 339
TB-9Sb, 474
TB-9Sb, arms uncovered by translocation and closest known genes (proximally or distally), 339
TB-9Sb, homozygous B-A translocation stocks and their inbred backgrounds, 321
TB-9Sb-4L6222, arms uncovered by translocation and closest known genes (proximally or distally), 338
TB-9Sb-4L6222, kernel and seedling factors for confirming the presence of the current basic set of B-A translocations, 320
TB-9Sb-4L6504, arms uncovered by translocation and closest known genes (proximally or distally), 338
TB-9Sd, 474
TB-9Sd, arms uncovered by translocation and closest known genes (proximally or distally), 338
TB-9Sd, homozygous B-A translocation stocks and their inbred backgrounds, 321
TB-9Sd, kernel and seedling factors for confirming the presence of the current basic set of B-A translocations, 320
TB-10Lb, arms uncovered by translocation and closest known genes (proximally or distally), 339
TB-10Ld, arms uncovered by translocation and closest known genes (proximally or distally), 340
TB-10L1, arms uncovered by translocation and closest known genes (proximally or distally), 339
TB-10L2, arms uncovered by translocation and closest known genes (proximally or distally), 340
TB-10L3, arms uncovered by translocation and closest known genes (proximally or distally), 339
TB-10L4, arms uncovered by translocation and closest known genes (proximally or distally), 339
TB-10L5, arms uncovered by translocation and closest known genes (proximally or distally), 339
TB-10L6, arms uncovered by translocation and closest known genes (proximally or distally), 339
TB-10L7, arms uncovered by translocation and closest known genes (proximally or distally), 339
TB-10L8, arms uncovered by translocation and closest known genes (proximally or distally), 339
TB-10L9, 474
TB-10L9, arms uncovered by translocation and closest known genes (proximally or distally), 339
TB-10L10, arms uncovered by translocation and closest known genes (proximally or distally), 339
TB-10L11, arms uncovered by translocation and closest known genes (proximally or distally), 339
TB-10L12, arms uncovered by translocation and closest known genes (proximally or distally), 339
TB-10L13, arms uncovered by translocation and closest known genes (proximally or distally), 340
TB-10L14, arms uncovered by translocation and closest known genes (proximally or distally), 339
TB-10L15, arms uncovered by translocation and closest known genes (proximally or distally), 340
TB-10L16, arms uncovered by translocation and closest known genes (proximally or distally), 339
TB-10L17, arms uncovered by translocation and closest known genes (proximally or distally), 339
TB-10L18, 334
TB-10L18, arms uncovered by translocation and closest known genes (proximally or distally), 339
TB-10L19, 474
TB-10L19, arms uncovered by translocation and closest known genes (proximally or distally), 339
TB-10L19, kernel and seedling factors for confirming the presence of the current basic set of B-A translocations, 320
TB-10L20, arms uncovered by translocation and closest known genes (proximally or distally), 339
TB-10L20, homozygous B-A translocation stocks and their inbred backgrounds, 321

TB-10L21, arms uncovered by translocation and closest known genes (proximally or distally), 340
TB-10L22, arms uncovered by translocation and closest known genes (proximally or distally), 339
TB-10L23, arms uncovered by translocation and closest known genes (proximally or distally), 340
TB-10L24, arms uncovered by translocation and closest known genes (proximally or distally), 340
TB-10L25, arms uncovered by translocation and closest known genes (proximally or distally), 339
TB-10L26, arms uncovered by translocation and closest known genes (proximally or distally), 339
TB-10L27, arms uncovered by translocation and closest known genes (proximally or distally), 340
TB-10L28, arms uncovered by translocation and closest known genes (proximally or distally), 339
TB-10L29, arms uncovered by translocation and closest known genes (proximally or distally), 340
TB-10L30, arms uncovered by translocation and closest known genes (proximally or distally), 340
TB-10L31, arms uncovered by translocation and closest known genes (proximally or distally), 339
TB-10L32, 474
TB-10L32, arms uncovered by translocation and closest known genes (proximally or distally), 340
TB-10L32, homozygous B-A translocations and their inbred backgrounds, 321
TB-10L33, arms uncovered by translocation and closest known genes (proximally or distally), 340
TB-10L34, arms uncovered by translocation and closest known genes (proximally or distally), 340
TB-10L35, arms uncovered by translocation and closest known genes (proximally or distally), 340
TB-10L36, arms uncovered by translocation and closest known genes (proximally or distally), 339
TB-10L37, arms uncovered by translocation and closest known genes (proximally or distally), 339
TB-10L38, arms uncovered by translocation and closest known genes (proximally or distally), 340
TB-10La, arms uncovered by translocation and closest known genes (proximally or distally), 340
TB-10La, homozygous B-A translocation stocks and their inbred backgrounds, 321
TB-10Sc, 474
TB-10Sc, arms uncovered by translocation and closest known genes (proximally or distally), 339
TB-10Sc, kernel and seedling factors for confirming the presence of the current basic set of B-A translocations, 320
Translocations as genetic markers, 361–363
Transmission electron microscopy (TEM), 118–133
  chemical (conventional) fixation, 118–121, 122, 128
  dehydration, 118, 119, 121
  embedding, 118, 119, 121, 127–128
  freeze-fracture, 132–133
  freezing methods, 118, 122–126, 128
  immunolocalization, 118, 128
  plastic histology and, 137
  plastic infiltration, 121–122
Transposition process, 220
Transposon-induced chromosome breakage, 258
Transposon tagging
  cloning genes, 647–651
  RFLP loci used, 653–658
  with *Mutator*, 243–248
  with *Spm*, 240–242
Transposon tagging with *Mutator*, 243–248
  crosses with newly arisen mutants, 246247
  DNA analysis, 247–248
  molecular markers, 245–246
  monitoring *Mutator* activity, 244–245
  Mutator system for tagging, 243–244
  non-autonomous genetic markers, 245
  Robertson's *Mutator* test, 245
  targeted and nontargeted approaches to tagging, 244
Transposon tagging with Spm, protocol, 240–242
  confirmation that the mutation is *Spm*-controlled, 242
  expected phenotype, 241
  expected segregation ratios, 241–242
  frequency of transposition of *Spm*, 241
Transverse veins, 26
Trinucleate, mutant genes, chromosomal deficiencies (TB) and cytoplasmic male sterility (cms) and their blocking effects, 474

Trisome, 326–327
Trisomic manipulation, 307
Trisomics, 312, 323
  partial, 392
  ring chromosomes, 503, 504
  r-X1 deficiency and, 351
  segmental, 377–378
Tryphine, 51
Tunicate (*Tu*) gene, 46, 72, 73, 498–499
Tunicate (*Tu1*) gene, 272
T-*waxy* stocks, 407–408
Two-dimensional gel electrophoresis of proteins, 425
Two-wavelength method, 396, 398

Ubisch bodies, 50
Unbranched tassel (*ub*) gene, 65
"Uncovered" phenotype, 325
Uninucleate, mutant genes, chromosomal deficiencies (TB) and cytoplasmic male sterility (cms) and their blocking effects, 474
Unrestricted mutagenesis, 656–657
Untranslated region (UTR), 637
Urea-based plant DNA miniprep, 526–527
Urediospores, 288–289, 290
USDA maize genome project, 429
UV nicking, 533

va1 (gene), 474
va2 (gene), 474
Va26 line, restoration patterns of inbred line of maize, 420
Vaccinia virus, 559
Vacuolation, mutant genes, chromosomal deficiencies (TB) and cytoplasmic male sterility (cms) and their blocking effects, 474
Vascular tissue, 18
Veins
  basipetal, 25
  cross, 26
  during leaf development, 25–26
  lateral, 25, 26
  primary intermediate, 25
  transverse, 26
Vestigial glumes (*Vg*) mutation, 46
Vp1 gene, 649
  cloned anthocyanin genes and their gene product, 283
vp5 gene, 322
vp9 (7S) gene, 259
Virtis Microhomogenizer, 299
Viviparous stocks, 412

w3 (2L) gene, 259
W22 line, 352
  androgenesis and cytoplasmic substitution, 391
  class and other zein characteristics, 641
  gametophyte factors, 501
  inbred line, 434
  inbred lines, male sterility, 393
W22-A (genomic clone), class and other zein characteristics, 642
W22-B (genomic clone), class and other zein characteristics, 642
W23 line, 344
  agroinfection, 601
  C-banding, 487, 488
  inbred, 17
  inbred lines, androgenesis and cytoplasmic substitution, 391
  inbred lines, male sterility, 392–393
  indeterminate gametophyte, 389
  as male parent of a newly arisen male sterile trait, 422
  pachytene chromosomes, 485–486
  restoration patterns of inbred line of maize, 420
  Rf1-Rf2 constitution of non-T-restoring inbred line, 421
  shoot meristem culture, hormones unnecessary, 718
W23R (pollen parent), 392
  frequencies of androgenesis in crosses of three inbred lines as male to W23 *ig ig* females, 391
W64 line, restoration patterns of inbred line of maize, 420
W64A (variety) line
  class and other zein characteristics, 641
  inbred lines, male sterility, 393
WA374 (pollen parent), 392
  frequencies of androgenesis in crosses of three inbred lines as male to W23 *ig ig* females, 391
Wall crenulation, epidermal cells, trait distinguishing juvenile and adult regions of the shoot, 12
Waxy endosperm (*wx*)
  frequencies and parents of monosomic types produced by *R/r-X1* x Mangelsdorf's tester cross, 353
  genetic marker, and monosomics, 353, 355
  isolation of RNA, 545–546
Waxy (*wx*) gene, 624–629
Wd gene, 439
wd mutation, 13
Wf9 line, 414
  as male parent of a newly arisen male sterile trait, 422

NCS mutations, 414
restoration patterns of inbred line of maize, 420
Rf1-Rf2 constitution of non-T-restoring inbred line, 421
White/Adams technique of the Zone System, 103
White kernel lethal stocks, 412
White Rice popcorn inbred line 4519–4, 500, 501
Whp gene, cloned anthocyanin genes and their gene product, 283
Windowing, 84
*wl**-217A(6L) gene, 259
Wx enzyme, 627
Wx gene, 277, 278
Wx mutations, 298–301

X-irradiation, 364

y1 gene, 649
Yellow-green-2 gene (*yg2*), 277, 278, 439
Yellow endosperm
frequencies and parents of monosomic types produced by *R*/*r-X1* x Mangelsdorf's tester cross, 353
genetic marker, and monosomics, 353, 355

Z4 (genomic clone), class and other zein characteristics, 641
Z7 (genomic clone), class and other zein characteristics, 641
Za1 (genomic clone), class and other zein characteristics, 641
zE19 (genomic clone), class and other zein characteristics, 641
zE25 (genomic clone), class and other zein characteristics, 641
Zea, DNA transmission, 414
Zea diploperennis, 286
Zea luxurians, gametophyte factors, 501
Zea mays, 289
Zea mays L., inbred lines, 423
Zea mays L. endosperm, establishment and culture of, 719–722
Zea mays L. ssp. *mays*, 66
Zea mays L. ssp. *mays*, vs. teosintes, 6670, 73, 74
Zea mays ssp. *huehuetenangensis*, gametophyte factors, 501
Zea mays ssp. *mexicana*, 66
gametophyte factors, 501
Zea mays ssp. *parviglumis*, 66
gametophyte factors, 501
Zea mays ssp. *parviglumis* race Balsas, 68
Zea mexicana, 286
Zea perennis, 286
Zein genes
characterization, and their regulation in endosperm, 639–644
Zeins
high-performance liquid chromatography of, 425
isoelectric focusing of, 425
$ZnSO_4 \cdot 7H_2O$, stock solutions (100×) and final quantity of components in a simplified medium for growth of maize endosperm, 720
Zpr10/(22) locus, genes affecting zein accumulation, 643
Zygotenes, 448, 449, 461, 462

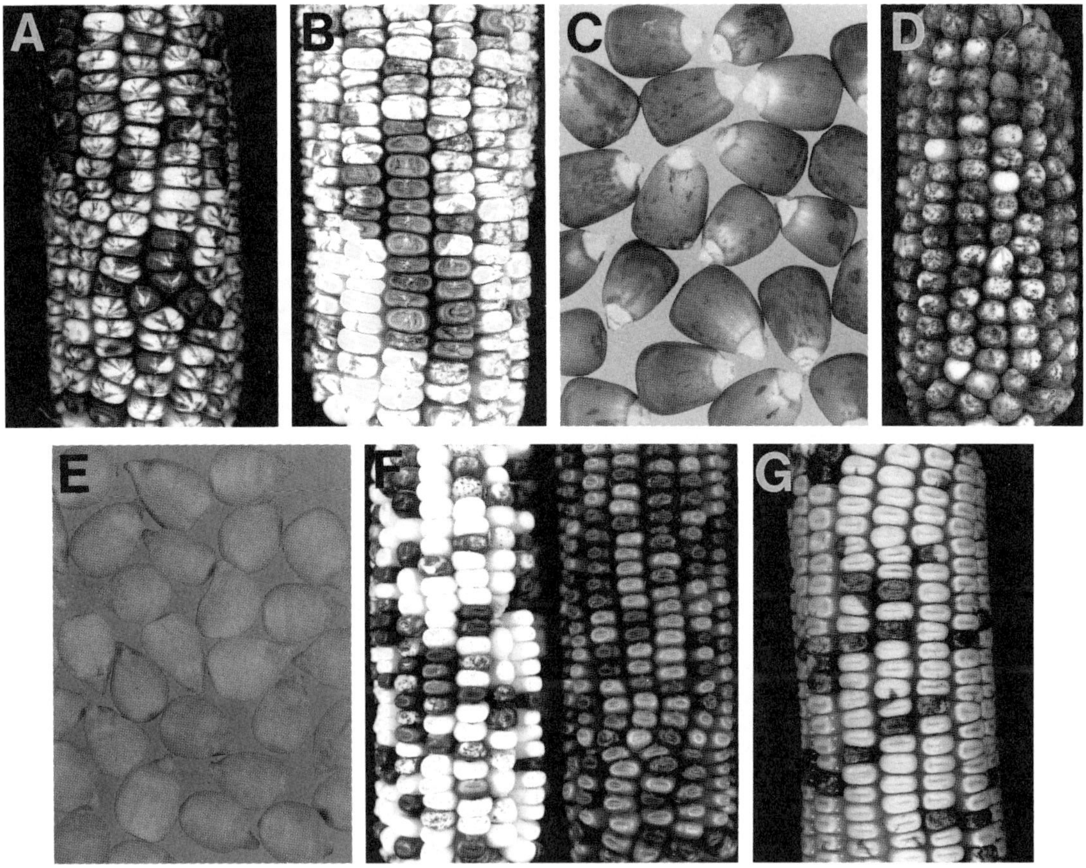

Figure 24.2 Gene tagging with *Ac*-Ear and Kernel Phenotypes. **A.** Medium variegated pericarp ear (*P-VV/P-WR*) conditioned by an insertion of *Ac* into the *P* gene. **B.** Strategy I ears showing a medium variegated pericarp phenotype (*P-VV/P-WR*). All F1 kernels receive the *r-sc:m3* allele from the pollen parent but only approximately one-half of the progeny receive an active *Ac* element which results in a spotted alcurone phenotype. The remaining kernels lack *Ac* activity and show a colorless alcurone phenotype. **C.** Strategy I kernels are selected as spotted aleurone kernels underlying red pericarp sectors. This phenotype is indicative of the loss of *Ac* from the donor locus (*P*) and its presence (*r-sc:m3* transactivation) elsewhere in the genome. **D.** Strategy II ears showing a homozygous variegated pericarp phenotype (*P-VV/P-VV*). All F1 kernels receive the *r-sc:m3* allele from the pollen parent and most kernels receive a single active *Ac* element from the maternally transmitted *P-VV* allele. **E.** Strategy II kernel selections with colorless or near colorless aleurone phenotype are indicative of an increase dosage of *Ac* (finely spotted) or loss of *Ac* (colorless) in kernel progeny. **F.** Ear and kernel phenotypes from red pericarp, spotted aleurone kernels (strategy I) show either colorless or full red pericarp, depending on which *P* allele segregates with the transposed *Ac* element. Both ears contain a fraction of kernels with spotted aleurones due to the presence of the *r-sc:m3* allele and transposed *Ac* element. **G.** Ear and kernel phenotypes from strategy II kernel selections. An increase in *Ac* copy number from one (*P-VV*) to two (*P-VV* and transposed *Ac*) result in a light-variegated pericarp phenotype. The proportion of fine-spotted, coarse-spotted, and colorless alcurones will reflect the linkage between the two *Ac* elements.